VOLUME ONE H

VITAMI
HORMC

Hormones

VOLUME ONE HUNDRED AND FIFTEEN

VITAMINS AND HORMONES

Hormones and Aging

Series Editor

GERALD LITWACK, PhD

Toluca Lake, North Hollywood, California

ELSEVIER

ACADEMIC PRESS

An imprint of Elsevier

Academic Press is an imprint of Elsevier
50 Hampshire Street, 5th Floor, Cambridge, MA 02139, United States
525 B Street, Suite 1650, San Diego, CA 92101, United States
The Boulevard, Langford Lane, Kidlington, Oxford OX5 1GB, United Kingdom
125 London Wall, London, EC2Y 5AS, United Kingdom

First edition 2021

ISBN: 978-0-323-85548-8
ISSN: 0083-6729

For information on all Academic Press publications
visit our website at https://www.elsevier.com/books-and-journals

Publisher: Zoe Kruze
Acquisitions Editor: Ashlie M. Jackman
Editorial Project Manager: Chris Hockaday
Production Project Manager: Denny Mansingh
Cover Designer: Victoria Pearson

Typeset by SPi Global, India

Contents

Contributors

Seemaab Ali
Department of Cancer Biology and Genetics, College of Medicine, The Ohio State University; The Ohio State University Comprehensive Cancer Center, Columbus, OH, United States

Francisco Amado
LAQV-REQUIMTE, Departamento de Química, Universidade de Aveiro, Aveiro, Portugal

Sabina Bhatta
Department of Biomedical Science, Kent State University, Cunningham Hall, Kent, OH, United States

Antonio Carlos Boschero
Obesity and Comorbidities Research Center, Institute of Biology, University of Campinas (UNICAMP), Campinas, Sao Paulo, Brazil

Renato Chaves Souto Branco
Obesity and Comorbidities Research Center, Institute of Biology, University of Campinas (UNICAMP), Campinas, Sao Paulo, Brazil

Gabriela Alves Bronczek
Obesity and Comorbidities Research Center, Institute of Biology, University of Campinas (UNICAMP), Campinas, Sao Paulo, Brazil

Zi-Jian Cai
CaiFortune Consulting, Suzhou, Jiangsu, PR China

Lei Cao
Department of Cancer Biology and Genetics, College of Medicine, The Ohio State University; The Ohio State University Comprehensive Cancer Center, Columbus, OH, United States

Xu Cao
From the Department of Orthopedic Surgery, The Johns Hopkins University School of Medicine, Baltimore, MD, United States

Daniel P. Cardinali
Faculty of Medical Sciences, Pontificia Universidad Católica Argentina, Buenos Aires, Argentina

Gemma Casadesus
Department of Biological Science, School of Arts and Sciences, Kent State University, Cunningham Hall, Kent, OH, United States

Mayank Choubey
Department of Zoology, Institute of Science, Banaras Hindu University, Varanasi, India; Department of Molecular Medicine, Columbia University Irvine Medical Center, New York, NY, United States

Sara De Vincentis
Unit of Endocrinology, Department of Biomedical, Metabolic and Neural Sciences, University of Modena and Reggio Emilia; Department of Medical Specialties, Azienda Ospedaliero-Universitaria of Modena, Ospedale Civile di Baggiovara, Modena, Italy

Maria Chiara Decaroli
Unit of Endocrinology, Department of Biomedical, Metabolic and Neural Sciences, University of Modena and Reggio Emilia; Department of Medical Specialties, Azienda Ospedaliero-Universitaria of Modena, Ospedale Civile di Baggiovara, Modena, Italy

Leonidas H. Duntas
Evgenideion Hospital, Unit of Endocrinology, Diabetes and Metabolism, University of Athens, Athens, Greece

Gustavo Duque
Australian Institute for Musculoskeletal Science (AIMSS), University of Melbourne and Western Health; Department of Medicine-Western Health, Melbourne Medical School, The University of Melbourne, St. Albans, VIC, Australia

D. González-Hedström
Departamento de Fisiología, Facultad de Medicina, Universidad Autónoma de Madrid; Pharmactive Biotech Products S.L. Parque Científico de Madrid. Avenida del Doctor Severo Ochoa, 37 Local 4J, 28108 Alcobendas, Madrid, Spain

Nicole Gorny
Department of Anatomy & Neurobiology, University of Maryland School of Medicine, Baltimore, MD, United States

M. Granado
Departamento de Fisiología, Facultad de Medicina, Universidad Autónoma de Madrid; CIBER Fisiopatología de la Obesidad y Nutrición. Instituto de Salud Carlos III, Madrid, Spain

Gail A. Greendale
Division of Geriatrics, Department of Medicine, David Geffen School of Medicine at UCLA, Los Angeles, CA, United States

Caroline Gurvich
Monash Alfred Psychiatry Research Centre, Monash University Central Clinical School and The Alfred Hospital, Melbourne, VIC, Australia

Kirsty Hamilton
Division of Systems Medicine, Ninewells Hospital and Medical School, University of Dundee, Dundee, United Kingdom

Jenni Harvey
Division of Systems Medicine, Ninewells Hospital and Medical School, University of Dundee, Dundee, United Kingdom

Christophe Hirtz
IRMB, Univ Montpellier, INSERM, CHU Montpellier, CNRS, Montpellier, France

Amit Jain
From the Department of Orthopedic Surgery, The Johns Hopkins University School of Medicine, Baltimore, MD, United States

Arun S. Karlamangla
Division of Geriatrics, Department of Medicine, David Geffen School of Medicine at UCLA, Los Angeles, CA, United States

Michy P. Kelly
Department of Anatomy & Neurobiology, University of Maryland School of Medicine, Baltimore, MD, United States

Amitabh Krishna
Department of Zoology, Institute of Science, Banaras Hindu University, Varanasi, India

Jayashri Kulkarni
Monash Alfred Psychiatry Research Centre, Monash University Central Clinical School and The Alfred Hospital, Melbourne, VIC, Australia

Mirian Ayumi Kurauti
Obesity and Comorbidities Research Center, Institute of Biology, University of Campinas (UNICAMP), Campinas, Sao Paulo; Department of Physiological Sciences, Biological Sciences Center, State University of Maringa (UEM), Maringa, Parana, Brazil

Makoto Kuro-o
Division of Anti-Aging Medicine, Center for Molecular Medicine, Jichi Medical University, Shimotsuke, Tochigi, Japan

Denise Lasigliè
Istituto Comprensivo "Franco Marro", Ministero dell'Istruzione Ministero dell'Università e della Ricerca (M.I.U.R), Villar Perosa, TO, Italy

Jessica Le
Monash Alfred Psychiatry Research Centre, Monash University Central Clinical School and The Alfred Hospital, Melbourne, VIC, Australia

T. Peter Li
From the Department of Orthopedic Surgery, The Johns Hopkins University School of Medicine, Baltimore, MD, United States

A. López-Calderón
Departamento de Fisiología, Facultad de Medicina, Universidad Complutense de Madrid, Madrid, Spain

Carine Marmentini
Obesity and Comorbidities Research Center, Institute of Biology, University of Campinas (UNICAMP), Campinas, Sao Paulo, Brazil

A.I. Martín
Departamento de Fisiología, Facultad de Medicina, Universidad Complutense de Madrid, Madrid, Spain

Megan Mey
Department of Biomedical Science, Kent State University, Cunningham Hall, Kent, OH, United States

Sneha Mishra
Cardiovascular and Muscle Research Laboratory, Department of Microbiology and Cell Biology, Indian Institute of Science, Bengaluru, India

Lavanya Murthy
Australian Institute for Musculoskeletal Science (AIMSS), University of Melbourne and Western Health; Department of Medicine-Western Health, Melbourne Medical School, The University of Melbourne, St. Albans, VIC, Australia

V. Papadopoulos
Department of Pharmacology and Pharmaceutical Sciences, School of Pharmacy, University of Southern California, Los Angeles, CA, United States

Yongsoon Park
Department of Food and Nutrition, Hanyang University, Seoul, South Korea

Luís Perpétuo
iBiMED, Department of Medical Sciences, University of Aveiro, Aveiro, Portugal

T. Priego
Departamento de Fisiología, Facultad de Medicina, Universidad Complutense de Madrid, Madrid, Spain

Nicholas J. Queen
Department of Cancer Biology and Genetics, College of Medicine, The Ohio State University; The Ohio State University Comprehensive Cancer Center, Columbus, OH, United States

Ashutosh Ranjan
Department of Zoology, Institute of Science, Banaras Hindu University, Varanasi, India

Venkatraman Ravi
Cardiovascular and Muscle Research Laboratory, Department of Microbiology and Cell Biology, Indian Institute of Science, Bengaluru, India

Vincenzo Rochira
Unit of Endocrinology, Department of Biomedical, Metabolic and Neural Sciences, University of Modena and Reggio Emilia; Department of Medical Specialties, Azienda Ospedaliero-Universitaria of Modena, Ospedale Civile di Baggiovara, Modena, Italy

Guillermo Romero
Department of Pharmacology & Chemical Biology, University of Pittsburgh, Pittsburgh, PA, United States

Guy Salama
Department of Medicine, Heart and Vascular Institute, University of Pittsburgh, Pittsburgh, PA, United States

Albert Shieh
Division of Geriatrics, Department of Medicine, David Geffen School of Medicine at UCLA, Los Angeles, CA, United States

Gabriela Moreira Soares
Obesity and Comorbidities Research Center, Institute of Biology, University of Campinas (UNICAMP), Campinas, Sao Paulo, Brazil

Nagalingam Ravi Sundaresan
Cardiovascular and Muscle Research Laboratory, Department of Microbiology and Cell Biology, Indian Institute of Science, Bengaluru, India

Veena Taneja
Department of Immunology and Rheumatology, Mayo Clinic, Rochester, MN, United States

Elizabeth H.X. Thomas
Monash Alfred Psychiatry Research Centre, Monash University Central Clinical School and The Alfred Hospital, Melbourne, VIC, Australia

Natalie Thomas
Monash Alfred Psychiatry Research Centre, Monash University Central Clinical School and The Alfred Hospital, Melbourne, VIC, Australia

Rui Vitorino
iBiMED, Department of Medical Sciences, University of Aveiro; LAQV-REQUIMTE, Departamento de Química, Universidade de Aveiro, Aveiro; UnIC, Departamento de Cirurgia e Fisiologia, Faculdade de Medicina da Universidade do Porto, Porto, Portugal

Pierre Marie Voisin
Pierre Fabre Oral Care, Castre, France

Bruce A. Watkins
Department of Nutrition, University of California, Davis, CA, United States

B.R. Zirkin
Department of Biochemistry and Molecular Biology, Johns Hopkins Bloomberg School of Public Health, Baltimore, MD, United States

Preface

Aging is a complicated process. Although there are some genes that are associated with aging, it is likely that pollution of the food chain, the atmosphere (the air we breathe probably contains nearly half the oxygen that is needed for longevity), water, and soil all play a major role. Ingestion or subjection to xenobiotics may play a role as well and some of our ingested medicines may fall into this category. Canned foods are usually extraordinarily high in sodium. The pollution of the oceans and rivers give rise to ingested fishes that have high levels of mercury, a toxin. In time, these factors, bacterial and viral infections, sources of foods and ingested xenobiotics result in organ degeneration and failure. Drinking large amounts of alcohol, drug abuse, and smoking must be counted as factors that give rise to liver and lung diseases. Unhealthy environmental conditions, such as working in mines or exposure to asbestos, can produce serious and life-threatening disease of the lungs. Anxiety and stress must also play a negative role in longevity. There are biochemical aspects of aging, such as alterations in genetic DNA and shortening of telomeres. Cancer is a major factor as is cardiovascular disease. Obesity is a surging problem in the world and it must account for the global epidemic of type 2 diabetes. Additionally, tissue caches of stem cells become depleted with aging and this must be a crucial determinant. Of course, there are genetic mutations that cause the development of disease. All of these factors, and more, are vectors for aging and death. Obviously, aging is accompanied by many metabolic alterations; not the least of these involve the endocrine system. Accordingly, the effects of aging on various hormonal activities are described in this volume. The content begins with some basic aspects, and then proceeds in the order of effects on the brain and then downward to the genital system. Virtually all of the 23 chapters included in this volume describe basic science approaches. The chapters are listed below in the sequence of their appearance.

L.H. Duntas, *Aging and the hypothalamic–pituitary–thyroid axis*; Z.J. Cai, *Hypothalamic aging and hormones*; L. Cao, S. Ali, and N.J. Queen, *Hypothalamic gene transfer of BDNF promotes healthy aging*; D.P. Cardinali, *Melatonin and healthy aging*; M. May, S. Bhatta, and G. Casadesus, *Luteinizing hormone and the aging brain*; K. Hamilton and J. Harvey, *Leptin regulation of hippocampal synaptic function in health and disease*; Y. Park and

B.A. Watkins, *Endocannabinoids and aging—inflammation, neuroplasticity, mood and pain*; L. Murthy and G. Duque, *Parathyroid hormones levels and aging: effect on balance*; M.A. Kurauti, G.M. Soares, C. Marmentini, G.A. Bronczek. R.C.S. Branco, and A.C. Boschero, *Insulin and aging*; D. Lasigliè, *Sirtuins and the prevention of immunoscenescence*; N. Gorny and M.P. Kelly, *Alterations in cyclic nucleotide signaling are implicated in healthy aging and age-related pathologies of the brain*; M. Kuro-O, *Aging and FGF23-klotho system*; M.C. Decaroli, S. DeVincentis, and V. Rochira, *Aging and sex hormones in males*; V. Taneja, *Sexual dimorphism, aging and immunity*; A.S. Karlamangla, A. Shieh, and G.A. Greendale, *Hormones and bone loss across the menopause transition*: G. Romero and G. Salama, *Relaxin abrogates genomic remodeling of the aged heart*; S. Mishra, V. Ravi, and N.R. Sundaresan, *Role of FoxO transcription factors in aging-associated cardiovascular diseases*; L. Perpétuo, P.M. Voisin, F. Amado, C. Hirtz, and R. Vitorino, *Ghrelin and adipokines: An overview of their physiological role, antimicrobial activity and impact on cardiovascular conditions*; C. Gurvich, J. Le, N. Thomas, E.H.X. Thomas, and J. Kulkarni, *Sex hormones and cognition in aging*; T. Priego, A. Martín, D. González-Hedström, M. Granado, A. López-Calderón, and D. Cardalini, *Role of hormones in sarcopenia*; T. Peter Li, A. Jain, and X. Cao, *Sources of lumbar back pain during aging and potential therapeutic targets*; V. Papadopoulos and B. Zirkin, *Leydig cell aging: Molecular mechanisms and treatments*; M. Choubey, A. Ranjan, and A. Krishna, *Adiponectin/AdipoRs signaling as a key player in testicular aging and associated-metabolic disorders*.

A new cover has been designed for this book. It will be used for successive volumes of *Vitamins and Hormones* with a different color background for each new volume. This is a time-saving device in that previous selections of figures from book chapters required extra effort from the cover designer.

I wish to thank Chris Hockaday and Ashlie Jackman in the Oxford, United Kingdom, office of Elsevier for their cooperation in the preparation of the volume for production. The production group in Chennai, India, as usual, excellently prepared the final volume.

Gerald Litwack, PhD
Toluca Lake, North Hollywood
CA, United States

CHAPTER ONE

Aging and the hypothalamic-pituitary-thyroid axis

Leonidas H. Duntas*
Evgenideion Hospital, Unit of Endocrinology, Diabetes and Metabolism, University of Athens, Athens, Greece
*Corresponding author: e-mail address: ledunt@otenet.gr

Contents

Abstract

The world's population is increasingly aging, this noted particularly in the Western world where there are ever greater numbers of centenarians and those over 85 years. Given the immense importance of the thyroid gland for optimal health and the fact that morphological and functional changes in the hypothalamic-pituitary-thyroid (HPT) axis take place as a natural adaptation to the aging process, a clear distinction must be made in older individuals between these and the onset of disease. However, this is problematic since, frequently, subtle differences exist between them, making diagnosis a challenging task, especially as concerns subclinical disease. The newly emerging interdisciplinary field of geroscience offers the prospect of being used as a platform to investigate the effect of disrupted HPT function on functional capacity and cognitive ability among the aged, as well as the risk or onset of age-associated diseases, thus enhancing healthspan and lifespan. Because optimal functioning of the thyroid gland is a prerequisite for longevity as well as for mental and physical wellbeing, this review summarizes the recent scientific data regarding HPT and aging while discussing alternative and personalized treatment approaches to maintaining a healthy thyroid as a means to ensuring a long, active, and healthy life.

Vitamins and Hormones, Volume 115
ISSN 0083-6729
https://doi.org/10.1016/bs.vh.2020.12.001

1. Introduction

The population of the Western world is aging, with those aged above 65 years projected to triple by 2050, reaching nearly 1.5 billion, as compared to 1950 (UN, 2010). Life expectancy has increased considerably over the past century, from 47 years in 1900 to over 78 years today, this due to successful management of the infectious and parasitic diseases that caused the death of infants and children at the beginning of the 20th century and improvement in social conditions and medical care during the last six to seven decades. On the other hand, the latter phenomenon has been accompanied by a steep increase in the incidence of chronic non-communicable diseases among the elderly, such as obesity, heart disease, diabetes, and cancer, which stem from radical changes in lifestyle and dietary habits in the previous century, which, in turn, affect the quality of life in this age group. Interventional studies targeting lifestyle and diet improvements may thus result in reducing susceptibility to and frequency of disease and age-related changes, thereby lessening functionality and increasing frailty and disability (National Institute of Aging, 2019).

Like all bodily systems, the endocrine system is affected by aging. This particularly applies to the thyroid gland in which numerous morphological and biochemical changes may be observed by the fifth decade, with the clinical presentation becoming progressively more nonspecific as individuals grow older and an array of physiological alterations possibly also affecting thyroid function test (TFT) results. Of note, mild or subclinical forms of thyroid dysfunction occur in more than 10% of individuals aged >80 years old, what is known as the "oldest old," although the threshold for treatment at this age is high, since their thyroid hormone requirements are generally lower (Leng & Razvi, 2019).

With regard to age-associated diseases (AADs), these naturally most severely affect the oldest old (Franceschi et al., 2018), who comprise the fastest growing age segment in many countries, their total being expected to have reached 14.1 million globally by 2040 (Lutz, Sanderson, & Scherbov, 2008).

In view of all the above, it is clear that in the years ahead, management of thyroid diseases will become not only ever more complex but also increasingly in demand as the global population ages. Emphasis should thus be placed on monitoring and maintaining a normally functioning thyroid gland and fine-tuning the activity of the hypothalamic-pituitary-thyroid (HPT) axis, with very careful attention paid to the individual's specific age and resultant needs and to what extent he/she is suffering from AADs. In this setting, the importance of optimal thyroid function for attainment of healthy longevity and "graceful aging" warrant persistent investigation.

The present review aims to evaluate the data of mainly the past 10 years, exploring how the HPT axis adapts to aging while analyzing and discussing some new concepts, including geroscience, as well as a number of challenges and dilemmas concerning the diagnosis and treatment of thyroid disease in older individuals.

2. Search strategy and selection criteria

The PubMed and Cochrane Library databases were searched for publications mainly from January 2010 to May 2020 using the terms "hypothalamic-pituitary thyroid axis in aging," "thyroid function in the elderly," "geroscience and thyroid" "thyroid disease and aging," "TRH in aging," "TSH in aging," "geroscience and hypothalamic-pituitary-thyroid axis." However, some older original studies on thyroid morphology and thyrotropin-releasing hormone (TRH) metabolism have also been considered.

3. Is aging a disease?

The interrelationship between aging and disease has long been under discussion, with pathological aging normally seen as separate from biological aging. Meanwhile, what is known as "successful aging," that is, aging without any detectable disease, is probably experienced by not more than 20% of the entire human population, with pathological aging being predominant (Gluckman, Buklijas, & Hanson, 2016). While many propose that aging is a disease, others are strongly against this notion, arguing that aging cannot be seen as a breakdown of the system but, instead, as a process that heightens the potency of different types of malfunctioning of the system as the individual ages. In view of this ambiguity, there is a need for continuous support from and investigation by the biomedical community as regards life-course epidemiology, in other words, the study of the long-term effects exerted on, inter alia, age-related disorders by physical and/or social exposures during gestation, childhood, adolescence, young adulthood, and later adult life. Such in-depth investigation will be carried out within the framework of the Developmental Origins of Health and Disease (DOHaD) concept.

4. The field of geroscience and the thyroid

The past decade has seen the emergence of a new interdisciplinary field, geroscience, devoted to research into biological aging and focusing on promoting among the elderly a healthy lifestyle and "optimal longevity," that is, living long while enjoying good health and quality of life (Kennedy et al., 2014; Vijk & Kennedy, 2016). The new field of geroscience has arisen to address, via a number of interventions, the progressive decline in physiological function that comes with aging, which leads to functional limitations (e.g., reduced mobility), increases the risk of AADs, disability and mortality, and, thus, impedes achievement of optimal longevity (Seals, Justice, & LaRocca, 2016). Most importantly, the novel interdisciplinary field of geroscience aims to investigate the relationship between the biology of aging and the pathophysiology of AADs, this integrated view proposing that aging is the predominant risk factor for AADs and that aging and AADs may have a common underlying basic mechanism (Sierra & Kohanski, 2017). The hypothesis thus is that manipulation of cellular aging processes may prolong both health and lifespan by delaying the manifestation of AADs (Fig. 1).

Identifying reliable biomarkers of aging is a major goal in geroscience. The first generation of epigenetic biomarkers were developed using only chronological age as a surrogate for biological age. However, the incorporation of clinical measures of phenotypic age that identify differences in lifespan and healthspan are expected to facilitate the development of more powerful epigenetic biomarkers of aging. In 2018, by applying an innovative two-step process, a new epigenetic biomarker of aging, DNA methylation (DNAm), PhenoAge, was developed (Levine et al., 2018). DNAm-based biomarkers of aging are able to predict chronological age, which results from hypo- and hypermethylation alterations across the genome induced by chronological time. Given that the thyroid plays a critical role in human physiology, thyroid function is highly likely to be intimately connected with the risk for AADs; therefore, TH, particularly free thyroxine (FT4), and the pituitary thyrotropin-stimulating hormone (TSH) could potentially serve as biomarkers of thyroid function over time (Fig. 2). Adapted to a model of geroscience, the latter could hence be capable of predicting decline in physiological thyroid function with aging, thereby enabling clinicians to intervene promptly in the case of thyroid dysfunction and thus enhance patients' health and longevity. The tight connection between thyroid functioning and optimal health is supported by studies in patients with a diagnosis

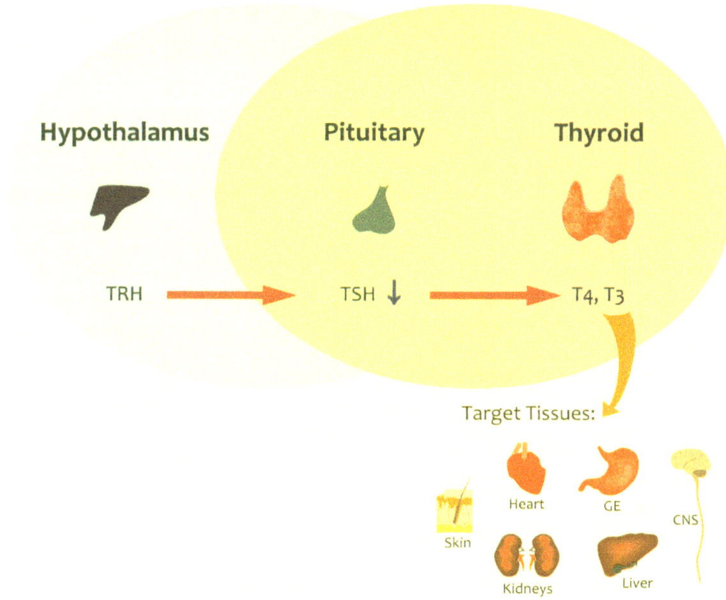

Fig. 1 Thyrotropin-releasing hormone (TRH) is the initiator of action of the hypothalamic-pituitary-thyroid (HPT) axis. TRH stimulates the synthesis and secretion of thyrotropin (TSH), which stimulates the synthesis and secretion of thyroxine from the thyroid follicular cells. Thyroxine is transported to the various tissues via the circulation, where it is converted to triiodothyronine and exerts its actions. Whereas TRH remains mainly unchanged in aging, TSH secretion declines slightly.

Fig. 2 According to the concept of geroscience, a decrease in FT4, or FT3 levels and consequently an increase in TSH (or vice versa) may be a predictable factor for comorbidities, while the existence of comorbidities may predict pathological changes in the HPT axis.

of hypothyroidism or hyperthyroidism which, when the subjects' TSH concentrations were within the reference range, did not have any clinically significant effect on long-term health outcomes, including mortality, atrial fibrillation, ischemic heart disease, stroke, and fractures (Gussekloo et al., 2004; Thayakaran et al., 2019). In contrast, when TSH concentration, adjusted to the patients' age, is observed to be outside the reference range, adverse health outcomes and shortening of lifespan are typically observed.

5. HPT axis morphology in aging

Aging is marked by subtle morphological and functional alterations in the hypothalamic-pituitary unit, as occurs in all biological systems. Changes in the structure of the hypothalamus (HT) observed in aged mice three decades ago were reported to be characterized by a gradual separation of the tight gap junctions between the tanycytes, the specialized ependymal cells lining the floor of the third ventricle (Morley & Korenman, 1992; Scott & Sladek, 1981). In addition, the nuclear volume of neurons in various regions of the hypothalamus declines, a finding of major importance for our understanding of the aging process.

Crucially, aging is accompanied by alterations in the brain neurotransmitters that stimulate or restrain pituitary hormone synthesis and secretion (Šošić-Jurjević et al., 2016).

The aging thyroid gland is characterized by microfollicular structure, relatively increased volume of follicular epithelium, decreased volume of luminal colloid, and increased basolateral expression of sodium-iodide symporter-immunopositivity (Veldhuis, 2013).

Aging is accompanied by progressive telomere shortening, which induces senescence and degeneration of cells. This phenomenon was studied with regard to its impact on the thyroid using normal thyroid tissue specimens from 46 human subjects aged between 0 and 98 years and it was observed that telomere shortening occurred after 50 years of age in thyroid and parathyroid tissues. The latter finding thus indicates that human thyroid and parathyroid tissues do not undergo the rapid reduction in telomere length early in life that has been reported for other human cell types, suggesting that the rate of telomere shortening has tissue-specific characteristics (Kammori et al., 2015).

6. Hypothalamic and pituitary physiology and rhythmicity

The key components of the hypothalamic-pituitary unit are hypothalamic TRH neurons, generated in the paraventricular nucleus, and TSH. Whereas prepro-thyrotropin releasing hormone messenger RNA levels in the HT do not change significantly with aging, the secretion rate of TRH approximately doubles (Pekary, Mirell, Turner, Walfish, & Hershman, 1987). An experimental study was conducted using young and aged male rats to explore the kinetics of the hypothalamic secretion of TRH to the anterior pituitary in aging mammals (Pekary et al., 1987). Though serum TSH levels typically increase when serum T4 levels decline, their values did not alter as the rats aged: this implies that in aging, there may be a compensation for the gradual loss of essential TRH stimulation of TSH release by a drop in T3 inhibition of TSH release in the pituitary. Also reported was a 66% decrease with aging in basal TRH release by incubated rat hypothalamic.

Hypothalamic TRH modulates the synthesis and secretion of TSH, which stimulates the synthesis, and secretion of T4 by the thyrocytes (Fig. 1).

A diurnal variation was also reported for the levels of TRH-like peptides (pGlu-X-Pro-NH(2), where "X" can be any amino acid residue) in some brain regions involved in mood regulation. These peptides have neuroprotective and antidepressant-like properties that may help stabilize chronobiologic systems, which are often abnormal in neuropsychiatric disease (Pekary, Stevens, & Sattin, 2006). TRH-like peptide levels, in general, are highly correlated with changes in TRH concentration, within and between brain regions: the observation that they are colocalized in large glutamatergic neurons innervating the rat limbic system has led to the hypothesis that they may also be involved in the mood changes and depression that often accompany hyper- and hypothyroidism in humans. However, the latter has not been confirmed, though the TRH-like peptides may be involved in maintaining autonomic, behavioral, and mood equilibria.

Sixty-five years ago, in 1955, Russfield and colleagues described enlargement of the hypophysis in young patients with hypothyroidism and a 10-fold "increase in amphophiles and hypertrophic amphophiles" (Russfield, 1955). Since then, it has been established that in humans, age alters the secretion of a number of pituitary hormones, this change starting with gradual erosion of physiological signaling mechanisms, resulting in lower incremental secretory-burst amplitude and blunted 24 h rhythmic secretion (Roelfsema & Veldhuis, 2013). The degree of these alterations is dependent on various factors and conditions, such as gender, body composition, stress, comorbidity, diet and caloric intake, medication use, physical frailty, immune status, level of exercise, and neurocognitive decline (Tarquini & Mazzoccoli, 2017). The rhythmic oscillations of signaling pathways, which regulate intermediate metabolism, are driven by molecular clockwork; the circadian timing system synchronizes behavioral cycles and anabolic/catabolic processes with environmental cues, mainly represented by light/darkness alternation (Roelfsema et al., 2010). Since metabolic pathways, such as bile acid synthesis, and autophagic and immune-inflammatory processes are driven by the biological clock, proper timing of hormone secretion is necessary to avoid dysmetabolism.

Pulsatile and circadian TSH secretions are predominantly controlled by TRH (Brabant, Prank, Hoang-Vu, Hesch, & von zur Mühlen, 1991). Physiologically, the lowest TSH concentrations are observed during the afternoon, followed by a rise during the evening, reaching a maximum in the first part of the night, with the acrophase (time of maximal concentration) of the diurnal rhythm in TSH concentrations occurring at 3 am. TSH levels physiologically decrease during sleep. In addition to this nychthemeral pattern, which is recurrent, variably sized bursts of TSH release occur during the 24-h period (Brabant et al., 1991). However, an age-related resetting of the pituitary threshold of TSH feedback suppression and loss of the TSH nocturnal surge with increasing age may also indicate a hypothalamic dysfunction, which is frequently seen progressively in older people and which becomes more evident in extreme senescence (Monzani et al., 1996).

More recent techniques have, however, yielded data that somewhat differ from those produced in older studies. According to automated deconvolution, percentage pulsatile TSH secretion was much lower in severely hypothyroid patients than that cited by Cluster or Desade (Adriaanse, Brabant, Prank, Endert, & Wiersinga, 1992). Meanwhile, relevant variables, such as pulsing regularity, secretion mode (shape of the pulse), and half-lives have propelled the continuing use both of serial blood sampling and of more sophisticated techniques for a better understanding of the functioning of the HPT axis.

7. Thyroid diseases and the hypothalamic-pituitary axis

Aging has been associated with impaired TSH secretion, the elderly frequently suffering from dysthyroidism. A study was therefore carried out to investigate whether there are differences among healthy young, middle-aged, and elderly people in 24 h secretory profiles of TRH, TSH, and FT4 determined in blood samples collected every 4 h for 24 h (Mazzoccoli et al., 2010). Fifteen healthy middle-aged subjects with mean age $\pm$ S.E. 44.1 ± 1.7 and 15 healthy elderly subjects with mean age $\pm$ S.E. 68.5 ± 1.2 were recruited to evaluate the circadian rhythmicity. Normal circadian rhythmicity for TRH and TSH was observed in the young and middle-aged subjects and for TSH in the older subjects. While elderly subjects exhibited lower TSH levels, no statistically significant difference in TRH and FT4 serum levels was detected between young, middle-aged, and elderly subjects.

On the other hand, in areas with adequate or high iodine intake, TSH levels have been found to be higher and positively associated with age, while, in contrast, mild or moderate iodine deficiency is accompanied by lower TSH levels and is negatively related to age (Diamanti-Kandarakis et al., 2017).

7.1 Overt and subclinical hyperthyroidism

Both identifying signs and symptoms of thyroid disease among the elderly and interpreting thyroid function tests are challenging. For example, because TSH secretion falls with aging, the finding of low serum TSH levels may or may not be indicative of hyperthyroidism, therefore, the diagnosis should be made following anamnesis, including detailed information on comorbidities and medications, combined with determination of serum FT4 and/or FT3 levels and ultrasound examination. A timely diagnosis is mandatory as both overt and subclinical hyperthyroidism increase cardiovascular mortality, frailty, and bone fractures, while they have also been associated with development of Alzheimer's disease (Aubert, 2017; Barbesino, 2019; George et al., 2019). Overt or subclinical hyperthyroidism in aging people calls for careful evaluation and timely treatment with thionamides or radioiodine therapy, must always be considered. Important factors influencing the decision to treat are the age of the patient and his/her general health condition (Boelaert, 2013).

7.2 Overt and subclinical hypothyroidism

Clinical diagnosis of hypothyroidism in elderly subjects is complicated by the fact that the TSH reference range needs to be very carefully evaluated according to the patient's age, while body weight and other variants, such as drugs, comorbidities, and general health condition, must be taken into account (Duntas & Yen, 2019). Clinical signs such as fatigue, cold intolerance, constipation, weakness, reduced appetite, hair loss and dry skin can easily be misdiagnosed and attributed to aging alone, as they can often be mild. The same can also apply to psychiatric disorders such as depression, which is common in the elderly (Kramer, von Muhlen, Kritz-Silverstein, & Barrett-Connor, 2009).

Thyroid hormone is a potent regulator of cholesterol metabolism, therefore, the possibility of thyroid dysfunction underlying cases of unexplained dyslipidemia should be considered (Duntas & Brenta, 2018). Once hypothyroidism has been confirmed, treatment requires caution and an individualized therapeutic approach.

Both cross-sectional and longitudinal studies have reported an increase in population of TSH distribution with age (Atzmon, Barzilai, Hollowell, Surks, & Gabriely, 2009). Meanwhile, data concerning adverse cardiovascular, metabolic, and cognitive effects of subclinical hypothyroidism in the elderly remain debatable (Gesing, Lewiński, & Karbownik-Lewińska, 2012; Hennessey & Espaillat, 2015). According to the TRUST study, treatment is not recommended in elderly subjects with subclinical hypothyroidism and $TSH < 10\,mU/L$ (Stott et al., 2017), and there is, indeed, some evidence that hypothyroidism may contribute to increased lifespan. It is thus generally considered that replacement therapy with L-thyroxine should not uniformly be recommended in elderly people, and particularly in those with subclinical hypothyroidism (Atzmon et al., 2009). However, in an observational study including 611 hospitalized patients, aged 60–80 years, with treated hypothyroidism, only TSH levels from 5 to 10 IU/L, and with no effect of FT4 levels, were associated with increased mortality (Akirov, Gimbel, Grossman, Shochat, & Shimon, 2017). The authors therefore conclude that the aim of treatment should be achievement of euthyroidism, according to age-related TSH reference levels, to improve survival. It is important to note that aging is typically associated with decreased secretion of TSH and thyroid hormones. However, reduced peripheral T4 degradation may explain the unchanged values of T4 in healthy elderly subjects, while serum T3 may show an age-specific decline, the extent of which could be associated with frailty, as recently reported (Bertoli et al., 2017; Jasim & Gharib, 2018).

7.3 Central hyperthyroidism

Secondary or central hyperthyroidism, a rare disorder due to TSH-secreting pituitary adenomas and forming part of the syndrome of "inappropriate secretion of TSH," is typically characterized by non-suppressed TSH and accompanied by elevated serum FT4 and FT3 levels (Beck-Peccoz, Lania, Beckers, Chatterjee, & Wemeau, 2013). Sick euthyroid syndrome is characterized by changes in the HPT axis (de Vries, Fliers, & Boelen, 2015), such as downregulation of TRH neurons in the paraventricular nucleus of the hypothalamus and reduced production of TSH from the thyrotrope. This condition, which is caused by altered feedback regulation during severe illness, needs skill for achievement of an accurate diagnosis and appropriate treatment.

Of interest is the fact that TRH has been applied as an anti-aging hormone following studies on the effects of TRH administration in mice, with particular focus on the gonadal-reproductive and kidney-urinary systems (Over, Mannan, Nsouli-Maktabi, Burman, & Jonklaas, 2010). Short-term acute and long-term chronic oral administration of TRH in drinking water to old mice over a period of 4 months resulted in a significant increase in testes weight, structure, and function (Pierpaoli, 2013). Similarly, TRH treatment protected the kidneys from amyloid and hyaline infiltration of tubuli and glomeruli, which is typically seen in aging mice.

8. Conclusions

The morphological and functional changes of the HPT axis as observed in the elderly should be evaluated as a progressive adaptation process rather than a disease state. However, the possibility of the existence of an inherent disease may complicate the diagnosis, while evaluation should be carried out according to the age adjusted reference levels of serum TSH levels. The question of whether aging-induced alterations of the HPT axis may lead to disease or if comorbidities might be responsible for pathological changes in the HPT is likely to be resolved with the development of the field of geroscience.

To conclude, aging should not be considered as a disease but as a stage of life during which the clinician's chief focus must be on the maintenance among the elderly, of optimal health and quality of life and promotion of mental, physical, and social activity (Mammen, 2019). The latter is a goal that can, in good part, be achievable through scrupulous differentiation between thyroid disease and healthy aging and via meticulous monitoring of the HPT axis and prudent use of TH in older adults.

References

Adriaanse, R., Brabant, G., Prank, K., Endert, E., & Wiersinga, W. M. (1992). Circadian changes in pulsatile TSH release in primary hypothyroidism. *Clinical Endocrinology*, *37*, 504–510.

Akirov, A., Gimbel, H., Grossman, A., Shochat, T., & Shimon, I. (2017). Elevated TSH in adults treated for hypothyroidism is associated with increased mortality. *European Journal of Endocrinology*, *176*(1), 57–66. https://doi.org/10.1530/EJE-16-0708.

Atzmon, G., Barzilai, N., Hollowell, J. G., Surks, M. I., & Gabriely, I. (2009). Extreme longevity is associated with increased serum thyrotropin. *The Journal of Clinical Endocrinology and Metabolism*, *94*(4), 1251–1254. https://doi.org/10.1210/jc.2008-2325.

Aubert, C. E., Bauer, D. C., da Costa, B. R., Feller, M., Rieben, C., Simonsick, E. M., et al. (2017). The association between subclinical thyroid dysfunction and dementia: The Health, Aging and Body Composition (Health ABC) Study. *Clinical Endocrinology*, *87*(5), 617–626. https://doi.org/10.1111/cen.13458.

Barbesino, G. (2019). Thyroid function changes in the elderly and their relationship to cardiovascular health: A mini-review. *Gerontology, 65*(1), 1–8. https://doi.org/10.1159/000490911.

Beck-Peccoz, P., Lania, A., Beckers, A., Chatterjee, K., & Wemeau, J. L. (2013). European thyroid association guidelines for the diagnosis and treatment of thyrotropin-secreting pituitary tumors. *The European Thyroid Journal, 2*(2), 76–82. https://doi.org/10.1159/000351007.

Bertoli, A., Valentini, A., Cianfarani, M. A., Gasbarra, E., Tarantino, U., & Federici, M. (2017). Low FT3: A possible marker of frailty in the elderly. *Clinical Interventions in Aging, 12,* 335–341. https://doi.org/10.2147/CIA.S125934 (eCollection 2017).

Boelaert, K. (2013). Thyroid dysfunction in the elderly. *Nature Reviews. Endocrinology, 9,* 194–204. https://doi.org/10.1038/nrendo.2013.30.

Brabant, G., Prank, K., Hoang-Vu, C., Hesch, R. D., & von zur Mühlen, A. (1991). Hypothalamic regulation of pulsatile thyrotopin secretion. *The Journal of Clinical Endocrinology and Metabolism, 72*(1), 145–150. https://doi.org/10.1210/jcem-72-1-145.

de Vries, E. M., Fliers, E., & Boelen, A. (2015). The molecular basis of the non-thyroidal illness syndrome. *Journal of Endocrinology, 225*(3), R67–R81. https://doi.org/10.1530/JOE-15-0133.

Diamanti-Kandarakis, E., Dattilo, M., Macut, D., Duntas, L., Gonos, E. S., Goulis, D. G., et al. (2017). MECHANISMS IN ENDOCRINOLOGY: Aging and anti-aging: A combo-endocrinology overview. *European Journal of Endocrinology, 176,* R283–R308. https://doi.org/10.1530/EJE-16-1061.

Duntas, L. H., & Brenta, G. (2018). A renewed focus on the association between thyroid hormones and lipid metabolism. *Frontiers in Endocrinology (Lausanne), 9,* 511. https://doi.org/10.3389/fendo.2018.00511 (eCollection 2018).

Duntas, L. H., & Yen, P. M. (2019). Diagnosis and treatment of hypothyroidism in the elderly. *Endocrine, 66*(1), 63–69. https://doi.org/10.1007/s12020-019-02067-9.

Franceschi, C., Garagnani, P., Morsiani, C., Conte, M., Santoro, A., Grignolio, A., et al. (2018). The continuum of aging and age-related diseases: common mechanisms but different rates. *Frontiers of Medicine (Lausanne), 12*(5), 61. https://doi.org/10.3389/fmed.2018.00061 (eCollection 2018).

George, K. M., Lutsey, P. L., Selvin, E., Palta, P., Windham, B. G., & Folsom, A. R. (2019). Association between thyroid dysfunction and incident dementia in the atherosclerosis risk in communities neurocognitive study. *Journal of Endocrinology and Metabolism, 9,* 82–89. https://doi.org/10.14740/jem588.

Gesing, A., Lewiński, A., & Karbownik-Lewińska, M. (2012). The thyroid gland and the process of aging; what is new? *Thyroid Research, 5,* 16. https://doi.org/10.1186/1756-6614-5-16.

Gluckman, P. D., Buklijas, T., & Hanson, M. (2016). The epigenome and developmental origins of health and disease. In *The developmental origins of health and disease (DOHaD) concept: Past, present, and future* (pp. 1–15).

Gussekloo, J., van Exel, E., de Craen, A. J., Meinders, A. E., Frölich, M., & Westendorp, R. G. (2004). Thyroid status, disability and cognitive function, and survival in old age. *Journal of the American Medical Association, 292,* 2591–2599. https://doi.org/10.1001/jama.292.21.2591.

Hennessey, J. V., & Espaillat, R. (2015). Diagnosis and management of subclinical hypothyroidism in elderly adults: A review of the literature. *Journal of the American Geriatrics Society, 63*(8), 1663–1673. https://doi.org/10.1111/jgs.13532.

Jasim, S., & Gharib, H. (2018). Thyroid and aging. *Endocrine Practice, 24*(4), 369–374. https://doi.org/10.4158/EP171796.RA.

Kammori, M., Sugishita, Y., Okamoto, T., Kobayashi, M., Yamazaki, K., Yamada, E., et al. (2015). Telomere shortening in breast cancer correlates with the pathological features of tumor progression. *Oncology Reports, 34*(2), 627–632. https://doi.org/10.3892/or.2015.4063.

Kennedy, B. K., Berger, S. L., Brunet, A., Campisi, J., Cuervo, A. M., Epel, E. S., et al. (2014). Geroscience: Linking aging to chronic disease. *Cell*, *6*(159), 709–713. https://doi.org/10.1016/j.cell.2014.10.039.

Kramer, C. K., von Muhlen, D., Kritz-Silverstein, D., & Barrett-Connor, E. (2009). Treated hypothyroidism, cognitive function, and depressed mood in old age: The Rancho Bernardo Study. *European Journal of Endocrinology*, *161*, 917–921.

Leng, O., & Razvi, S. (2019). Hypothyroidism in the older population. *Thyroid Research*, *12*, 2. https://doi.org/10.1186/s13044-019-0063-3 (eCollection 2019).

Levine, M. E., Lu, A. T., Quach, A., Chen, B. H., Assimes, T. L., Bandinelli, S., et al. (2018). An epigenetic biomarker of aging for lifespan and healthspan. *Aging (Albany NY)*, *10*, 573–591. https://doi.org/10.18632/aging.101414.

Lutz, W., Sanderson, W., & Scherbov, S. (2008). The coming acceleration of global population ageing. *Nature*, *451*, 716–719. https://doi.org/10.1038/nature0651.

Mammen, J. S. (2019). Interpreting elevated TSH in older adults. *Current Opinion in Endocrine and Metabolic Research*, *5*, 68–73. https://doi.org/10.1016/j.coemr.2019.04.001.

Mazzoccoli, G., Pazienza, V., Piepoli, A., Muscarella, L. A., Inglese, M., De Cata, A., et al. (2010). Hypothalamus-hypophysis-thyroid axis function in healthy aging. *Journal of Biological Regulators and Homeostatic Agents*, *24*(4), 433–439.

Monzani, F., Del Guerra, P., Caraccio, N., Del Corso, L., Casolaro, A., Mariotti, S., et al. (1996). Age-related modifications in the regulation of the hypothalamic-pituitary-thyroid axis. *Hormone Research*, *46*(3), 107–112. https://doi.org/10.1159/000185005.

Morley, J., & Korenman, S. J. (1992). *Endocrinology and metabolism in the elderly* (p. XV). Blackwell.

National Institute on Aging. (2019). *Understanding the dynamics of aging process*. https://www.nia.nih.gov/about/aging-strategic-directions-research/understanding-dynamics-aging.

Over, R., Mannan, S., Nsouli-Maktabi, H., Burman, K. D., & Jonklaas, J. (2010). Age and thyrotropin response to hypothyroxinemia. *The Journal of Clinical Endocrinology and Metabolism*, *95*, 3675–3683. https://doi.org/10.1210/jc.2010-0281.

Pekary, A. E., Mirell, C. J., Turner, L. F., Jr., Walfish, P. G., & Hershman, J. M. (1987). J. Hypothalamic secretion of thyrotropin releasing hormone declines in aging rats. *Journal of Gerontology*, *42*(4), 447–450. https://doi.org/10.1093/geronj/42.4.447.

Pekary, A. E., Stevens, S. A., & Sattin, A. (2006). Circadian rhythms of trh-like peptide levels in rat. *Brain*, *1125*(1), 67–76. https://doi.org/10.1016/j.brainres.2006.10.003.

Pierpaoli, W. (2013). Aging-reversing properties of thyrotropin-releasing hormone. *Current Aging Science*, *6*(1), 92–98. https://doi.org/10.2174/18746098113060100l2.

Roelfsema, F., Pereira, A. M., Adriaanse, R., Endert, E., Fliers, E., Romijn, J. A., et al. (2010). Thyrotropin secretion in mild and severe primary hypothyroidism is distinguished by amplified burst mass and basal secretion with increased spikiness and approximate entropy. *The Journal of Clinical Endocrinology and Metabolism*, *95*, 928–934.

Roelfsema, F., & Veldhuis, J. D. (2013). Thyrotropin secretion patterns in health and disease. *Endocrine Reviews*, *34*(5), 619–657. https://doi.org/10.1210/er.2012-1076.

Russfield, A. B. (1955). Histology of the human hypophysis in thyroid disease-hypothyroidism, hyperthyroidism, and cancer. *The Journal of Clinical Endocrinology and Metabolism*, *15*, 1393–1408. https://doi.org/10.1210/jcem-15-11-1393.

Scott, D. E., & Sladek, J. R., Jr. (1981). Age related changes in the endocrine hypothalamus: I. Tanycytes and the blood-brain-cerebrospinal fluid barrier. *Neurobiology of Aging*, *2*(2), 89–94. https://doi.org/10.1016/0197-4580(81)90004-x.

Seals, D. R., Justice, J. N., & LaRocca, T. J. (2016). Physiological geroscience: Targeting function to increase healthspan and achieve optimal longevity. *The Journal of Physiology*, *594*(8), 2001–2024. https://doi.org/10.1113/jphysiol.2014.282665.

Sierra, F., & Kohanski, R. (2017). Geroscience and the trans-NIH Geroscience Interest Group, GSIG. *Geroscience*, *39*, 1–5. https://doi.org/10.1007/s11357 016-9954-6.

Šošić-Jurjević, B., Ajdžanović, V., Filipović, B., Trifunović, S., Jarić, I., Ristić, N., et al. (2016). Functional morphology of pituitary-thyroid and -adrenocortical axes in middle-aged male rats treated with vitex agnus castus essential oil. *Acta Histochemica, 118*(7), 736–745. https://doi.org/10.1016/j.acthis.2016.07.007.

Stott, D. J., Gussekloo, J., Kearney, P. M., Rodondi, N., Westendorp, R. G., Mooijaart, S., et al. (2017). Study protocol; thyroid hormone replacement for untreated older adults with subclinical hypothyroidism—A randomised placebo controlled trial (TRUST). *BMC Endocrine Disorders, 17*, 6.

Tarquini, R., & Mazzoccoli, G. (2017). Clock genes, metabolism, and cardiovascular risk. *Heart Failure Clinics, 13*(4), 645–655. https://doi.org/10.1016/j.hfc.2017.05.001.

Thayakaran, R., Adderley, N. J., Sainsbury, C., Torlinska, B., Boelaert, K., Šumilo, D., et al. (2019). Thyroid replacement therapy, thyroid stimulating hormone concentrations, and long term health outcomes in patients with hypothyroidism: Longitudinal study. *BMJ, 366*, l4892. https://doi.org/10.1136/bmj.l4892.

United Nations. (2010). *World population prospects: The 2010 revision.* http://esa.un.org/unpd/wpp.

Veldhuis, J. D. (2013). Changes in pituitary function with ageing and implications for patient care. *Nature Reviews. Endocrinology, 9*(4), 205–215. https://doi.org/10.1038/nrendo.2013.38.

Vijk, J., & Kennedy, B. K. (2016). The essence of aging. *Gerontology, 62*(4), 381–385. https://doi.org/10.1159/000439348.

CHAPTER TWO

Hypothalamic aging and hormones

Zi-Jian Cai*
CaiFortune Consulting, Suzhou, Jiangsu, PR China
*Corresponding author: e-mail address: hrsh8@126.com

Contents

Abstract

It is the heterogeneous changes of hypothalamic functions that determine the chronological sequence of aging in mammals. Recently, it was hypothesized by Cai the decrease in slow-wave sleep (SWS) resulting from skin aging as responsible for the degeneration of hypothalamic suprachiasmatic nucleus (SCN). It was soon hypothesized by the European people in television that the increase in body fat as responsible for the

Vitamins and Hormones, Volume 115
ISSN 0083-6729
https://doi.org/10.1016/bs.vh.2020.12.002

degeneration of male preoptic sexually dimorphic nucleus (SDN-POA), via the aromatase converting testosterone to estradiol as proposed by Cohen. It is the hypothalamic paraventricular nucleus (PVN) that remains unchanged in neuron number during aging for psychological stress. In this chapter, it is briefly reviewed more manifestations of hypothalamic related mammalian aging processes, including (1) the aging of ovary by lipid, estradiol and hypothalamus; (2) the aging of muscle, stomach, intestine, thymus, and the later aging of brain, regulated by growth hormone/insulin-like growth factor 1(GH/IGF1); (3) the cardiovascular hypertension from PVN activation, the bone and other peripheral aging by psychological stress, and that of kidney by vasopressin. It is classified these aging processes by the primary regulation from one of the three hypothalamic nuclei, although still necessary to investigate and supplement their secondary regulation by the hypothalamic nuclei in future. It is the hypothalamic structural changes that shift the functional balance among these three hypothalamic systems toward aging.

Highlights

- The hypothalamic nuclei undergo changes in heterogeneity during aging.
- The decrease in SWS and testosterone degenerates SCN and SDN-POA respectively.
- The SCN and GH/IGF1 are related to aging of muscle, stomach and intestine.
- The PVN and stress are related to aging of bone, kidney and hypertension.
- The mechanisms for ovary and lipid aging require more investigations.

1. The introduction

The hypothalamus controls the hormones and determines the chronological sequence of aging in the ontogenetic life of mammals (Aguilera, 2011; Hofman, 1997). The hypothalamic nuclei are heterogeneous in degeneration during aging in mammals (Engelberth et al., 2014; Hofman, 1997; Hsieh, Hsu, Yang, Hsu, & Peng, 1996; Panzica, García-Ojeda, Viglietti-Panzica, Thompson, & Ottinger, 1996; Zhou & Swaab, 1999). The paraventricular nucleus (PVN), responsible for stress response, is functional and maintains its neuron number throughout the whole lifespan in humans (Hofman, 1997; Zhou & Swaab, 1999). Whereas, the suprachiasmatic nucleus (SCN), controlling the circadian rhythm, is decreased in number of neurons during aging in humans and marmosets (Engelberth et al., 2014; Hofman, 1997; Zhou & Swaab, 1999). Besides, the male sexually dimorphic nucleus in preoptic area (SDN-POA) declines sharply in cell number during aging in humans and rats (Hofman, 1997; Hsieh et al., 1996; Panzica et al., 1996; Zhou & Swaab, 1999). Evidently, the hypothalamic nuclei manifest heterogeneous in degeneration during aging.

Swaab and Bao early suggested that the "use or lose" could be the mechanism underlying the heterogeneity in degeneration of hypothalamic nuclei (Swaab & Bao, 2011). Recently, Cai hypothesized (Cai, 2016a, 2018a) that the continuous decrement in duration of slow-wave sleep (SWS) demonstrated in many senescent observations (Ehlers & Kupfer, 1989; Espiritu, 2008; Hirshkowitz, Moore, Hamilton, Rando, & Karacan, 1992) would cause the senescent degeneration of SCN (Engelberth et al., 2014; Hofman, 1997; Zhou & Swaab, 1999), which resulted from the reduction of emotional responses and memories from skin aging (Cai, 1991, 1995, 2016a, 2016b, 2018a). Soon, on July 4, 2016, the European people in television speculated that there was an additional senescent pathway for male reproduction from the common knowledge of adipose accumulation and reduction of spermatogenesis in middle age/early old age to analogously elicit the degeneration of hypothalamic preoptic area (POA) (Cai, 2017, 2018a). Finally, the PVN with its neuronal number unchanged in aging may cause many cellular and molecular changes of aging from chronic stress (Cai, 2018a; Swaab & Bao, 2011).

In this chapter, based on these new hypothalamic senescent hypotheses, it is briefly reviewed more manifestations of hypothalamic related mammalian aging processes in addition to the aging of skin and adipose, and grossly classified the aging processes as primarily driven by one of the three hypothalamic nuclei. Whereas, the secondary regulation of these aging processes by these hypothalamic nuclei should be investigated and supplemented more in future.

2. The senescent degeneration of reproductive system

2.1 The senescent degeneration of male reproductive system

On July 4, 2016, the European people in television speculated that there was a senescent pathway for male reproduction from the common knowledge of adipose accumulation and reduction of spermatogenesis in aging to elicit the degeneration of hypothalamic preoptic area (POA) (Cai, 2017, 2018a). Indeed, it was observed that the male sexually dimorphic nucleus in preoptic area (SDN-POA) declined in cell number after aging (Hofman, 1997; Hsieh et al., 1996; Panzica et al., 1996; Zhou & Swaab, 1999). Besides, it was even shown that the higher estradiol could result in the neuronal loss in the preoptic sexually dimorphic nucleus (SDN-POA) in male rats (Hsieh et al., 1996), while the testosterone reverse the neuronal degeneration within SDN-POA in male quails (Panzica et al., 1996), demonstrating that the reproductive hormonal levels directly able to regulate the neuronal degeneration within the male SDN-POA.

The morphogenetic development and growth in animals is controlled by both induction and termination (Cai, 2015). Induction initiates the morphogenetic development while termination controls the final shape of the generated organ during morphogenesis and growth (Cai, 2015). In vertebrates, the fat depot sizes reach a peak by middle age or early old age, followed by a substantial decline in advanced old age (Cartwright, Tchkonia, & Kirkland, 2007). Fat tissue growth occurs by increase in size and number of fat cells. Beginning at middle age, the subcutaneous fat loses earlier and consequently the intra-abdominal fat increases relatively in the course toward aging (Cartwright et al., 2007), resulting in insulin resistance and increased risk of atherosclerosis and diabetes, even in lean subjects (Cartwright et al., 2007). Nutrients, hormonal effectors such as insulin, leptin, IGF-1, glucocorticoids, and so on are responsible for the regulation of adipose size and redistribution in development and aging (Carrascosa, Ros, Andrés, Fernández-Agulló, & Arribas, 2009; Cartwright et al., 2007).

The enzyme aromatase in vertebrates converts the testosterone to estradiol and decreases the testosterone levels (Cohen, 2001; Yamada & Harada, 1990). The aromatase is expressed in the differentiated adipose cells (Yamada & Harada, 1990). Accordingly, Cohen suggested that the increase in adipose tissue in middle age be associated with an increase in the enzyme aromatase and lead to diminished testosterone levels during male aging (Cohen, 2001; Yamada & Harada, 1990). Aromatase is a remarkable and evolutionarily conserved enzyme present in amphioxus (Callard et al., 2011), fishes (Böhne, Heule, Boileau, & Salzburger, 2013; Zhang, Zhang, Lu, Zhang, & Zhang, 2014), amphibians (Nakamura, 2013), reptiles (Girondot, Ben Hassine, Sellos, Godfrey, & Guillon, 2010), and mammals including humans (Cohen, 2001; Yamada & Harada, 1990), and is suggested as the key enzyme in the epigenetic determination of sex as male or female during ontogeny in many species of vertebrates (Nakamura, 2010), as demonstrated in some fishes (Böhne et al., 2013), amphibians (Nakamura, 2013) and reptiles (Girondot et al., 2010). Obviously, the increase in aromatase from increment of fat size in middle age is sufficient to disturb the male reproductive system via converting the testosterone to estradiol.

The disturbance on the male reproductive system from adipose and aromatase in aging can even be assayed in vivo. In one report, it was demonstrated that the free and bioavailable estradiol levels decreased modestly with age as did the ratio of free testosterone to free estradiol, the latter testifying to the age-related increase in aromatization of testosterone (Vermeulen, Kaufman, Goemaere, & van Pottelberg, 2002). The estradiol levels were highly, significantly and positively related to the body fat mass (Vermeulen et al., 2002). In another paper comparing BMI in relation to sperm count, it was shown that the overweight and obesity were associated with an increased prevalence of azoospermia or oligozoospermia (Sermondade et al., 2013).

In this regard, due to the regulation of neuronal numbers within the hypothalamic SDN-POA by the testosterone and estradiol (Hsieh et al., 1996; Panzica et al., 1996), the increment of adipose and aromatase converting the testosterone to estradiol in aging (Cohen, 2001; Vermeulen et al., 2002) would cause the neuronal degeneration within the male SDN-POA. These manifest the major feature of senescent degeneration of male reproductive system.

2.2 The senescent degeneration of female reproductive system

The aging of female reproductive system in mammals is characterized by the decrease in follicle numbers and decay in oocyte quality to gradual decline in fertility and final occurrence of sterility (Broekmans, Knauff, te Velde, Macklon, & Fauser, 2007). Concurrently, changes in peripheral and hypothalamic endocrinal secretions occur during perimenopause, notably as the increase in estradiol, elevation of follicle-stimulating hormone (FSH) and change in amplitude of pulsatile luteinizing hormone (LH) (Prior, 2005; Reame et al., 1996).

Currently, most theories regarding the causes of ovarian aging have considered the intrinsic decay in oocyte quality or ovarian follicle as the key mechanisms. (1) Consistent with the reactive oxygen species (ROS) in causing aging, some authors have considered the advanced glycation end products (AGEs) as relevant to protein damage, ROS, and increase of inflammatory reactions of ovarian tissues (Pertynska-Marczewska & Diamanti-Kandarakis, 2017; Tatone et al., 2008). (2) Consistent with the telomere theory in causing aging, many authors have considered the decrease in length of telomeres as a parsimonious explanation for the effects of reproductive aging on oocyte quality (Kalmbach, Antunes, Kohlrausch, & Keefe, 2015; Keefe, Marquard, & Liu, 2006). (3) Consistent with the role of mitochondria in efficient utilizing the nutrients and energy, many authors have considered the mitochondrial DNA (mtDNA) copy number as a potential biomarker of embryo viability and the use of mitochondrial nutrients and mitochondrial transfer as a potential treatment for poor ovarian function and response (May-Panloup et al., 2016; Wang, Zhang, Jiang, & Seli, 2017). (4) Recently, some authors have argued the traditional dogma of oocytes not able to renew themselves in postnatal or adult life, and considered the aging and degradation of ovarian stem cell nests rather than ovarian germ cells themselves as the main factor leading to the decline of the reproductive function (Ye et al., 2017; Zhang, Adhikari, Zheng, & Liu, 2013).

Obviously, among the four categories of aging theories on female reproductive system, the (1), (2) and (3) are only relevant to the intrinsic cellular degeneration of oocytes. Whereas, the (4) requires the environmental influences on the ovarian stem cell nests, so that the relevant authors pointed to the hormonal interactions between the female reproductive system and immune system in plausible causing the aging of ovarian stem cell nests (Ye et al., 2017; Zhang et al., 2013).

Whereas, the characteristic increases in estradiol, FSH, pulsatile LH and other hypothalamic regulated hormones during perimenopause could also be involved in the aging mechanisms of female reproductive system (Prior, 2005; Reame et al., 1996). The increment of fat size in middle age can increase the estradiol from testosterone (Cohen, 2001; Vermeulen et al., 2002) and decrease the male reproductive ability (Hsieh et al., 1996; Panzica et al., 1996; Zhou & Swaab, 1999), which can herein be speculated from such enlightenment to correspondingly increase the female reproductive ability in perimenopause (Prior, 2005; Reame et al., 1996), resulting in acceleration and precocity of oocytes in contrast to gradual degeneration of male reproductive system in aging.

3. The special aging of skin and its causing SCN aging via SWS

3.1 The special aging of skin

Skin aging is characterized in appearance as gray in hairs, increase in wrinkles, deposition of pigments and so on. Skin is constantly exposed to sunshine and oxygen, which affects the skin aging more than the influence from hypothalamic neuroendocrine system. It has been demonstrated that the dysfunction of skin collagen (Zhuang, Hou, Zhao, Zhang, & Li, 2009) and elastin (Imokawa, 2008) may be responsible for the generation of skin wrinkles from photoaging. It has also been demonstrated that oxidative accumulations are intimately associated with skin aging (Kammeyer &

Luiten, 2015; Rinnerthaler, Bischof, Streubel, Trost, & Richter, 2015). Obviously, both sunshine and oxygen are the environmental causes resulting in the aging of skin.

In addition to the damage from environmental sunshine and oxygen, the skin also undergoes aging genetically by shortening the length of telomeres. It has been shown that the telomerase activity and telomere length may be relevant to skin aging (Buckingham & Klingelhutz, 2011). Particularly, it was reported that the telomerase reversed the hair follicle stem cell defects in epidermis (Siegl-Cachedenier, Flores, Klatt, & Blasco, 2007). Obviously, genetic shortening telomere length is also a concurrent mechanism causing skin aging.

One of the important consequences of skin aging is the change in electrodermal activities. It was reported that the electrodermal activities decreased in the older subjects than younger ((Gavazzeni, Wiens, & Fischer, 2008); Powell, Milligan, & Furchtgott, 1980). In parallel, it has also been demonstrated that the psychological stress can elicit significant changes in electrodermal activities in humans (Panconesi & Hautmann, 1996; Rodriguez-Vallecillo & Woodbury-Fariña, 2014).

It is noted that the skin aging is also affected by the hypothalamic regulation of psychological stress (Dunn & Koo, 2013), while the biological rhythms in skin are as well under the regulation of central hypothalamic SCN (Matsui, Pelle, Dong, & Pernodet, 2016). However, as compared to the constant exposure to sunshine (Imokawa, 2008; Zhuang et al., 2009) and oxygen (Kammeyer & Luiten, 2015; Rinnerthaler et al., 2015), plus the intracellular decrease in length of telomeres (Buckingham & Klingelhutz, 2011; Siegl-Cachedenier et al., 2007), the hypothalamic regulation of skin aging (Dunn & Koo, 2013) could be secondary.

3.2 The hypothetic aging pathway from skin to hypothalamic SCN via SWS

Cai recently suggested a new peripheral hypothesis to explain the aging of hypothalamic SCN (Cai, 2016a, 2018a). On the one hand, the continuous decrement in duration of SWS demonstrated in many senescent observations (Ehlers & Kupfer, 1989; Espiritu, 2008; Hirshkowitz et al., 1992) would cause the continuous functional degeneration of SCN in hypothalamus during aging (Cai, 2016a, 2018a; Engelberth et al., 2014; Hofman, 1997; Zhou & Swaab, 1999). On the other hand, the skin underwent aging from exposure to the environmental sunshine (Imokawa, 2008; Zhuang et al., 2009) and oxygen (Kammeyer & Luiten, 2015; Rinnerthaler et al., 2015) as well as from genetic shortening of the telomere (Buckingham & Klingelhutz, 2011; Siegl-Cachedenier et al., 2007). The skin senescence was repeatedly reported to result in reduction of the electrodermal activities (Gavazzeni et al., 2008; Powell et al., 1980), which would reduce the emotional responses and memories in brain manifesting as electrodermal changes (Panconesi & Hautmann, 1996; Rodriguez-Vallecillo & Woodbury-Fariña, 2014), and would in turn reduce the requirement for slow wave sleep (SWS) (Cai, 1991, 1995, 2016a, 2016b, 2018a). In these two respects, from skin senescence to slow wave sleep (SWS), Cai formulated a new degenerative aging pathway to explain the change of hypothalamic SCN during aging (Cai, 2016a, 2018a).

The aging-related disorganization in expression of core clock genes in various brain regions supports this hypothetic pathway. It is common knowledge that the hippocampus is responsible for learning and memory (Cai, 1990, 2018b). It was demonstrated that the expression of Clock, Bmal1 and Per2 genes in hippocampus lost circadian rhythm earlier than those in hypothalamic SCN (Duncan, Prochot, Cook, Tyler Smith, & Franklin, 2013; Wyse & Coogan, 2010), implicating that the hippocampal circadian rhythm relevant to sleep and memories actually decreased earlier in response to the reduction of emotional responses, while such dysfunctions later began to affect the hypothalamic SCN (Cai, 2016a, 2018a). If the degeneration of hypothalamic SCN during aging results from the changes in the secretion of growth hormone (GH), there would not have shown this pattern of disorganization in expression of core clock genes earlier in the hippocampus and then in the hypothalamic SCN (Duncan et al., 2013; Wyse & Coogan, 2010).

It is noted that some peripheral clocks such as those in heart were shown to preserve their activity even after SCN having lost some circadian functions during aging (Bonaconsa et al., 2014; Yamazaki et al., 2002). Obviously, these special peripheral circadian clocks must have originated earlier in evolution than SCN.

4. The peripheral aging under the regulation of the GH/IGF1 axis

Because the secretion of growth hormone (GH) occurs in SWS in primates (Cai, 1991, 1995), the growth hormone (GH)/insulin-like growth factor 1 (IGF1) axis must partly be regulated by the circadian SCN. Those aging components under the regulation of GH/IGF1 axis would in turn be under the regulation of the circadian SCN.

4.1 The aging of skeletal muscle

Aging is associated with sarcopenia as the loss of skeletal muscle mass, accompanied by a decrease in muscle strength and resistance. Functional denervation is a manifestation of aging sarcopenia (Badawi & Nishimune, 2018; Rudolf, Khan, Labeit, & Deschenes, 2014).

Currently, most theories regarding the aging mechanisms of skeletal muscle fall into three categories. (1) The neuromuscular junctions of sarcopenia are not stable structures and become impaired during aging due to the selective degeneration of specific active zone proteins (Badawi & Nishimune, 2018; Rudolf et al., 2014), which may cause or worsen sarcopenia. Exercise can achieve a remarkable reversibility of the degeneration (Badawi & Nishimune, 2018; Rudolf et al., 2014). (2) Recent studies have identified that not only mitochondrial metabolic dysfunction but also mitochondrial dynamics and mitochondrial calcium uptake could be involved in the degeneration of skeletal muscle mass (Del Campo, Jaimovich, & Tevy, 2016; Hepple, 2016). (3) The down regulation of GH/IGF1 axis during aging may cause or worsen sarcopenia, while modulation of GH/IGF-1 axis may counteract sarcopenia in older adults (Giovannini, Marzetti, Borst, & Leeuwenburgh, 2008; Perrini et al., 2010). Obviously, skeletal sarcopenia is under the regulation of GH/IGF-1 axis in aging.

Among the three categories of aging mechanisms of skeletal muscle, the (2) is intrinsic and cellular, while (1) and (3) require environmental coordination, and (3) is evidenced to be regulated by GH/IGF-1 axis (Giovannini et al., 2008; Perrini et al., 2010). Considering the direct influence of GH/IGF-1 axis on muscular function, it is likely that the GH/IGF-1 axis may be the primary hypothalamic regulation of muscular aging. More investigations are required for clarification of this issue.

4.2 The aging of stomach and intestine

The aging processes of digestive stomach and intestine are correlated to the GH/IGF-1 axis and SCN respectively. Stomach is special during aging in that ghrelin sourcing from stomach can feedback to regulate the GH/IGF-1 axis (Muller et al., 2002). This is also a candidate of feedback on GH/IGF-1 axis by the effects of stress and obesity via stomach, while the biological importance of this feedback on aging requires further investigation.

On the other hand, there is also the circadian timing system in the gastrointestinal tract (GIT) culminating in the rhythmic expression and activity of a set of clock genes and related hormones (Konturek, Brzozowski, & Konturek, 2011). The clocks in the GIT are responsible for the periodic activity of its various segments and transit along the GIT (Konturek et al., 2011). Certainly, the central clock on SCN communicates and interacts with the gut clocks (Konturek et al., 2011), while more investigations are required to further clarify the details of their interactions during aging.

4.3 The aging of thymus

Thymic involution is the genetically programmed aging process via which the thymus undergoes a progressive reduction in size due to the loss of thymic epithelial cells and decrease in thymopoiesis (Bar-Dayan, Afek, Bar-Dayan, Goldberg, & Kopolovic, 1999; Taub & Longo, 2005).

On the one hand, both estrogen and testosterone were reported to enhance the induction of apoptotic thymic involution (Hareramadas & Rai, 2006; Ventevogel & Sempowski, 2013; Yao & Hou, 2004). On the other hand, both GH and ghrelin were evidenced as effective for the restoration of thymic function in aged individuals (Taub, Murphy, & Longo, 2010; Ventevogel & Sempowski, 2013). Accordingly, the decline of GH and increase of sex steroids around puberty may in synergy cause the apoptotic thymic involution in young adults (Hirokawa, Utsuyama, & Kobayashi, 2001). Even though both decline of GH and increase of sex steroids may be considered as the primary causes of thymic involution, the longer restorative effect of GH can last after menopause (Taub et al., 2010; Ventevogel & Sempowski, 2013) when sex steroids are decreased.

Of special interest for application is that both GH and ghrelin have been considered as potential in use for the restoration of thymic function in aged individuals (Taub et al., 2010; Ventevogel & Sempowski, 2013), which deserves more investigations.

5. The peripheral aging under the regulation of PVN

5.1 The PVN and psychological stress during aging

The hypothalamic PVN is responsible for psychological stress, functional throughout the lifespan and maintains its neuron number in senescence in humans (Hofman, 1997; Zhou & Swaab, 1999). The activity of the corticotrophin releasing hormone (CRH) neurons in the hypothalamic PVN, as the basis for the activity of the hypothalamo-pituitary-adrenal (HPA) axis, is activated during aging in a sex-dependent way (Swaab & Bao, 2011). The maintained psychological response of stress in aging has been demonstrated to cause decrease in proteolytic clearance within peripheral cells (Koga, Kaushik, & Cuervo, 2011). Besides, it has been reviewed that the psychological stress might lead to aging via many other molecular processes in cells, such as the change in length of telomere, increase in DNA damage, alternation in immunological inflammation, and so on (Moreno-Villanueva & Bürkle, 2015).

Because of the complex and vague manifestations of molecular and cellular changes in peripheral cells following psychological stress during aging, in this chapter it is further reviewed the aging changes in various peripheral organs primarily driven by PVN, manifesting more clearly.

5.2 The cardiovascular hypertension from PVN activation during aging

Aging is associated with increase in total and low-density lipoprotein (LDL) cholesterol, and triglycerides, and decrease in high-density lipoprotein (HDL) cholesterol (O'Connor et al., 1999). It has been demonstrated that the endogenous nocturnal GH secretion may be correlated with total, LDL, and HDL cholesterol levels independently of total or abdominal fat (O'Connor et al., 1999). Besides, it has as well been shown that the estrogen and cholesterol-lowering therapies could reduce LDL cholesterol and raise HDL cholesterol (Koh et al., 1999). Obviously, both GH/IGF-1 axis and estrogen are related to the cholesterol levels as the risky factor of hypertension.

The research on hypertension has even been extended to examine the relevant changes within the hypothalamus. It was shown that the downregulation in binding of GABAA receptor occurred in the hypothalamic PVN of spontaneously hypertensive rats and might play a role in genetic hypertension (Kunkler & Hwang, 1995). Besides, it was recently demonstrated that the increase in activation of hypothalamic PVN could enhance the sympathetic outflow and increase the blood pressure, which might be a major contributor to the pathogenesis of hypertension (Carmichael & Wainford, 2015).

Because the change in cholesterol levels, regulated by GH/IGF-1 axis (O'Connor et al., 1999) and estrogen (Koh et al., 1999), is just the risky factor of hypertension, the direct activation of hypothalamic PVN, via enhancing the sympathetic outflow and increasing the blood pressure (Carmichael & Wainford, 2015), should be the primary hypothalamic regulation on hypertension.

It is noted that, besides the skin aging as the good cause to reduce the duration of SWS and cause the degeneration of SCN (Cai, 2016a, 2018a), hypertension may be another cause able to reduce the duration of SWS (Cai, 2018a), which is nonetheless more heterogeneously distributed in population (Cai, 2018a), secondary and requiring more investigations.

5.3 The aging of bone

The humans and mammals undergo bodily growth until after puberty. The height of humans and size of mammals are controlled without further increment after mature, which is an important process related to the termination of morphological development (Cai, 2015).

The elongation of cartilage of long bones is stimulated by hypothalamic GH/IGF-1 axis before puberty (Isaksson, Ohlsson, Nilsson, Isgaard, & Lindahl, 1991). However, after puberty the hypothalamic stimulation on secretion of sex hormones is increased. The estrogen can counteract against the growth hormone to control the growth of longitudinal bone in both females and males (Bachrach & Smith, 1996; Chagin et al., 2004), so that the height of humans or size of mammals is controlled without increment after mature.

Bone remodeling and aging both occur afterward. Because the role of GH on bone growth is inhibited by estrogen after puberty (Bachrach & Smith, 1996; Chagin et al., 2004), the effects of other hormones on bone remodeling and aging would be more predominant. It has been demonstrated that IGF-1 plays important roles in bone growth and aging (Yakar, Werner, & Rosen, 2018), while the hypothalamic oxytocin is essential for optimal skeletal remodeling (Colaianni et al., 2014).

Recently, it has been reviewed the roles of sympathetic nervous system in causing the osteoporotic diseases and aging (He, Jiang, & Dai, 2011), as well as the chronic psychological stress in causing the osteoporosis and bone loss (Azuma, Adachi, Hayashi, & Kubo, 2015). Considering the roles of IGF-1 on bone growth and aging mainly as peripheral (Yakar, Werner, & Rosen, 2018), and oxytocin mainly for skeletal remodeling (Colaianni et al., 2014), the sympathetic nervous system (He et al., 2011) and chronic psychological stress (Azuma et al., 2015) would become the primary regulator of bone aging by hypothalamus.

5.4 The aging of kidney

Typical age-related changes in kidney include loss of parenchymal mass, progressive glomerulosclerosis, tubulopathy, interstitial fibrosis, and the development of afferent-efferent arteriolar shunts (Cowen, Hodak, & Verbalis, 2013). Concomitant with the loss of diluting capacity, the aging kidney also loses the ability to maximally conserve body water during states of dehydration (Cowen et al., 2013).

The renal function of kidney is under the central neuroendocrine control of vasopressin, which is synthesized by the cell bodies of hypothalamic supraoptic nuclei (SON) for osmotic sensation and PVN for stress during aging (Cowen et al., 2013; Goudsmit, Neijmeijer-Leloux, & Swaab, 1992; Prager-Khoutorsky & Bourque, 2015). An increase in secretion of vasopressin for each unit change in plasma osmolality has been observed during senescence (Cowen et al., 2013; Goudsmit et al., 1992), as well a decrease in normal renal response to vasopressin (Cowen et al., 2013; Goudsmit et al., 1992). Briefly, the aging of kidney is under the control of vasopressin and SON/PVN.

5.5 The aging of skin

The skin aging is affected by the hypothalamic regulation of psychological stress (Dunn & Koo, 2013). Whereas, because of the constant exposure to sunshine (Imokawa, 2008; Zhuang et al., 2009) and oxygen (Kammeyer & Luiten, 2015; Rinnerthaler et al., 2015), and the intracellular decrease in length of telomeres (Buckingham & Klingelhutz, 2011; Siegl-Cachedenier et al., 2007), this PVN regulation of skin aging (Dunn & Koo, 2013) could only be considered as secondary.

6. The aging later in brain and later off leptin

6.1 The later aging of brain

Brain also undergoes aging. The Alzheimer's disease is characterized as the early accumulation of amyloid-beta due to genetic mutation (Sompol et al., 2008; Wang, Su, Perry, Smith, & Zhu, 2007), which also appears in aging later. Recently, it has been demonstrated that sleep could help biophysical clearance of amyloid-beta from the adult brain (Kang et al., 2009; Xie et al., 2013).

However, the functional decline of episodic memory, the most vulnerable brain function to aging, usually begins after 60 years old in humans (Nyberg, 2017), while even begins later in a portion of them (Nyberg, 2017),

In this regard, the hypothalamic aging would result from the aging mechanisms other than the forebrain and the toxicity of amyloid-beta.

Because of the later onset of brain aging (Nyberg, 2017), the peripheral influences on brain aging would be considered as important. Of particular interest, the homeostasis of cholesterol and lipid has been shown to be involved in modulation of synaptic function and plasticity vulnerable to dementia (Grösgen, Grimm, Friess, & Hartmann, 2010; Koudinov & Koudinova, 2003). Because it has been reported that sleep helped the biophysical clearance of amyloid-beta from brain (Kang et al., 2009; Xie et al., 2013), and that the endogenous nocturnal GH secretion was correlated with total, LDL, and HDL cholesterol levels (O'Connor et al., 1999), the later aging of brain is regulated by the hypothalamic SCN for sleep and GH/IGF-1 axis for cholesterol.

6.2 The leptin resistance during aging

Because of the importance of lipid in regulating the aging processes as mentioned above, it is necessary to consider the role of leptin in aging. Leptin is a catabolic adipose signal acting on hypothalamus, suppressing food intake and inducing hypermetabolism (Pétervári et al., 2014). It is well evidenced that the central leptin resistance occurs during aging (Balaskó, Soós, Székely, & Pétervári, 2014; Pétervári et al., 2014), downregulating the leptin roles in aging. Besides, it has also been shown the interaction present between leptin and GH/IGF1 axis (Isozaki, Tsushima, Miyakawa, Demura, & Seki, 1999).

However, because the fat depot sizes increase to peak by middle age or early old age (Cartwright et al., 2007), while the consequent increment of aromatase converting the testosterone to estradiol (Cohen, 2001; Vermeulen et al., 2002) directly causes the senescent neuronal degeneration within the hypothalamic SDN-POA, the roles of leptin in causing aging before menopause would be considered as relatively indirect or secondary when compared to that mediated by aromatase. The central leptin resistance in aged mammals later than menopause is stable as downregulation of the leptin roles in causing further aging, and should be coherent with other changes in senescent animals, which requires more investigations.

7. The preliminary classification

From all of the above reviews in this chapter, it is classified the hypothalamic related aging processes preliminarily into the followings:

(1) Processes regulated by SCN and GH/IGF1 axis: As primary there include the aging of skeletal muscle, stomach, intestine, thymus and brain. As secondary there include the aging of bone, hypertension, and so on others requiring further investigation.

(2) Processes regulated by PVN and HPA axis: As primary there include the aging from psychological stress, the aging of bone, kidney and hypertension. As secondary there include the aging of skin, stomach, and so on others requiring more investigations.

(3) Processes regulated by SDN-POA and sex hormones: As primary there include the aging of ovary and male reproductive system. As secondary there include the aging of bone, thymus, hypertension, and so on others requiring further investigation. The testosterone and estradiol can feedback to regulate the neuronal loss of SDN-POA during aging.

(4) Specialties: The skin aging primarily results from sunshine, oxygen and intracellular decrease in length of telomeres, which in turn feedbacks to regulate the neuronal loss of SCN during aging via SWS. Hypertension can likewise feedback on SCN via SWS but is more heterogeneously distributed in population. The aging of adipose tissue also seems to be very genetically programmed at the cellular level (Cartwright et al., 2007), which requires further investigation.

The classification of aging of various organs by their primary regulation from one of the three hypothalamic nuclei can make it easy and intuitive to comprehend the complex aging processes of the whole organism of mammals. With such classification, it becomes intuitive that some organs are relevant in aging more to nutrition and SCN, GH/IGF1 axis, such as skeletal muscle, stomach, intestine, thymus and brain; some other organs vulnerable to stress, such as bone, kidney, skin and hypertension; while the reproductive organs mainly related in aging to SDN-POA and sex hormones. Obviously, this classification can prospectively promote and guide the investigations of aging mechanisms.

The SCN, GH/IGF1 axis and nutritional functions are required for normal survival and repair of mammals while aging manifests as degeneration of these functions as mentioned above. Likewise, SDN-POA and sex hormones are required for normal reproduction of mammals while aging manifests as degeneration of these functions. In contrast, PVN, HPA axis and stress functions are required for normal activities and avoidance of mammals while aging manifests as maintenance of these functions. In this regard, aging manifests as the shift of balance among these three hypothalamic systems, with degeneration of SCN, GH/IGF1 axis, nutritional system and SDN-POA, reproductive system, but maintenance of PVN, HPA axis, and stress system (Table 1).

Table 1 The three hypothalamic aging systems.

	I	II	III
Hypothalamic structures/hormones	SCN and GH/IGF1 axis	SDN-POA and sex hormones	PVN, HPA axis and stress hormones
Hypothalamic structural changes in aging	SCN degenerates in aging	Male SDN-POA degenerates in aging	PVN remains unchanged in aging
Senescent causes of hypothalamic structural changes	Decrease in SWS from special skin aging, and so on	Testosterone decrease by increased fat and aromatase in aging	Unchanged
Functional balance and aging	Function maintained as anti-aging	Function maintained as anti-aging	Function maintained as aging
Related primary regulation on aging of organs	Muscle, stomach, intestine, thymus, brain and so on	Female ovary, male reproductive system and so on	The aging of bone, kidney, hypertension, from psychological stress, and so on
Related secondary regulation on aging of organs	Bone, hypertension, and so on others requiring further investigation	Bone, thymus, hypertension, and so on others requiring further investigation	Skin, stomach, and so on others requiring more investigations

In this chapter, it is also discussed a portion of secondary regulation on these aging processes by hypothalamic nuclei, while it is required more investigations and supplements.

Besides, it is necessary to point out that the cellular aging of ovary, the aging of adipose tissue, stomach, intestine and brain, the peripheral aging following psychological stress, and the functional restoration of thymus in aging are important but still vague, which requires more investigations.

8. The conclusions

It was recently hypothesized by Cai that the decrease in SWS resulting from skin aging would cause the degeneration of SCN. It was soon hypothesized by the European people in television that the increase in body fat in aging would cause the degeneration of male preoptic SDN-POA, which was mediated via the aromatase and testosterone as proposed by Cohen. The neuron number in paraventricular nucleus (PVN) remains unchanged during aging.

In this chapter, it is further briefly reviewed more manifestations of hypothalamic related mammalian aging processes. Besides, it is grossly classified the aging processes according to the primary regulation by one of the three hypothalamic nuclei as SCN, PVN and SDN-POA. (1) Processes primarily regulated by SCN and GH/IGF1 axis include the aging of skeletal muscle, stomach, intestine, thymus and brain, relying more on nutritional supply. (2) Processes primarily regulated by PVN and HPA axis are vulnerable more to stress, including the aging from psychological stress, the aging of bone, kidney and hypertension. (3) Processes primarily regulated by SDN-POA are mostly relevant to sex hormones, including the aging of ovary and male reproductive system, while the testosterone and estradiol feedback regulating the neuronal loss of male SDN-POA during aging. (4) The skin aging primarily results from sunshine, oxygen and intracellular decrease in length of telomeres, and feedback regulates the neuronal loss of SCN in aging via SWS, while the aging of adipose tissue may likewise also be programmed at the cellular level. Shift of balance occurs during aging among the three hypothalamic systems, with degeneration of SCN, GH/IGF1 axis, nutritional system and SDN-POA, reproductive system, but maintenance of PVN, HPA axis and stress system.

In this chapter, it is also discussed a portion of secondary regulations on these aging processes by these hypothalamic nuclei, which requires more investigations and supplements.

The conflict of interest

The author declares no conflict of interest nor financial support for this work.

References

Aguilera, G. (2011). HPA axis responsiveness to stress: Implications for healthy aging. *Experimental Gerontology*, *46*(2–3), 90–95. https://doi.org/10.1016/j.exger.2010.08.023.

Azuma, K., Adachi, Y., Hayashi, H., & Kubo, K. Y. (2015). Chronic psychological stress as a risk factor of osteoporosis. *Journal of University of Occupational and Environmental Health*, *37*(4), 245–253. https://doi.org/10.7888/juoeh.37.245.

Bachrach, B. E., & Smith, E. P. (1996). The role of sex steroids in bone growth and development: Evolving new concepts. *The Endocrinologist*, *6*, 362–368. https://doi.org/10.1097/00019616-199609000-00003.

Badawi, Y., & Nishimune, H. (2018). Presynaptic active zones of mammalian neuromuscular junctions: Nanoarchitecture and selective impairments in aging. *Neuroscience Research*, *127*, 78–88. https://doi.org/10.1016/j.neures.2017.11.014.

Balaskó, M., Soós, S., Székely, M., & Pétervári, E. (2014). Leptin and aging: Review and questions with particular emphasis on its role in the central regulation of energy balance. *Journal of Chemical Neuroanatomy, 61–62*, 248–255. https://doi.org/10.1016/j.jchemneu.2014.08.006.
Bar-Dayan, Y., Afek, A., Bar-Dayan, Y., Goldberg, I., & Kopolovic, J. (1999). Proliferation, apoptosis and thymic involution. *Tissue and Cell, 31*(4), 391–396. https://doi.org/10.1054/tice.1999.0001.
Böhne, A., Heule, C., Boileau, N., & Salzburger, W. (2013). Expression and sequence evolution of aromatase cyp19a1 and other sexual development genes in East African cichlid fishes. *Molecular Biology and Evolution, 30*(10), 2268–2285. https://doi.org/10.1093/molbev/mst124.
Bonaconsa, M., Malpeli, G., Montaruli, A., Carandente, F., Grassi-Zucconi, G., & Bentivoglio, M. (2014). Differential modulation of clock gene expression in the suprachiasmatic nucleus, liver and heart of aged mice. *Experimental Gerontology, 55*, 70–79. https://doi.org/10.1016/j.exger.2014.03.011.
Broekmans, F. J., Knauff, E. A., te Velde, E. R., Macklon, N. S., & Fauser, B. C. (2007). Female reproductive ageing: Current knowledge and future trends. *Trends in Endocrinology and Metabolism, 18*(2), 58–65. https://doi.org/10.1016/j.tem.2007.01.004.
Buckingham, E. M., & Klingelhutz, A. J. (2011). The role of telomeres in the ageing of human skin. *Experimental Dermatology, 20*(4), 297–302. https://doi.org/10.1111/j.1600-0625.2010.01242.x.
Cai, Z.-J. (1990). The neural mechanism of declarative memory consolidation and retrieval: A hypothesis. *Neuroscience and Biobehavioral Reviews, 14*, 295–304. https://doi.org/spi6;10.1111/10.1016/S0149-7634(05)80039-9.
Cai, Z.-J. (1991). The functions of sleep: Further analysis. *Physiology & Behavior, 50*, 53–60. https://doi.org/10.1111/10.1016/0031-9384(91)90497-C.
Cai, Z.-J. (1995). An integrative analysis to sleep functions. *Behavioural Brain Research, 69*, 187–194. https://doi.org/10.1016/0166-4328(95)00005-E.
Cai, Z.-J. (2015). Termination of organogenesis as intrinsic constraint on animal development and evolution: A theory. *Open Access Library Journal, 2*, e1646. https://doi.org/10.4236/oalib.1101646.
Cai, Z.-J. (2016a). A hypothetic aging pathway from skin to hypothalamic suprachiasmatic nucleus via slow wave sleep. *Sleep Science, 9*, 212–215. https://doi.org/10.1016/j.slsci.2016.09.004.
Cai, Z.-J. (2016b). Progressions of sleep, memory and depression applicable to psychoanalysis: A review. *Current Psychiatry Reviews, 12*, 240–245. https://doi.org/10.2174/1573400512666160610083505.
Cai, Z.-J. (2017). The adipose cause of senescence of male hypothalamic preoptic nucleus: An European view. *Research & Reviews: Neuroscience, 1*(2), 21–26.
Cai, Z.-J. (2018a). The peripheral hypotheses of hypothalamic aging. *Open Access Library Journal, 5*, e4445. https://doi.org/10.4236/oalib.1104445.
Cai, Z.-J. (2018b). The limbic-reticular coupling theory of memory processing in the brain and its greater compatibility over other theories. *Dementia & Neuropsychologia, 12*(2), 105–113. https://doi.org/10.1590/1980-57642018dn12-020002.
Callard, G. V., Tarrant, A. M., Novillo, A., Yacci, P., Ciaccia, L., Vajda, S., et al. (2011). Evolutionary origins of the estrogen signaling system: Insights from amphioxus. *The Journal of Steroid Biochemistry and Molecular Biology, 127*(3–5), 176–188. https://doi.org/10.1016/j.jsbmb.2011.03.022.
Carmichael, C. Y., & Wainford, R. D. (2015). Hypothalamic signaling mechanisms in hypertension. *Current Hypertension Reports, 17*(5), 39. https://doi.org/10.1007/s11906-015-0550-4.

Carrascosa, J. M., Ros, M., Andrés, A., Fernández-Agulló, T., & Arribas, C. (2009). Changes in the neuroendocrine control of energy homeostasis by adiposity signals during aging. *Experimental Gerontology*, *44*(1–2), 20–25. https://doi.org/10.1016/j.exger.2008.05.005.
Cartwright, M. J., Tchkonia, T., & Kirkland, J. L. (2007). Aging in adipocytes: Potential impact of inherent, depot-specific mechanisms. *Experimental Gerontology*, *42*(6), 463–471. https://doi.org/10.1016/j.exger.2007.03.003.
Chagin, A. S., Lindberg, M. K., Andersson, N., Moverare, S., Gustafsson, J. A., Sävendahl, L., et al. (2004). Estrogen receptor-beta inhibits skeletal growth and has the capacity to mediate growth plate fusion in female mice. *Journal of Bone and Mineral Research*, *19*(1), 72–77. https://doi.org/10.1359/JBMR.0301203.
Cohen, P. G. (2001). Aromatase, adiposity, aging and disease. The hypogonadal-metabolic-atherogenic-disease and aging connection. *Medical Hypotheses*, *56*(6), 702–708. https://doi.org/10.1054/mehy.2000.1169.
Colaianni, G., Tamma, R., Di Benedetto, A., Yuen, T., Sun, L., Zaidi, M., et al. (2014). The oxytocin-bone axis. *Journal of Neuroendocrinology*, *26*(2), 53–57. https://doi.org/10.1111/jne.12120.
Cowen, L. E., Hodak, S. P., & Verbalis, J. G. (2013). Age-associated abnormalities of water homeostasis. *Endocrinology and Metabolism Clinics of North America*, *42*(2), 349–370. https://doi.org/10.1016/j.ecl.2013.02.005.
Del Campo, A., Jaimovich, E., & Tevy, M. F. (2016). Mitochondria in the aging muscles of flies and mice: New perspectives for old characters. *Oxidative Medicine and Cellular Longevity*, *2016*, 9057593. https://doi.org/10.1155/2016/9057593.
Duncan, M. J., Prochot, J. R., Cook, D. H., Tyler Smith, J., & Franklin, K. M. (2013). Influence of aging on Bmal1 and Per2 expression in extra-SCN oscillators in hamster brain. *Brain Research*, *1491*, 44–53. https://doi.org/10.1016/j.brainres.2012.11.008.
Dunn, J. H., & Koo, J. (2013). Psychological stress and skin aging: A review of possible mechanisms and potential therapies. *Dermatology Online Journal*, *19*(6), 18561.
Ehlers, C. L., & Kupfer, D. J. (1989). Effects of age on delta and REM sleep parameters. *Electroencephalography and Clinical Neurophysiology*, *72*(2), 118–125. https://doi.org/10.1016/0013-4694(89)90172-7.
Engelberth, R. C., Silva, K. D., Azevedo, C. V., Gavioli, E. C., dos Santos, J. R., Soares, J. G., et al. (2014). Morphological changes in the suprachiasmatic nucleus of aging female marmosets (*Callithrix jacchus*). *BioMed Research International*, *2014*, 243825. https://doi.org/10.1155/2014/243825.
Espiritu, J. R. (2008). Aging-related sleep changes. *Clinics in Geriatric Medicine*, *24*(1), 1–14. https://doi.org/10.1016/j.cger.2007.08.007.
Gavazzeni, J., Wiens, S., & Fischer, H. (2008). Age effects to negative arousal differ for self-report and electrodermal activity. *Psychophysiology*, *45*(1), 148–151. https://doi.org/10.1111/j.1469-8986.2007.00596.x.
Giovannini, S., Marzetti, E., Borst, S. E., & Leeuwenburgh, C. (2008). Modulation of GH/IGF-1 axis: Potential strategies to counteract sarcopenia in older adults. *Mechanisms of Ageing and Development*, *129*(10), 593–601. https://doi.org/10.1016/j.mad.2008.08.001.
Girondot, M., Ben Hassine, S., Sellos, C., Godfrey, M., & Guillon, J. M. (2010). Modeling thermal influence on animal growth and sex determination in reptiles: Being closer to the target gives new views. *Sexual Development*, *4*(1–2), 29–38. https://doi.org/10.1159/000280585.
Goudsmit, E., Neijmeijer-Leloux, A., & Swaab, D. F. (1992). The human hypothalamo-neurohypophyseal system in relation to development, aging and Alzheimer's disease. *Progress in Brain Research*, *93*, 237–247. https://doi.org/10.1016/S0079-6123(08)64575-6.

Grösgen, S., Grimm, M. O., Friess, P., & Hartmann, T. (2010). Role of amyloid beta in lipid homeostasis. *Biochimica et Biophysica Acta, 1801*(8), 966–974. https://doi.org/10.1016/j.bbalip.2010.05.002.

Hareramadas, B., & Rai, U. (2006). Cellular mechanism of estrogen-induced thymic involution in wall lizard: Caspase-dependent action. *Journal of Experimental Zoology Part A: Comparative Experimental Biology, 305*(5), 396–409. https://doi.org/10.1002/jez.a.260.

He, J. Y., Jiang, L. S., & Dai, L. Y. (2011). The roles of the sympathetic nervous system in osteoporotic diseases: A review of experimental and clinical studies. *Ageing Research Reviews, 10*(2), 253–263. https://doi.org/10.1016/j.arr.2011.01.002.

Hepple, R. T. (2016). Impact of aging on mitochondrial function in cardiac and skeletal muscle. *Free Radical Biology and Medicine, 98*, 177–186. https://doi.org/10.1016/10.1016/j.brainres.2010.03.113.

Hirokawa, K., Utsuyama, M., & Kobayashi, S. (2001). Hypothalamic control of thymic function. *Cellular and Molecular Biology (Noisy-le-Grand, France), 47*(1), 97–102. https://doi.org/10.1016/10.1247/csf.26.49.

Hirshkowitz, M., Moore, C. A., Hamilton, C. R., 3rd., Rando, K. C., & Karacan, I. (1992). Polysomnography of adults and elderly: Sleep architecture, respiration, and leg movement. *Journal of Clinical Neurophysiology, 9*(1), 56–62. https://doi.org/10.1097/00004691-199201000-00006.

Hofman, M. A. (1997). Lifespan changes in the human hypothalamus. *Experimental Gerontology, 32*(4–5), 559–575. https://doi.org/10.1016/s0531-5565(96)00162-3.

Hsieh, Y. L., Hsu, C., Yang, S. L., Hsu, H. K., & Peng, M. T. (1996). Estradiol modulation of neuron loss in the medial division of medial preoptic nucleus in rats during aging. *Gerontology, 42*(1), 18–24. https://doi.org/10.1159/000213766.

Imokawa, G. (2008). Recent advances in characterizing biological mechanisms underlying UV-induced wrinkles: A pivotal role of fibroblast-derived elastase. *Archives of Dermatological Research, 300*(Suppl. 1), S7–20. https://doi.org/10.1007/s00403-007-0798-x.

Isaksson, O. G., Ohlsson, C., Nilsson, A., Isgaard, J., & Lindahl, A. (1991). Regulation of cartilage growth by growth hormone and insulin-like growth factor I. *Pediatric Nephrology, 5*(4), 451–453. https://doi.org/10.1007/BF01453680.

Isozaki, O., Tsushima, T., Miyakawa, M., Demura, H., & Seki, H. (1999). Interaction between leptin and growth hormone (GH)/IGF-I axis. *Endocrine Journal, 46*(Suppl), S17–S24. https://doi.org/10.1507/endocrj.46.Suppl_S17.

Kalmbach, K. H., Antunes, D. M., Kohlrausch, F., & Keefe, D. L. (2015). Telomeres and female reproductive aging. *Seminars in Reproductive Medicine, 33*(6), 389–395. https://doi.org/10.1055/s-0035-1567823.

Kammeyer, A., & Luiten, R. M. (2015). Oxidation events and skin aging. *Ageing Research Reviews, 21*, 16–29. https://doi.org/10.1016/j.arr.2015.01.001.

Kang, J. E., Lim, M. M., Bateman, R. J., Lee, J. J., Smyth, L. P., Cirrito, J. R., et al. (2009). Amyloid-beta dynamics are regulated by orexin and the sleep-wake cycle. *Science, 326*(5955), 1005–1007. https://doi.org/10.1126/science.1180962.

Keefe, D. L., Marquard, K., & Liu, L. (2006). The telomere theory of reproductive senescence in women. *Current Opinion in Obstetrics and Gynecology, 18*(3), 280–285. https://doi.org/10.1097/01.gco.0000193019.05686.49.

Koga, H., Kaushik, S., & Cuervo, A. M. (2011). Protein homeostasis and aging: The importance of exquisite quality control. *Ageing Research Reviews, 10*(2), 205–215. https://doi.org/10.1016/j.arr.2010.02.001.

Koh, K. K., Cardillo, C., Bui, M. N., Hathaway, L., Csako, G., Waclawiw, M. A., et al. (1999). Vascular effects of estrogen and cholesterol-lowering therapies in hypercholesterolemic postmenopausal women. *Circulation, 99*(3), 354–360. https://doi.org/10.1161/01.CIR.99.3.354.

Konturek, P. C., Brzozowski, T., & Konturek, S. J. (2011). Gut clock: Implication of circadian rhythms in the gastrointestinal tract. *Journal of Physiology and Pharmacology*, *62*(2), 139–150. http://www.jpp.krakow.pl/journal/archive/04_11/pdf/139_04_11_article.pdf.

Koudinov, A. R., & Koudinova, N. V. (2003). Cholesterol, synaptic function and Alzheimer's disease. *Pharmacopsychiatry*, *36*(Suppl. 2), S107–S112. https://doi.org/10.1055/s-2003-43055.

Kunkler, P. E., & Hwang, B. H. (1995). Lower GABAA receptor binding in the amygdala and hypothalamus of spontaneously hypertensive rats. *Brain Research Bulletin*, *36*(1), 57–61. https://doi.org/10.1016/0361-9230(94)00164-V.

Matsui, M. S., Pelle, E., Dong, K., & Pernodet, N. (2016). Biological rhythms in the skin. *International Journal of Molecular Sciences*, *17*(6), E801. https://doi.org/10.3390/ijms17060801.

May-Panloup, P., Boucret, L., Chao de la Barca, J. M., Desquiret-Dumas, V., Ferré-L'Hotellier, V., Morinière, C., et al. (2016). Ovarian ageing: The role of mitochondria in oocytes and follicles. *Human Reproduction Update*, *22*(6), 725–743. https://doi.org/10.1093/humupd/dmw028.

Moreno-Villanueva, M., & Bürkle, A. (2015). Molecular consequences of psychological stress in human aging. *Experimental Gerontology*, *68*, 39–42. https://doi.org/10.1016/j.exger.2014.12.003.

Muller, E. E., Rigamonti, A. E., Colonna Vde, G., Locatelli, V., Berti, F., & Cella, S. G. (2002). GH-related and extra-endocrine actions of GH secretagogues in aging. *Neurobiology of Aging*, *23*(5), 907–919. https://doi.org/10.1016/s0197-4580(02)00026-x.

Nakamura, M. (2010). The mechanism of sex determination in vertebrates-are sex steroids the key-factor? *Journal of Experimental Zoology Part A Ecological Genetics and Physiology*, *313*(7), 381–398. https://doi.org/10.1002/jez.616.

Nakamura, M. (2013). Is a sex-determining gene(s) necessary for sex-determination in amphibians? Steroid hormones may be the key factor. *Sexual Development*, 7(1–3), 104–114. https://doi.org/10.1159/000339661.

Nyberg, L. (2017). Functional brain imaging of episodic memory decline in ageing. *Journal of Internal Medicine*, *281*(1), 65–74. https://doi.org/10.1111/joim.12533.

O'Connor, K. G., Harman, S. M., Stevens, T. E., Jayme, J. J., Bellantoni, M. F., Busby-Whitehead, M. J., et al. (1999). Interrelationships of spontaneous growth hormone axis activity, body fat, and serum lipids in healthy elderly women and men. *Metabolism*, *48*(11), 1424–1431. https://doi.org/10.1016/S0026-0495(99)90154-3.

Panconesi, E., & Hautmann, G. (1996). Psychophysiology of stress in dermatology: The psychobiologic pattern of psychosomatics. *Dermatologic Clinics*, *14*(3), 399–421. https://doi.org/10.1016/S0733-8635(05)70368-5.

Panzica, G. C., García-Ojeda, E., Viglietti-Panzica, C., Thompson, N. E., & Ottinger, M. A. (1996). Testosterone effects on vasotocinergic innervation of sexually dimorphic medial preoptic nucleus and lateral septum during aging in male quail. *Brain Research*, *712*(2), 190–198. https://doi.org/10.1016/0006-8993(95)01386-5.

Perrini, S., Laviola, L., Carreira, M. C., Cignarelli, A., Natalicchio, A., & Giorgino, F. (2010). The GH/IGF1 axis and signaling pathways in the muscle and bone: Mechanisms underlying age-related skeletal muscle wasting and osteoporosis. *Journal of Endocrinology*, *205*(3), 201–210. https://doi.org/10.1677/JOE-09-0431.

Pertynska-Marczewska, M., & Diamanti-Kandarakis, E. (2017). Aging ovary and the role for advanced glycation end products. *Menopause*, *24*(3), 345–351. https://doi.org/10.1097/GME.0000000000000755.

Pétervári, E., Rostás, I., Soós, S., Tenk, J., Mikó, A., Füredi, N., et al. (2014). Age versus nutritional state in the development of central leptin resistance. *Peptides*, *56*, 59–67. https://doi.org/10.1016/j.peptides.2014.03.011.

Powell, D. A., Milligan, W. L., & Furchtgott, E. (1980). Peripheral autonomic changes accompanying learning and reaction time performance in older people. *Journal of Gerontology*, *35*(1), 57–65. https://doi.org/10.1093/geronj/35.1.57.

Prager-Khoutorsky, M., & Bourque, C. W. (2015). Mechanical basis of osmosensory transduction in magnocellular neurosecretory neurones of the rat supraoptic nucleus. *Journal of Neuroendocrinology*, *27*(6), 507–515. https://doi.org/10.1111/jne.12270.

Prior, J. C. (2005). Ovarian aging and the perimenopausal transition: The paradox of endogenous ovarian hyperstimulation. *Endocrine*, *26*(3), 297–300. https://doi.org/10.1385/ENDO:26:3:297.

Reame, N. E., Kelche, R. P., Beitins, I. Z., Yu, M. Y., Zawacki, C. M., & Padmanabhan, V. (1996). Age effects of follicle-stimulating hormone and pulsatile luteinizing hormone secretion across the menstrual cycle of premenopausal women. *The Journal of Clinical Endocrinology & Metabolism*, *81*(4), 1512–1518. https://doi.org/10.1210/jc.81.4.1512.

Rinnerthaler, M., Bischof, J., Streubel, M. K., Trost, A., & Richter, K. (2015). Oxidative stress in aging human skin. *Biomolecules*, *5*(2), 545–589. https://doi.org/10.3390/biom5020545.

Rodriguez-Vallecillo, E., & Woodbury-Fariña, M. A. (2014). Dermatological manifestations of stress in normal and psychiatric populations. *Psychiatric Clinics of North America*, *37*(4), 625–651. https://doi.org/10.1016/j.psc.2014.08.009.

Rudolf, R., Khan, M. M., Labeit, S., & Deschenes, M. R. (2014). Degeneration of neuromuscular junction in age and dystrophy. *Frontiers in Aging Neuroscience*, *6*, 99. https://doi.org/10.3389/fnagi.2014.00099.

Sermondade, N., Faure, C., Fezeu, L., Shayeb, A. G., Bonde, J. P., Jensen, T. K., et al. (2013). BMI in relation to sperm count: An updated systematic review and collaborative meta-analysis. *Human Reproduction Update*, *19*(3), 221–231. https://doi.org/10.1093/humupd/dms050.

Siegl-Cachedenier, I., Flores, I., Klatt, P., & Blasco, M. A. (2007). Telomerase reverses epidermal hair follicle stem cell defects and loss of long-term survival associated with critically short telomeres. *The Journal of Cell Biology*, *179*(2), 277–290. https://doi.org/10.1083/jcb.200704141.

Sompol, P., Ittarat, W., Tangpong, J., Chen, Y., Doubinskaia, I., Batinic-Haberle, I., et al. (2008). A neuronal model of Alzheimer's disease: An insight into the mechanisms of oxidative stress–mediated mitochondrial injury. *Neuroscience*, *153*(1), 120–130. https://doi.org/10.1016/j.neuroscience.2008.01.044.

Swaab, D. F., & Bao, A. M. (2011). (Re-)activation of neurons in aging and dementia: Lessons from the hypothalamus. *Experimental Gerontology*, *46*(2–3), 178–184. https://doi.org/10.1016/j.exger.2010.08.028.

Tatone, C., Amicarelli, F., Carbone, M. C., Monteleone, P., Caserta, D., Marci, R., et al. (2008). Cellular and molecular aspects of ovarian follicle ageing. *Human Reproduction Update*, *14*(2), 131–142. https://doi.org/10.1093/humupd/dmm048.

Taub, D. D., & Longo, D. L. (2005). Insights into thymic aging and regeneration. *Immunological Reviews*, *205*, 72–93. https://doi.org/10.1111/j.0105-2896.2005.00275.x.

Taub, D. D., Murphy, W. J., & Longo, D. L. (2010). Rejuvenation of the aging thymus: Growth hormone-mediated and ghrelin-mediated signaling pathways. *Current Opinion in Pharmacology*, *10*(4), 408–424. https://doi.org/10.1016/j.coph.2010.04.015.

Ventevogel, M. S., & Sempowski, G. D. (2013). Thymic rejuvenation and aging. *Current Opinion in Immunology*, *25*(4), 516–522. https://doi.org/10.1016/j.coi.2013.06.002.

Vermeulen, A., Kaufman, J. M., Goemaere, S., & van Pottelberg, I. (2002). Estradiol in elderly men. *The Aging Male*, *5*(2), 98–102. https://doi.org/10.1080/tam.5.2.98.102.

Wang, X. L., Su, B., Perry, G., Smith, M. A., & Zhu, X. W. (2007). Insights into amyloid-β induced mitochondrial dysfunction in Alzheimer disease. *Free Radical Biology and Medicine*, *43*(12), 1569–1573. https://doi.org/10.1016/j.freeradbiomed.2007.09.007.

Wang, T., Zhang, M., Jiang, Z., & Seli, E. (2017). Mitochondrial dysfunction and ovarian aging. *American Journal of Reproductive Immunology*, 77(5), e12651. https://doi.org/10.1111/aji.12651.

Wyse, C. A., & Coogan, A. N. (2010). Impact of aging on diurnal expression patterns of CLOCK and BMAL1 in the mouse brain. *Brain Research*, *1337*, 21–31. https://doi.org/10.1016/j.brainres.2010.03.113.

Xie, L., Kang, H., Xu, Q., Chen, M. J., Liao, Y., Thiyagarajan, M., et al. (2013). Sleep drives metabolite clearance from the adult brain. *Science*, *342*(6156), 373–377. https://doi.org/10.1126/science.1241224.

Yakar, S., Werner, H., & Rosen, C. J. (2018). Insulin-like growth factors: Actions on the skeleton. *Journal of Molecular Endocrinology*, *61*(1), T115–T137. https://doi.org/10.1530/JME-17-0298.

Yamada, K., & Harada, N. (1990). Expression of estrogen synthetase (P-450 aromatase) during adipose differentiation of 3T3-L1 cells. *Biochemical and Biophysical Research Communications*, *169*(2), 531–536. https://doi.org/10.1016/0006-291X(90)90363-R.

Yamazaki, S., Straume, M., Tei, H., Sakaki, Y., Menaker, M., & Block, G. D. (2002). Effects of aging on central and peripheral mammalian clocks. *Proceedings of the National Academy of Sciences of the United States of America*, *99*(16), 10801–10806. https://doi.org/10.1073/pnas.152318499.

Yao, G., & Hou, Y. (2004). Thymic atrophy via estrogen-induced apoptosis is related to Fas/FasL pathway. *International Immunopharmacology*, *4*(2), 213–221. https://doi.org/10.1016/j.intimp.2003.12.005.

Ye, H., Zheng, T., Li, W., Li, X., Fu, X., Huang, Y., et al. (2017). Ovarian stem cell nests in reproduction and ovarian aging. *Cellular Physiology and Biochemistry*, *43*(5), 1917–1925. https://doi.org/10.1159/000484114.

Zhang, H., Adhikari, D., Zheng, W., & Liu, K. (2013). Combating ovarian aging depends on the use of existing ovarian follicles, not on putative oogonial stem cells. *Reproduction*, *146*(6), R229–R233. https://doi.org/10.1530/REP-13-0202.

Zhang, Y., Zhang, S., Lu, H., Zhang, L., & Zhang, W. (2014). Genes encoding aromatases in teleosts: Evolution and expression regulation. *General and Comparative Endocrinology*, *205*, 151–158. https://doi.org/10.1016/j.ygcen.2014.05.008.

Zhou, J. N., & Swaab, D. F. (1999). Activation and degeneration during aging: A morphometric study of the human hypothalamus. *Microscopy Research and Technique*, *44*(1), 36–48. https://doi.org/10.1002/(SICI)1097-0029(19990101)44:1 < 36::AID-JEMT5 > 3.0.CO;2-F.

Zhuang, Y., Hou, H., Zhao, X., Zhang, Z., & Li, B. (2009). Effects of collagen and collagen hydrolysate from jellyfish (*Rhopilema esculentum*) on mice skin photoaging induced by UV irradiation. *Journal of Food Science*, 74(6), H183–H188. https://doi.org/10.1111/j.1750-3841.2009.01236.x.

CHAPTER THREE

Hypothalamic gene transfer of BDNF promotes healthy aging

Lei Cao[a,b,*], Seemaab Ali[a,b], and Nicholas J. Queen[a,b]

[a]Department of Cancer Biology and Genetics, College of Medicine, The Ohio State University, Columbus, OH, United States

[b]The Ohio State University Comprehensive Cancer Center, Columbus, OH, United States

*Corresponding author: e-mail address: lei.cao@osumc.edu

Contents

Abstract

The aging process and age-related diseases all involve metabolic decline and impaired ability to cope with adversity. Environmental enrichment (EE)—a housing environment which recapitulates aspects of active lifestyle—exerts a wide range of health benefits in laboratory rodents. Brain-derived neurotrophic factor (BDNF) in the hypothalamus orchestrates autonomic and neuroendocrine processes, serving as one key brain mediator of EE-induced resistance to obesity, cancer, and autoimmunity. Recombinant adeno-associated virus (AAV)-mediated hypothalamic BDNF gene transfer alleviates obesity, diabetes, and metabolic syndromes in both diet-induced and genetic models. One recent study by our lab demonstrates the efficacy and safety of a built-in autoregulatory system to control transgene BDNF expression, mimicking the body's natural feedback systems in middle-age mice. Twelve-month old mice were treated with

Vitamins and Hormones, Volume 115
ISSN 0083-6729
https://doi.org/10.1016/bs.vh.2020.12.003

autoregulatory BDNF vector and monitored for 7 months. BDNF gene transfer prevented age-associated metabolic decline by: reducing adiposity, preventing the decline of brown fat activity, increasing adiponectin while reducing leptin and insulin in circulation, improving glucose tolerance, increasing energy expenditure, alleviating hepatic steatosis, and suppressing inflammatory genes in the hypothalamus and adipose tissues. Furthermore, BDNF treatment reduced anxiety-like and depression-like behaviors. This chapter summarizes this work and discusses potential roles that hypothalamic BDNF might play in promoting healthy aging.

1. Metabolic problems of aging

1.1 Targeting Healthspan

The number of Americans over the age of 65 is currently over 50 million (*National Population Totals and Components of Change: 2010-2019,* 2019). By 2030, 1 in 5 Americans is projected to be over the age of 65 (Colby & Ortman, 2014). The United States' aging demographic shift mirrors the global phenomenon of increasing lifespan (He, Goodkind, & Kowal, 2016). Elderly populations worldwide are expected only to continue to grow during this epidemiologic transition.

With increased life expectancy comes an accompanied risk for chronic morbidities, including heart disease, neurological disorders, and metabolic dysfunction (Ferrucci, Giallauria, & Guralnik, 2008; Mather, Jacobsen, & Pollard, 2015; Needham, 2014). These conditions have significant sequelae—leading to poor quality of life and potential mortality, especially in older adults. Depending on the onset and severity of late-life chronic disease, these increases in median lifespan may improve or worsen the lives of the elderly. In the face of aging demographic trends, this is conceptualized best in the framework of either an expansion or a compression of lifetime morbidity accompanying increased longevity (Crimmins, 2015; Fries, 1980; Gore, Kingston, Johnson, Kirkwood, & Jagger, 2018).

With morbidity-related disability, researchers have repeatedly found population level trends of a progressive reduction in functional impairment in older Americans towards the end of the 20th century (Freedman et al., 2004; Schoeni, Freedman, & Martin, 2008). More recently, this has not been the case. The fraction of years spent without disability has not gone down, despite the epidemiologic and demographic forces decreasing mortality (Crimmins, 2015; Crimmins & Beltran-Sanchez, 2011; Seeman, 2010; Zhang, Saito, & Crimmins, 2019). These findings underscore the importance of determining when, where, and how age-related morbidities begin.

Recent analyses show that, while global disease burden may be declining when measured since 1990, high burdens of age-related diseases exist among both longer-lived and shorter-lived populations (Chang, Skirbekk, Tyrovolas, Kassebaum, & Dieleman, 2019). This suggests that other predictors of age acceleration, and not necessarily chronological age, will predict the incidence of age-related disease. Research and clinical efforts are now focused on extending and improving healthspan, or the length of healthy, disability-free life (Olshansky, 2018).

1.2 Obesity in older adults

As demographic trends show increasing aged populations, obesity, defined as abnormal or excessive fat accumulation, is also increasing globally. Current estimates put obesity rates in the United States at 42% in adults (Hales, Carroll, Fryar, & Ogden, 2020). Overnutrition, defined as excessive caloric intake exceeding caloric expenditure, and sedentary lifestyles are the primary drivers of obesity (Hruby & Hu, 2015). Without significant interventions, nearly 1 in 2 adults in the US are projected to be obese by 2030 (Ward et al., 2019). The convergence of aging and obesity is also becoming apparent. Obesity disproportionally affects middle-aged and older adults, with the prevalence of obesity being highest in middle-aged adults, aged 40–59 (Hales et al., 2020).

Obesity is indirectly measured using the body mass index (BMI), calculated from an individual's weight in kilograms divided by the square of their height in meters. A BMI of greater than 25 is considered overweight, whereas a BMI greater than 30 is obese. Obesity is a significant health concern because of its associated risk for other chronic diseases, including but not limited to hypertension, arteriosclerosis, stroke, steatohepatitis, diabetes and cancer (Attlee, Kassem, Hashim, & Obaid, 2015; Pi-Sunyer, 2009). Above 25, every 5-unit increment of BMI is associated with a 30% increases in all-cause mortality, a 40% higher vascular mortality, and a 120% increase in diabetes-related mortality (Prospective Studies et al., 2009).

1.3 Pathobiology of adipose tissue in obesity and aging

In addition to its role in energy storage, adipose tissues comprise a dynamic and adaptive organ system, responsive to immune, nervous system, and nutrition signals. Dysfunction in adipose tissue processes during aging is now thought to be a driver of the aging process (Palmer & Kirkland, 2016). Aging is associated with changes in the distribution of adipose tissue, its immune and inflammatory signaling, and its cellular composition. Aging in both men and women leads to progressively increasing distribution of adipose tissue in the abdominal region and deposition in ectopic sites (Kuk, Saunders, Davidson, & Ross, 2009). Adipose tissue around abdominal viscera, or visceral adipose tissue, correlates strongly with health risks of obesity, compared to adipose distribution in the rest of the body (Ibrahim, 2010; Kissebah & Krakower, 1994; Matsuzawa, 2008; Tchernof & Despres, 2013).

Expansion of adipose tissue in obesity is associated with immune cell infiltration which promotes chronic low-grade inflammation (Choe, Huh, Hwang, Kim, & Kim, 2016). Aging is also associated with a systemic pro-inflammatory state termed "inflammaging," related to reduced autophagy, increased oxidative stress, and subsequent activation of inflammasome proteins (Salminen, Kaarniranta, & Kauppinen, 2012). Adipocytes become senescent with advancing age and acquire a senescence-associated secretory phenotype (SASP), in which they release high levels of interleukin 6 (IL-6), IL-1β, monocyte chemoattractant protein (MCP1 or CCL2), and other chemokines and matrix metalloproteases (Stout, Justice, Nicklas, & Kirkland, 2017). IL-6 from adipose tissue is a major contributor to systemic inflammation in old age (Starr, Evers, & Saito, 2009). These chronic inflammation states are closely linked to the health hazards of obesity, including the development of insulin resistance and cardiovascular disease (Ellulu, Patimah, Khaza'ai, Rahmat, & Abed, 2017; Ferrucci & Fabbri, 2018; Rodriguez-Hernandez, Simental-Mendia, Rodriguez-Ramirez, & Reyes-Romero, 2013).

Interventions reducing adiposity may be beneficial in addressing this pathophysiology. Meta-analyses show exercise is an effective intervention for reducing visceral adiposity in overweight adults (Vissers et al., 2013). Obese patients undergoing gastric bypass surgery for weight loss show significantly reduced premature aging markers (Hohensinner et al., 2018). Coupled with dramatically lower BMI in these patients, reductions in plasma concentrations of CRP, IL6, and PAI-1 all point to reduced circulating SASP load 2 years after surgery. Given that interventions targeting obesity may reduce health risks and address the adipose drivers of aging, our research focus has been on the exploration of anti-obesity mechanisms and the development of treatments for obesity in mouse models. Here we explain central nervous system (CNS) mechanisms, as well as environmental and genetic approaches to preventing and treating obesity.

2. CNS energy regulation and the hypothalamic-sympathoneural-adipocyte axis

2.1 Hypothalamic control of energy balance

The hypothalamus is the CNS neuroendocrine coordinator for the body. The hypothalamus regulates many physiological processes critical to life, including circadian rhythms, social drives, reproduction, homeostasis, thermoregulation, energy balance, and stress responses (Burbridge, Stewart, & Placzek, 2016). These processes are coordinated across discrete neuronal nuclei which are interconnected with the autonomic nervous system and are responsive to endocrine signals. The feeding and satiety circuitry of the hypothalamus is involved in the development of obesity, and is situated across the medial basal hypothalamus (MBH), containing the arcuate nucleus (ARC) and the ventromedial nucleus of the hypothalamus (VMH), and the paraventricular nucleus (PVN). Anorexigenic pro-opiomelanocortin (POMC) neurons in the hypothalamus signal for satiety and in opposition to orexigenic neuropeptide Y (NPY)/agouti-related peptide (AgRP) neurons which drive feeding. The circuit is partly integrated in the PVN, which receives inputs from a wide swath of the hypothalamus and the rest of the CNS (Luo et al., 2019). These effects are signaled by hormones including leptin, insulin, and ghrelin (Austin & Marks, 2009; Gaspar & Velloso, 2018).

2.2 Brain-derived neurotrophic factor (BDNF)

BDNF is a widely expressed neurotrophin with key roles in synaptic plasticity within the CNS. As an immediate early gene responding to neuronal activity, BDNF transcription is controlled by calcium responsive elements and is significantly induced by calcium influx in neurons (Finkbeiner, 2000). The mouse *Bdnf* gene contains eight 5′ noncoding exons each with distinct promoter regions, which are alternatively spliced with exon IX, containing the one coding region of the *Bdnf* gene. (Liu et al., 2006). Human *Bdnf* shares a similar structure with eleven noncoding exons which splice to a single coding exon (Pruunsild, Kazantseva, Aid, Palm, & Timmusk, 2007). Translated proBDNF is processed by proteases intracellularly to produce mature BDNF, although it is also released in its pro-form in some contexts to directly bind the p75 neurotrophin receptor or to be processed extracellularly into mature BDNF (Lessmann, Gottmann,

& Malcangio, 2003). Intracellular mature BDNF is generally trafficked via the activity-regulated synaptic packaging pathway to be stored both pre- and post-synaptically at dendrites for exocytosis (Leschik et al., 2019). BDNF can also act via autocrine and paracrine release (Mowla et al., 2001). BDNF signals through the neurotrophin receptor tropomyosin receptor kinase B (TrkB), whose full-length isoform is expressed predominantly on neurons (TrkB.FL). A truncated isoform lacking receptor tyrosine kinase activity (TrkB.T1) is expressed predominantly on astrocytes (Holt et al., 2019).

Hypothalamic BDNF is critical for maintaining energy balance (Xu et al., 2003; Xu & Xie, 2016). BDNF is implicated in the development of human obesity. Genome-wide association studies have found that BDNF is one of 18 genetic loci associated with BMI (Jiao et al., 2011; Speliotes et al., 2010). Haploinsufficiency for BDNF or its receptor TrkB is associated with obesity and hyperphagia (Gray et al., 2006; Han et al., 2008; Yeo et al., 2004).

2.3 Environmental enrichment (EE)

Age-related changes to an individual's health arise from a combination of person's physical age, their genetics, and their environment, and yet we still know very little about how these interact to yield either healthy "successful" aging or unsuccessful aging. There are stereotyped features of aging that we see in humans and animal models, but it is becoming clear that they result from complex interactions between an organism's biology and its environment (Bektas, Schurman, Sen, & Ferrucci, 2018; de Magalhaes, Wuttke, Wood, Plank, & Vora, 2012; Kennedy et al., 2014; Lopez-Otin, Blasco, Partridge, Serrano, & Kroemer, 2013).

EE serves as a tool for studying cognitive and physical interventions in laboratory animals. Compared to standard lab environments (SE), EE housing exposes animals to species-specific, multimodal cognitive, social, motor, and sensory stimuli in novel and complex ways. For rodents such as the wild-type C57BL/6 strain of mice, this includes larger housing enclosures, supplemented with tunnels, igloos, huts, retreats, toys, platform structures, and running wheels (Slater & Cao, 2015). This model provides an effective approach for studying environment interactions with biology, such as those contributing to the process of aging. The mechanisms of EE appear to be multifactorial, as exercise and non-exercise enrichment both yield distinct and synergistic inductions of neurogenesis and neurotrophic factor production (Birch, McGarry, & Kelly, 2013; Cao et al., 2010; Steiner, Zurborg, Horster, Fabel, & Kempermann, 2008; Zhao, Jou, Wolff, Sun, & Gage, 2014). Our previous work has demonstrated that this EE housing promotes a healthy metabolic phenotype in young and aged mice (Cao et al., 2011; During et al., 2015; Foglesong et al., 2016; McMurphy et al., 2018; Queen, Boardman, et al., 2020).

2.4 Hypothalamic-sympathoneural-adipocyte axis

EE provides a physiologically relevant manipulation of dysfunctional adipose tissue during aging. We previously identified a novel neuroendocrine axis, named the hypothalamic-sympathoneural-adipocyte (HSA) axis, which is involved in many of the metabolic effects of EE (Cao et al., 2009, 2010, 2011; During et al., 2015; Liu et al., 2014; Siu et al., 2017). In this axis, an increase in the hypothalamic expression of BDNF leads to elevated sympathetic tone to adipose tissue. Sympathetic norepinephrine release in adipose tissue results in adipose tissue remodeling, including the induction of thermogenically-active beige cells and the suppression of leptin. Activation of this axis in rodents reduces adiposity, improves resistance to genetic and diet-induced obesity, and increases insulin sensitivity (Cao et al., 2009, 2010, 2011; During et al., 2015; Liu et al., 2014; Siu et al., 2017). For animal welfare considerations, it is notable that sympathetic activation generally indicates a stress response. EE is considered a model of eustress or positive stress, associated with beneficial health outcomes (Milsum, 1985; Sanchis-Gomar et al., 2012).

Metabolic parameters such as adipose hormones, blood glucose levels and related hormones, and blood lipids are variably affected by EE in mice, depending on the model. Reductions in leptin, insulin, insulin-like growth factor 1 (IGF-1), glucose, and cholesterol are associated with EE in young male mice with diet-induced obesity (Cao et al., 2011). In healthy males, EE also lowers leptin and IGF-1, and increases levels of adiponectin and corticosterone (Cao et al., 2010). Of these serum markers, leptin is the most reliable to change in the context of EE. Leptin is a hormone produced by adipocytes which is released into circulation, with blood levels correlating with overall adiposity (Frederich et al., 1995). Leptin acts on receptors in the hypothalamus to regulate food intake (van Swieten, Pandit, Adan, & van der Plasse, 2014).

2.5 EE promotes healthy aging

Intrigued by the metabolic benefits of EE observed in young animals and the importance of metabolic decline in systemic aging, we investigated the ability of EE to promote healthy aging. In middle-aged female mice (10-month old), we observed that a short-term EE exposure of 6-weeks induced hypothalamic *Bdnf* expression and thereby activated the HSA axis, leading to profound adipose remodeling mirroring those observed in young counterparts, such as lower adiposity and reduced circulating level of leptin (McMurphy et al., 2018). EE-housed animals had reduced blood glucose excursions in response to a glucose bolus associated with elevated adipose glucose uptake. Adipose tissue levels of leptin mRNA were reduced, and adipose β3-adrenergic receptor (*Adrb3*) expression was increased. These changes were associated with activation of sympathetic response in adipose tissue, including increased peroxisome proliferator-activated receptor gamma coactivator 1α (PGC-1α) mRNA expression and protein levels. PGC-1α is the master regulator of mitochondrial biogenesis and regulates thermogenesis and oxidative metabolism (Barbera et al., 2001; Puigserver et al., 1998). Importantly, these improvements in adipose tissue metabolism were not replicated by voluntary running alone, despite dramatic reductions in adiposity in running wheel animals (McMurphy et al., 2018). EE and running wheel treatments displayed distinct peripheral tissue changes, with voluntary running strongly affecting muscle metabolism and increasing muscle size, and EE affecting adipose tissue more strongly, as described.

Longer EE for 12 months, initiated at middle age, maintained an EE adipose tissue pattern of gene expression relative to SE housed mice. The long-term effects of EE also include reduced adiposity, improved glucose tolerance, and prevention of age-related brown adipose tissue (BAT) dysfunction. In addition to adipose remodeling, EE suppressed age-related liver steatosis, reduced hepatic glucose production, and increased glucose uptake contributing to the improved glycemic control. The EE-induced liver modulation was associated with a suppression of protein kinase Cε. Moreover, EE improved motor abilities and reduced anxiety-like behavior (McMurphy et al., 2018). Of note, EE downregulated the expression of inflammatory genes in the brain, adipose tissue, and liver. EE initiated at 18-month of age resulted in a trend of positive impact on mean lifespan (~45 days). These data suggest that a physically, mentally, and socially active environment can promote healthy aging and the HSA axis might serve as one of the underpinning mechanisms.

2.6 Clinical relevance of EE studies in mouse models

Attempts to translate EE-mediated therapy originated from animal models remain in their infancy. Three clinical trials by Woo and Leon are among the first examples of direct translation of EE in animal models of autism spectrum disorder (ASD) to clinical studies (Aronoff, Hillyer, & Leon, 2016; Woo, Donnelly, Steinberg-Epstein, & Leon, 2015; Woo & Leon, 2013). These randomized clinical trials have shown that EE in the form of Sensory Enrichment Therapy is capable of ameliorating symptoms of ASD (Aronoff et al., 2016). EE has also been shown to improve the symptoms of children with Rett syndrome (Downs et al., 2018). These studies provide a clear example of how preclinical studies in animal models of brain disorders can directly inform clinical investigations and trials.

Conversely, EE is a laboratory condition that recapitulates some aspects of an active lifestyle, and therefore can facilitate mechanistic studies. These mechanistic studies in animals are vital to understand conserved mechanisms that may contribute to healthy aging. For example, numerous clinical data have shown that behavioral interventions (cognitive stimulation, exercise, etc.) and lifestyle (education level, work position, cognitive and social activities) have important benefits for healthy aging of the brain (Sampedro-Piquero & Begega, 2017). Our research on EE's impact on healthy aging supports the notion that EE is a successful laboratory model to facilitate a dynamic cycle of "bench to bedside and back again" (Hill-Yardin & Hannan, 2013). Recently, we summarize the benefits of EE across the lifespan, and address the "translatability challenge" of such research in a review article (Queen, Hassan, & Cao, 2020), suggesting several human analogs of laboratory EE.

3. Molecular therapy of obesity with an autoregulatory BDNF vector

Our research on EE and the HSA axis has promoted us to develop a molecular therapy for obesity and cancer. Recombinant adeno-associated virus (rAAV) vectors are attractive vehicles for gene therapy because they can transduce both dividing and postmitotic tissues with low immunogenicity and long-lasting transgene expression (Mingozzi & High, 2011). In a series of studies, we have shown rAAV-mediated gene delivery of BDNF to the hypothalamus reproduces many health benefits of EE, including the anti-obesity and anticancer phenotypes (Cao et al., 2009, 2010, 2011; Cao & During, 2012; Foglesong, Queen, Huang, Widstrom, & Cao, 2019).

As rAAV-mediated transgene expression is sustained long-term and BDNF plays important role in weight regulation, it is essential to regulate the therapeutic gene expression in order to avoid serious side effects such as pathological cachexia. Thus, we developed a built-in autoregulatory system to control therapeutic gene expression, mimicking the body's natural feedback systems (U.S. Patent 9,265,843 B2). This autoregulatory system involves a single rAAV vector harboring two expression cassettes, one driving BDNF by a general constitutive chicken β-actin (CBA) promoter, and the other driving a specific microRNA targeting BDNF under the control of a promoter from the agouti-related protein (AGRP) (Fig. 1) (Cao et al., 2009). Because rAAV-mediated BDNF expression leads to weight loss and reduced adiposity, this physiological event will activate AGRP promoter activity to maintain homeostasis. In turn, the AGRP promoter drives microRNA expression to inhibit BDNF expression (miR-Bdnf). Therefore, the therapeutic gene expression is coupled to the physiological changes induced by the expression of the introduced therapeutic gene.

The efficacy of this autoregulatory approach was first tested in a mouse model of type 2 diabetes (*db/db*), where high-dose gene therapy was delayed until mice were extremely obese (Cao et al., 2009). rAAV vectors were stereotaxically delivered to the hypothalamus and metabolic parameters were monitored. Control mice receiving a rAAV vector encoding yellow fluorescent protein (YFP) continued to gain weight. In contrast, when mice were given an unregulated vector (BDNF-miR-scr, i.e., a BDNF-encoding

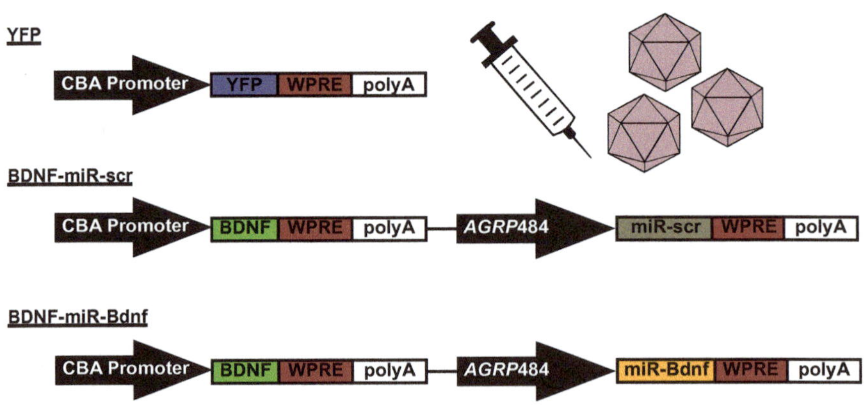

Fig. 1 Schematic of the rAAV vectors. *Adapted from Cao, L., Lin, E. J., Cahill, M. C., Wang, C., Liu, X., & During, M. J. (2009). Molecular therapy of obesity and diabetes by a physiological autoregulatory approach.* Nature Medicine, 15*(4), 447–454. doi:10.1038/nm.1933.*

vector with a scrambled microRNA targeting no known genes), their weight dropped precipitously by approximately 45% in 3 weeks (from 55–30 g). Mice receiving the autoregulatory BDNF vector (BDNF-miR-Bdnf) displayed a marked decrease in body weight, followed by stabilization between 3 and 4 weeks after rAAV injection. Weight loss was maintained for the 11-week duration of experiment with narrow fluctuation, indicating efficient autoregulation of the BDNF transgene expression. The gene therapy sufficiently alleviated obesity and diabetes.

Next, we examined the efficacy and safety of this autoregulatory BDNF gene therapy in melanocortin 4 receptor (MC4R)-deficient mice, a model representing the most common monogenic form of human obesity (Farooqi & O'Rahilly, 2006). Melanocortin-4 receptor (MC4R) mutations have been found in up to 6% of severe early-onset obesity (Farooqi et al., 2003), and conventional treatment are either ineffective long-term or contraindicated. BDNF is immediately downstream of MC4R in the key pathways for controlling energy homeostasis (Nicholson, Peter, Lecourt, Barde, & Hofbauer, 2007; Xu et al., 2003). MC4R heterozygous mice, which best resemble MC4R obese patients, were randomized to receive either autoregulatory BDNF vector or YFP control, and monitored for 30 weeks. Autoregulatory BDNF treatment completely prevented the development of obesity and metabolic syndrome, without adverse cardiovascular function or behavioral disturbances, providing compelling evidence to support further development as a gene therapy for MC4R-deficient obese patients (Siu et al., 2017).

BDNF deficiency is also associated with the obesity seen in the Prader-Willi syndrome and WAGR syndrome (Han et al., 2008; Han, Muehlbauer, Cui, Newgard, & Haqq, 2010). We are currently assessing this autoregulatory BDNF gene therapy in a mouse model of Prader-Willi syndrome.

4. Hypothalamic gene transfer of BDNF promotes healthy aging

Our recent study showed that implementing EE after middle age promoted healthy aging in wild type mice. Many of the observed EE-aging phenotypes were associated with activation of the HSA axis (McMurphy et al., 2018). Thus, we asked whether genetically activating the HSA axis by overexpressing BDNF in the hypothalamus could mimic the EE-induced benefits on health in aged mice. BDNF signaling, insulin, IGF-1, and serotonin have been proposed to be important determinants of health during aging because of their cooperative influence on energy metabolism, stress response and cardiovascular function (Mattson, Maudsley, & Martin, 2004). Decreased BDNF levels are associated with vulnerable neuronal populations in several neurodegenerative disorders including Alzheimer's, Parkinson's and Huntington's diseases. In addition, reduced BDNF signaling has been documented during normal aging. These data demonstrate the need for further therapeutic research on components of the BDNF signaling pathway (Tapia-Arancibia, Aliaga, Silhol, & Arancibia, 2008). Administration of exogenous BDNF and/or stimulation of its receptor expression can offset some physiologic or pathologic age-related changes in the CNS (Tapia-Arancibia et al., 2008). Moreover, BDNF signaling in the brain is thought to mediate at least some of the anti-aging effects of an intermittent fasting regiment (Duan et al., 2003; Lee, Duan, & Mattson, 2002), although data on the hypothalamus are not reported. It is also unclear how BDNF signaling in neurons is transferred to the periphery to improve the healthspan of many different organ systems. Thus, we investigated the long-term effects of hypothalamic gene transfer of BDNF in middle-aged mice using an autoregulatory rAAV vector.

We conducted a long-term study to comprehensively characterize the metabolic and behavioral implications of hypothalamic BDNF overexpression on normal aging (McMurphy et al., 2019). Female C56BL/6 mice, 12-months of age, were randomized to receive autoregulatory BDNF vector or YFP vector and were monitored for 7 months. YFP-treated mice gradually gained weight. In contrast, BDNF gene transfer completely prevented aging-related weight gain, and stable body weight was maintained throughout the 7-month duration of the study. Food intake was decreased proportionally to body weight in BDNF-treated mice, suggesting that hypophagia is unlikely to be the primary cause of weight loss. At 13-weeks post-surgery, BDNF-treated mice performed better in a glucose tolerance test. Oxygen consumption in BDNF-treated mice was significantly increased in both dark and light phases measured by indirect calorimetry. Interestingly, physical activity was significantly decreased in BDNF-treated mice while food intake was no different within the metabolic chambers. The higher oxygen consumption concurrent with lower physical activity indicates that BDNF overexpression elevates the resting metabolic rate.

Hypothalamic BDNF gene transfer resulted in a lean phenotype. Significant and large reductions in absolute and relative mass were observed across all subcutaneous and visceral fat depots in the BDNF-treated mice. The leanness of BDNF mice was associated with a serum biomarker profile characterized by a sharp drop of leptin and insulin and a rise of adiponectin and IGF-1.

BDNF treatment led to profound adipose remodeling consistent with the activation of the HSA axis. BAT dissipates energy via releasing chemical energy from mitochondria in the form of heat. This process is primarily mediated by uncoupling protein-1 (UCP1), which is a specific BAT marker (Enerback et al., 1997). BAT plays important roles in thermogenesis and energy homeostasis and its activity declines with aging (Enerback, 2010). In this study, the BATs of 19-month old YFP mice were visibly pale whereas the BAT in BDNF mice appeared darker. H&E staining revealed the BAT of BDNF mice maintained typical morphology of young mice and was devoid of white adipocyte infiltration often associated with aging. The morphological changes were associated with robust regulation of BAT gene expression, featuring upregulation of thermogenic markers *Ucp1* and *Ppargc1a* (encoding PGC-1α). These data indicate the preservation of proper BAT functions against aging-related loss.

BDNF treatment induced a gene expression signature in WAT depots similar to that observed in mice after long-term EE, including upregulation of *Adrb3*, *Adipoq* (encoding adiponectin), *Glut4* (encoding glucose transporter 4), *Insr* (encoding insulin receptor), *Hsl* (encoding hormone-sensitive lipase), *Srebp1c* (encoding sterol regulatory element binding protein 1c) and *Pparg* (encoding peroxisome proliferator-activated receptor γ). *Fh1* (encoding mitochondrial fumarate hydratase) and *Parp1* (encoding poly ADP-ribose polymerase 1) are associated with caloric restriction-induced metabolic adaption (Mitchell et al., 2016); both were upregulated in BDNF-treated mice. Leptin and inflammation-modulatory genes (*Il1b*, *Il6*, and *Saa3*) were downregulated. Moreover, clusters of beige cells were observed in the retroperitoneal WAT, consistent with the induction of beige cell gene expression program.

Aging is associated with hepatic steatosis (Sheedfar, Di Biase, Koonen, & Vinciguerra, 2013). Lipid accumulation was evident in the livers of YFP mice at 19 months of age. In contrast, BDNF treatment diminished liver steatosis and decreased the hepatic triglyceride levels by approximately 80% compared to YFP mice. The major transcription factors that regulate de novo lipogenesis enzymes, *Chrebp1a* (encoding carbohydrate-responsive element-binding protein 1a) and *Srebp1c*, were significantly downregulated in the liver of BDNF mice, consistent with reduced steatosis (Horton, Goldstein, & Brown, 2002; Iizuka, Bruick, Liang, Horton, & Uyeda, 2004).

In addition to metabolic adaptations, a battery of behavioral assessments was performed 15–17 weeks post rAAV injection. BDNF treatment reduced anxiety-like behaviors as measured during cold-induced defecation and novelty suppressed feeding tests. The treatment additionally decreased depression-like behaviors as measured during tail suspension and forced swimming tests. Together, these data suggest long-term hypothalamic BDNF gene transfer does not elicit adverse behavioral changes.

Sustained hypothalamic BDNF overexpression was confirmed at the end of the study, 193 days after AAV injection. Of note, inflammation-modulatory genes were collectively downregulated in BDNF overexpressing hypothalamus including *Il1b*, *Il6*, *Il2rg* (encoding interkeukin-2 receptor γ), *Ly6d* (encoding lymphocyte antigen 6 family member D), *Cxcl10* (encoding C-X-C motif chemokine 10), and *Nfkbia* (encoding NFκB inhibitor α). These changes were not observed in the amygdala or hippocampus.

5. Conclusions and future directions

Our data demonstrate that hypothalamic gene transfer of BDNF prevents aging-associated metabolic decline including weight gain, loss of BAT function, hepatic steatosis, and impaired glucose tolerance. The physiological autoregulatory BDNF expression vector achieves a sustainable plateau of weight loss and leanness in aged mice maintained on a normal diet (McMurphy et al., 2019) (Fig. 2). Hypothalamic overexpression of BDNF largely reproduces the metabolic benefits of EE, suggesting a brain-fat axis—the HSA axis—as one pathway mediating several of the anti-aging effects of a socially, physically, cognitively active lifestyle (McMurphy et al., 2018). These results also raise more questions to tackle in future investigations.

Chronic inflammation is implicated in the etiology of disease across the lifespan (Furman et al., 2019). A conceptual model has been proposed in which hypothalamic microinflammation is a common basis of metabolic syndrome and aging (Tang, Purkayastha, & Cai, 2015). Chronic over-nutrition induces inflammation-like changes in the hypothalamus (including TNF-α and interleukins) and interrupts the central regulation of energy balance, glucose homeostasis and core features of metabolic syndromes (De Souza et al., 2005; Kleinridders et al., 2009; Purkayastha, Zhang, & Cai, 2011; Thaler et al., 2012; Zhang et al., 2008). Notably, low-grade

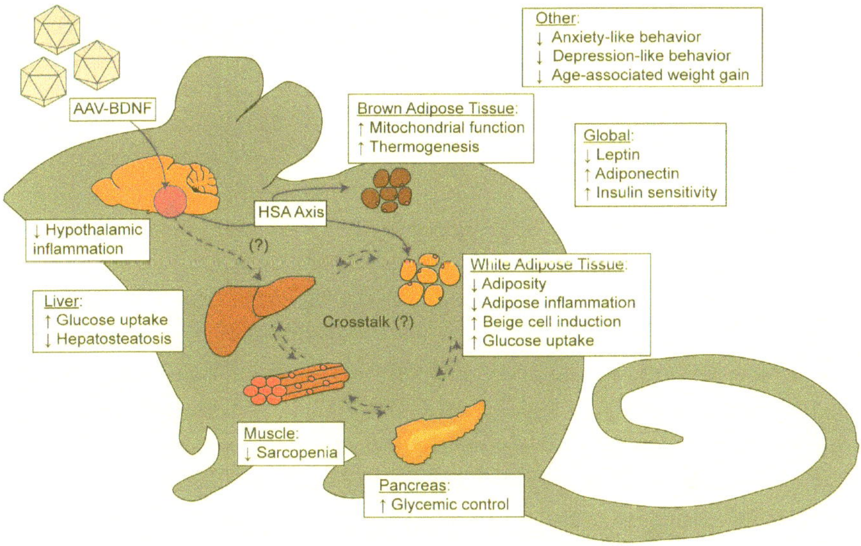

Fig. 2 Hypothalamic BDNF gene transfer promotes healthy aging.

inflammation is also a hallmark of aging, and systemic inflammation is negatively correlated with human longevity (Franceschi et al., 2005; Harris et al., 1999; Khabour & Barnawi, 2010). Recent studies have demonstrated that hypothalamic microinflammation promotes systemic aging (Tang & Cai, 2013; Zhang et al., 2013). We observed a suppression of NFκB pathway genes following hypothalamic BDNF treatment in young obese mice and in aged mice of normal weight. Whether this effect contributes to the observed systemic metabolic modulations requires further investigation. Of note, the collective downregulation of the inflammation-modulatory genes was observed specifically in the hypothalamus but not in the amygdala or hippocampus of the same mouse, suggesting the suppression is likely downstream of BDNF signaling instead of feedback from systemic metabolic improvement. AAV1 vectors predominantly transduce neurons (Wang, Wang, Clark, & Sferra, 2003) and therefore BDNF was primarily overexpressed in neurons in this study (McMurphy et al., 2019). However, the current study is unable to distinguish between the autocrine and paracrine effects of BDNF, since BDNF can be secreted from transduced neurons. We are currently using AAV1 to deliver a dominant-negative form of the TrkB receptor (Cao et al., 2011) to the hypothalamus of aged mice. Antagonizing BDNF signaling specifically in the hypothalamic neurons will identify the roles of neuronal BDNF signaling in hypothalamic microinflammation and systemic aging.

Microglial cells are the resident immune cells of the CNS. In their "quiescent" state, these cells continuously survey their territory using thin processes with multiple branches. Microglia can be activated and undergo phenotypic and morphological changes when homeostatic disturbance is detected. Microglia can produce proinflammatory mediators, are capable of phagocytosis, and present antigens (Guillemot-Legris & Muccioli, 2017; Jha, Lee, & Suk, 2016). Recent studies have demonstrated the role of microglia in orchestrating the hypothalamic inflammatory response to dietary excess and mediating obesity susceptibility (Aguzzi, Barres, & Bennett, 2013; Valdearcos et al., 2014, 2017). Notably, aging can trigger dysfunction of microglia (Perry, Nicoll, & Holmes, 2010; Streit, Xue, Tischer, & Bechmann, 2014; Zhang et al., 2013). We recently report that long-term EE (initiated at 10-months of age for 7.5 months) results in morphological changes of microglia characterized by hypertrophy and ramification without increases in microglial cell density (Ali et al., 2019). Moreover, we show that the CSF1R inhibitor PLX5622 (PLX)—which depletes microglia—in combination with EE substantially improves metabolic outcomes in middle-aged mice over PLX or EE alone. Chronic PLX treatment depleted 75% of microglia from the hypothalamus and reduced inflammatory markers without affecting the upregulation of BDNF by EE (Ali et al., 2020). These data intrigue us to test the hypothesis that EE modulates microglia function via hypothalamic BDNF signaling and subsequently contributing to the metabolic adaptations in aging.

Hypothalamic BDNF gene delivery leads to robust remodeling in both BAT and WAT of aged mice. Shared features across all adipose depots include suppression of leptin, stimulation of adiponectin, and upregulation of insulin receptor and mitochondrial genes. Interestingly, gene expression profiling shows upregulation of lipolytic and lipogenic genes as well as genes involved in lipid flux and fatty acid oxidation in BDNF-treated mice (McMurphy et al., 2019). It will be interesting to investigate whether hypothalamic BDNF modulates aging-associated adipose senescence and functional decline.

BDNF gene transfer improves glucose tolerance in aged mice on a normal diet. Our previous studies show improvement in glycemic control in obesity and diabetes models but not in young mice of normal weight. This improved glycemic control is associated with decreased adiposity, an improved adipokine profile (lower leptin, higher adiponectin), and alleviation of hepatic steatosis in aged mice. Here, adipose remodeling may contribute to the improved glucose tolerance directly by elevation of adipose glucose uptake or indirectly by crosstalk to the liver. Of note, the drastic shrinkage of adipose tissue and hypoleptinemia in BDNF-treated mice were not associated with lipodystrophy. On the contrary, the hepatic triglyceride level was decreased by 5-fold, and the transcription factors critical for de novo lipogenesis were decreased in the livers of BDNF-treated mice (McMurphy et al., 2019). Meek and colleagues report that infusion of BDNF to either the lateral cerebral ventricle or the VMH in a model of uncontrolled insulin-deficient diabetes attenuates diabetic hyperglycemia via an insulin-independent inhibition of hepatic glucose production (Meek et al., 2013). Future studies will elucidate whether a hypothalamic BDNF-liver axis directly modulates liver glucose and lipid metabolism, thereby contributing to glycemic control in aged animals. Moreover, how hypothalamic BDNF might affect the complex crosstalk among metabolic tissues such as muscle, pancreas, adipose tissue, and liver warrants investigation (Fig. 2).

With regard to the effects on behavior, most studies link a reduction in BDNF with anxiety (Rios et al., 2001), impaired spatial learning and discrimination (Gorski, Balogh, Wehner, & Jones, 2003), and depression (Karege et al., 2005; Shimizu et al., 2003). However, far less evidence is available regarding the role of hypothalamic BDNF in emotionality. To our knowledge, our study is the first assessing hypothalamic BDNF-induced behavioral adaptations in aging, characterized by reduced anxiety- and depression-like behaviors. In this study (McMurphy et al., 2019), the BDNF transgene was expressed mainly in the ARC and VMH nuclei of the hypothalamus, which could increase the BDNF protein level in the adjacent dorsomedial hypothalamus (DMH). The DMH is a brain region not only involved in physiological functions such as metabolism and stress responses to environmental threats, but is also critically involved in behavioral regulation, particularly fear, anxiety and panic-like disorders (Shekhar, Sims, & Bowsher, 1993; Silva et al., 2014). Additionally, we have reported that hypothalamic BDNF modulates the hypothalamic-pituitary-adrenal (HPA) axis mediating the EE's regulation of T cell immunity in young mice (Xiao et al., 2016, 2019). The body's stress response system becomes less agile and dysfunctional with advanced age. It will be interesting to investigate whether hypothalamic BDNF's modulation of the HPA axis contributes to the reduction of depression-like behaviors in aged animals. Another possibility is that hypothalamic BDNF overexpression leads to global improvement of metabolism and in turn indirectly influences other limbic structures, such as the prefrontal cortex, hippocampus, nucleus accumbens, ventral striatum, and amygdala (Russo, Murrough, Han, Charney, & Nestler, 2012), and thereby affects brain functions and behaviors (de Noronha et al., 2016).

More recently, we have expanded research from investigating the metabolic role of EE to characterizing EE-induced immune phenotypes. EE reduces the CD4/CD8 T cell ratio in the spleen and lymph node and enhances tumor-infiltrating CD8 T cells, contributing to the EE-induced anticancer effect via both the sympathetic nervous system (SNS) and the HPA axis (Xiao et al., 2016). Additionally, EE regulates thymocyte maturation and selection, CD4/CD8-lineage choice, and T cell export; such regulation was found to be mediated by the HPA axis and be dependent on thymocyte glucocorticoid receptor (GR). In a model of multiple sclerosis—experimental autoimmune encephalomyelitis (EAE)—EE alleviates symptoms, inhibits spinal cord inflammation through regulation of type 1 T-helper cells, and prevents EAE-induced thymic disturbance (Xiao et al., 2019). Moreover, we have observed that EE enhances NK cell maturation in the spleen and promotes NK cell reconstitution after NK cell depletion (manuscript in revision at Journal of Immunology). Interestingly, our mechanistic studies have shown that these diverse immunomodulatory effects of EE are mediated by hypothalamic BDNF in young animals (Fig. 3). The aging process has been shown by many to influence immune function (Cai & Khor, 2019; Nikolich-Žugich, 2018; Pereira & Akbar, 2016; Weiskopf, Weinberger, & Grubeck-Loebenstein, 2009). However, how EE affects the aging-immune interplay is understudied, and the role of hypothalamic BDNF in this process is unknown—and worthy of attention.

Although the current data indicate benefits of hypothalamic BDNF gene transfer in middle-age mice, we are mindful of potential adverse effects. While hypothalamic BDNF gene transfer clearly modulates adipose tissue, it is important to note that fat stores help organisms sustain metabolic function in times of stress. For the elderly, a large reduction of these stores might be contraindicated. For example, in contrast to the increased risks in younger patients—and perhaps contrary to intuition—morbidity and mortality are reduced for older adults with overweight and mildly obese BMIs (Chang, Beason, Hunleth, & Colditz, 2012; Hainer & Aldhoon-Hainerova, 2013). While the mechanisms are still under investigation, most recent studies cite the BMI measure itself as misleading; this is especially true in older patients, for whom adiposity, body fat distribution, and preservation of lean mass are better predictors of survival (Abramowitz et al., 2018; Chang et al., 2012; Lee et al., 2018). Similarly, how hypothalamic BDNF gene transfer affects lean mass remains to be investigated. As such, it is possible that sustained

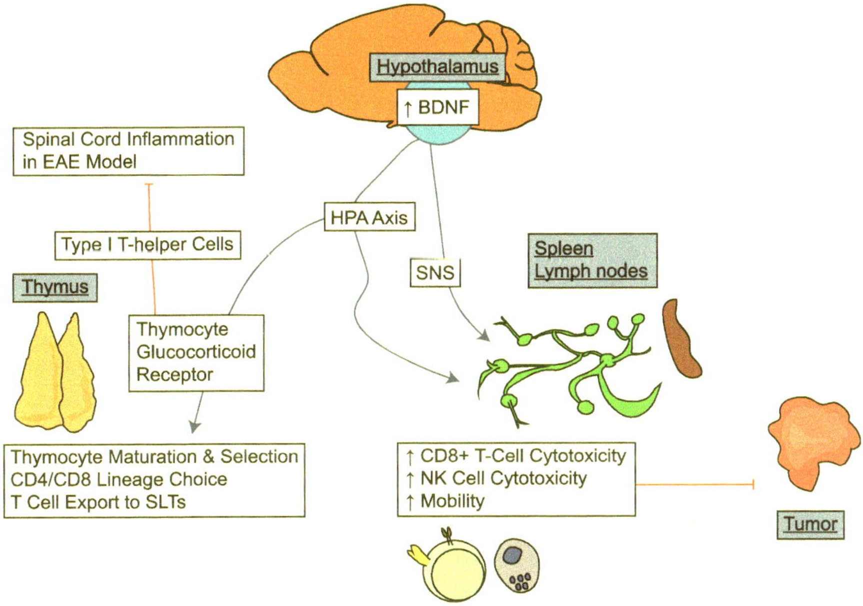

Fig. 3 Hypothalamic BDNF modulates immune functions in young animals.

low adiposity and elevated basal metabolic rate resulting from BDNF overexpression might not be beneficial at old age. Our data demonstrate that implementing EE in 18-month old mice positively affects mean lifespan (McMurphy et al., 2018). Thus, it will be interesting to identify mechanisms independent of hypothalamic BDNF with regard to the EE-induced anti-aging benefits.

Acknowledgments

We thank past and present lab members for their data and discussions. This work was supported by NIH grants CA163640, CA166590, and AG041250 to L. Cao.

Declaration of interests

L.C. is an inventor of US patent 9,265,843 B2 on the autoregulatory BDNF vector. All the other authors declare no conflict of financial interests.

References

Abramowitz, M. K., Hall, C. B., Amodu, A., Sharma, D., Androga, L., & Hawkins, M. (2018). Muscle mass, BMI, and mortality among adults in the United States: A population-based cohort study. *PLoS One*, *13*(4), e0194697. https://doi.org/10.1371/journal.pone.0194697.

Aguzzi, A., Barres, B. A., & Bennett, M. L. (2013). Microglia: Scapegoat, saboteur, or something else? *Science*, *339*(6116), 156–161. https://doi.org/10.1126/science.1227901.

Ali, S., Liu, X., Queen, N. J., Patel, R. S., Wilkins, R. K., Mo, X., et al. (2019). Long-term environmental enrichment affects microglial morphology in middle age mice. *Aging (Albany NY), 11*(8), 2388–2402. https://doi.org/10.18632/aging.101923.

Ali, S., Mansour, A. G., Huang, W., Queen, N. J., Mo, X., Anderson, J. M., et al. (2020). CSF1R inhibitor PLX5622 and environmental enrichment additively improve metabolic outcomes in middle-aged female mice. *Aging (Albany NY), 12*, 2101–2122. https://doi.org/10.18632/aging.102724.

Aronoff, E., Hillyer, R., & Leon, M. (2016). Environmental enrichment therapy for autism: Outcomes with increased access. *Neural Plasticity, 2016*, 2734915. https://doi.org/10.1155/2016/2734915.

Attlee, A., Kassem, H., Hashim, M., & Obaid, R. S. (2015). Physical status and feeding behavior of children with autism. *Indian Journal of Pediatrics, 82*(8), 682–687. https://doi.org/10.1007/s12098-015-1696-4.

Austin, J., & Marks, D. (2009). Hormonal regulators of appetite. *International Journal of Pediatric Endocrinology, 2009*, 141753. https://doi.org/10.1155/2009/141753.

Barbera, M. J., Schluter, A., Pedraza, N., Iglesias, R., Villarroya, F., & Giralt, M. (2001). Peroxisome proliferator-activated receptor alpha activates transcription of the brown fat uncoupling protein-1 gene. A link between regulation of the thermogenic and lipid oxidation pathways in the brown fat cell. *The Journal of Biological Chemistry, 276*(2), 1486–1493. https://doi.org/10.1074/jbc.M006246200.

Bektas, A., Schurman, S. H., Sen, R., & Ferrucci, L. (2018). Aging, inflammation and the environment. *Experimental Gerontology, 105*, 10–18. https://doi.org/10.1016/j.exger.2017.12.015.

Birch, A. M., McGarry, N. B., & Kelly, A. M. (2013). Short-term environmental enrichment, in the absence of exercise, improves memory, and increases NGF concentration, early neuronal survival, and synaptogenesis in the dentate gyrus in a time-dependent manner. *Hippocampus, 23*(6), 437–450. https://doi.org/10.1002/hipo.22103.

Burbridge, S., Stewart, I., & Placzek, M. (2016). Development of the neuroendocrine hypothalamus. *Comprehensive Physiology, 6*(2), 623–643. https://doi.org/10.1002/cphy.c150023.

Cai, D., & Khor, S. (2019). "Hypothalamic microinflammation" paradigm in aging and metabolic diseases. *Cell Metabolism, 30*(1), 19–35. https://doi.org/10.1016/j.cmet.2019.05.021.

Cao, L., Choi, E. Y., Liu, X., Martin, A., Wang, C., Xu, X., et al. (2011). White to brown fat phenotypic switch induced by genetic and environmental activation of a hypothalamic-adipocyte axis. *Cell Metabolism, 14*(3), 324–338. https://doi.org/10.1016/j.cmet.2011.06.020.

Cao, L., & During, M. J. (2012). What is the brain-cancer connection? *Annual Review of Neuroscience, 35*, 331–345. https://doi.org/10.1146/annurev-neuro-062111-150546.

Cao, L., Lin, E. J., Cahill, M. C., Wang, C., Liu, X., & During, M. J. (2009). Molecular therapy of obesity and diabetes by a physiological autoregulatory approach. *Nature Medicine, 15*(4), 447–454. https://doi.org/10.1038/nm.1933.

Cao, L., Liu, X., Lin, E. J., Wang, C., Choi, E. Y., Riban, V., et al. (2010). Environmental and genetic activation of a brain-adipocyte BDNF/leptin axis causes cancer remission and inhibition. *Cell, 142*(1), 52–64. https://doi.org/10.1016/j.cell.2010.05.029.

Chang, S. H., Beason, T. S., Hunleth, J. M., & Colditz, G. A. (2012). A systematic review of body fat distribution and mortality in older people. *Maturitas, 72*(3), 175–191. https://doi.org/10.1016/j.maturitas.2012.04.004.

Chang, A. Y., Skirbekk, V. F., Tyrovolas, S., Kassebaum, N. J., & Dieleman, J. L. (2019). Measuring population ageing: an analysis of the Global Burden of Disease Study 2017. *The Lancet Public Health, 4*(3), e159–e167. https://doi.org/10.1016/S2468-2667(19)30019-2.

Choe, S. S., Huh, J. Y., Hwang, I. J., Kim, J. I., & Kim, J. B. (2016). Adipose tissue remodeling: Its role in energy metabolism and metabolic disorders. *Frontiers in Endocrinology (Lausanne)*, 7, 30. https://doi.org/10.3389/fendo.2016.00030.

Colby, S. L., & Ortman, J. M. (2014). Projections of the size and composition of the U.S. population: 2014 to 2060. In *Current Population Reports* United States Census Bureau.

Crimmins, E. M. (2015). Lifespan and healthspan: past, present, and promise. *Gerontologist, 55*(6), 901–911. https://doi.org/10.1093/geront/gnv130.

Crimmins, E. M., & Beltran-Sanchez, H. (2011). Mortality and morbidity trends: is there compression of morbidity? *The Journals of Gerontology. Series B, Psychological Sciences and Social Sciences, 66*(1), 75–86. https://doi.org/10.1093/geronb/gbq088.

de Magalhaes, J. P., Wuttke, D., Wood, S. H., Plank, M., & Vora, C. (2012). Genome-environment interactions that modulate aging: powerful targets for drug discovery. *Pharmacological Reviews, 64*(1), 88–101. https://doi.org/10.1124/pr.110.004499.

de Noronha, S. R., Campos, G. V., Abreu, A. R., de Souza, A. A., Chianca, D. A., Jr., & de Menezes, R. C. (2016). High fat diet induced-obesity facilitates anxiety-like behaviors due to GABAergic impairment within the dorsomedial hypothalamus in rats. *Behavioural Brain Research, 316*, 38–46. https://doi.org/10.1016/j.bbr.2016.08.042.

De Souza, C. T., Araujo, E. P., Bordin, S., Ashimine, R., Zollner, R. L., Boschero, A. C., et al. (2005). Consumption of a fat-rich diet activates a proinflammatory response and induces insulin resistance in the hypothalamus. *Endocrinology, 146*(10), 4192–4199. https://doi.org/10.1210/en.2004-1520.

Downs, J., Rodger, J., Li, C., Tan, X., Hu, N., Wong, K., et al. (2018). Environmental enrichment intervention for Rett syndrome: an individually randomised stepped wedge trial. *Orphanet Journal of Rare Diseases, 13*(1), 3. https://doi.org/10.1186/s13023-017-0752-8.

Duan, W., Guo, Z., Jiang, H., Ware, M., Li, X. J., & Mattson, M. P. (2003). Dietary restriction normalizes glucose metabolism and BDNF levels, slows disease progression, and increases survival in huntingtin mutant mice. *Proceedings of the National Academy of Sciences of the United States of America, 100*(5), 2911–2916. https://doi.org/10.1073/pnas.0536856100 0536856100.

During, M. J., Liu, X., Huang, W., Magee, D., Slater, A., McMurphy, T., et al. (2015). Adipose VEGF links the white-to-brown fat switch with environmental, genetic, and pharmacological stimuli in male mice. *Endocrinology, 156*(6), 2059–2073. https://doi.org/10.1210/en.2014-1905.

Ellulu, M. S., Patimah, I., Khaza'ai, H., Rahmat, A., & Abed, Y. (2017). Obesity and inflammation: the linking mechanism and the complications. *Archives of Medical Science, 13*(4), 851–863. https://doi.org/10.5114/aoms.2016.58928.

Enerback, S. (2010). Human brown adipose tissue. *Cell Metabolism, 11*(4), 248–252. https://doi.org/10.1016/j.cmet.2010.03.008.

Enerback, S., Jacobsson, A., Simpson, E. M., Guerra, C., Yamashita, H., Harper, M. E., et al. (1997). Mice lacking mitochondrial uncoupling protein are cold-sensitive but not obese. *Nature, 387*(6628), 90–94. https://doi.org/10.1038/387090a0.

Farooqi, I. S., Keogh, J. M., Yeo, G. S., Lank, E. J., Cheetham, T., & O'Rahilly, S. (2003). Clinical spectrum of obesity and mutations in the melanocortin 4 receptor gene. *The New England Journal of Medicine, 348*(12), 1085–1095. https://doi.org/10.1056/NEJMoa022050348/12/1085.

Farooqi, S., & O'Rahilly, S. (2006). Genetics of obesity in humans. *Endocrine Reviews, 27*(7), 710–718. doi:er.2006-0040 [pii]. https://doi.org/10.1210/er.2006-0040.

Ferrucci, L., & Fabbri, E. (2018). Inflammageing: chronic inflammation in ageing, cardiovascular disease, and frailty. *Nature Reviews. Cardiology, 15*(9), 505–522. https://doi.org/10.1038/s41569-018-0064-2.

Ferrucci, L., Giallauria, F., & Guralnik, J. M. (2008). Epidemiology of aging. *Radiologic Clinics of North America, 46*(4), 643–652. v https://doi.org/10.1016/j.rcl.2008.07.005.
Finkbeiner, S. (2000). Calcium regulation of the brain-derived neurotrophic factor gene. *Cellular and Molecular Life Sciences, 57*(3), 394–401. https://doi.org/10.1007/PL00000701.
Foglesong, G. D., Huang, W., Liu, X., Slater, A. M., Siu, J., Yildiz, V., et al. (2016). Role of hypothalamic VGF in energy balance and metabolic adaption to environmental enrichment in mice. *Endocrinology, 157*(3), 983–996. https://doi.org/10.1210/en.2015-1627.
Foglesong, G. D., Queen, N. J., Huang, W., Widstrom, K. J., & Cao, L. (2019). Enriched environment inhibits breast cancer progression in obese models with intact leptin signaling. *Endocrine-Related Cancer, 26*(5), 483–495. https://doi.org/10.1530/ERC-19-0075.
Franceschi, C., Olivieri, F., Marchegiani, F., Cardelli, M., Cavallone, L., Capri, M., et al. (2005). Genes involved in immune response/inflammation, IGF1/insulin pathway and response to oxidative stress play a major role in the genetics of human longevity: the lesson of centenarians. *Mechanisms of Ageing and Development, 126*(2), 351–361. https://doi.org/10.1016/j.mad.2004.08.028.
Frederich, R. C., Hamann, A., Anderson, S., Lollmann, B., Lowell, B. B., & Flier, J. S. (1995). Leptin levels reflect body lipid content in mice: evidence for diet-induced resistance to leptin action. *Nature Medicine, 1*(12), 1311–1314. https://doi.org/10.1038/nm1295-1311.
Freedman, V. A., Crimmins, E., Schoeni, R. F., Spillman, B. C., Aykan, H., Kramarow, E., et al. (2004). Resolving inconsistencies in trends in old-age disability: report from a technical working group. *Demography, 41*(3), 417–441. https://doi.org/10.1353/dem.2004.0022.
Fries, J. F. (1980). Aging, natural death, and the compression of morbidity. *The New England Journal of Medicine, 303*(3), 130–135. https://doi.org/10.1056/NEJM198007173030304.
Furman, D., Campisi, J., Verdin, E., Carrera-Bastos, P., Targ, S., Franceschi, C., et al. (2019). Chronic inflammation in the etiology of disease across the life span. *Nature Medicine, 25*(12), 1822–1832. https://doi.org/10.1038/s41591-019-0675-0.
Gaspar, J. M., & Velloso, L. A. (2018). Hypoxia inducible factor as a central regulator of metabolism—Implications for the development of obesity. *Frontiers in Neuroscience, 12*, 813. https://doi.org/10.3389/fnins.2018.00813.
Gore, P. G., Kingston, A., Johnson, G. R., Kirkwood, T. B. L., & Jagger, C. (2018). New horizons in the compression of functional decline. *Age and Ageing, 47*(6), 764–768. https://doi.org/10.1093/ageing/afy145.
Gorski, J. A., Balogh, S. A., Wehner, J. M., & Jones, K. R. (2003). Learning deficits in forebrain-restricted brain-derived neurotrophic factor mutant mice. *Neuroscience, 121*(2), 341–354. Retrieved from https://www.ncbi.nlm.nih.gov/pubmed/14521993.
Gray, J., Yeo, G. S., Cox, J. J., Morton, J., Adlam, A. L., Keogh, J. M., et al. (2006). Hyperphagia, severe obesity, impaired cognitive function, and hyperactivity associated with functional loss of one copy of the brain-derived neurotrophic factor (BDNF) gene. *Diabetes, 55*(12), 3366–3371. https://doi.org/10.2337/db06-0550.
Guillemot-Legris, O., & Muccioli, G. G. (2017). Obesity-induced neuroinflammation: Beyond the hypothalamus. *Trends in Neurosciences, 40*(4), 237–253. https://doi.org/10.1016/j.tins.2017.02.005.
Hainer, V., & Aldhoon-Hainerova, I. (2013). Obesity paradox does exist. *Diabetes Care, 36*(Suppl 2), S276–S281. https://doi.org/10.2337/dcS13-2023.
Hales, C., Carroll, M., Fryar, C., & Ogden, C. (2020). Prevalence of obesity and severe obesity among adults: United States, 2017–2018. In *NCHS Data Brief, no 360: Centers for Disease Control and Prevention*, National Center for Health Statistics.
Han, J. C., Liu, Q. R., Jones, M., Levinn, R. L., Menzie, C. M., Jefferson-George, K. S., et al. (2008). Brain-derived neurotrophic factor and obesity in the WAGR syndrome. *The New England Journal of Medicine, 359*(9), 918–927. https://doi.org/10.1056/NEJMoa0801119.

Han, J. C., Muehlbauer, M. J., Cui, H. N., Newgard, C. B., & Haqq, A. M. (2010). Lower brain-derived neurotrophic factor in patients with prader-willi syndrome compared to obese and lean control subjects. *The Journal of Clinical Endocrinology and Metabolism*, *95*(7), 3532–3536. https://doi.org/10.1210/jc.2010-0127.

Harris, T. B., Ferrucci, L., Tracy, R. P., Corti, M. C., Wacholder, S., Ettinger, W. H., Jr., et al. (1999). Associations of elevated interleukin-6 and C-reactive protein levels with mortality in the elderly. *The American Journal of Medicine*, *106*(5), 506–512. Retrieved from https://www.ncbi.nlm.nih.gov/pubmed/10335721.

He, W., Goodkind, D., & Kowal, P. (2016). *An aging world: 2015. international population reports*. U.S. Census Bureau.

Hill-Yardin, E. L., & Hannan, A. J. (2013). Translating preclinical environmental enrichment studies for the treatment of autism and other brain disorders: Comment on Woo and Leon (2013). *Behavioral Neuroscience*, *127*(4), 606–609. https://doi.org/10.1037/a0033319.

Hohensinner, P. J., Kaun, C., Ebenbauer, B., Hackl, M., Demyanets, S., Richter, D., et al. (2018). Reduction of premature aging markers after gastric bypass surgery in morbidly obese patients. *Obesity Surgery*, *28*(9), 2804–2810. https://doi.org/10.1007/s11695-018-3247-3.

Holt, L. M., Hernandez, R. D., Pacheco, N. L., Torres Ceja, B., Hossain, M., & Olsen, M. L. (2019). Astrocyte morphogenesis is dependent on BDNF signaling via astrocytic TrkB. T1. *eLife*, *8*, e44667. https://doi.org/10.7554/eLife.44667.

Horton, J. D., Goldstein, J. L., & Brown, M. S. (2002). SREBPs: activators of the complete program of cholesterol and fatty acid synthesis in the liver. *The Journal of Clinical Investigation*, *109*(9), 1125–1131. https://doi.org/10.1172/JCI15593.

Hruby, A., & Hu, F. B. (2015). The epidemiology of obesity: A big picture. *PharmacoEconomics*, *33*(7), 673–689. https://doi.org/10.1007/s40273-014-0243-x.

Ibrahim, M. M. (2010). Subcutaneous and visceral adipose tissue: Structural and functional differences. *Obesity Reviews*, *11*(1), 11–18. https://doi.org/10.1111/j.1467-789X.2009.00623.x.

Iizuka, K., Bruick, R. K., Liang, G., Horton, J. D., & Uyeda, K. (2004). Deficiency of carbohydrate response element-binding protein (ChREBP) reduces lipogenesis as well as glycolysis. *Proceedings of the National Academy of Sciences of the United States of America*, *101*(19), 7281–7286. https://doi.org/10.1073/pnas.0401516101.

Jha, M. K., Lee, W. H., & Suk, K. (2016). Functional polarization of neuroglia: Implications in neuroinflammation and neurological disorders. *Biochemical Pharmacology*, *103*, 1–16. https://doi.org/10.1016/j.bcp.2015.11.003.

Jiao, H., Arner, P., Hoffstedt, J., Brodin, D., Dubern, B., Czernichow, S., et al. (2011). Genome wide association study identifies KCNMA1 contributing to human obesity. *BMC Medical Genomics*, *4*, 51. https://doi.org/10.1186/1755-8794-4-51.

Karege, F., Bondolfi, G., Gervasoni, N., Schwald, M., Aubry, J. M., & Bertschy, G. (2005). Low brain-derived neurotrophic factor (BDNF) levels in serum of depressed patients probably results from lowered platelet BDNF release unrelated to platelet reactivity. *Biological Psychiatry*, *57*(9), 1068–1072. https://doi.org/10.1016/j.biopsych.2005.01.008.

Kennedy, B. K., Berger, S. L., Brunet, A., Campisi, J., Cuervo, A. M., Epel, E. S., et al. (2014). Geroscience: Linking aging to chronic disease. *Cell*, *159*(4), 709–713. https://doi.org/10.1016/j.cell.2014.10.039.

Khabour, O. F., & Barnawi, J. M. (2010). Association of longevity with IL-10 -1082 G/A and TNF-alpha-308 G/A polymorphisms. *International Journal of Immunogenetics*, *37*(4), 293–298. https://doi.org/10.1111/j.1744-313X.2010.00925.x.

Kissebah, A. H., & Krakower, G. R. (1994). Regional adiposity and morbidity. *Physiological Reviews*, *74*(4), 761–811. https://doi.org/10.1152/physrev.1994.74.4.761.

Kleinridders, A., Schenten, D., Konner, A. C., Belgardt, B. F., Mauer, J., Okamura, T., et al. (2009). MyD88 signaling in the CNS is required for development of fatty acid-induced leptin resistance and diet-induced obesity. *Cell Metabolism, 10*(4), 249–259. https://doi.org/10.1016/j.cmet.2009.08.013.

Kuk, J. L., Saunders, T. J., Davidson, L. E., & Ross, R. (2009). Age-related changes in total and regional fat distribution. *Ageing Research Reviews, 8*(4), 339–348. https://doi.org/10.1016/j.arr.2009.06.001.

Lee, J., Duan, W., & Mattson, M. P. (2002). Evidence that brain-derived neurotrophic factor is required for basal neurogenesis and mediates, in part, the enhancement of neurogenesis by dietary restriction in the hippocampus of adult mice. *Journal of Neurochemistry, 82*(6), 1367–1375. (doi:1085 [pii].

Lee, D. H., Keum, N., Hu, F. B., Orav, E. J., Rimm, E. B., Willett, W. C., et al. (2018). Predicted lean body mass, fat mass, and all cause and cause specific mortality in men: Prospective US cohort study. *BMJ, 362*, k2575. https://doi.org/10.1136/bmj.k2575.

Leschik, J., Eckenstaler, R., Endres, T., Munsch, T., Edelmann, E., Richter, K., et al. (2019). Prominent postsynaptic and dendritic exocytosis of endogenous BDNF vesicles in BDNF-GFP knock-in mice. *Molecular Neurobiology, 56*(10), 6833–6855. https://doi.org/10.1007/s12035-019-1551-0.

Lessmann, V., Gottmann, K., & Malcangio, M. (2003). Neurotrophin secretion: Current facts and future prospects. *Progress in Neurobiology, 69*(5), 341–374. https://doi.org/10.1016/s0301-0082(03)00019-4.

Liu, Q. R., Lu, L., Zhu, X. G., Gong, J. P., Shaham, Y., & Uhl, G. R. (2006). Rodent BDNF genes, novel promoters, novel splice variants, and regulation by cocaine. *Brain Research, 1067*(1), 1–12. https://doi.org/10.1016/j.brainres.2005.10.004.

Liu, X., McMurphy, T., Xiao, R., Slater, A., Huang, W., & Cao, L. (2014). Hypothalamic gene transfer of BDNF inhibits breast cancer progression and metastasis in middle age obese mice. *Molecular Therapy, 22*(7), 1275–1284. https://doi.org/10.1038/mt.2014.45.

Lopez-Otin, C., Blasco, M. A., Partridge, L., Serrano, M., & Kroemer, G. (2013). The hallmarks of aging. *Cell, 153*(6), 1194–1217. https://doi.org/10.1016/j.cell.2013.05.039.

Luo, F., Mu, Y., Gao, C., Xiao, Y., Zhou, Q., Yang, Y., et al. (2019). Whole-brain patterns of the presynaptic inputs and axonal projections of BDNF neurons in the paraventricular nucleus. *Journal of Genetics and Genomics, 46*(1), 31–40. https://doi.org/10.1016/j.jgg.2018.11.004.

Mather, M., Jacobsen, L. A., & Pollard, K. M. (2015). Aging in the United States. *Population Bulletin, 70*(2), 12.

Matsuzawa, Y. (2008). The role of fat topology in the risk of disease. *International Journal of Obesity, 32*(Suppl 7), S83–S92. https://doi.org/10.1038/ijo.2008.243.

Mattson, M. P., Maudsley, S., & Martin, B. (2004). A neural signaling triumvirate that influences ageing and age-related disease: insulin/IGF-1, BDNF and serotonin. *Ageing Research Reviews, 3*(4), 445–464. https://doi.org/10.1016/j.arr.2004.08.001.

McMurphy, T., Huang, W., Liu, X., Siu, J. J., Queen, N. J., Xiao, R., et al. (2019). Hypothalamic gene transfer of BDNF promotes healthy aging in mice. *Aging Cell, 18*(2), e12846. https://doi.org/10.1111/acel.12846.

McMurphy, T., Huang, W., Queen, N. J., Ali, S., Widstrom, K. J., Liu, X., et al. (2018). Implementation of environmental enrichment after middle age promotes healthy aging. *Aging (Albany NY), 10*(7), 1698–1721. https://doi.org/10.18632/aging.101502.

Meek, T. H., Wisse, B. E., Thaler, J. P., Guyenet, S. J., Matsen, M. E., Fischer, J. D., et al. (2013). BDNF action in the brain attenuates diabetic hyperglycemia via insulin-independent inhibition of hepatic glucose production. *Diabetes, 62*(5), 1512–1518. https://doi.org/10.2337/db12-0837.

Milsum, J. H. (1985). A model of the eustress system for health/illness. *Behavioral Science, 30*(4), 179–186. https://doi.org/10.1002/bs.3830300402.

Mingozzi, F., & High, K. A. (2011). Therapeutic in vivo gene transfer for genetic disease using AAV: progress and challenges. *Nature Reviews. Genetics, 12*(5), 341–355. https://doi.org/10.1038/nrg2988.

Mitchell, S. J., Madrigal-Matute, J., Scheibye-Knudsen, M., Fang, E., Aon, M., Gonzalez-Reyes, J. A., et al. (2016). Effects of sex, strain, and energy intake on hallmarks of aging in mice. *Cell Metabolism, 23*(6), 1093–1112. https://doi.org/10.1016/j.cmet.2016.05.027.

Mowla, S. J., Farhadi, H. F., Pareek, S., Atwal, J. K., Morris, S. J., Seidah, N. G., et al. (2001). Biosynthesis and post-translational processing of the precursor to brain-derived neurotrophic factor. *The Journal of Biological Chemistry, 276*(16), 12660–12666. https://doi.org/10.1074/jbc.M008104200.

National Population Totals and Components of Change: 2010-2019. (2019). United States Census Bureau. Retrieved from https://www.census.gov/data/tables/time-series/demo/popest/2010s-national-total.html.

Needham, S. L. (2014). Toward priorities for aging research. *Rejuvenation Research, 17*(2), 154–156. https://doi.org/10.1089/rej.2013.1508.

Nicholson, J. R., Peter, J. C., Lecourt, A. C., Barde, Y. A., & Hofbauer, K. G. (2007). Melanocortin-4 receptor activation stimulates hypothalamic brain-derived neurotrophic factor release to regulate food intake, body temperature and cardiovascular function. *Journal of Neuroendocrinology, 19*(12), 974–982. doi:JNE1610 [pii]. https://doi.org/10.1111/j.1365-2826.2007.01610.x.

Nikolich-Žugich, J. (2018). The twilight of immunity: Emerging concepts in aging of the immune system. *Nature Immunology, 19*(1), 10–19.

Olshansky, S. J. (2018). From Lifespan to Healthspan. *JAMA, 320*(13), 1323–1324. https://doi.org/10.1001/jama.2018.12621.

Palmer, A. K., & Kirkland, J. L. (2016). Aging and adipose tissue: Potential interventions for diabetes and regenerative medicine. *Experimental Gerontology, 86*, 97–105. https://doi.org/10.1016/j.exger.2016.02.013.

Pereira, B. I., & Akbar, A. N. (2016). Convergence of innate and adaptive immunity during human aging. *Frontiers in Immunology, 7*, 445.

Perry, V. H., Nicoll, J. A., & Holmes, C. (2010). Microglia in neurodegenerative disease. *Nature Reviews. Neurology, 6*(4), 193–201. https://doi.org/10.1038/nrneurol.2010.17.

Pi-Sunyer, X. (2009). The medical risks of obesity. *Postgraduate Medicine, 121*(6), 21–33. https://doi.org/10.3810/pgm.2009.11.2074.

Prospective Studies, C., Whitlock, G., Lewington, S., Sherliker, P., Clarke, R., Emberson, J., et al. (2009). Body-mass index and cause-specific mortality in 900 000 adults: Collaborative analyses of 57 prospective studies. *Lancet, 373*(9669), 1083–1096. https://doi.org/10.1016/S0140-6736(09)60318-4.

Pruunsild, P., Kazantseva, A., Aid, T., Palm, K., & Timmusk, T. (2007). Dissecting the human BDNF locus: Bidirectional transcription, complex splicing, and multiple promoters. *Genomics, 90*(3), 397–406. https://doi.org/10.1016/j.ygeno.2007.05.004.

Puigserver, P., Wu, Z., Park, C. W., Graves, R., Wright, M., & Spiegelman, B. M. (1998). A cold-inducible coactivator of nuclear receptors linked to adaptive thermogenesis. *Cell, 92*(6), 829–839. https://doi.org/10.1016/s0092-8674(00)81410-5.

Purkayastha, S., Zhang, G., & Cai, D. (2011). Uncoupling the mechanisms of obesity and hypertension by targeting hypothalamic IKK-beta and NF-kappaB. *Nature Medicine, 17*(7), 883–887. https://doi.org/10.1038/nm.2372.

Queen, N. J., Boardman, A. A., Patel, R. S., Siu, J. J., Mo, X., & Cao, L. (2020). Environmental enrichment improves metabolic and behavioral health in the BTBR mouse model of autism. *Psychoneuroendocrinology, 111*, 104476. https://doi.org/10.1016/j.psyneuen.2019.104476.

Queen, N. J., Hassan, Q. N., II, & Cao, L. (2020). Improvements to healthspan through environmental enrichment and lifestyle interventions: Where are we now? *Frontiers in Neuroscience*, *14*, 605. https://doi.org/10.3389/fnins.2020.00605.

Rios, M., Fan, G., Fekete, C., Kelly, J., Bates, B., Kuehn, R., et al. (2001). Conditional deletion of brain-derived neurotrophic factor in the postnatal brain leads to obesity and hyperactivity. *Molecular Endocrinology*, *15*(10), 1748–1757. https://doi.org/10.1210/mend.15.10.0706.

Rodriguez-Hernandez, H., Simental-Mendia, L. E., Rodriguez-Ramirez, G., & Reyes-Romero, M. A. (2013). Obesity and inflammation: Epidemiology, risk factors, and markers of inflammation. *International Journal of Endocrinology*, *2013*, 678159. https://doi.org/10.1155/2013/678159.

Russo, S. J., Murrough, J. W., Han, M. H., Charney, D. S., & Nestler, E. J. (2012). Neurobiology of resilience. *Nature Neuroscience*, *15*(11), 1475–1484. https://doi.org/10.1038/nn.3234.

Salminen, A., Kaarniranta, K., & Kauppinen, A. (2012). Inflammaging: Disturbed interplay between autophagy and inflammasomes. *Aging (Albany NY)*, *4*(3), 166–175. https://doi.org/10.18632/aging.100444.

Sampedro-Piquero, P., & Begega, A. (2017). Environmental enrichment as a positive behavioral intervention across the lifespan. *Current Neuropharmacology*, *15*(4), 459–470. https://doi.org/10.2174/1570159X14666160325115909.

Sanchis-Gomar, F., Garcia-Gimenez, J. L., Perez-Quilis, C., Gomez-Cabrera, M. C., Pallardo, F. V., & Lippi, G. (2012). Physical exercise as an epigenetic modulator: Eustress, the "positive stress" as an effector of gene expression. *Journal of Strength and Conditioning Research*, *26*(12), 3469–3472. https://doi.org/10.1519/JSC.0b013e31825bb594.

Schoeni, R. F., Freedman, V. A., & Martin, L. G. (2008). Why is late-life disability declining? *The Milbank Quarterly*, *86*(1), 47–89. https://doi.org/10.1111/j.1468-0009.2007.00513.x.

Seeman, I. (2010). Data on disability. *American Journal of Public Health*, *100*(8), 1367. https://doi.org/10.2105/AJPH.2010.193276.

Sheedfar, F., Di Biase, S., Koonen, D., & Vinciguerra, M. (2013). Liver diseases and aging: Friends or foes? *Aging Cell*, *12*(6), 950–954. https://doi.org/10.1111/acel.12128.

Shekhar, A., Sims, L. S., & Bowsher, R. R. (1993). GABA receptors in the region of the dorsomedial hypothalamus of rats regulate anxiety in the elevated plus-maze test. II. Physiological measures. *Brain Research*, *627*(1), 17–24. Retrieved from http://www.ncbi.nlm.nih.gov/pubmed/8293298.

Shimizu, E., Hashimoto, K., Okamura, N., Koike, K., Komatsu, N., Kumakiri, C., et al. (2003). Alterations of serum levels of brain-derived neurotrophic factor (BDNF) in depressed patients with or without antidepressants. *Biological Psychiatry*, *54*(1), 70–75. Retrieved from https://www.ncbi.nlm.nih.gov/pubmed/12842310.

Silva, M. S., Pereira, B. A., Cespedes, I. C., Nascimento, J. O., Bittencourt, J. C., & Viana, M. B. (2014). Dorsomedial hypothalamus CRF type 1 receptors selectively modulate inhibitory avoidance responses in the elevated T-maze. *Behavioural Brain Research*, *271*, 249–257. https://doi.org/10.1016/j.bbr.2014.06.018.

Siu, J. J., Queen, N. J., Liu, X., Huang, W., McMurphy, T., & Cao, L. (2017). Molecular therapy of melanocortin-4-receptor obesity by an autoregulatory BDNF vector. *Molecular Therapy. Methods & Clinical Development*, 7, 83–95. https://doi.org/10.1016/j.omtm.2017.09.005.

Slater, A. M., & Cao, L. (2015). A protocol for housing mice in an enriched environment. *Journal of Visualized Experiments*, (100), e52874. https://doi.org/10.3791/52874.

Speliotes, E. K., Willer, C. J., Berndt, S. I., Monda, K. L., Thorleifsson, G., Jackson, A. U., et al. (2010). Association analyses of 249,796 individuals reveal 18 new loci associated with body mass index. *Nature Genetics*, *42*(11), 937–948. https://doi.org/10.1038/ng.686.

Starr, M. E., Evers, B. M., & Saito, H. (2009). Age-associated increase in cytokine production during systemic inflammation: Adipose tissue as a major source of IL-6. *The Journals of Gerontology. Series A, Biological Sciences and Medical Sciences*, *64*(7), 723–730. https://doi.org/10.1093/gerona/glp046.

Steiner, B., Zurborg, S., Horster, H., Fabel, K., & Kempermann, G. (2008). Differential 24 h responsiveness of Prox1-expressing precursor cells in adult hippocampal neurogenesis to physical activity, environmental enrichment, and kainic acid-induced seizures. *Neuroscience*, *154*(2), 521–529. https://doi.org/10.1016/j.neuroscience.2008.04.023.

Stout, M. B., Justice, J. N., Nicklas, B. J., & Kirkland, J. L. (2017). Physiological aging: Links among adipose tissue dysfunction, diabetes, and frailty. *Physiology (Bethesda)*, *32*(1), 9–19. https://doi.org/10.1152/physiol.00012.2016.

Streit, W. J., Xue, Q. S., Tischer, J., & Bechmann, I. (2014). Microglial pathology. *Acta Neuropathologica Communications*, *2*, 142. https://doi.org/10.1186/s40478-014-0142-6.

Tang, Y., & Cai, D. (2013). Hypothalamic inflammation and GnRH in aging development. *Cell Cycle*, *12*(17), 2711–2712. https://doi.org/10.4161/cc.26054.

Tang, Y., Purkayastha, S., & Cai, D. (2015). Hypothalamic microinflammation: A common basis of metabolic syndrome and aging. *Trends in Neurosciences*, *38*(1), 36–44. https://doi.org/10.1016/j.tins.2014.10.002.

Tapia-Arancibia, L., Aliaga, E., Silhol, M., & Arancibia, S. (2008). New insights into brain BDNF function in normal aging and Alzheimer disease. *Brain Research Reviews*, *59*(1), 201–220. doi:S0165-0173(08)00077-5 [pii]. https://doi.org/10.1016/j.brainresrev.2008.07.007.

Tchernof, A., & Despres, J. P. (2013). Pathophysiology of human visceral obesity: an update. *Physiological Reviews*, *93*(1), 359–404. https://doi.org/10.1152/physrev.00033.2011.

Thaler, J. P., Yi, C. X., Schur, E. A., Guyenet, S. J., Hwang, B. H., Dietrich, M. O., et al. (2012). Obesity is associated with hypothalamic injury in rodents and humans. *The Journal of Clinical Investigation*, *122*(1), 153–162. https://doi.org/10.1172/JCI59660.

Valdearcos, M., Douglass, J. D., Robblee, M. M., Dorfman, M. D., Stifler, D. R., Bennett, M. L., et al. (2017). Microglial inflammatory signaling orchestrates the hypothalamic immune response to dietary excess and mediates obesity susceptibility. *Cell Metabolism*, *26*(1), 185–197. e183 https://doi.org/10.1016/j.cmet.2017.05.015.

Valdearcos, M., Robblee, M. M., Benjamin, D. I., Nomura, D. K., Xu, A. W., & Koliwad, S. K. (2014). Microglia dictate the impact of saturated fat consumption on hypothalamic inflammation and neuronal function. *Cell Reports*, *9*(6), 2124–2138. https://doi.org/10.1016/j.celrep.2014.11.018.

van Swieten, M. M., Pandit, R., Adan, R. A., & van der Plasse, G. (2014). The neuroanatomical function of leptin in the hypothalamus. *Journal of Chemical Neuroanatomy*, *61–62*, 207–220. https://doi.org/10.1016/j.jchemneu.2014.05.004.

Vissers, D., Hens, W., Taeymans, J., Baeyens, J. P., Poortmans, J., & Van Gaal, L. (2013). The effect of exercise on visceral adipose tissue in overweight adults: A systematic review and meta-analysis. *PLoS One*, *8*(2), e56415. https://doi.org/10.1371/journal.pone.0056415.

Wang, C., Wang, C. M., Clark, K. R., & Sferra, T. J. (2003). Recombinant AAV serotype 1 transduction efficiency and tropism in the murine brain. *Gene Therapy*, *10*(17), 1528–1534. https://doi.org/10.1038/sj.gt.3302011.

Ward, Z. J., Bleich, S. N., Cradock, A. L., Barrett, J. L., Giles, C. M., Flax, C., et al. (2019). Projected U.S. state-level prevalence of adult obesity and severe obesity. *The New England Journal of Medicine*, *381*(25), 2440–2450. https://doi.org/10.1056/NEJMsa1909301.

Weiskopf, D., Weinberger, B., & Grubeck-Loebenstein, B. (2009). The aging of the immune system. *Transplant International*, *22*(11), 1041–1050.

Woo, C. C., Donnelly, J. H., Steinberg-Epstein, R., & Leon, M. (2015). Environmental enrichment as a therapy for autism: A clinical trial replication and extension. *Behavioral Neuroscience*, *129*(4), 412–422. https://doi.org/10.1037/bne0000068.

Woo, C. C., & Leon, M. (2013). Environmental enrichment as an effective treatment for autism: A randomized controlled trial. *Behavioral Neuroscience*, *127*(4), 487–497. https://doi.org/10.1037/a0033010.

Xiao, R., Bergin, S. M., Huang, W., Mansour, A. G., Liu, X., Judd, R. T., et al. (2019). Enriched environment regulates thymocyte development and alleviates experimental autoimmune encephalomyelitis in mice. *Brain, Behavior, and Immunity*, *75*, 137–148. https://doi.org/10.1016/j.bbi.2018.09.028.

Xiao, R., Bergin, S. M., Huang, W., Slater, A. M., Liu, X., Judd, R. T., et al. (2016). Environmental and genetic activation of hypothalamic BDNF modulates T-cell immunity to exert an anticancer phenotype. *Cancer Immunology Research*, *4*(6), 488–497. https://doi.org/10.1158/2326-6066.CIR-15-0297.

Xu, B., Goulding, E. H., Zang, K., Cepoi, D., Cone, R. D., Jones, K. R., et al. (2003). Brain-derived neurotrophic factor regulates energy balance downstream of melanocortin-4 receptor. *Nature Neuroscience*, *6*(7), 736–742. https://doi.org/10.1038/nn1073.

Xu, B., & Xie, X. (2016). Neurotrophic factor control of satiety and body weight. *Nature Reviews. Neuroscience*, *17*(5), 282–292. https://doi.org/10.1038/nrn.2016.24.

Yeo, G. S., Connie Hung, C. C., Rochford, J., Keogh, J., Gray, J., Sivaramakrishnan, S., et al. (2004). A de novo mutation affecting human TrkB associated with severe obesity and developmental delay. *Nature Neuroscience*, 7(11), 1187–1189. https://doi.org/10.1038/nn1336.

Zhang, G., Li, J., Purkayastha, S., Tang, Y., Zhang, H., Yin, Y., et al. (2013). Hypothalamic programming of systemic ageing involving IKK-beta, NF-kappaB and GnRH. *Nature*, *497*(7448), 211–216. https://doi.org/10.1038/nature12143.

Zhang, Y. S., Saito, Y., & Crimmins, E. M. (2019). Changing impact of obesity on active life expectancy of older Americans. *The Journals of Gerontology. Series A, Biological Sciences and Medical Sciences*, *74*(12), 1944–1951. https://doi.org/10.1093/gerona/glz133.

Zhang, X., Zhang, G., Zhang, H., Karin, M., Bai, H., & Cai, D. (2008). Hypothalamic IKKbeta/NF-kappaB and ER stress link overnutrition to energy imbalance and obesity. *Cell*, *135*(1), 61–73. https://doi.org/10.1016/j.cell.2008.07.043.

Zhao, C., Jou, J., Wolff, L. J., Sun, H., & Gage, F. H. (2014). Spine morphogenesis in newborn granule cells is differentially regulated in the outer and middle molecular layers. *The Journal of Comparative Neurology*, *522*(12), 2756–2766. https://doi.org/10.1002/cne.23581.

CHAPTER FOUR

Melatonin and healthy aging

Daniel P. Cardinali*
Faculty of Medical Sciences, Pontificia Universidad Católica Argentina, Buenos Aires, Argentina
*Corresponding author: e-mail addresses: daniel_cardinali@uca.edu.ar; danielcardinali@fibertel.com.ar

Contents

Abstract

Preservation of a robust circadian rhythmicity (particulsarly of the sleep/wake cycle), a proper nutrition and adequate physical exercise are key elements for healthy aging. Aging comes along with circadian alteration, e.g. a disrupted sleep and inflammation, that leads to metabolic disorders. In turn, sleep cycle disturbances cause numerous pathophysiological changes that accelerates the aging process. In the central nervous system, sleep disruption impairs several functions, among them, the clearance of waste molecules. The decrease of plasma melatonin, a molecule of unusual phylogenetic conservation present in all known aerobic organisms, plays a particular role as far as the endocrine sequels of aging. Every day, the late afternoon/nocturnal increase of melatonin synchronizes both the central circadian pacemaker located in the hypothalamic suprachiasmatic nuclei as well as myriads of peripheral cellular circadian clocks. This is called the "chronobiotic effect" of melatonin, the methoxyindole being the prototype of the endogenous family of chronobiotic agents. In addition, melatonin exerts a significant cytoprotective action by buffering free radicals and reversing inflammation via down regulation of proinflammatory cytokines, suppression of low degree inflammation and prevention of insulin resistance. Because of these properties melatonin has been advocated to be a potential therapeutic tool in COVID 19 pandemic. Melatonin administration to aged animals counteracts a significant number of senescence-related changes. In humans, melatonin is effective both as a chronobiotic and a cytoprotective agent to maintain a healthy aging. Circulating melatonin levels are consistently reduced in the metabolic syndrome, ischemic and non-ischemic cardiovascular diseases and neurodegenerative disorders like the Alzheimer's and Parkinson's

Vitamins and Hormones, Volume 115
ISSN 0083-6729
https://doi.org/10.1016/bs.vh.2020.12.004

diseases. The potential therapeutic value of melatonin has been suggested by a limited number of clinical trials generally employing melatonin in the 2–10 mg/day range. However, from animal studies the cytoprotective effects of melatonin need higher doses to become apparent (i.e. in the 100 mg/day range). Hence, controlled studies employing melatonin doses in this range are urgently needed.

1. Introduction

Aging is the leading risk factor for a variety of chronic diseases, including cancer, metabolic, cardiovascular, and neurodegenerative diseases, which result in poor quality of life and an increase in morbidity and mortality worldwide. Among several other comorbidities, the aged population is more prone to suffer from coronavirus infection and the association of aging with a higher vulnerability to SARS-CoV-2 infection is a subject of major importance in COVID-19 pandemic (Esteve, Permanyer, Boertien, & Vaupel, 2020). Chronic, endemic disorders such as the metabolic syndrome (MS) and Alzheimer's disease (AD) are at the core of pathological aging and their prevention is presently a fundamental goal for governmental and non-governmental health organizations. The prevalence of MS varies from 15% to 30% depending on the region of the world considered (O'Neill & O'Driscoll, 2015). An increase of 1.5 to 2.5 times in cardiovascular mortality occurs when MS is present. Likewise, the global prevalence of dementia among people aged $\geq$60 years is 5% to 7% (Prince et al., 2013). In 2010, approximately 35.6 million people lived with dementia, and this number is expected to double every 20 years. These are a few examples that the aging of the world population is a recognized challenge for current health system.

Several interrelated processes, such as circadian desynchronization, free radical-mediated damage, mitochondrial dysfunction, low degree of inflammation ("inflammaging") and immunosenescence have been identified as major pathophysiological mechanisms in aging (Welz & Benitah, 2019). Environmental and nutritional factors related to different lifestyles are variables of significant importance for healthy aging. In this context, melatonin, an unusual phylogenetically conserved compound present in all known aerobic phyla, has a promising significance. Melatonin is the prototype of the drugs that influence the circadian apparatus (i.e. chronobiotics) (Arendt & Skene, 2005; Dawson & Armstrong, 1996) and plays a major function in the coordination of circadian rhythmicity. In addition, melatonin is a potent cytoprotective agent having strong antioxidant, anti-inflammatory and immunoregulating activity (Cardinali, 2019b). Because of these properties melatonin has been advocated to be a potential therapeutic tool in COVID 19 pandemic (Cardinali, Brown, & Pandi-Perumal, 2020; Reiter, Ma, & Sharma, 2020).

The pineal gland is the demonstrable source of circulating melatonin in humans, the decrease in plasma melatonin being one of the characteristics of the advancing age (Claustrat & Leston, 2015). This Chapter is focused on the use of melatonin as a chronobiotic and cytoprotector to contribute to healthy aging.

2. Melatonin as a chronobiotic

Organisms populating the Earth are under the steady influence of daily and seasonal changes resulting from the planet's rotation and orbit around the sun. This periodic pattern is most prominently manifested by the light–dark cycle and has led to the establishment of endogenous circadian timing systems that synchronize biological functions to the environment. This is the basis of predictive homeostasis evolving as an adaptation to anticipate predictable changes in the environment, such as light and darkness, temperature, food availability, or predator activity (Riede, van der Vinne, & Hut, 2017). Therefore, the circadian clock is one of the most indispensable biological functions for living organisms that acts like a multifunctional timer to adjust the homeostatic system, including sleep and wakefulness, hormonal secretions, immune function and various other bodily functions, to the 24-h cycle.

Research in animals and humans has shown that only a few environmental cues, like light–dark cycles, are effective entraining agents ("zeitgebers") for the circadian oscillator. An entraining agent can actually reset, or phase shift, the internal clock. Depending on when an organism is exposed to such an entraining agent, circadian rhythms can be advanced, delayed, or not shifted at all. Therefore, adjusting the daily activity pattern to the appropriate time of day involves a rhythmic variation in the influence of the zeitgeber as a resetting factor (Duffy, Zitting, & Chinoy, 2015).

In mammals, the circadian system is composed of many individual, tissue-specific cellular clocks. To generate coherent physiological and behavioral responses, the phases of this multitude of cellular clocks are orchestrated by a master circadian pacemaker residing in the suprachiasmatic nuclei (SCN) of the hypothalamus (Buijs et al., 2017). At a molecular level, circadian clocks are based on clock genes, some of which encode proteins able to feedback and inhibit their own transcription. These cellular oscillators consist of interlocked transcriptional and post-translational feedback loops that involve a small number of core clock genes (about 12 genes identified currently) (Welz & Benitah, 2019). The negative and positive transcriptional/translational feedback loops to form the core clockwork, have been characterized in rodents by transgenic gene deletion methodology. Clock gene expression oscillates because of the delay in the feedback loops, regulated in part by phosphorylation of the clock proteins that control their stability, nuclear re-entry, and transcription complex formation (Welz & Benitah, 2019).

As noted, the trillions of cellular clocks in primates are synchronized by a few thousand neurons located in the SCN. It is remarkable that such a small group of neurons display the properties of a central clock. Indeed, these "neuronal oligarchies," like the human ones, control trillions of cells in the body by: (a) taking control of the major communication channels (the endocrine and autonomic nervous systems); (b) concentrating the relevant information in a private way (i.e., light information arriving via the retino-hypothalamic tract) (Cardinali, 2017). Thus, it is not surprising that anatomical studies have showed that the SCN projects to at least three different neuronal targets: endocrine neurons, autonomic neurons of the paraventricular nucleus (PVN) of the hypothalamus, and other diencephalic structures that transmit the circadian signal to other brain regions (Buijs, Guzman Ruiz, Mendez, & Rodriguez, 2019). Through autonomic nervous system projections involving the superior cervical ganglia the SCN controls the release of a major internal synchronizer from the pineal gland, namely melatonin (Cardinali, 2017).

The current view of the physiological regulation of the circadian rhythm of sleep/wakefulness (the main circadian rhythm in the body) holds that it is regulated by two components, namely, a circadian ($\sim$24 h) component and a homeostatic component (Borbely, Daan, Wirz-Justice, & Deboer, 2016). Melatonin plays a major role in the circadian component that regulates the timing of sleep. The circadian rhythm in synthesis and secretion of pineal melatonin is closely associated with the sleep rhythm in both normal and blind subjects (Emens & Eastman, 2017). The onset of nighttime melatonin secretion is initiated approximately 2 h in advance to individual's habitual bedtime and has been shown to correlate with the onset evening sleepiness. Several studies implicate endogenous melatonin in the physiological regulation of the circadian mechanisms ruling sleep propensity (Auld, Maschauer, Morrison, Skene, & Riha, 2017; Gobbi & Comai, 2018).

Aging has been associated with a significant reduction in sleep efficiency and continuity and coincides with a significant reduction in amplitude of the melatonin rhythm and consequently of many circadian rhythms in the body (Duffy et al., 2015). Increase in early morning awakenings and difficulty in falling sleep have been frequent reported in the elderly. An impaired melatonin secretion is associated with the sleep disorders that are encountered in elderly insomniacs. Indeed, aging may be a process resulting from the relative circadian desynchrony produced by melatonin deficiency and melatonin can be effective for improving the quality of life in the elderly via its recognized chronobiotic capacity (Cardinali, 2019b).

Melatonin plays a major function in the coordination of circadian rhythmicity. Melatonin secretion is an "arm" of the biologic clock in the sense that it responds to signals from the SCN and that the timing of the melatonin rhythm indicates the status of the clock, both in terms of phase (i.e., internal clock time relative to external clock time) and amplitude. From another point of view, melatonin is also a chemical code of night: the longer the night, the longer the duration of its secretion. In most mammalian species, this pattern of secretion serves as a time cue for seasonal rhythms (Clarke & Caraty, 2013).

Pineal melatonin production is controlled by a complex neural system originating in the hypothalamic PVN and terminating in the high levels of the thoracic spinal cord—the superior cervical ganglion sympathetic system. The postganglionic sympathetic nerve terminals of the superior cervical ganglion release norepinephrine into the pineal gland that triggers melatonin synthesis by its interaction with β- (mainly) and α-adrenoceptors on the membrane of pineal cells. Melatonin, due to its high diffusibility, is not stored inside the pineal and is released as soon as it is produced (Tan, Xu, Zhou, & Reiter, 2018). The structures which regulate circadian rhythms have been described as the SCN-melatonin loop (Tan et al., 2018). This loop includes melanopsin-containing retinal ganglion cells, the retino-hypothalamic tract, SCN, PVN, intermediolateral cell column, the sympathetic cervical ganglia, the pineal gland, and the melatonin rhythm which feedback impacts the SCN.

As a result, the melatonin production, and consequently its cerebrospinal fluid (CSF) and blood levels are circadian in nature and tightly synchronized with the environmental light/dark cycle. Indeed, the circadian pineal production of melatonin is restricted to the dark phase of the light/dark cycle in all mammalian species. It is noteworthy that melatonin is always produced during the night independent of the daily pattern of activity/rest of the species, indicating its strong relationship with the external photoperiod. Additionally, melatonin is produced during the night provided there is no light in the environment.

Light at night, mainly in the blue range, activates melanopsin-containing retinal ganglion cells, a special retinal system that inhibits the pineal sympathetic norepinephrine release and consequently decreases or abolishes melatonin production. Given the regularity of the daily melatonin production that is associated with high and low or absent blood concentrations during the night and day, respectively, melatonin is able to synchronize the circadian rhythms of several organs and their functions. In vitro experiments have shown that the artificial day and night melatonin profile can function as a pacemaker for the daily rhythmic functions of the cells (Hardeland et al., 2011).

Like the effects induced by the external zeitgeber light, effects by the internal zeitgeber melatonin on the circadian clock are time-dependent. Daily timed administration of melatonin to rats shifts the phase of the circadian clock, and this phase shifting may explain the effect of melatonin on sleep in humans. Indirect support for such a physiological role derived from clinical studies on blind subjects (who show free running of their circadian rhythms) treated with melatonin (Skene & Arendt, 2007). More direct support for this hypothesis was provided by the demonstration that the phase response curve for melatonin was opposite (i.e., about 180 degrees out of phase) to that of light (Lewy, 2010).

The direct effects in regions containing high densities of melatonin receptors, such as the circadian pacemaker SCN, or the pars tuberalis (PT), a site of particular relevance for photoperiodically controlled reproduction (Clarke & Caraty, 2013) strongly supported the premier significance of melatonin's physiological role. Although the control of circadian and seasonal rhythmicity represents melatonin's main physiological function, the actions of the methoxyindole are by no means restricted to areas of high receptor density.

The pleiotropy of melatonin has to be analyzed at different levels, from the sites of synthesis and local dynamics, distribution of receptors and other binding sites in target organs, cell-specific differences in signaling as related to the presence of G protein variants, and intracellular effects—with a particular focus on mitochondrial actions—to numerous secondary changes induced by influencing other hormones, neurotransmitters, neurotrophins and further signal molecules (Hardeland et al., 2011). In functional terms, melatonin exerts a host of effects that can be under the control of the

SCN and, in seasonal breeders, the premammillary hypothalamus and the PT, and may also have direct effects in numerous peripheral organs. In particular, melatonin is involved in sleep initiation, vasomotor control, adrenal function, antiexcitatory actions, immunomodulation including anti-inflammatory properties, antioxidant actions, and energy metabolism, influencing mitochondrial electron flux, the mitochondrial permeability transition pore, and mitochondrial biogenesis (Hardeland, 2018a, 2018b; Tan & Reiter, 2019).

The chronobiotic action of melatonin is mediated via the melatonin receptors, which have been identified both in the CNS and in the periphery (Dubocovich et al., 2010). MT_1 and MT_2 receptors all belonging to the superfamily of membrane receptors associated with G proteins (G-protein coupled receptors, GPCR) have been cloned. More recently, another member, GPR50, was included in the melatonin receptor subfamily. GPR50 shows high sequence homology to MT_1 and MT_2 but does not bind to melatonin or any other known ligand. An interesting feature of these receptors is their capacity to form homo- and heteromers between each other and also with other GPCRs (Cecon, Oishi, & Jockers, 2018) among them the serotonin 5-HT_{2C}. The heteromers display functional properties different from those of the corresponding homomers. For example, acting on the MT_2/5-HT_{2C} heteromer, melatonin binding induces the activation of Gq signaling through a transactivation of the serotonergic receptor caused by conformational changes of the MT_2, which is normally not coupled to a Gq.

Circulating melatonin is loosely bound to albumin and in the liver, it is first hydroxylated and then conjugated with sulfate and glucuronide (Claustrat & Leston, 2015). In human urine, 6-sulfatoxymelatonin has been identified as the main metabolite. In the brain melatonin is metabolized into kynurenine derivatives. It is of interest that the well-documented antioxidant properties of melatonin are shared by some of its metabolites including cyclic 3-hydroxymelatonin, N^1-acetyl-N^2-formyl-5-methoxykynuramine (AFMK) and, with highest potency, N^1-acetyl-5-methoxykynuramine (AMK). Thus an "antioxidant cascade" is triggered by melatonin administration to experimental animals and humans (Galano & Reiter, 2018).

As mentioned, in mammals, circulating melatonin is derived almost exclusively from the pineal gland. In addition, melatonin is synthesized locally in most cells, tissues and organs, including lymphocytes, bone marrow, thymus, gastrointestinal tract, skin and eyes, where it can play an autocrine or paracrine role (Acuña-Castroviejo et al., 2014). Indeed, there is now strong evidence that melatonin is produced in every animal cell that has mitochondria (Tan & Reiter, 2019). In both animals and humans, melatonin participates in diverse physiological functions that indicate not only the duration of the night, but also improve the elimination of free radicals and the immune response, showing relevant cytoprotective properties (Hardeland, 2018a).

Melatonin is a powerful chronobiotic with very slight hypnotic capacity. Daily doses of 2–5 mg melatonin, timed to advance the phase of the internal clock by interaction with MT_1 receptors in the SCN, maintains synchronization of the circadian rhythms to a 24-h cycle in sighted persons who are living in conditions likely to induce a free-running rhythm (Lewy, 2010). Melatonin synchronizes the rhythm in persons after a short period of free-running. In blind subjects with free-running rhythms, it has been possible to stabilize, or entrain, the sleep/wake cycle to a 24-h period by giving melatonin, with resulting improvements in sleep and mood (Skene & Arendt, 2007). In normal aged subjects and in demented patients with disturbed synchronization of the sleep/wake cycle melatonin administration is helpful to reduce the variation of onset time of sleep. The phase shifting effects of melatonin are also sufficient to explain its effectiveness as a treatment for circadian-related sleep disorders, such as jet lag or delayed phase sleep syndrome (Burgess & Emens, 2018) (Fig. 1).

3. Use of melatonin in sleep aging

The relationship between the sleep/wake cycle and aging is bidirectional (Mazzotti et al., 2014). Inadequate sleep, both in terms of its duration and quality, can affect health and thus adversely contribute to the aging process. On the other hand, sleep/wake cycle difficulties typically increase with age as a result of flattening and misalignment of circadian rhythms like that of melatonin secretion, as well as of the sleep-disturbing influences of aging-related disorders and diseases (Hardeland, 2015). The most convincing examples are that of immunosenescence, which also affects the brain (Cardinali, Esquifino, Srinivasan, & Pandi-Perumal, 2008; Hardeland, 2018a), as well as the almost exponential increase of hydroxyl radical generation seen in the senescent brain (Poeggeler, Reiter, Tan, Chen, & Manchester, 1993). The association of age with a higher vulnerability to SARS-CoV-2 infection is a subject of major importance nowadays. Several factors, including higher stress due to social isolation, diminished melatonin levels with age, and higher exposure of individuals to light at the evening, which reduces

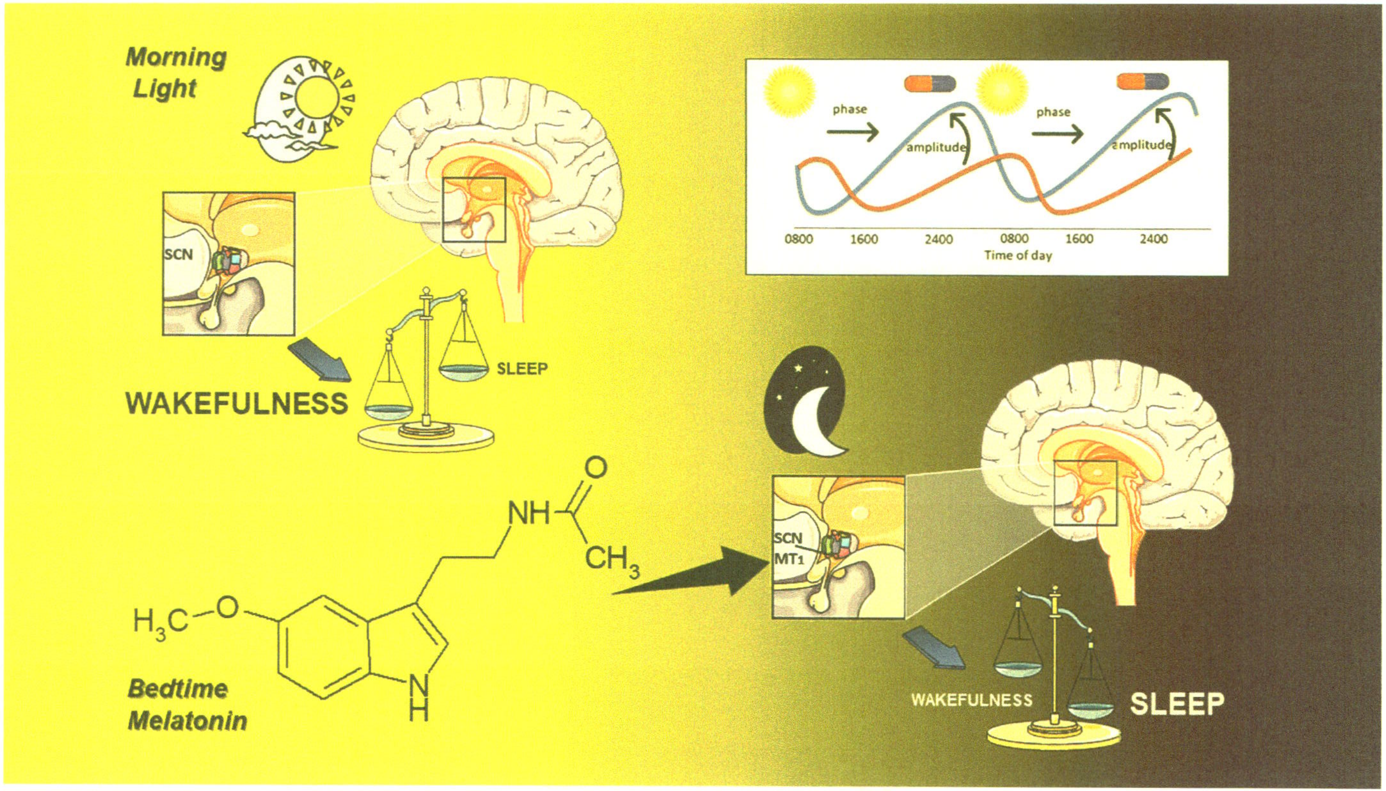

Fig. 1 Chronobiotic activity of melatonin. The combined use of melatonin and bright light to augment the amplitude and synchronize endogenous circadian rhythmicity in aging. Bright light in the morning together with melatonin at bedtime potentiate to augment the amplitude of sleep/wake cycle and other circadian rhythms. Melatonin activity is achieved via MT1 receptors located in SCN. Insert: In delayed phase sleep syndrome melatonin augments the amplitude and phase-advanced the sleep/wake rhythm.

melatonin levels and disrupts circadian rhythmicity are relevant for maintaining the circadian health in aged individuals in COVID 19 pandemic (Cardinali, Brown, Reiter, & Pandi-Perumal, 2020).

There is now considerable evidence that, with advancing age, there are increases in sleep/wake disorders. Epidemiological research has shown that more than 50% of adults above the age of 65 years suffer from some form of a chronic sleep-related complaint (Foley et al., 1995). It is important to distinguish the changes due to aging per se vs changes due to aging-related disease and secondary to the use of concomitant medications.

Several meta-analyses support the view that the chronobiotic/hypnotic properties of melatonin are useful in aged patients with primary sleep disorders to decrease sleep onset latency and to increase total sleep time, with little if any effect on sleep efficiency (Auld et al., 2017; Ferracioli-Oda, Qawasmi, & Bloch, 2013; Li et al., 2019). Several expert consensus reports also support such a role of melatonin in adult insomnia (Geoffroy, Micoulaud Franchi, Lopez, & Schroder, 2019; Palagini et al., 2020; Vecchierini, Kilic-Huck, & Quera-Salva, 2020; Wilson et al., 2010).

Many studies have shown that sleep/wake disruption is an important contributor to neuropathology. Sleep deprivation (Ooms et al., 2014) or NREM disruption (Ju et al., 2017) have been shown in healthy subjects to increase CSF levels of amyloid β $(A\beta)_{1-42}$ and $A\beta_{1-40}$, respectively. This reduction was reported to be prevented by a single night of total sleep deprivation. Similar findings were obtained in mice, in which sleep deprivation caused increases of Aβ peptides in the interstitial fluid of the brain (Kang et al., 2009). A "glymphatic" hypothesis (Iliff et al., 2012) holds that an active, lymphatic-like movements, occurs in brain extracellular space driven by perivascular astrocytes, which are strongly enriched in aquaporin-4 and by changes in the vascular lumen. Water release via aquaporin-4 would be responsible for an actively driven fluid exchange between para-arterial and paravenous spaces. The exchange of solutes between the CSF and the interstitial fluid occurs mostly during NREM sleep when the cortical interstitial space increases by >60% and provides a low resistance path for the movement of CSF and interstitial fluid in the brain parenchyma (Iliff et al., 2012). This is affected in the aging human brain. Various neurological disease states such as stroke, traumatic brain injury, and AD have been interpreted in terms of the contribution by glymphatic dysfunction (Boespflug & Iliff, 2018). The administration of melatonin to AD transgenic mice augments the glymphatic clearance of Aβ (Pappolla et al., 2018).

The prevalence of primary insomnia varies up to 10% in the general population and up to 25–30% in the elderly, for whom the treatment of insomnia is a clear medical necessity. The direct and indirect costs of insomnia represent a substantial social economic burden. Benzodiazepines (BZD) and other BZD receptor agonists (Z-drugs like zolpidem, zaleplon, zopiclone) are the most commonly prescribed drugs for the treatment of insomnia in the elderly. Several meta-analyses pondering the risks and benefits of these therapeutic options in older patients have reported statistically significant improvements in sleep but have also reported a statistically significant risk of life-threatening adverse events (Winkler, Auer, Doering, & Rief, 2014). In fact, these drugs are only approved by regulatory agencies for treatment for older adults no more than a few weeks in length due to safety reasons. Adverse effects have been reported in >40% of users of both BZD and Z drugs.

Health authorities in Europe are undertaking policies and recommendations to reduce the consumption of BZD and Z drugs. However, the campaigns have not been generally successful, despite national guidelines and recommendations, and the use of these drugs has continued to increase. The clearer strategy to reduce chronic BZD use is to reduce medication; abrupt cessation can only be justified if a serious adverse effect occurs during treatment. There is no clear evidence regarding the optimal rate of taper, and times vary from 4 weeks to several months.

In 1986 we demonstrated the interaction of melatonin with central BZD receptors (Acuña-Castroviejo, Lowenstein, Rosenstein, & Cardinali, 1986) and in 1997 we published the first demonstration of the reduction of BZD consumption in melatonin-treated elderly subjects (Fainstein, Bonetto, Brusco, & Cardinali, 1997). Melatonin competes with BZD and Z drugs at their site of action. A facilitator role of melatonin on GABA neurotransmission explains the anxiolytic, antihyperalgesic and antinociceptive effects of melatonergic agents (Cardinali, Golombek, Rosenstein, Brusco, & Vigo, 2016). Several clinical studies now support the efficacy of melatonin to reduce BZD use in chronically treated patients (Baandrup et al., 2018). In a pharmaco-epidemiologic study aimed to evaluate the impact of anti-BZD/Z drugs campaigns and the availability of alternative pharmacotherapy (melatonin) on the consumption of BZD and Z drugs in several European countries, the results indicated that campaigns failed unless they were associated with the availability of melatonin in the market (Clay, Falissard, Moore, & Toumi, 2013). Melatonin has thus become an effective treatment for the preservation of a healthy sleep in the elderly. Melatonin has been used as a sleep aid for decades without any serious adverse effects being reported. Moreover, it has often been used in critically ill patients to improve sleep and wellbeing, both of which are also beneficial to SARS-CoV-2 infected patients. It is a molecule with an uncommonly high safety profile and can be administered via numerous routes including orally. It is inexpensive, stable without refrigeration, and would be particularly useful in underdeveloped countries where access to high-quality health care may be lacking.

4. Cytoprotection by melatonin in animal models of MS and AD

The cytoprotective activity of melatonin exceeds that mediated via receptors. Almost every cell in the human body contains melatonin, in quantities much higher than those circulating in blood derived from the pineal gland (Acuña-Castroviejo et al., 2014). The mitochondrial capacity to synthesize melatonin is now confirmed, but for reasons that remain unexplained, intracellular melatonin does not get to the extracellular space.

To modify intracellular melatonin levels, doses much higher than those employed as a chronobiotic are needed (Reiter et al., 2018; Venegas et al., 2013). Most studies on neuroprotective and anti-inflammatory effects in animals employ pharmacological doses, which clearly exceed the saturation of the receptor.

In both the cytoplasm and the cell nucleus, melatonin has important antioxidant and scavenging effects on free radicals, which are largely independent of receptors (Manchester et al., 2015). These effects are exerted in three ways: (a) melatonin is a free radical scavenger; (b) melatonin is metabolized to compounds with high antioxidant activity; (c) melatonin is an indirect antioxidant, which stimulates the synthesis of antioxidant enzymes and inhibits that of prooxidant enzymes. Melatonin has a proven superiority to vitamin C and E in protection against oxidative damage and in the elimination of free radicals (Galano, Tan, & Reiter, 2011). In addition, melatonin potentiates the effects of other antioxidants, such as vitamin C and Trolox. Several antiapoptotic and cytoprotective effects of melatonin are exerted under conditions of ischemia (unrelated to free radicals) and can be attributed to the stabilizing activity of the mitochondrial membrane (Reiter et al., 2018).

Melatonin is also an immunological modulator that shows proinflammatory and anti-inflammatory properties (Carrillo-Vico, Lardone, Alvarez-Sanchez,

Rodriguez-Rodriguez, & Guerrero, 2013; Hardeland, 2018a). The anti-inflammatory actions are of medicinal interest, since they are observed in high-grade inflammation such as sepsis, ischemia/reperfusion and brain injury, as well as in the low-grade inflammation seen in MS, neurodegenerative disorders and aging. Melatonin has significant anti-inflammatory properties presumably by inhibiting the binding of nuclear factor κB (NF κB) to DNA, thus decreasing the synthesis of proinflammatory cytokines, by inhibiting cyclooxygenase (Cox) (Cardinali, Ritta, Fuentes, Gimeno, & Gimeno, 1980) in particular Cox-2 (Deng, Tang, Tseng, & Wu, 2006), and by suppressing the expression of the inducible gene of nitric oxide synthase (Costantino, Cuzzocrea, Mazzon, & Caputi, 1998). In addition, other pathways of secondary signaling are involved (Hardeland, 2018a).

Treatment with melatonin of rats can reduce obesity, type 2 diabetes and hepatic steatosis. In several animal models of hyperadiposity, the injection of melatonin could normalize most of the observed alterations and correct the altered biochemical proinflammatory profile (for reference see Cardinali, 2019a). In addition, melatonin is effective in animal models of ischemic and nonischemic heart failure, an important comorbidity of MS.

Collectively, the results in animal studies indicate that the administration of melatonin effectively counteracts some of the disrupting effects seen in diet-induced obesity in animals insulin resistance, dyslipidemia and obesity, and the consequences of ischemic and non-ischemic heart disease. Interestingly, a recent study examining the subcellular distribution of the melatonin in the heart of rats indicates that at a dose of 40 mg/kg b.w, maximal concentration of melatonin was reached in the nucleus and mitochondrion. The authors concluded that doses of melatonin $\geq$112 mg/day are required for therapeutic purpose in a 70 kg adult (Acuña-Castroviejo, Noguera-Navarro, Reiter, & Escames, 2018).

Concerning experimental models of AD, cell line studies regarding AD and melatonin have delineated important melatonin mediated mechanisms in AD prevention. For a comprehensive review on melatonin activity to reverse disrupted signaling mechanisms in neurodegeneration, including proteostasis dysfunction, disruption of autophagic integrity, and anomalies in the insulin, Notch, and Wnt/β-catenin signaling pathways see reference Shukla, Govitrapong, Boontem, Reiter, and Satayavivad (2017).

In transgenic models of AD the obtained results are compatible with the view that melatonin regulates Aß metabolism mainly at the initial phases of the pathological process. From the doses of melatonin used in these different transgenic models, the HED of melatonin for a 75 kg adult can be calculated by normalization of body surface area (Reagan-Shaw, Nihal, & Ahmad, 2008). Theoretical human equivalent doses ranged from 2- to 3-orders of magnitude greater than those employed in humans.

A summary of the cytoprotective effects of melatonin in healthy aging is depicted in Fig. 2.

5. Use of melatonin in MS and AD patients

Type 2 diabetic patients have low circulating levels of melatonin (Tutuncu et al., 2005). In addition, allelic variants for melatonin receptors were associated with an increase in fasting blood glucose levels and/or an increased risk of type 2 diabetes (see for reference (Vallim, Amaral, Cipolla-Neto, & D'Almeida, 2019)) and with the polycystic ovarian syndrome (PCOS) (Song et al., 2015; Spinedi & Cardinali, 2018). Patients with coronary artery disease show a decrease in melatonin secretion and among elderly hypertensive patients, nocturnal urinary melatonin excretion was inversely associated with the non-dipper pattern of hypertensive disease (Obayashi et al., 2013). Melatonin administration ($\leq$5 mg/day) proved capable of reducing nocturnal blood pressure in hypertensives and attenuated age-dependent disturbances of cardiovascular rhythms (see for reference (Nduhirabandi & Maarman, 2018; Pandi-Perumal et al., 2017)).

Treatment with melatonin ($\leq$5 mg/day) improves MS in obese and PCOS patients as well as in bipolar and schizophrenic patients receiving second generation antipsychotics The administration of melatonin normalizes MS in elderly hypertensive patients and improves the enzyme profile in patients with alcoholic hepatic steatosis. The combination of melatonin and zinc acetate, when used alone or in combination with metformin, improved the glycemic control in type 2 diabetic patients (see for reference Cardinali, 2019a). Additional information concerning a glucose tolerance-reducing property of melatonin in humans came from the detection of melatonin receptor polymorphisms. To date, several single nucleotide polymorphisms located near or inside the gene encoding MTNR1B with an association with type 2 diabetes mellitus have been identified (see for reference Vallim et al., 2019). Some of these melatonin receptor polymorphisms are dysfunctional because of their incapability of binding melatonin, and others were found to be unable to interact with G_i proteins. Thus, the absence of melatonin signaling is presumably diabetogenic (Vallim et al., 2019).

Concerning AD, CSF melatonin levels decrease even in preclinical stages of the disease when the patients do not manifest any cognitive impairment,

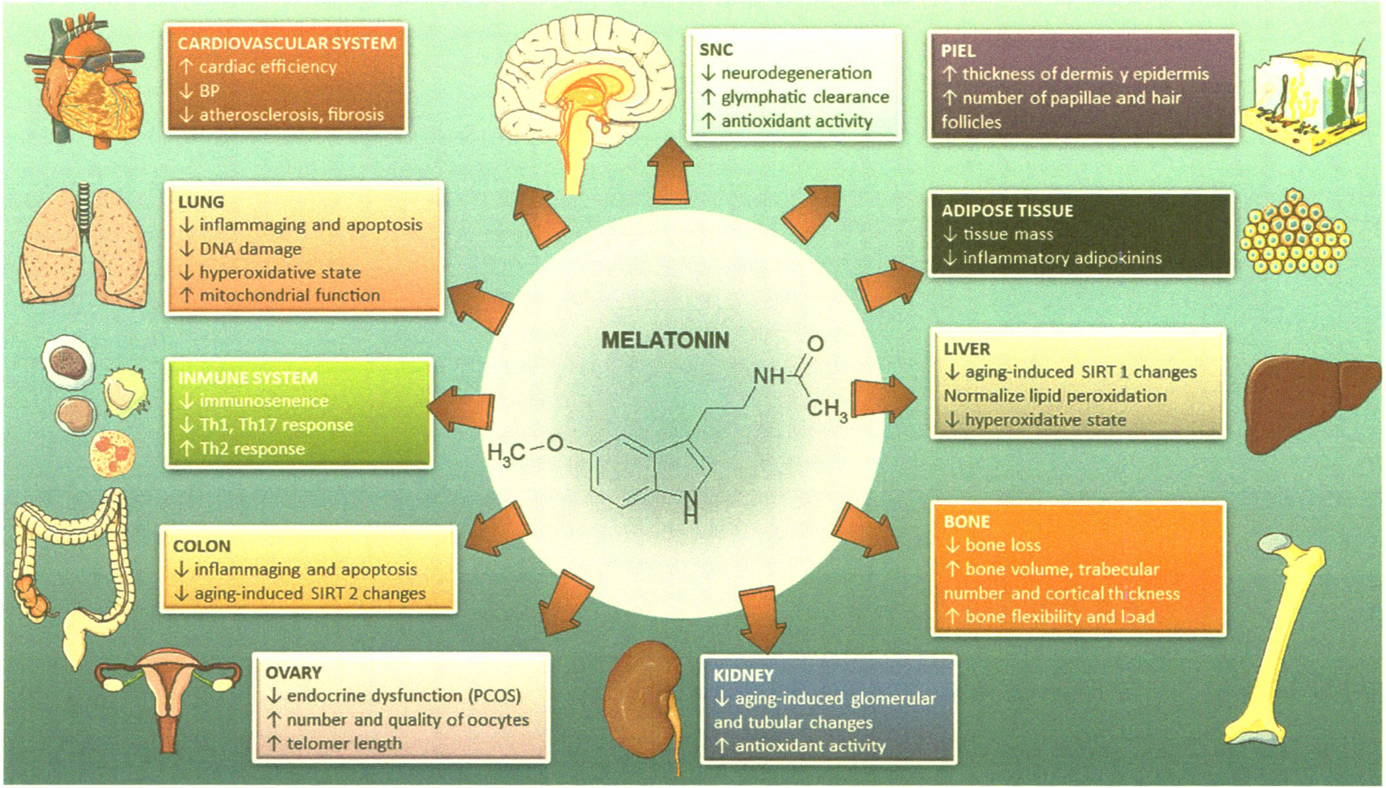

Fig. 2 Melatonin as a cytoprotector in healthy aging. The anti-aging potential of melatonin is shown in different organs.

suggesting that the reduction in CSF melatonin may be an early trigger and marker for AD (Liu, Zhou, van Heerikhuize, Hofman, & Swaab, 1999). Although it is not known whether the relative melatonin deficiency is either a consequence or a cause of neurodegeneration, it seems clear that the loss in melatonin aggravates the disease and that early circadian disruption can be an important deficit to be considered. Significant differences were observed in melatonin levels between mild cognitive impairment (MCI) and AD patients with a negative correlation between neuropsychological assessment of dementia and melatonin levels (Sirin et al., 2015).

Two meta-analyses endorsed the view that melatonin therapy is effective in improving sleep in AD patients (Xu et al., 2015; Zhang et al., 2016). Moreover, the melatonergic agonist ramelteon was reported as effective to treat delirium, an acute state of mental confusion that can lead to many adverse sequelae in intensive care unit elder patients (Furuya et al., 2012).

An analysis of published data of the use of melatonin in the early stages of cognitive decline consistently showed that administration of melatonin, every night before retiring, improves the quality of sleep and cognitive performance in this phase of the disease (see for reference Cardinali, 2019c). Patients treated with melatonin showed a significantly better performance in various neuropsychological tests. They also had lower scores in the Beck Depression Inventory concomitantly with improvement in the quality of sleep and wakefulness. Therefore, melatonin treatment can be effective in the early stages of the neurodegenerative disease (Cardinali, 2019c).

6. Conclusions

Melatonin exhibits both hypnotic and chronobiotic properties and has been used therapeutically for the treatment of insomnia related to age, as well as other primary and secondary insomnia (Leger, Laudon, & Zisapel, 2004; Zhdanova et al., 2001). Several meta-analyses and consensus reports support this role.

As discussed in this article, studies using 2–5 mg of melatonin/day are not adequate to provide an adequate comparison with data on the protection of MS or AD derived from animal studies. Hence clinical trials with doses in the 40–100 mg/day range are urgently needed.

It should be noted that melatonin is remarkably non-toxic, and its safety is very high. The lethal dose 50 for the intraperitoneal injection of melatonin was determined for rats (1168 mg/kg) and mice (1131 mg/kg), but the oral administration of melatonin (tested up to 3200 mg/kg in rats) could not be determined and for melatonin subcutaneous injection (tested up to 1600 mg/kg in rats and mice) (Sugden, 1983). In humans, melatonin has a high safety profile and, in general, is very well tolerated [see for reference (Cardinali, 2019a)].

Unfortunately, the pharmaceutical industry is refractive to support melatonin studies because of the lack of protective patents for a natural compound. Hence, only with the involvement of governmental and non-profit organizations such a goal can be achieved. At present, the only option for the attending physician interested in the use of melatonin as a cytoprotective is the off-label indication of the drug.

In conclusion, from studies in animals, several potentially useful effects of melatonin, such as those in MS and AD, require high doses of melatonin to be evident. If melatonin is expected to be effective in improving health, especially in the elderly, it is likely that the low doses of melatonin commonly used are not very beneficial so far (Cardinali, 2019b). The question of whether melatonin has a therapeutic value in the prevention or treatment of MS or AD deserves further analysis. Multicenter double-blind studies are needed to explore and further investigate the potential and utility of melatonin. The doses of melatonin used should be re-evaluated in view of the equivalent human doses of melatonin derived from preclinical data. Unfortunately, of the 139 clinical trials related to melatonin in an initial state (recruitment and non-recruitment) listed in PubMed (ClinicalTrials.gov Search results 05/05/2020) none is directed to address this query.

Acknowledgments

DPC is an Emeritus Superior Investigator from the Argentine National Research Council (CONICET) and Emeritus Professor, University of Buenos Aires.

Conflict of interest

The author declares that there are no commercial or financial relationships that could be construed as a potential conflict of interest.

References

Acuña-Castroviejo, D., Escames, G., Venegas, C., Diaz-Casado, M. E., Lima-Cabello, E., Lopez, L. C., et al. (2014). Extrapineal melatonin: sources, regulation, and potential functions. *Cellular and Molecular Life Sciences*, *71*(16), 2997–3025.

Acuña-Castroviejo, D., Lowenstein, P. R., Rosenstein, R., & Cardinali, D. P. (1986). Diurnal variations of benzodiazepine binding in rat cerebral cortex: Disruption by pinealectomy. *Journal of Pineal Research*, *3*(2), 101–109.

Acuña-Castroviejo, D., Noguera-Navarro, M. T., Reiter, R. J., & Escames, G. (2018). Melatonin actions in the heart: More than a hormone. *Melatonin Research*, *1*(1), 21–26.

Arendt, J., & Skene, D. J. (2005). Melatonin as a chronobiotic. *Sleep Medicine Reviews*, *9*(1), 25–39.
Auld, F., Maschauer, E. L., Morrison, I., Skene, D. J., & Riha, R. L. (2017). Evidence for the efficacy of melatonin in the treatment of primary adult sleep disorders. *Sleep Medicine Reviews*, *34*, 10–22.
Baandrup, L., Ebdrup, B. H., Rasmussen, J. O., Lindschou, J., Gluud, C., & Glenthoj, B. Y. (2018). Pharmacological interventions for benzodiazepine discontinuation in chronic benzodiazepine users. *The Cochrane Database of Systematic Reviews*, *3*, CD011481.
Boespflug, E. L., & Iliff, J. J. (2018). The emerging relationship between interstitial fluid-cerebrospinal fluid exchange, amyloid-beta, and sleep. *Biological Psychiatry*, *83*(4), 328–336.
Borbely, A. A., Daan, S., Wirz-Justice, A., & Deboer, T. (2016). The two-process model of sleep regulation: A reappraisal. *Journal of Sleep Research*, *25*(2), 131–143.
Buijs, R. M., Guzman Ruiz, M. A., Mendez, H. R., & Rodriguez, C. B. (2019). The suprachiasmatic nucleus; a responsive clock regulating homeostasis by daily changing the setpoints of physiological parameters. *Autonomic Neuroscience*, *21*, 843–850.
Buijs, F. N., Guzman-Ruiz, M., Leon-Mercado, L., Basualdo, M. C., Escobar, C., Kalsbeek, A., et al. (2017). Suprachiasmatic nucleus interaction with the Arcuate nucleus; essential for organizing physiological rhythms. *eNeuro*, *4*(2).
Burgess, H. J., & Emens, J. S. (2018). Drugs used in circadian sleep-wake rhythm disturbances. *Sleep Medicine Clinics*, *13*(2), 231–241.
Cardinali, D. P. (2017). *Autonomic nervous system. Basic and clinical aspects*. Switzerland: Springer International Publishing.
Cardinali, D. P. (2019a). Are melatonin doses employed clinically adequate for melatonin-induced cytoprotection? *Melatonin Research*, *2*(2), 106–132.
Cardinali, D. P. (2019b). Melatonin as a chronobiotic/cytoprotector: Its role in healthy aging. *Biological Rhythm Research*, *50*(1), 28–45.
Cardinali, D. P. (2019c). Melatonin: Clinical perspectives in neurodegeneration. *Frontiers in Endocrinology (Lausanne)*, *10*, 480.
Cardinali, D. P., Brown, G. M., & Pandi-Perumal, S. R. (2020). Can melatonin be a potential "silver bullet" in treating COVID-19 patients? *Diseases*. https://doi.org/10.3390/diseases8040044.
Cardinali, D. P., Esquifino, A. I., Srinivasan, V., & Pandi-Perumal, S. R. (2008). Melatonin and the immune system in aging. *Neuroimmunomodulation*, *15*(4-6), 272–278.
Cardinali, D. P., Brown, G. M., Reiter, R. J., & Pandi-Perumal, S. R. (2020). Elderly as a high-risk group during Covid-19 pandemic: Effect of circadian misalignment, sleep dysregulation and melatonin administration. *Sleep & Vigilance*. https://doi.org/10.1007/s41782-020-00111-7.
Cardinali, D. P., Golombek, D., Rosenstein, R. E., Brusco, L. I., & Vigo, D. E. (2016). Assessing the efficacy of melatonin to curtail benzodiazepine/Z drug abuse. *Pharmacological Research*, *109*, 12–23.
Cardinali, D. P., Ritta, M. N., Fuentes, A. M., Gimeno, M. F., & Gimeno, A. L. (1980). Prostaglandin E release by rat medial basal hypothalamus in vitro. Inhibition by melatonin at submicromolar concentrations. *European Journal of Pharmacology*, *67*, 151–153.
Carrillo-Vico, A., Lardone, P. J., Alvarez-Sanchez, N., Rodriguez-Rodriguez, A., & Guerrero, J. M. (2013). Melatonin: Buffering the immune system. *International Journal of Molecular Sciences*, *14*(4), 8638–8683.
Cecon, E., Oishi, A., & Jockers, R. (2018). Melatonin receptors: Molecular pharmacology and signalling in the context of system bias. *British Journal of Pharmacology*, *75*(16), 3263–3280.
Clarke, I. J., & Caraty, A. (2013). Kisspeptin and seasonality of reproduction. *Advances in Experimental Medicine and Biology*, *784*, 411–430.

Claustrat, B., & Leston, J. (2015). Melatonin: Physiological effects in humans. *Neurochirurgie, 61*(2–3), 77–84.

Clay, E., Falissard, B., Moore, N., & Toumi, M. (2013). Contribution of prolonged-release melatonin and anti-benzodiazepine campaigns to the reduction of benzodiazepine and Z-drugs consumption in nine European countries. *European Journal of Clinical Pharmacology, 69*(4), 1–10.

Costantino, G., Cuzzocrea, S., Mazzon, E., & Caputi, A. P. (1998). Protective effects of melatonin in zymosan-activated plasma-induced paw inflammation. *European Journal of Pharmacology, 363*(1), 57–63.

Dawson, D., & Armstrong, S. M. (1996). Chronobiotics—Drugs that shift rhythms. *Pharmacology & Therapeutics, 69*(1), 15–36.

Deng, W. G., Tang, S. T., Tseng, H. P., & Wu, K. K. (2006). Melatonin suppresses macrophage cyclooxygenase-2 and inducible nitric oxide synthase expression by inhibiting p52 acetylation and binding. *Blood, 108*(2), 518–524.

Dubocovich, M. L., Delagrange, P., Krause, D. N., Sugden, D., Cardinali, D. P., & Olcese, J. (2010). International Union of Basic and Clinical Pharmacology. LXXV. Nomenclature, classification, and pharmacology of G protein-coupled melatonin receptors. *Pharmacological Reviews, 62*(3), 343–380.

Duffy, J. F., Zitting, K. M., & Chinoy, E. D. (2015). Aging and circadian rhythms. *Sleep Medicine Clinics, 10*(4), 423–434.

Emens, J. S., & Eastman, C. I. (2017). Diagnosis and treatment of Non-24-h sleep-wake disorder in the blind. *Drugs, 77*(6), 637–650.

Esteve, A., Permanyer, I., Boertien, D., & Vaupel, J. W. (2020). National age and coresidence patterns shape COVID-19 vulnerability. *Proceedings of the National Academy of Sciences of the United States of America.* https://doi.org/10.1073/pnas.2008764117.

Fainstein, I., Bonetto, A., Brusco, L. I., & Cardinali, D. P. (1997). Effects of melatonin in elderly patients with sleep disturbance. A pilot study. *Current Therapeutic Research, 58*, 990–1000.

Ferracioli-Oda, E., Qawasmi, A., & Bloch, M. H. (2013). Meta-analysis: Melatonin for the treatment of primary sleep disorders. *PLoS One, 8*(5), e63773.

Foley, D. J., Monjan, A. A., Brown, S. L., Simonsick, E. M., Wallace, R. B., & Blazer, D. G. (1995). Sleep complaints among elderly persons: An epidemiologic study of three communities. *Sleep, 18*, 425–432.

Furuya, M., Kunishige, K., Miyaoka, T., Wake, R., Liaury, K., Sadakuni, F., et al. (2012). Augmentation with ramelteon to achieve remission in geriatric major depression. *Psychiatry and Clinical Neurosciences, 66*(1), 81–82.

Galano, A., & Reiter, R. J. (2018). Melatonin and its metabolites vs oxidative stress: From individual actions to collective protection. *Journal of Pineal Research, 65*(1), e12514.

Galano, A., Tan, D. X., & Reiter, R. J. (2011). Melatonin as a natural ally against oxidative stress: A physicochemical examination. *Journal of Pineal Research, 51*(1), 1–16.

Geoffroy, P. A., Micoulaud Franchi, J. A., Lopez, R., & Schroder, C. M. (2019). The use of melatonin in adult psychiatric disorders: Expert recommendations by the French institute of medical research on sleep (SFRMS). *Encephale, 45*(5), 413–423.

Gobbi, G., & Comai, S. (2018). Sleep well. Untangling the role of melatonin MT1 and MT2 receptor in sleep. *Journal of Pineal Research, 66*, e12544.

Hardeland, R. (2015). Melatonin and circadian oscillators in aging—A dynamic approach to the multiply connected players. *Interdisciplinary Topics in Gerontology, 40*, 128–140.

Hardeland, R. (2018a). Melatonin and inflammation-story of a double-edged blade. *Journal of Pineal Research, 65*(4), e12525.

Hardeland, R. (2018b). Recent findings in melatonin research and their relevance to the CNS. *Central Nervous System Agents in Medicinal Chemistry, 18*(2), 102–114.

Hardeland, R., Cardinali, D. P., Srinivasan, V., Spence, D. W., Brown, G. M., & Pandi-Perumal, S. R. (2011). Melatonin—A pleiotropic, orchestrating regulator molecule. *Progress in Neurobiology*, *9*, 3350–3384.

Iliff, J. J., Wang, M., Liao, Y., Plogg, B. A., Peng, W., Gundersen, G. A., et al. (2012). A paravascular pathway facilitates CSF flow through the brain parenchyma and the clearance of interstitial solutes, including amyloid beta. *Science Translational Medicine*, *4*(147), 147ra111.

Ju, Y. S., Ooms, S. J., Sutphen, C., Macauley, S. L., Zangrilli, M. A., Jerome, G., et al. (2017). Slow wave sleep disruption increases cerebrospinal fluid amyloid-beta levels. *Brain*, *140*(8), 2104–2111.

Kang, J. E., Lim, M. M., Bateman, R. J., Lee, J. J., Smyth, L. P., Cirrito, J. R., et al. (2009). Amyloid-beta dynamics are regulated by orexin and the sleep-wake cycle. *Science*, *326*(5955), 1005–1007.

Leger, D., Laudon, M., & Zisapel, N. (2004). Nocturnal 6-sulfatoxymelatonin excretion in insomnia and its relation to the response to melatonin replacement therapy. *The American Journal of Medicine*, *116*(2), 91–95.

Lewy, A. (2010). Clinical implications of the melatonin phase response curve. *The Journal of Clinical Endocrinology and Metabolism*, *95*(7), 3158–3160.

Li, T., Jiang, S., Han, M., Yang, Z., Lv, J., Deng, C., et al. (2019). Exogenous melatonin as a treatment for secondary sleep disorders: A systematic review and meta-analysis. *Frontiers in Neuroendocrinology*, *52*(1), 22–28.

Liu, R. Y., Zhou, J. N., van Heerikhuize, J., Hofman, M. A., & Swaab, D. F. (1999). Decreased melatonin levels in postmortem cerebrospinal fluid in relation to aging, Alzheimer's disease, and apolipoprotein E-epsilon4/4 genotype. *The Journal of Clinical Endocrinology and Metabolism*, *84*(1), 323–327.

Manchester, L. C., Coto-Montes, A., Boga, J. A., Andersen, L. P., Zhou, Z., Galano, A., et al. (2015). Melatonin: An ancient molecule that makes oxygen metabolically tolerable. *Journal of Pineal Research*, *59*(4), 403–419.

Mazzotti, D. R., Guindalini, C., Moraes, W. A., Andersen, M. L., Cendoroglo, M. S., Ramos, L. R., et al. (2014). Human longevity is associated with regular sleep patterns, maintenance of slow wave sleep, and favorable lipid profile. *Frontiers in Aging Neuroscience*, *6*, 134.

Nduhirabandi, F., & Maarman, G. J. (2018). Melatonin in heart failure: A promising therapeutic strategy? *Molecules*, *23*(7), E1819.

Obayashi, K., Saeki, K., Iwamoto, J., Okamoto, N., Tomioka, K., Nezu, S., et al. (2013). Nocturnal urinary melatonin excretion is associated with non-dipper pattern in elderly hypertensives. *Hypertension Research*, *36*(8), 736–740.

O'Neill, S., & O'Driscoll, L. (2015). Metabolic syndrome: A closer look at the growing epidemic and its associated pathologies. *Obesity Reviews*, *16*(1), 1–12.

Ooms, S., Overeem, S., Besse, K., Rikkert, M. O., Verbeek, M., & Claassen, J. A. (2014). Effect of 1 night of total sleep deprivation on cerebrospinal fluid beta-amyloid 42 in healthy middle-aged men: A randomized clinical trial. *JAMA Neurology*, *71*(8), 971–977.

Palagini, L., Manni, R., Aguglia, E., Amore, M., Brugnoli, R., Girardi, P., et al. (2020). Expert opinions and consensus recommendations for the evaluation and management of insomnia in clinical practice: Joint statements of five Italian scientific societies. *Frontiers in Psychiatry*. https://doi.org/10.3389/fpsyt.2020.00558.

Pandi-Perumal, S. R., BaHammam, A. S., Ojike, N. I., Akinseye, O. A., Kendzerska, T., Butoo, K., et al. (2017). Melatonin and human cardiovascular disease. *Journal of Cardiovascular Pharmacology and Therapeutics*, *22*, 122–133.

Pappolla, M. A., Matsubara, E., Vidal, R., Pacheco-Quinto, J., Poeggeler, B., Zagorski, M., et al. (2018). Melatonin treatment enhances A lymphatic clearance in a transgenic mouse model of amyloidosis. *Current Alzheimer Research*, *15*(7), 637–642.

Poeggeler, B., Reiter, R. J., Tan, D. X., Chen, L. D., & Manchester, L. C. (1993). Melatonin, hydroxyl radical-mediated oxidative damage, and aging: A hypothesis. *Journal of Pineal Research, 14*(4), 151–168.

Prince, M., Bryce, R., Albanese, E., Wimo, A., Ribeiro, W., & Ferri, C. P. (2013). The global prevalence of dementia: A systematic review and metaanalysis. *Alzheimer's & Dementia, 9*(1), 63–75.

Reagan-Shaw, S., Nihal, M., & Ahmad, N. (2008). Dose translation from animal to human studies revisited. *The FASEB Journal, 22*(3), 659–661.

Reiter, R. J., Ma, Q., & Sharma, R. (2020). Treatment of ebola and other infectious diseases: Melatonin "goes viral". *Melatonin Research, 3*(1), 43–57. https://doi.org/10.32794/mr11250047.

Reiter, R. J., Tan, D. X., Rosales-Corral, S., Galano, A., Jou, M. J., & Acuña-Castroviejo, D. (2018). Melatonin mitigates mitochondrial meltdown: Interactions with SIRT3. *International Journal of Molecular Sciences, 19*(8), E2439.

Riede, S. J., van der Vinne, V., & Hut, R. A. (2017). The flexible clock: Predictive and reactive homeostasis, energy balance and the circadian regulation of sleep-wake timing. *The Journal of Experimental Biology, 220*(Pt. 5), 738–749.

Shukla, M., Govitrapong, P., Boontem, P., Reiter, R. J., & Satayavivad, J. (2017). Mechanisms of melatonin in alleviating Alzheimer's disease. *Current Neuropharmacology, 15*(7), 1010–1031.

Sirin, F. B., Kumbul, D. D., Vural, H., Eren, I., Inanli, I., Sutcu, R., et al. (2015). Plasma 8-isoPGF2alpha and serum melatonin levels in patients with minimal cognitive impairment and Alzheimer disease. *Turkish Journal of Medical Sciences, 45*(5), 1073–1077.

Skene, D. J., & Arendt, J. (2007). Circadian rhythm sleep disorders in the blind and their treatment with melatonin. *Sleep Medicine, 8*(6), 651–655.

Song, X., Sun, X., Ma, G., Sun, Y., Shi, Y., Du, Y., et al. (2015). Family association study between melatonin receptor gene polymorphisms and polycystic ovary syndrome in Han Chinese. *European Journal of Obstetrics, Gynecology, and Reproductive Biology, 195*, 108–112.

Spinedi, E., & Cardinali, D. P. (2018). The polycystic ovary syndrome and the metabolic syndrome: A possible chronobiotic-cytoprotective adjuvant therapy. *International Journal of Endocrinology, 2018*, 1349868.

Sugden, D. (1983). Psychopharmacological effects of melatonin in mouse and rat. *The Journal of Pharmacology and Experimental Therapeutics, 227*(3), 587–591.

Tan, D. X., & Reiter, R. J. (2019). Mitochondria: The birth place, the battle ground and the site of melatonin metabolism. *Melatonin Research*, 244–266.

Tan, D. X., Xu, B., Zhou, X., & Reiter, R. J. (2018). Pineal calcification, melatonin production, aging, associated health consequences and rejuvenation of the pineal gland. *Molecules, 23*(2), E301.

Tutuncu, N. B., Batur, M. K., Yildirir, A., Tutuncu, T., Deger, A., Koray, Z., et al. (2005). Melatonin levels decrease in type 2 diabetic patients with cardiac autonomic neuropathy. *Journal of Pineal Research, 39*(1), 43–49.

Vallim, J. R. D. S., Amaral, F. G. D., Cipolla-Neto, J., & D'Almeida, V. (2019). Rhythmic changes in Fabry disease: Inversion and non-oscillatory pattern in 6-sulfatoxymelatonin daily profile. *Chronobiology International, 36*, 1–11.

Vecchierini, M., Kilic-Huck, U., & Quera-Salva, M. (2020). Melatonin (MEL) and its use in neurological diseases and insomnia: Recommendations of the French Medical and Research Sleep Society. *Revue Neurologique*. https://doi.org/10.1016/j.neurol.2020.06.009.

Venegas, C., Garcia, J. A., Doerrier, C., Volt, H., Escames, G., Lopez, L. C., et al. (2013). Analysis of the daily changes of melatonin receptors in the rat liver. *Journal of Pineal Research, 54*(3), 313–321.

Welz, P. S., & Benitah, S. A. (2019). Molecular connections between circadian clocks and aging. *Journal of Molecular Biology*, *S0022-2836*(19). 30746-6.

Wilson, S. J., Nutt, D. J., Alford, C., Argyropoulos, S. V., Baldwin, D. S., Bateson, A. N., et al. (2010). British Association for Psychopharmacology consensus statement on evidence-based treatment of insomnia, parasomnias and circadian rhythm disorders. *Journal of Psychopharmacology*, *24*(11), 1577–1601.

Winkler, A., Auer, C., Doering, B. K., & Rief, W. (2014). Drug treatment of primary insomnia: A meta-analysis of polysomnographic randomized controlled trials. *CNS Drugs*, *28*(9), 799–816.

Xu, J., Wang, L. L., Dammer, E. B., Li, C. B., Xu, G., Chen, S. D., et al. (2015). Melatonin for sleep disorders and cognition in dementia: A meta-analysis of randomized controlled trials. *American Journal of Alzheimer's Disease and Other Dementias*, *30*(5), 439–447.

Zhang, W., Chen, X. Y., Su, S. W., Jia, Q. Z., Ding, T., Zhu, Z. N., et al. (2016). Exogenous melatonin for sleep disorders in neurodegenerative diseases: A meta-analysis of randomized clinical trials. *Neurological Sciences*, *37*(1), 57–65.

Zhdanova, I. V., Wurtman, R. J., Regan, M. M., Taylor, J. A., Shi, J. P., & Leclair, O. U. (2001). Melatonin treatment for age-related insomnia. *The Journal of Clinical Endocrinology and Metabolism*, *86*(10), 4727–4730.

CHAPTER FIVE

Luteinizing hormone and the aging brain

Megan Mey[a], Sabina Bhatta[a], and Gemma Casadesus[b,*]
[a]Department of Biomedical Science, Kent State University, Cunningham Hall, Kent, OH, United States
[b]Department of Biological Science, School of Arts and Sciences, Kent State University, Cunningham Hall, Kent, OH, United States
*Corresponding author: e-mail addresses: gcasades@kent.edu; gcasades@aim.com

Contents

Abstract

Fluctuations in luteinizing hormone (LH) release contribute to the development and maintenance of the reproductive system and become dysregulated during aging. Of note, increasing evidence supports extra-gonadal roles for LH within the CNS, particularly as it relates to cognition and plasticity in aging and age-related degenerative diseases such as Alzheimer's disease (AD). However, despite increasing evidence that supports a link between this hormone and CNS function, the mechanisms underlying LH action within the brain and how they influence cognition and plasticity during the lifespan is poorly understood and, in fact, often in conflict. This chapter aims to provide an up-to-date review of the literature addressing the role of LH signaling in the context of CNS aging and disease and put forward a unifying hypothesis that may explain currently conflicting theories regarding the role of LHCGR signaling in CNS function and dysfunction in aging and disease.

List of Abbreviations

AD	Alzheimer's disease
BBB	blood brain barrier
CamKII	calcium-calmodulin kinase II
cAMP	cyclic AMP
CNS	central nervous system
FSH	follicle stimulating hormone

Vitamins and Hormones, Volume 115
ISSN 0083-6729
https://doi.org/10.1016/bs.vh.2020.12.005

GnRH	gonadotropin releasing hormone
GnRHR	gonadotropin releasing hormone receptor
GSK3β	glycogen synthase kinase 3 beta
hCG	human chorionic gonadotropin
HPG axis	hypothalamic-Pituitary-Gonadal axis
ICV	intracerebroventricular
LA	leuprolide acetate
LH	luteinizing hormone
LHCGR	luteinizing hormone receptor
OVX	ovariectomy
PKA	protein kinase A
PLC	phospholipase C

1. Introduction

The role of circulating luteinizing hormone (LH) and its receptor in are well understood in the context of steroid production and reproductive function as well as in changes that the hypothalamic-pituitary-gonadal (HPG) axis undergoes during our lifespan. Furthermore, it is also well known that the loss of any components within this axis such as during menopause, leads to significant dysregulation of the system, affects homeostasis, and results in functional decline throughout the body, including the brain (McEwen, Akama, Spencer-Segal, Milner, & Waters, 2012; Greendale, Derby, & Maki, 2011). These age-related changes, which are more prominent in women, are thought to contribute to the higher incidence of age-related neurodegenerative diseases such as Alzheimer's disease (Pike, 2017).

Until the last decade the relationship between menopause hormones and CNS alterations has been primarily focused on understanding the role of estrogen loss and replacement in this process. In this regard, classic studies have conclusively demonstrated the beneficial impact of estrogen replacement after ovariectomy in rodents (McEwen et al., 2012 for review). However, the lack of ability of estrogen replacement to improve function in animals (Blair, Bhatta, McGee, & Casadesus, 2015; Bohacek & Daniel, 2010) and humans (Henderson & Sherwin, 2007) when there is a delay between menopause and treatment onset, unveils the complex nature of this system and the likely involvement of multiple HPG axis hormones on brain function.

In connection with the above, emerging studies during the last decade support a new role for LH and its receptor in the CNS. Importantly, the age-related upregulation of LH, more prominent in women than men due to menopause, have been linked to cognitive dysfunction, neuronal plasticity impairment, and increased Alzheimer's disease risk (Bhatta, Blair, & Casadesus, 2018; Blair et al., 2015; Burnham & Thornton, 2015 for review). At a cellular level, knowledge of broad LH receptor expression (AL-Hader, Lei, & Rao, 1997; Apaja, Harju, Aatsinki, Petäjä-Repo, & Rajaniemi, 2004; Bukovsky et al., 2003; Gallo, Johnson, Kalra, Whitmoyer, & Sawyer, 1972; Lei, Rao, Kornyei, Licht, & Hiatt, 1993; Mak et al., 2007; Yang, Nasipak, & Kelley, 2007; You, Kim, Hsu, El Halawani, & Foster, 2000; Yuan, Peng, Liu, Feng, & Qiu, 2019) and early reports of LH action on neuronal electrophysiology (Gallo et al., 1972) of behavior go as far back as four decades and support receptor functionality within the brain. However, despite multiple reports of LHCGR functionality in the CNS over the years, the precise mechanisms underlying LH's regulation of neuronal function and plasticity and how its signaling dysregulation during reproductive and neuronal aging impacts these functions remain largely unknown. Furthermore, current findings seem to support both detrimental and beneficial roles for LHCGR activation in age-related CNS function. Therefore, this chapter aims to provide an up to date survey of the work that supports a role for LH in the aging CNS as it pertains to cognition and neuronal plasticity. This chapter also aims to provide critical insight on novel findings that can reconcile current conflicting hypotheses for LHCGR action in the CNS (Fig. 1).

1.1 HPG axis regulation and LH release across the lifespan

LH is one of three gonadotropins that regulate gonadal function and are essential for normal growth, sexual development and reproduction. Pituitary-derived LH and follicle stimulating hormone (FSH) and placental-derived chorionic gonadotropin (hCG) are heterodimeric proteins containing two chains. A common alpha chain is shared between the three gonadotropins and a beta chain is unique to each and determines the specific functions of each hormone and the affinity to their corresponding receptor. It is important to note, however, that LH and hCG share 84% homology and share the same receptor, hence the common use of hCG to induce ovulation during assisted reproductive therapy. LH and FSH are under the control of gonadotropin releasing hormone (GnRH), which is released from the hypothalamus in response to circulating levels of gonadal steroids through the kisspeptin system (reviewed in (Lehman, Coolen, & Goodman, 2010;

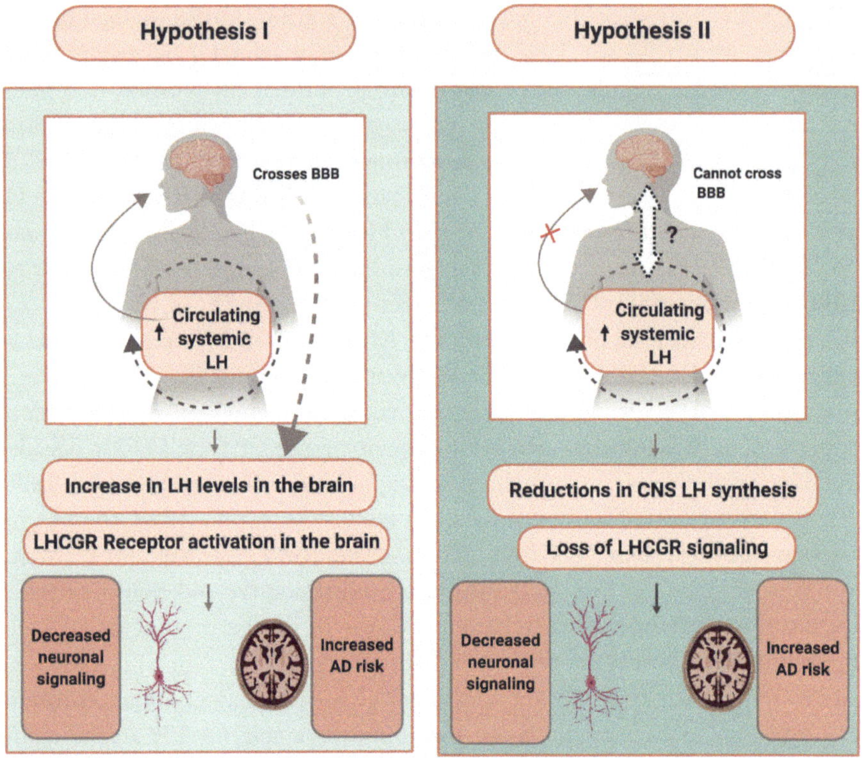

Fig. 1 Hypotheses for LH involvement in age-related cognitive decline and Alzheimer's disease risk. Conflicting hypotheses explain the mechanism by which increased circulating LH could be influencing associated cognitive decline, and increased risk of Alzheimer's disease. One hypothesis assumes that LH crosses the BBB, binding LHCGR contributing to cognitive decline. The second hypothesis considers an inverse relationship between circulating and brain LH, suggesting that increased circulating LH leads to decreased synthesis of CNS LH and therefore decreased LHCGR activation contributing to cognitive decline.

Marques, Skorupskaite, George, & Anderson, 2000). This feedback loop system is termed the hypothalamic-pituitary-gonadal (HPG) axis. Specifically, during reproductive stages, HPG axis action involves the activation of the gonadotropin-releasing hormone receptor (GnRHR) in the pituitary by steroid induced hypothalamic GnRH release. This, in turn, leads to the release of gonadotropins into the bloodstream that reach the gonads and initiate the production of sex steroids. These sex steroids complete the HPG axis by providing negative feedback to shut down further GnRH.

LH has different functions in males and females and its release patterns is therefore also widely different between sexes. In females, LH binds its receptor in ovarian follicles to promote their maturation. Additionally, the mid-cycle surge of LH triggers ovulation and corpus luteum production of progesterone to ensure optimal maturation of the uterine endometrium and implantation of the fertilized egg. In males, LH stimulates production of testosterone by the testes (Rahman & Rao, 2009). Because of its different role a between males and females, its levels remain stable from puberty on in males while in females, LH fluctuates widely with menstrual cycle until reproductive senescence. To this end, with increasing age, males gradually produce less testosterone leading to gradual decreases in negative feedback onto hypothalamic GnRH neurons and gradually rising LH levels (Daniel, Hulst, & Berbling, 2006; Neaves, Johnson, Porter, Parker, & Petty, 1984). In females, however, the rise in circulating is more sudden. The progressive exhaustion of the follicle pool and eventual complete loss of steroid negative feedback onto the pituitary gland during the fourth and fifth decade of life, leads to large increases in gonadotropin hormone synthesis and release (Couzinet, Meduri, & Leese, 2001) that denote the onset of menopause.

1.2 Age-related changes in circulating LH and cognitive decline

The concept that levels of LH or hCG may have an impact on the brain that is reflected in behavior is not new (Berkowitz et al., 1981; Toth et al., 1994). For example, during pregnancy changes in serum hCG are reflected by changes in sleep pattern (Berkowitz et al., 1981). Other studies have suggested that LHCGR may be involved in wakefulness following a decrease in activity after female rats were injected intraperitoneally with hCG (Toth et al., 1994). LHR activation in the CNS of *Xenopus laevis* is known to affect courtship songs (Yang et al., 2007), and LH receptor activation has been linked to increased neurogenesis in the dentate gyrus of the hippocampus in relation to pheromone-related behaviors (Mak et al., 2007). Furthermore, administration of LH or hCG in rats reduced locomotor activity levels and taste neophobia as well as food intake (Emanuele, Oslapas, Connick, Kirsteins, & Lawrence, 1981; Lukacs, Hiatt, Lei, & Rao, 1995). However, more recently, the relationship between circulating LH and behavior has extended to the realm of memory function and has also been implicated in the development of Alzheimer's disease (AD).

During the last two decades, multiple clinical studies have shown a correlation between high circulating LH and cognitive decline and AD development (Hogervorst, Combrinck, & Smith, 2003; Short, Bowen, O'Brien, & Graff-Radford, 2001) as well as AD pathology (Bowen et al., 2002; Verdile et al., 2014, 2008). In rodents, earlier studies demonstrated that over-expression of LH in transgenic mice (Casadesus et al., 2007), or exogenous treatment with human LH (Wahjoepramono et al., 2011) without any estrogen manipulations, impaired working memory and/or increased AD pathology. Furthermore, in animal studies, administration of the analog hCG, leads to cognitive decline or increased presence of AD-related neuronal markers (Barron, Verdile, Taddei, Bates, & Martins, 2010; Berry, Tomidokoro, Ghiso, & Thornton, 2008; Burnham, Sundby, Laman-Maharg, & Thornton, 2017). The relationship persists even with the presence of estrogen replacement, suggesting an independent action of the LH peptide on function (Berry et al., 2008).

The clinical associations summarized above have led to a more direct interrogation of the involvement of LH in age-related conditions, namely by suppressing high post-menopausal LH levels. To this end, LH (and gonadal steroids) can be suppressed by GnRH super agonists or antagonists, an approach is commonly used to suppress steroid synthesis in patients with steroid-positive cancers. Interestingly, while such treatments have been linked to cognitive impairments in reproductively active patients (Henderson & Sherwin, 2007). In the context of elderly populations, the reverse seems to be true. To this end, a small phase II trial using leuprolide acetate, a GnRHR negative agonist, to investigate benefits of circulating LH suppression in women with AD showed stabilization of cognition in a subgroup taking acetylcholinesterase inhibitors (Bowen, Perry, Xiong, Smith, & Atwood, 2015). Interestingly, although not all studies concur (Nead, Sinha, & Nguyen, 2017), another study also showed that gonadotropin suppression in prostate cancer patients may shield against AD risk (Bowen et al., 2015).

The findings observed in clinical cohorts have been extensively re-created in rodents. To this end, the use of different types of GnRHR negative modulators to downregulate high circulating ovariectomy/orchiectomy-induced LH levels have consistently shown benefits on learning and memory and associated molecular mechanisms (Bohm-Levine, Goldberg, Mariani, Frankfurt, & Thornton, 2020; Bryan et al., 2010; Telegdy, Adamik, Tanaka, & Schally, 2010). For example, Bryan et al. (2010) demonstrated that improvement in spatial memory in ovariectomized female mice was associated with increased auto-phosphorylation of CAMKII and subsequent phosphorylation of the AMPA GluR1 receptor subunit, required for induction and maintenance of LTP (Hayashi et al., 2000). Similarly, downregulation of LH using comparable approaches have shown increased spine density (Blair et al., 2015). More recently, a study using antide, a GnRHR antagonist, demonstrated that, the neuronal trophic factor, BDNF may be required for such cognitive improvements (Bohm-Levine et al., 2020). Interestingly, the benefits of LH suppression in ovariectomized "menopausal" females extended beyond the window of estrogen replacement effectiveness (Blair et al., 2015), suggesting that some of the neuronal benefits of estrogen replacement are at least partially due to its negative feedback onto pituitary LH synthesis.

Such benefits extend to animal models of AD. To this end early studies showed that LH suppression in aged gonadally-intact AD mouse models (Tg2576) improved function and marginally but significantly reduced AD pathology (Casadesus et al., 2007). More recently, Palm et al. (2014) demonstrated that suppression of circulating LH in ovariectomized 3XADtg mice improved cognitive function and regulated important signaling pathways associated with cognition. However, improvements in cognition in this study were not matched by reductions in amyloid and tau pathology (Palm et al., 2014) suggesting that the neuroprotective effects of lowering circulating LH may occur independently of the regulation of AD pathology. Given that LH suppression has also been shown to mitigate neuronal injury (Mariani, Mojziszek, Curley, & Thornton, 2019), it is likely that the impact of this manipulation on trophic factors such as BNDF (Bohm-Levine et al., 2020; Palm et al., 2014) play as an important role in mediating such benefits.

1.3 LHCGR activation effects on CNS function

The studies cited above suggest that increased levels of circulating ligand during aging and its binding and activation of the LHCGR receptor yield a negative impact in the aging CNS. However, the concept that LHCGR is a negative regulator of memory is difficult to reconcile based on the canonical signaling cascades associated with LHCGR activation. LHCGR is known to signal through Gs and Gq, these cascades are both linked to learning and memory formation, specifically LTP mediated synaptic plasticity (Lisman, Yasuda, & Raghavachari, 2012). For example, LHR is a G-protein coupled receptor that signals through Gs, ultimately leading to increased cAMP, CREB, and extracellular signal-related protein kinase (ERK) (reviewed in Bhatta et al., 2018) or the activating protein kinase A (PKA) (Menon & Menon, 2012). ERK and PKA are critical cascades in long-term potentiation, memory and structural plasticity (Hardingham, Arnold, & Bading, 2001; Hebert & Dash, 2002; Impey et al., 1998). Gq signaling through LHCGR has been shown to drive GSK3β inhibition and β-catenin activation (Breen et al., 2013; Palm et al., 2014), which is also known to be beneficial for cognition, plasticity and is involved in AD neuroprotection (Kleppisch, Voigt, Allmann, & Offermanns, 2001; McQuail et al., 2013; Saudargiene, Cobb, & Graham, 2015).

In addition to the signaling nature implications of the LHCGR, the assertion that a large peptide molecule such as LH can freely cross the blood brain barrier without a transporter must be carefully considered (Ondo, Mical, & Porter, 1972). Also important to consider when formulating a hypothesis involving LH's impact in the aging brain, is the ability of both LH and hCG to dose-dependently downregulate its own receptor (Peegel, Randolph, Midgley, & Menon, 1994). In light of this, studies testing direct activation of LHCGR through central intracerebroventricular (ICV) delivery of LH or LH analogs like hCG, have yielded contradictory results. For example, hCG ICV treatment directly to the dorsal hippocampus led to impaired memory function (Burnham et al., 2017) and AD-related phenotype acceleration in and AD mouse model (Barron et al., 2010). ICV hCG administration has also been shown to depress activity and result with motor and cognitive deficits in rats (Lukacs et al., 1995). On the other hand, a recent study using much lower doses of hCG, the LH analog, was able to rescue ovariectomy-related cognitive deficits (Blair, Bhatta, & Casadesus, 2019). These results were paralleled by increased phosphorylation of ERK and increased expression of synaptophysin, a synaptic marker that lies downstream of ERK activation (Giachello et al., 2010; Prekeris & Terrian, 1997). Furthermore, similar neuroprotective findings have been reported for LHCGR activation against hypoxic neuronal injury (Movsas, Weiner, Greenberg, Holtzman, & Galindo, 2017), suggesting that some of this conflict may lie in the doses delivered in the cited studies.

Additional studies are required to identify the role of LHCGR signaling within the brain, its impact on behavior, and its dysregulation during aging. Understanding the dynamics of LHCGR in presence of its ligand could potentially be able to reconcile the contradictory findings discussed above. To this end, it well understood that that LHCGR is quickly internalized and downregulated in the presence of high levels of its ligand (Peegel et al., 1994). Thus, under the hypothesis that LH crosses the blood brain barrier to bind LHCGR, high levels of LH within the brain could downregulate LHCGR expression and lead to cognitive deficits. Reducing circulating LH with GnRHR antagonists could in turn allow LHCGR levels to normalize and resume its normal function. Alternatively, one could also speculate that supra-physiological levels of LH or hCG delivered directly into the brain or delivered to gonadectomized animals (already with high levels of circulating LH) could have similar downregulating effects onto LHCGR expression, which could also impact signaling and function negatively. This would explain why studies using high doses of hCG (Barron et al., 2010; Lukacs et al., 1995) would have detrimental effects on function but studies using lower doses (Blair et al., 2019; Movsas et al., 2017) would have beneficial effects. Thus, precisely characterizing central LHCGR dynamics under different treatment conditions seems to be imperative to understand the nature of the effects of this hormone and its receptor's potential as a therapeutic target for age-related neurodegenerative disease such as AD.

2. Central LH synthesis in aging and CNS function as a unifying hypothesis

The synthesis and expression of LH within the brain has been reported previously (Emanuele et al., 1981; Glass & McClusky, 1987). Importantly, early studies (Emanuele et al., 1981) and later work (Blair et al., 2019; Palm et al., 2014), indicate an inverse relationship between brain derived LH and pituitary-derived (circulating) LH. To this end, in cycling adult female rats, hypothalamic LH decreased during proestrus, a period where LH release from the pituitary is maximal (Emanuele et al., 1981). Similarly, LH treatment into the median eminence (ME) of the hypothalamus in intact or castrated males and females significantly decreased circulating LH levels.

More recently, such inverse relationship has been directly linked to menopause and AD related functional loss as evidenced by the fact that LH mRNA levels are reduced in both the hippocampus and the cortex in AD brains (Palm et al., 2014), while LH circulating levels are reported to be are increased compared to cognitively normal individuals (Short et al., 2001). Similarly, Palm et al. (2014), demonstrated that brain LH was decreased in ovariectomized 3xTg AD female mice and negatively correlated to circulating LH. Importantly, lower LH immunoreactivity within the brain was reversed when ovariectomy-related high circulating LH levels were normalized and this paralleled cognitive improvements and increases in BDNF transcription, inhibition of GSK3β and upregulation of β-catenin, all important in neuronal plasticity. More recently, another study has shown a strong negative correlation between brain and circulating LH levels in ovariectomized animals that when normalized through low central re-administration of hCG led to cognitive improvements and

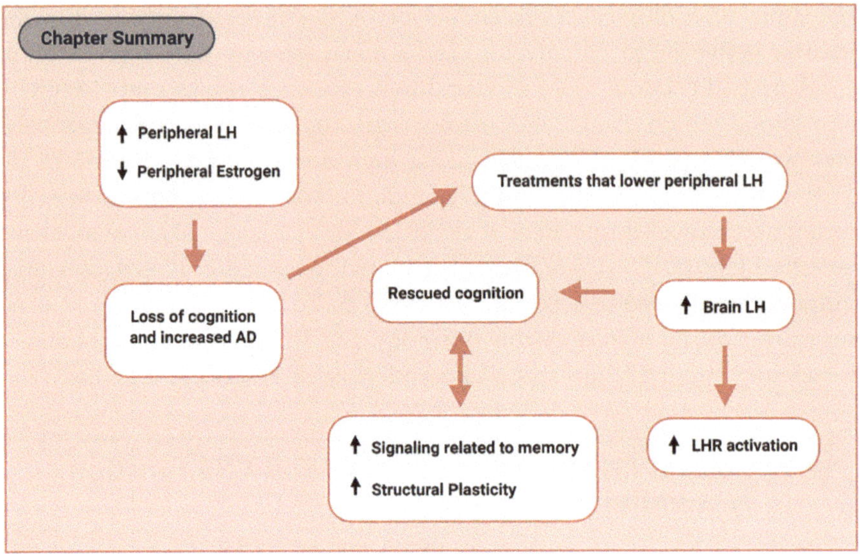

Fig. 2 Luteinizing hormone action in learning and memory. Increased peripheral LH acquired by age related dysregulation of HPG axis is paralleled by decreased synthesis of LH in the CNS and cognitive deficit following impaired plasticity. Treatments that down regulate circulating LH increase LH synthesis in the CNS and rescue cognition. Increased synthesis of CNS LH could be stimulating LHR in the brain. The downstream effects of LHR are associated with improved synaptic plasticity which may underly the rescued cognition following circulating LH downregulation.

activation of plasticity-associated signaling (Blair et al., 2019). Taken together, while the mechanism underlying this inverse relationship between circulating and local CNS LH levels is beyond the scope of this chapter, this inverse relationship suggests that, perhaps, chronic upregulation of pituitary LH associated with aging, leads to downregulation of central LH synthesis. This, in turn, results in loss of LHCGR signaling and CNS impairment. Improtantly, these data may provide a reconciling link between data demonstrating that peripheral downregulation of LH is beneficial and the cognition-enhancing (not impairing) signaling pathways that underly LHCGR activation (Fig. 2).

3. Conclusion

The tight inter-relationship between hormonal systems and the brain now extends to the understanding of LH signaling in the CNS. This is particularly true for cognition within reproductive senescence-related aging and AD. However, the mechanism of action of LH within the brain continue to be poorly understood. Work stemming from studies using GNRHR-related mechanisms to deplete circulating LH rather than by direct LHCGR inhibition confound the potential conclusions that can be made in relation to the role of this receptor in mediating function. Additionally, conflicting results in studies using analogs such as hCG to drive LHCGR activation further complicate our ability to interpret the impact of LHCGR activation or inhibition on CNS function during reproductive senescence. These conflicts may stem from the nature of LHCGR regulation dynamics upon changing levels of ligand and the lack of clarity as to whether LH or hCG cross the blood brain barrier. Therefore, an in-depth characterization of central LHCGR trafficking and transcription under different treatment and hormonal status conditions seems to be imperative to be able to understand the function of this hormone and its receptor on CNS across the lifespan. Lastly, the potential role of brain LH in mediating LHR receptor activation and its inverse relationship with pituitary-derived LH must be mechanistically probed in depth to develop a comprehensive understanding of this hormone role in the age-related cognitive decline and AD development. However, such understanding poses the potential key to unify the current conflicts in the literature.

References

AL-Hader, A. A., Lei, Z. M., & Rao, C. V. (1997). Novel expression of functional luteinizing hormone/chorionic gonadotropin receptors in cultured glial cells from neonatal rat brains. *Biology of Reproduction*, *56*(2), 501–507. https://doi.org/10.1095/biolreprod56.2.501.

Apaja, P. M., Harju, K. T., Aatsinki, J. T., Petäjä-Repo, U. E., & Rajaniemi, H. J. (2004). Identification and structural characterization of the neuronal luteinizing hormone receptor associated with sensory systems. *Journal of Biological Chemistry*, *279*(3), 1899–1906. https://doi.org/10.1074/jbc.M311395200.

Barron, A. M., Verdile, G., Taddei, K., Bates, K. A., & Martins, R. N. (2010). Effect of chronic hCG administration on Alzheimer's-related cognition and a beta accumulation in PS1KI mice. *Endocrinology*, *151*(11), 5380–5388. https://doi.org/10.1210/en.2009-1168.

Berkowitz, R. S., Osathanondh, R., Goldstein, D. P., Martin, P. M., Mallampati, S. R., & Datta, S. (1981). Cerebrospinal fluid human chorionic gonadotropin levels in normal pregnancy and choriocarcinoma. *Surgery, Gynecology & Obstetrics*, *153*(5), 687–689.

Berry, A., Tomidokoro, Y., Ghiso, J., & Thornton, J. (2008). Human chorionic gonadotropin (a luteinizing hormone homologue) decreases spatial memory and increases brain amyloid-beta levels in female rats. *Hormones and Behavior*, *54*(1), 143–152. https://doi.org/10.1016/j.yhbeh.2008.02.006.

Bhatta, S., Blair, J. A., & Casadesus, G. (2018). Luteinizing hormone involvement in aging female cognition: Not all is estrogen loss. *Frontiers in Endocrinology*, *9*, 544. https://doi.org/10.3389/fendo.2018.00544.

Blair, J. A., Bhatta, S., & Casadesus, G. (2019). CNS luteinizing hormone receptor activation rescues ovariectomy-related loss of spatial memory and neuronal plasticity. *Neurobiology of Aging*, *78*, 111–120. https://doi.org/10.1016/j.neurobiolaging.2019.02.002.

Blair, J. A., Bhatta, S., McGee, H., & Casadesus, G. (2015). Luteinizing hormone: Evidence for direct action in the CNS. *Hormones and Behavior, 76*, 57–62. https://doi.org/10.1016/j.yhbeh.2015.06.020.

Bohacek, J., & Daniel, J. M. (2010). The beneficial effects of estradiol on attentional processes are dependent on timing of treatment initiation following ovariectomy in middle-aged rats. *Psychoneuroendocrinology, 35*(5), 694–705. https://doi.org/10.1016/j.psyneuen.2009.10.010.

Bohm-Levine, N., Goldberg, A. R., Mariani, M., Frankfurt, M., & Thornton, J. (2020). Reducing luteinizing hormone levels after ovariectomy improves spatial memory: Possible role of brain-derived neurotrophic factor. *Hormones and Behavior, 118*, 104590. https://doi.org/10.1016/j.yhbeh.2019.104590.

Bowen, R. L., Perry, G., Xiong, C., Smith, M. A., & Atwood, C. S. (2015). A clinical study of lupron depot in the treatment of women with Alzheimer's disease: Preservation of cognitive function in patients taking an acetylcholinesterase inhibitor and treated with high dose lupron over 48 weeks. *Journal of Alzheimer's Disease: JAD, 44*(2), 549–560. https://doi.org/10.3233/JAD-141626.

Bowen, R. L., Smith, M. A., Harris, P. L. R., Kubat, Z., Martins, R. N., Castellani, R. J., et al. (2002). Elevated luteinizing hormone expression colocalizes with neurons vulnerable to Alzheimer's disease pathology. *Journal of Neuroscience Research, 70*(3), 514–518. https://doi.org/10.1002/jnr.10452.

Breen, S. M., Andric, N., Ping, T., Xie, F., Offermans, S., Gossen, J. A., et al. (2013). Ovulation involves the luteinizing hormone-dependent activation of G(q/11) in granulosa cells. *Molecular Endocrinology (Baltimore, Md.), 27*(9), 1483–1491. https://doi.org/10.1210/me.2013-1130.

Bryan, K. J., Mudd, J. C., Richardson, S. L., Chang, J., Lee, H., Zhu, X., et al. (2010). Downregulation of serum gonadotropins is as effective as Estrogen replacement at improving menopause-associated cognitive deficits. *Journal of Neurochemistry, 112*(4), 870–881. https://doi.org/10.1111/j.1471-4159.2009.06502.x.

Bukovsky, A., Indrapichate, K., Fujiwara, H., Cekanova, M., Ayala, M. E., Dominguez, R., et al. (2003). Multiple luteinizing hormone receptor (LHR) protein variants, interspecies reactivity of anti-LHR mAb clone 3B5, subcellular localization of LHR in human placenta, pelvic floor and brain, and possible role for LHR in the development of abnormal pregnancy, pelvic floor disorders and Alzheimer's disease. *Reproductive Biology and Endocrinology, 1*(1), 46. https://doi.org/10.1186/1477-7827-1-46.

Burnham, V., Sundby, C., Laman-Maharg, A., & Thornton, J. (2017). Luteinizing hormone acts at the hippocampus to dampen spatial memory. *Hormones and Behavior, 89*, 55–63. https://doi.org/10.1016/j.yhbeh.2016.11.007.

Burnham, V. L., & Thornton, J. E. (2015). Luteinizing hormone as a key player in the cognitive decline of Alzheimer's disease. *Hormones and Behavior, 76*, 48–56. https://doi.org/10.1016/j.yhbeh.2015.05.010.

Casadesus, G., Milliken, E. L., Webber, K. M., Bowen, R. L., Lei, Z., Rao, C. V., et al. (2007). Increases in luteinizing hormone are associated with declines in cognitive performance. *Molecular and Cellular Endocrinology, 269*(1–2), 107–111. https://doi.org/10.1016/j.mce.2006.06.013.

Couzinet, B., Meduri, G., & Leese, M. (2001). Postmenopausal ovary is not a major androgen-producing gland. *The Journal of Clinical Endocrinology and Metabolism, 86*, 5060–5066. Oxford Academic https://academic.oup.com/jcem/article/86/10/5060/2849419.

Daniel, J. M., Hulst, J. L., & Berbling, J. L. (2006). Estradiol replacement enhances working memory in middle-aged rats when initiated immediately after ovariectomy but not after a long-term period of ovarian hormone deprivation. *Endocrinology, 147*(1), 607–614. https://doi.org/10.1210/en.2005-0998.

Emanuele, N., Oslapas, R., Connick, E., Kirsteins, L., & Lawrence, A. M. (1981). Hypothalamic LH may play a role in control of pituitary LH release. *Neuroendocrinology*, *33*(1), 12–17. https://doi.org/10.1159/000123194.

Gallo, R. V., Johnson, J. H., Kalra, S. P., Whitmoyer, D. I., & Sawyer, C. H. (1972). Effects of luteinizing hormone on multiple-unit activity in the rat Hippocampus. *Neuroendocrinology*, *9*(3), 149–157. https://doi.org/10.1159/000122046.

Giachello, C. N. G., Fiumara, F., Giacomini, C., Corradi, A., Milanese, C., Ghirardi, M., et al. (2010). MAPK/Erk-dependent phosphorylation of synapsin mediates formation of functional synapses and short-term homosynaptic plasticity. *Journal of Cell Science*, *123*(6), 881–893. https://doi.org/10.1242/jcs.056846.

Glass, J. D., & McClusky, M. E. (1987). Immunoreactive luteinizing hormone-containing neurons in the brain of the white-footed mouse, Peromyscus leucopus. *Experientia*, *43*, 188–190.

Greendale, G. A., Derby, C. A., & Maki, P. M. (2011). Perimenopause and cognition. *Obstetrics and Gynecology Clinics of North America*, *38*(3), 519–535. https://doi.org/10.1016/j.ogc.2011.05.007.

Hardingham, G. E., Arnold, F. J., & Bading, H. (2001). Nuclear calcium signaling controls CREB-mediated gene expression triggered by synaptic activity. *Nature Neuroscience*, *4*(3), 261–267. https://doi.org/10.1038/85109.

Hayashi, Y., Shi, S. H., Esteban, J. A., Piccini, A., Poncer, J. C., & Malinow, R. (2000). Driving AMPA receptors into synapses by LTP and CaMKII: Requirement for GluR1 and PDZ domain interaction. *Science (New York, N.Y.)*, *287*(5461), 2262–2267. https://doi.org/10.1126/science.287.5461.2262.

Hebert, A. E., & Dash, P. K. (2002). Extracellular signal-regulated kinase activity in the entorhinal cortex is necessary for long-term spatial memory. *Learning & Memory (cold Spring Harbor, N.Y.)*, *9*(4), 156–166. https://doi.org/10.1101/lm.48502.

Henderson, V. W., & Sherwin, B. B. (2007). Surgical versus natural menopause: Cognitive issues. *Menopause (New York, N.Y.)*, *14*(3 Pt 2), 572–579. https://doi.org/10.1097/gme.0b013e31803df49c.

Hogervorst, E., Combrinck, M., & Smith, A. D. (2003). Testosterone and gonadotropin levels in men with dementia. *Neuro Endocrinology Letters*, *24*(3–4), 203–208.

Impey, S., Obrietan, K., Wong, S. T., Poser, S., Yano, S., Wayman, G., et al. (1998). Cross talk between ERK and PKA is required for Ca2+ stimulation of CREB-dependent transcription and ERK nuclear translocation. *Neuron*, *21*(4), 869–883. https://doi.org/10.1016/s0896-6273(00)80602-9.

Kleppisch, T., Voigt, V., Allmann, R., & Offermanns, S. (2001). G(alpha)q-deficient mice lack metabotropic glutamate receptor-dependent long-term depression but show normal long-term potentiation in the hippocampal CA1 region. *The Journal of Neuroscience: The Official Journal of the Society for Neuroscience*, *21*(14), 4943–4948.

Lehman, M. N., Coolen, L. M., & Goodman, R. L. (2010). Minireview: Kisspeptin/neurokinin B/dynorphin (KNDy) cells of the arcuate nucleus: A central node in the control of gonadotropin-releasing hormone secretion. *Endocrinology*, *151*(8), 3479–3489. https://doi.org/10.1210/en.2010-0022.

Lei, Z. M., Rao, C. V., Kornyei, J. L., Licht, P., & Hiatt, E. S. (1993). Novel expression of human chorionic gonadotropin/luteinizing hormone receptor gene in brain. *Endocrinology*, *132*(5), 2262–2270. https://doi.org/10.1210/endo.132.5.8477671.

Lisman, J., Yasuda, R., & Raghavachari, S. (2012). Mechanisms of CaMKII action in long-term potentiation. *Nature Reviews. Neuroscience*, *13*(3), 169–182. https://doi.org/10.1038/nrn3192.

Lukacs, H., Hiatt, E. S., Lei, Z. M., & Rao, C. V. (1995). Peripheral and intracerebroventricular administration of human chorionic gonadotropin alters several hippocampus-associated behaviors in cycling female rats. *Hormones and Behavior*, *29*(1), 42–58. https://doi.org/10.1006/hbeh.1995.1004.

Mak, G. K., Enwere, E. K., Gregg, C., Pakarainen, T., Poutanen, M., Huhtaniemi, I., et al. (2007). Male pheromone-stimulated neurogenesis in the adult female brain: Possible role in mating behavior. *Nature Neuroscience, 10*(8), 1003–1011. https://doi.org/10.1038/nn1928.

Mariani, M. M., Mojziszek, K., Curley, E., & Thornton, J. E. (2019). Lowering luteinizing hormone (LH) reverses spatial memory deficits associated with neurotoxin infusion into the hippocampus of ovx rats. *Hormones and Behavior, 119*, 104631. https://doi.org/10.1016/j.yhbeh.2019.104631.

Marques, P., Skorupskaite, K., George, J. T., & Anderson, R. A. (2000). Physiology of GNRH and gonadotropin secretion (Updated 2018 Jun 19). In K. R. Feingold, B. Anawalt, A. Boyce, et al. (Eds.), *Endotext [Internet]*. South Dartmouth (MA): MDText.com, Inc. Available from https://www.ncbi.nlm.nih.gov/sites/books/NBK279070/.

McEwen, B. S., Akama, K. T., Spencer-Segal, J. L., Milner, T. A., & Waters, E. M. (2012). Estrogen effects on the brain: Actions beyond the hypothalamus via novel mechanisms. *Behavioral Neuroscience, 126*(1), 4–16. https://doi.org/10.1037/a0026708.

McQuail, J. A., Davis, K. N., Miller, F., Hampson, R. E., Deadwyler, S. A., Howlett, A. C., et al. (2013). Hippocampal Gαq/11 but not Gαo-coupled receptors are altered in aging. *Neuropharmacology, 70*, 63–73. https://doi.org/10.1016/j.neuropharm.2013.01.009.

Menon, K. M. J., & Menon, B. (2012). Structure, function and regulation of gonadotropin receptors—A perspective. *Molecular and Cellular Endocrinology, 356*(1–2), 88–97. https://doi.org/10.1016/j.mce.2012.01.021.

Movsas, T. Z., Weiner, R. L., Greenberg, M. B., Holtzman, D. M., & Galindo, R. (2017). Pretreatment with human chorionic gonadotropin protects the neonatal brain against the effects of hypoxic-ischemic injury. *Frontiers in Pediatrics, 5*, 232, Published 2017 Nov 3. https://doi.org/10.3389/fped.2017.00232.

Nead, K. T., Sinha, S., & Nguyen, P. L. (2017). Androgen deprivation therapy for prostate cancer and dementia risk: A systematic review and meta-analysis. *Prostate Cancer and Prostatic Diseases, 20*(3), 259–264. https://doi.org/10.1038/pcan.2017.10.

Neaves, W. B., Johnson, L., Porter, J. C., Parker, C. R., & Petty, C. S. (1984). Leydig cell numbers, daily sperm production, and serum gonadotropin levels in aging men. *The Journal of Clinical Endocrinology & Metabolism, 59*(4), 756–763. https://doi.org/10.1210/jcem-59-4-756.

Ondo, J. G., Mical, R. S., & Porter, J. C. (1972). Passage of radioactive substances from CSF to hypophysial portal blood. *Endocrinology, 91*(5), 1239–1246. https://doi.org/10.1210/endo-91-5-1239.

Palm, R., Chang, J., Blair, J., Garcia-Mesa, Y., Lee, H., Castellani, R. J., et al. (2014). Down-regulation of serum gonadotropins but not estrogen replacement improves cognition in aged-ovariectomized 3xTg AD female mice. *Journal of Neurochemistry, 130*(1), 115–125. https://doi.org/10.1111/jnc.12706.

Peegel, H., Randolph, J., Midgley, A. R., & Menon, K. M. (1994). In situ hybridization of luteinizing hormone/human chorionic gonadotropin receptor messenger ribonucleic acid during hormone-induced down-regulation and the subsequent recovery in rat corpus luteum. *Endocrinology, 135*(3), 1044–1051. https://doi.org/10.1210/endo.135.3.8070346.

Pike, C. J. (2017). Sex and the development of Alzheimer's disease. *Journal of Neuroscience Research, 95*(1–2), 671–680. https://doi.org/10.1002/jnr.23827.

Prekeris, R., & Terrian, D. M. (1997). Brain myosin V is a synaptic vesicle-associated motor protein: Evidence for a Ca2+−dependent interaction with the synaptobrevin-synaptophysin complex. *The Journal of Cell Biology, 137*(7), 1589–1601. https://doi.org/10.1083/jcb.137.7.1589.

Rahman, N. A., & Rao, C. V. (2009). Recent progress in luteinizing hormone/human chorionic gonadotrophin hormone research. *Molecular Human Reproduction, 15*(11), 703–711. https://doi.org/10.1093/molehr/gap067.

Saudargiene, A., Cobb, S., & Graham, B. P. (2015). A computational study on plasticity during theta cycles at Schaffer collateral synapses on CA1 pyramidal cells in the hippocampus. *Hippocampus*, *25*(2), 208–218. https://doi.org/10.1002/hipo.22365.

Short, R., Bowen, R., O'Brien, P., & Graff-Radford, N. (2001). Elevated gonadotropin levels in patients with Alzheimer disease. Mayo Clinic proceedings. *Mayo Clinic*, *76*, 906–909. https://doi.org/10.4065/76.9.906.

Telegdy, G., Adamik, A., Tanaka, M., & Schally, A. V. (2010). Effects of the LHRH antagonist Cetrorelix on affective and cognitive functions in rats. *Regulatory Peptides*, *159*(1–3), 142–147. https://doi.org/10.1016/j.regpep.2009.08.005.

Toth, P., Lukacs, H., Hiatt, E. S., Reid, K. H., Iyer, V., & Rao, C. V. (1994). Administration of human chorionic gonadotropin affects sleep-wake phases and other associated behaviors in cycling female rats. *Brain Research*, *654*(2), 181–190. https://doi.org/10.1016/0006-8993(94)90478-2.

Verdile, G., Laws, S. M., Henley, D., Ames, D., Bush, A. I., Ellis, K. A., et al. (2014). Associations between gonadotropins, testosterone and β amyloid in men at risk of Alzheimer's disease. *Molecular Psychiatry*, *19*(1), 69–75. https://doi.org/10.1038/mp.2012.147.

Verdile, G., Yeap, B. B., Clarnette, R. M., Dhaliwal, S., Burkhardt, M. S., Chubb, S. A. P., et al. (2008). Luteinizing hormone levels are positively correlated with plasma amyloid-beta protein levels in elderly men. *Journal of Alzheimer's Disease: JAD*, *14*(2), 201–208. https://doi.org/10.3233/jad-2008-14208.

Wahjoepramono, E. J., Wijaya, L. K., Taddei, K., Bates, K. A., Howard, M., Martins, G., et al. (2011). Direct exposure of Guinea pig CNS to human luteinizing hormone increases cerebrospinal fluid and cerebral beta amyloid levels. *Neuroendocrinology*, *94*(4), 313–322. https://doi.org/10.1159/000330812.

Yang, E.-J., Nasipak, B. T., & Kelley, D. B. (2007). Direct action of gonadotropin in brain integrates behavioral and reproductive functions. *Proceedings of the National Academy of Sciences of the United States of America*, *104*(7), 2477–2482. https://doi.org/10.1073/pnas.0608391104.

You, S., Kim, H., Hsu, C.-C., El Halawani, M. E., & Foster, D. N. (2000). Three different Turkey luteinizing hormone receptor (tLH-R) isoforms I: Characterization of alternatively spliced tLH-R isoforms and their regulated expression in diverse Tissues1. *Biology of Reproduction*, *62*(1), 108–116. https://doi.org/10.1095/biolreprod62.1.108.

Yuan, L.-J., Peng, C., Liu, B.-H., Feng, J.-B., & Qiu, G.-F. (2019). Identification and characterization of a luteinizing hormone receptor (LHR) homolog from the Chinese mitten crab Eriocheir sinensis. *International Journal of Molecular Sciences*, *20*(7), 1736. https://doi.org/10.3390/ijms20071736.

Further reading

Alzheimer's Association. (2014). 2014 Alzheimer's disease facts and figures. *Alzheimer's & Dementia: The Journal of the Alzheimer's Association*, *10*(2), e47–e92. https://doi.org/10.1016/j.jalz.2014.02.001.

Banks, W. A., Kastin, A. J., Akerstrom, V., & Jaspan, J. B. (1993). Radioactively iodinated cyclo(his-pro) crosses the blood-brain barrier and reverses ethanol-induced narcosis. *The American Journal of Physiology*, *264*(5 Pt 1), E723–E729. https://doi.org/10.1152/ajpendo.1993.264.5.E723.

Bernichtein, S., Alevizaki, M., & Huhtaniemi, I. (2008). Is the adrenal cortex a target for gonadotropins? *Trends in Endocrinology and Metabolism: TEM*, *19*(7), 231–238. https://doi.org/10.1016/j.tem.2008.06.003.

Chakravarti, S., Collins, W. P., Forecast, J. D., Newton, J. R., Oram, D. H., & Studd, J. W. (1976). Hormonal profiles after the menopause. *British Medical Journal*, *2*(6039), 784–787.

Dudley, E. C., Hopper, J. L., Taffe, J., Guthrie, J. R., Burger, H. G., & Dennerstein, L. (1998). Using longitudinal data to define the perimenopause by menstrual cycle characteristics. *Climacteric: The Journal of the International Menopause Society*, *1*(1), 18–25. https://doi.org/10.3109/13697139809080677.

Garcia-Segura, L. M., Azcoitia, I., & DonCarlos, L. L. (2001). Neuroprotection by estradiol. *Progress in Neurobiology*, *63*(1), 29–60. https://doi.org/10.1016/s0301-0082(00)00025-3.

Lai, W., Li, X., Zhu, H., Zhu, X., Tan, H., Feng, P., et al. (2020). Plasma luteinizing hormone level affects the brain activity of patients with polycystic ovary syndrome. *Psychoneuroendocrinology*, *112*, 104535. https://doi.org/10.1016/j.psyneuen.2019.104535.

Möller, C., Vrenken, H., Jiskoot, L., Versteeg, A., Barkhof, F., Scheltens, P., et al. (2013). Different patterns of gray matter atrophy in early- and late-onset Alzheimer's disease. *Neurobiology of Aging*, *34*(8), 2014–2022. https://doi.org/10.1016/j.neurobiolaging.2013.02.013.

Odle, J., Jacobi, S. K., Boyd, R. D., Bauman, D. E., Anthony, R. V., Bazer, F. W., et al. (2017). The potential impact of animal science research on global maternal and child nutrition and health: A landscape review. *Advances in Nutrition (Bethesda, Md.)*, *8*(2), 362–381. https://doi.org/10.3945/an.116.013896.

Ohnishi, T., Matsuda, H., Tabira, T., Asada, T., & Uno, M. (2001). Changes in brain morphology in Alzheimer disease and normal aging: Is Alzheimer disease an exaggerated aging process? *AJNR. American Journal of Neuroradiology*, *22*(9), 1680–1685.

Rodrigues, M. A., Verdile, G., Foster, J. K., Hogervorst, E., Joesbury, K., Dhaliwal, S., et al. (2008). Gonadotropins and cognition in older women. *Journal of Alzheimer's Disease*, *13*(3), 267–274. https://doi.org/10.3233/jad-2008-13304.

Stefanska, A., Cembrowska, P., Kubacka, J., Kuligowska-Prusinska, M., & Sypniewska, G. (2019). Gonadotropins and their association with the risk of prediabetes and type 2 diabetes in middle-aged postmenopausal women. *Disease Markers*, *2019*, 2384069. https://doi.org/10.1155/2019/2384069.

Stenman, U.-H., Alfthan, H., & Hotakainen, K. (2004). Human chorionic gonadotropin in cancer. *Clinical Biochemistry*, *37*(7), 549–561. https://doi.org/10.1016/j.clinbiochem.2004.05.008.

Weiss, G., Skurnick, J. H., Goldsmith, L. T., Santoro, N. F., & Park, S. J. (2004). Menopause and hypothalamic-pituitary sensitivity to estrogen. *JAMA*, *292*(24), 2991–2996. https://doi.org/10.1001/jama.292.24.2991.

Weller, J., & Budson, A. (2018). Current understanding of Alzheimer's disease diagnosis and treatment. *F1000Research*, 7, 1161. https://doi.org/10.12688/f1000research.14506.1.

Wiacek, M., Hagner, W., & Zubrzycki, I. Z. (2011). Measures of menopause driven differences in levels of blood lipids, follicle-stimulating hormone, and luteinizing hormone in women aged 35 to 60 years: National Health and nutrition examination survey III and National Health and nutrition examination survey 1999–2002 study. *Menopause*, *18*(1), 60–66. https://doi.org/10.1097/gme.0b013e3181e7060b.

CHAPTER SIX

Leptin regulation of hippocampal synaptic function in health and disease

Kirsty Hamilton and Jenni Harvey*
Division of Systems Medicine, Ninewells Hospital and Medical School, University of Dundee, Dundee, United Kingdom
*Corresponding author: e-mail address: j.z.harvey@dundee.ac.uk

Contents

Abstract

It is widely accepted that the metabolic hormone leptin regulates food intake and body weight via activation of hypothalamic leptin receptors. However, as leptin receptors are also highly expressed in other brain regions, such as the hippocampus, alterations in leptin responsiveness also impacts on key functions of the hippocampus, like learning and memory. Within the hippocampus, high levels of leptin receptors are expressed at excitatory synapses, and in accordance with a synaptic localization, leptin potently regulates synaptic transmission at both Schaffer collateral (SC) and temporoammonic (TA) inputs to CA1 pyramidal neurons. Increasing evidence from cellular and behavioral studies examining leptin action at CA1 synapses support the notion that leptin is a potential cognitive enhancer. However, the capacity of leptin to regulate synaptic efficacy at SC-CA1 and TA-CA1 synapses declines in an age-dependent manner. Moreover, clinical evidence that supports a link between circulating leptin levels and the risk of the age-related neurodegenerative disorder, Alzheimer's disease (AD) is accumulating. Consequently, it has been proposed that the leptin system is a potential therapeutic

Vitamins and Hormones, Volume 115
ISSN 0083-6729
https://doi.org/10.1016/bs.vh.2020.12.006

target in AD, and that boosting the hippocampal actions of leptin may be beneficial in the treatment of AD. Here we review recent progress in our understanding of the neuronal and hippocampal synaptic functions that are regulated by leptin and how alterations in the leptin system influence age-related CNS-related disorders like AD.

1. Leptin

It is now more than 50 years since it was first proposed that a circulating hormone or factor was able to regulate food intake and body weight, and this hypothesis was supported by parabiosis experiments (Hervey, 1959). Later studies identified this circulatory factor as leptin, which is derived from leptos, the Greek word for thin. Leptin is a 16 kDa peptide hormone that is made by the *ob* gene (Halaas et al., 1998). It is primarily made and released from white adipose tissue and the circulating plasma levels of leptin are directly proportional to body fat content (Campfield, Smith, Guisez, Devos, & Burn, 1995). One of the main targets for leptin is the hypothalamus, and leptin that is made in the periphery can reach the hypothalamus via saturable transport across the blood brain barrier (Banks, Kastin, Huang, Jaspan, & Maness, 1996), or in the cerebrospinal fluid. Within the hypothalamus, leptin plays a pivotal role in controlling food intake and body weight. After feeding, leptin levels rise and leptin acts to signal satiety, whereas food restriction or fasting results in a decrease in leptin (Schwartz, Peskind, Raskind, Boyko, & Porte, 1996).

In addition to controlling feeding behavior, leptin also regulates neuronal brain development, with lack of leptin associated with abnormal or delayed development of various brain regions. High levels of leptin receptor expression have been detected in brain regions, such as the hippocampus and cortex, which are influenced by leptin during developmental stages. Moreover, in support of leptin's role in developmental processes, significant elevations in leptin occurs during the critical period of development (P5–8) (Ahima, Prabakaran, & Flier, 1998). The hippocampus is also a key target for leptin beyond development stages with leptin-driven effects observed in adulthood and during aging. In this respect, increasing evidence indicates that leptin is a potential cognitive enhancer, as treatment with leptin significantly enhances key cellular events linked to hippocampus-dependent learning and memory, including trafficking of AMPA receptors to and away from synapses as well as synaptic plasticity (Irving & Harvey, 2014; McGregor & Harvey, 2019). Conversely, lack or insensitivity to leptin is associated with memory impairments (Li et al., 2002).

2. Leptin receptor and its signaling pathways

Leptin mediates its biological actions via activation of leptin receptors (LEPR). Leptin receptors are encoded by the LEPR gene and six isoforms of the receptor exist (LEPRa-f) that are generated by alternative splicing. All isoforms are transmembrane receptors, with the exception of LEPRe which is thought to act as a carrier for leptin as it circulates freely in the plasma. The remaining receptor isoforms can be sub-divided into two main types: the short isoforms (LEPRa,c,d and f) and long form (LEPRb). These receptors differ not only in terms of the length of their C-terminal domain, but also in terms of their signaling capacity. The long form, LEPRb, contains specific tyrosine residues within its extended C-terminal domain, that enable the full complement of signaling pathways to be activated following leptin binding (Banks, Davis, Bates, & Myers, 2000; Fig. 1). In contrast, the short isoforms have limited signaling ability due to their truncated C-terminal regions.

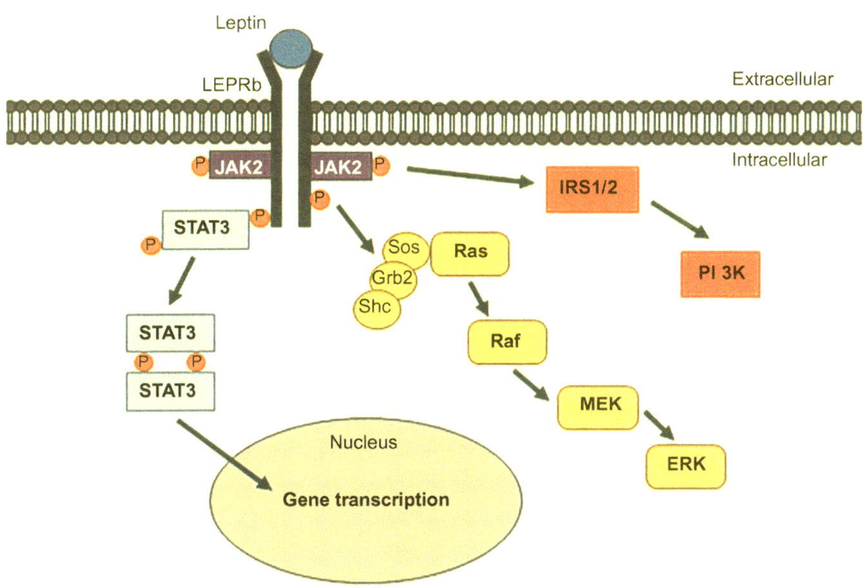

Fig. 1 Leptin receptor signaling. Binding of leptin to the long form of the leptin receptor (LEPRb) results in janus tyrosine kinase 2 (JAK2) activation and subsequent phosphorylation of specific tyrosine residues within the C-terminal domain of the receptor. This promotes phosphorylation and dimerization of STAT3 which translocates to the nucleus to regulate gene transcription. Binding of leptin to LEPRb also results in activation of canonical PI 3-kinase (PI 3K) and ERK signaling cascades.

The short LEPR isoforms are thought to play an important role in the blood to brain transport of leptin. The cerebrospinal fluid is also implicated in transporting leptin to the CNS as a high density of LEPRa are expressed at the choroid plexus.

LEPRs show greatest homology to the class I cytokine receptor superfamily, and there are parallels in the signaling pathways that are activated by LEPRs and other members of this receptor family (Ihle, 1995). Following leptin binding to LEPRb, janus tyrosine kinase 2 (JAK2) associates with the receptor which in turn drives phosphorylation of specific tyrosine residues within the receptor C-terminal region. Phosphorylation of tyrosine residues enables recruitment of STAT3 (signal transducer and activator of transcription 3) which is then phosphorylated by JAK2, leading to STAT3 dimerization and translocation to the nucleus where it regulates gene transcription (Fig. 1). In addition to JAK2-STAT3-driven regulation of nuclear gene transcription, PI 3-kinase and p42/p44 MAPK (ERK) are also key signaling cascades activated by leptin receptors (Fig. 1). Activation of JAK2-STAT3, ERK and PI 3-kinase have been implicated in leptin regulation of hypothalamic feeding behavior, and hippocampal synaptic function.

2.1 Expression of leptin and leptin receptors in the CNS

Numerous studies have used a combination of immunocytochemical, histological and in situ hybridization techniques to characterize the cellular distribution of leptin and LEPR expression in the CNS. These studies have found good evidence for leptin mRNA and protein expression in the hypothalamus, cortex and hippocampus (Hâkansson, Brown, Ghilardi, Skoda, & Meister, 1998; Mercer et al., 1996; Schwartz et al., 1996), however the expression of leptin within these brain regions varies among different neuronal populations. Detection of leptin protein in the brain has also fuelled suggestions that leptin may be produced and released locally in the CNS (Morash, Li, Murphy, Wilkinson, & Ur, 1999), and this hypothesis has been reinforced experimentally by studies demonstrating net leptin efflux into the jugular vein (Eikelis et al., 2004; Wiesner et al., 1999).

In accordance with leptin's role in regulating energy homeostasis, LEPRs are highly expressed in specific hypothalamic brain regions that control food intake and body weight (Fei et al., 1997; Hâkansson, Hulting, & Meister, 1996). However wide-spread expression of LEPRs has been reported in several other brain regions such as the brain stem, cerebellum and hippocampus (Hâkansson et al., 1998; Scott et al., 2009). Evaluation of the sub-cellular localization of LEPRs has revealed high levels of expression at excitatory and inhibitory synapses in hypothalamic and hippocampal brain regions (Shanley, O'Malley, Irving, Ashford, & Harvey, 2002), suggesting that the ability of leptin to regulate synaptic transmission may be key for its physiological functions.

One of the key central targets for leptin is the hypothalamus. Leptin receptors are highly expressed in several nuclei within this brain region, including areas specifically associated with the homeostatic control of energy balance (Ahima & Flier, 2000) such as the ventromedial nucleus (VMN), dorsomedial nucleus (DMN) and arcuate nucleus (ARC; Elmquist, Bjørbaek, Ahima, Flier, & Saper, 1998; Håkansson et al., 1996). Peripheral administration of leptin results in the activation of the ARC, DMN and VMN, suggesting a role for leptin in energy homeostatic mechanisms. Indeed, significant reductions in food intake and body weight are observed after intra-cerebroventricular (icv) administration of leptin into mice (Ahima & Flier, 2000). Within the ARC, two specific neuronal populations are highly responsive to leptin. The orexigenic circuit comprises neuropeptide-Y (NPY) and agouti-related protein (AGRP)-containing neurons which act to stimulate appetite. Opposing this action, the anorexigenic circuit consists of proopiomelanocortin (POMC) and cocaine- and amphetamine-regulated transcript (CART)-containing neurons which suppress appetite. These ARC neurons respond to the circulating levels of leptin and in turn relay information on the status of energy stores to various second order neurons, which ultimately control information flow to spinal neurons and peripheral organs (Spiegelman & Flier, 2001).

3. Leptin regulation of hippocampal synaptic function

It is well established that the forebrain is involved in higher cognitive function, and the hippocampus, specifically plays a pivotal part in learning and remembering events. The ability of hippocampal excitatory synapses to alter their strength in response to changes in the frequency of synaptic activation, is known as synaptic plasticity. Moreover, activity-dependent forms of hippocampal synaptic plasticity such as long-term potentiation (LTP) are thought to be cellular mechanisms underlying hippocampus-dependent learning and memory (Bliss & Collingridge, 1993). Increasing evidence indicates not only that leptin can regulate hippocampal synaptic function but that leptin has potential cognitive enhancing actions (Harvey, Solovyova, & Irving, 2006). Studies performed in leptin-insensitive rodents (*db/db* mice; *fa/fa* rats) indicate that these rodents have impairments in their ability to perform spatial memory tasks and impaired LTP (Li et al., 2002; Winocur et al., 2005). Moreover, direct administration of leptin into the hippocampus enhances the magnitude of hippocampal LTP (Wayner, Armstrong, Phelix, & Oomura, 2004). Cellular studies performed in acute hippocampal slices have replicated these findings such that exposure to leptin promotes conversion of short-term potentiation (STP) into LTP (Shanley, Irving, & Harvey, 2001).

It is well established that activation of NMDA receptors is crucial for the induction of LTP at hippocampal Schaffer collateral (SC) to CA1 synapses (Bliss & Collingridge, 1993), and that hormones such as insulin and estrogens, facilitate LTP induction by enhancing NMDA receptor function (Moult & Harvey, 2008). In accordance with this, treatment with leptin facilitates NMDA-evoked currents in *Xenopus* oocytes expressing NMDA receptor subunits and LEPRb (Shanley et al., 2001). In digital imaging studies, leptin enhances Ca^{2+} influx via NMDA receptor channels, via a process involving activation of PI 3-kinase and ERK signaling (Shanley et al., 2002). Thus, it has been proposed that leptin, that is either made and released in the periphery is transported to the CNS, or it is released near synapses during periods of high frequency synaptic stimulation, and it modulates the threshold for induction of LTP, via enhancement of NMDA receptor function.

3.1 Regulation of SC-CA1 synapses

Previous studies have demonstrated that application of leptin to acute hippocampal slices from 2 to 3 week-old rats results in a short-lived depression of excitatory synaptic transmission that returns to baseline levels after leptin washout (Shanley et al., 2001; Xu et al., 2008). By contrast, leptin has a completely opposing action at adult SC-CA1 hippocampal synapses, as a persistent increase in synaptic transmission (leptin-induced LTP) is observed (Moult et al., 2010; Fig. 2). Activation of leptin receptors is required for leptin-induced LTP at adult SC-CA1 synapses as leptin failed to influence synaptic transmission in slices from Zucker *fa/fa* rats (Moult et al., 2010). A postsynaptic expression mechanism underlies this process as leptin-induced LTP was not accompanied by any change in paired pulse ratio or coefficient of variance (Moult et al., 2010). Synaptic activation of NMDA receptors plays a role as pharmacological blockade of NMDA receptors prevented leptin-induced LTP, and leptin had no effect on synaptic efficacy when synaptic stimulation was stopped during addition of leptin.

Animal age	SC-CA1 input	TA-CA1 input
Juvenile	Transient depression in synaptic transmission - Involves inhibition of AMPAR-rnediated synaptic transmission	LTP - Involves GluA2-locking AMPAR insertion via PI 3K signalling - Dependent on GluN2B-containing NMDARs
Adult	LTP - Involves GluA2-locking AMPAR insertion via PTEN inhibition - NMDAR-dependent	LTD - Involves AMPAR internalisation via JAK2/STAT3 signalling - Dependent on GluN2A-containing NMDARs

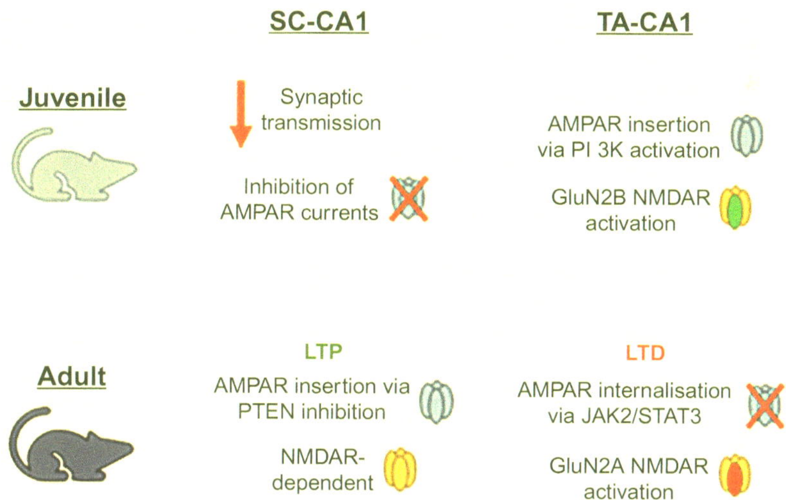

Fig. 2 Differential effects of leptin at SC-CA1 and TA-CA1 synapses in juvenile and adult hippocampi. In juvenile hippocampi, leptin administration results in a transient depression of synaptic transmission at SC-CA1 synapses via inhibition of AMPA receptor (AMPAR)-mediated synaptic transmission. At TA-CA1 synapses in juvenile hippocampi, leptin treatment results in long-term potentiation (LTP) which involves activation of GluN2B-containing NMDA receptors (NMDARs) and GluA2-lacking AMPAR insertion via PI 3-kinase (PI 3K) signaling. In contrast, leptin administration in adult hippocampi results in LTP at SC-CA1 synapses which is dependent on NMDAR activation and insertion of GluA2-lacking AMPARs via PTEN inhibition. At TA-CA1 synapses in adult hippocampi, leptin treatment results in long-term depression (LTD) which is dependent on AMPAR internalization via JAK2/STAT3 signaling and activation of GluN2A-containing NMDARs.

NMDA receptor activation is known to trigger movement of AMPA receptors to hippocampal synapses during LTP (Collingridge, Isaac, & Wang, 2004). Moreover, changes in the molecular composition of AMPA receptors occurs following induction of LTP, as insertion of GluA2-lacking AMPA receptors accompanies induction of hippocampal LTP (Isaac, Ashby, & McBain, 2007). Recent evidence indicates that leptin also regulates trafficking of AMPA receptors to and away from hippocampal synapses as exposure of cultured hippocampal neurons to leptin results in a significant increase in the surface expression of the AMPA receptor subunit, GluA1 (Malekizadeh et al., 2017; Moult et al., 2010). This effect is due to enhanced exocytosis of GluA1 as the effects of leptin were blocked by inhibitors of exocytosis. Leptin also increases the synaptic density of AMPA receptors as in dual labeling studies an increase in the co-localization of surface GluA1 and the postsynaptic marker, PSD-95 was observed (Moult et al., 2010). Leptin-driven changes in AMPA receptor trafficking also contributes to leptin-induced LTP at adult SC-CA1 synapses as an increase in the rectification of synaptic AMPA receptor currents was evident after leptin treatment, and pharmacological blockade of GluA2-lacking AMPA receptors with philanthotoxin reversed leptin-induced LTP (Moult et al., 2010).

The cellular mechanisms underlying leptin-driven insertion of GluA2-lacking AMPA receptors into synapses has been studied in detail. Initial studies suggested a potential role for PI 3-kinase signaling as an increase in intracellular phosphoinositide 3,4,5,-trisphosphate (PIP_3) accompanied leptin-driven effects on surface GluA1. However, the phosphatase PTEN also regulates PIP_3 levels, and inhibition of PTEN with bisperoxovanadium (BpV) mirrors and occludes the ability of leptin to increase the synaptic expression of GluA1 and to increase the efficacy of excitatory synaptic transmission (Moult et al., 2010). These findings indicate that leptin-driven inhibition of PTEN underlies the leptin-dependent alterations in AMPA receptor trafficking and synaptic efficacy at adult SC-CA1 synapses (Fig. 2).

3.2 Regulation of TA-CA1 synapses

In addition to the classical SC input, hippocampal CA1 pyramidal neurons are also directly innervated by the temporoammonic (TA) pathway which extends from layer III of the entorhinal cortex (EC) and directly innervates stratum lacunosum-moleculare (*s l.m.*). Numerous lines of evidence indicate that the TA input, like the SC, plays an important role in hippocampus-dependent memory processes (Remondes & Schuman, 2002, 2004).

Recent evidence indicates that TA-CA1 synapses are also a target for leptin, but that TA-CA1 and SC-CA1 synapses are differentially regulated by this hormone. For instance, leptin induces a novel form of NMDA receptor-dependent LTP at juvenile TA-CA1 synapses, whereas at juvenile SC-CA1 synapses a synaptic depression is observed in response to leptin (Moult & Harvey, 2011; Fig. 2). In accordance with other studies (Liu et al., 2004; Massey et al., 2004), distinct NMDA receptor subunits are implicated in different forms of leptin-driven synaptic plasticity as the novel form of LTP evoked by leptin at TA-CA1 synapses requires selective activation of GluN2B, but not GluN2A, subunits (Luo, McGregor, Irving, & Harvey, 2015). Moreover, leptin-driven activation of PI 3-kinase is key for insertion of GluA2-lacking AMPA receptors and induction of leptin-induced LTP at juvenile TA-CA1 synapses (Luo et al., 2015).

Differential regulation of synaptic efficacy at the two inputs to CA1 neurons is also evident in adult hippocampus. Thus, in contrast to leptin-induced LTP at adult SC-CA1 synapses, application of leptin results in induction of a novel form of NMDA receptor-dependent LTD at adult TA-CA1 synapses (McGregor, Clements, Farah, Irving, & Harvey, 2018). Activation of canonical JAK2-STAT3 signaling is pivotal for leptin-driven removal of AMPA receptors and induction of LTD at adult TA-CA1 synapses (Fig. 2). The expression mechanisms underlying leptin-induced LTD and activity-dependent LTD at adult TA-CA1 synapses are analogous, as there is occlusion between both processes (McGregor et al., 2018). Increasing evidence suggests that TA-CA1 synapses play a key role in controlling spatial memory and consolidation of memory (Remondes & Schuman, 2004; Vago, Bevan, & Kesner, 2007). TA-CA1 synapses are also implicated in episodic memory processes via their ability to integrate place cell information (Nakashiba, Young, McHugh, Buhl, & Tonegawa, 2008). Consequently, the ability of leptin to regulate excitatory synaptic efficacy at TA-CA1 synapses suggests that leptin is likely to influence spatial and episodic memory processes.

4. Leptin and the aging brain

It is well established that adipose tissue, which is the main site for leptin production, undergoes significant change with age. Clinical evidence indicates that as people get older redistribution of peripheral fat stores is common as build-up of visceral fat but a decline in subcutaneous fat occurs. Age-related changes also occur at the level of adipocytes, with alterations in the density and function of adipocytes taking place which can in turn influence the production and release of adipocyte-derived hormones, such as leptin.

Age-dependent alterations in metabolic systems are well documented as numerous studies indicate that there is a significant decline in the ability of metabolic hormones to perform their peripheral functions with age. Moreover, development of resistance to leptin, enhanced adiposity, and increased body weight are all linked to aging in humans (Mattson & Arumugam, 2018; Fig. 3). Age-related changes in metabolic function are not restricted to peripheral tissues as there is accumulating evidence that the sensitivity of central neurons to metabolic hormones also deteriorates during the aging process. Thus, age-dependent reductions in the sensitivity to leptin as well as alterations in the leptin system have been observed in several brain regions. Indeed, comparison between young and aged rats reveals a significant decrease in the ability of a leptin infusion to reduce food intake in aged animals, suggesting altered responsiveness of hypothalamic neurons to leptin (Scarpace, Matheny, Moore, & Tümer, 2000). This finding is also supported by the observation that leptin-driven stimulation of STAT3 signaling is attenuated in aged animals (Scarpace, Matheny, & Tümer, 2001; Fig. 3).

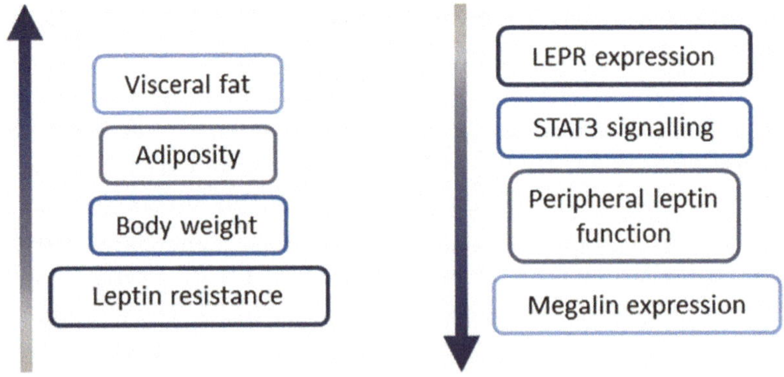

Fig. 3 Age-dependent changes in the leptin system. Many changes in leptin signaling and production have been reported to occur during the aging process. An increase in visceral fat stores, adiposity and body weight have been linked to aging in humans, thereby causing alterations in leptin production. Reduced megalin expression associated with age likely contributes to a decrease in blood to brain transport of leptin. Development of leptin resistance, decreased leptin receptor (LEPR) expression in several brain regions and reduced STAT3 signaling are also associated with aging. Several studies have therefore shown alterations in leptin production, transport and signaling occur with age.

Reductions in receptor-mediated uptake of leptin into the hypothalamus has also been observed in aged animals, which may in part be due to decreased expression of leptin receptors in this brain region (Fernández-Galaz et al., 2001). Recent studies indicate that the leptin-responsiveness of astrocytes also varies with age. Indeed, leptin treatment of hypothalamic astrocytes derived from aged animals, leads to increased release of the pro-inflammatory cytokines, tumor necrosis factor α (TNF-α) and interleukin 1β (IL-1β). In contrast, however leptin has no effect on the release of TNF-α and IL-1β in neonatal hypothalamic astrocytes (Santos, Bobermin, Souza, & Quincozes-Santos, 2018).

Aging is also associated with impairments in the ability of leptin to access the brain. It is known that effective transport of leptin from the periphery to the brain is required for normal brain function, and recent evidence suggests that megalin, a receptor that is expressed at the choroid plexus epithelium, plays a part in assisting the blood to brain transport of leptin (Dietrich et al., 2008). However, marked downregulation of megalin expression occurs during the aging process, which is likely to lead to reduced transport of leptin into the brain with age (Dietrich et al., 2008; Fig. 3). Alterations in the density of leptin receptors has been identified in adult and aged hippocampus and cortex (Caron, Sachot, Prevot, & Bouret, 2010), which is also likely to contribute to age-related changes in leptin sensitivity in these brain regions. Cellular studies support this possibility as the ability of leptin to induce LTP at SC-CA1 synapses in hippocampal slices from aged (12–14 month) rats is significantly attenuated relative to adult (3–4 month; Moult & Harvey, 2011). Indeed, the magnitude of leptin-induced LTP in aged tissue is around 50% less than that observed in adulthood (Moult & Harvey, 2011). Similar age-related changes in leptin sensitivity are evident at hippocampal TA-CA1 synapses. Thus, leptin induces a novel form of NMDA-dependent LTD at adult TA-CA1 synapses, however this effect of leptin is completely absent at aged TA-CA1 synapses (McGregor et al., 2018). In accord with other studies, the failure of leptin to induce LTD in aged hippocampus suggests possible development of leptin resistance. However, this is unlikely to be the case for all central functions of leptin, as in parallel studies application of leptin to aged slices resulted in the induction of robust LTP at SC-CA1 synapses (Moult & Harvey, 2011). This suggests that the ability of leptin to induce LTP at TA-CA1 synapses is unaffected by age. In view of these findings, it is more likely that neuron-specific, or indeed synapse-specific alterations in leptin sensitivity occur during the aging process, rather than development of leptin resistance per se across the whole brain.

5. Leptin and Alzheimer's disease risk

It is well established that aging is the biggest risk factor for development of neurodegenerative disorders like Alzheimer's disease (AD). Other key factors that influence risk of developing AD include diet and lifestyle. Indeed, increasing evidence indicates that impaired metabolic function increases the risk of AD (Kim & Feldman, 2015; Stranahan & Mattson, 2011). Clinical evidence supports a link between lifestyle factors and AD as weight gain in midlife and a sedentary lifestyle have been identified as key features that elevate AD risk. This in turn has fuelled speculation of a link between obesity and AD (Cai, 2013). Indeed, obesity-linked type II diabetes is associated with around twofold greater risk of AD than non-diabetics. Moreover, clinical studies have shown correlations between midlife obesity and development of AD later in life (Hassing et al., 2009; Kivipelto et al., 2005; Xu et al., 2011). Accumulation of fat mass is a key feature of the obese state, and as adipocytes are the main source of leptin, this results in significant elevations in the circulating levels of leptin, which ultimately leads to development of leptin resistance.

Thus, if the obese state elevates AD risk, this suggests that development of leptin resistance and/or dysfunctions in the leptin system may contribute to this risk. This possibility is supported by prospective data from the Framingham Heart Study which has identified that non-obese individuals with high circulating leptin levels have a much lower AD risk compared to individuals with low plasma leptin (Lieb et al., 2009). Loss of weight is a common feature of AD, and weight loss has been identified as an indicator of mortality in AD patients (White, Pieper, & Schmader, 1998). In addition to weight loss, significant reductions in the levels of circulating leptin are observed in AD patients compared to non-AD individuals with similar BMI (Bigalke et al., 2011; Power, Collins, & Neill, 2001). Thus, in humans both low and elevated levels of leptin (and subsequent leptin resistance) are associated with a greater likelihood of developing AD. Overall this suggests that the risk of AD is significantly elevated in individuals with plasma leptin levels that fall out-with the normal physiological range of this hormone (Friedman, 2014).

6. Neuroprotective actions of leptin

Numerous studies indicate that leptin is a potent neuroprotective hormone as treatment with leptin limits the extent of neuronal cell death induced in peripheral and central neurons by various toxic stimuli (Davis, Mudd, & Hawkins, 2014; Doherty, Oldreive, & Harvey, 2008; Guo, Jiang, Xu, Duan, & Mattson, 2008). Leptin also increases neuronal viability in several models of CNS-driven disease, including rodent models of ischemia where exposure to leptin significantly reduces overall brain injury (Zhang & Chen, 2008; Zhang et al., 2013). In MPTP and 6-OH dopamine models of Parkinson's disease, dopamine neuron survival is also significantly elevated after treatment with leptin (Doherty et al., 2008; Weng et al., 2008). The protective effects of leptin extend to other neurodegenerative disorders. In AD, toxic accumulation of amyloid β (Aβ) is a key pathological trait and Aβ build-up is implicated in neuronal cell death. Several lines of evidence indicate that leptin reduces toxic neuronal accumulation of Aβ by either directly reducing Aβ generation by inhibiting β secretase activity (Fewlass et al., 2004) or by elevating the expression of insulin degrading enzyme thereby promoting degradation of Aβ (Marwarha, Dasari, Prasanthi, Schommer, & Ghribi, 2010; Fig. 4). The uptake and clearance of Aβ is also influenced by leptin as this hormone is reported to enhance ApoE-dependent uptake of Aβ into human neuroblastoma cells, indicating that leptin promotes neuronal clearance of toxic Aβ (Fewlass et al., 2004). In addition to influencing the brain levels of Aβ, numerous studies have shown that leptin directly prevents the neurotoxic apoptotic effects of Aβ. Indeed, the scale of cell death induced in hippocampal or cortical neurons by chronic exposure to oligomeric Aβ is significantly attenuated by leptin in a range of cellular toxicity assays (Doherty, Beccano-Kelly, Yan, Gunn-Moore, & Harvey, 2013; Martins et al., 2013). The protective actions of leptin against Aβ also extend to the hypothalamus, as leptin reduces Aβ-driven elevations in intracellular calcium and subsequent generation of superoxide in a hypothalamic cell line (Gomes et al., 2014).

Another key pathological hallmark of AD is neurofibrillary tangles containing elevated levels of hyper-phosphorylated tau protein. Increasing evidence indicates that modification of tau is a likely causative factor in the progression of AD, as the degree of tau phosphorylation and its neuronal accumulation display a very strong association with the cognitive impairments that are observed in AD patients. Activation of the serine/threonine protein kinase, glycogen synthase kinase 3 beta (GSK3β) is known to play a key role in phosphorylating tau protein. Consequently, inhibition of GSK3β function is an important route for limiting phosphorylation of tau. Several studies have examined if the hormone leptin regulates tau phosphorylation as it is well established that leptin receptor activation triggers stimulation of PI 3-kinase which in turn leads to inhibition of GSK3β. Indeed, treatment of

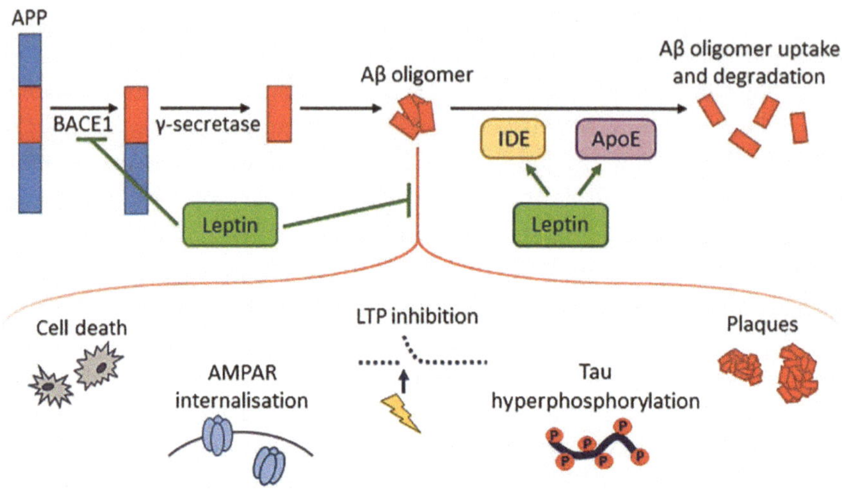

Fig. 4 Leptin prevents the acute and chronic synapto-toxic effects of amyloid β. Cleavage of the amyloid precursor protein (APP) by BACE1 and γ-secretase results in formation of Aβ protein that us more prone to aggregation into oligomers. Aβ oligomers have many synapto-toxic effects including enhanced cell death, AMPA receptor (AMPAR) internalization, hyperphosphorylation of tau, plaque formation and inhibition of LTP induced by high frequency stimulation. Aβ oligomers are degraded via ApoE-dependent uptake and insulin degrading enzyme (IDE). Leptin reduces BACE1 activity and promotes IDE- and ApoE-dependent degradation thereby reducing Aβ production and burden. Leptin also prevents the aberrant synapto-toxic effects of Aβ oligomers.

TgCRND8 mice that overexpress mutant APP, with leptin is reported to significantly reduce phosphorylated tau levels via a process that involves GSK3β inhibition (Greco et al., 2010). Moreover, leptin-driven activation of a key enzyme regulator of energy homeostasis, namely AMP-activated protein kinase (AMPK) has been linked to inhibition of GSK3β and subsequent reduction in tau phosphorylation in human neuroblastoma, SH-SY5Y cells (Greco et al., 2008). Chronic exposure to Aβ which itself promotes phosphorylation of tau, is also markedly decreased in cortical neurons after treatment with leptin (Doherty et al., 2013; Fig. 4). Conversely, development of neuronal leptin resistance and/or insensitivity to leptin has been linked to elevated levels of phosphorylated tau. Thus, studies in Zucker *fa/fa* rats as well as *ob/ob* and *db/db* mice, have revealed significantly elevated neuronal levels of phosphorylated tau compared to wild-type littermates, suggesting that lack or insensitivity to leptin influences the phosphorylation status of tau (Doherty et al., 2013; El Khoury et al., 2016; Gratuze et al., 2017). These observations are also supported by the finding that dietary-linked development of leptin resistance results in upregulated levels of phosphorylated tau (Platt, Beckett, Kohler, Niedowicz, & Murphy, 2016). Thus, there is good evidence that the protective actions of leptin are not restricted to preventing the aberrant actions of Aβ, as leptin also influences the expression and phosphorylation status of tau. Recent studies suggest that phosphorylation of tau is not only a key pathological feature of AD, but that tau phosphorylation is also linked to the synaptic abnormalities that occur in the early stages of the disease. Indeed, it is known that tau is concentrated in axons in the healthy brain (Avila et al., 2004), however in AD, phosphorylation of tau promotes its targeting to synapses where it disrupts synaptic function (Hoover et al., 2010). Thus, it is feasible that leptin by inhibiting GSK3β, limits tau phosphorylation and thereby restricts toxic accumulation of tau at synapses.

6.1 Leptin prevents the acute synapto-toxic effects of Aβ

Increasing evidence indicates that acute exposure to oligomeric Aβ has rapid detrimental effects on neuronal function. Indeed, numerous studies have reported that acute treatment with Aβ interferes with excitatory synaptic transmission at central synapses. In hippocampal CA1 neurons, Aβ induces a persistent synaptic depression that involves endocytosis and synaptic removal of AMPA receptors and subsequent loss of dendritic spines (Hsieh et al., 2006; Fig. 4). Cellular studies examining the molecular identity of the AMPA receptors involved identified that GluA1 subunits are removed from hippocampal synapses after acute treatment with oligomeric Aβ (Doherty et al., 2013; Li et al., 2009). Moreover, recent evidence indicates that leptin has the ability to counter this synapto-toxic effect of Aβ as prior treatment of neurons with low nanomolar concentrations of leptin blocked Aβ-driven removal of GluA1 from hippocampal synapses (Doherty et al., 2013; Malekizadeh et al., 2017; Fig. 4). Studies using selective pharmacological tools to dissect the potential signaling cascades involved have found that leptin receptor activation of PI 3-kinase is crucial for this protective effect (Doherty et al., 2013).

Activity-dependent synaptic plasticity is also influenced by Aβ as short-term application of either synthetic oligomeric Aβ or Aβ oligomers isolated from AD patients to hippocampal slices not only blocks the induction of NMDA-dependent LTP but also facilitates LTD induction (Li et al., 2009; Shankar et al., 2008). Several studies have shown that exposure to various hormones and growth factors, such as insulin or BDNF, can prevent the acute effects of Aβ on activity-dependent synaptic plasticity

(Criscuolo, Fabiani, Bonadonna, Origlia, & Domenici, 2015; Lee, Kuo, Huang, & Hsu, 2009). In a similar manner, application of leptin to acute brain slices protects against the rapid aberrant effects of oligomeric Aβ on synaptic plasticity evoked at hippocampal SC-CA1 synapses (Doherty et al., 2013; Malekizadeh et al., 2017). Thus, treatment of slices with low concentrations of leptin completely reverses the Aβ-induced block of NMDA-dependent hippocampal LTP (Fig. 4). Moreover, the ability of Aβ to promote the induction of LTD is also significantly attenuated in hippocampal slices treated with leptin (Doherty et al., 2013). Thus, there is now compelling evidence that leptin protects against both the acute and chronic actions of Aβ in various synapto-toxic models of AD.

Although acute exposure to Aβ leads to significant synaptic impairments that can be reversed by leptin treatment, it is not clear if the beneficial effects of leptin at synapses translate into improved hippocampal function in vivo. It is known that activity-dependent changes in synaptic efficacy, such as LTP and LTD, are key cellular events that occur at hippocampal synapses and these events are implicated in hippocampus-dependent learning and memory (Bliss & Collingridge, 1993). Indeed, several studies have shown that administration of leptin to wild-type animals leads to enhanced cognitive performance in a variety of behavioral assays of hippocampus-dependent learning and memory (Malekizadeh et al., 2017; Wayner et al., 2004), and significant impairments in hippocampal-specific memory tasks have been observed in rodents that are leptin-insensitive (e.g. Zucker *fa/fa* rats and *db/db* mice; Li et al., 2002, Winocur et al., 2005). However, do the potential cognitive enhancing actions of leptin extend to influencing behavior in models of AD? Accumulating evidence shows that leptin does have beneficial cognitive effects in models of AD. For instance, bilateral injection of leptin into the hippocampus of SAMP8 mice that have memory deficits due to overexpression of toxic Aβ, results in improved performance in various hippocampus-dependent memory tasks (Farr, Banks, & Morley, 2006). Similarly, the ability of tgCRND8 mice to perform novel object recognition and fear conditioning tasks is significantly improved following treatment with leptin (Greco et al., 2010). More recent studies in rats indicate that Aβ-driven deficits in hippocampal synaptic plasticity and spatial memory are alleviated after ICV administration with leptin (Tong et al., 2015), which provides further evidence in support of the cognitive benefits of leptin in models of AD.

6.2 Protective actions of leptin at TA-CA1 synapses?

Increasing evidence indicates the TA input to CA1 pyramidal neurons is an important area that is targeted in AD pathology. Indeed, aging per se is associated with key morphological changes at TA-CA1 synapses. Moreover, tau dilations that mirror the pathological alterations that occur in AD, are observed in the TA pathway (Braak & Braak, 1997; Yassa, Muftuler, & Stark, 2010). Reduction in the levels of myelination specifically in the TA input, in addition to significant loss of synapses in the hippocampal *s l.m.* region have been reported in rodent models of tauopathy (Maurin et al., 2014, 2013). Impairments in activity-dependent LTP at hippocampal TA-CA1 synapses are observed in rodent models of AD that overexpress mutant human tau (Booth et al., 2016). Although it is evident that leptin can regulate synaptic processes at TA-CA1 synapses, it remains to be seen if leptin has any neuroprotective effects at the TA pathway in models of AD and other tauopathies.

7. Therapeutic potential of leptin

Although numerous studies in animal models provide compelling support for the use of leptin as a therapy in human cases of AD, there has, as yet, been no clinical evaluation of the potential benefits of using leptin in AD trials. However, clinical studies using leptin as a replacement therapy in congenital cases of leptin deficiency clearly demonstrate that this hormone poses no safety risk for use in humans (Paz-Filho et al., 2008; Paz-Filho, Mastronardi, & Licinio, 2015). Questions regarding the use of leptin will need to be addressed before its use in future AD clinical trials. For example, it will be necessary to identify AD patients that will most likely benefit from a leptin-based therapy, as individuals with resistance to leptin may not respond to this treatment. Limiting potential side effects is also an important consideration with any drug therapy and this will also be pertinent in the use of a leptin-based therapy, in view of the wide-spread expression of leptin receptors in the brain and periphery. One way of circumventing the risk of side effects may be to modify the whole leptin molecule, to potentially enhance the potency of leptin at the same time as reducing the likelihood of potential side effects. Indeed, several groups have identified that small fragments derived from the whole leptin peptide are bioactive, and are capable of reproducing the antiobesity actions of leptin by targeting leptin receptors in the hypothalamus (Grasso, Leinung, Ingher, & Lee, 1997; Rozhavskaya-Arena, Lee, Leinung, & Grasso, 2000). In these studies, two specific leptin fragments (leptin$_{22-56}$ and leptin$_{116-130}$) have been well characterized and found to mirror the actions of leptin. Recent preclinical studies have extended these findings to demonstrate that one of the leptin fragments, namely leptin$_{116-130}$, mirrors the full spectrum of anti-AD actions of whole leptin (Malekizadeh et al., 2017). Thus, in a variety of cellular assays, leptin$_{116-130}$, but not leptin$_{22-56}$, replicates the cognitive enhancing effects of leptin, as indicated by the ability of leptin$_{116-130}$ to facilitate hippocampal LTP, increase trafficking of GluA1 to synapses, and enhance performance in episodic-like memory tasks in rodents (Malekizadeh et al., 2017). In addition, leptin$_{116-130}$ displayed powerful protective actions, as treatment with this leptin fragment not only counteracted the harmful effects of Aβ at excitatory synapses, but also limited the extent of neuronal cell death induced by chronic treatment with toxic Aβ (Malekizadeh et al., 2017). In view of these findings, it is feasible that further evaluation of leptin$_{116-130}$ could ultimately lead to the identification of small leptin-like mimetics that could be developed into potential therapeutic agents for use in AD.

References

Ahima, R. S., & Flier, J. S. (2000). Leptin. *Annual Review of Physiology*, *62*, 413–437.

Ahima, R. S., Prabakaran, D., & Flier, J. S. (1998). Postnatal leptin surge and regulation of circadian rhythm of leptin by feeding. Implications for energy homeostasis and neuroendocrine function. *The Journal of Clinical Investigation*, *101*(5), 1020–1027.

Avila, J., Pérez, M., Lucas, J. J., Gómez-Ramos, A., Santa María, I., Moreno, F., et al. (2004). Assembly in vitro of tau protein and its implications in Alzheimer's disease. *Current Alzheimer Research*, *1*(2), 97–101.

Banks, A. S., Davis, S. M., Bates, S. H., & Myers, M. G. (2000). Activation of downstream signals by the long form of the leptin receptor. *Journal of Biological Chemistry*, *275*, 14563–14572.

Banks, W. A., Kastin, A. J., Huang, W., Jaspan, J. B., & Maness, L. M. (1996). Leptin enters the brain by a saturable system independent of insulin. *Peptides*, *17*(2), 305–311.

Bigalke, B., Schreitmüller, B., Sopova, K., Paul, A., Stransky, E., Gawaz, M., et al. (2011). Adipocytokines and CD34 progenitor cells in Alzheimer's disease. *PLoS One*, *6*, e20286.

Bliss, T. V., & Collingridge, G. L. (1993). A synaptic model of memory: Long-term potentiation in the hippocampus. *Nature*, *361*, 31–39.

Booth, C., Witton, J., Nowacki, J., Tsaneva-Atanasova, K., Jones, M. W., Randall, A. D., et al. (2016). Alterations to intrinsic pyramidal neuron properties and temporoammonic synaptic plasticity underlie deficits in hippocampal network function in a mouse model of tauopathy. *Journal of Neuroscience*, *36*, 350–363.

Braak, H., & Braak, E. (1997). Frequency of stages of Alzheimer-related lesions in different age categories. *Neurobiology of Aging*, *18*(4), 351–357.

Cai, D. (2013). Neuroinflammation and neurodegeneration in overnutrition-induced diseases. *Trends in Endocrinology and Metabolism*, *24*, 40–47.

Campfield, L., Smith, F. J., Guisez, Y., Devos, R., & Burn, P. (1995). Recombinant mouse OB protein: Evidence for a peripheral signal linking adiposity and central neural networks. *Science*, *269*, 546–549.

Caron, E., Sachot, C., Prevot, V., & Bouret, S. G. (2010). Distribution of leptin-sensitive cells in the postnatal and adult mouse brain. *Journal of Comparative Neurology*, *518*, 459–476.

Collingridge, G. L., Isaac, J. T. R., & Wang, Y. T. (2004). Receptor trafficking and synaptic plasticity. *Nature Reviews Neuroscience*, *5*, 952–962.

Criscuolo, C., Fabiani, C., Bonadonna, C., Origlia, N., & Domenici, L. (2015). BDNF prevents amyloid-dependent impairment of LTP in the entorhinal cortex by attenuating p38 MAPK phosphorylation. *Neurobiology of Aging*, *36*, 1303–1309.

Davis, C., Mudd, J., & Hawkins, M. (2014). Neuroprotective effects of leptin in the context of obesity and metabolic disorders. *Neurobiology of Disease*, *72*, 61–71.

Dietrich, M. O., Spuch, C., Antequera, D., Rodal, I., de Yébenes, J. G., Molina, J. A., et al. (2008). Megalin mediates the transport of leptin across the blood-CSF barrier. *Neurobiology of Aging*, *29*, 902–912.

Doherty, G. H., Beccano-Kelly, D., Yan, S. D., Gunn-Moore, F. J., & Harvey, J. (2013). Leptin prevents hippocampal synaptic disruption and neuronal cell death induced by amyloid β. *Neurobiology of Aging*, *34*, 226–237.

Doherty, G. H., Oldreive, C., & Harvey, J. (2008). Neuroprotective actions of leptin on central and peripheral neurons in vitro. *Neuroscience*, *154*, 1297–1307.

Eikelis, N., Lambert, G., Wiesner, G., Kaye, D., Schlaich, M., Morris, M., et al. (2004). Extra-adipocyte leptin release in human obesity and its relation to sympathoadrenal function. *American Journal of Physiology - Endocrinology and Metabolism*, *286*(5), E744–E752.

El Khoury, N. B., Gratuze, M., Petry, F., Papon, M. A., Julien, C., Marcouiller, F., et al. (2016). Hypothermia mediates age-dependent increase of tau phosphorylation in db/db mice. *Neurobiology of Disease*, *88*, 55–65.

Elmquist, J. K., Bjørbaek, C., Ahima, R. S., Flier, J. S., & Saper, C. B. (1998). Distributions of leptin receptor mRNA isoforms in the rat brain. *Journal of Comparative Neurology*, *395*, 535–547.

Farr, S., Banks, W., & Morley, J. E. (2006). Effects of leptin on memory processing. *Peptides*, *27*, 1420–1425.

Fei, H., Okano, H. J., Li, C., Lee, G., Zhao, C., Darnell, R., et al. (1997). Anatomic localization of alternatively spliced leptin receptors (Ob-R) in mouse brain and other tissues. *Proceedings of the National Academy of Sciences of the United States of America*, *94*(13), 7001–7005.

Fernández-Galaz, C., Fernández-Agulló, T., Campoy, F., Arribas, C., Gallardo, N., Andrés, A., et al. (2001). Decreased leptin uptake in hypothalamic nuclei with ageing in Wistar rats. *Journal of Endocrinology*, *171*, 23–32.

Fewlass, D. C., Noboa, K., Pi-Sunyer, F. X., Johnston, J. M., Yan, S. D., & Tezapsidis, N. (2004). Obesity-related leptin regulates Alzheimer's abeta. *FASEB Journal*, *18*, 1870–1878.

Friedman, J. (2014). 20 years of leptin: Leptin at 20: An overview. *The Journal of Endocrinology*, *223*, T1–T8.

Gomes, S., Martins, I., Fonseca, A. C., Oliveira, C. R., Resende, R., & Pereira, C. M. (2014). Protective effect of leptin and ghrelin against toxicity induced by amyloid-β oligomers in a hypothalamic cell line. *Journal of Neuroendocrinology*, *26*, 176–185.

Grasso, P., Leinung, M. C., Ingher, S. P., & Lee, D. W. (1997). In vivo effects of leptin-related synthetic peptides on body weight and food intake in female ob/ob mice: Localization of leptin activity to domains between amino acid residues 106-140. *Endocrinology*, *138*, 1413–1418.

Gratuze, M., El Khoury, N. B., Turgeon, A., Julien, C., Marcouiller, F., Morin, F., et al. (2017). Tau hyperphosphorylation in the brain of ob/ob mice is due to hypothermia: Importance of thermoregulation in linking diabetes and Alzheimer's disease. *Neurobiology of Disease, 98*, 1–8.
Greco, S. J., Bryan, K. J., Sarkar, S., Zhu, X., Smith, M. A., Ashford, J. W., et al. (2010). Leptin reduces pathology and improves memory in a transgenic mouse model of Alzheimer's disease. *Journal of Alzheimers Disease, 19*, 1155–1167.
Greco, S. J., Sarkar, S., Johnston, J. M., Zhu, X., Su, B., Casadesus, G., et al. (2008). Leptin reduces Alzheimer's disease-related tau phosphorylation in neuronal cells. *Biochemical and Biophysical Research Communications, 376*, 536–541.
Guo, Z., Jiang, H., Xu, X., Duan, W., & Mattson, M. P. (2008). Leptin-mediated cell survival signaling in hippocampal neurons mediated by JAK STAT3 and mitochondrial stabilization. *Journal of Biological Chemistry, 283*, 1754–1763.
Håkansson, M. L., Brown, H., Ghilardi, N., Skoda, R. C., & Meister, B. (1998). Leptin receptor immunoreactivity in chemically defined target neurons of the hypothalamus. *Journal of Neuroscience, 18*, 559–572.
Håkansson, M. L., Hulting, A. L., & Meister, B. (1996). Expression of leptin receptor mRNA in the hypothalamic arcuate nucleus—Relationship with NPY neurones. *Neuroreport*, 7(18), 3087–3092.
Halaas, J. L., Gajiwala, K. S., Maffei, M., Cohen, S. L., Chait, B. T., Rabinowitz, D., et al. (1998). Weight-reducing effects of the plasma protein encoded by the obese gene. *Science, 269*, 543–546.
Harvey, J., Solovyova, N., & Irving, A. (2006). Leptin and its role in hippocampal synaptic plasticity. *Progress in Lipid Research, 45*, 369–378.
Hassing, L. B., Dahl, A. K., Thorvaldsson, V., Berg, S., Gatz, M., Pedersen, N. L., et al. (2009). Overweight in midlife and risk of dementia: A 40-year follow-up study. *International Journal of Obesity, 33*, 893–898.
Hervey, G. R. (1959). The effects of lesions in the hypothalamus in parabiotic rats. *Journal of Physiology, 145*, 336–352.
Hoover, B. R., Reed, M. N., Su, J., Penrod, R. D., Kotilinek, L. A., Grant, M. K., et al. (2010). Tau mislocalization to dendritic spines mediates synaptic dysfunction independently of neurodegeneration. *Neuron, 68*, 1067–1081.
Hsieh, H., Boehm, J., Sato, C., Iwatsubo, T., Tomita, T., Sisodia, S., et al. (2006). AMPAR removal underlies Abeta-induced synaptic depression and dendritic spine loss. *Neuron, 52*, 831–843.
Ihle, J. N. (1995). Cytokine receptor signalling. *Nature, 377*, 5910–5914.
Irving, A. J., & Harvey, J. (2014). Leptin regulation of hippocampal synaptic function in health and disease. *Philosophical Transactions of the Royal Society of London. Series B, Biological Sciences, 369*, 20130155.
Isaac, J. T. R., Ashby, M. C., & McBain, C. J. (2007). The role of the GluR2 subunit in AMPA receptor function and synaptic plasticity. *Neuron, 54*(6), 859–871.
Kim, B., & Feldman, E. L. (2015). Insulin resistance as a key link for the increased risk of cognitive impairment in the metabolic syndrome. *Experimental & Molecular Medicine, 47*, e149.
Kivipelto, M., Ngandu, T., Fratiglioni, L., Viitanen, M., Kåreholt, I., Winblad, B., et al. (2005). Obesity and vascular risk factors at midlife and the risk of dementia and Alzheimer disease. *Archives of Neurology, 62*, 1556–1560.
Lee, C. C., Kuo, Y. M., Huang, C. C., & Hsu, K. S. (2009). Insulin rescues amyloid beta-induced impairment of hippocampal long-term potentiation. *Neurobiology of Aging, 30*, 377–387.
Li, X. L., Aou, S., Oomura, Y., Hori, N., Fukunaga, K., & Hori, T. (2002). Impairment of long-term potentiation and spatial memory in leptin receptor-deficient rodents. *Neuroscience, 113*, 607–615.

Li, S., Hong, S., Shepardson, N. E., Walsh, D. M., Shankar, G. M., & Selkoe, D. (2009). Soluble oligomers of amyloid Beta protein facilitate hippocampal long-term depression by disrupting neuronal glutamate uptake. *Neuron*, *62*, 788–801.

Lieb, W., Beiser, A. S., Vasan, R. S., Tan, Z. S., Au, R., Harris, T. B., et al. (2009). Association of plasma leptin levels with incident Alzheimer disease and MRI measures of brain aging. *JAMA*, *302*, 2565–2572.

Liu, L., Wong, T. P., Pozza, M. F., Lingenhoehl, K., Wang, Y., Sheng, M., et al. (2004). Role of NMDA receptor subtypes in governing the direction of hippocampal synaptic plasticity. *Science*, *304*, 1021–1024.

Luo, X., McGregor, G., Irving, A. J., & Harvey, J. (2015). Leptin induces a novel form of NMDA receptor-dependent LTP at hippocampal temporoammonic- CA1 synapses. *eNeuro*, *2*, 1–17.

Malekizadeh, Y., Holiday, A., Redfearn, D., Ainge, J. A., Doherty, G., & Harvey, J. (2017). A leptin fragment mirrors the cognitive enhancing and neuroprotective actions of leptin. *Cerebral Cortex*, *27*, 4769–4782.

Martins, I., Gomes, S., Costa, R. O., Otvos, L., Oliveira, C. R., & Resende, R. (2013). Leptin and ghrelin prevent hippocampal dysfunction induced by Aβ oligomers. *Neuroscience*, *241*, 41–51.

Marwarha, G., Dasari, B., Prasanthi, J. R., Schommer, J., & Ghribi, O. (2010). Leptin reduces the accumulation of Ab and phosphorylated tau induced by 27-hydroxycholesterol in rabbit organotypic slices. *Journal of Alzheimers Disease*, *19*, 1007–1019.

Massey, P. V., Johnson, B. E., Moult, P. R., Auberson, Y. P., Brown, M. W., Molnar, E., et al. (2004). Differential roles of NR2A and NR2B-containing NMDA receptors in cortical long-term potentiation and long-term depression. *Journal of Neuroscience*, *24*, 7821–7828.

Mattson, M. P., & Arumugam, T. V. (2018). Hallmarks of brain aging: Adaptive and pathological modification by metabolic states. *Cell Metabolism*, *27*, 1176–1199.

Maurin, H., Chong, S.-A., Kraev, I., Davies, H., Kremer, A., Seymour, C. M., et al. (2014). Early structural and functional defects in synapses and myelinated axons in stratum lacunosum moleculare in two preclinical models for tauopathy. *PLoS One*, *9*, e87605.

Maurin, H., Seymour, C. M., Lechat, B., Borghgraef, P., Devijver, H., Jaworski, T., et al. (2013). Tauopathy differentially affects cell adhesion molecules in mouse brain: Early down-regulation of nectin-3 in stratum lacunosum moleculare. *PLoS One*, *8*(5), e63589.

McGregor, G., Clements, L., Farah, A., Irving, A. J., & Harvey, J. (2018). Age-dependent regulation of excitatory synaptic transmission at hippocampal temporoammonic-CA1 synapses by leptin. *Neurobiology of Aging*, *69*, 76–93.

McGregor, G., & Harvey, J. (2019). Leptin regulation of synaptic function at hippocampal TA-CA1 and SC-CA1 synapses: Implications for health and disease. *Neurochemical Research*, *44*, 650–660.

Mercer, J. G., Hoggard, N., Williams, L. M., Lawrence, C. B., Hannah, L. T., & Trayhurn, P. (1996). Localization of leptin receptor mRNA and the long form splice variant (Ob-Rb) in mouse hypothalamus and adjacent brain regions by in situ hybridization. *FEBS Letters*, *387*, 113–116.

Morash, B., Li, A., Murphy, P. R., Wilkinson, M., & Ur, E. (1999). Leptin gene expression in the brain and pituitary gland. *Endocrinology*, *140*, 5995–5998.

Moult, P. R., Cross, A., Santos, S. D., Carvalho, A.-L., Lindsay, Y., Connolly, C. N., et al. (2010). Leptin regulates AMPA receptor trafficking via PTEN inhibition. *Journal of Neuroscience*, *30*, 4088–4101.

Moult, P. R., & Harvey, J. (2008). Hormonal regulation of hippocampal dendritic morphology and synaptic plasticity. *Cell Adhesion & Migration*, *2*, 269–275.

Moult, P. R., & Harvey, J. (2011). NMDA receptor subunit composition determines the polarity of leptin-induced synaptic plasticity. *Neuropharmacology*, *61*, 924–936.

Nakashiba, T., Young, J. Z., McHugh, T. J., Buhl, D. L., & Tonegawa, S. (2008). Transgenic inhibition of synaptic transmission reveals role of CA3 output in hippocampal learning. *Science*, *319*(5867), 1260–1264.

Paz-Filho, G. J., Babikian, T., Asarnow, R., Delibasi, T., Esposito, K., Erol, H. K., et al. (2008). Leptin replacement improves cognitive development. *PLoS One*, *3*, e3098.

Paz-Filho, G. J., Mastronardi, C. A., & Licinio, J. (2015). Leptin treatment: Facts and expectations. *Metabolism*, *64*(1), 146–156.

Platt, T. L., Beckett, T. L., Kohler, K., Niedowicz, D. M., & Murphy, M. P. (2016). Obesity, diabetes, and leptin resistance promote tau pathology in a mouse model of disease. *Neuroscience*, *315*, 162–174.

Power, D. A., Collins, R., & Neill, O. (2001). Circulating leptin levels and weight loss in Alzheimer's disease patients. *Dementia and Geriatric Cognitive Disorders*, *12*, 167–170.

Remondes, M., & Schuman, E. M. (2002). Direct cortical input modulates plasticity and spiking in CA1 pyramidal neurons. *Nature*, *416*, 736–740.

Remondes, M., & Schuman, E. M. (2004). Role for a cortical input to hippocampal area CA1 in the consolidation of a long-term memory. *Nature*, *430*, 699–703.

Rozhavskaya-Arena, M., Lee, D. W., Leinung, M. C., & Grasso, P. (2000). Design of a synthetic leptin agonist: Effects on energy balance, glucose homeostasis, and thermoregulation. *Endocrinology*, *141*, 2501–2507.

Santos, C. L., Bobermin, L. D., Souza, D. O., & Quincozes-Santos, A. (2018). Leptin stimulates the release of pro-inflammatory cytokines in hypothalamic astrocyte cultures from adult and aged rats. *Metabolic Brain Disease*, *33*, 2059–2063.

Scarpace, P. J., Matheny, M., Moore, R. L., & Tümer, N. (2000). Impaired leptin responsiveness in aged rats. *Diabetes*, *49*, 431–435.

Scarpace, P. J., Matheny, M., & Tümer, N. (2001). Hypothalamic leptin resistance is associated with impaired leptin signal transduction in aged obese rats. *Neuroscience*, *104*, 1111–1117.

Schwartz, W. Z., Peskind, E., Raskind, M., Boyko, E. J., & Porte, D. (1996). Cerebrospinal fluid leptin levels: Relationship to plasma levels and to adiposity in humans. *Nature Medicine*, *2*, 589–593.

Scott, M. M., Lachey, J. L., Sternson, S. M., Lee, C. E., Elias, C. F., Friedman, J. M., et al. (2009). Leptin targets in the mouse brain. *The Journal of Comparative Neurology*, *514*(5), 518–532.

Shankar, G. M., Li, S., Mehta, T. H., Garcia-Munoz, A., Shepardson, N. E., Smith, I., et al. (2008). Amyloid-beta protein dimers isolated directly from Alzheimer's brains impair synaptic plasticity and memory. *Nature Medicine*, *14*, 837–842.

Shanley, L. J., Irving, A. J., & Harvey, J. (2001). Leptin enhances NMDA receptor function and modulates hippocampal synaptic plasticity. *Journal of Neuroscience*, *21*, RC186.

Shanley, L. J., O'Malley, D., Irving, A. J., Ashford, M. L., & Harvey, J. (2002). Leptin inhibits epileptiform-like activity in rat hippocampal neurones via PI 3-kinase-driven activation of BK channels. *Journal of Physiology*, *545*, 933–944.

Spiegelman, B. M., & Flier, J. S. (2001). Obesity and the regulation of energy balance. *Cell*, *104*, 531–543.

Stranahan, A. M., & Mattson, M. P. (2011). Bidirectional metabolic regulation of neurocognitive function. *Neurobiology of Learning and Memory*, *96*, 507–516.

Tong, J. Q., Zhang, J., Hao, M., Yang, J., Han, Y. F., Liu, X. J., et al. (2015). Leptin attenuates the detrimental effects of b-amyloid on spatial memory and hippocampal later-phase long term potentiation in rats. *Hormone and Behaviour*, *73*, 125–130.

Vago, D. R., Bevan, A., & Kesner, R. P. (2007). The role of the direct perforant path input to the CA1 subregion of the dorsal hippocampus in memory retention and retrieval. *Hippocampus*, *17*, 977–987.

Wayner, M. J., Armstrong, D. L., Phelix, C. F., & Oomura, Y. (2004). Orexin-a (Hypocretin-1) and leptin enhance LTP in the dentate gyrus of rats in vivo. *Peptides*, *25*, 991–996.

Weng, Z., Signore, A. P., Gao, Y., Wang, S., Zhang, F., Hastings, T., et al. (2008). Leptin protects against 6-hydroxydopamine-induced dopaminergic cell death via mitogen-activated protein kinase signaling. *Journal of Biological Chemistry*, *282*, 34479–34491.

White, H., Pieper, C., & Schmader, K. (1998). The association of weight change in Alzheimer's disease with severity of disease and mortality: A longitudinal analysis. *Journal of the American Geriatric Society*, *46*, 1223–1227.

Wiesner, G., Vaz, M., Collier, G., Seals, D., Kaye, D., Jennings, G., et al. (1999). Leptin is released from the human brain: Influence of adiposity and gender1. *The Journal of Clinical Endocrinology & Metabolism*, *84*(7), 2270–2274.

Winocur, G., Greenwood, C. E., Piroli, G. G., Grillo, C. A., Reznikov, L. R., Reagan, L. P., et al. (2005). Memory impairment in obese Zucker rats: An investigation of cognitive function in an animal model of insulin resistance and obesity. *Behavioural Neuroscience*, *119*, 1389–1395.

Xu, W. L., Atti, A. R., Gatz, M., Pedersen, N. L., Johansson, B., & Fratiglioni, L. (2011). Midlife overweight and obesity increase late-life dementia risk: A population-based twin study. *Neurology*, *76*, 1568–1574.

Xu, L., Rensing, N., Yang, X., Zhang, H. X., Thio, L. L., Rothman, S. M., et al. (2008). Leptin inhibits 4-aminopyridine–And pentylenetetrazole-induced seizures and AMPAR-mediated synaptic transmission in rodents. *Journal of Clinical Investigation*, *118*(1), 272–280.

Yassa, M., Muftuler, L. T., & Stark, C. E. L. (2010). Ultrahigh-resolution microstructural diffusion tensor imaging reveals perforant path degradation in aged humans in vivo. *Proceedings of the National Academy of Sciences of the United States of America*, *107*, 12687–12691.

Zhang, F., & Chen, J. (2008). Leptin protects hippocampal CA1 neurons against ischemic injury. *Journal of Neurochemistry*, *107*, 578–587.

Zhang, J., Deng, Z., Liao, J., Song, C., Liang, C., Xue, H., et al. (2013). Leptin attenuates cerebral ischemia injury through the promotion of energy metabolism via the PI3K/Akt pathway. *Journal of Cerebral Blood Flow and Metabolism*, *33*, 567–574.

CHAPTER SEVEN

Endocannabinoids and aging—Inflammation, neuroplasticity, mood and pain

Yongsoon Park[a] and Bruce A. Watkins[b,*]
[a]Department of Food and Nutrition, Hanyang University, Seoul, South Korea
[b]Department of Nutrition, University of California, Davis, CA, United States
*Corresponding author: e-mail address: bawatkins@ucdavis.edu

Contents

Abstract

Aging is associated with changes in hormones, slowing of metabolism, diminished physiological processes, chronic inflammation and high exposure to oxidative stress factors, generally described as the biological cost of living. Lifestyle interventions of diet and exercise can improve the quality of life during aging and lower diet-related chronic disease. The endocannabinoid system (ECS) has important effects on systemic metabolism and physiological systems, including the central and peripheral nervous systems. Exercise can reduce the loss of muscle mass and improve strength, and increase the levels of endocannabinoids (eCB) in brain and blood. Although the ECS exerts controls on multiple systems throughout life it affords benefits to natural aging. The eCB are synthesized from polyunsaturated fatty acids (PUFA) and the primary ones are produced from arachidonic acid (n-6 PUFA) and others from the n-3 PUFA, namely eicosapentaenoic and docosahexaenoic acids. The eCB ligands bind to their receptors, CB1 and CB2, with effects on appetite stimulation, metabolism, immune functions,

Vitamins and Hormones, Volume 115
ISSN 0083-6729
https://doi.org/10.1016/bs.vh.2020.12.007

and brain physiology and neuroplasticity. Dietary families of PUFA are a primary factor that can influence the types and levels of eCB and as a consequence, the downstream actions when the ligands bind to their receptors. Furthermore, the association of eCB with the synthesis of oxylipins (OxL) is a connection between the physiological actions of eCB and the lipid derived immunological OxL mediators of inflammation. OxL are ubiquitous and influence neuroinflammation and inflammatory processes. The emerging actions of eCB on neuroplasticity, well-being and pain are important to aging. Herein, we present information about the ECS and its components, how exercise and diet affects specific eCB, their role in neuroplasticity, neuroinflammation, pain, mood, and relationship to OxL. Poor nutrition status and low nutrient intakes observed with many elderly are reasons to examine the role of dietary PUFA actions on the ECS to improve health.

Abbreviations

anandamide or AEA	*N*-arachidonoylethanolamine
2-AG	2-arachidonoylglycerol
BDNF	brain-derived neurotrophic factor
CNS	central nervous system
CREB	cyclic AMP response element binding protein
eCB	endocannabinoids
ECS	endocannabinoid system
OEA	*N*-oleylethanolamine
PEA	*N*-palmitoylethanolamine
PUFA	polyunsaturated fatty acids

1. Endocannabinoids

The endocannabinoids (eCB) exert multiple actions on systemic biology, macronutrient metabolism, brain and the peripheral nervous system. Exercise is one recent discovery that induces the production of eCB as an important aspect of health throughout life. Recent studies reveal that exercise is a factor that induces the production of eCB in the brain and the eCB play a role in neuroplasticity and neuroinflammation. These actions have important implications for aging. Moreover, the endocannabinoid system (ECS) supports normal physiological functions in the brain and peripheral nervous system (Nikhra, 2019; Ohno-Shosaku & Kano, 2014). Further, the ECS acts as a neuromodulator and immunomodulatory (Lopez et al., 2018), and the oxylipins (OxL) likely participate in the later. In general, the activation of the ECS receptors by the eCB promote cell proliferation, neurogenesis both in a post- and pre-synaptic capacity (Sallaberry & Astern, 2018).

The ECS is comprised of the G-protein-coupled cannabinoid receptors CB1 (type 1) and CB2 (type 2) and the endogenous ligands agonists known as eCB. The eCBs are lipid mediators including esters, amides, and ethers of arachidonic acid, which mimic the effects of Δ9-tetrahydrocannabinol primarily by binding to and activating CB1 and CB2 receptors (Dyall et al., 2016; Howlett et al., 2011). The eCB function in a modulatory fashion on the brain and peripheral nervous system to influence neurobiology and neurophysiology, such as neuroplasticity, mood, cognition, reward, pain, sleep, and appetite (Greco, Gasperi, Maccarrone, & Tassorelli, 2010; Hill et al., 2010; Patel, Davison, Pittman, & Sharkey, 2010; Racz, Nent, Erxlebe, & Zimmer, 2015; Tantimonaco et al., 2014; Viscomi et al., 2009). An overview of the eCB, their receptors and actions on the brain, peripheral nervous and other systems are shown in Fig. 1.

The most studied eCBs are *N*-arachidonoylethanolamine (anandamide, AEA), an *N*-acylethanolamine (NAE), and 2-arachidonoylglycerol (2-AG), a monoacylglycerol (Fig. 1). The synthesis and degradation of eCB is achieved by phospholipases and lipases associated with intracellular membranes. AEA is biosynthesized from arachidonic acid by N-acylphosphatidyl ethanolamine

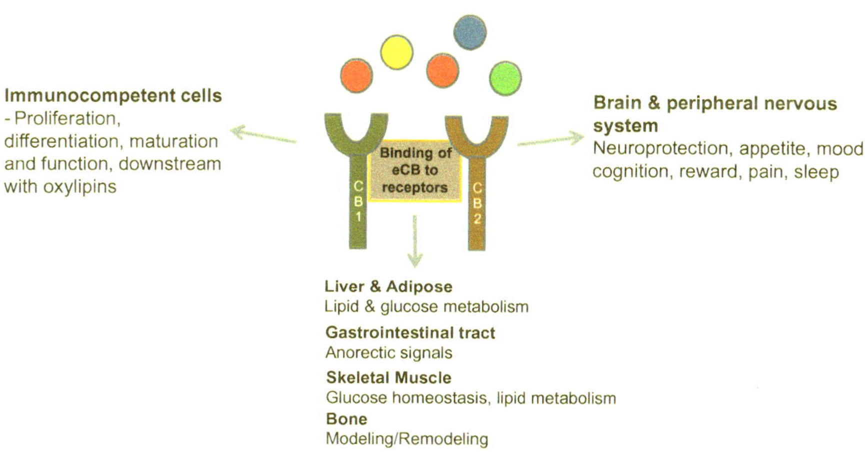

Fig. 1 Endocannabinoids (eCB) include anandamide (red) and 2-AG (yellow), both derived from arachidonic acid, and *N*-docosahexaenoylethanolamine (DHEA) (blue) and *N*-eicosapentaenoylethanolamine (EPEA) (green) are synthesized from DHA and EPA, respectively. The eCB ligands bind to receptors CB1 and CB2 resulting in activation of the endocannabinoid system. The formation of oxylipins from multiple enzyme systems (cyclooxygenase COX 1 and COX2, and lipoxygenase LOX) share common substrates (PUFA) and participate in inflammation.

(NAPE)-specific phospholipase D (NAPE-PLD), while 2-AG is biosynthesized by diacylglycerol lipase alpha (DAGLα) and diacylglycerol lipase beta (DAGLβ). The degrading enzymes for AEA and 2-AG include fatty acid amide hydrolase (FAAH) and monoacylglycerol lipase (MAGL), respectively (Kano, Ohno-Shosaku, Hashimotodani, Uchigashima, & Watanabe, 2009).

As AEA and 2-AG are derived from arachidonic acid, an n-6 PUFA, *N*-docosahexaenoylethanolamine (DHEA) and *N*-eicosapentaenoylethanolamine (EPEA) are derived from the n-3 PUFA, docosahexaenoic acid (DHA) and eicosapentaenoic acid (EPA), respectively (Fig. 1). In contrast, the synthesis of 2-docosahexaenoylglycerol (2-DHG) and 2-eicosapentaenoylglycerol (2-EPG) is not well characterized in the literature, but it is likely they are produced via the same pathways as 2-AG (Dyall, 2017).

Both DHEA and EPEA have been shown to activate CB1 and CB2 (Brown et al., 2010). Furthermore, the amounts of these eCB can be influenced by dietary n-3 PUFA that affect the concentrations of the substrates as well as arachidonic acid. Kim, Carlson, Kuchel, Newman, and Watkins (2016), Kim, Carlson, and Watkins (2014), Kim, Li, and Watkins (2013) observed that n-3 PUFA have been effective in lowering substrate for the synthesis of AEA and 2-AG. For example, a dietary source of DHA can significantly increase DHEA in mice (Kim et al., 2016) and postmenopausal women, including an increased ratio of DHEA/AEA (Watkins et al., 2016). A recent clinical trial on major depressive disorder participants revealed that supplementation of EPA increased EPEA levels compared to the level of DHEA, while EPA + DHA supplementation increased the DHEA levels more so than EPA treatment alone (Yang et al., 2019). In a study in which a high fat diet was fed to obese Zucker rats, increasing amounts of krill oil (rich in EPA and DHA) in the diet reduced the levels of arachidonic acid based on the amounts of diacylglycerol (DAG) species in tissues as well as for AEA and 2-AG in tissues, and also resulted in reductions of *N*-oleylethanolamine (OEA) and *N*-palmitoylethanolamine (PEA) (Piscitelli et al., 2011). In addition to eCB derived from n-3 PUFA, an n-3 PUFA containing diet increased the mRNA levels of eCB synthesis enzymes (NAPE and DAGLα), and CB1 and CB2 in muscle of mice (Hutchins-Wiese, Li, Hannon, & Watkins, 2012).

The eCB operate on demand to influence synaptic transmission in a Ca2+ dependent manner. Different from the classical vesicle-based system of synaptic transmission, AEA and 2-AG act through a retrograde mechanism from postsynaptic neurons to target presynaptic cannabinoid (CB) receptors

(Hashimotodani, Ohno-Shosaku, & Kano, 2007; Ohno-Shosaku, Maejima, & Kano, 2001). The eCB are synthesized by postsynaptic neurons and bind to CB receptors located in presynaptic terminals of GABA or glutamate neurons to inhibit their activity through Gi/o signaling (Silvestri & Di Marzo, 2013). The eCB activate cannabinoid Gi/o protein coupled CB receptors, which regulates calcium and potassium channels and suppresses docking of new synaptic vesicles, impairing the release of neurotransmitters (Blankman & Cravatt, 2013). Thus the eCB are active participants in neuronal control and cellular functions.

The CB1 receptor is highly expressed in brain areas that control emotion, cognition, memory, motor, and nociception, and that includes the anatomical regions of the cortex, limbic system, hippocampus, cerebellum, and several nuclei of the basal ganglia (Howlett, Blume, & Dalton, 2010; Pertwee et al., 2010). Deletion of the CB1 receptor gene and the fatty acid amide hydrolase (FAAH) enzyme results in suppressed proliferation of progenitor cells and dentate gyrus (Aguado et al., 2006; Jin et al., 2004; Kim et al., 2006), consequently the evidence supports the role of eCB in progenitor cell proliferation in nervous tissues. Furthermore, CB1 is also expressed in peripheral cells and tissues, including adipose tissue, liver, and skeletal muscle.

The CB2 receptor was initially described in the peripheral nervous and immune systems (Fernandez-Ruiz et al., 2007) but is also expressed within the central nervous system (CNS) (Patel et al., 2010) where it plays a relevant role in coping with insults to the brain (Viscomi et al., 2009). Thus, the binding of the eCB to CB receptors is involved in neuroplastic events altering neural connections to improve, strengthen, or delete them while reorganizing neural pathways and synapse (Hill et al., 2011; Hill, Titterness, et al., 2010). Therefore, the eCB and CB2 receptor are integral components of maintaining and supporting the normal physiology and responsiveness of the brain to various stimuli.

Besides binding to CB1 and CB2, known as classical receptors, eCB also bind to vanilloid (TRPV, related transient receptor potential) and possibly other receptors that activate key signaling pathways in the CNS and peripheral tissues (Cascio, 2013). By triggering signaling cascades associated with Ca2+ homeostasis, the AEA-TRPV1 interaction is involved in the control of the CNS including nociception (Greco et al., 2010), and of basic biological processes like the induction of apoptosis (Maccarrone, Lorenzon, Bari, Melino, & Finazzi-Agro, 2000; Stock et al., 2012). Thus, the eCB and its receptors are a vital participant in pain reception and other stimuli.

Both AEA and 2-AG bind to nuclear peroxisome proliferator-activated receptors (PPARs) α and γ, thus regulating lipid and glucose metabolism (Gasperi et al., 2007), as well as inflammatory responses (Rockwell, Snider, Thompson, Heuvel, & Kaminski, 2006). Concerning eCB-like compounds, both OEA and PEA seem to act through the activation of PPAR-α, thus regulating appetite and body weight (OEA) or reducing inflammatory responses (PEA) (Okamoto, Morishita, Tsuboi, Tonai, & Ueda, 2004).

Generally, the eCB act by a two-step process. First, by a membrane transporter facilitated cellular uptake and second, by intracellular degradation facilitated by FAAH for AEA or MAGL for 2-AG (Kano et al., 2009). Although the primary fate of AEA and 2-AG is enzymatic hydrolysis, eCB might also be oxidized by cyclooxygenase-2 (COX2) and lipoxygenase isozymes, thus producing oxidized eCBs that are involved in regulating brain synaptic transmission and other biological processes (Nomura et al., 2011), or in the production of prostanoids and the collective, oxylipins (Kim et al., 2013, 2014, 2016; Kim, Li, & Watkins, 2011).

Interestingly the eCB and oxylipins share some common aspects of metabolism and actions on neuroinflammation. The eCB and oxylipins are synthesized from a mutual source of substrate polyunsaturated fatty acids (PUFA), both from the n-6 PUFA and n-3 PUFA families, for their formation. Thus diet is a principle factor influencing the production of eCB and oxylipins, and to alter their production and downstream actions on metabolism, physiology, health, and disease. Furthermore, the eCB and oxylipins influence pain and well-being. The emerging relationships between exercise and elaboration of eCB in brain creates research opportunities to better understand health and rewards of exercise throughout the life cycle and specifically for aging adults as the interplay of exercise and well-being are explored.

1.1 The eCB and aging

Aging is associated with significant metabolic and physiological changes described as the biological cost of living. Furthermore, with these changes come much greater risk of chronic disease and lowered resistance to infectious agents often associated with diet-related chronic disease. Improving diet, nutrient status, and incorporating exercise as another lifestyle factor can improve the quality of life for advanced age. Although chronological aging is set in time, the impact on biological aging is malleable with proper diet and lifestyle. As with many hormones, metabolic pathways and systems decline, the ECS is diminished with aging. Furthermore, the ability to handle oxidative stress and chronic inflammation associated with disease is compromised as we age and these effects are also manifest in the brain and peripheral nervous system with lowered homeostasis and greater neurodegeneration (Wyss-Coray, 2016).

Excessive neuroinflammation can proceed to a chronic condition of neurodegeneration and several neurodegenerative diseases. Aging and neuronal changes can manifest to declines of motor coordination, sensory abilities, attention and cognitive performance. Cerebral atrophy affects the cortex, thalamus and accumbens but not the brain stem, however, neurogenesis occurs in the glial cells and somewhat in hippocampus (Nikhra, 2019). Cannabinoids afford peripheral analgesic, anti-inflammatory, immunosuppressive and neuroprotective benefits. Thus the ECS supports some important molecular and physiological processes during aging (Di Marzo, Stella, & Zimmer, 2015; Nikhra, 2019). Stress also contributes to neuroinflammation which impairs neurogenesis and synaptic function; and cognitive decline leads to changes in synaptic plasticity (Bilkei-Gorzo, 2012; Di Marzo et al., 2015). Further, greater oxidative stress in the brain is associated with risk for neurodegenerative disorders and cognitive decline. The damaging macromolecules and condition of oxidative stress compromises neuronal functions causing neuroinflammation and contributing to neurodegeneration in aging brain. Increasingly, the ECS is a target for therapeutics at several molecular hubs in the CNS for Alzheimer's disease, Parkinson's disease and other neurodegenerative disorders (Di Marzo et al., 2015). The evidence also suggests that the ECS acts to slow down the progression of brain aging, and cannabinoids afford neuroprotective antioxidant capacity due to their metal-ion chelating property (Nikhra, 2019). Thus the ECS may control levels of reactive oxygen species indirectly through the regulation of glial activity as both astrocytes and microglia express CB1 and CB2 receptors. Moreover expression of CB2 receptor exceeds that for the expression of CB1 receptors in microglia and the activation of glial CB2 receptors attenuates glial activation and prevents neurodegeneration (Glisky, 2007). The glial cells produce eCB such as AEA, 2-AG and palmitoyl-ethanol-amide, and express their degradation enzymes.

2. Neuroinflammation, disease, eCB and oxylipins

Neuroinflammation is a localized form of inflammation that occurs in the peripheral nervous system and the CNS to restore damaged neurons and glia, with microglia and astrocytes the predominant effectors, but excess neuroinflammation may inhibit neuronal regeneration (Dyall, 2017). Neuroinflammation is a major cause of several neurologic and neuropsychiatric diseases such as Alzheimer's disease, Parkinson's disease, multiple sclerosis, depression, bipolar disorder, and autism, as well as postoperative neurocognitive disorders.

During neuroinflammation there is a general up-regulation of ECS activity, with predominantly anti-inflammatory effects (Bilkei-Gorzo, 2012). The eCB are secreted through microglial extracellular membrane vesicles and these extracellular vesicles carry AEA on their surface, which stimulates CB1 receptors on neurons and inhibits presynaptic transmission. In addition to microglia, astrocytes respond to CNS damage by producing cytokines and inflammatory mediators, thereby mediating crosstalk between neuroinflammatory and neural systems in the process of "reactive astrogliosis" (Pekny & Pekna, 2014).

Acute exercise markedly increases pro-inflammatory cytokines, such as IL-1β, IL-6, IL-8, and tumor necrosis factor (TNF)-α (Scott et al., 2011), which return to basal levels within 24h from the end of exercise in leukocytes (Zaldivar et al., 2006). However, it is well-recognized that regular exercise exerts general positive effects in preventing and reducing systemic inflammation by: downregulating the release of pro-inflammatory cytokines from monocytes and macrophages; decreasing the number of circulating activated immune cells (Timmerman, Flynn, Coen, Markofski, & Pence, 2008); and increasing the amount of blood T regulatory cells (Wang et al., 2012). In addition, a key role is played by skeletal muscle that, during contraction, releases myokines like IL-6, IL-8, and IL-15 (Pedersen, 2009), and among them, IL-6 promotes an anti-inflammatory environment that suppresses the long-term effects of elevated pro-inflammatory cytokines (Pedersen, 2012), which could be mediated through eCB (Pedersen, 2012). Thus an orchestrated balance in synthesis of eCB and oxylipins from PUFA substrate would afford a mediating control on inflammatory processes and functioning of cells in the brain and peripheral nervous system for optimum physiological outcomes.

AEA suppresses pro-inflammatory cytokines, and inhibiting MAGL affords some antinociceptive and anti-inflammatory actions by increasing levels of 2-AG, suggesting that eCB degradative enzymes, FAAH and MAGL, are logical targets for drug development (Donvito et al., 2018). Stella, Schweitzer, and Piomelli (1997) reported that the concentration of 2-AG was significantly higher than AEA in the basal state of brain.

The predominant receptor in the CNS is CB1 (Pertwee et al., 2010), which appears to be a target for neuroinflammatory diseases such as Multiple Sclerosis, a demyelinating disease of the CNS (Chiurchiù, Leuti, & Maccarrone, 2018). Neuropathic pain is mediated in microglia of the spinal cord and brain, and associated with the release of pro-inflammatory cytokines, such as interleukins and TNFα (Baron, Binder, & Wasner, 2010), and a target for treating inflammatory neuropathic pain (Donvito et al., 2018).

Whether the CB1 receptor plays a tonic role in nociception remains to be established. However, CB1 (−/−) mice do not display apparent altered nociception (Nadal, La Porta, Bura, & Maldonado, 2013), and show a reduction in sucrose preference and home cage activity (Racz et al., 2015). In addition, selective CB1 receptor deletion in peripheral nociceptors leads to enhanced pain behavior as well as diminished anti-nociceptive effects of cannabinoid receptor agonists, suggesting an important role of peripheral CB1 receptors in neuropathic pain (Agarwal et al., 2007). Racz et al. (2008) reported that CB2 (−/−) mice also show exacerbated pain behavior, while CB2 overexpression in the CNS attenuates neuropathic pain, suggesting that CB2 receptor deletion also leads to varying effects on neuropathic pain.

Beside the eCB, the oxylipins are well known lipid-derived mediators of inflammation and immune cell functions (Lundstrom et al., 2013). The multiple oxylipin lipid mediators are derived from arachidonic acid, EPA and DHA, and produced by the actions of COX and LOX (Dyall, 2017). Dietary intake of n-3 PUFA are associated with not only changes in AEA, EPEA and DHEA, but also in the levels of proinflammatory oxylipins, and those from EPEA and DHEA exert strong anti-inflammatory and immune-modulating properties (Balvers et al., 2010; Balvers et al., 2012). In mouse peritoneal macrophages, EPEA and DHEA were found to reduce the lipopolysaccharide-induced production of the proinflammatory chemokines (Balvers et al., 2012). In addition, postmenopausal women supplemented with n-3 PUFA (EPA and DHA) showed increased levels of oxylipins derived from EPA but decreased those derived from arachidonic acid (Watkins et al., 2016). Furthermore, profiling of oxylipins during exercise would be important and in joint inflammatory diseases such as found in osteoarthritic donors, confirmed differences in many oxylipins compared to normal donor subjects in infrapatellar fat pad specimens (Gierman et al., 2013). Kim et al. 2016 reported the reduction in classic

5-lipoxygenase-dependent proinflammatory mediators in mice fed a DHA supplemented semi-purified diet, suggesting a significant reduction in inflammatory tone. It is well known that n-3 PUFA produces the prostanoid derived from n-3 PUFA via the COX enzymes and inflammatory mediators, and COX enzymes are also affected by eCB (Kim and Watkins (2014). Thus, it is conceivable that both changes in eCB level and types as well as reduced inflammatory mediators such as cytokines and oxylipins could be responsible for controlling pain and improving mood after exercise. However, the extent that EPEA and DHEA affect the nervous system and eCB signaling is not yet known (Watkins, 2018).

In summary, lower levels of neuroinflammation are important for brain physiology, however, excess is deleterious to health (Fig. 2). Exercise does increase eCB in brain (Fig. 2). The evidence suggests that eCB and oxylipins participate in events associated with neuroinflammation and changes in the production of both eCB and oxylipins via dietary n-3 PUFA appear to decrease inflammatory actions in the brain and peripheral nervous system (Fig. 3).

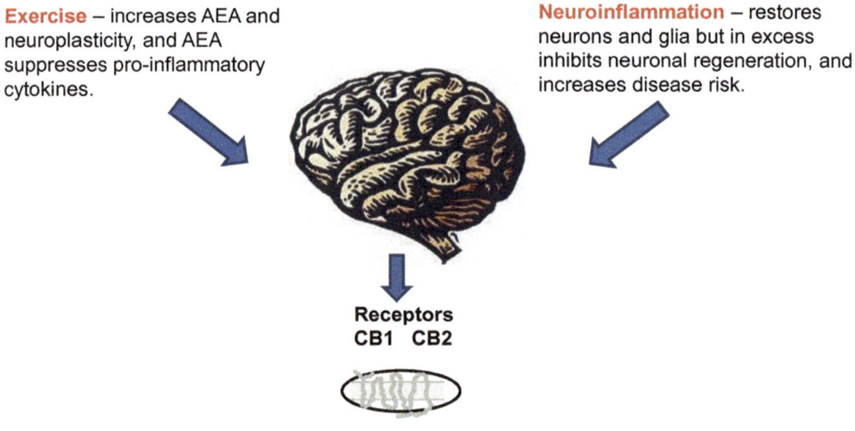

Fig. 2 Activation of the cannabinoid receptors by its ligand anandamide (AEA) influence synaptic transmission in the brain and peripheral nervous system. The eCB are active participants in neuronal controls and cellular functions. The CB1 receptor is highly expressed in brain and controls emotion, cognition, memory and nociception. The CB2 receptor is found in the peripheral nervous system, immune cells and the brain. In general the binding of eCB to its receptors are part of the events that induce neuroplasticity. Other factors influenced by the ECS include the release of brain neurotrophic factors and cytokines, the latter are upregulated during neuroinflammation.

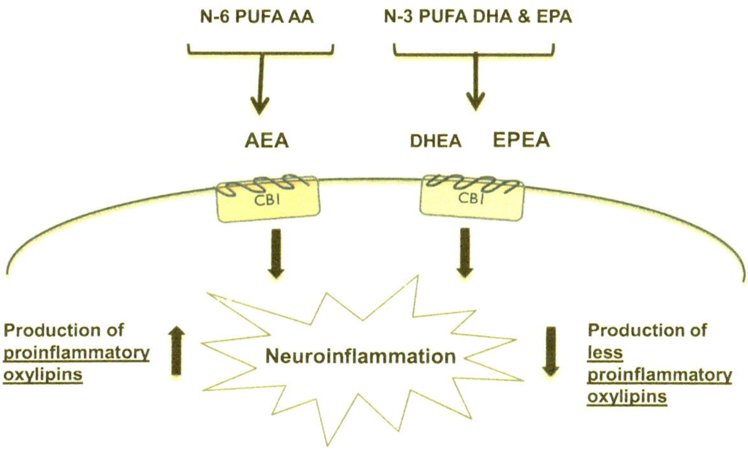

Fig. 3 During neuroinflammation an up-regulation of the ECS and anti-inflammatory effects occur in the brain. The eCB AEA and 2-AG (derived from n-6 PUFA arachidonic acid or AA) are secreted to stimulate CB1 on neurons and inhibit presynaptic transmission. It is not entirely clear how the n-3 PUFA (EPA and DHA) derived eCB DHEA and EPEA influence the specific neuroinflammatory events; however, they appear to reduce the production of pro-inflammatory chemokines or oxylipins. Both groups of eCB, those derived from n-6 and n-3 PUFA, result in the synthesis and formation of oxylipins that impact inflammatory processes. In general, lower neuroinflammation is important for brain physiology but excess can lead to disease.

3. Exercise effects on eCB: Differences and similarities in human and animal studies

At present, twelve human studies reported the effect of exercise on blood levels of eCB. The majority of these investigations showed a significant increase in plasma AEA following exercise (Brellenthin, Crombie, Hillard, & Koltyn, 2017; Feuerecker et al., 2012; Heyman et al., 2012; Meyer, Crombie, Cook, Hillard, & Koltyn, 2019; Raichlen, Foster, Gerdeman, Seillier, & Giuffrida, 2012; Sparling, Giuffrida, Piomelli, Rosskopf, & Dietrich, 2003), while 2-AG remained unchanged (Feuerecker et al., 2012; Heyman et al., 2012; Meyer et al., 2019; Raichlen et al., 2012; Sparling et al., 2003; Stone et al., 2018). In addition, two AEA congeners, OEA and PEA underwent the same exercise-induced changes as AEA (Heyman et al., 2012; Meyer et al., 2019; Stone et al., 2018).

Raichlen et al. (2012) compared the effect of exercise on plasma eCB levels in humans to that in another cursorial mammal (the dog) and in a non-cursorial species (the ferret). The authors reported that humans and dogs share significantly increased exercise-induced plasma AEA, but not 2-AG, levels following 30-min high-intensity endurance running, but not following low-intensity 30-min walking in these taxa, whereas eCB signaling did not significantly increase in ferrets following exercise at any intensity (Raichlen et al., 2012). Increased AEA levels following exercise have also been detected centrally in rodents (Galdino et al., 2014; King-Himmelreich et al., 2017; Raichlen, Foster, Seillier, Giuffrida, & Gerdeman, 2013; Thompson, Argueta, Garland, & DiPatrizio, 2017). Like human studies, plasma 2-AG levels were not changed by exercise in rats (Biedermann et al., 2016; Jennings, Okine, Olango, Roche, & Finn, 2016). Hill, Karatsoreos, Hillard, and McEwen (2010) and Hill, Titterness, et al. (2010) also reported that running wheels elevated the total content of AEA but not 2-AG in hippocampus, without any change in FAAH maximal hydrolytic activity, reflecting a potential increase in AEA biosynthesis. Moreover, exercise in rodents cannot be the same activity as voluntary exercise in humans. Unlike no changes in 2-AG after voluntary exercise in humans, treadmill exercise until fatigue (Galdino et al., 2014) and restraint stress (Hill et al., 2011) increased not only AEA but also 2-AG in mice.

Yasari et al. (2012) supported the premise that exercise induced AEA biosynthesis because eight-weeks of regular running had no effect on hepatic expression of FAAH, CB1, and CB2 in mice. Swenson, Hamilton, Robison, and Thanos (2019) also showed that exercise had no effect on CB1 receptor expression in female and male mice. In contrast, regular exercise in human subjects combined with caloric restriction decreased FAAH gene expression of abdominal adipose tissue, indicating that exercise could reduce AEA degradation (You, Disanzo, Wang, Yang, & Gong, 2011). AEA, OEA and PEA are degraded by FAAH, while 2-AG degraded by MAGL, supporting the premise that AEA, OEA and PEA share similar patterns of changes following exercise, whereas changes of 2-AG is different.

Endocannabinoid signaling is induced by stress and glucocorticoid hormones, particularly cortisol (Hill, Titterness, et al., 2010). Hill, Karatsoreos, et al. (2010) demonstrated that systemic administration of corticosterone resulted in an increase in the tissue concentrations of AEA in amygdala, hippocampus and hypothalamus of brain in rats, while 2-AG appeared less responsive. The rapid increase in AEA following corticosterone administration, suggests that glucocorticoids act through a non-genomic pathway (Hill, Karatsoreos, et al., 2010), compatible with acute adaptations to a physiological stressful condition such as exercise. Heyman et al. (2012) observed a positive correlation between serum cortisol and plasma AEA levels during exercise and recovery in healthy subjects, thus the peripheral increase in AEA during exercise and recovery could be partly triggered by exercise-induced changes in circulating cortisol. Furthermore, a transient increase in insulin also rapidly leads to decreases in plasma AEA, to a greater extent than plasma 2-AG levels, in lean and obese subjects (Di Marzo et al., 2009), possibly by stimulating the expression of FAAH to degrade AEA (Murdolo et al., 2007), which the FAAH has a stronger action on AEA than 2-AG. In addition, insulin production decreases during acute exercise in healthy subjects because of inhibition of β-cell activity by the sympathetic system (Heyman et al., 2007). Therefore, it is possible that the decrease in circulating insulin levels during exercise and the persistence of low levels of insulin during recovery also contribute to the increase in AEA plasma concentrations without affecting 2-AG levels.

Interestingly, Galdino et al. (2014) showed that acute aerobic exercise in rats was accompanied by significant elevation of the plasma levels of AEA, OEA, PEA and 2-AG, which resulted in CB1- and CB2-mediated mechanical and thermal antinociception. Therefore, studies are needed to confirm that exercise induced increase of eCB does not directly act through a genomic-pathway and on the signaling pathway for pain.

There has been no study investigating the effect of exercise on eCB induced by or synthesized from n-3 PUFA. However, Hutchins-Wiese et al. (2012) reported that a semi-purified diet enriched with n-3 PUFA increased the expression of NAPE-PLD, DAGLα, FAAH, CB1, and CB2 mRNA coupled with PUFA preservation during hind limb suspension, which caused peripheral neuron excitement, suggesting that n-3 PUFA with unloading skeletal muscle could modulate inflammation through eCB and the ECS.

A dependence on the increase of AEA level during exercise intensity is suggested indirectly. Plasma levels of AEA increased after performing 30–45 min of moderate exercise (running and cycling), but not of lower or significantly higher exercise intensities where the maximal heart rate was 44% and 92%, respectively (Feuerecker et al., 2012; Heyman et al., 2012; Raichlen et al., 2012; Sparling et al., 2003). Feuerecker et al. (2012) studied the effects of longer moderate exercise bouts consisting of hiking in a normoxic environment (4–4.5 h) or in a hypoxic environment (3.5–5.5 h). Both types of exercise caused an average of two- and threefold increase in plasma AEA levels, respectively, while 2-AG levels were not significantly changed.

In addition, Stone et al. (2018) reported that cycling but not dancing increased plasma level of OEA in post-menopausal women during fasting. In summary, a key research question is what is the role of increased AEA during specific exercise levels?

On the one hand, Thompson et al. (2017) reported that female mice had higher AEA but lower 2-AG than male mice at baseline, and after wheel exercise, the level of AEA increased in both female and male control mice. On the other hand, wheel exercise decreased the level of 2-AG in female control mice, but increased 2-AG in male control mice. Sex differences are not unexpected, given that previous studies on these mice revealed sex differences in the response of wheel running to CB1 agonists and antagonists (Keeney et al., 2008; Keeney, Meek, Middleton, Holness, & Garland, 2012). However, Swenson et al. (2019) showed that exercise had no effect on CB1 receptor expression in female and male mice, but at baseline, female mice had lower CB1 than male mice. In addition, adolescent females are especially more vulnerable than males to disruption of CB1 signaling by repeated exposure to Δ9-tetrahydrocannabinol (Burston, Wiley, Craig, Selley, & Sim-Selley, 2010), and studies in humans indicate that females are more sensitive than males to the effects of cannabis (Craft, Marusich, & Wiley, 2013). It is not clear if differences in CB1 between male and females at baseline is related to the levels of AEA and 2-AG.

There has been no study comparing the effect of exercise on eCB between young and old humans, but studies in rodents show that the ECS is susceptible to age-related deficits (Di Marzo et al., 2015). Old mice had lower AEA and 2-AG levels in brain than young mice (Leishman, Mackie, & Bradshaw, 2019; Piyanoya et al., 2015). CB1 receptor levels decrease, along with the activity NAPE-PLD and DAGL with increasing age (Di Marzo et al., 2015). Furthermore, decreases in DAGL, coupled with elevated MAGL, leads to specific decreases in 2-AG levels in the hippocampus of aging mice (Piyanoya et al., 2015). These age differences likely occur in humans.

At present, there are many gaps and limitations in the evidence regarding the eCB and ECS in exercise levels, sex, and the human. Future research must explain these relationships and further describe the role on well-being and physiological processes that benefit the human during aging. The importance of diet and specifically PUFA as substrate for eCB will be a key aspect of this research. Many of these aforementioned aspects of exercise effects on eCB are presented in Fig. 4.

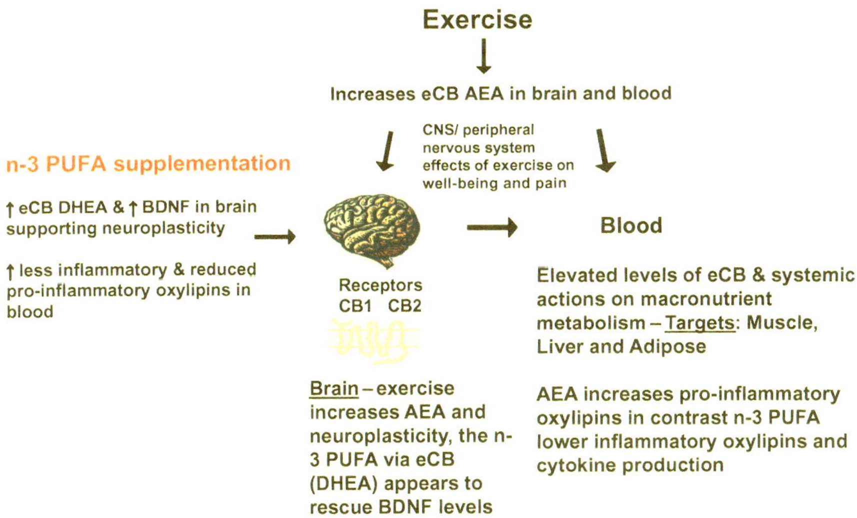

Fig. 4 Exercise has many effects on the brain and peripheral nervous system as well as actions on systemic macronutrient metabolism. Brain and blood levels of AEA increase after exercise in rodents and humans. The eCB support neuroplasticity and the n-3 PUFA derived eCB appears to support greater production of brain-derived neurotrophic factor (BDNF) in brain. The type of n-6 and n-3 PUFA can direct the biosynthesis of different eCB and oxylipins. The n-6 PUFA lead to greater pro-inflammatory and the n-3 PUFA less inflammatory oxylipins. Generally, the eCB are associated with anabolic processes and stimulation of appetite, restoring macronutrient storage of glycogen synthesis and fat accretion following digestion and absorption. It is interesting to point out the positive correlation between serum cortisol and plasma AEA levels during exercise and recovery in healthy subjects.

3.1 Aging and the role for eCB and exercise in neuroplasticity, neurogenesis and BDNF

Neuroplasticity is described as the ability of the nervous system to respond to intrinsic or extrinsic stimuli by reorganizing its structure, function, and connections (Cramer et al., 2011). The role of neurotrophins, in particular, brain-derived neurotrophic factor (BDNF) has been shown to modulate neuroplasticity such as neuronal branching, dendrite formation, and synaptic plasticity (Gomez-Palacio-Schjetnan & Escobar, 2013). BDNF exerts its neurotrophic effects by activating the tropomyosin-related kinase receptor B (TrkB), which signals the downstream activation of the transcription factor cAMP response element binding protein (CREB). Activation of CREB turns on the expression of genes involved in neuronal survival and plasticity.

Abnormalities in neurotrophins expression in the brain underlie the pathology of most neurological disorders. BDNF also binds to p75NTR albeit with a lower affinity, to the receptor, which generally is known to promote proteolysis and apoptosis (Levy et al., 2018). BDNF is abundantly expressed in the mammalian brain, with the highest concentrations found in the hippocampus and cortex (Ernfors, Wetmore, Olson, & Persson, 1990).

Neurogenesis occurs mainly in two brain areas, the subventricular zone and the subgranular zone within the dentate gyrus of the hippocampus. These regions are packed with neural stem/progenitor cells that originate neuroblasts, and migrate toward their destination, where they differentiate into mature neurons and are integrated into the neuronal circuity (Vilar & Mira, 2016). BDNF is expressed in both subventricular zone and subgranular zone neurogenic niches, and able to affect early steps of postnatal neurogenesis, such as cell fate, cell proliferation and neuronal differentiation of the subventricular zone, and in dentate gyrus cultures (Ferreira, Ribeiro, Rodrigues, Sebastiao, & Xapelli, 2018). In addition, the volume of hippocampal dentate gyrus region was decreased in BDNF +/− mice and BDNF knock-out mice (Lee, Duan, & Mattson, 2002; Magarinos et al., 2011; Rauskolb et al., 2010), while BDNF treatment increased neurogenesis, such as number of cells or volume of the hippocampal dentate gyrus region (Alonso, Medina, & Pozzo-Miller, 2004; Scharfman et al., 2005; Ying et al., 2002). A summary of the actions of exercise on specific aspects of brain physiology and BDNF is illustrated in Fig. 5.

Besides the process of neurogenesis, another mechanism of neuroplasticity is the modification of mature neuronal morphology involving synaptogenesis (Levy et al., 2018). BDNF can act on tyrosine kinase B (Trk B) receptor signaling, resulting in activation of synaptic proteins. BDNF +/− mice revealed that BDNF was required for several forms of long-term potentiation mediating the dynamic and adaptive changes in synaptic strength (Aarse, Herlitze, & Manahan-Vaughan, 2016), and BDNF infusion in the rat hippocampus induced long-term potentiation (LTP), and triggered synaptic strengthening (Bramham, 2007; Ying et al., 2002). In addition, structural MRI studies showed the reduced LTP and BDNF expression in individuals during a depressive episode in comparison to patients in remission (Miller & Hen, 2015). These studies suggest that BDNF is an important factor in the regulation of neuroplasticity by affecting both neurogenesis and synaptic plasticity.

Exercise has also shown beneficial effects on neuroplasticity (Fig. 5). A meta-analysis suggested that acute aerobic, but not strength exercise,

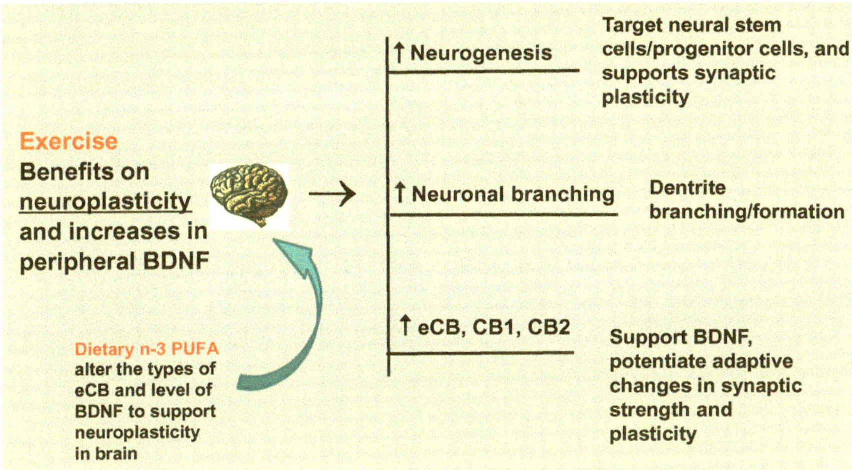

Fig. 5 Exercise has potent effects on the brain to support changes in neuroplasticity and neurogenesis both important throughout the life cycle. These actions are likely critical to improve cognition and neuromuscular control in the elderly. Neurogenesis and adaptive changes in synaptic plasticity are vital to the ability of the nervous system to respond to intrinsic and extrinsic stimuli. The production of BDNF induced by exercise, supports the collective actions on neuroplasticity. The eCB are participants in the effects of exercise on brain, and CB2 appears to have a preponderant role in modulating BDNF in the brain. Acute exercise appears to increase oxylipins and these compounds and down-stream products influence inflammatory pain and neuropathic pain. Furthermore, dietary n-3 PUFA may provide beneficial effects on neurodegenerative diseases such as Alzheimer's disease, depression, and Parkinson's disease.

increases basal peripheral BDNF concentrations, although this effect was only transient (Knaepen, Goekint, Heyman, & Meeusen, 2010). In preclinical studies, swimming showed antidepressant-like effects on hedonia in stressed rodents along with a normalization of a stress-induced BDNF mRNA expression decrease (Jiang et al., 2014). Neuroplasticity happens throughout the growth period and the entire life cycle to adapt to our environment and survive, thus BDNF is an essential aspect of health and living and perhaps well-being, however, these relationships must be verified in future studies.

An important aspect of BDNF actions appears to be its relationship with the ECS. For example, neural stem/progenitor cells present in the neurogenic niches express all elements of the ECS including CB1 and CB2 receptors (Rodrigues et al., 2017). Ferreira et al. (2018) reported that the CB2 receptor has a preponderant role in modulating BDNF actions on dentate gyrus and BDNF-mediated postnatal neurogenesis of subventricular zone influenced by both CB1 and CB2 receptors, thus suggesting that BDNF is required for CB actions to occur. In addition, genetic deletion of CB1 receptor was shown to promote a decrease in BDNF expression (Aso et al., 2008), while induction of BDNF expression contributed to the protective effect of CB1 receptor activity against excitotoxicity (Khaspekov et al., 2004; Marsicano et al., 2003).

The principal active component of cannabis, Δ^9-tetrahydrocannabinol, was shown to promote upregulation of BDNF expression (Butovsky et al., 2005), which rescues the cognitive deficits promoted by administration of Δ^9-tetrahydrocannabinol (Segal-Gavish et al., 2017). Clinical data also suggests that acute and chronic intermittent exposure to Δ^9-tetrahydrocannabinol alters serum levels of BDNF in humans (D'Souza, Pittman, Perry, & Simen, 2009). The CREB could be the common linkage because it is an important regulator of BDNF-induced gene expression, and is reported to control several steps of the neurogenic process in the adult hippocampus (Nakagawa et al., 2002) and superior vena cava (Giachino et al., 2005). Consistently, cannabinoids have been shown to induce CREB phosphorylation (Estrada & Isokawa, 2009), and also to promote changes in BDNF and CREB gene expression (Grigorenko et al., 2002). There has been no study showing the association between eCB derived from n-3 PUFA and BDNF, however, Rashid, Katakura, Kharebava, Kevala, and Kim (2013) reported that DHEA induces differentiation of neural stem cells via protein kinase A (PKA)/CREB, suggesting DHEA could increase BDNF.

Exercise increased BDNF, which could be responsible for increasing neurogenesis, improving memory, and affording antidepressant effects of exercise (dos Santos et al., 2018; Rasmussen et al., 2009). Thus, the research on exercise and eCB is extended to BDNF, a possible molecule involved in exercise-induced brain plasticity (Heyman et al., 2012). Exercise resulted in elevated AEA (PEA and OEA but not 2-AG) and BDNF, and AEA and BDNF concentrations were positively correlated at the end of exercise and after the 15 min recovery. These findings suggest that AEA increment during exercise might be one of the factors involved in exercise-induced increase in peripheral BDNF levels and that AEA high levels during recovery might delay the return of BDNF to basal levels in young male cyclists (Heyman et al., 2012). In this study, the positive correlation with serum cortisol and plasma AEA levels might suggest that cortisol resulted in an increase in AEA after acute exercise in men (Heyman et al., 2012).

The BDNF/Trk system has been proposed as a principle element in the pathophysiology of several mood disorders, including drug addiction, depression, and anxiety. Accordingly, BDNF signaling in the stress system of the brain, including the hippocampus and the hypothalamo-pituitary-adrenocortical (HPA) axis, is thought to be a potential target for antidepressant action (Martinowich, Manji, & Lu, 2007). Recent meta-analyses showed that serum BDNF levels were lower in depressed patients, suggesting the use of serum BDNF levels as a reliable biomarker for mood disorders (Kishi, Yoshimura, Ikuta, & Iwata, 2018; Molendijk et al., 2014; Polyakova et al., 2015). The effect of antidepressants on BDNF in the treatment of major depressive disorder have been widely studied, showing that BDNF expression was increased in post-mortem brains of depressed patients treated with antidepressant drugs as compared to non-treated patients (Chen, Dowlatshahi, MacQueen, Wang, & Young, 2001; Dunham, Deakin, Miyajima, Payton, & Toro, 2009). In this way, although indirectly, BDNF is a likely candidate to elicit benefits of well-being and presumably via the actions of exercise on the brain and coupled to serotonin secretion. Further, n-3 PUFA may improve mood via DHEA and through a connection to increasing BDNF, but not all of these aspects were evaluated in one study on mood outcomes.

Neurotrophins are small proteins that are crucial for neuronal differentiation, growth, survival, and plasticity. Nerve growth factor (NGF), brain-derived neurotrophic factor (BDNF), neurotrophin-3 (NT-3), and neurotrophin-4/5 (NT-4/5) are members of the mammalian neurotrophin family. The impact of these molecules on the nervous system is mediated by the tropomyosin receptor kinase (Trk) receptors and membrane-bound receptor tyrosine kinases that activate a number of cell signaling pathways which are linked to growth, differentiation, and survival (Patapoutian & Reichardt, 2001). In particular, NGF has been proven to be strongly involved in regulating plasticity by impacting upon almost all of the neuroplasticity processes and the effects of antidepressant therapies, including neuronal survival, differentiation, and proliferation (Crutcher, 1986). Moreover, Balogun and Cheema (2014) reported that dietary n-3 PUFA (rich in EPA and DHA) dependent regulation of NGF and CREB at the post-translational level, suggesting that maintaining the levels of neurotrophins including NGF throughout the life span could be important to prevent neurological disorders.

4. The role of N-3 PUFA in brain and aging

Epidemiologic studies suggest that lower per capita fish/seafood consumption, which is used as a surrogate measurement of n-3 PUFA dietary intake (EPA and DHA together, e.g., fish oil), is associated with a higher prevalence of bipolar spectrum disorders and depression (Hibbeln, 1998; Noaghiul & Hibbeln, 2003). Further, it has been shown that dietary PUFA have a modulatory effect on BDNF (Hadjighassem et al., 2015). Ferreira et al. (2014) showed that n-3 PUFA intake was associated with serum BDNF levels in adolescents, corroborating the suggestion that the relationship between mental disorders and BDNF could be mediated by the intake of nutrients such as n-3 PUFA. In line with the positive effect of n-3 PUFA on BDNF regulation, is the finding that an alpha-linolenic acid deficient diet decreased striatal BDNF content in mice (Fan et al., 2016; Ferreira et al., 2013; Pietropaolo et al., 2014). An n-3 PUFA deficiency particularly during pregnancy and lactation will make a lasting harmful impact on brain neurogenesis and apoptosis of the adult offspring and that these harmful effects may be attributable to the epigenetic changes in BDNF (Fan et al., 2016).

In addition, an increase in DHA, and a concomitant decrease in arachidonic acid on a high n-3 PUFA diet reveals that an increase in brain DHA is compensated by a decrease in brain arachidonic acid and increase in the protein expression of BDNF (Balogun & Cheema, 2014; Jiang, Shi, Wang, & Yang, 2009). Furthermore, studies reported that although EPA is far less abundant in the brain than DHA, it is highly neuroactive, and may be more effective on BDNF than DHA in the treatment of psychiatric disorders (Dong, Xu, Kalueff, & Song, 2018) and major depression (Bot et al., 2011). In a rat model of Alzheimer's disease, IL-1β triggering glial activation and increasing dietary intake of n-6 PUFA, induces neuroinflammation and amyloid precursor protein expression, as well as suppresses BDNF function (Dong et al., 2018). However, EPA supplementation treatment restored the n-3 and n-6 PUFA concentrations in the hippocampus, and normalized glial cell activity and BDNF signaling, thereby inducing anti-inflammatory actions in IL-1β-injected animals. These findings suggest that n-3 PUFA may provide beneficial effects on neurodegenerative diseases such as Alzheimer's disease, depression, and Parkinson's disease (Bousquet et al., 2009; Dong et al., 2018).

Gu et al. (2018) reported that endogenous n-3 PUFA (EPA and DHA) could ameliorate depression-like behavior via an anti-inflammatory mechanism in Fat-1 mice because pro-inflammatory cytokines IL-1, IL-17, and TNF-α were inhibited, while anti-inflammatory cytokines IL-10, IL-4 and TGF-β1 were increased. These anti-inflammatory effects are accompanied by normalizing astrocyte function, shown through the decreased expression of p75, but increased BDNF, which may contribute to the improvement of depression-like changes.

Recently, supplementation of n-3 PUFA, EPA and DHA, was shown to increase BDNF expression with beneficial effects on emotional status in male and female depressed rodents (Choi & Park, 2017; Ferreira et al., 2013; Jin & Park, 2015; Park, Moon, & Kim, 2012). Previous studies consistently show that n-3 PUFA helps in the maintenance of learning and memory performance and prevention of age-related synaptic plasticity changes via potentiating the activation of CREB and its downstream effectors BDNF/TrkB via the ERK and Akt pathways (Balogun & Cheema, 2014; Gao et al., 2016). In addition, Wu, Ying, and Gomez-Pinilla (2013) reported that the combination of DHA and physical exercise increased BDNF expression, synaptic plasticity, and cognition in rats after traumatic brain injury, as compared with DHA or physical exercise alone, indicating synergistic effect of diet and exercise (Fig. 5). In summary, n-3 PUFA have been found to influence the cellular and physiological processes involved in psychiatric/mood disorders, such as the fatty acid composition of cellular membranes, and the modulation of behavioral systems, neurodevelopment, and oxidative stress in specific brain areas (Wu et al., 2013).

4.1 Oxylipins, inflammation and disease

Oxylipins are bioactive oxidation compounds derived from dietary n-6 and n-3 PUFA or the linoleic acid and α-linolenic elongation/desaturation pathway products, respectively. The oxylipins are not stored due to their potency, and are instead synthesized de novo in a tightly regulated manner and rapidly degraded (Dyall, 2017). Oxylipins act as autocrine and paracrine lipid mediators by binding to cell surface G protein-coupled receptors, CB1 and CB2, or to multiple intracellular and nuclear receptors such as PPAR-γ involved in specific neuroinflammatory pathways (Montagner et al., 2016). Membrane phospholipid PUFA are first released by phospholipase A2 (PLA2) in response to cell activation from various stress-related stimuli including inflammation (Barquissau et al., 2017). Three main pathways are involved in the production of numerous and diverse oxylipins: (1) cyclooxygenase (COX) and subsequent synthases, (2) lipoxygenase (LOX), and (3) cytochrome P450 mixed function oxidase enzymes (CYP) (Barquissau et al., 2017). These canonical pathways produce the classic mediators, with those produced from 20 carbon PUFA, such as arachidonic acid and EPA, called eicosanoids, whereas those from 22 carbon PUFA, such as DHA, called docosanoids. AEA and 2-AG are produced from membrane-bound phospholipid arachidonic acid, with synthesis occurring at the post-synaptic terminal via increased levels of intracellular calcium with both made in response to demand and rapidly degraded to arachidonic acid or oxygenated to further bioactive oxylipins (Fonseca, Costa, Almada, Correia-da-Silva, & Teixeira, 2013).

The COX enzymes (COX1 and COX2) catalyze the initial oxygenation of non-esterified fatty acids to produce prostaglandin H (PGH), a short-lived intermediate, which is further metabolized into prostanoids, such as other prostaglandin series (PGD, PGE, PGF), prostacyclins (PGI), thromboxanes (TX), and lipoxins (LX), hydroxy and hydroperoxy fatty acids (Alhouayek & Muccioli, 2014). COX catalyzes the initial oxygenation of non-esterified fatty acids to produce prostaglandin H (PGH), a short-lived intermediate, which is further metabolized into prostanoids, such as other prostaglandin series (PGD, PGE, PGF), prostacyclins (PGI), thromboxanes (TX), and lipoxins (LX), hydroxy and hydroperoxy fatty acids (Alhouayek & Muccioli, 2014). COX-2 was traditionally thought to be responsible for contributing to inflammation and neuroinflammation by converting arachidonic acid to PG and TX. However, increasing dietary n-3 PUFA changes the lipid profile of cell membranes (remodeling Fig. 6) and alters the balance of n-6 and n-3 PUFA competing as substrates for COX, consequently changing the series of PG synthesized, which ultimately leads to different cellular responses to mitogenic and inflammatory stimuli.

LOX enzymes are traditionally classified based on the position of the hydroxyl and hydroperoxyl fatty acids they produce from arachidonic acid to LX, LT, HX, hydroxy-eicosatetraenoic acid (HETE), hydroperoxy-eicosateraenoic acid (HpETE), and the "specialized lipid mediators" (Dyall, 2017). The specialized lipid mediators include the resolvins (Rv), protectins (PD) and maresins (MaR) derived from n-3 PUFA. D-series resolvins (RvD), PD and MaR produced from DHA, whereas E-series resolvins (RvE) from EPA, which demonstrate similar anti-inflammatory and pro-resolving properties (Serhan, Dalli, Colas, Winkler, & Chiang, 2015). These metabolites are active lipid mediators with specific effects on resolving neuroinflammation in different body systems including the brain (Sun et al., 2018). While the RvD1 target is toward activating the nuclear factor kappa-light-chain-enhancer (NF-κB) pathways, RvE1 appears to be regulated by miRNAs instead (Rey et al., 2016). A further class of metabolites generated from n-3 PUFA by LOX is the electrophilic fatty acid oxo-derivatives (EFOX), with 7-oxo-DHA, 7-oxo-DPA and 5-oxo-EPA produced from DHA, DPA n-3 and EPA, respectively (Cipollina, Salvatore, Muldoon, Freeman, & Schopfer, 2014). EFOXs display a wide range of anti-inflammatory actions,

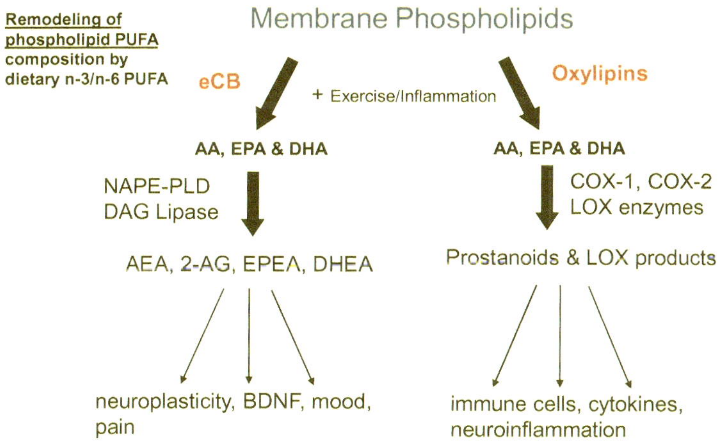

Fig. 6 Both biosynthetic pathways for endocannabinoids and oxylipins share common dietary or endogenously formed n-6 and n-3 PUFA. Thus, these PUFA dictate the types of eCB and oxylipins produced and as a consequence their actions. The ECS is regulated by AA eCB agonists, and EPA may decrease eCB agonists in brain as well as other tissues and cells. As dietary n-3 PUFA increase, the amounts of AA are replaced in the composition of membrane phospholipid PUFA which lowers synthesis of AEA. Under conditions of exercise when AEA increases it is not known if n-3 PUFA supplementation replaces AEA with EPEA and DHEA eCB; however, such does occur under other conditions. Further, it is not clear if the EPEA and DHEA mediate the same effects as AEA or afford more positive benefits on mood and well-being during exercise. The current evidence suggests that replacement of AA in membrane phospholipids with EPA and DHA will reduce the synthesis of proinflammatory oxylipins and increase BDNF.

including acting as agonists nuclear receptors, such as the PPAR and inhibiting cytokine production in activated macrophages (Groeger et al., 2010). In neutrophils isolated from healthy human subjects who were supplemented with EPA and DHA for 4 months, increased formation of 7-oxo-DHA and 5-oxo-EPA was observed (Cipollina et al., 2014).

The third oxidative pathway involves CYP epoxygenases and ω-hydrolases, which metabolize PUFA to lipid mediators (Dyall, 2017). Regio- and stereoisomers of epoxy-eicosatetraenoic acids (EET) and HETE are synthesized from arachidonic acid, whereas those derived from EPA include epoxy-eicosatetraenoic acids (EETeTR) and hydroxy-eicosapentaenoic acids (HEPE) and epoxy-docosapentaenoic acids (EDP), and HDoHE from DHA (Arnold et al., 2010). EPA is the preferred substrate for most isoforms of CYP, with metabolism of DHA and arachidonic acid occurring at similar rates (Arnold et al., 2010).

4.2 Physiologic impact of oxylipins and downstream biomarkers

Oxylipins have diverse actions and differ by structural and biochemical sources. The AA-derived mediators from the COX and LOX pathways tend to be associated with pro-inflammatory responses while those from EPA and DHA are either less inflammatory or anti-inflammatory mediators (Lundstrom et al., 2013). Further, arachidonic acid metabolites from the CYP epoxygenases are themselves anti-inflammatory agents, while the n-3 PUFA epoxides are even more potent anti-inflammatory agents (Ulu et al., 2013; Wagner, Inceoglu, & Hammock, 2011; Zhang et al., 2013).

Watkins et al. reported that supplementation with n-3 PUFA increased n-3 PUFA-derived oxylipins such as 14,15-DiHETE, 17,18-DiHETE, 17(18)-EpETE, 5-HEPE, and 19,20-DiHDoPE but decreased n-6 PUFA derived oxylipins such as 14,15-DiHETrE, 11,12-DiHETrE, and 8,9-DiHETrE, (CYP450/soluble epoxide hydrolase metabolites) in serum of postmenopausal women (Watkins et al., 2016). The consequence of DHA tissue enrichment resulting in an increase of DHEA in postmenopausal women supports a rationale for supplementing with DHA to improve endocannabinoid tone and control age-related inflammation by altering oxylipins (Watkins et al., 2016). DHEA as well as the metabolites produced from its oxidative metabolism are known to produce novel molecules possessing anti-inflammatory and organ protective properties (Yang et al., 2011). In addition, FAT-1 transgenic mice endogenously synthesizing n-3 PUFA had higher serum levels of cytokines such as TNF-α, IL-6, IL-1b, and MCP-1, and also oxylipin, 12-HETE than FAT-2 transgenic mice endogenously synthesizing n-6 PUFA (Kaliannan et al., 2019).

Nociceptors are activated or sensitized by inflammatory mediators such as bradykinin, prostaglandins, nerve growth factor, proinflammatory cytokines (TNF-α, IL-1β), and proinflammatory chemokines that directly bind and stimulate G-protein-coupled receptors, ionotropic receptors, and tyrosine kinase receptors (Basbaum, Bautista, Scherrer, & Julius, 2009). Specialized proresolution mediators, including lipoxins, resolvins, protectins, and maresins, are biosynthesized from n-3 PUFA, DHA and EPA, two major components of fish oil. Specialized proresolution mediators have multiple beneficial actions for treating inflammation-associated disease conditions (Ji, Nackley, Huh, Terrando, & Maixner, 2018). Lipoxin A4, resolvin D1, resolvin E1, resolvin D2, neuroprotection D1, and maresin 1, at very low doses (10–500, ng range), reduced inflammatory pain, postoperative pain, and neuropathic pain in animal models (Ji et al., 2018).

Neutrophils isolated from healthy human subjects who were supplemented with EPA and DHA for 4 months, showed increased formation of 7-oxo-DHA and 5-oxo-EPA (Cipollina et al., 2014). These results support the precept that DHA can form oxygenated metabolites with electrophilic properties and that these metabolites can transduce anti-inflammatory actions through activation of the Nrf2/ARE pathway. DHA also undergoes lipid peroxidation and produces oxylipin metabolites including 4-HHE. There is evidence that these metabolites may play a role in modulating oxidative homeostasis in cells through the NF-κB and the Nrf2/ARE pathways (Sun et al., 2018).

As presented, both n-6 and n-3 PUFA can serve as substrate for oxylipins which influence inflammation and participate in neuroinflammation (Figs. 3 and 4). The oxylipins are factors that participate with eCB in normal brain physiology (Fig. 4) to include homeostasis (neuroplasticity) and pain (Fig. 5).

4.3 Metabolic aspects of eCB and oxylipins: Exercise from brain to muscle, cognition, mood, pain, energy homeostasis and immune function

Exercise is based on a complex network of connections extending from brain, the peripheral nerves to muscle, and influenced by the ECS (Tantimonaco et al., 2014). Exercise effects on brain functions via eCB and CB1 being the most abundant G-protein coupled receptors in the brain, illustrates the participation in neuronal plasticity, thus eCB are involved in several brain responses that greatly overlap with the positive effects of exercise and is able to cross the blood-brain barrier, and exercise increases eCB plasma levels (Pertwee et al., 2010). The CB1 co-localized with type-2 dopamine receptors (Pertwee et al., 2010), which enhances AEA production (Giuffrida et al., 1999), demonstrates that the eCB system controls motor activity by interacting with the dopaminergic system in the basal ganglia network. The CB1 is also implicated in the regulation of contraction frequency, since eCB are released upon type-1 metabotropic glutamate receptor activation induced long-term depression in the isolated lamprey spinal cord during rhythmic locomotion, in a CB1-dependent manner (Kyriakatos & El Manira, 2007). In CB1 −/− mice, alterations in corticospinal motor neuron generation and subcerebral connectivity caused defective skilled motor function (Díaz-Alonso et al., 2012).

Activation of CB1 in nerve terminals (where the receptor is highly expressed) inhibits acetylcholine release in vertebrate striated neuromuscular junctions (Newman et al., 2007). In isolated fast and slow muscles of the frog, CB1 was reported to reduce tension evoked by caffeine, which is known to act on sarcoplasmic calcium release (Huerta et al., 2009). ATP is necessary for muscle contraction, and for major calcium-handling proteins like sarcoplasmic reticulum Ca^{2+}-ATPase, whose uncoupling causes muscle fatigue (James, Walter, & Seebacher, 2011; Seebacher, Pollard, & James, 2012). AEA and CB1 agonists have been shown to increase the locomotor activity of mice and rats (Pandolfo, Pamplona, Prediger, & Takahashi, 2007; Wiley, 2003), and moreover, the activation and the blockade of CB1 decreased running activity in mice selectively bred for high voluntary wheel running, as well as in their wild-type littermates (Keeney et al., 2008, 2012).

Previous studies support that the ECS at least in part explain exercise effects on functions of the brain and peripheral tissues related to mood, reward, pain, appetite, immune functions, and energy metabolism (Tantimonaco et al., 2014; Watkins, 2018). Increased hippocampal neurogenesis correlates with both high AEA levels and increases of CB1 activity, which positively regulates BDNF activity (Butovsky et al., 2005; Hill, Titterness, et al., 2010). BDNF is a mood controller, and reduction of BDNF in the hippocampus critically contributes to stress vulnerability and depression (Yau et al., 2012). Voluntary running was reported to positively correlate with hippocampal BDNF levels in a rodent model of depression and anxiety (Yau et al., 2012).

Cognitive processes are also influenced by reward circuitry and by dopamine signaling (Chiew & Braver, 2011), which are crucial in processing environmental rewarding stimuli in voluntary exercise (De Chiara et al., 2010; Dubreucq et al., 2013). Reinforcing properties of exercise mediated by the ECS in mouse striatum alters sensitivity of CB1 in controlling γ-aminobutyric acid (GABA) transmission (De Chiara et al., 2010) in voluntary exercise. The synaptic response to the selective CB1 agonist HU210 was normal after a single day of exposure to a running wheel or to sucrose consumption, and potentiated after 7 days of treatment. GABA functions slowly returned to normal after treatment discontinuation (De Chiara et al., 2010), thus confirming that the GABA receptor adapts to CB1 signaling in response to running activity.

Increased eCB concentrations in blood, observed during exercise, might lead to central reduction in pain perception. The eCB congeners, PEA and *N*-stearoylethanolamine, substrates of FAAH, have been found to increase after low-force exercise in women with chronic widespread or neck/shoulder pain, in association with reduced pain intensity (Ghafouri et al., 2013). Systemic and intrathecal injections of specific antagonists of CB1 and CB2, as well inhibitors of eCB transport and hydrolysis, showed that eCB and their congeners contribute to exercise analgesic effects, either centrally or peripherally.

Exercise increases energy expenditure, includes compensatory mechanisms activated in hypothalamus and peripheral levels that impact food intake (Blundell et al., 2012). Energy expenditure as an anabolic response and specifically appetite is influenced by eCB. High-intensity training increases the orexigenic neuropeptide Y (NPY) (de Rijke, Hillebrand, Verhagen, Roeling, & Adan, 2005); similarly, eCB increases NPY release at hypothalamus of rats (Gamber, Macarthur, & Westfall, 2005) and increases feeding behavior in rodents (Kirkham, Williams, Fezza, & Di Marzo, 2002). In addition, CB1 −/− (much like FAAH −/−) mice are lean, hypophagic, and resistant to diet-induced obesity (Soria-Gomez et al., 2007; Wiley et al., 2005), and voluntary exercise on a running wheel enhances the CB1 inverse agonist-dependent reduction of appetite in mice (Zhou & Shearman, 2004), suggesting that eCB effects on feeding are CB1-dependent. Furthermore, exercise exerts a positive effect on peripheral energy homeostasis, while eCB mainly exerts a negative influence (Tantimonaco et al., 2014). AEA-treated soleus muscles from lean rats showed decreased basal and insulin-induced glucose uptake (Lindborg, Teachey, Jacob, & Henriksen, 2010), and a CB1 blockade in L6 rat myotubes increased glucose uptake through phosphatidylinositol 3-kinase/protein kinase B (Akt) (Esposito et al., 2008) and ERK1/2 kinases (Lipina et al., 2010). However, AEA was also able to induce basal glucose uptake on its own at high concentrations (10 μM), perhaps due to 5′ AMP-activated protein kinase (AMPK) activation (Eckardt et al., 2009). AEA (5 μM) was also able to increase the expression of PPAR-γ co-activator-1α, a transcription factor engaged to regulate mitochondrial biogenesis, glucose uptake, and fat oxidation (Valdeolivas et al., 2013). The latter data suggested a possible eCB positive effect on muscles through alternative pathways that may be targeted by an eCB-like compound, for example, OEA and PEA.

Acute exercise was shown to increase oxylipins involved to some degree in regulating the transient immune and physiological responses (Markworth et al., 2013). The oxylipin response to exercise is a new area of scientific endeavor, and the data from this study indicate that prolonged and intensive aerobic exercise causes large increases in a diverse range of plasma oxylipins generated from linoleic acid, dihomo-γ-linolenic, arachidonic acid, α-linolenic, EPA and DHA substrates through COX, LOX, and CYP enzyme systems. Post-exercise plasma concentrations were highest for oxylipins generated from n-6 PUFA substrate arachidonic acid to include at least 15 varieties of HETEs and DiHETrEs from COX and CYP enzymes, might be linked to both oxidative stress and inflammation (Nieman et al., 2019). Systemic and intrathecal injections of specific antagonists of CB1 and CB2, as well inhibitors of eCB transport and hydrolysis, showed that eCB and their congeners contribute to exercise analgesic effects, either centrally or peripherally. In summary, both eCB and oxylipins are involved in the physiological aspects of exercise outcomes and impact on neuroinflammation, and their amounts are influenced by dietary PUFA (Fig. 6).

5. Conclusions

At present, there are many gaps and limitations in the evidence regarding the eCB and ECS in exercise levels, as well as sex differences, aging and the specific mechanisms in the human. Future research much result in a better understanding of these relationships to delineate the role of eCB on well-being. The importance of diet and specifically the amounts and types of PUFA, which serve as substrate for eCB, will be a critical aspect of new research.

The eCB act in specific biochemical pathways, by a membrane transporter facilitated cellular uptake and by intracellular degradation by FAAH for AEA and MAGL for 2-AG (Kano et al., 2009). In this case the primary fate of AEA and 2-AG is enzymatic hydrolysis, however, eCB might also be oxidized by cyclooxygenase-2 and lipoxygenase isozymes, thus producing oxidized eCB that are involved in regulating brain synaptic transmission and other biological processes (Nomura et al., 2011), or via the synthesis of prostanoids and the collective, oxylipins (Kim et al., 2013, 2014, 2016; Kim & Watkins, 2014).

As described herein, dietary PUFA such as EPA and DHA help in the maintenance of learning and memory performance. Moreover, these PUFA by virtue of the eCB produced help in the prevention of age-related synaptic plasticity changes by potentiating the activation of CREB and its downstream effectors BDNF/TrkB via the ERK and Akt pathways (Balogun & Cheema, 2014; Gao et al., 2016). For example, Wu et al. (2013) reported that the combination of DHA and physical exercise increased BDNF expression, synaptic plasticity, and cognition in rats after traumatic brain injury, as compared with DHA or physical exercise alone, suggesting synergistic effects of dietary n-3 PUFA and exercise. Furthermore, the n-3 PUFA influence the cellular and physiological processes involved in psychiatric/mood disorders by changing the fatty acid composition of cellular membranes, the modulation of behavioral systems, neurodevelopment, and oxidative stress in specific brain areas (Wu et al., 2013).

BDNF is a likely candidate to elicit benefits of well-being presumably via the actions of exercise on the brain, as well as from eCB (ECS), and perhaps linked to serotonin secretion. Increased hippocampal neurogenesis correlates with both high AEA levels and increases in CB1 activity, which positively regulate BDNF activity (Butovsky et al., 2005; Hill, Titterness, et al., 2010). BDNF is a mood controller, and reducing BDNF in the hippocampus contributes to stress vulnerability and depression (Yau et al., 2012). Voluntary running was shown to positively correlate with hippocampal BDNF levels in a rodent model of depression and anxiety (Yau et al., 2012).

The n-3 PUFA may improve mood via DHEA and likely the link to increasing BDNF but not all of these aspects were evaluated on mood outcomes. Recently, Balogun and Cheema (2014) reported that dietary n-3 PUFA (rich in EPA and DHA) afford dependent regulation of NGF and CREB at a post-translational level, suggesting that maintaining the levels of neurotrophins including NGF throughout the life span could be important to prevent neurological disorders. Other studies consistently show that n-3 PUFA help in the maintenance of learning and memory performance and prevention of age-related synaptic plasticity changes by potentiating the activation of CREB and its downstream effectors BDNF/TrkB via the ERK and Akt pathways (Balogun & Cheema, 2014; Gao et al., 2016).

Cognitive processes are also influenced by reward circuitry and by dopamine signaling (Chiew & Braver, 2011), which are crucial in processing environmental reward stimuli in voluntary exercise (De Chiara et al., 2010; Dubreucq et al., 2013). Reinforcing properties of exercise mediated by the ECS was reported in mouse striatum, where voluntary exercise alters sensitivity of CB1 in controlling γ-aminobutyric acid (GABA) transmission. Further, exercise improves memory via the ECS in mice (Ferreira-Vieira, Bastos, Pereira, Moreira, & Massensini, 2014).

The oxylipins are synthesized from both n-6 and n-3 PUFA to affect inflammation and participate in neuroinflammation. The oxylipins are factors that participate with eCB in normal brain physiology to include homeostasis and pain. Acute exercise was shown to increase oxylipins involved to some degree in regulating the transient immune and physiological responses (Markworth et al., 2013). The oxylipin response to exercise is a new area of scientific endeavor, and the recent data indicate that prolonged intensive aerobic exercise causes large increases in a diverse range of plasma oxylipins generated from linoleic acid, dihomo-γ-linolenic, arachidonic acid, α-linolenic, EPA, and DHA substrates through the COX, LOX, and CYP enzyme systems.

Thus both eCB and oxylipins are involved in the physiological aspects of exercise outcomes and impact on neuroinflammation. These compounds also appear to be associated with well-being. The collective actions of eCB, oxylipins and dietary PUFA are important research targets to understand healthy aging and the control of pain.

The evidence presented in this review merit future research to explain the roles of eCB and oxylipins in exercise and their potential health benefits. Still many specifics about how eCB and oxylipins act in the brain and peripheral nervous system must be elucidated. The research must explore these relationships relative to the effects of exercise and their consequences on well-being and pain. This review also extended the work on eCB and oxylipins to include dietary PUFA substrates that can direct the biosynthesis of these compounds to influence inflammation and specifically in neuroinflammation.

One intriguing area of research needed is a complete analysis of all eCB prior, during and after exercise in men and women, and types of exercise that involve both mind and body such as Tai Chi as well as differences between men and women. Moreover, studies should describe the type of research models to elucidate the relationships between eCB and oxylipins in neuroinflammation with emphasis on aging.

Future research should include how eCB and oxylipins influence BDNF levels and actions on the brain and peripheral nervous system. It will be critical to describe how other biomarkers and cytokines participate in the actions of eCB and oxylipins as the work continues on exercise and especially in the aging adult.

Acknowledgment

The authors received no funding for writing this review and declare no conflict of interest.

References

Aarse, J., Herlitze, S., & Manahan-Vaughan, D. (2016). The requirement of BDNF for hippocampal synaptic plasticity is experience-dependent. *Hippocampus*, *26*(6), 739–751. https://doi.org/10.1002/hipo.22555.

Agarwal, N., Pacher, P., Tegeder, I., Amaya, F., Constantin, C. E., Brenner, G. J., et al. (2007). Cannabinoids mediate analgesia largely via peripheral type 1 cannabinoid receptors in nociceptors. *Nature Neuroscience*, *10*(7), 870–879. https://doi.org/10.1038/nn1916.

Aguado, T., Palazuelos, J., Monory, K., Stella, N., Cravatt, B., Lutz, B., et al. (2006). The endocannabinoid system promotes astroglial differentiation by acting on neural progenitor cells. *The Journal of Neuroscience, 26*(5), 1551–1561. https://doi.org/10.1523/JNEUROSCI.3101-05.2006.

Alhouayek, M., & Muccioli, G. G. (2014). COX-2-derived endocannabinoid metabolites as novel inflammatory mediators. *Trends in Pharmacological Sciences, 35*(6), 284–292. https://doi.org/10.1016/j.tips.2014.03.001.

Alonso, M., Medina, J. H., & Pozzo-Miller, L. (2004). ERK1/2 activation is necessary for BDNF to increase dendritic spine density in hippocampal CA1 pyramidal neurons. *Learning & Memory, 11*(2), 172–178. https://doi.org/10.1101/lm.67804.

Arnold, C., Markovic, M., Blossey, K., Wallukat, G., Fischer, R., Dechend, R., et al. (2010). Arachidonic acid-metabolizing cytochrome P450 enzymes are targets of omega-3 fatty acids. *Journal of Biological Chemistry, 285*(43), 32720–32733. https://doi.org/10.1074/jbc.M110.118406.

Aso, E., Ozaita, A., Valdizán, E. M., Ledent, C., Pazos, Á., Maldonado, R., et al. (2008). BDNF impairment in the hippocampus is related to enhanced despair behavior in CB1 knockout mice. *Journal of Neurochemistry, 105*(2), 565–572. https://doi.org/10.1111/j.1471-4159.2007.05149.x.

Balogun, K. A., & Cheema, S. K. (2014). The expression of neurotrophins is differentially regulated by omega-3 polyunsaturated fatty acids at weaning and postweaning in C57BL/6 mice cerebral cortex. *Neurochemistry International, 66*, 33–42. https://doi.org/10.1016/j.neuint.2014.01.007.

Balvers, M. G., Verhoeckx, K. C., Bijlsma, S., Rubingh, C. M., Meijerink, J., Wortelboer, H. M., et al. (2012). Fish oil and inflammatory status alter the n-3 to n-6 balance of the endocannabinoid and oxylipin metabolomes in mouse plasma and tissues. *Metabolomics, 8*(6), 1130–1147. https://doi.org/10.1007/s11306-012-0421-9.

Balvers, M. G., Verhoeckx, K. C., Meijerink, J., Bijlsma, S., Rubingh, C. M., Wortelboer, H. M., et al. (2012). Time-dependent effect of in vivo inflammation on eicosanoid and endocannabinoid levels in plasma, liver, ileum and adipose tissue in C57BL/6 mice fed a fish-oil diet. *International Immunopharmacology, 13*(2), 204–214. https://doi.org/10.1016/j.intimp.2012.03.022.

Balvers, M. G., Verhoeckx, K. C., Plastina, P., Wortelboer, H. M., Meijerink, J., & Witkamp, R. F. (2010). Docosahexaenoic acid and eicosapentaenoic acid are converted by 3T3-L1 adipocytes to N-acyl ethanolamines with anti-inflammatory properties. *Biochimica et Biophysica Acta, 1801*(10), 1107–1114. https://doi.org/10.1016/j.bbalip.2010.06.006.

Baron, R., Binder, A., & Wasner, G. (2010). Neuropathic pain: Diagnosis, pathophysiological mechanisms, and treatment. *Lancet Neurology, 9*(8), 807–819. https://doi.org/10.1016/S1474-4422(10)70143-5.

Barquissau, V., Ghandour, R. A., Ailhaud, G., Klingenspor, M., Langin, D., Amri, E.-Z., et al. (2017). Control of adipogenesis by oxylipins, GPCRs and PPARs. *Biochimie, 136*, 3–11. https://doi.org/10.1016/j.biochi.2016.12.012.

Basbaum, A. I., Bautista, D. M., Scherrer, G., & Julius, D. (2009). Cellular and molecular mechanisms of pain. *Cell, 139*(2), 267–284. https://doi.org/10.1016/j.cell.2009.09.028.

Biedermann, S. V., Auer, M. K., Bindila, L., Ende, G., Lutz, B., Weber-Fahr, W., et al. (2016). Restricted vs. unrestricted wheel running in mice: Effects on brain, behavior and endocannabinoids. *Hormones and Behavior, 86*, 45–54. https://doi.org/10.1016/j.yhbeh.2016.09.007.

Bilkei-Gorzo, A. (2012). The endocannabinoid system in normal and pathological brain ageing. *Philosophical Transactions of the Royal Society of London. Series B, Biological Sciences, 367*(1607), 3326–3341. https://doi.org/10.1098/rstb.2011.0388.

Blankman, J. L., & Cravatt, B. F. (2013). Chemical probes of endocannabinoid metabolism. *Pharmacological Reviews*, *65*(2), 849–871. https://doi.org/10.1124/pr.112.006387.

Blundell, J. E., Caudwell, P., Gibbons, C., Hopkins, M., Naslund, E., King, N., et al. (2012). Role of resting metabolic rate and energy expenditure in hunger and appetite control: A new formulation. *Disease Models & Mechanisms*, *5*(5), 608–613. https://doi.org/10.1242/dmm.009837.

Bot, M., Pouwer, F., Assies, J., Jansen, E. H., Beekman, A. T., & De Jonge, P. (2011). Supplementation with eicosapentaenoic omega-3 fatty acid does not influence serum brain-derived neurotrophic factor in diabetes mellitus patients with major depression: A randomized controlled pilot study. *Neuropsychobiology*, *63*(4), 219–223. https://doi.org/10.1159/000321804.

Bousquet, M., Gibrat, C., Saint-Pierre, M., Julien, C., Calon, F., & Cicchetti, F. (2009). Modulation of brain-derived neurotrophic factor as a potential neuroprotective mechanism of action of omega-3 fatty acids in a parkinsonian animal model. *Progress in Neuro-Psychopharmacology & Biological Psychiatry*, *33*(8), 1401–1408. https://doi.org/10.1016/j.pnpbp.2009.07.018.

Bramham, C. R. (2007). Control of synaptic consolidation in the dentate gyrus: Mechanisms, functions, and therapeutic implications. In *Vol. 163. Dentate Gyrus: A Comphrehensive Guide to Structure, Function, and Clinical Implications* (pp. 453–471). https://doi.org/10.1016/S0079-6123(07)63025-8.

Brellenthin, A. G., Crombie, K. M., Hillard, C. J., & Koltyn, K. F. (2017). Endocannabinoid and mood responses to exercise in adults with varying activity levels. *Medicine and Science in Sports and Exercise*, *49*(8), 1688–1696. https://doi.org/10.1249/Mss.0000000000001276.

Brown, I., Cascio, M. G., Wahle, K. W., Smoum, R., Mechoulam, R., Ross, R. A., et al. (2010). Cannabinoid receptor-dependent and -independent anti-proliferative effects of omega-3 ethanolamides in androgen receptor-positive and -negative prostate cancer cell lines. *Carcinogenesis*, *31*(9), 1584–1591. https://doi.org/10.1093/carcin/bgq151.

Burston, J. J., Wiley, J. L., Craig, A. A., Selley, D. E., & Sim-Selley, L. J. (2010). Regional enhancement of cannabinoid CB1 receptor desensitization in female adolescent rats following repeated Delta 9-tetrahydrocannabinol exposure. *British Journal of Pharmacology*, *161*(1), 103–112. https://doi.org/10.1111/j.1476-5381.2010.00870.x.

Butovsky, E., Juknat, A., Goncharov, I., Elbaz, J., Eilam, R., Zangen, A., et al. (2005). In vivo up-regulation of brain-derived neurotrophic factor in specific brain areas by chronic exposure to Delta(9)-tetrahydrocannabinol. *Journal of Neurochemistry*, *93*(4), 802–811. https://doi.org/10.1111/j.1471-4159.2005.03074.x.

Cascio, M. G. (2013). PUFA-derived endocannabinoids: An overview. *The Proceedings of the Nutrition Society*, *72*(4), 451–459. https://doi.org/10.1017/S0029665113003418.

Chen, B., Dowlatshahi, D., MacQueen, G. M., Wang, J. F., & Young, L. T. (2001). Increased hippocampal BDNF immunoreactivity in subjects treated with antidepressant medication. *Biological Psychiatry*, *50*(4), 260–265. https://doi.org/10.1016/S0006-3223(01)01083-6.

Chiew, K. S., & Braver, T. S. (2011). Positive affect versus reward: Emotional and motivational influences on cognitive control. *Frontiers in Psychology*, *2*, 279. https://doi.org/10.3389/fpsyg.2011.00279.

Chiurchiù, V., Leuti, A., & Maccarrone, M. (2018). Bioactive lipids and chronic inflammation: Managing the fire within. *Frontiers in Immunology*, *9*, 38. https://doi.org/10.3389/fimmu.2018.00038.

Choi, J.-E., & Park, Y. (2017). EPA and DHA, but not ALA, have antidepressant effects with 17β-estradiol injection via regulation of a neurobiological system in ovariectomized rats. *Journal of Nutritional Biochemistry*, *49*, 101–109. https://doi.org/10.1016/j.jnutbio.2017.07.012.

Cipollina, C., Salvatore, S. R., Muldoon, M. F., Freeman, B. A., & Schopfer, F. J. (2014). Generation and dietary modulation of anti-inflammatory electrophilic omega-3 fatty acid derivatives. *PLoS One*, *9*(4), e94836. https://doi.org/10.1371/journal.pone.0094836.

Craft, R. M., Marusich, J. A., & Wiley, J. L. (2013). Sex differences in cannabinoid pharmacology: A reflection of differences in the endocannabinoid system? *Life Sciences*, *92*(8–9), 476–481. https://doi.org/10.1016/j.lfs.2012.06.009.

Cramer, S. C., Sur, M., Dobkin, B. H., O'Brien, C., Sanger, T. D., Trojanowski, J. Q., et al. (2011). Harnessing neuroplasticity for clinical applications. *Brain*, *134*, 1591–1609. https://doi.org/10.1093/brain/awr039.

Crutcher, K. A. (1986). The role of growth-factors in neuronal development and plasticity. *CRC Critical Reviews in Clinical Neurobiology*, *2*(3), 297–333. Retrieved from <Go to ISI>://WOS:A1986F053600002.

De Chiara, V., Errico, F., Musella, A., Rossi, S., Mataluni, G., Sacchetti, L., et al. (2010). Voluntary exercise and sucrose consumption enhance cannabinoid CB1 receptor sensitivity in the striatum. *Neuropsychopharmacology*, *35*(2), 374–387. https://doi.org/10.1038/npp.2009.141.

de Rijke, C. E., Hillebrand, J. J. G., Verhagen, L. A. W., Roeling, T. A. P., & Adan, R. A. H. (2005). Hypothalamic neuropeptide expression following chronic food restriction in sedentary and wheel-running rats. *Journal of Molecular Endocrinology*, *35*(2), 381–390. https://doi.org/10.1677/jme.1.01808.

Di Marzo, V., Cote, M., Matias, I., Lemieux, I., Arsenault, B. J., Cartier, A., et al. (2009). Changes in plasma endocannabinoid levels in viscerally obese men following a 1 year lifestyle modification programme and waist circumference reduction: Associations with changes in metabolic risk factors. *Diabetologia*, *52*(2), 213–217. https://doi.org/10.1007/s00125-008-1178-6.

Di Marzo, V., Stella, N., & Zimmer, A. (2015). Endocannabinoid signalling and the deteriorating brain. *Nature Reviews. Neuroscience*, *16*(1), 30–42. https://doi.org/10.1038/nrn3876.

Díaz-Alonso, J., Aguado, T., Wu, C.-S., Palazuelos, J., Hofmann, C., Garcez, P., et al. (2012). The CB1 cannabinoid receptor drives corticospinal motor neuron differentiation through the Ctip2/Satb2 transcriptional regulation axis. *Journal of Neuroscience*, *32*(47), 16651–16665. https://doi.org/10.1523/JNEUROSCI.0681-12.2012.

Dong, Y. L., Xu, M., Kalueff, A. V., & Song, C. (2018). Dietary eicosapentaenoic acid normalizes hippocampal omega-3 and 6 polyunsaturated fatty acid profile, attenuates glial activation and regulates BDNF function in a rodent model of neuroinflammation induced by central interleukin-1 beta administration. *European Journal of Nutrition*, *57*(5), 1781–1791. https://doi.org/10.1007/s00394-017-1462-7.

Donvito, G., Nass, S. R., Wilkerson, J. L., Curry, Z. A., Schurman, L. D., Kinsey, S. G., et al. (2018). The endogenous cannabinoid system: A budding source of targets for treating inflammatory and neuropathic pain. *Neuropsychopharmacology*, *43*(1), 52–79. https://doi.org/10.1038/npp.2017.204.

dos Santos, F. V., Targa, A. D. S., Hammerschmidt, I., Zanata, S. M., Maia, F. G., Visentainer, J. V., et al. (2018). Fish oil supplementation reverses behavioral and neurochemical alterations induced by swimming exercise in rats. *Physiology & Behavior*, *194*, 95–102. https://doi.org/10.1016/j.physbeh.2018.05.001.

D'Souza, D., Pittman, B., Perry, E., & Simen, A. (2009). Preliminary evidence of cannabinoid effects on brain-derived neurotrophic factor (BDNF) levels in humans. *Psychopharmacology*, *202*(4), 569–578. https://doi.org/10.1007/s00213-008-1333-2.

Dubreucq, S., Durand, A., Matias, I., Benard, G., Richard, E., Soria-Gomez, E., et al. (2013). Ventral tegmental area cannabinoid type-1 receptors control voluntary exercise performance. *Biological Psychiatry*, *73*(9), 895–903. https://doi.org/10.1016/j.biopsych.2012.10.025.

Dunham, J. S., Deakin, J. F. W., Miyajima, F., Payton, A., & Toro, C. T. (2009). Expression of hippocampal brain-derived neurotrophic factor and its receptors in Stanley consortium brains. *Journal of Psychiatric Research*, *43*(14), 1175–1184. https://doi.org/10.1016/j.jpsychires.2009.03.008.

Dyall, S. C. (2017). Interplay between n-3 and n-6 long-chain polyunsaturated fatty acids and the endocannabinoid system in brain protection and repair. *Lipids*, *52*(11), 885–900. https://doi.org/10.1007/s11745-017-4292-8.

Dyall, S. C., Mandhair, H. K., Fincham, R. E. A., Kerr, D. M., Roche, M., & Molina-Holgado, F. (2016). Distinctive effects of eicosapentaenoic and docosahexaenoic acids in regulating neural stem cell fate are mediated via endocannabinoid signalling pathways. *Neuropharmacology*, *107*, 387–395. https://doi.org/10.1016/j.neuropharm.2016.03.055.

Eckardt, K., Sell, H., Taube, A., Koenen, M., Platzbecker, B., Cramer, A., et al. (2009). Cannabinoid type 1 receptors in human skeletal muscle cells participate in the negative crosstalk between fat and muscle. *Diabetologia*, *52*(4), 664–674. https://doi.org/10.1007/s00125-008-1240-4.

Ernfors, P., Wetmore, C., Olson, L., & Persson, H. (1990). Identification of cells in rat brain and peripheral tissues expressing mRNA for members of the nerve growth factor family. *Neuron*, *5*(4), 511–526. https://doi.org/10.1016/0896-6273(90)90090-3.

Esposito, I., Proto, M. C., Gazzerro, P., Laezza, C., Miele, C., Alberobello, A. T., et al. (2008). The cannabinoid CB1 receptor antagonist rimonabant stimulates 2-deoxyglucose uptake in skeletal muscle cells by regulating the expression of phosphatidylinositol-3-kinase. *Molecular Pharmacology*, *74*(6), 1678–1686. https://doi.org/10.1124/mol.108.049205.

Estrada, N. M., & Isokawa, M. (2009). Metabolic demand stimulates CREB Signaling in the limbic cortex: Implication for the induction of hippocampal synaptic plasticity by intrinsic stimulus for survival. *Frontiers in Systems Neuroscience*, *3*, 5. https://doi.org/10.3389/neuro.06.005.2009.

Fan, C. N., Fu, H. C., Dong, H., Lu, Y. Y., Lu, Y. F., & Qi, K. M. (2016). Maternal n-3 polyunsaturated fatty acid deprivation during pregnancy and lactation affects neurogenesis and apoptosis in adult offspring: Associated with DNA methylation of brain-derived neurotrophic factor transcripts. *Nutrition Research*, *36*(9), 1013–1021. https://doi.org/10.1016/j.nutres.2016.06.005.

Fernandez-Ruiz, J., Romero, J., Velasco, G., Tolon, R. M., Ramos, J. A., & Guzman, M. (2007). Cannabinoid CB2 receptor: A new target for controlling neural cell survival? *Trends in Pharmacological Sciences*, *28*(1), 39–45. https://doi.org/10.1016/j.tips.2006.11.001.

Ferreira, C. F., Bernardi, J. R., Bosa, V. L., Schuch, I., Goldani, M. Z., Kapczinski, F., et al. (2014). Correlation between n-3 polyunsaturated fatty acids consumption and BDNF peripheral levels in adolescents. *Lipids in Health and Disease*, *13*, 44. https://doi.org/10.1186/1476-511x-13-44.

Ferreira, C. F., Bernardi, J. R., Krolow, R., Arcego, D. M., Fries, G. R., de Aguiar, B. W., et al. (2013). Vulnerability to dietary n-3 polyunsaturated fatty acid deficiency after exposure to early stress in rats. *Pharmacology, Biochemistry, and Behavior*, *107*, 11–19. https://doi.org/10.1016/j.pbb.2013.03.006.

Ferreira, F. F., Ribeiro, F. F., Rodrigues, R. S., Sebastiao, A. M., & Xapelli, S. (2018). Brain-derived neurotrophic factor (BDNF) role in cannabinoid-mediated neurogenesis. *Frontiers in Cellular Neuroscience*, *12*, 441. https://doi.org/10.3389/fncel.2018.00441.

Ferreira-Vieira, T. H., Bastos, C. P., Pereira, G. S., Moreira, F. A., & Massensini, A. R. (2014). A role for the endocannabinoid system in exercise-induced spatial memory enhancement in mice. *Hippocampus*, *24*(1), 79–88. https://doi.org/10.1002/hipo.22206.

Feuerecker, M., Hauer, D., Toth, R., Demetz, F., Hölzl, J., Thiel, M., et al. (2012). Effects of exercise stress on the endocannabinoid system in humans under field conditions. *European Journal of Applied Physiology*, *112*(7), 2777–2781. https://doi.org/10.1007/s00421-011-2237-0.

Fonseca, B., Costa, M., Almada, M., Correia-da-Silva, G., & Teixeira, N. (2013). Endogenous cannabinoids revisited: A biochemistry perspective. *Prostaglandins & Other Lipid Mediators*, *102*, 13–30. https://doi.org/10.1016/j.prostaglandins.2013.02.002.

Galdino, G., Romero, T. R. L., Silva, J. F. P., Aguiar, D. C., de Paula, A. M., Cruz, J. S., et al. (2014). The endocannabinoid system mediates aerobic exercise-induced antinociception in rats. *Neuropharmacology*, 77, 313–324. https://doi.org/10.1016/j.neuropharm.2013.09.022.

Gamber, K. M., Macarthur, H., & Westfall, T. C. (2005). Cannabinoids augment the release of neuropeptide Y in the rat hypothalamus. *Neuropharmacology*, *49*(5), 646–652. https://doi.org/10.1016/j.neuropharm.2005.04.017.

Gao, H., Yan, P., Zhang, S., Huang, H., Huang, F., Sun, T., et al. (2016). Long-term dietary alpha-linolenic acid supplement alleviates cognitive impairment correlate with activating hippocampal CREB signaling in natural aging rats. *Molecular Neurobiology*, *53*(7), 4772–4786. https://doi.org/10.1007/s12035-015-9393-x.

Gasperi, V., Fezza, F., Pasquariello, N., Bari, M., Oddi, S., Agro, A. F., et al. (2007). Endocannabinoids in adipocytes during differentiation and their role in glucose uptake. *Cellular and Molecular Life Sciences*, *64*(2), 219–229. https://doi.org/10.1007/s00018-006-6445-4.

Ghafouri, N., Ghafouri, B., Larsson, B., Stensson, N., Fowler, C. J., & Gerdle, B. (2013). Palmitoylethanolamide and stearoylethanolamide levels in the interstitium of the trapezius muscle of women with chronic widespread pain and chronic neck-shoulder pain correlate with pain intensity and sensitivity. *Pain*, *154*(9), 1649–1658. https://doi.org/10.1016/j.pain.2013.05.002.

Giachino, C., De Marchis, S., Giampietro, C., Parlato, R., Perroteau, I., Schutz, G., et al. (2005). cAMP response element-binding protein regulates differentiation and survival of newborn neurons in the olfactory bulb. *Journal of Neuroscience*, *25*(44), 10105–10118. https://doi.org/10.1523/Jneurosci.3512-05.2005.

Gierman, L., Wopereis, S., van El, B., Verheij, E., Werff-van der Vat, B., Bastiaansen-Jenniskens, Y., et al. (2013). Metabolic profiling reveals differences in concentrations of oxylipins and fatty acids secreted by the infrapatellar fat pad of donors with end-stage osteoarthritis and normal donors. *Arthritis and Rheumatism*, *65*(10), 2606–2614. https://doi.org/10.1002/art.38081.

Giuffrida, A., Parsons, L. H., Kerr, T. M., de Fonseca, F. R., Navarro, M., & Piomelli, D. (1999). Dopamine activation of endogenous cannabinoid signaling in dorsal striatum. *Nature Neuroscience*, *2*(4), 358–363. https://doi.org/10.1038/7268.

Glisky, E. L. (2007). Changes in cognitive function in human aging. In D. R. Riddle (Ed.), *Brain Aging: Models, Methods, and Mechanisms*. Boca Raton (FL).

Gomez-Palacio-Schjetnan, A., & Escobar, M. L. (2013). Neurotrophins and synaptic plasticity. *Current Topics in Behavioral Neurosciences*, *15*, 117–136. https://doi.org/10.1007/7854_2012_231.

Greco, R., Gasperi, V., Maccarrone, M., & Tassorelli, C. (2010). The endocannabinoid system and migraine. *Experimental Neurology*, *224*(1), 85–91. https://doi.org/10.1016/j.expneurol.2010.03.029.

Grigorenko, E., Kittler, J., Clayton, C., Wallace, D., Zhuang, S., Bridges, D., et al. (2002). Assessment of cannabinoid induced gene changes: Tolerance and neuroprotection. *Chemistry and Physics of Lipids*, *121*(1–2), 257–266. https://doi.org/10.1016/s0009-3084(02)00161-5.

Groeger, A. L., Cipollina, C., Cole, M. P., Woodcock, S. R., Bonacci, G., Rudolph, T. K., et al. (2010). Cyclooxygenase-2 generates anti-inflammatory mediators from omega-3 fatty acids. *Nature Chemical Biology*, *6*(6), 433–441. https://doi.org/10.1038/nchembio.367.

Gu, M., Li, Y., Tang, H., Zhang, C., Li, W., Zhang, Y., et al. (2018). Endogenous omega (N)-3 fatty acids in fat-1 mice attenuated depression-like behavior, imbalance between microglial M1 and M2 phenotypes, and dysfunction of neurotrophins induced by lipopolysaccharide administration. *Nutrients*, *10*(10), 1351. https://doi.org/10.3390/nu10101351.

Hadjighassem, M., Kamalidehghan, B., Shekarriz, N., Baseerat, A., Molavi, N., Mehrpour, M., et al. (2015). Oral consumption of α-linolenic acid increases serum BDNF levels in healthy adult humans. *Nutrition Journal*, *14*(1), 20. https://doi.org/10.1186/s12937-015-0012-5.

Hashimotodani, Y., Ohno-Shosaku, T., & Kano, M. (2007). Endocannabinoids and synaptic function in the CNS. *The Neuroscientist*, *13*(2), 127–137. https://doi.org/10.1177/1073858406296716.

Heyman, E., Delamarche, P., Berthon, P., Meeusen, R., Briard, D., Vincent, S., et al. (2007). Alteration in sympathoadrenergic activity at rest and during intense exercise despite normal aerobic fitness in late pubertal adolescent girls with type 1 diabetes. *Diabetes & Metabolism*, *33*(6), 422–429. https://doi.org/10.1016/j.diabet.2007.10.003.

Heyman, E., Gamelin, F. X., Goekint, M., Piscitelli, F., Roelands, B., Leclair, E., et al. (2012). Intense exercise increases circulating endocannabinoid and BDNF levels in humans-Possible implications for reward and depression. *Psychoneuroendocrinology*, *37*(6), 844–851. https://doi.org/10.1016/j.psyneuen.2011.09.017.

Hibbeln, J. R. (1998). Fish consumption and major depression. *Lancet*, *351*(9110), 1213. https://doi.org/10.1016/S0140-6736(05)79168-6.

Hill, M. N., Karatsoreos, I. N., Hillard, C. J., & McEwen, B. S. (2010). Rapid elevations in limbic endocannabinoid content by glucocorticoid hormones in vivo. *Psychoneuroendocrinology*, *35*(9), 1333–1338. https://doi.org/10.1016/j.psyneuen.2010.03.005.

Hill, M. N., McLaughlin, R. J., Pan, B., Fitzgerald, M. L., Roberts, C. J., Lee, T. T., et al. (2011). Recruitment of prefrontal cortical endocannabinoid signaling by glucocorticoids contributes to termination of the stress response. *The Journal of Neuroscience*, *31*(29), 10506–10515. https://doi.org/10.1523/JNEUROSCI.0496-11.2011.

Hill, M. N., Titterness, A. K., Morrish, A. C., Carrier, E. J., Lee, T. T. Y., Gil-Mohapel, J., et al. (2010). Endogenous cannabinoid signaling is required for voluntary exercise-induced enhancement of progenitor cell proliferation in the hippocampus. *Hippocampus*, *20*(4), 513–523. https://doi.org/10.1002/hipo.20647.

Howlett, A. C., Blume, L. C., & Dalton, G. D. (2010). CB1 cannabinoid receptors and their associated proteins. *Current Medicinal Chemistry*, *17*(14), 1382–1393. https://doi.org/10.2174/092986710790980023.

Howlett, A. C., Reggio, P. H., Childers, S. R., Hampson, R. E., Ulloa, N. M., & Deutsch, D. G. (2011). Endocannabinoid tone versus constitutive activity of cannabinoid receptors. *British Journal of Pharmacology*, *163*(7), 1329–1343. https://doi.org/10.1111/j.1476-5381.2011.01364.x.

Huerta, M., Ortiz-Mesina, M., Trujillo, X., Sanchez-Pastor, E., Vasquez, C., Castro, E., et al. (2009). Effects of cannabinoids on caffeine contractures in slow and fast skeletal muscle fibers of the frog. *Journal of Membrane Biology*, *229*(2), 91–99. https://doi.org/10.1007/s00232-009-9174-4.

Hutchins-Wiese, H. L., Li, Y., Hannon, K., & Watkins, B. A. (2012). Hind limb suspension and long-chain omega-3 PUFA increase mRNA endocannabinoid system levels in skeletal muscle. *Journal of Nutritional Biochemistry*, *23*(8), 986–993. https://doi.org/10.1016/j.jnutbio.2011.05.005.

James, R. S., Walter, I., & Seebacher, F. (2011). Variation in expression of calcium-handling proteins is associated with inter-individual differences in mechanical performance of rat (Rattus norvegicus) skeletal muscle. *Journal of Experimental Biology*, *214*(21), 3542–3548. https://doi.org/10.1242/jeb.058305.

Jennings, E. M., Okine, B. N., Olango, W. M., Roche, M., & Finn, D. P. (2016). Repeated forced swim stress differentially affects formalin-evoked nociceptive behaviour and the endocannabinoid system in stress normo-responsive and stress hyper-responsive rat strains. *Progress in Neuro-Psychopharmacology & Biological Psychiatry, 64*, 181–189. https://doi.org/10.1016/j.pnpbp.2015.05.008.

Ji, R. R., Nackley, A., Huh, Y., Terrando, N., & Maixner, W. (2018). Neuroinflammation and central sensitization in chronic and widespread pain. *Anesthesiology, 129*(2), 343–366. https://doi.org/10.1097/ALN.0000000000002130.

Jiang, P., Dang, R. L., Li, H. D., Zhang, L. H., Zhu, W. Y., Xue, Y., et al. (2014). The impacts of swimming exercise on hippocampal expression of neurotrophic factors in rats exposed to chronic unpredictable mild stress. *Evidence-based Complementary and Alternative Medicine, 2014*, 729827. https://doi.org/10.1155/2014/729827.

Jiang, L. H., Shi, Y., Wang, L. S., & Yang, Z. R. (2009). The influence of orally administered docosahexaenoic acid on cognitive ability in aged mice. *Journal of Nutritional Biochemistry, 20*(9), 735–741. https://doi.org/10.1016/j.jnutbio.2008.07.003.

Jin, Y., & Park, Y. (2015). N-3 polyunsaturated fatty acids and 17β-estradiol injection induce antidepressant-like effects through regulation of serotonergic neurotransmission in ovariectomized rats. *Journal of Nutritional Biochemistry, 26*(9), 970–977. https://doi.org/10.1016/j.jnutbio.2015.04.005.

Jin, K., Xie, L., Kim, S. H., Parmentier-Batteur, S., Sun, Y., Mao, X. O., et al. (2004). Defective adult neurogenesis in CB1 cannabinoid receptor knockout mice. *Molecular Pharmacology, 66*(2), 204–208. https://doi.org/10.1124/mol.66.2.204.

Kaliannan, K., Li, X.-Y., Wang, B., Pan, Q., Chen, C.-Y., Hao, L., et al. (2019). Multi-omic analysis in transgenic mice implicates omega-6/omega-3 fatty acid imbalance as a risk factor for chronic disease. *Communications Biology, 2*(1), 1–18. https://doi.org/10.1038/s42003-019-0521-4.

Kano, M., Ohno-Shosaku, T., Hashimotodani, Y., Uchigashima, M., & Watanabe, M. (2009). Endocannabinoid-mediated control of synaptic transmission. *Physiological Reviews, 89*(1), 309–380. https://doi.org/10.1152/physrev.00019.2008.

Keeney, B. K., Meek, T. H., Middleton, K. M., Holness, L. F., & Garland, T. (2012). Sex differences in cannabinoid receptor-1 (CB1) pharmacology in mice selectively bred for high voluntary wheel-running behavior. *Pharmacology, Biochemistry, and Behavior, 101*(4), 528–537. https://doi.org/10.1016/j.pbb.2012.02.017.

Keeney, B. K., Raichlen, D. A., Meek, T. H., Wijeratne, R. S., Middleton, K. M., Gerdeman, G. L., et al. (2008). Differential response to a selective cannabinoid receptor antagonist (SR141716: Rimonabant) in female mice from lines selectively bred for high voluntary wheel-running behaviour. *Behavioural Pharmacology, 19*(8), 812–820. https://doi.org/10.1097/FBP.0b013e32831c3b6b.

Khaspekov, L. G., Verca, M. S. B., Frumkina, L. E., Hermann, H., Marsicano, G., & Lutz, B. (2004). Involvement of brain-derived neurotrophic factor in cannabinoid receptor-dependent protection against excitotoxicity. *European Journal of Neuroscience, 19*(7), 1691–1698. https://doi.org/10.1111/j.1460-9568.2004.03285.x.

Kim, J., Carlson, M. E., Kuchel, G. A., Newman, J. W., & Watkins, B. A. (2016). Dietary DHA reduces downstream endocannabinoid and inflammatory gene expression and epididymal fat mass while improving aspects of glucose use in muscle in C57BL/6J mice. *International Journal of Obesity, 40*(1), 129–137. https://doi.org/10.1038/ijo.2015.135.

Kim, J., Carlson, M. E., & Watkins, B. A. (2014). Docosahexaenoyl ethanolamide improves glucose uptake and alters endocannabinoid system gene expression in proliferating and differentiating C2C12 myoblasts. *Frontiers in Physiology, 5*, 100. https://doi.org/10.3389/fphys.2014.00100.

Kim, J., Li, Y., & Watkins, B. A. (2011). Endocannabinoid signaling and energy metabolism: A target for dietary intervention. *Nutrition*, *27*(6), 624–632. https://doi.org/10.1016/j.nut.2010.11.003.

Kim, J., Li, Y., & Watkins, B. A. (2013). Fat to treat fat: Emerging relationship between dietary PUFA, endocannabinoids, and obesity. *Prostaglandins & Other Lipid Mediators*, *104*, 32–41. https://doi.org/10.1016/j.prostaglandins.2012.11.005.

Kim, J., & Watkins, B. A. (2014). Cannabinoid receptor antagonists and fatty acids alter endocannabinoid system gene expression and COX activity. *The Journal of Nutritional Biochemistry*, *25*(8), 815–823. https://doi.org/10.1016/j.jnutbio.2014.03.012.

Kim, S. H., Won, S. J., Mao, X. O., Ledent, C., Jin, K., & Greenberg, D. A. (2006). Role for neuronal nitric-oxide synthase in cannabinoid-induced neurogenesis. *The Journal of Pharmacology and Experimental Therapeutics*, *319*(1), 150–154. https://doi.org/10.1124/jpet.106.107698.

King-Himmelreich, T. S., Moser, C. V., Wolters, M. C., Schmetzer, J., Moller, M., Schreiber, Y., et al. (2017). AMP-activated kinase and the endogenous endocannabinoid system might contribute to antinociceptive effects of prolonged moderate caloric restriction in mice. *Molecular Pain*, *13*, 1–12. https://doi.org/10.1177/1744806917703111.

Kirkham, T. C., Williams, C. M., Fezza, F., & Di Marzo, V. (2002). Endocannabinoid levels in rat limbic forebrain and hypothalamus in relation to fasting, feeding and satiation: Stimulation of eating by 2-arachidonoyl glycerol. *British Journal of Pharmacology*, *136*(4), 550–557. https://doi.org/10.1038/sj.bjp.0704767.

Kishi, T., Yoshimura, R., Ikuta, T., & Iwata, N. (2018). Brain-derived neurotrophic factor and major depressive disorder: Evidence from meta-analyses. *Frontiers in Psychiatry*, *8*, 308. https://doi.org/10.3389/fpsyt.2017.00308.

Knaepen, K., Goekint, M., Heyman, E. M., & Meeusen, R. (2010). Neuroplasticity—Exercise-induced response of peripheral brain-derived neurotrophic factor. *Sports Medicine*, *40*(9), 765–801. https://doi.org/10.2165/11534530-000000000-00000.

Kyriakatos, A., & El Manira, A. (2007). Long-term plasticity of the spinal locomotor circuitry mediated by endocannabinoid and nitric oxide signaling. *Journal of Neuroscience*, *27*(46), 12664–12674. https://doi.org/10.1523/Jneurosci.3174-07.2007.

Lee, J., Duan, W., & Mattson, M. P. (2002). Evidence that brain-derived neurotrophic factor is required for basal neurogenesis and mediates, in part, the enhancement of neurogenesis by dietary restriction in the hippocampus of adult mice. *Journal of Neurochemistry*, *82*(6), 1367–1375. https://doi.org/10.1046/j.1471-4159.2002.01085.x.

Leishman, E., Mackie, K., & Bradshaw, H. B. (2019). Elevated levels of arachidonic acid-derived lipids including prostaglandins and endocannabinoids are present throughout ABHD12 knockout brains: Novel insights into the neurodegenerative phenotype. *Frontiers in Molecular Neuroscience*, *12*, 142. https://doi.org/10.3389/fnmol.2019.00142.

Levy, M. J. F., Boulle, F., Steinbusch, H. W., van den Hove, D. L. A., Kenis, G., & Lanfumey, L. (2018). Neurotrophic factors and neuroplasticity pathways in the pathophysiology and treatment of depression. *Psychopharmacology*, *235*(8), 2195–2220. https://doi.org/10.1007/s00213-018-4950-4.

Lindborg, K. A., Teachey, M. K., Jacob, S., & Henriksen, E. J. (2010). Effects of in vitro antagonism of endocannabinoid-1 receptors on the glucose transport system in normal and insulin-resistant rat skeletal muscle. *Diabetes, Obesity & Metabolism*, *12*(8), 722–730. https://doi.org/10.1111/j.1463-1326.2010.01227.x.

Lipina, C., Stretton, C., Hastings, S., Hundal, J. S., Mackie, K., Irving, A. J., et al. (2010). Regulation of MAP kinase-directed mitogenic and protein kinase B-mediated signaling by cannabinoid receptor type 1 in skeletal muscle cells. *Diabetes*, *59*(2), 375–385. https://doi.org/10.2337/db09-0979.

Lopez, A., Aparicio, N., Pazos, M. R., Grande, M. T., Barreda-Manso, M. A., Benito-Cuesta, I., et al. (2018). Cannabinoid CB2 receptors in the mouse brain: Relevance for Alzheimer's disease. *Journal of Neuroinflammation, 15*(1), 158. https://doi.org/10.1186/s12974-018-1174-9.

Lundstrom, S. L., Yang, J., Brannan, J. D., Haeggstrom, J. Z., Hammock, B. D., Nair, P., et al. (2013). Lipid mediator serum profiles in asthmatics significantly shift following dietary supplementation with omega-3 fatty acids. *Molecular Nutrition & Food Research, 57*(8), 1378–1389. https://doi.org/10.1002/mnfr.201200827.

Maccarrone, M., Lorenzon, T., Bari, M., Melino, G., & Finazzi-Agro, A. (2000). Anandamide induces apoptosis in human cells via vanilloid receptors. Evidence for a protective role of cannabinoid receptors. *Journal of Biological Chemistry, 275*(41), 31938–31945. https://doi.org/10.1074/jbc.M005722200.

Magarinos, A. M., Li, C. J., Toth, J. G., Bath, K. G., Jing, D., Lee, F. S., et al. (2011). Effect of brain-derived neurotrophic factor haploinsufficiency on stress-induced remodeling of hippocampal neurons. *Hippocampus, 21*(3), 253–264. https://doi.org/10.1002/hipo.20744.

Markworth, J. F., Vella, L., Lingard, B. S., Tull, D. L., Rupasinghe, T. W., Sinclair, A. J., et al. (2013). Human inflammatory and resolving lipid mediator responses to resistance exercise and ibuprofen treatment. *American Journal of Physiology. Regulatory, Integrative and Comparative Physiology, 305*(11), R1281–R1296. https://doi.org/10.1152/ajpregu.00128.2013.

Marsicano, G., Goodenough, S., Monory, K., Hermann, H., Eder, M., Cannich, A., et al. (2003). CB1 cannabinoid receptors and on-demand defense against excitotoxicity. *Science, 302*(5642), 84–88. https://doi.org/10.1126/science.1088208.

Martinowich, K., Manji, H., & Lu, B. (2007). New insights into BDNF function in depression and anxiety. *Nature Neuroscience, 10*(9), 1089–1093. https://doi.org/10.1038/nn1971.

Meyer, J. D., Crombie, K. M., Cook, D. B., Hillard, C. J., & Koltyn, K. F. (2019). Serum endocannabinoid and mood changes after exercise in major depressive disorder. *Medicine and Science in Sports and Exercise, 51*(9), 1909–1917. https://doi.org/10.1249/Mss.0000000000002006.

Miller, B. R., & Hen, R. (2015). The current state of the neurogenic theory of depression and anxiety. *Current Opinion in Neurobiology, 30*, 51–58. https://doi.org/10.1016/j.conb.2014.08.012.

Molendijk, M. L., Spinhoven, P., Polak, M., Bus, B. A. A., Penninx, B. W. J. H., & Elzinga, B. M. (2014). Serum BDNF concentrations as peripheral manifestations of depression: Evidence from a systematic review and meta-analyses on 179 associations (N = 9484). *Molecular Psychiatry, 19*(7), 791–800. https://doi.org/10.1038/mp.2013.105.

Montagner, A., Polizzi, A., Fouché, E., Ducheix, S., Lippi, Y., Lasserre, F., et al. (2016). Liver PPARα is crucial for whole-body fatty acid homeostasis and is protective against NAFLD. *Gut, 65*(7), 1202–1214. https://doi.org/10.1136/gutjnl-2015-310798.

Murdolo, G., Kempf, K., Hammarstedt, A., Herder, C., Smith, U., & Jansson, P. A. (2007). Insulin differentially modulates the peripheral endocannabinoid system in human subcutaneous abdominal adipose tissue from lean and obese individuals. *Journal of Endocrinological Investigation, 30*(8), RC17–21. https://doi.org/10.1007/BF03347440.

Nadal, X., La Porta, C., Bura, S. A., & Maldonado, R. (2013). Involvement of the opioid and cannabinoid systems in pain control: New insights from knockout studies. *European Journal of Pharmacology, 716*(1–3), 142–157. https://doi.org/10.1016/j.ejphar.2013.01.077.

Nakagawa, T., Ono-Kishino, M., Sugaru, E., Yamanaka, M., Taiji, M., & Noguchi, H. (2002). Brain-derived neurotrophic factor (BDNF) regulates glucose and energy metabolism in diabetic mice. *Diabetes/Metabolism Research and Reviews, 18*(3), 185–191. https://doi.org/10.1002/dmrr.290.

Newman, Z., Malik, P., Wu, T. Y., Ochoa, C., Watsa, N., & Lindgren, C. (2007). Endocannabinoids mediate muscarine-induced synaptic depression at the vertebrate neuromuscular junction. *European Journal of Neuroscience*, *25*(6), 1619–1630. https://doi.org/10.1111/j.1460-9568.2007.05422.x.

Nieman, D. C., Gillitt, N. D., Chen, G. Y., Zhang, Q. B., Sakaguchi, C. A., & Stephan, E. H. (2019). Carbohydrate intake attenuates post-exercise plasma levels of cytochrome P450-generated oxylipins. *PLoS One*, *14*(3), e0213676. https://doi.org/10.1371/journal.pone.0213676.

Nikhra, V. (2019). The role of ECS in regulation of neuronal aging, neuroinflammation, neurodegeneration and cognitive decline. *EC Endocrinology and Metabolic Research*, *4*(6), 261–273.

Noaghiul, S., & Hibbeln, J. R. (2003). Cross-national comparisons of seafood consumption and rates of bipolar disorders. *American Journal of Psychiatry*, *160*(12), 2222–2227. https://doi.org/10.1176/appi.ajp.160.12.2222.

Nomura, D. K., Morrison, B. E., Blankman, J. L., Long, J. Z., Kinsey, S. G., Marcondes, M. C. G., et al. (2011). Endocannabinoid hydrolysis generates brain prostaglandins that promote neuroinflammation. *Science*, *334*(6057), 809–813. https://doi.org/10.1126/science.1209200.

Ohno-Shosaku, T., & Kano, M. (2014). Endocannabinoid-mediated retrograde modulation of synaptic transmission. *Current Opinion in Neurobiology*, *29*, 1–8. https://doi.org/10.1016/j.conb.2014.03.017.

Ohno-Shosaku, T., Maejima, T., & Kano, M. (2001). Endogenous cannabinoids mediate retrograde signals from depolarized postsynaptic neurons to presynaptic terminals. *Neuron*, *29*(3), 729–738. https://doi.org/10.1016/s0896-6273(01)00247-1.

Okamoto, Y., Morishita, J., Tsuboi, K., Tonai, T., & Ueda, N. (2004). Molecular characterization of a phospholipase D generating anandamide and its congeners. *Journal of Biological Chemistry*, *279*(7), 5298–5305. https://doi.org/10.1074/jbc.M306642200.

Pandolfo, P., Pamplona, F. A., Prediger, R. D. S., & Takahashi, R. N. (2007). Increased sensitivity of adolescent spontaneously hypertensive rats, an animal model of attention deficit hyperactivity disorder, to the locomotor stimulation induced by the cannabinoid receptor agonist WIN 55,212-2. *European Journal of Pharmacology*, *563*(1–3), 141–148. https://doi.org/10.1016/j.ejphar.2007.02.013.

Park, Y., Moon, H. J., & Kim, S. H. (2012). N-3 polyunsaturated fatty acid consumption produces neurobiological effects associated with prevention of depression in rats after the forced swimming test. *The Journal of Nutritional Biochemistry*, *23*(8), 924–928. https://doi.org/10.1016/j.jnutbio.2011.04.018.

Patapoutian, A., & Reichardt, L. F. (2001). Trk receptors: Mediators of neurotrophin action. *Current Opinion in Neurobiology*, *11*(3), 272–280. https://doi.org/10.1016/S0959-4388(00)00208-7.

Patel, K. D., Davison, J. S., Pittman, Q. J., & Sharkey, K. A. (2010). Cannabinoid CB(2) receptors in health and disease. *Current Medicinal Chemistry*, *17*(14), 1393–1410. https://doi.org/10.2174/092986710790980041.

Pedersen, B. K. (2009). The diseasome of physical inactivity—And the role of myokines in muscle—Fat cross talk. *The Journal of Physiology*, *587*(Pt. 23), 5559–5568. https://doi.org/10.1113/jphysiol.2009.179515.

Pedersen, B. K. (2012). Muscular interleukin-6 and its role as an energy sensor. *Medicine and Science in Sports and Exercise*, *44*(3), 392–396. https://doi.org/10.1249/MSS.0b013e31822f94ac.

Pekny, M., & Pekna, M. (2014). Astrocyte reactivity and reactive astrogliosis: Costs and benefits. *Physiological Reviews*, *94*(4), 1077–1098. https://doi.org/10.1152/physrev.00041.2013.

Pertwee, R. G., Howlett, A., Abood, M. E., Alexander, S., Di Marzo, V., Elphick, M., et al. (2010). International union of basic and clinical pharmacology. LXXIX. Cannabinoid receptors and their ligands: Beyond CB1 and CB2. *Pharmacological Reviews, 62*(4), 588–631. https://doi.org/10.1124/pr.110.003004.

Pietropaolo, S., Goubran, M. G., Joffre, C., Aubert, A., Lemaire-Mayo, V., Crusio, W. E., et al. (2014). Dietary supplementation of omega-3 fatty acids rescues fragile X phenotypes in Fmr1-Ko mice. *Psychoneuroendocrinology, 49*, 119–129. https://doi.org/10.1016/j.psyneuen.2014.07.002.

Piscitelli, F., Carta, G., Bisogno, T., Murru, E., Cordeddu, L., Berge, K., et al. (2011). Effect of dietary krill oil supplementation on the endocannabinoidome of metabolically relevant tissues from high-fat-fed mice. *Nutrition and Metabolism, 8*(1), 51. https://doi.org/10.1186/1743-7075-8-51.

Piyanoya, A., Lomazzo, E., Bindila, L., Lerner, R., Albayram, O., Ruhl, T., et al. (2015). Age-related changes in the endocannabinoid system in the mouse hippocampus. *Mechanisms of Ageing and Development, 150*, 55–64. https://doi.org/10.1016/j.mad.2015.08.005.

Polyakova, M., Stuke, K., Schuemberg, K., Mueller, K., Schoenknecht, P., & Schroeter, M. L. (2015). BDNF as a biomarker for successful treatment of mood disorders: A systematic & quantitative meta-analysis. *Journal of Affective Disorders, 174*, 432–440. https://doi.org/10.1016/j.jad.2014.11.044.

Racz, I., Nadal, X., Alferink, J., Banos, J. E., Rehnelt, J., Martin, M., et al. (2008). Crucial role of CB2 cannabinoid receptor in the regulation of central immune responses during neuropathic pain. *Journal of Neuroscience, 28*(46), 12125–12135. https://doi.org/10.1523/Jneurosci.3400-08.2008.

Racz, I., Nent, E., Erxlebe, E., & Zimmer, A. (2015). CB1 receptors modulate affective behaviour induced by neuropathic pain. *Brain Research Bulletin, 114*, 42–48. https://doi.org/10.1016/j.brainresbull.2015.03.005.

Raichlen, D. A., Foster, A. D., Gerdeman, G. L., Seillier, A., & Giuffrida, A. (2012). Wired to run: Exercise-induced endocannabinoid signaling in humans and cursorial mammals with implications for the 'runner's high'. *Journal of Experimental Biology, 215*(8), 1331–1336. https://doi.org/10.1242/jeb.063677.

Raichlen, D. A., Foster, A. D., Seillier, A., Giuffrida, A., & Gerdeman, G. L. (2013). Exercise-induced endocannabinoid signaling is modulated by intensity. *European Journal of Applied Physiology, 113*(4), 869–875. https://doi.org/10.1007/s00421-012-2495-5.

Rashid, M. A., Katakura, M., Kharebava, G., Kevala, K., & Kim, H. Y. (2013). N-docosahexaenoylethanolamine is a potent neurogenic factor for neural stem cell differentiation. *Journal of Neurochemistry, 125*(6), 869–884. https://doi.org/10.1111/jnc.12255.

Rasmussen, P., Brassard, P., Adser, H., Pedersen, M. V., Leick, L., Hart, E., et al. (2009). Evidence for a release of brain-derived neurotrophic factor from the brain during exercise. *Experimental Physiology, 94*(10), 1062–1069. https://doi.org/10.1113/expphysiol.2009.048512.

Rauskolb, S., Zagrebelsky, M., Dreznjak, A., Deogracias, R., Matsumoto, T., Wiese, S., et al. (2010). Global deprivation of brain-derived neurotrophic factor in the CNS reveals an area-specific requirement for dendritic growth. *Journal of Neuroscience, 30*(5), 1739–1749. https://doi.org/10.1523/Jneurosci.5100-09.2010.

Rey, C., Nadjar, A., Buaud, B., Vaysse, C., Aubert, A., Pallet, V., et al. (2016). Resolvin D1 and E1 promote resolution of inflammation in microglial cells in vitro. *Brain, Behavior, and Immunity, 55*, 249–259. https://doi.org/10.1016/j.bbi.2015.12.013.

Rockwell, C. E., Snider, N. T., Thompson, J. T., Heuvel, J. P. V., & Kaminski, N. E. (2006). Interleukin-2 suppression by 2-arachidonyl glycerol is mediated through peroxisome proliferator-activated receptor γ independently of cannabinoid receptors 1 and 2. *Molecular Pharmacology, 70*(1), 101–111. https://doi.org/10.1124/mol.105.019117.

Rodrigues, R. S., Ribeiro, F. F., Ferreira, F., Vaz, S. H., Sebastiao, A. M., & Xapelli, S. (2017). Interaction between cannabinoid type 1 and type 2 receptors in the modulation of subventricular zone and dentate gyrus neurogenesis. *Frontiers in Pharmacology*, *8*, 516. https://doi.org/10.3389/fphar.2017.00516.

Sallaberry, A., & Astern, L. (2018). The endocannabinoid system, our universal regulator. *Journal of Young Investigators*, *34*, 48–55.

Scharfman, H., Goodman, J., Macleod, A., Phani, S., Antonelli, C., & Croll, S. (2005). Increased neurogenesis and the ectopic granule cells after intrahippocampal BDNF infusion in adult rats. *Experimental Neurology*, *192*(2), 348–356. https://doi.org/10.1016/j.expneurol.2004.11.016.

Scott, J. P. R., Sale, C., Greeves, J. P., Casey, A., Dutton, J., & Fraser, W. D. (2011). Effect of exercise intensity on the cytokine response to an acute bout of running. *Medicine and Science in Sports and Exercise*, *43*(12), 2297–2306. https://doi.org/10.1249/MSS.0b013e31822113a9.

Seebacher, F., Pollard, S. R., & James, R. S. (2012). How well do muscle biomechanics predict whole-animal locomotor performance? The role of Ca2+ handling. *Journal of Experimental Biology*, *215*(11), 1847–1853. https://doi.org/10.1242/jeb.067918.

Segal-Gavish, H., Gazit, N., Barhum, Y., Ben-Zur, T., Taler, M., Hornfeld, S. H., et al. (2017). BDNF overexpression prevents cognitive deficit elicited by adolescent cannabis exposure and host susceptibility interaction. *Human Molecular Genetics*, *26*(13), 2462–2471. https://doi.org/10.1093/hmg/ddx139.

Serhan, C. N., Dalli, J., Colas, R. A., Winkler, J. W., & Chiang, N. (2015). Protectins and maresins: New pro-resolving families of mediators in acute inflammation and resolution bioactive metabolome. *Biochimica et Biophysica Acta*, *1851*(4), 397–413. https://doi.org/10.1016/j.bbalip.2014.08.006.

Silvestri, C., & Di Marzo, V. (2013). The endocannabinoid system in energy homeostasis and the etiopathology of metabolic disorders. *Cell Metabolism*, *17*(4), 475–490. https://doi.org/10.1016/j.cmet.2013.03.001.

Soria-Gomez, E., Matias, I., Rueda-Orozco, P. E., Cisneros, M., Petrosino, S., Navarro, L., et al. (2007). Pharmacological enhancement of the endocannabinoid system in the nucleus accumbens shell stimulates food intake and increases c-Fos expression in the hypothalamus. *British Journal of Pharmacology*, *151*(7), 1109–1116. https://doi.org/10.1038/sj.bjp.0707313.

Sparling, P. B., Giuffrida, A., Piomelli, D., Rosskopf, L., & Dietrich, A. (2003). Exercise activates the endocannabinoid system. *Neuroreport*, *14*(17), 2209–2211. https://doi.org/10.1097/00001756-200312020-00015.

Stella, N., Schweitzer, P., & Piomelli, D. (1997). A second endogenous cannabinoid that modulates long-term potentiation. *Nature*, *388*(6644), 773–778. https://doi.org/10.1038/42015.

Stock, K., Kumar, J., Synowitz, M., Petrosino, S., Imperatore, R., Smith, E. S., et al. (2012). Neural precursor cells induce cell death of high-grade astrocytomas through stimulation of TRPV1. *Nature Medicine*, *18*(8), 1232–1238. https://doi.org/10.1038/nm.2827.

Stone, N. L., Millar, S. A., Herrod, P. J. J., Barrett, D. A., Ortori, C. A., Mellon, V. A., et al. (2018). An analysis of endocannabinoid concentrations and mood following singing and exercise in healthy volunteers. *Frontiers in Behavioral Neuroscience*, *12*, 269. https://doi.org/10.3389/fnbeh.2018.00269.

Sun, G. Y., Simonyi, A., Fritsche, K. L., Chuang, D. Y., Hannink, M., Gu, Z., et al. (2018). Docosahexaenoic acid (DHA): An essential nutrient and a nutraceutical for brain health and diseases. *Prostaglandins, Leukotrienes, and Essential Fatty Acids*, *136*, 3–13. https://doi.org/10.1016/j.plefa.2017.03.006.

Swenson, S., Hamilton, J., Robison, L., & Thanos, P. K. (2019). Chronic aerobic exercise: Lack of effect on brain CB1 receptor levels in adult rats. *Life Sciences*, *230*, 84–88. https://doi.org/10.1016/j.lfs.2019.05.058.

Tantimonaco, M., Ceci, R., Sabatini, S., Catani, M. V., Rossi, A., Gasperi, V., et al. (2014). Physical activity and the endocannabinoid system: An overview. *Cellular and Molecular Life Sciences*, *71*(14), 2681–2698. https://doi.org/10.1007/s00018-014-1575-6.

Thompson, Z., Argueta, D., Garland, T., & DiPatrizio, N. (2017). Circulating levels of endocannabinoids respond acutely to voluntary exercise, are altered in mice selectively bred for high voluntary wheel running, and differ between the sexes. *Physiology & Behavior*, *170*, 141–150. https://doi.org/10.1016/j.physbeh.2016.11.041.

Timmerman, K. L., Flynn, M. G., Coen, P. M., Markofski, M. M., & Pence, B. D. (2008). Exercise training-induced lowering of inflammatory (CD14+CD16+) monocytes: A role in the anti-inflammatory influence of exercise? *Journal of Leukocyte Biology*, *84*(5), 1271–1278. https://doi.org/10.1189/jlb.0408244.

Ulu, A., Harris, T. R., Morisseau, C., Miyabe, C., Inoue, H., Schuster, G., et al. (2013). Anti-inflammatory effects of omega-3 polyunsaturated fatty acids and soluble epoxide hydrolase inhibitors in angiotensin-II-dependent hypertension. *Journal of Cardiovascular Pharmacology*, *62*(3), 285–297. https://doi.org/10.1097/FJC.0b013e318298e460.

Valdeolivas, S., Pazos, M., Bisogno, T., Piscitelli, F., Iannotti, F., Allara, M., et al. (2013). The inhibition of 2-arachidonoyl-glycerol (2-AG) biosynthesis, rather than enhancing striatal damage, protects striatal neurons from malonate-induced death: A potential role of cyclooxygenase-2-dependent metabolism of 2-AG. *Cell Death & Disease*, *4*(10), e862. https://doi.org/10.1038/cddis.2013.387.

Vilar, M., & Mira, H. (2016). Regulation of neurogenesis by neurotrophins during adulthood: Expected and unexpected roles. *Frontiers in Neuroscience*, *10*, 16. https://doi.org/10.3389/fnins.2016.00026.

Viscomi, M. T., Oddi, S., Latini, L., Pasquariello, N., Florenzano, F., Bernardi, G., et al. (2009). Selective CB2 receptor agonism protects central neurons from remote axotomy-induced apoptosis through the PI3K/Akt pathway. *Journal of Neuroscience*, *29*(14), 4564–4570. https://doi.org/10.1523/Jneurosci.0786-09.2009.

Wagner, K., Inceoglu, B., & Hammock, B. D. (2011). Soluble epoxide hydrolase inhibition, epoxygenated fatty acids and nociception. *Prostaglandins & Other Lipid Mediators*, *96*(1–4), 76–83. https://doi.org/10.1016/j.prostaglandins.2011.08.001.

Wang, J., Song, H., Tang, X., Yang, Y., Vieira, V. J., Niu, Y., et al. (2012). Effect of exercise training intensity on murine T-regulatory cells and vaccination response. *Scandinavian Journal of Medicine & Science in Sports*, *22*(5), 643–652. https://doi.org/10.1111/j.1600-0838.2010.01288.x.

Watkins, B. A. (2018). Endocannabinoids, exercise, pain, and a path to health with aging. *Molecular Aspects of Medicine*, *64*, 68–78. https://doi.org/10.1016/j.mam.2018.10.001.

Watkins, B. A., Kim, J., Kenny, A., Pedersen, T. L., Pappan, K. L., & Newman, J. W. (2016). Circulating levels of endocannabinoids and oxylipins altered by dietary lipids in older women are likely associated with previously identified gene targets. *Biochimica et Biophysica Acta*, *1861*(11), 1693–1704. https://doi.org/10.1016/j.bbalip.2016.07.007.

Wiley, J. L. (2003). Sex-dependent effects of delta 9-tetrahydrocannabinol on locomotor activity in mice. *Neuroscience Letters*, *352*(2), 77–80. https://doi.org/10.1016/j.neulet.2003.08.050.

Wiley, J. L., Burston, J. J., Leggett, D. C., Alekseeva, O. O., Razdan, R. K., Mahadevan, A., et al. (2005). CB1 cannabinoid receptor-mediated modulation of food intake in mice. *British Journal of Pharmacology*, *145*(3), 293–300. https://doi.org/10.1038/sj.bjp.0706157.

Wu, A., Ying, Z., & Gomez-Pinilla, F. (2013). Exercise facilitates the action of dietary DHA on functional recovery after brain trauma. *Neuroscience*, *248*, 655–663. https://doi.org/10.1016/j.neuroscience.2013.06.041.

Wyss-Coray, T. (2016). Ageing, neurodegeneration and brain rejuvenation. *Nature*, *539*(7628), 180–186.

Yang, R., Fredman, G., Krishnamoorthy, S., Agrawal, N., Irimia, D., Piomelli, D., et al. (2011). Decoding functional metabolomics with docosahexaenoyl ethanolamide

(DHEA) identifies novel bioactive signals. *Journal of Biological Chemistry, 286*(36), 31532–31541. https://doi.org/10.1074/jbc.M111.237990.

Yang, B., Lin, L., Bazinet, R. P., Chien, Y.-C., Chang, J. P.-C., Satyanarayanan, S. K., et al. (2019). Clinical efficacy and biological regulations of ω–3 PUFA-derived endocannabinoids in major depressive disorder. *Psychotherapy and Psychosomatics, 88*(4), 215–224.

Yasari, S., Prud'homme, D., Tesson, F., Jankowski, M., Gutkowska, J., Levy, E., et al. (2012). Effects of exercise training on molecular markers of lipogenesis and lipid partitioning in fructose-induced liver fat accumulation. *Journal of Nutrition and Metabolism, 2012*, 181687. https://doi.org/10.1155/2012/181687.

Yau, S. Y., Lau, B. W. M., Zhang, E. D., Lee, J. C. D., Li, A., Lee, T. M. C., et al. (2012). Effects of voluntary running on plasma levels of neurotrophins, hippocampal cell proliferation and learning and memory in stressed rats. *Neuroscience, 222*, 289–301. https://doi.org/10.1016/j.neuroscience.2012.07.019.

Ying, S. W., Futter, M., Rosenblum, K., Webber, M. J., Hunt, S. P., Bliss, T. V. P., et al. (2002). Brain-derived neurotrophic factor induces long-term potentiation in intact adult hippocampus: Requirement for ERK activation coupled to CREB and upregulation of Arc synthesis. *Journal of Neuroscience, 22*(5), 1532–1540. https://doi.org/10.1523/Jneurosci.22-05-01532.2002.

You, T. J., Disanzo, B. L., Wang, X. W., Yang, R. Z., & Gong, D. W. (2011). Adipose tissue endocannabinoid system gene expression: Depot differences and effects of diet and exercise. *Lipids in Health and Disease, 10*, 194. https://doi.org/10.1186/1476-511x-10-194.

Zaldivar, F., Wang-Rodriguez, J., Nemet, D., Schwindt, C., Galassetti, P., Mills, P. J., et al. (2006). Constitutive pro- and anti-inflammatory cytokine and growth factor response to exercise in leukocytes. *Journal of Applied Physiology, 100*(4), 1124–1133. https://doi.org/10.1152/japplphysiol.00562.2005.

Zhang, G. D., Panigrahy, D., Mahakian, L. M., Yang, J., Liu, J. Y., Lee, K. S. S., et al. (2013). Epoxy metabolites of docosahexaenoic acid (DHA) inhibit angiogenesis, tumor growth, and metastasis. *Proceedings of the National Academy of Sciences of the United States of America, 110*(16), 6530–6535. https://doi.org/10.1073/pnas.1304321110.

Zhou, D., & Shearman, L. P. (2004). Voluntary exercise augments acute effects of CB1-receptor inverse agonist on body weight loss in obese and lean mice. *Pharmacology, Biochemistry, and Behavior*, 77(1), 117–125. https://doi.org/10.1016/j.pbb.2003.10.015.

CHAPTER EIGHT

Parathyroid hormone levels and aging: Effect on balance

Lavanya Murthy[a,b] **and Gustavo Duque**[a,b,*]

[a]Australian Institute for Musculoskeletal Science (AIMSS), University of Melbourne and Western Health, St. Albans, VIC, Australia

[b]Department of Medicine-Western Health, Melbourne Medical School, The University of Melbourne, St. Albans, VIC, Australia

*Corresponding author: e-mail address: gustavo.duque@unimelb.edu.au

Contents

Abstract

Falls result in multiple adverse health outcomes including mortality in older persons. Impaired balance is known to increase the risk of falls. Numerous factors play a role in the etiology of balance disturbance. Among the calciotropic hormones, the role of vitamin D in falls risk has been extensively studied. There is evolving interest in the role of parathyroid hormone (PTH) in this area. Elevated PTH hormone levels have been associated with age-related syndromes such as frailty, osteoporosis, and sarcopenia. Among the existing studies, most have focused on muscle strength with few studies evaluating its role in balance disturbances. In this chapter, we will summarize the aspects of PTH and aging, its biological effects on muscle, and the known associations between PTH and balance.

1. Aging, frailty and sarcopenia

Aging is associated with a progressive decline of the musculoskeletal system resulting in reduced physical and functional capacity, owing to loss in muscle mass, strength and function, and coupled with deteriorating bone density and structure (Harridge & Lazarus, 2017). Frailty is a highly prevalent syndrome in older persons that can be defined as a state of decline and

Vitamins and Hormones, Volume 115
ISSN 0083-6729
https://doi.org/10.1016/bs.vh.2020.12.009

vulnerability in late life, characterized by weakness and decreased physiologic reserve (Topinková, 2008). Sarcopenia (loss of muscle mass and strength), osteopenia/osteoporosis, slow gait velocity, poor balance, reduced physical activity are among important features of the frailty syndrome (Fried et al., 2001; Rockwood & Mitnitski, 2007). The increased vulnerability in this population contributes to adverse outcomes including procedural complications, falls, institutionalization, poor psychosocial well-being, reduced quality of life, disability, and death (Fried et al., 2001; Landi et al., 2012; Yu et al., 2014).

Sarcopenia is defined based on the revised European Working Group on Sarcopenia in Older People (EWGSOP2) recommendations. Following its meeting in 2018, the EWGSOP2 defined low muscle strength as the "key characteristic of sarcopenia." EWGSOP2 also uses "detection of low muscle quantity and quality to confirm the sarcopenia diagnosis and identifies poor physical performance as indicative of severe sarcopenia" (Cruz-Jentoft et al., 2019). The combined presence of sarcopenia and osteopenia or osteoporosis in an individual is termed osteosarcopenia (OS). Such individuals form a subgroup in frailty who are at higher risk of severe adverse outcomes (Binkley & Buehring, 2009; Huo et al., 2015a, 2015b; Pasco et al., 2017).

Multiple factors including nutritional, immunological, metabolic and age-related changes in hormones play a major role in the development of sarcopenia by reducing muscle mass and function (Morley, Baumgartner, Roubenoff, Mayer, & Nair, 2001; Roubenoff & Hughes, 2000). Similarly, there are multiple risk factors for falls and fractures in older age, and poor balance is important among them.

2. Falls and balance

The incidence of falls increases with age and is estimated to occur in around 50% of persons ≥80 years of age (Chang et al., 2004; Tinetti, Speechley, & Ginter, 1988). Falls are associated with increased morbidity and mortality. They can result in catastrophic outcomes including injuries (i.e. hip fracture), permanent disability, loss of independence, and institutionalization (Bergen, Stevens, & Burns, 2016; Salkeld et al., 2000). The associated fear of falling also induces significant anxiety among older persons (Austin, Devine, Dick, Prince, & Bruce, 2007), and restricts their future physical activities (Deshpande et al., 2008).

Falls in older individuals are usually a result of impairments in multiple domains which cumulatively affect the ability of the individual to compensate and maintain postural stability (Tinetti et al., 1988). Cardiovascular disorders such as arrhythmias or syncopal events, muscle weakness, visual impairment, cognitive impairment, Parkinson's disease, psychotropic medication use, balance problems, advancing age, etc. are all well known to be among the risk factors for falls in older individuals. Aging is also related to a decline in the ability to rapidly develop joint torque in the lower limb muscles in the setting of sudden postural disturbances, which may affect balance recovery (Lang et al., 2008; Maki & McIlroy, 1996; Thelen, Schultz, Alexander, & Ashton-Miller, 1996). In addition, reduction in the muscle cross-section, increased muscle adiposity, and delay in the onset of muscle activation are all known to contribute to impaired posture control in this age group.

"Static balance" and "dynamic balance" are the two components of balance performance and is considered an indicator of functional status. Poor balance influences a large number of older individuals and results in threefold increase in risk of falls due to instability during ambulation (dynamic balance) or while standing (static balance) (Bracchi et al., 2002; Ju et al., 2018; Rubenstein & Josephson, 2006). To be able to maintain normal balance without swaying requires a harmonious relationship between afferent sensory inputs and motor control to ensure that the gravity line is maintained even in the event of disturbances (de França et al., 2019; Pizzigalli, Micheletti Cremasco, Mulasso, & Rainoldi, 2016).

3. Aging and parathyroid hormone (PTH) levels

PTH levels increase with age. Secondary hyperparathyroidism in older persons may be the result of renal insufficiency, low dietary calcium intake, and low vitamin D levels (Freaney, McBrinn, & McKenna, 1993; Lips, 2001; Souberbielle et al., 2001). It has been demonstrated that there is reduced renal hydroxylation of vitamin D with aging. Low vitamin D levels together with reduced calcitriol and renal calcium loss, results in a reduction in calcium absorption in the gut. This in turn leads to a negative external calcium balance which can increase PTH levels (Prince et al., 1995). It has also been shown that circulating PTH levels can increase with age independently of vitamin D levels (Carrivick, Walsh, Brown, Wardrop, & Hadlow, 2015).

Both PTH levels and free calcitriol index have been noted to increase at 10 years post menopause along with increase in both resorption (Prince et al., 1995). A cross-sectional study of postmenopausal women showed

PTH levels increased with age and serum 1,25-dihydroxyvitamin D [1,25 $(OH)_2D$]. Also, 25(OH)D and plasma ionized calcium were noted to be inversely associated with PTH levels. 25(OH)D levels below 40 nmol/L resulted in stepwise rise in PTH levels (Need, Horowitz, Morris, & Nordin, 2000). In the MINOS study, elevated PTH > 45 pg/ml was shown to have an inverse association with "age-related muscle loss" independent of vitamin D levels.

4. The biological effects of PTH on muscle

PTH affects the musculoskeletal system through several mechanisms. Its role in mineral homeostasis is well known. It regulates calcium metabolism and bone resorption which in turn impacts bone mass (Swarthout, D'Alonzo, Selvamurugan, & Partridge, 2002). Its spectrum impact also extends to skeletal muscle in that elevated PTH levels are also manifest as muscle atrophy and weakness (Houston et al., 2007). PTH affects protein and amino acid metabolism in skeletal muscle in rats (Garber, 1983). It has been studied to impact skeletal muscle through increasing protein catabolism, decreasing the number of type 2 muscle fibers and mitochondrial oxygen uptake (Lips, 2001). It has also been reported that PTH administration in rats impairs energy production, transfer and utilization in skeletal muscle (Baczynski et al., 1985). A study in mice showed that the presence of PTH receptors (PTHR) in skeletal muscle fibers enable PTH mediated uptake and retention of vitamin D (Abboud et al., 2017). There is also evidence that PTH may have a negative impact on muscle structure and function by increasing the free intracellular calcium in muscle tissue (Begum, Sussman, & Draznin, 1992; McCarty, 2002).

An indirect role of PTH on muscle mass and strength has been hypothesized via elevated interleukin-6 (IL-6) levels which are known to be associated with sarcopenia in later age (Pahor & Kritchevsky, 1998; Visser et al., 2002). PTH has been studied to bring about increased production of IL-6 and IL-6sR in rat liver thereby contributing to a rise in their in vivo circulating levels (Mitnick et al., 2001).

Kir et al. (2016) studied that elevated PTH has a potential role in reducing insulin-like growth factor 1 (IGF-1) and enhancing the expression of Murf-1, Atrogin-1 and Myostatin (which regulate muscle atrophy), thereby mediating muscle wasting in chronic kidney disease. In addition, fat-PTH receptor knockout mice maintained their muscle mass and strength which was indicative of a possible PTH-PTHR signaling axis and fat-skeletal muscle crosstalk.

Clinically, the Longitudinal Aging Study Amsterdam showed that lower 25-OHD and higher PTH levels increase the risk of sarcopenia in older men and women (Visser, Deeg, & Lips, 2003).

5. What is known about the role of PTH in balance?

Among the metabolic/endocrine causes, the multiple adverse effects of low vitamin D levels have been extensively studied. It is well known that falls and fractures in older individuals is associated with low vitamin D levels (Duque, Daly, Sanders, & Kiel, 2017). Low vitamin D levels have been associated with frailty (Hirani et al., 2013; Shardell et al., 2009), muscle weakness, and slow gait speed (Annweiler et al., 2009, 2017). However, recent evidence also shows a possible U-shaped association between vitamin D levels and adverse health outcomes since higher concentrations (100–112 nmol/L) have also been related to an increased number of falls events (Gallagher, 2016).

There is evolving interest in the role of PTH in frailty and sarcopenia and its outcomes. Recent evidence demonstrating an inconsistent role of vitamin D in falls have allowed us to question whether PTH has a role in mediating this relationship. However, most studies in this area, have only looked at the combined effect of vitamin D and secondary hyperparathyroidism on muscle strength and performance. Also, the focus has been on muscle mass and strength with limited evidence available on balance association.

It has recently been studied in community dwelling individuals with history of falls, an independent association exists between elevated PTH levels and the clinical manifestations of frailty and osteosarcopenia. Individuals with high PTH levels have been shown to suffer more incident falls (Fig. 1) (Murthy et al., 2018; Suriyaarachchi et al., 2018). The Longitudinal Aging Study Amsterdam showed In a study based on ambulant nursing home residents and hostel residents, lower serum 25-hydroxyvitamin D and higher serum PTH levels were associated with increased number of falls (Stein et al., 1999). PTH, but not vitamin D, has been studied to predict time to first fall (Sambrook et al., 2004). It has been demonstrated that high PTH levels are associated with slower gait velocity, independent of vitamin D (Montero-Odasso et al., 2016) and longer time to complete repeated sit to stand test compared to healthy age-matched individuals (Morris et al., 2012). A recent prospective study also observed a trend between higher PTH levels and loss of hip flexor and knee extensor strength (Verreault et al., 2002).

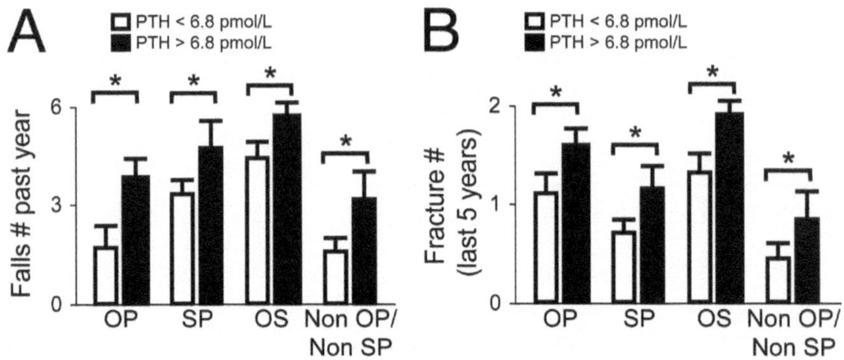

Fig. 1 A cross sectional study in 400 subjects (mean age = 79, 65% women) in Western Sydney showed that subjects in the osteosarcopenic (OS) group had a higher prevalence of self-reported falls and fractures than those in the other three subgroups. In every subgroup, the number of falls and fractures was higher among those with high PTH. The highest number of falls and fractures occurred in the OS group with high PTH. *$P < 0.01$. SP, sarcopenic; OP, osteopenic/osteoporotic. *Adapted from Suriyaarachchi, P., Gomez, F., Curcio, C. L., Boersma, D., Murthy, L., Grill, V., et al. (2018). High parathyroid hormone levels are associated with osteosarcopenia in older individuals with a history of falling.* Maturitas, 113, *21–25.*

The beneficial effects of surgical correction of primary hyperparathyroidism has been demonstrated. In their study, Deutch et al., evaluated changes in muscular strength and endurance, work capacity, and subjective fatigue following surgical treatment of primary hyperparathyroidism, with patients surgically treated for nontoxic goiter serving as controls. The results showed improved knee extension force after parathyroidectomy as a result of increased force generation capacity of the muscle (Deutch, Jensen, Christiansen, & HessovD, 2000). Morris et al. demonstrated a meaningful improvement in 6-min walking distance following surgery for primary hyperparathyroidism (Morris et al., 2010). Small clinical studies have shown that patients with hyperparathyroidism have impaired muscle function that can be restored after treatment (Joborn, Joborn, Rastad, Äkerström, & Ljunghall, 1988; Kristoffersson, Boström, & Söderberg, 1992).

Few studies have evaluated the association of PTH levels with balance. Bird, El Haber, Batchelor, Hill, and Wark (2018) conducted an observational cohort study of 61 pairs of female twins aged >45 years and examined "environmental and heritable features of gait and balance." The average age of the participants was 59.4 ± 8.2 years and included both peri-menopausal as well as post-menopausal women. The results showed that elevated PTH levels and low vitamin D levels were associated with gait instability as

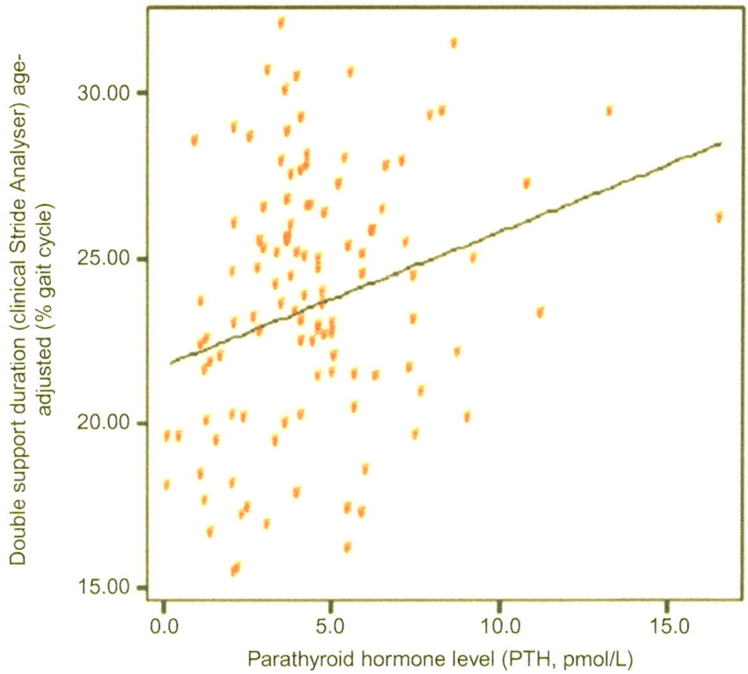

Fig. 2 Association between parathyroid hormone levels (PTH) and double support duration of the clinical stride analyzer. *Adapted from Bird, M.-L., El Haber, N., Batchelor, F., Hill, K., & Wark, J. D. (2018). Vitamin D and parathyroid hormone are associated with gait instability and poor balance performance in mid-age to older aged women.* Gait & Posture, 59, *71–75.*

demonstrated by increase in double support duration (Fig. 2). Only vitamin D was associated with dynamic balance measured by the Step test (Hill, Bernhardt, McGann, Maltese, & Berkovits, 1996). However, the study did not assess static balance.

De Franca et al. performed a retrospective, cross sectional study on older persons with history of falls. Static balance (posturography) was assessed by using the Balance Rehabilitation Unit (BRU; Medicaa, Uruguay). Dynamic balance was evaluated by the timed "up and go" (TUG) (Shumway-Cook, Brauer, & Woollacott, 2000) and four-square step (FSS) tests (Dite & Temple, 2002), and by gait parameters, showed that rising PTH levels were associated with increase in the time to complete the TUG test, worsening the performance as well as increase in deviation in center of pressure (COP) with eyes closed on hard surface, decreasing the stability. The study did not evaluate the association of PTH levels with number of falls (de França et al., 2019).

Frailty, poor balance, falls, OS, and low functional status are not known to be typical manifestations of primary normocalcemic hyperparathyroidism. However, identifying nontraditional manifestations and their impact in general health and quality of life also emerges as a current need, improving the identification of these individuals and preventing severe injuries. The above evidence suggests a role for PTH in balance performance in older individuals. The exact mechanism of how PTH mediates its effects on muscle strength and performance or balance is unclear. Further longitudinal studies aiming to study the independent association of PTH with falls risk and balance are still required.

6. Conclusion

PTH plays an important role in bone and muscle biology and function. It also plays a role in the regulation of gait and balance through recently established biological mechanisms associated to its role in muscle strength and function, which are independent of other causes of elevated PTH such as low serum vitamin D, adenomas, or renal diseases. Whether correction of serum levels of PTH have a beneficial impact on gait and balance remains unknown.

References

Abboud, M., Rybchyn, M., Liu, J., Ning, Y., Gordon-Thomson, C., Brennan-Speranza, T., et al. (2017). The effect of parathyroid hormone on the uptake and retention of 25-hydroxyvitamin D in skeletal muscle cells. *The Journal of Steroid Biochemistry and Molecular Biology*, *173*, 173–179.

Annweiler, C., Henni, S., Walrand, S., Montero-Odasso, M., Duque, G., & Duval, G. T. (2017). Vitamin D and walking speed in older adults: Systematic review and meta-analysis. *Maturitas*, *106*, 8–25.

Annweiler, C., Schott-Petelaz, A. M., Berrut, G., Kressig, R. W., Bridenbaugh, S., Herrmann, F. R., et al. (2009). Vitamin D deficiency-related quadriceps weakness: Results of the epidemiologie De l'Osteoporose cohort. *Journal of the American Geriatrics Society*, *57*(2), 368–369.

Austin, N., Devine, A., Dick, I., Prince, R., & Bruce, D. (2007). Fear of falling in older women: A longitudinal study of incidence, persistence, and predictors. *Journal of the American Geriatrics Society*, *55*(10), 1598–1603.

Baczynski, R., Massry, S. G., Magott, M., El-Belbessi, S., Kohan, R., & Brautbar, N. (1985). Effect of parathyroid hormone on energy metabolism of skeletal muscle. *Kidney International*, *28*(5), 722–727.

Begum, N., Sussman, K., & Draznin, B. (1992). Calcium-induced inhibition of phosphoserine phosphatase in insulin target cells is mediated by the phosphorylation and activation of inhibitor 1. *Journal of Biological Chemistry*, *267*(9), 5959–5963.

Bergen, G., Stevens, M. R., & Burns, E. R. (2016). Falls and fall injuries among adults aged $\geq$ 65 years—United States, 2014. *Morbidity and Mortality Weekly Report*, *65*(37), 993–998.

Binkley, N., & Buehring, B. (2009). Beyond FRAX®: It's time to consider "sarco-osteopenia". *Journal of Clinical Densitometry*, *4*(12), 413–416.

Bird, M.-L., El Haber, N., Batchelor, F., Hill, K., & Wark, J. D. (2018). Vitamin D and parathyroid hormone are associated with gait instability and poor balance performance in mid-age to older aged women. *Gait & Posture*, *59*, 71–75.

Bracchi, E., Rizzo, S., Longari, F., Bernardini, M., Bizzotti, C., & Frenguelli, A. (2002). Balance disorders in the elderly. *Acta Otorhinolaryngologica Italica: Organo Ufficiale Della Societa Italiana di Otorinolaringologia e Chirurgia Cervico-facciale*, *22*(5), 263–267.

Carrivick, S. J., Walsh, J. P., Brown, S. J., Wardrop, R., & Hadlow, N. C. (2015). Brief report: Does PTH increase with age, independent of 25-hydroxyvitamin D, phosphate, renal function, and ionized calcium? *The Journal of Clinical Endocrinology and Metabolism*, *100*(5), 2131–2134.

Chang, J. T., Morton, S. C., Rubenstein, L. Z., Mojica, W. A., Maglione, M., Suttorp, M. J., et al. (2004). Interventions for the prevention of falls in older adults: Systematic review and meta-analysis of randomised clinical trials. *BMJ*, *328*(7441), 680.

Cruz-Jentoft, A. J., Bahat, G., Bauer, J., Boirie, Y., Bruyère, O., Cederholm, T., et al. (2019). Sarcopenia: Revised European consensus on definition and diagnosis. *Age and Ageing*, *48*(1), 16–31.

de França, N. A. G., Murthy, L. S., Phu, S., Liberts, E., Vogrin, S., Martini, L. A., et al. (2019). High parathyroid hormone levels are associated with poor balance in older persons: A cross-sectional study. *Maturitas*, *121*, 57–62.

Deshpande, N., Metter, E. J., Lauretani, F., Bandinelli, S., Guralnik, J., & Ferrucci, L. (2008). Activity restriction induced by fear of falling and objective and subjective measures of physical function: A prospective cohort study. *Journal of the American Geriatrics Society*, *56*(4), 615–620.

Deutch, S. R., Jensen, M. B., Christiansen, P. M., & HessovD, I. (2000). Muscular performance and fatigue in primary hyperparathyroidism. *World Journal of Surgery*, *24*(1), 102–107.

Dite, W., & Temple, V. A. (2002). A clinical test of stepping and change of direction to identify multiple falling older adults. *Archives of Physical Medicine and Rehabilitation*, *83*(11), 1566–1571.

Duque, G., Daly, R. M., Sanders, K., & Kiel, D. P. (2017). Vitamin D, bones and muscle: Myth versus reality. *Australasian Journal on Ageing*, *36*, 8–13.

Freaney, R., McBrinn, Y., & McKenna, M. J. (1993). Secondary hyperparathyroidism in elderly people: Combined effect of renal insufficiency and vitamin D deficiency. *The American Journal of Clinical Nutrition*, *58*(2), 187–191.

Fried, L. P., Tangen, C. M., Walston, J., Newman, A. B., Hirsch, C., Gottdiener, J., et al. (2001). Frailty in older adults: Evidence for a phenotype. *The Journals of Gerontology Series A: Biological Sciences and Medical Sciences*, *56*(3), M146–M157.

Gallagher, J. C. (2016). Vitamin D and falls—The dosage conundrum. *Nature Reviews. Endocrinology*, *12*(11), 680–684.

Garber, A. J. (1983). Effects of parathyroid hormone on skeletal muscle protein and amino acid metabolism in the rat. *The Journal of Clinical Investigation*, *71*(6), 1806–1821.

Harridge, S. D., & Lazarus, N. R. (2017). Physical activity, aging, and physiological function. *Physiology (Bethesda)*, *32*(2), 152–161.

Hill, K. D., Bernhardt, J., McGann, A. M., Maltese, D., & Berkovits, D. (1996). A new test of dynamic standing balance for stroke patients: Reliability, validity and comparison with healthy elderly. *Physiotherapy Canada*, *48*(4), 257–262.

Hirani, V., Naganathan, V., Cumming, R. G., Blyth, F., Le Couteur, D. G., Handelsman, D. J., et al. (2013). Associations between frailty and serum 25-hydroxyvitamin D and 1,25-dihydroxyvitamin D concentrations in older Australian men: The concord health and ageing in men project. *The Journals of Gerontology Series A: Biological Sciences and Medical Sciences*, *68*(9), 1112–1121.

Houston, D. K., Cesari, M., Ferrucci, L., Cherubini, A., Maggio, D., Bartali, B., et al. (2007). Association between vitamin D status and physical performance: The InCHIANTI study. *The Journals of Gerontology. Series A, Biological Sciences and Medical Sciences*, *62*(4), 440–446.

Huo, Y. R., Suriyaarachchi, P., Gomez, F., Curcio, C. L., Boersma, D., Gunawardene, P., et al. (2015a). Comprehensive nutritional status in sarco-osteoporotic older fallers. *The Journal of Nutrition, Health & Aging*, *19*(4), 474–480.

Huo, Y. R., Suriyaarachchi, P., Gomez, F., Curcio, C. L., Boersma, D., Muir, S. W., et al. (2015b). Phenotype of osteosarcopenia in older individuals with a history of falling. *Journal of the American Medical Directors Association*, *16*(4), 290–295.

Joborn, C., Joborn, H., Rastad, J., Åkerström, G., & Ljunghall, S. (1988). Maximal isokinetic muscle strength in patients with primary hyperparathyroidism before and after parathyroid surgery. *British Journal of Surgery*, *75*(1), 77–80.

Ju, J., Jiang, Y., Zhou, P., Li, L., Ye, X., Wu, H., et al. (2018). Evaluation of the reliability and validity for X16 balance testing scale for the elderly. *BMC Geriatrics*, *18*(1), 112.

Kir, S., Komaba, H., Garcia, A. P., Economopoulos, K. P., Liu, W., Lanske, B., et al. (2016). PTH/PTHrP receptor mediates cachexia in models of kidney failure and cancer. *Cell Metabolism*, *23*(2), 315–323.

Kristoffersson, A., Boström, A., & Söderberg, T. (1992). Muscle strength is improved after parathyroidectomy in patients with primary hyperparathyroidism. *British Journal of Surgery*, *79*(2), 165–168.

Landi, F., Liperoti, R., Russo, A., Giovannini, S., Tosato, M., Capoluongo, E., et al. (2012). Sarcopenia as a risk factor for falls in elderly individuals: Results from the ilSIRENTE study. *Clinical Nutrition*, *31*(5), 652–658.

Lang, T., Koyama, A., Li, C., Li, J., Lu, Y., Saeed, I., et al. (2008). Pelvic body composition measurements by quantitative computed tomography: Association with recent hip fracture. *Bone*, *42*(4), 798–805.

Lips, P. (2001). Vitamin D deficiency and secondary hyperparathyroidism in the elderly: Consequences for bone loss and fractures and therapeutic implications. *Endocrine Reviews*, *22*(4), 477–501.

Maki, B. E., & McIlroy, W. E. (1996). Postural control in the older adult. *Clinics in Geriatric Medicine*, *12*(4), 635–658.

McCarty, M. F. (2002). Vitamin D status and muscle strength. *The American Journal of Clinical Nutrition*, *76*(6), 1454–1455.

Mitnick, M. A., Grey, A., Masiukiewicz, U., Bartkiewicz, M., Rios-Velez, L., Friedman, S., et al. (2001). Parathyroid hormone induces hepatic production of bioactive interleukin-6 and its soluble receptor. *American Journal of Physiology. Endocrinology and Metabolism*, *280*(3), E405–E412.

Montero-Odasso, M., Sakurai, R., Muir-Hunter, S., Islam, A., Doherty, T., Duque, G., et al. (2016). Serum parathyroid hormone but not vitamin D is associated with impaired gait in community-dwelling older adults. *Journal of the American Geriatrics Society*, *64*(12), 2606–2608.

Morley, J. E., Baumgartner, R. N., Roubenoff, R., Mayer, J., & Nair, K. S. (2001). Sarcopenia. *Journal of Laboratory and Clinical Medicine*, *137*(4), 231–243.

Morris, G. S., Grubbs, E. G., Hearon, C. M., Gantela, S., Lee, J. E., Evans, D. B., et al. (2010). Parathyroidectomy improves functional capacity in "asymptomatic" older patients with primary hyperparathyroidism: A randomized control trial. *Annals of Surgery*, *251*(5), 832–837.

Morris, G., Landry, C., Grubbs, E., Jimenez, C., Busaidy, N., & Perrier, N. (2012). Greater than age-predicted functional deficits in older patients with primary hyperparathyroidism. *Endocrine Practice*, *18*(4), 450–455.

Murthy, L., Dreyer, P., Suriyaarachchi, P., Gomez, F., Curcio, C. L., Boersma, D., et al. (2018). Association between high levels of parathyroid hormone and frailty: The Nepean osteoporosis and frailty (NOF) study. *The Journal of Frailty & Aging*, 7(4), 253–257.

Need, A. G., Horowitz, M., Morris, H. A., & Nordin, B. C. (2000). Vitamin D status: Effects on parathyroid hormone and 1, 25-dihydroxyvitamin D in postmenopausal women. *The American Journal of Clinical Nutrition*, *71*(6), 1577–1581.

Pahor, M., & Kritchevsky, S. (1998). Research hypotheses on muscle wasting, aging, loss of function and disability. *The Journal of Nutrition, Health & Aging*, *2*(2), 97–100.

Pasco, J. A., Mohebbi, M., Holloway, K. L., Brennan-Olsen, S. L., Hyde, N. K., & Kotowicz, M. A. (2017). Musculoskeletal decline and mortality: Prospective data from the Geelong osteoporosis study. *Journal of Cachexia, Sarcopenia and Muscle*, *8*(3), 482–489.

Pizzigalli, L., Micheletti Cremasco, M., Mulasso, A., & Rainoldi, A. (2016). The contribution of postural balance analysis in older adult fallers: A narrative review. *Journal of Bodywork and Movement Therapies*, *20*(2), 409–417.

Prince, R. L., Dick, I., Devine, A., Price, R. I., Gutteridge, D. H., Kerr, D., et al. (1995). The effects of menopause and age on calcitropic hormones: A cross-sectional study of 655 healthy women aged 35 to 90. *Journal of Bone and Mineral Research*, *10*(6), 835–842.

Rockwood, K., & Mitnitski, A. (2007). Frailty in relation to the accumulation of deficits. *The Journals of Gerontology Series A: Biological Sciences and Medical Sciences*, *62*(7), 722–727.

Roubenoff, R., & Hughes, V. A. (2000). Sarcopenia: Current concepts. *The Journals of Gerontology Series A: Biological Sciences and Medical Sciences*, *55*(12), M716–M724.

Rubenstein, L. Z., & Josephson, K. R. (2006). Falls and their prevention in elderly people: What does the evidence show? *The Medical Clinics of North America*, *90*(5), 807–824.

Salkeld, G., Ameratunga, S. N., Cameron, I., Cumming, R., Easter, S., Seymour, J., et al. (2000). Quality of life related to fear of falling and hip fracture in older women: A time trade off study commentary: Older people's perspectives on life after hip fractures. *BMJ*, *320*(7231), 341–346.

Sambrook, P., Chen, J., March, L., Cameron, I., Cumming, R., Lord, S., et al. (2004). Serum parathyroid hormone predicts time to fall independent of vitamin D status in a frail elderly population. *The Journal of Clinical Endocrinology & Metabolism*, *89*(4), 1572–1576.

Shardell, M., Hicks, G. E., Miller, R. R., Kritchevsky, S., Andersen, D., Bandinelli, S., et al. (2009). Association of low vitamin D levels with the frailty syndrome in men and women. *The Journals of Gerontology Series A: Biological Sciences and Medical Sciences*, *64*(1), 69–75.

Shumway-Cook, A., Brauer, S., & Woollacott, M. (2000). Predicting the probability for falls in community-dwelling older adults using the Timed Up & Go Test. *Physical Therapy*, *80*(9), 896–903.

Souberbielle, J.-C., Cormier, C., Kindermans, C., Gao, P., Cantor, T., Forette, F.o., et al. (2001). Vitamin D status and redefining serum parathyroid hormone reference range in the elderly. *The Journal of Clinical Endocrinology & Metabolism*, *86*(7), 3086–3090.

Stein, M. S., Wark, J. D., Scherer, S. C., Walton, S. L., Chick, P., Di Carlantonio, M., et al. (1999). Falls relate to vitamin D and parathyroid hormone in an Australian nursing home and hostel. *Journal of the American Geriatrics Society*, *47*(10), 1195–1201.

Suriyaarachchi, P., Gomez, F., Curcio, C. L., Boersma, D., Murthy, L., Grill, V., et al. (2018). High parathyroid hormone levels are associated with osteosarcopenia in older individuals with a history of falling. *Maturitas*, *113*, 21–25.

Swarthout, J. T., D'Alonzo, R. C., Selvamurugan, N., & Partridge, N. C. (2002). Parathyroid hormone-dependent signaling pathways regulating genes in bone cells. *Gene*, *282*(1–2), 1–17.

Thelen, D. G., Schultz, A. B., Alexander, N. B., & Ashton-Miller, J. A. (1996). Effects of age on rapid ankle torque development. *The Journals of Gerontology Series A: Biological Sciences and Medical Sciences*, *51*(5), M226–M232.

Tinetti, M. E., Speechley, M., & Ginter, S. F. (1988). Risk factors for falls among elderly persons living in the community. *New England Journal of Medicine*, *319*(26), 1701–1707.

Topinková, E. (2008). Aging, disability and frailty. *Annals of Nutrition and Metabolism*, *52*(Suppl. 1), 6–11.

Verreault, R., Semba, R. D., Volpato, S., Ferrucci, L., Fried, L. P., & Guralnik, J. M. (2002). Low serum vitamin D does not predict new disability or loss of muscle strength in older women. *Journal of the American Geriatrics Society*, *50*(5), 912–917.

Visser, M., Deeg, D. J. H., & Lips, P. (2003). Low vitamin D and high parathyroid hormone levels as determinants of loss of muscle strength and muscle mass (sarcopenia): The longitudinal aging study Amsterdam. *The Journal of Clinical Endocrinology & Metabolism*, *88*(12), 5766–5772.

Visser, M., Pahor, M., Taaffe, D. R., Goodpaster, B. H., Simonsick, E. M., Newman, A. B., et al. (2002). Relationship of interleukin-6 and tumor necrosis factor-α with muscle mass and muscle strength in elderly men and women: The health ABC study. *The Journals of Gerontology Series A: Biological Sciences and Medical Sciences*, *57*(5), M326–M332.

Yu, S., Appleton, S., Adams, R., Chapman, I., Wittert, G., Visvanathan, T., et al. (2014). The impact of low muscle mass definition on the prevalence of sarcopenia in older Australians. *BioMed Research International*, *2014*.

CHAPTER NINE

Insulin and aging

Mirian Ayumi Kurauti[a,b,*], Gabriela Moreira Soares[a], Carine Marmentini[a], Gabriela Alves Bronczek[a], Renato Chaves Souto Branco[a], and Antonio Carlos Boschero[a]

[a]Obesity and Comorbidities Research Center, Institute of Biology, University of Campinas (UNICAMP), Campinas, Sao Paulo, Brazil

[b]Department of Physiological Sciences, Biological Sciences Center, State University of Maringa (UEM), Maringa, Parana, Brazil

*Corresponding author: e-mail address: mirian.kurauti@hotmail.com

Contents

Abstract

Aging is characterized by a progressive loss of physiological function leading to increase in the vulnerability to death. This deterioration process occurs in all living organisms and is the primary risk factor for pathological conditions including obesity, type 2 diabetes mellitus, Alzheimer's disease and cardiovascular diseases. Most of the age-related diseases have been associated with impairment of action of an important hormone, namely insulin. It is well-known that this hormone is a critical mediator of metabolism, growth, proliferation and differentiation. Insulin action depends on two processes that determine its circulating levels, insulin secretion and clearance, and insulin sensitivity in its target tissues. Aging has deleterious effects on these three mechanisms, impairing insulin action, thereby increasing the risk for diseases and death. Thus, improving insulin action may be an important strategy to have a healthier and longer life.

Vitamins and Hormones, Volume 115
ISSN 0083-6729
https://doi.org/10.1016/bs.vh.2020.12.010

1. Introduction

Insulin is an important hormone that regulates metabolism, growth, proliferation and differentiation. The release of this hormone, by the pancreas, is enhanced at postprandial state, when the nutrients are abundant in the circulation, increasing its blood concentration (insulinemia). This increase favors the insulin action on its target tissues, such as liver, skeletal muscle, adipose tissue and brain. The activation of insulin signaling pathway, in these tissues, is essential for a proper handle of the excess of nutrients, primarily glucose, at postprandial state.

Aging seems to impair the insulin action through several mechanism leading to glucose intolerance and type 2 diabetes mellitus (T2DM), besides others age-related diseases (Kalyani & Egan, 2013; Watson & Craft, 2003), thereby increasing morbidity and mortality. Indeed, improved insulin action is associated with a healthier and longer life (Wijsman et al., 2011).

Insulin action depends on: (1) its circulating levels, controlled by insulin secretion and clearance; and (2) insulin sensitivity, determined by the insulin signaling pathway in the target tissues. Therefore, in this chapter we will firstly discuss how aging impacts on insulin secretion, clearance and sensitivity. Finally, we will discuss the function of insulin on the healthy longevity.

2. Insulin secretion and aging

Glucose is the main source of energy for the brain maintenance. Once metabolized, it generates adenosine triphosphate (ATP), required for the synthesis of neurotransmitters and maintenance of neuronal and non-neuronal cells functions (Mergenthaler, Lindauer, Dienel, & Meisel, 2013). Thus, the regulation of glucose metabolism is essential for the brain physiology and an impairment on this regulation, impacts on the function of brain, as well as, on the function of others important organs.

In this scenario, insulin shows as a master regulator of glucose metabolism. This hormone is synthesized and secreted by pancreatic β-cells, which sense the blood glucose concentration and respond with a release of an appropriated insulin amount to maintain blood glucose at normal levels (Schmitz, Rungby, Edge, & Juhl, 2008). The lack of insulin or its receptor causes a severe hyperglycemia (Wang, Zhou, Palyha, & Mu, 2019).

Insulin secretion is the first step for insulin action, contributing to regulate its circulating levels. It was extensively reported that aging impairs insulin secretion and to understand the mechanism involved in this alteration, it is important to describe briefly the physiology of this process.

2.1 Mechanism of insulin secretion

Although glucose is the major stimulus for insulin secretion, some amino acids and fatty acids can also stimulate β-cells to release insulin. In addition to these nutritionals stimulus for insulin secretion, this process is also regulated by paracrine, endocrine and neuronal factors through activation of several intracellular signaling pathways (Skelin Klemen, Dolenšek, Slak Rupnik, & Stožer, 2017). Here, we summarized the mechanism of glucose-stimulated insulin secretion (GSIS), illustrated in the Fig. 1.

To initiate insulin secretion, glucose enters in the pancreatic β-cells through specifics glucose transporters (GLUTs) in the plasma membrane, GLUT2 in rodents and GLUT1 in humans (McCulloch et al., 2011). In the cytosol, glucose is rapidly phosphorylated by glucokinase (GK) to generate glucose-6-phosphate (G6P), which is metabolized to pyruvate via glycolysis. Pyruvate enters the mitochondria through specifics pyruvate carriers (Patterson et al., 2014), feeding the tricarboxylic acid cycle (TCA). The reduction of nicotinamide and flavin adenine dinucleotides (NAD and FAD, respectively), generates NADH and FADH, which donate protons to export to the intermembrane space of mitochondria through the electron transport chain. This creates a proton gradient necessary to ATP production by ATP synthase. The subsequent elevation of cytosolic ATP:ADP ratio inhibits ATP-sensitive K^+ (K_{ATP}) channels leading to membrane depolarization, which is also modulated by several additional ions channels (Braun et al., 2008; Malaisse, Boschero, Kawazu, & Hutton, 1978). When the membrane potential reaches to a threshold (− 55 mV to −50 mV), voltage-dependents Na^+ and Ca^{2+} channels are activated, firing action potentials. The associated Ca^{2+} influx increases its intracellular concentration, triggering the exocytosis of insulin granules (Boschero & Malaisse, 1978; Rorsman, Braun, & Zhang, 2012). In this final step of insulin secretion, Ca^{2+} seems to modulate the function of several exocytosis-related proteins including syntaxin 1, vesicle-associated membrane protein-2 (VAMP2) and synaptosomal-associated protein of 25 kDa (SNAP25) (Jacobsson et al., 1994).

The mechanism of insulin secretion is complex and goes beyond this brief description, but the effects of aging on this mechanism is much more complicated. In the next section, we will discuss about the effects of aging on insulin secretion.

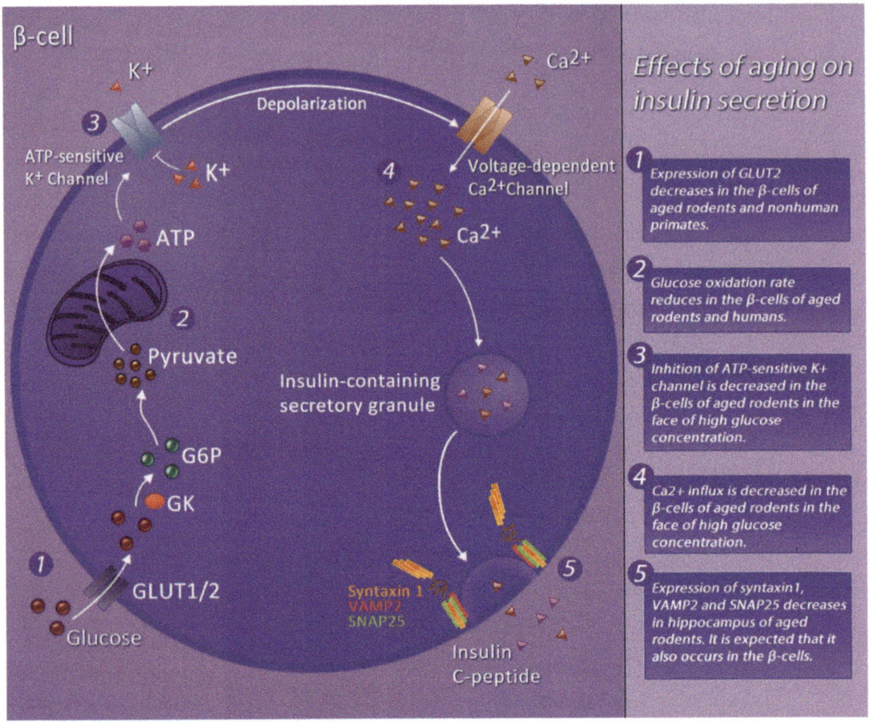

Fig. 1 Mechanism of GSIS and the effects of aging on this process. Glucose enters in the β-cell through the GLUT1/2 and is metabolized, increasing the ATP levels. High ATP:ADP ratio inhibits ATP-sensitive K^{+} channels leading to membrane depolarization and, consequently, activation of voltage-dependents Ca^{2+} channels. Ca^{2+} influx increases its intracellular levels, triggering the exocytosis of insulin granules through modulation of exocytosis-related proteins including syntaxin 1, vesicle-associated membrane protein-2 (VAMP2) and synaptosomal-associated protein of 25 kDa (SNAP25). The effects of aging on insulin secretion are summarized in the boxes with the numbers indicating in the figure, the steps of GSIS impaired with age. GK, glucokinase; G6P, glucose-6-phosphate.

2.2 Effects of aging on insulin secretion

Glucose tolerance decreases with age, explaining the high prevalence of T2DM in the elderly. The mechanism of the impairment of glucose tolerance is not entirely understood but, in general, glucose intolerance may be a result of impaired insulin sensitivity and/or reduced insulin secretion, triggering the development of T2DM. Indeed, most of the studies, but not all (Barbieri, Rizzo, Manzella, & Paolisso, 2001; Bourey et al., 1993), have reported insulin resistance and decreased insulin secretion in older compared with younger individuals (Fink, Kolterman, Griffin, & Olefsky, 1983; Møller, Gormsen, Fuglsang, & Gjedsted, 2003). However, it is very difficult to attribute these alterations to the aging process *per se*, since several risk factors for glucose intolerance are commonly associated with age, such as, increased adiposity, decreased physical activities, use of different medications and coexisting illness (Chang & Halter, 2003). Therefore, the degree of these age-related factors may explain the discrepant effects of aging on insulin sensitivity (discussed later in this chapter) and insulin secretion.

Regarding to insulin secretion, most of the studies have found impaired insulin secretion in older *vs* younger individuals (Coordt, Ruhe, & McDonald, 1995; Møller et al., 2003). The mechanisms involved with this deleterious effect of aging on insulin secretion seems to include, decline in β-cells function and mass.

Beta-cells dysfunction is characterized by inadequate GSIS, which occurs in several steps, as described previously in this chapter. Earlier studies have found impaired GSIS in aged rodents (Muzumdar et al., 2004; Wang, Halban, & Rowe, 1988) and humans (Pacini, Beccaro, Valerio, Nosadini, & Crepaldi, 1990) and this effect may be due to alterations in various steps of GSIS (see the boxes in Fig. 1).

Glucose transport is the first step of GSIS and may be altered with age, since it was found a decreased GLUT2 expression in aged rodents (Ohneda et al., 1993) and nonhuman primates (Kramer, Moeller, Hachey, Mansfield, & Wachtman, 2009). However, in humans, GLUT1 and GLUT3 are higher expressed, compared with GLUT 2 (McCulloch et al., 2011). Thus, the effects of aging, on this first step of GSIS, may be different in humans and rodents, and to clarify this mechanism, further investigations are necessary.

Inside the β-cells, glucose undergoes mitochondrial oxidation to increase ATP:ADP ratio, which is essential to the membrane depolarization and insulin secretion. It was observed a decreased glucose oxidation in aged rodents (Reaven & Reaven, 1980) and humans (Bonadonna, Groop, Simonson, & DeFronzo, 1994), an effect probably related to a lower activity of a key enzyme that links glycolysis and oxidative phosphorylation, the mitochondrial glycerophosphate dehydrogenase (mGPDH) (Azhar, Ho, Reaven, & Reaven, 1983). Indeed, several studies have supported this hypothesis, showing attenuated glucose oxidation concomitant with decreased insulin secretion in islets from aged rodents (Elahi, Muller, Andersen, Tobin, & Andres, 1985; Sartin, Chaudhuri, Farina, & Adelman, 1986). However, it was also reported no difference (Ammon et al., 1984) and even a higher glucose oxidative rate in islets from older compared with younger rodents (Ruhe, Curry, & McDonald, 1993), raising doubts regarding the effect of aging on glucose oxidation in β-cells.

The increase of ATP:ADP ratio reduces K^+ membrane permeability by inhibiting K_{ATP} channels. The inhibition of these channels seems to be decreased in aged rats, since it was observed an inadequate K^+ efflux when glucose concentration raises (3–16.7 mM). Also, Ca^{2+} influx significantly decreased in these old rats and only 60–70% of insulin was released, compared with young rats (Ammon, Fahmy, Mark, Wahl, & Youssif, 1987). These data support the hypothesis that aging reduces glucose oxidation, disrupting the following steps of GSIS. It is known that aging induces an abnormal production of reactive oxygen species (ROS), leading to mitochondrial dysfunction (Dai, Chiao, Marcinek, Szeto, & Rabinovitch, 2014), which is currently considered an important mechanism whereby aging impairs β-cells function, disrupting the physiological stimulus-secretion coupling in these cells (De Tata, 2014).

In the exocytosis of insulin granules, aging may decrease the expression of several proteins involved with this process (syntaxin 1, VAMP2 and SNAP25). It was observed that, in the hippocampus of aged rats, these proteins were decreased. In this study, the lower expression of these exocytosis-related proteins was associated with an impairment on neurotransmitter release (VanGuilder, Yan, Farley, Sonntag, & Freeman, 2010). Thus, it is expected that aging may also impair insulin release from β-cells by a similar mechanism. However, further studies are necessary to confirm this hypothesis.

In addition to β-cells dysfunction, aging also impairs the regulation of β-cells mass, which is determined by death and turnover of these cells. A gradual reduction in β-cells proliferation and increase in the frequency of apoptotic β-cells were observed in islets from 4, 14 and 24-month-old rats (Gu et al., 2012). In human pancreas, β-cells proliferation also decreased with age while β-cells apoptosis did not change significantly (Reers et al., 2009). In addition to proliferation, differentiation from non-β-cells progenitors is also a mechanism for β-cell turnover, but this mechanism was not changed in the pancreas of aged humans. Therefore, aging seems to affect mainly the β-cells proliferation, impairing the capacity of pancreas to adapt to a higher demand of insulin secretion that arises upon high-fat diet, low-dose streptozotocin and partial pancreatomy (Rankin & Kushner, 2009; Tschen, Dhawan, Gurlo, & Bhushan, 2009). The loss of this β-cells adaptative response may explain the increased risk for glucose intolerance and T2DM in aged subjects.

The declined β-cell proliferation, observed in the pancreas of aged humans (Reers et al., 2009), was directly associated with a lower expression of pancreatic duodenal homeobox-1 (PDX-1), a transcription factor that plays a crucial role in β-cell proliferation. In addition, aging also increases the expression of cell cycle proteins inhibitors, such as the tumor suppressor protein $p16^{INK4a}$, a cyclin-dependent kinase inhibitor (CDKI) that emerges as a key control point for cell cycle entry of β-cells. In fact, $p16^{INK4a}$ knockout mice display a significantly increase in β-cells proliferation (Krishnamurthy et al., 2004). Thus, the higher expression of $p16^{INK4a}$ in these cells, may be an important mechanism by which aging decreases β-cells proliferation (De Tata, 2014).

However, the mechanisms responsible for the decline in insulin secretion with age are still a matter of debate. Although most of the studies have found decrease, others have reported no difference (Bourey et al., 1993; Elahi, Muller, McAloon-Dyke, Tobin, & Andres, 1993) and even increase (Tricò, Natali, Arslanian, Mari, & Ferrannini, 2018) in insulin secretion during the aging process. Similarly, the insulinemia after glucose administration have been observed to be lower (Muller, Elahi, Tobin, & Andres, 1996), similar (Broughton, James, Alberti, & Taylor, 1991) and even higher (Gumbiner et al., 1989; Nolan et al., 1973) in elderly *vs* young subjects.

To explain these discrepant data about insulin secretion and insulinemia, we must consider several variables in these different studies, for example, the age range of younger and older experimental groups and the age-related risk factors for glucose intolerance and T2DM, including increased adiposity, decreased physical activities, use of diverse medications and coexisting illness.

Regarding to the age range of experimental groups, increased insulin secretion and hyperinsulinemia is more likely to be found in early stages of aging (Kurauti et al., 2018; Pyörälä, Miettinen, Laakso, & Pyörälä, 1998) whereas decreased insulin secretion, accompanied by similar or lower insulinemia, is frequently observed in late stages of aging (Garcia et al., 1997; Ribeiro et al., 2012). It seems that with advancing age, the ability of β-cells to compensate the increase in insulin resistance is impaired, predisposing glucose intolerance and T2DM in aged subjects. In fact, an interesting model proposes that multiples age-related risk factors, in addition to the effects of aging *per se*, predispose to develop insulin resistance and impair β-cells function and turnover. Thus, it is suggested that, in late stages of aging, compensatory hyperinsulinemia does not occurs, resulting in higher prevalence of glucose intolerance and T2DM (Chang & Halter, 2003).

Irrespective of the discrepant data regarding to insulin secretion and insulinemia in old age, it is intuitive to assume that circulant insulin levels must accompany insulin secretion. Indeed, in our recently study (Kurauti et al., 2018), we observed that hyperinsulinemia was associated with hypersecretion of insulin in the isolated islets from 10-month-old *vs* 3-month-old mice. Conversely, others reported no difference in the insulinemia, which was accompanied by a lower insulin secretion in the elderly compared with young adults (Pacini et al., 1990). In this study, it was also observed that aging reduces insulin clearance, counteracting the declined insulin secretion. Therefore, we must be aware that insulinemia is determined not only by insulin secretion, but also by insulin clearance, an important process that we will discuss in the next section.

3. Insulin clearance and aging

Although insulin secretion seems to decline during the aging process, age-related hyperinsulinemia has been extensively reported in aged rodents and humans (Burchfiel et al., 1998; Chow et al., 2019; Fink, Kolterman, & Olefsky, 1984). Considering all the described deleterious effects of aging on insulin secretion, is expected that insulin clearance also declines with advancing age, as reported previously (Minaker, Rowe, Tonino, & Pallotta, 1982; Pacini et al., 1990). To understand the effect of aging on insulin clearance, it is important to review, briefly, the mechanism of this process.

3.1 Mechanism of insulin clearance

Following the secretion, insulin enters into the portal circulation whereby reaches to the liver, the primary site of insulin clearance. About 50% of secreted insulin is removed during its first passage through this organ. The remaining unprocessed insulin is delivered to the peripheral tissues through the systemic circulation. Insulin is also removed by these peripheral tissues, such as skeletal muscle, adipose tissue and specially kidney, the major site of insulin clearance from systemic circulation, that contributes with about 50% of peripheral insulin removal (Duckworth, Bennett, & Hamel, 1998).

Considering the important function of hepatic insulin clearance upon homeostatic regulation of peripheral insulin levels, we will focus in this process. Basically, hepatic insulin clearance occurs in three steps illustrated in the Fig. 2: (A) binding of insulin to its receptor (IR); (B) internalization of insulin-IR complex; and (C) degradation of insulin by specifics enzymes. In the first step of insulin clearance, the binding of insulin to its receptor induces a conformational change, leading to autophosphorylation at tyrosine residues. It activates IR, which in turn recruits and phosphorylates regulatory proteins, modulating metabolic and mitogenic functions in the target

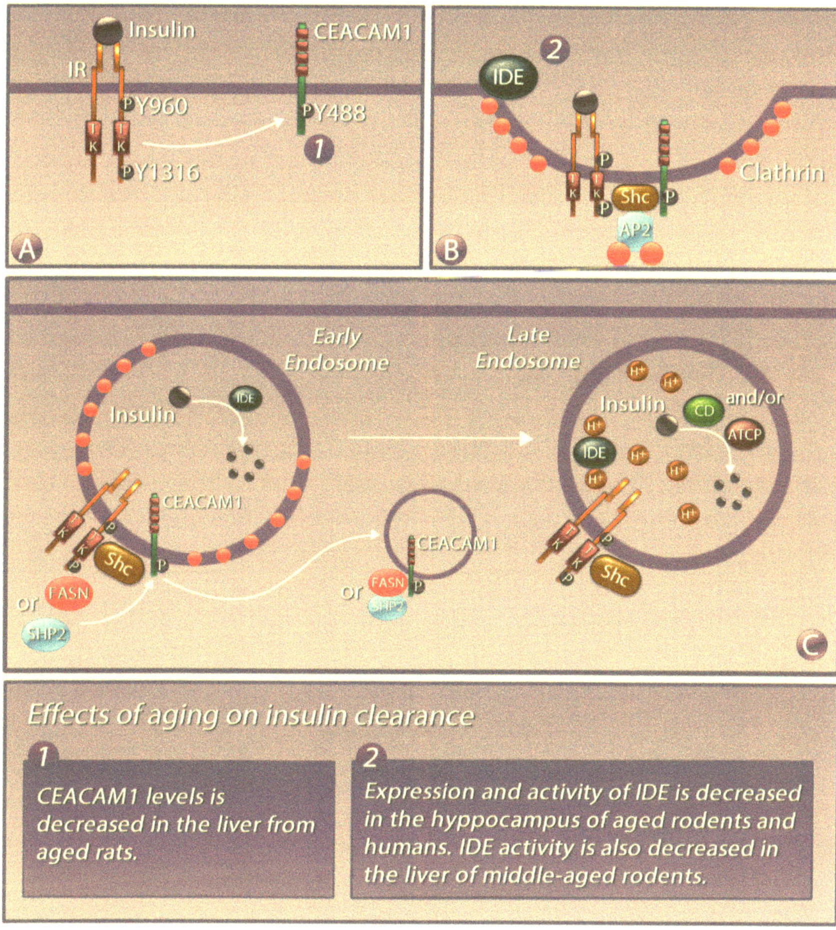

Fig. 2 Mechanism of hepatic insulin clearance and the effects of aging on this process. (A) Insulin binds to its receptor (IR) which phosphorylates the carcinoembryonic antigen-related cell adhesion molecule-1 (CEACAM1). (B) Phosphorylated CEACAM1 associates with insulin-IR complex, via Shc (SH2-containing adapter protein), targeting this complex to clathrin-coated pits/vesicles by interaction with the adaptor protein-2 (AP2) complex. (C) In the early endosome, CEACAM1 binds to others proteins, such as SH2-containing phosphatase-2 (SHP2) and fatty acid synthase (FASN), destabilizing insulin-IR complex and insulin can be cleaved by insulin-degrading enzyme (IDE). As the endosome matures, it is acidified, favoring insulin-IR complex to dissociate. Finally, acidic proteases, such as acidic thiol protease (AcTP) and/or aspartic acid protease cathepsin D (CD) degrade insulin in the late endosome. The effects of aging on insulin clearance is summarized in the boxes with the numbers indicating the age-targets proteins.

cells (discussed later in this chapter). In addition, an important protein that promotes receptor-mediated insulin internalization, namely carcinoembryonic antigen-related cell adhesion molecule-1 (CEACAM1), can also be phosphorylated by IR (Najjar et al., 1995). After phosphorylation at specific tyrosine residue (Tyr 488), CEACAM1 associates with insulin-IR complex, via Shc (SH2-containing adapter protein), targeting this complex to clathrin-coated pits/vesicles by interaction with the adaptor protein-2 (AP2) complex (Najjar & Perdomo, 2019; Okabayashi et al., 1996), thereby triggering the endocytosis process (Fig. 2B).

In early endosome (Fig. 2C), insulin-IR complex is destabilized, in part by the dissociation from CEACAM1 that occurs when it binds to others proteins, including SH2-containing phosphatase-2 (SHP2) and fatty acid synthase (FASN) (Najjar & Perdomo, 2019). Generally, IR returns to the plasma membrane, via retro-endocytosis, and insulin is cleaved by the major enzyme responsible for its degradation, the insulin-degrading enzyme (IDE). As the endosome matures (Fig. 2C), it undergoes to acidification induced by a specific H^+-ATPase, also known as vacuolar ATPase (Maxson & Grinstein, 2014). This acid environment, in late endosome, favors insulin to dissociate from its receptor, while inactivates IDE (Grasso, Satriano, & Milardi, 2015). Therefore, acidic proteases, such as acidic thiol protease (AcTP) and/or aspartic acid protease cathepsin D (CD), seem to be more important for insulin degradation than IDE, in the late endosome (Authier, Rachubinski, Posner, & Bergeron, 1994). Nevertheless, IDE have been described as the major enzyme related with insulin degradation and clearance (Duckworth et al., 1998).

IDE, also known as insulysin or insulinase, is a zinc metallopeptidase ubiquitously expressed. The liver and kidneys seem to display the highest IDE expression, supporting the major role of these organs on insulin degradation and clearance (Hulse, Ralat, & Wei-Jen, 2009). Although initially described as an enzyme responsible for insulin catabolism (Mirsky & Broh-Kahn, 1949; Tomizawa & Halsey, 1959), further studies demonstrate that IDE also degrades other peptides including amyloid-β, amylin and glucagon (Farris et al., 2003). All of these substrates display specifics amino acids sequences that share a tendency to aggregate, forming fibrillary amyloid-like structures, which in turn are toxic for the cells and, probably, the cause of amyloidogenic disorders, such as Alzheimer's disease (Selkoe & Hardy, 2016) and T2DM (Abedini & Schmidt, 2013). Thus, IDE have been proposed as an important target to treat these illness (Kurochkin, Guarnera, & Berezovsky, 2018; Tang, 2016). In T2DM, this enzyme seems to have an important role in the regulation of insulinemia and insulin sensitivity. This hypothesis is supported by studies with IDE knockout mice which display hyperinsulinemia, leading to insulin resistance with age that results from reductions in IR levels in skeletal muscle, adipose tissue and liver (Abdul-Hay et al., 2011). Similarly, mice with liver-specific deletion of IDE (L-IDE-KO) also displayed glucose intolerance and insulin resistance which were associated with lower IR levels in the liver. However, hepatic insulin clearance did not decrease and the circulant insulin levels of L-IDE-KO mice was similar to the levels found in control mice (Villa-Pérez et al., 2018), raising doubts on IDE function in the regulation of insulinemia.

Several studies have investigated insulin clearance in aged subjects; however, the molecular mechanism of aging in this process is poorly understood. Next, we will discuss what is known about the effect of aging on insulin clearance.

3.2 Effects of aging on insulin clearance

Impaired insulin clearance has been associated with metabolic diseases, such as obesity and T2DM (Bergman, Piccinini, Kabir, Kolka, & Ader, 2019; Erdmann et al., 2008). Since aging is a risk factor to these pathological conditions, it is expected that insulin clearance diminishes with advancing age, explaining, at least in part, the age-related hyperinsulinemia (Fink, Revers, Kolterman, & Olefsky, 1985). Corroborating this hypothesis, most of the studies have shown reduced insulin clearance in elderly compared with young individuals (Fink et al., 1985; Iozzo et al., 1999; Minaker et al., 1982; Pacini et al., 1990). However, it is not a consensus that aging *per se* impairs insulin clearance, since several studies have found no difference between young and old subjects, when the clearance of this hormone was measured in humans (Gumbiner et al., 1989) and in rodents (Kurauti et al., 2018). In these studies, insulin secretion was increased and the higher levels of insulin that reached the liver might have blunted the reduction of hepatic insulin clearance, as previously proposed (Fink et al., 1985). In fact, hepatic insulin clearance seems to be even increased in 12-month-old *vs* 2-month-old rats and this effect was due to the higher insulin levels in the portal vein of the older rats (Matveyenko, Veldhuis, & Butler, 2008).

To explain these discrepant findings, we may look at the insulin clearance in obesity. As observed in aging, not all obese individuals display impaired insulin clearance (Polonsky et al., 1988) and this might be explained by their metabolic profile. In the last decade, obese individuals have been characterized in metabolically healthy obese (MHO) and metabolically abnormal obese (MAO). Despite the excessive body fatness, MHO individuals display a favorable cardio-metabolic risk profile, with higher insulin sensitivity, compared with MAO individuals (Primeau et al., 2011). Comparing these two obese groups, it was observed that insulin clearance was significantly lower in MAO group, contributing to hyperinsulinemia found in these individuals (Marini et al., 2014). Thus, it is possible that in metabolically health aged subjects, insulin clearance might be maintained at normal levels (Broughton et al., 1991), whereas in aged subjects that display metabolic abnormalities, such as insulin resistance, insulin clearance might be impaired (Ehrhardt et al., 2019). However, not all studies have supported this hypothesis (Ahrén & Pacini, 1998) and further investigations is necessary to clarify this matter.

Although several studies have demonstrated a lower insulin clearance in aged subjects, few studies have investigated the mechanism whereby aging reduces the clearance of this hormone. To be cleared, insulin is transferred across the endothelium to the target cells and this process can limit the rate of insulin clearance. In the liver, the sinusoidal endothelial cells line the wall of the hepatic sinusoid and are perforated with transcellular pores (50–150 nm), namely fenestrations, which facilitate the transfer of several molecules, including insulin. With advancing age, the diameter of the pores and frequency of fenestrations are diminished in the liver sinusoidal endothelial cells and this seems to be an important mechanism by which aging impairs hepatic insulin clearance (Mohamad et al., 2016).

After the transfer of insulin across the hepatic sinusoid, this hormone can bind to its receptor in the hepatocyte, triggering the endocytosis process via phosphorylation of CEACAM1 which, in turn, stabilizes the insulin-IR complex and inhibits SHP-2, sustaining the phosphorylation of insulin receptor substrate-1 (IRS1), thereby prolonging the transduction of IRS1/Phosphatidylinositol-3 kinase (PI3K)/AKT pathway, activated by insulin (Horst, Najjar, Wagener, & Tiegs, 2018). It has been described in Zucker rats (an obesity model) (Heinrich et al., 2017), and also in obese humans subjects (Lee, 2011), that CEACAM1 is downregulated, compromising insulin sensitivity and clearance. Therefore, age-related insulin resistance, may be also associated with decrease in CEACAM1 levels, explaining the impaired insulin sensitivity and clearance in aged subjects, as we observed in 18-month-old *vs* 3-month-old rats and mice (unpublished data: C. Marmentini; G. M. Soares; G. A. Bronczek; S. Piovani; C. E. Mareze-Costa; E. M. Carneiro; A. C. Boschero; and M. A. Kurauti).

IDE was also described as a downstream target of IR signaling pathway. Inhibition of PI3K abolished IDE upregulation induced by insulin, as reported in hippocampal neurons of rats (Zhao et al., 2004). Considering that old age is associated with impaired insulin sensitivity, downregulation of IDE may also be a possible mechanism whereby aging reduces insulin degradation and clearance. Corroborating this hypothesis, decreased IDE expression and activity have been associated with advancing age in the brain, especially in hippocampus (Caccamo, Oddo, Sugarman, Akbari, & LaFerla, 2005; Hwang et al., 2005). This effect of aging is a possible explanation to the higher risk for Alzheimer's disease in aged subjects. In the liver, we also found lower IDE activity in 10-month-old *vs* 3-month-old mice; however, IDE expression was increased in this tissue (Kurauti et al., 2018). The hypothesis is that decreased IDE activity was compensated by its higher expression, explaining the similar level of insulin clearance, observed in old *vs* control group. It is possible that with advancing age, this compensatory effect may be lost. Indeed, it was observed a reduced IDE expression in the liver from 12-month-old rats (Kochkina et al., 2015). However, further studies are necessary to confirm this hypothesis.

It is not clear whether insulin resistance is a cause or consequence of the impaired insulin clearance. However, both alterations are frequently observed in aged subjects. Therefore, it is important to discuss the mechanisms by which aging impairs the insulin sensitivity.

4. Insulin sensitivity and aging

Insulin sensitivity is determined by the insulin signal transduction in its target cells. Thus, it is necessary to review the insulin signaling pathway before discussing the effects of aging on insulin sensitivity.

4.1 Insulin signaling pathway

Insulin signaling (IS) pathway has a crucial role on cell metabolism, growth, proliferation and differentiation. The activation of this pathway initiates when insulin binds to its receptor on the cell surface. Although IR is expressed in several tissues, here, we will focus on IS pathway in the liver, skeletal muscle, adipose tissue and brain, which represent the major sites of insulin action. The well-characterized components of IS pathway are illustrated in the Fig. 3.

IR is constituted by two extracellular α subunits and two transmembrane β subunits joined by disulfide bonds (Boucher, Kleinridders, & Kahn, 2014). Insulin interacts with α subunit and induces a conformational

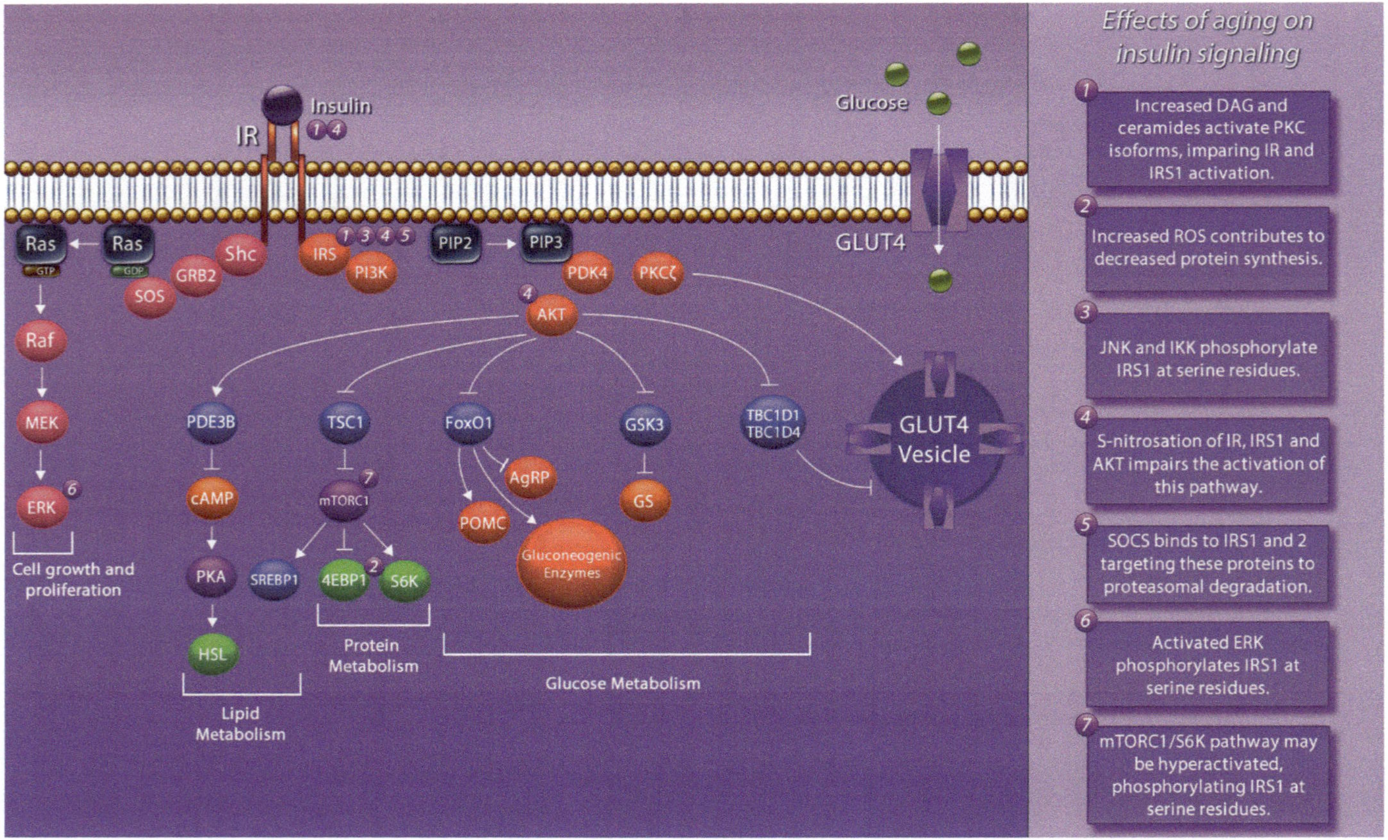

Fig. 3 See figure legend on opposite page.

change, leading to autophosphorylation of β subunits at tyrosine residues. It activates IR that recruits and phosphorylates scaffolds proteins to regulate mitogenic and metabolic functions (Belfiore et al., 2017).

Insulin stimulates mitogenesis via activation of a mitogen-activated protein kinase (MAPK) pathway (Fig. 3) that initiates after the interaction of growth factor receptor-bound protein-2 (GRB2) with IR, which occurs via Shc. GRB2 is associated with SOS (son of sevenless) that activates a GTP-binding protein Ras (Lemmon & Schlessinger, 2010). It recruits Raf (MAPKKK), to the plasma membrane which in turn phosphorylates MEK (MAPKK). Subsequently, MEK phosphorylates the extracellular signal-regulated kinase (ERK or MAPK) that induces several nuclear transcription factors to enhance the expression of proteins involved with growth and proliferation (Zhang & Liu, 2002).

In addition to mitogenesis, insulin also regulates carbohydrate, protein and lipid metabolism via activation of IRS/PI3K/AKT pathway (Fig. 3). After insulin binding, IR recruits and phosphorylates, at tyrosine residues, IRS proteins, which in turn recruits downstream proteins, such as PI3K. This kinase phosphorylates phosphatidylinositol-4,5-bisphosphate (PIP2) generating phosphatidylinositol-3,4,5-trisphosphate (PIP3), at the plasma membrane. Phosphoinositide-dependent protein kinase-4 (PDK4) binds to PIP3, recruiting and phosphorylating, at tyrosine residues, the protein kinase B (PKB or AKT), which needs to be phosphorylated at serine residues by the mammalian target of rapamycin complex 2 (mTORC2) to reach its

Fig. 3 Insulin signaling pathway and the effects of aging on this process. Insulin binds to its receptor (IR) enhancing cell growth and proliferation via activation of the ERK/MAPK pathway. Insulin-IR complex also activates IRS/PI3K/AKT pathway that plays a crucial role in the glucose, protein and lipid metabolism. A brief description of these insulin signaling pathways is presented in the main text. The effects of aging on insulin signaling pathways is summarized in the boxes with the numbers indicating the age-targets proteins. Shc, SH2-containing adapter protein; GRB2, growth factor receptor-bound protein-2; SOS, son of sevenless; Ras, a GTP-binding protein; Raf, a MAPKKK; MEK, a MAPKK; ERK, a MAPK; IRS, insulin receptor substrates; PI3K, phosphoinositide 3-kinase; PIP2, phosphatidylinositol-4,5-bisphosphate; PIP3, phosphatidylinositol-3,4,5-trisphosphate; PDK4, phosphoinositide-dependent protein kinase-4; AKT or PKB, protein kinase B; TBC1D1/D4, TBC1 domain family member 1/4; GLUT4, glucose transporter-4; GSK3, glycogen synthase kinase-3; GS, glycogen synthase; FoxO1, forkhead box O1; AgRP, agouti-related protein; POMC, proopiomelanocortin; TSC1, tuberous sclerosis complex-1; mTORC1, mammalian target of rapamycin complex 1; S6K, protein S6 kinase; 4EBP1, eukaryotic translation initiation factor 4E binding protein 1; SREBP1, steroid regulatory element-binding protein-1c; PDE3B, phosphodiesterase 3B; cAMP, cyclic AMP; PKA, protein kinase A; HSL, homone-sensitive lipase; PKCζ, atypical protein kinase C ζ.

fully activity (Taniguchi, Emanuelli, & Kahn, 2006). This proximal IS pathway, is similar among insulin-responsive cells, but the phosphorylation of AKT, induces the activation of specific proteins from distal IS pathways, leading to distinct regulatory effects on carbohydrate, protein and lipid metabolism in the liver, skeletal muscle, adipose tissue and brain.

In the liver, AKT substrates includes glycogen synthase kinase-3 (GSK3, that regulates glycogenesis), transcription factor forkhead box O1 (FoxO1, that regulates gluconeogenic genes expression) and multiple regulators of mammalian target of rapamycin complex 1 (mTORC1) activity, which in turn upregulates protein synthesis and lipogenic genes expression. The stimulation of glycogenesis is the major physiological function of insulin in the liver. AKT phosphorylates and inactivates GSK3, a negative regulator of glycogen synthase (GS), stimulating glycogenesis. In addition, AKT also phosphorylates and activates the protein phosphatase 1 (PP1), which dephosphorylates and inactivates glycogen phosphorylase (GP), inhibiting glycogenolysis. It is important to know that insulin suppresses hepatic glucose production, not only by inhibiting glycogenolysis, but also by downregulating gluconeogenic gene expression via phosphorylation and inactivation of FoxO1 via AKT (Petersen & Shulman, 2018).

Skeletal muscle is responsible for approximately 70% of glucose uptake in the whole body (Ferrannini et al., 1985) and the postprandial insulin is the major stimulus for this process. AKT phosphorylates and inactivates proteins involved with glucose uptake, such as the AS160 (AKT substrate of 160 kDa), also known as TBC1D4, and the related TBC1D1. The inactivation of these proteins triggers the translocation of GLUT4-storage vesicles (GSV) to the plasma membrane (Jaldin-Fincati, Pavarotti, Frendo-Cumbo, Bilan, & Klip, 2017), allowing glucose to enter in the muscle cells. In addition to AKT, PI3K also phosphorylates and activates an atypical protein kinase C ζ (PKCζ), which contributes to the translocation of GSV, through the activation of exocytosis-related proteins, such as VAMP2 (Braiman et al., 2001).

Protein synthesis is also an important process stimulated by insulin, in the skeletal muscle. This effect is mediated via phosphorylation and inactivation of a tuberous sclerosis complex-1 (TSC1) by AKT. The inactivation of TSC inhibits its negative regulatory effect upon mTORC1 and this complex can phosphorylate the eukaryotic translation initiation factor 4E (eIF-4E)-binding protein 1 (4E-BP1). It inhibits 4E-BP1, allowing the recruitment of other eukaryotic translation initiation factors. The ribosomal protein S6 kinases (S6Ks) constitutes another important target of mTORC1.

S6Ks modulates the functions of eukaryotic translation initiation factors and coordinates the regulation of ribosome biogenesis, contributing to an efficient translation (Ma & Blenis, 2009).

In adipose tissue, insulin stimulates glucose uptake by similar mechanisms described for the skeletal muscle (Leney & Tavaré, 2009). The pyruvate, generated in the glycolysis, is directed to de novo fatty acid synthesis, which is also stimulated by PI3K/AKT/mTORC1 pathway. This pathway upregulates the steroid regulatory element-binding protein-1c (SREBP1c), a transcription factor that enhances the expression of several lipogenic genes, including fatty acid synthase (FAS) and acetyl-CoA carboxylase (ACC) (Caron, Richard, & Laplante, 2015). Also, insulin has antilipolytic effect that seems to be mediated by AKT dependent (Czech, Tencerova, Pedersen, & Aouadi, 2013) and independent (DiPilato et al., 2015) pathways. AKT dependent pathway is well-characterized and is mediated by phosphorylation and activation of phosphodiesterase 3B (PDE3B), which degrades intracellular cAMP, suppressing the allosteric activation of protein kinase A (PKA). This inhibits the phosphorylation of perilipin 1 (PLN1) and hormone-sensitive lipase (HSL), thereby suppressing lipolysis.

Metabolic function of insulin is not limited to its action on the liver, skeletal muscle and adipose tissue. Actually, this hormone also plays an important role on the regulation of energy metabolism through activation of its receptor in the hypothalamus (Obici, Feng, Karkanias, Baskin, & Rossetti, 2002). In this brain area, the activation of IS pathway modulates central and peripheral metabolic functions. Through PI3K/AKT pathway, insulin inactivates FoxO1 inhibiting the transcription of the agouti-related protein (AgRP), an orexigenic peptide, while enhances the transcription of proopiomelanocortin (POMC), an anorexigenic peptide (Kitamura et al., 2006), thereby regulating food intake and energy balance. Hepatic glucose production is a peripheral metabolic function regulated by insulin in the hypothalamus. Insulin acts mainly in the AgRP-expressing neurons, inhibiting electrical activity through the activation of K_{ATP} channels, via PI3K. It reduces the release of the AgRP, NPY and other neurotransmitters, regulating the innervation to the liver (through the vagus nerve), thereby suppressing hepatic glucose production (Könner et al., 2007).

Additionally, hippocampus, a critical brain area for learning and memory, is also an important target of insulin. Similar to peripheral tissues, insulin activates the IRS/PI3K/AKT pathway inducing the translocation of GSV to the plasma membrane. However, due to the higher expression of GLUT1 and GLUT3 in neuronal cells, the function of GLUT4 in the hippocampus remains to be determined. It has been suggested that the insulin-stimulated translocation of GSV contributes to a rapid fluctuation of glucose levels in the hippocampus that occurs during learning task (McNay, Fries, & Gold, 2000). Acute administration of physiological insulin concentration, enhances spatial working memory, via PI3K. When insulin signaling is blocked, in the hippocampus of rats, cognitive performance is impaired (McNay et al., 2010). In fact, insulin has been proposed as a regulatory factor of neuronal functions, not only by stimulating glucose uptake in the neurons, but also by modulating synaptic plasticity, neurotransmission and neurogenesis (Steculorum, Solas, & Brüning, 2014).

IR is expressed in many other tissues where insulin exerts important physiological functions. In the heart, besides regulating glucose uptake, insulin also induces long-chain fatty acid (LCFA) uptake by stimulating the translocation of the fatty acid transporter (FAT/CD36) to the plasma membrane via PI3K (Luiken et al., 2002). Similar to skeletal muscle, cardiac muscle atrophy is prevented by the activation of IRS/PI3K/AKT pathway. In the vascular endothelium, insulin increases oxide nitric (NO) production by enhancing endothelial nitric oxide synthase (eNOS) via AKT (Zeng et al., 2000). NO relaxes vascular smooth muscle, increasing the blood flow into the circulatory system. In the kidneys, insulin has an anti-natriuretic property, increasing sodium absorption by regulating the activities of specifics renal sodium channels (Horita et al., 2011). Considering these effects of insulin in the cardiovascular and renal systems, this hormone can be also a regulatory factor of blood pressure.

In summary, IS pathway has an important function in several tissues and impairments on this pathway can lead to deleterious effects on metabolic, neural and cardiovascular functions. Next, we will present how aging can impair insulin sensitivity.

4.2 Effects of aging on insulin sensitivity

Impaired IS pathway reduces insulin sensitivity leading to insulin resistance, which is defined as a subnormal biological response to a given insulin concentration (Moller & Flier, 1991). Insulin resistance is characterized by a dysfunctional IRS/PI3K/AKT pathway in the target tissues, turning them unable to coordinate a normal glucose-lowering response involving, mainly, glycogen synthesis, suppression of hepatic glucose production and cellular uptake of glucose. Thus, insulin resistance is considered a hallmark of metabolic diseases, such as obesity and T2DM (Czech, 2017), both age-related pathological conditions. Indeed, insulin resistance has been frequently described in elderly (Morley, 2008) and the deleterious effects on IS pathway that may occur with advancing age, are summarized in the boxes of Fig. 3.

Aging is closely related to changes in energy balance. Studies have shown that resting metabolic rate and total energy expenditure is declined in aged subjects (Poehlman et al., 1992). These effects may contribute to increase adiposity, which is frequently associated with aging (Speakman & Westerterp, 2010). Typically, fat mass increases between the third and seventh decades of life, and is redistributed throughout the body with advancing aging. While subcutaneous fat decreases, visceral fat increases (Folsom et al., 1993), elevating the risk for insulin resistance. Visceral fat accumulation has an important function in the maintenance of chronic low-grade inflammation by increasing circulant levels of pro-inflammatory cytokines (Huffman & Barzilai, 2009), which have important implications on the development of local and systemic insulin resistance. Impaired IS pathway in the visceral fat depots increases lipolysis, contributing to the release of large amount of free fatty acids (FFA) in the circulation. These fatty acids can be taken by other organs, such as liver and skeletal muscle, increasing the ectopic fat accumulation that contributes to insulin resistance by activating inflammatory pathways and increasing bioactive lipids, including diacylglycerol (DAG) and sphingolipids (Sears & Perry, 2015). Indeed, higher content of these two bioactive lipids was found in old compared with young obese subjects (Søgaard et al., 2019).

In the liver, DAG activates PKCε, an PKC isoform highly expressed in hepatocytes. This kinase phosphorylates IR, at specific threonine residue (Thr1160), impairing the tyrosine kinase activity of this receptor (Petersen et al., 2016). PKCε has been described as a critical mediator of fat-induced hepatic insulin resistance in nonalcoholic fatty liver disease, in rodents and humans (Jornayvaz & Shulman, 2012; Kumashiro et al., 2011), a disorder that is commonly associated with aging (Bertolotti et al., 2014). Similarly, in the skeletal muscle, DAG activates another novel PKC isoform, the PKCθ, which in turn phosphorylates IRS1, at specific serine residues (Ser1101), inhibiting its activation by IR (Li et al., 2004). High activity of PKCθ has been observed in skeletal muscle in obesity and T2DM (Szendroedi et al., 2014), both age-related diseases.

Sphingolipids, such as ceramides, also seems to mediate insulin resistance in the skeletal muscle from obese subjects (Adams et al., 2004). It was reported that ceramides activate PKCζ, a downstream effector of PI3K, that also phosphorylates IRS1 at several serine residues, as described in rodents (Lee, Lynn, Kim, & Quon, 2008). These phosphorylations impair IRS1/PI3K/AKT pathway, leading to insulin resistance.

Although increased adiposity is an important factor in the development of insulin resistance during aging, it has been demonstrated that lean aged subjects may be also insulin resistant compared with young subjects, matched for lean body mass and fat mass (Petersen et al., 2003). In this study, insulin resistance was associated with increased ectopic fat accumulation and decreased mitochondrial oxidative and phosphorylation activity, mainly in the skeletal muscle. It suggests that mitochondrial dysfunction predispose to age-related ectopic fat accumulation and insulin resistance in this tissue. In fact, reduction in the number and/or function of this organelle is tightly involved with aging process and its related disorders, such as insulin resistance (Shou, Chen, & Xiao, 2020).

In addition to ectopic fat accumulation, mitochondrial dysfunction also increases ROS production, causing oxidative damage in lipids, proteins and nucleic acids, mainly mitochondrial DNA (mtDNA). Oxidative damage in the mtDNA impairs the respiratory chain complexes creating a vicious cycle with an increase in ROS production and a subsequent accumulation of more mtDNA damages (Bonomini, Rodella, & Rezzani, 2015). During aging, mitochondrial dysfunction seems to occur in several tissues, such as skeletal muscle and liver (Trounce, Byrne, & Marzuki, 1989; Yen, Chen, King, Yeh, & Wei, 1989), inducing insulin resistance not only by increasing ectopic fat accumulation but also by increasing oxidative stress.

During the aging process, oxidative stress seems to occur mainly due to mitochondrial dysfunction; however, decrease in the capacity of antioxidant systems has been also reported in several tissues from aged rodents (Ji, 1993), and in the skeletal muscle from elderly subjects (Fulle et al., 2005), contributing to exacerbate ROS oxidative damages. At physiological levels, hydrogen peroxide (H_2O_2) facilitates the activation of IS pathway (Tiganis, 2011). However, high levels of ROS, as occurs with aging, activate inflammatory pathways that suppress the activation of IRS1/PI3K/AKT pathway (Gual, Le Marchand-Brustel, & Tanti, 2005). Also, oxidative stress may induce endoplasmic reticulum (ER) stress, which also impairs insulin sensitivity through activation of inflammatory pathways (Salvadó, Palomer, Barroso, & Vázquez-Carrera, 2015).

In the skeletal muscle, oxidative stress also inhibits protein synthesis and stimulates protein degradation contributing to the sarcopenia observed with advancing age (Szentesi, Csernoch, Dux, & Keller-Pintér, 2019). Since skeletal muscle represents the major site for insulin-stimulated glucose uptake (Ferrannini et al., 1985), the loss of this tissue represents an important mechanism to impair glucose-lowering effects of insulin, during aging.

Irrespective of the mechanism whereby aging impairs insulin sensitivity, it is clear that inflammation is a common process induced by visceral and ectopic fat accumulation, mitochondrial dysfunction, oxidative stress and ER stress. The chronic low-grade inflammation, commonly found in the elderly, contributes to the development of insulin resistance (Franceschi & Campisi, 2014). This low-grade inflammation is characterized by a chronic production of proinflammatory factors, notably tumor necrosis factor α (TNFα), interleukin-1β (IL-1 β) and interleukin-6 (IL-6) (Michaud et al., 2013).

TNFα and IL-1β act on the insulin-responsive cells through their specifics receptors activating the c-Jun NH2-terminal kinase (JNK) and the inhibitor of nuclear factor κB kinase (IKK). These kinases phosphorylate IRS1 at specifics serine residues suppressing the activation of PI3K/AKT pathway (Aguirre, Uchida, Yenush, Davis, & White, 2000; Gao et al., 2002). IKK also phosphorylates the inhibitor of nuclear factor κB (IκB) that dissociates from the nuclear factor κB (NF-κB), which in turn is translocated to the nucleus. The NF-κB activates the gene transcription of several proinflammatory mediators, including cytokines, chemokines, cycloxigenase-2 (COX-2) and the inducible oxide nitric synthase (iNOS) (Pahl, 1999). Enhanced iNOS transcription increases the synthesis of NO, which can covalently attach to cysteine residues of target proteins forming *S*-nitrosothiol adducts in a reversible posttranslational modification, termed S-nitrosation. Increased iNOS expression and S-nitrosation of the IR, IRS1 and AKT was observed in the skeletal muscle from insulin-resistant aged mice (28-month-old vs 3-month-old) (Ropelle et al., 2013). In this study, using iNOS-null mice, pharmacological inhibition of iNOS and acute exercise, which reduces iNOS expression, it was reported that each approach protects these aged mice against insulin resistance, providing evidence that iNOS-mediated protein S-nitrosation may be an important mechanism whereby aging impairs IS pathway.

Activation of JNK and IKK pathways are important mechanisms by which aging may impair insulin sensitivity. These inflammatory pathways are also activated by FFA, DAG, ceramides, ROS and ER stress (Shoelson, Lee, & Goldfine, 2006). Other proinflammatory cytokines that are increased in old age, can also activate JNK and IKK; however, these cytokines, in particular IL-6, induces insulin resistance mainly by increasing the expression of the suppressor of cytokine signaling (SOCS) through the activation of the Janus kinase (JAK)/signal transducer and activator of transcription (STAT) pathway (Kim, Bachmann, & Chen, 2009). It has been reported that SOCS binds to IR, thereby inhibiting its tyrosine kinase activity. Also, SOCS targets IRS1 and IRS2 to proteasomal degradation, reducing their protein levels (Howard & Flier, 2006). Therefore, IL-6 may be also an important mechanism that induces insulin resistance, during the aging process.

Considering the mechanisms whereby aging impairs insulin sensitivity, indeed, inflammation seems to be an important process that leads to insulin resistance in the liver, skeletal muscle, adipose tissue, brain and other insulin-target organs by impairing IRS/PI3K/AKT pathway, thereby leading to alterations in metabolic, neural and cardiovascular functions. It is important to keep in mind that the Ras/Raf/MEF/ERK pathway is not impaired, in fact it even increases with age (Williamson, Gallagher, Harber, Hollon, & Trappe, 2003) contributing to downregulation of IRS/PI3K/AKT pathway. It is known that ERK can phosphorylate IRS1 at specific serine residue (Ser616) inhibiting its activation by IR (De Fea & Roth, 1997). In addition, ERK also upregulates mTORC1/S6K pathway which in turn also inhibits the activation of IRS1 (Slack, 2017; Ueno et al., 2005). Thus, hyperactivity of mTORC1 is closely associated with insulin resistance-related diseases, such as T2DM and Alzheimer's disease (Bové, Martínez-Vicente, & Vila, 2011; Guillén & Benito, 2018).

Although it is widely known that aging is associated with insulin resistance, these findings are not universal. It was reported that centenarians (90–100 years-old) have a preserved insulin sensitivity compared with old-aged individuals (>65 years-old) (Barbieri et al., 2001), suggesting that maintenance of insulin sensitivity may be associated with a longer life. In the next section, we will discuss the importance of insulin action in the healthy longevity.

5. Insulin action in the healthy longevity

As described in this chapter, insulin action may be impaired through several mechanisms, during aging. Alterations in insulin secretion, clearance and sensitivity are commonly found in the elderly, leading to impairments on insulin action in the liver, skeletal muscle, adipose tissue, brain and other insulin-target organs. Defects on insulin action in these organs are strongly associated with various adverse outcomes that increase the risk for age-related diseases and death. Therefore, improving insulin action may be an effective strategy to achieve a healthier and longer life.

In addition to the studies of centenarians, which have a preserved insulin sensitivity (Barbieri et al., 2001), studies of familial longevity also demonstrate the association between the maintenance of insulin sensitivity and longevity. It was reported that offspring from long-lived siblings displayed lower circulant levels of glucose and insulin, which were associated with enhanced insulin sensitivity compared with their age-matched controls (Wijsman et al., 2011). It is suggested that lower insulinemia combined with an increased insulin sensitivity may extend the lifespan in humans.

In fact, hyperinsulinemia combined with insulin resistance tend to increase the risk for age-related diseases. Although it remains unclear which alteration comes first, during the aging process, it is hypothesized that hyperinsulinemia may trigger the insulin resistance, thereby shortening the lifespan. Genetic ablation of insulin gene (*Ins2*$^{+/-}$) in mice reduced the circulant levels of this hormone which was associated with a preserved insulin sensitivity compared to controls as they age (Templeman et al., 2017). Moreover, lowered insulinemia in the *Ins2*$^{+/-}$ mice induced a significant lifespan extension, suggesting that lower circulant insulin levels may be crucial to live a healthier and longer life.

During the aging process, insulin levels may be increased by an enhanced insulin secretion and/or decreased insulin clearance, as discussed in this chapter. Irrespective of the mechanism whereby hyperinsulinemia emerges with age, this higher insulin levels may impair insulin sensitivity. Thus, although insulin resistance may be associated with harmful effects of aging, it may be also a primarily protective mechanism against some dangerous threat to life, as previously hypothesized (Barzilai & Ferrucci, 2012).

It is well-documented in invertebrates that genetic downregulation of insulin/insulin-like growth factor-1 (IGF1) signaling (IIS) pathway extends lifespan. In *Caenorhabditis elegans*, mutations in *daf-2* and *age-1*, homologs to mammalian genes coding for insulin/IGF1 receptor and PI3K, respectively, leads to defects in the function of these IIS components, thereby increasing longevity in these nematodes (Friedman & Johnson, 1988; Kenyon, Chang, Gensch, Rudner, & Tabtiang, 1993). This effect on lifespan seems to occur by activating transcription factors that are inhibited by IIS pathway, such as Daf-16 (homolog to mammalian FoxO), which regulates genes involved with stress responses, enhancing the cellular resistance to external and internal stresses (Altintas, Park, & Lee, 2016). Since many targets of Daf-16/FoxO are conserved among species, this IIS-mediated longevity seems to occur in other invertebrates (*Drosophila melanogaster*) and mammals (Fontana, Partridge, & Longo, 2010; Selman et al., 2008).

It is difficult to separate the effects of insulin and IGF1 in the regulation of the longevity, since both hormones activate the IRS/PI3K/AKT pathway. However, using whole body insulin receptor knockout (IRKO) and fat-specific insulin receptor knockout (FIRKO) mice, it was concluded that insulin has a critical role in the regulation of longevity, since these mice displayed an increase in lifespan compared to wild-type mice (Blüher, Kahn, & Kahn, 2003; Nelson, Strong, Bokov, Diaz, & Ward, 2012). In particular, the long-lived FIRKO mice displayed a decreased fat mass, which was associated with an increased mitochondrial activity and decreased oxidative stress, in their adipose tissue, compared to controls as they age. Despite insulin resistance in the adipose tissue, whole body insulin sensitivity improves with age, compared to controls. Thus, fat-specific downregulation of insulin action may have an important function in the regulation of longevity (Katic et al., 2007).

Similar to FIRKO mice, dietary restriction in rodents also decreases fat mass while increases lifespan (Das, Gabriely, & Barzilai, 2004). However, in these animals, insulin sensitivity is increased in the whole body including adipose tissue. The mechanism by which dietary restriction increases lifespan is not entirely understood, but it is hypothesized that nutrients-sensing pathways may be involved. Indeed, under starved conditions AMP-activated protein kinase (AMPK) is activated, thereby inhibiting mTORC1 (Pan & Finkel, 2017). It is well-characterized that genetic or pharmacological inhibition of mTORC1 increases lifespan from yeast to mammals (Fontana et al., 2010). It suggests that the inhibition of mTORC1 increases lifespan by stimulating an important mechanism that protects cells from damages, namely autophagy (Pan & Finkel, 2017). In fact, this mechanism is decreased with advancing age and hyperactivity of mTORC1 is associated with age-related diseases including T2DM, Alzheimer's disease and cancer (Saxton & Sabatini, 2017). Therefore, inhibiting mTORC1 may be also an important strategy to increase lifespan.

It appears to be controversial that increased insulin sensitivity may extend lifespan, since insulin signaling activates mTORC1. However, it is important to keep in mind that to live a longer life, increased insulin sensitivity must be accompanied by a lower insulin levels, which have been found in centenarians, offspring from long-lived parents, $Ins^{+/-}$ mice, FIRKO and dietary restriction models. Thus, even with high insulin sensitivity, the lower circulant levels of insulin counteract the likely hyperactivity of IS pathway, preventing the negative effects of mTORC1 hyperactivation.

In summary, it is undeniable that insulin is an important regulator of longevity; however, it is not the only factor involved with aging. How important insulin is for longevity is not completely established, since IGF1 also activates IRS/PI3K/AKT pathway. Therefore, further investigations are necessary to clarify this matter, what certainly will contribute to the search for effective strategies to prevent the harmful effects of aging, leading to a healthier and longer life.

6. Conclusion

Aging may impair insulin action by affecting the three processes that determines the action of this hormone. Insulin secretion may be higher at the beginning, decreasing with age due to the decline in β-cell function and mass. Insulin clearance may also impair with advancing age, contributing to hyperinsulinemia commonly found in elderly. Finally, aging may induce insulin resistance by several mechanism, including hyperinsulinemia, thereby increasing the risk for age-related diseases and death. Therefore, protecting insulin secretion, clearance and sensitivity from the harmful effects of aging may maintain a lower insulinemia and a preserved insulin sensitivity, two keystones for a healthier and longer live.

Acknowledgments

The authors thanks to the São Paulo Research Foundation (FAPESP, grant numbers: 2015/12611-0, 2018/24368-0, 2018/15032-9 and 2019/00728-0) for financial support.

References

Abdul-Hay, S. O., Kang, D., McBride, M., Li, L., Zhao, J., & Leissring, M. A. (2011). Deletion of insulin-degrading enzyme elicits antipodal, age-dependent effects on glucose and insulin tolerance. *PLoS One*, *6*(6), e20818. https://doi.org/10.1371/journal.pone.0020818.

Abedini, A., & Schmidt, A. M. (2013). Mechanisms of islet amyloidosis toxicity in type 2 diabetes. *FEBS Letters*, *587*(8), 1119–1127. https://doi.org/10.1016/j.febslet.2013.01.017.

Adams, J. M., Pratipanawatr, T., Berria, R., Wang, E., DeFronzo, R. A., Sullards, M. C., et al. (2004). Ceramide content is increased in skeletal muscle from obese insulin-resistant humans. *Diabetes*, *53*(1), 25–31. https://doi.org/10.2337/diabetes.53.1.25.

Aguirre, V., Uchida, T., Yenush, L., Davis, R., & White, M. F. (2000). The c-Jun NH(2)-terminal kinase promotes insulin resistance during association with insulin receptor substrate-1 and phosphorylation of Ser(307). *The Journal of Biological Chemistry*, *275*(12), 9047–9054. https://doi.org/10.1074/jbc.275.12.9047.

Ahrén, B., & Pacini, G. (1998). Age-related reduction in glucose elimination is accompanied by reduced glucose effectiveness and increased hepatic insulin extraction in man. *The Journal of Clinical Endocrinology and Metabolism*, *83*(9), 3350–3356. https://doi.org/10.1210/jcem.83.9.5107.

Altintas, O., Park, S., & Lee, S. J. (2016). The role of insulin/IGF-1 signaling in the longevity of model invertebrates, *C. elegans* and *D. melanogaster*. *BMB Reports*, *49*(2), 81–92. https://doi.org/10.5483/bmbrep.2016.49.2.261.

Ammon, H. P., Amm, U., Eujen, R., Hoppe, E., Trier, G., & Verspohl, E. J. (1984). The role of old age in the effects of glucose on insulin secretion, pentosephosphate shunt activity, pyridine nucleotides and glutathione of rat pancreatic islets. *Life Sciences*, *34*(3), 247–257. https://doi.org/10.1016/0024-3205(84)90596-4.

Ammon, H. P., Fahmy, A., Mark, M., Wahl, M. A., & Youssif, N. (1987). The effect of glucose on insulin release and ion movements in isolated pancreatic islets of rats in old age. *The Journal of Physiology*, *384*, 347–354. https://doi.org/10.1113/jphysiol.1987.sp016458.

Authier, F., Rachubinski, R. A., Posner, B. I., & Bergeron, J. J. (1994). Endosomal proteolysis of insulin by an acidic thiol metalloprotease unrelated to insulin degrading enzyme. *The Journal of Biological Chemistry*, *269*(4), 3010–3016.

Azhar, S., Ho, H., Reaven, E. P., & Reaven, G. M. (1983). Evidence of an age-related decline in mitochondrial glycerophosphate dehydrogenase activity of isolated rat islets. *Metabolism*, *32*(11), 1019–1021. https://doi.org/10.1016/0026-0495(83)90070-7.

Barbieri, M., Rizzo, M. R., Manzella, D., & Paolisso, G. (2001). Age-related insulin resistance: Is it an obligatory finding? The lesson from healthy centenarians. *Diabetes/Metabolism Research and Reviews*, *17*(1), 19–26. https://doi.org/10.1002/dmrr.178.

Barzilai, N., & Ferrucci, L. (2012). Insulin resistance and aging: A cause or a protective response? *The Journals of Gerontology. Series A, Biological Sciences and Medical Sciences*, *67*(12), 1329–1331. https://doi.org/10.1093/gerona/gls145.

Belfiore, A., Malaguarnera, R., Vella, V., Lawrence, M. C., Sciacca, L., Frasca, F., et al. (2017). Insulin receptor isoforms in physiology and disease: An updated view. *Endocrine Reviews*, *38*(5), 379–431. https://doi.org/10.1210/er.2017-00073.

Bergman, R. N., Piccinini, F., Kabir, M., Kolka, C. M., & Ader, M. (2019). Hypothesis: Role of reduced hepatic insulin clearance in the pathogenesis of type 2 diabetes. *Diabetes*, *68*(9), 1709–1716. https://doi.org/10.2337/db19-0098.

Bertolotti, M., Lonardo, A., Mussi, C., Baldelli, E., Pellegrini, E., Ballestri, S., et al. (2014). Nonalcoholic fatty liver disease and aging: Epidemiology to management. *World Journal of Gastroenterology*, *20*(39), 14185–14204. https://doi.org/10.3748/wjg.v20.i39.14185.

Blüher, M., Kahn, B. B., & Kahn, C. R. (2003). Extended longevity in mice lacking the insulin receptor in adipose tissue. *Science*, *299*(5606), 572–574. https://doi.org/10.1126/science.1078223.

Bonadonna, R. C., Groop, L. C., Simonson, D. C., & DeFronzo, R. A. (1994). Free fatty acid and glucose metabolism in human aging: Evidence for operation of the Randle cycle. *The American Journal of Physiology*, *266*(3 Pt 1), E501–E509. https://doi.org/10.1152/ajpendo.1994.266.3.E501.

Bonomini, F., Rodella, L. F., & Rezzani, R. (2015). Metabolic syndrome, aging and involvement of oxidative stress. *Aging and Disease*, *6*(2), 109–120. https://doi.org/10.14336/AD.2014.0305.

Boschero, A. C., & Malaisse, W. J. (1978). Effect of glucose on 22Na+ efflux in pancreatic islets. *Archives Internationales de Physiologie et de Biochimie*, *86*(3), 479–485. https://doi.org/10.3109/13813457809055918.

Boucher, J., Kleinridders, A., & Kahn, C. R. (2014). Insulin receptor signaling in normal and insulin-resistant states. *Cold Spring Harbor Perspectives in Biology*, *6*(1). https://doi.org/10.1101/cshperspect.a009191.

Bourey, R. E., Kohrt, W. M., Kirwan, J. P., Staten, M. A., King, D. S., & Holloszy, J. O. (1993). Relationship between glucose tolerance and glucose-stimulated insulin response in 65-year-olds. *Journal of Gerontology*, *48*(4), M122–M127. https://doi.org/10.1093/geronj/48.4.m122.

Bové, J., Martínez-Vicente, M., & Vila, M. (2011). Fighting neurodegeneration with rapamycin: Mechanistic insights. *Nature Reviews. Neuroscience*, *12*(8), 437–452. https://doi.org/10.1038/nrn3068.

Braiman, L., Alt, A., Kuroki, T., Ohba, M., Bak, A., Tennenbaum, T., et al. (2001). Activation of protein kinase C zeta induces serine phosphorylation of VAMP2 in the GLUT4 compartment and increases glucose transport in skeletal muscle. *Molecular and Cellular Biology*, *21*(22), 7852–7861. https://doi.org/10.1128/MCB.21.22.7852-7861.2001.

Braun, M., Ramracheya, R., Bengtsson, M., Zhang, Q., Karanauskaite, J., Partridge, C., et al. (2008). Voltage-gated ion channels in human pancreatic beta-cells: Electrophysiological characterization and role in insulin secretion. *Diabetes*, *57*(6), 1618–1628. https://doi.org/10.2337/db07-0991.

Broughton, D. L., James, O. W., Alberti, K. G., & Taylor, R. (1991). Peripheral and hepatic insulin sensitivity in healthy elderly human subjects. *European Journal of Clinical Investigation*, *21*(1), 13–21. https://doi.org/10.1111/j.1365-2362.1991.tb01352.x.

Burchfiel, C. M., Sharp, D. S., Curb, J. D., Rodriguez, B. L., Abbott, R. D., Arakaki, R., et al. (1998). Hyperinsulinemia and cardiovascular disease in elderly men: The Honolulu heart program. *Arteriosclerosis, Thrombosis, and Vascular Biology*, *18*(3), 450–457. https://doi.org/10.1161/01.atv.18.3.450.

Caccamo, A., Oddo, S., Sugarman, M. C., Akbari, Y., & LaFerla, F. M. (2005). Age- and region-dependent alterations in Abeta-degrading enzymes: Implications for Abeta-induced disorders. *Neurobiology of Aging*, *26*(5), 645–654. https://doi.org/10.1016/j.neurobiolaging.2004.06.013.

Caron, A., Richard, D., & Laplante, M. (2015). The roles of mTOR complexes in lipid metabolism. *Annual Review of Nutrition*, *35*, 321–348. https://doi.org/10.1146/annurev-nutr-071714-034355.

Chang, A. M., & Halter, J. B. (2003). Aging and insulin secretion. *American Journal of Physiology. Endocrinology and Metabolism*, *284*(1), E7–12. https://doi.org/10.1152/ajpendo.00366.2002.

Chow, H. M., Shi, M., Cheng, A., Gao, Y., Chen, G., Song, X., et al. (2019). Age-related hyperinsulinemia leads to insulin resistance in neurons and cell-cycle-induced senescence. *Nature Neuroscience*, *22*(11), 1806–1819. https://doi.org/10.1038/s41593-019-0505-1.

Coordt, M. C., Ruhe, R. C., & McDonald, R. B. (1995). Aging and insulin secretion. *Proceedings of the Society for Experimental Biology and Medicine*, *209*(3), 213–222. https://doi.org/10.3181/00379727-209-43879b.

Czech, M. P. (2017). Insulin action and resistance in obesity and type 2 diabetes. *Nature Medicine*, *23*(7), 804–814. https://doi.org/10.1038/nm.4350.

Czech, M. P., Tencerova, M., Pedersen, D. J., & Aouadi, M. (2013). Insulin signalling mechanisms for triacylglycerol storage. *Diabetologia*, *56*(5), 949–964. https://doi.org/10.1007/s00125-013-2869-1.

Dai, D. F., Chiao, Y. A., Marcinek, D. J., Szeto, H. H., & Rabinovitch, P. S. (2014). Mitochondrial oxidative stress in aging and healthspan. *Longevity & Healthspan*, *3*, 6. https://doi.org/10.1186/2046-2395-3-6.

Das, M., Gabriely, I., & Barzilai, N. (2004). Caloric restriction, body fat and ageing in experimental models. *Obesity Reviews*, *5*(1), 13–19. https://doi.org/10.1111/j.1467-789x.2004.00115.x.

De Fea, K., & Roth, R. A. (1997). Modulation of insulin receptor substrate-1 tyrosine phosphorylation and function by mitogen-activated protein kinase. *The Journal of Biological Chemistry*, *272*(50), 31400–31406. https://doi.org/10.1074/jbc.272.50.31400.

De Tata, V. (2014). Age-related impairment of pancreatic Beta-cell function: Pathophysiological and cellular mechanisms. *Frontiers in Endocrinology*, *5*, 138. https://doi.org/10.3389/fendo.2014.00138.

DiPilato, L. M., Ahmad, F., Harms, M., Seale, P., Manganiello, V., & Birnbaum, M. J. (2015). The role of PDE3B phosphorylation in the inhibition of lipolysis by insulin. *Molecular and Cellular Biology*, *35*(16), 2752–2760. https://doi.org/10.1128/MCB.00422-15.

Duckworth, W. C., Bennett, R. G., & Hamel, F. G. (1998). Insulin degradation: Progress and potential. *Endocrine Reviews*, *19*(5), 608–624.

Ehrhardt, N., Cui, J., Dagdeviren, S., Saengnipanthkul, S., Goodridge, H. S., Kim, J. K., et al. (2019). Adiposity-independent effects of aging on insulin sensitivity and clearance in mice and humans. *Obesity (Silver Spring)*, *27*(3), 434–443. https://doi.org/10.1002/oby.22418.

Elahi, D., Muller, D. C., Andersen, D. K., Tobin, J. D., & Andres, R. (1985). The effect of age and glucose concentration on insulin secretion by the isolated perfused rat pancreas. *Endocrinology, 116*(1), 11–16. https://doi.org/10.1210/endo-116-1-11.

Elahi, D., Muller, D. C., McAloon-Dyke, M., Tobin, J. D., & Andres, R. (1993). The effect of age on insulin response and glucose utilization during four hyperglycemic plateaus. *Experimental Gerontology, 28*(4–5), 393–409. https://doi.org/10.1016/0531-5565(93)90066-m.

Erdmann, J., Kallabis, B., Oppel, U., Sypchenko, O., Wagenpfeil, S., & Schusdziarra, V. (2008). Development of hyperinsulinemia and insulin resistance during the early stage of weight gain. *American Journal of Physiology. Endocrinology and Metabolism, 294*(3), E568–E575. https://doi.org/10.1152/ajpendo.00560.2007.

Farris, W., Mansourian, S., Chang, Y., Lindsley, L., Eckman, E. A., Frosch, M. P., et al. (2003). Insulin-degrading enzyme regulates the levels of insulin, amyloid beta-protein, and the beta-amyloid precursor protein intracellular domain in vivo. *Proceedings of the National Academy of Sciences of the United States of America, 100*(7), 4162–4167. https://doi.org/10.1073/pnas.0230450100.

Ferrannini, E., Bjorkman, O., Reichard, G. A., Pilo, A., Olsson, M., Wahren, J., et al. (1985). The disposal of an oral glucose load in healthy subjects. A quantitative study. *Diabetes, 34*(6), 580–588. https://doi.org/10.2337/diab.34.6.580.

Fink, R. I., Kolterman, O. G., Griffin, J., & Olefsky, J. M. (1983). Mechanisms of insulin resistance in aging. *The Journal of Clinical Investigation, 71*(6), 1523–1535. https://doi.org/10.1172/jci110908.

Fink, R. I., Kolterman, O. G., & Olefsky, J. M. (1984). The physiological significance of the glucose intolerance of aging. *Journal of Gerontology, 39*(3), 273–278. https://doi.org/10.1093/geronj/39.3.273.

Fink, R. I., Revers, R. R., Kolterman, O. G., & Olefsky, J. M. (1985). The metabolic clearance of insulin and the feedback inhibition of insulin secretion are altered with aging. *Diabetes, 34*(3), 275–280.

Folsom, A. R., Kaye, S. A., Sellers, T. A., Hong, C. P., Cerhan, J. R., Potter, J. D., et al. (1993). Body fat distribution and 5-year risk of death in older women. *JAMA, 269*(4), 483–487.

Fontana, L., Partridge, L., & Longo, V. D. (2010). Extending healthy life span—From yeast to humans. *Science, 328*(5976), 321–326. https://doi.org/10.1126/science.1172539.

Franceschi, C., & Campisi, J. (2014). Chronic inflammation (inflammaging) and its potential contribution to age-associated diseases. *The Journals of Gerontology. Series A, Biological Sciences and Medical Sciences, 69*(Suppl 1), S4–S9. https://doi.org/10.1093/gerona/glu057.

Friedman, D. B., & Johnson, T. E. (1988). A mutation in the age-1 gene in Caenorhabditis elegans lengthens life and reduces hermaphrodite fertility. *Genetics, 118*(1), 75–86.

Fulle, S., Di Donna, S., Puglielli, C., Pietrangelo, T., Beccafico, S., Bellomo, R., et al. (2005). Age-dependent imbalance of the antioxidative system in human satellite cells. *Experimental Gerontology, 40*(3), 189–197. https://doi.org/10.1016/j.exger.2004.11.006.

Gao, Z., Hwang, D., Bataille, F., Lefevre, M., York, D., Quon, M. J., et al. (2002). Serine phosphorylation of insulin receptor substrate 1 by inhibitor kappa B kinase complex. *The Journal of Biological Chemistry, 277*(50), 48115–48121. https://doi.org/10.1074/jbc.M209459200.

Garcia, G. V., Freeman, R. V., Supiano, M. A., Smith, M. J., Galecki, A. T., & Halter, J. B. (1997). Glucose metabolism in older adults: A study including subjects more than 80 years of age. *Journal of the American Geriatrics Society, 45*(7), 813–817. https://doi.org/10.1111/j.1532-5415.1997.tb01507.x.

Grasso, G., Satriano, C., & Milardi, D. (2015). A neglected modulator of insulin-degrading enzyme activity and conformation: The pH. *Biophysical Chemistry, 203–204*, 33–40. https://doi.org/10.1016/j.bpc.2015.05.010.

Gu, Z., Du, Y., Liu, Y., Ma, L., Li, L., Gong, Y., et al. (2012). Effect of aging on islet beta-cell function and its mechanisms in Wistar rats. *Age (Dordrecht, Netherlands), 34*(6), 1393–1403. https://doi.org/10.1007/s11357-011-9312-7.

Gual, P., Le Marchand-Brustel, Y., & Tanti, J. F. (2005). Positive and negative regulation of insulin signaling through IRS-1 phosphorylation. *Biochimie, 87*(1), 99–109. https://doi.org/10.1016/j.biochi.2004.10.019.

Guillén, C., & Benito, M. (2018). mTORC1 Overactivation as a key aging factor in the progression to type 2 diabetes mellitus. *Frontiers in Endocrinology, 9*, 621. https://doi.org/10.3389/fendo.2018.00621.

Gumbiner, B., Polonsky, K. S., Beltz, W. F., Wallace, P., Brechtel, G., & Fink, R. I. (1989). Effects of aging on insulin secretion. *Diabetes, 38*(12), 1549–1556.

Heinrich, G., Muturi, H. T., Rezaei, K., Al-Share, Q. Y., DeAngelis, A. M., Bowman, T. A., et al. (2017). Reduced hepatic carcinoembryonic antigen-related cell adhesion molecule 1 level in obesity. *Frontiers in Endocrinology, 8*, 54. https://doi.org/10.3389/fendo.2017.00054.

Horita, S., Seki, G., Yamada, H., Suzuki, M., Koike, K., & Fujita, T. (2011). Insulin resistance, obesity, hypertension, and renal sodium transport. *International Journal of Hypertension, 2011*, 391762. https://doi.org/10.4061/2011/391762.

Horst, A. K., Najjar, S. M., Wagener, C., & Tiegs, G. (2018). CEACAM1 in liver injury, metabolic and immune regulation. *International Journal of Molecular Sciences, 19*(10), 3110. https://doi.org/10.3390/ijms19103110.

Howard, J. K., & Flier, J. S. (2006). Attenuation of leptin and insulin signaling by SOCS proteins. *Trends in Endocrinology and Metabolism, 17*(9), 365–371. https://doi.org/10.1016/j.tem.2006.09.007.

Huffman, D. M., & Barzilai, N. (2009). Role of visceral adipose tissue in aging. *Biochimica et Biophysica Acta, 1790*(10), 1117–1123. https://doi.org/10.1016/j.bbagen.2009.01.008.

Hulse, R. E., Ralat, L. A., & Wei-Jen, T. (2009). Structure, function, and regulation of insulin-degrading enzyme. *Vitamins and Hormones, 80*, 635–648. https://doi.org/10.1016/S0083-6729(08)00622-5.

Hwang, D. Y., Cho, J. S., Kim, C. K., Shim, S. B., Jee, S. W., Lee, S. H., et al. (2005). Aging-related correlation of insulin-degrading enzyme with gamma-secretase-generated products involving insulin and glucose levels in transgenic mice. *Neurochemical Research, 30*(9), 1171–1177. https://doi.org/10.1007/s11064-005-7952-7.

Iozzo, P., Beck-Nielsen, H., Laakso, M., Smith, U., Yki-Järvinen, H., & Ferrannini, E. (1999). Independent influence of age on basal insulin secretion in nondiabetic humans. European Group for the Study of insulin resistance. *The Journal of Clinical Endocrinology and Metabolism, 84*(3), 863–868. https://doi.org/10.1210/jcem.84.3.5542.

Jacobsson, G., Bean, A. J., Scheller, R. H., Juntti-Berggren, L., Deeney, J. T., Berggren, P. O., et al. (1994). Identification of synaptic proteins and their isoform mRNAs in compartments of pancreatic endocrine cells. *Proceedings of the National Academy of Sciences of the United States of America, 91*(26), 12487–12491. https://doi.org/10.1073/pnas.91.26.12487.

Jaldin-Fincati, J. R., Pavarotti, M., Frendo-Cumbo, S., Bilan, P. J., & Klip, A. (2017). Update on GLUT4 vesicle traffic: A cornerstone of insulin action. *Trends in Endocrinology and Metabolism, 28*(8), 597–611. https://doi.org/10.1016/j.tem.2017.05.002.

Ji, L. L. (1993). Antioxidant enzyme response to exercise and aging. *Medicine and Science in Sports and Exercise, 25*(2), 225–231.

Jornayvaz, F. R., & Shulman, G. I. (2012). Diacylglycerol activation of protein kinase Cε and hepatic insulin resistance. *Cell Metabolism, 15*(5), 574–584. https://doi.org/10.1016/j.cmet.2012.03.005.

Kalyani, R. R., & Egan, J. M. (2013). Diabetes and altered glucose metabolism with aging. *Endocrinology and Metabolism Clinics of North America, 42*(2), 333–347. https://doi.org/10.1016/j.ecl.2013.02.010.

Katic, M., Kennedy, A. R., Leykin, I., Norris, A., McGettrick, A., Gesta, S., et al. (2007). Mitochondrial gene expression and increased oxidative metabolism: Role in increased lifespan of fat-specific insulin receptor knock-out mice. *Aging Cell*, *6*(6), 827–839. https://doi.org/10.1111/j.1474-9726.2007.00346.x.

Kenyon, C., Chang, J., Gensch, E., Rudner, A., & Tabtiang, R. (1993). A C. elegans mutant that lives twice as long as wild type. *Nature*, *366*(6454), 461–464. https://doi.org/10.1038/366461a0.

Kim, J. H., Bachmann, R. A., & Chen, J. (2009). Interleukin-6 and insulin resistance. *Vitamins and Hormones*, *80*, 613–633. https://doi.org/10.1016/S0083-6729(08)00621-3.

Kitamura, T., Feng, Y., Kitamura, Y. I., Chua, S. C., Xu, A. W., Barsh, G. S., et al. (2006). Forkhead protein FoxO1 mediates Agrp-dependent effects of leptin on food intake. *Nature Medicine*, *12*(5), 534–540. https://doi.org/10.1038/nm1392.

Kochkina, E. G., Plesneva, S. A., Vasilev, D. S., Zhuravin, I. A., Turner, A. J., & Nalivaeva, N. N. (2015). Effects of ageing and experimental diabetes on insulin-degrading enzyme expression in male rat tissues. *Biogerontology*, *16*(4), 473–484. https://doi.org/10.1007/s10522-015-9569-9.

Könner, A. C., Janoschek, R., Plum, L., Jordan, S. D., Rother, E., Ma, X., et al. (2007). Insulin action in AgRP-expressing neurons is required for suppression of hepatic glucose production. *Cell Metabolism*, *5*(6), 438–449. https://doi.org/10.1016/j.cmet.2007.05.004.

Kramer, J., Moeller, E. L., Hachey, A., Mansfield, K. G., & Wachtman, L. M. (2009). Differential expression of GLUT2 in pancreatic islets and kidneys of new and old world nonhuman primates. *American Journal of Physiology. Regulatory, Integrative and Comparative Physiology*, *296*(3), R786–R793. https://doi.org/10.1152/ajpregu.90694.2008.

Krishnamurthy, J., Torrice, C., Ramsey, M. R., Kovalev, G. I., Al-Regaiey, K., Su, L., et al. (2004). Ink4a/Arf expression is a biomarker of aging. *The Journal of Clinical Investigation*, *114*(9), 1299–1307. https://doi.org/10.1172/JCI22475.

Kumashiro, N., Erion, D. M., Zhang, D., Kahn, M., Beddow, S. A., Chu, X., et al. (2011). Cellular mechanism of insulin resistance in nonalcoholic fatty liver disease. *Proceedings of the National Academy of Sciences of the United States of America*, *108*(39), 16381–16385. https://doi.org/10.1073/pnas.1113359108.

Kurauti, M. A., Ferreira, S. M., Soares, G. M., Vettorazzi, J. F., Carneiro, E. M., Boschero, A. C., et al. (2018). Hyperinsulinemia is associated with increasing insulin secretion but not with decreasing insulin clearance in an age-related metabolic dysfunction mice model. *Journal of Cellular Physiology*, *234*(6), 9802–9809. https://doi.org/10.1002/jcp.27667.

Kurochkin, I. V., Guarnera, E., & Berezovsky, I. N. (2018). Insulin-degrading enzyme in the fight against Alzheimer's disease. *Trends in Pharmacological Sciences*, *39*(1), 49–58. https://doi.org/10.1016/j.tips.2017.10.008.

Lee, W. (2011). The CEACAM1 expression is decreased in the liver of severely obese patients with or without diabetes. *Diagnostic Pathology*, *6*, 40. https://doi.org/10.1186/1746-1596-6-40.

Lee, S., Lynn, E. G., Kim, J. A., & Quon, M. J. (2008). Protein kinase C-zeta phosphorylates insulin receptor substrate-1, -3, and -4 but not -2: Isoform specific determinants of specificity in insulin signaling. *Endocrinology*, *149*(5), 2451–2458. https://doi.org/10.1210/en.2007-1595.

Lemmon, M. A., & Schlessinger, J. (2010). Cell signaling by receptor tyrosine kinases. *Cell*, *141*(7), 1117–1134. https://doi.org/10.1016/j.cell.2010.06.011.

Leney, S. E., & Tavaré, J. M. (2009). The molecular basis of insulin-stimulated glucose uptake: Signalling, trafficking and potential drug targets. *The Journal of Endocrinology*, *203*(1), 1–18. https://doi.org/10.1677/JOE-09-0037.

Li, Y., Soos, T. J., Li, X., Wu, J., Degennaro, M., Sun, X., et al. (2004). Protein kinase C theta inhibits insulin signaling by phosphorylating IRS1 at Ser(1101). *The Journal of Biological Chemistry*, *279*(44), 45304–45307. https://doi.org/10.1074/jbc.C400186200.

Luiken, J. J., Koonen, D. P., Willems, J., Zorzano, A., Becker, C., Fischer, Y., et al. (2002). Insulin stimulates long-chain fatty acid utilization by rat cardiac myocytes through cellular redistribution of FAT/CD36. *Diabetes*, *51*(10), 3113–3119. https://doi.org/10.2337/diabetes.51.10.3113.

Ma, X. M., & Blenis, J. (2009). Molecular mechanisms of mTOR-mediated translational control. *Nature Reviews. Molecular Cell Biology*, *10*(5), 307–318. https://doi.org/10.1038/nrm2672.

Malaisse, W. J., Boschero, A. C., Kawazu, S., & Hutton, J. C. (1978). The stimulus secretion coupling of glucose-induced insulin release. XXVII. Effect of glucose on K+ fluxes in isolated islets. *Pflügers Archiv*, *373*(3), 237–242. https://doi.org/10.1007/bf00580830.

Marini, M. A., Frontoni, S., Succurro, E., Arturi, F., Fiorentino, T. V., Sciacqua, A., et al. (2014). Differences in insulin clearance between metabolically healthy and unhealthy obese subjects. *Acta Diabetologica*, *51*(2), 257–261. https://doi.org/10.1007/s00592-013-0511-9.

Matveyenko, A. V., Veldhuis, J. D., & Butler, P. C. (2008). Adaptations in pulsatile insulin secretion, hepatic insulin clearance, and beta-cell mass to age-related insulin resistance in rats. *American Journal of Physiology. Endocrinology and Metabolism*, *295*(4), E832–E841. https://doi.org/10.1152/ajpendo.90451.2008.

Maxson, M. E., & Grinstein, S. (2014). The vacuolar-type H^+-ATPase at a glance—More than a proton pump. *Journal of Cell Science*, *127*(Pt 23), 4987–4993. https://doi.org/10.1242/jcs.158550.

McCulloch, L. J., van de Bunt, M., Braun, M., Frayn, K. N., Clark, A., & Gloyn, A. L. (2011). GLUT2 (SLC2A2) is not the principal glucose transporter in human pancreatic beta cells: Implications for understanding genetic association signals at this locus. *Molecular Genetics and Metabolism*, *104*(4), 648–653. https://doi.org/10.1016/j.ymgme.2011.08.026.

McNay, E. C., Fries, T. M., & Gold, P. E. (2000). Decreases in rat extracellular hippocampal glucose concentration associated with cognitive demand during a spatial task. *Proceedings of the National Academy of Sciences of the United States of America*, *97*(6), 2881–2885. https://doi.org/10.1073/pnas.050583697.

McNay, E. C., Ong, C. T., McCrimmon, R. J., Cresswell, J., Bogan, J. S., & Sherwin, R. S. (2010). Hippocampal memory processes are modulated by insulin and high-fat-induced insulin resistance. *Neurobiology of Learning and Memory*, *93*(4), 546–553. https://doi.org/10.1016/j.nlm.2010.02.002.

Mergenthaler, P., Lindauer, U., Dienel, G. A., & Meisel, A. (2013). Sugar for the brain: The role of glucose in physiological and pathological brain function. *Trends in Neurosciences*, *36*(10), 587–597. https://doi.org/10.1016/j.tins.2013.07.001.

Michaud, M., Balardy, L., Moulis, G., Gaudin, C., Peyrot, C., Vellas, B., et al. (2013). Proinflammatory cytokines, aging, and age-related diseases. *Journal of the American Medical Directors Association*, *14*(12), 877–882. https://doi.org/10.1016/j.jamda.2013.05.009.

Minaker, K. L., Rowe, J. W., Tonino, R., & Pallotta, J. A. (1982). Influence of age on clearance of insulin in man. *Diabetes*, *31*(10), 851–855. https://doi.org/10.2337/diab.31.10.851.

Mirsky, I. A., & Broh-Kahn, R. H. (1949). The inactivation of insulin by tissue extracts; the distribution and properties of insulin inactivating extracts. *Archives of Biochemistry*, *20*(1), 1–9.

Mohamad, M., Mitchell, S. J., Wu, L. E., White, M. Y., Cordwell, S. J., Mach, J., et al. (2016). Ultrastructure of the liver microcirculation influences hepatic and systemic insulin activity and provides a mechanism for age-related insulin resistance. *Aging Cell*, *15*(4), 706–715. https://doi.org/10.1111/acel.12481.

Moller, D. E., & Flier, J. S. (1991). Insulin resistance—Mechanisms, syndromes, and implications. *The New England Journal of Medicine*, *325*(13), 938–948. https://doi.org/10.1056/NEJM199109263251307.

Møller, N., Gormsen, L., Fuglsang, J., & Gjedsted, J. (2003). Effects of ageing on insulin secretion and action. *Hormone Research*, *60*(Suppl. 1), 102–104. https://doi.org/10.1159/000071233.

Morley, J. E. (2008). Diabetes and aging: Epidemiologic overview. *Clinics in Geriatric Medicine*, *24*(3), 395–405. v https://doi.org/10.1016/j.cger.2008.03.005.

Muller, D. C., Elahi, D., Tobin, J. D., & Andres, R. (1996). Insulin response during the oral glucose tolerance test: The role of age, sex, body fat and the pattern of fat distribution. *Aging (Milano)*, *8*(1), 13–21. https://doi.org/10.1007/bf03340110.

Muzumdar, R., Ma, X., Atzmon, G., Vuguin, P., Yang, X., & Barzilai, N. (2004). Decrease in glucose-stimulated insulin secretion with aging is independent of insulin action. *Diabetes*, *53*(2), 441–446.

Najjar, S. M., & Perdomo, G. (2019). Hepatic insulin clearance: Mechanism and physiology. *Physiology (Bethesda)*, *34*(3), 198–215. https://doi.org/10.1152/physiol.00048.2018.

Najjar, S. M., Philippe, N., Suzuki, Y., Ignacio, G. A., Formisano, P., Accili, D., et al. (1995). Insulin-stimulated phosphorylation of recombinant pp120/HA4, an endogenous substrate of the insulin receptor tyrosine kinase. *Biochemistry*, *34*(29), 9341–9349. https://doi.org/10.1021/bi00029a009.

Nelson, J. F., Strong, R., Bokov, A., Diaz, V., & Ward, W. (2012). Probing the relationship between insulin sensitivity and longevity using genetically modified mice. *The Journals of Gerontology. Series A, Biological Sciences and Medical Sciences*, *67*(12), 1332–1338. https://doi.org/10.1093/gerona/gls199.

Nolan, S., Stephan, T., Chae, S., Vidalon, C., Gegick, C., Khurana, R. C., et al. (1973). Age-related insulin patterns in normal glucose tolerance. *Journal of the American Geriatrics Society*, *21*(3), 106–111. https://doi.org/10.1111/j.1532-5415.1973.tb00854.x.

Obici, S., Feng, Z., Karkanias, G., Baskin, D. G., & Rossetti, L. (2002). Decreasing hypothalamic insulin receptors causes hyperphagia and insulin resistance in rats. *Nature Neuroscience*, *5*(6), 566–572. https://doi.org/10.1038/nn0602-861.

Ohneda, M., Johnson, J. H., Inman, L. R., Chen, L., Suzuki, K., Goto, Y., et al. (1993). GLUT2 expression and function in beta-cells of GK rats with NIDDM. Dissociation between reductions in glucose transport and glucose-stimulated insulin secretion. *Diabetes*, *42*(7), 1065–1072. https://doi.org/10.2337/diab.42.7.1065.

Okabayashi, Y., Sugimoto, Y., Totty, N. F., Hsuan, J., Kido, Y., Sakaguchi, K., et al. (1996). Interaction of Shc with adaptor protein adaptins. *The Journal of Biological Chemistry*, *271*(9), 5265–5269. https://doi.org/10.1074/jbc.271.9.5265.

Pacini, G., Beccaro, F., Valerio, A., Nosadini, R., & Crepaldi, G. (1990). Reduced beta-cell secretion and insulin hepatic extraction in healthy elderly subjects. *Journal of the American Geriatrics Society*, *38*(12), 1283–1289. https://doi.org/10.1111/j.1532-5415.1990.tb03449.x.

Pahl, H. L. (1999). Activators and target genes of Rel/NF-kappaB transcription factors. *Oncogene*, *18*(49), 6853–6866. https://doi.org/10.1038/sj.onc.1203239.

Pan, H., & Finkel, T. (2017). Key proteins and pathways that regulate lifespan. *The Journal of Biological Chemistry*, *292*(16), 6452–6460. https://doi.org/10.1074/jbc.R116.771915.

Patterson, J. N., Cousteils, K., Lou, J. W., Manning Fox, J. E., MacDonald, P. E., & Joseph, J. W. (2014). Mitochondrial metabolism of pyruvate is essential for regulating glucose-stimulated insulin secretion. *The Journal of Biological Chemistry*, *289*(19), 13335–13346. https://doi.org/10.1074/jbc.M113.521666.

Petersen, K. F., Befroy, D., Dufour, S., Dziura, J., Ariyan, C., Rothman, D. L., et al. (2003). Mitochondrial dysfunction in the elderly: Possible role in insulin resistance. *Science*, *300*(5622), 1140–1142. https://doi.org/10.1126/science.1082889.

Petersen, M. C., Madiraju, A. K., Gassaway, B. M., Marcel, M., Nasiri, A. R., Butrico, G., et al. (2016). Insulin receptor Thr1160 phosphorylation mediates lipid-induced hepatic insulin resistance. *The Journal of Clinical Investigation*, *126*(11), 4361–4371. https://doi.org/10.1172/JCI86013.

Petersen, M. C., & Shulman, G. I. (2018). Mechanisms of insulin action and insulin resistance. *Physiological Reviews*, *98*(4), 2133–2223. https://doi.org/10.1152/physrev.00063.2017.

Poehlman, E. T., Berke, E. M., Joseph, J. R., Gardner, A. W., Katzman-Rooks, S. M., & Goran, M. I. (1992). Influence of aerobic capacity, body composition, and thyroid hormones on the age-related decline in resting metabolic rate. *Metabolism*, *41*(8), 915–921. https://doi.org/10.1016/0026-0495(92)90177-c.

Polonsky, K. S., Given, B. D., Hirsch, L., Shapiro, E. T., Tillil, H., Beebe, C., et al. (1988). Quantitative study of insulin secretion and clearance in normal and obese subjects. *The Journal of Clinical Investigation*, *81*(2), 435–441. https://doi.org/10.1172/JCI113338.

Primeau, V., Coderre, L., Karelis, A. D., Brochu, M., Lavoie, M. E., Messier, V., et al. (2011). Characterizing the profile of obese patients who are metabolically healthy. *International Journal of Obesity*, *35*(7), 971–981. https://doi.org/10.1038/ijo.2010.216.

Pyörälä, M., Miettinen, H., Laakso, M., & Pyörälä, K. (1998). Hyperinsulinemia and the risk of stroke in healthy middle-aged men: The 22-year follow-up results of the Helsinki policemen study. *Stroke*, *29*(9), 1860–1866. https://doi.org/10.1161/01.str.29.9.1860.

Rankin, M. M., & Kushner, J. A. (2009). Adaptive beta-cell proliferation is severely restricted with advanced age. *Diabetes*, *58*(6), 1365–1372. https://doi.org/10.2337/db08-1198.

Reaven, G. M., & Reaven, P. D. (1980). Effect of age on glucose oxidation by isolated rat islets. *Diabetologia*, *18*(1), 69–71. https://doi.org/10.1007/bf01228306.

Reers, C., Erbel, S., Esposito, I., Schmied, B., Büchler, M. W., Nawroth, P. P., et al. (2009). Impaired islet turnover in human donor pancreata with aging. *European Journal of Endocrinology*, *160*(2), 185–191. https://doi.org/10.1530/EJE-08-0596.

Ribeiro, R. A., Batista, T. M., Coelho, F. M., Boschero, A. C., Lopes, G. S., & Carneiro, E. M. (2012). Decreased β-cell insulin secretory function in aged rats due to impaired Ca(2+) handling. *Experimental Physiology*, *97*(9), 1065–1073. https://doi.org/10.1113/expphysiol.2012.064790.

Ropelle, E. R., Pauli, J. R., Cintra, D. E., da Silva, A. S., De Souza, C. T., Guadagnini, D., et al. (2013). Targeted disruption of inducible nitric oxide synthase protects against aging, S-nitrosation, and insulin resistance in muscle of male mice. *Diabetes*, *62*(2), 466–470. https://doi.org/10.2337/db12-0339.

Rorsman, P., Braun, M., & Zhang, Q. (2012). Regulation of calcium in pancreatic α- and β-cells in health and disease. *Cell Calcium*, *51*(3–4), 300–308. https://doi.org/10.1016/j.ceca.2011.11.006.

Ruhe, R. C., Curry, D. L., & McDonald, R. B. (1993). Stimulus-secretion coupling and insulin secretion in aging: Evaluation of glucose oxidation and ATP-sensitive potassium channel responsiveness. *Mechanisms of Ageing and Development*, *71*(3), 161–167. https://doi.org/10.1016/0047-6374(93)90081-2.

Salvadó, L., Palomer, X., Barroso, E., & Vázquez-Carrera, M. (2015). Targeting endoplasmic reticulum stress in insulin resistance. *Trends in Endocrinology and Metabolism*, *26*(8), 438–448. https://doi.org/10.1016/j.tem.2015.05.007.

Sartin, J. L., Chaudhuri, M., Farina, S., & Adelman, R. C. (1986). Regulation of insulin secretion by glucose during aging. *Journal of Gerontology*, *41*(1), 30–35. https://doi.org/10.1093/geronj/41.1.30.

Saxton, R. A., & Sabatini, D. M. (2017). mTOR signaling in growth, metabolism, and disease. *Cell*, *169*(2), 361–371. https://doi.org/10.1016/j.cell.2017.03.035.

Schmitz, O., Rungby, J., Edge, L., & Juhl, C. B. (2008). On high-frequency insulin oscillations. *Ageing Research Reviews*, 7(4), 301–305. https://doi.org/10.1016/j.arr.2008.04.002.

Sears, B., & Perry, M. (2015). The role of fatty acids in insulin resistance. *Lipids in Health and Disease*, *14*, 121. https://doi.org/10.1186/s12944-015-0123-1.

Selkoe, D. J., & Hardy, J. (2016). The amyloid hypothesis of Alzheimer's disease at 25 years. *EMBO Molecular Medicine*, *8*(6), 595–608. https://doi.org/10.15252/emmm.201606210.

Selman, C., Lingard, S., Choudhury, A. I., Batterham, R. L., Claret, M., Clements, M., et al. (2008). Evidence for lifespan extension and delayed age-related biomarkers in insulin receptor substrate 1 null mice. *The FASEB Journal*, *22*(3), 807–818. https://doi.org/10.1096/fj.07-9261com.

Shoelson, S. E., Lee, J., & Goldfine, A. B. (2006). Inflammation and insulin resistance. *The Journal of Clinical Investigation*, *116*(7), 1793–1801. https://doi.org/10.1172/JCI29069.

Shou, J., Chen, P. J., & Xiao, W. H. (2020). Mechanism of increased risk of insulin resistance in aging skeletal muscle. *Diabetology and Metabolic Syndrome*, *12*, 14. https://doi.org/10.1186/s13098-020-0523-x.

Skelin Klemen, M., Dolenšek, J., Slak Rupnik, M., & Stožer, A. (2017). The triggering pathway to insulin secretion: Functional similarities and differences between the human and the mouse β cells and their translational relevance. *Islets*, *9*(6), 109–139. https://doi.org/10.1080/19382014.2017.1342022.

Slack, C. (2017). Ras signaling in aging and metabolic regulation. *The Journal of Nutrition, Health & Aging*, *4*(3), 195–205. https://doi.org/10.3233/NHA-160021.

Søgaard, D., Baranowski, M., Larsen, S., Taulo Lund, M., Munk Scheuer, C., Vestergaard Abildskov, C., et al. (2019). Muscle-saturated bioactive lipids are increased with aging and influenced by high-intensity interval training. *International Journal of Molecular Sciences*, *20*(5), 1240. https://doi.org/10.3390/ijms20051240.

Speakman, J. R., & Westerterp, K. R. (2010). Associations between energy demands, physical activity, and body composition in adult humans between 18 and 96 y of age. *The American Journal of Clinical Nutrition*, *92*(4), 826–834. https://doi.org/10.3945/ajcn.2009.28540.

Steculorum, S. M., Solas, M., & Brüning, J. C. (2014). The paradox of neuronal insulin action and resistance in the development of aging-associated diseases. *Alzheimers Dement*, *10*(1 Suppl), S3–11. https://doi.org/10.1016/j.jalz.2013.12.008.

Szendroedi, J., Yoshimura, T., Phielix, E., Koliaki, C., Marcucci, M., Zhang, D., et al. (2014). Role of diacylglycerol activation of PKCθ in lipid-induced muscle insulin resistance in humans. *Proceedings of the National Academy of Sciences of the United States of America*, *111*(26), 9597–9602. https://doi.org/10.1073/pnas.1409229111.

Szentesi, P., Csernoch, L., Dux, L., & Keller-Pintér, A. (2019). Changes in redox signaling in the skeletal muscle with aging. *Oxidative Medicine and Cellular Longevity*, *2019*, 4617801. https://doi.org/10.1155/2019/4617801.

Tang, W. J. (2016). Targeting insulin-degrading enzyme to treat type 2 diabetes mellitus. *Trends in Endocrinology and Metabolism*, *27*(1), 24–34. https://doi.org/10.1016/j.tem.2015.11.003.

Taniguchi, C. M., Emanuelli, B., & Kahn, C. R. (2006). Critical nodes in signalling pathways: Insights into insulin action. *Nature Reviews. Molecular Cell Biology*, 7(2), 85–96. https://doi.org/10.1038/nrm1837.

Templeman, N. M., Flibotte, S., Chik, J. H. L., Sinha, S., Lim, G. E., Foster, L. J., et al. (2017). Reduced circulating insulin enhances insulin sensitivity in old mice and extends lifespan. *Cell Reports*, *20*(2), 451–463. https://doi.org/10.1016/j.celrep.2017.06.048.

Tiganis, T. (2011). Reactive oxygen species and insulin resistance: The good, the bad and the ugly. *Trends in Pharmacological Sciences*, *32*(2), 82–89. https://doi.org/10.1016/j.tips.2010.11.006.

Tomizawa, H. H., & Halsey, Y. D. (1959). Isolation of an insulin-degrading enzyme from beef liver. *The Journal of Biological Chemistry*, *234*(2), 307–310.

Tricò, D., Natali, A., Arslanian, S., Mari, A., & Ferrannini, E. (2018). Identification, pathophysiology, and clinical implications of primary insulin hypersecretion in nondiabetic adults and adolescents. *JCI Insight*, *3*(24), e124912. https://doi.org/10.1172/jci.insight.124912.

Trounce, I., Byrne, E., & Marzuki, S. (1989). Decline in skeletal muscle mitochondrial respiratory chain function: Possible factor in ageing. *Lancet*, *1*(8639), 637–639. https://doi.org/10.1016/s0140-6736(89)92143-0.

Tschen, S. I., Dhawan, S., Gurlo, T., & Bhushan, A. (2009). Age-dependent decline in beta-cell proliferation restricts the capacity of beta-cell regeneration in mice. *Diabetes*, *58*(6), 1312–1320. https://doi.org/10.2337/db08-1651.

Ueno, M., Carvalheira, J. B., Tambascia, R. C., Bezerra, R. M., Amaral, M. E., Carneiro, E. M., et al. (2005). Regulation of insulin signalling by hyperinsulinaemia: Role of IRS-1/2 serine phosphorylation and the mTOR/p70 S6K pathway. *Diabetologia*, *48*(3), 506–518. https://doi.org/10.1007/s00125-004-1662-6.

VanGuilder, H. D., Yan, H., Farley, J. A., Sonntag, W. E., & Freeman, W. M. (2010). Aging alters the expression of neurotransmission-regulating proteins in the hippocampal synaptoproteome. *Journal of Neurochemistry*, *113*(6), 1577–1588. https://doi.org/10.1111/j.1471-4159.2010.06719.x.

Villa-Pérez, P., Merino, B., Fernández-Díaz, C. M., Cidad, P., Lobatón, C. D., Moreno, A., et al. (2018). Liver-specific ablation of insulin-degrading enzyme causes hepatic insulin resistance and glucose intolerance, without affecting insulin clearance in mice. *Metabolism*, *88*, 1–11. https://doi.org/10.1016/j.metabol.2018.08.001.

Wang, S. Y., Halban, P. A., & Rowe, J. W. (1988). Effects of aging on insulin synthesis and secretion. Differential effects on preproinsulin messenger RNA levels, proinsulin biosynthesis, and secretion of newly made and preformed insulin in the rat. *The Journal of Clinical Investigation*, *81*(1), 176–184. https://doi.org/10.1172/JCI113291.

Wang, Y., Zhou, H., Palyha, O., & Mu, J. (2019). Restoration of insulin receptor improves diabetic phenotype in T2DM mice. *JCI Insight*, *4*(15), e124945. https://doi.org/10.1172/jci.insight.124945.

Watson, G. S., & Craft, S. (2003). The role of insulin resistance in the pathogenesis of Alzheimer's disease: Implications for treatment. *CNS Drugs*, *17*(1), 27–45. https://doi.org/10.2165/00023210-200317010-00003.

Wijsman, C. A., Rozing, M. P., Streefland, T. C., le Cessie, S., Mooijaart, S. P., Slagboom, P. E., et al. (2011). Familial longevity is marked by enhanced insulin sensitivity. *Aging Cell*, *10*(1), 114–121. https://doi.org/10.1111/j.1474-9726.2010.00650.x.

Williamson, D., Gallagher, P., Harber, M., Hollon, C., & Trappe, S. (2003). Mitogen-activated protein kinase (MAPK) pathway activation: Effects of age and acute exercise on human skeletal muscle. *The Journal of Physiology*, *547*(Pt 3), 977–987. https://doi.org/10.1113/jphysiol.2002.036673.

Yen, T. C., Chen, Y. S., King, K. L., Yeh, S. H., & Wei, Y. H. (1989). Liver mitochondrial respiratory functions decline with age. *Biochemical and Biophysical Research Communications*, *165*(3), 944–1003. https://doi.org/10.1016/0006-291x(89)92701-0.

Zeng, G., Nystrom, F. H., Ravichandran, L. V., Cong, L. N., Kirby, M., Mostowski, H., et al. (2000). Roles for insulin receptor, PI3-kinase, and Akt in insulin-signaling pathways related to production of nitric oxide in human vascular endothelial cells. *Circulation*, *101*(13), 1539–1545. https://doi.org/10.1161/01.cir.101.13.1539.

Zhang, W., & Liu, H. T. (2002). MAPK signal pathways in the regulation of cell proliferation in mammalian cells. *Cell Research*, *12*(1), 9–18. https://doi.org/10.1038/sj.cr.7290105.

Zhao, L., Teter, B., Morihara, T., Lim, G. P., Ambegaokar, S. S., Ubeda, O. J., et al. (2004). Insulin-degrading enzyme as a downstream target of insulin receptor signaling cascade: Implications for Alzheimer's disease intervention. *The Journal of Neuroscience*, *24*(49), 11120–11126. https://doi.org/10.1523/JNEUROSCI.2860-04.2004.

CHAPTER TEN

Sirtuins and the prevention of immunosenescence

Denise Lasigliè*

Istituto Comprensivo "Franco Marro", Ministero dell'Istruzione Ministero dell'Università e della Ricerca (M.I.U.R), Villar Perosa, TO, Italy

*Corresponding author: e-mail address: deniselasiglie@gmail.com

Contents

Abstract

Aging of hematopoietic stem cells (HSCs) has been largely described as one underlying cause of senescence of the immune-hematopoietic system (immunosenescence). A set of well-defined hallmarks characterizes aged HSCs contributing to unbalanced hematopoiesis and aging-associated functional alterations of both branches of the immune system. In this chapter, the contribution of sirtuins, a family of conserved NAD^+ dependent deacetylases with key roles in metabolism, genome integrity, aging and lifespan, to immunosenescence, will be addressed. In particular, the role of SIRT6 will be deeply analyzed highlighting a multifaceted part of this deacetylase in HSCs aging as well as in the immunosenescence of dendritic cells (DCs). These and other emerging data are currently paving the way for future design and development of rejuvenation means aiming at rescuing age-related changes in immune function in the elderly and combating age-associated hematopoietic diseases.

Vitamins and Hormones, Volume 115
ISSN 0083-6729
https://doi.org/10.1016/bs.vh.2020.12.011

1. Immunosenescence

As a complex and multifaceted process, senescence regards many molecular and cellular mechanisms in different organs and systems, with a decline in health partially attributed to defects in immunity occurring with natural age advancement (Katz et al., 2004). Vaccine failure, increased susceptibility to anemia, infection, autoimmune and myelogenous diseases as well as prevalence of cancer were reported in old age reflecting age-dependent decreases of immune competence (Dorshkind, Montecino-Rodriguez, & Signer, 2009; Goronzy & Weyand, 2013). Immunosenescence was shown to involve various cells types in primary and secondary lymphoid organs as well as circulating and tissue resident adaptive and innate immune cells. Cell intrinsic defects, impairment at the systemic level and in lymphoid and nonlymphoid organs, lifelong exposure to pathogens and antigens as well as genetic predisposition were reported to be at the core of the age-related immune dysfunction (Thompson, Smithey, Surh, & Nikolich-Zugich, 2017). Though severe age-associated changes were thought to regard *in primis* adaptive immunity (Weng, 2006), subsequent deeper analyses uncovered a profound impact of senescence on the innate branch of the immune system (Solana, Pawelec, & Tarazona, 2006). Besides, along with age, pathophysiological changes occur in hematopoietic stem cells (HSCs) and progenitors cells (HPCs) pools as well as in their associated bone marrow (BM) niche (Geiger, de Haan, & Florian, 2013; Morrison & Scadden, 2014; Rossi, Jamieson, & Weissman, 2008; Wagner, Horn, Bork, & Ho, 2008) contributing to defective HSCs repopulation capacity and homing, increased mobilization and myeloid skewed differentiation potentials (Latchney & Calvi, 2017), with aged HSCs dictating hierarchically several age-related changes occurring during immunosenescence (Miller & Allman, 2005). Indeed, recent growing evidence sustains the aging process being initiated at the very top of the hematopoietic hierarchy (Geiger et al., 2013; Wang, Geiger, & Rudolph, 2011). In the following paragraphs, I will summarize the principal effects of senescence on hematopoiesis, myeloid commitment and innate immune cells functions, with a particular accent on classical dendritic cells (cDCs) biology. This chapter is not intended to deeply review the effects of aging on the immune-hematopoietic system and/or HSCs, rather, to evaluate sirtuins contribution, in particular SIRT6, to immunosenescence, a less explored aspect of sirtuins biology.

2. Hematopoiesis in young and old age

Cells of the immune-hematopoietic system are continuously consumed and replenished by new ones produced by HSCs in the BM through a strictly organized lineage commitment (Rossi et al., 2008; Weissman & Shizuru, 2008). In steady-state HSCs are kept in a quiescent state and daily hematopoiesis is guaranteed by downstream hematopoietic progenitors cells (HPCs) proliferation and differentiation (Sun et al., 2014). However, in response to hematopoietic emergency (infection, inflammation, bleeding, tissue injury, altered nutrition) (stress-induced hematopoiesis) and upon external environmental cues (cytokines, chemokines, growth factors, (neuro) hormones, metabolites, pathogen associated molecular pattern (PAMPs)), HSCs are able to increase their proliferation, mobilization, differentiation and modify their fate decisions (Kaushansky, 2006; Méndez-Ferrer, Lucas, Battista, & Frenette, 2008; Morrison & Scadden, 2014; Wilkinson & Yamazaki, 2018; Zhao et al., 2014). In the classical immunophenotype-based three like hierarchical model, which was largely established by Weissman's group (Kondo, Weissman, & Akashi, 1997; Manz, Miyamoto, Akashi, & Weissman, 2002), HSCs sit at the apex of a developmental hierarchy and lineage commitment requires progressive loss of multi-potency and self-renewal. In such a vision it proceeds in a stepwise manner with quiescent long-term HSCs (LT-HSCs), which possess self-renewal and multi-lineage differentiation properties, generating themselves and giving rise to cycling short-term HSCs (ST-HSCs), which in turn generate multi-potent progenitors (MPPs) with reduced self-renewal ability (see Fig. 1). Downstream lineage commitment goes on with a first bifurcation occurring between common lymphoid progenitors (CLPs) (with lymphoid potential) (Kondo et al., 1997) and common myeloid progenitors (CMPs) (with myeloid, erythroid and megakaryocytic potential) (Akashi, Traver, Miyamoto, & Weissman, 2000). Then CMPs bifurk giving rise to megacaryocyte-erythroid progenitors (MEPs) and granulocyte-macrophage progenitors (GMPs). CLPs give rise to T, B, Natural Killer cells (NK cells) and eventually DCs. MEPs differentiate into platelets and erythrocytes, whereas GMPs generate granulocytes, monocytes/macrophages and DCs (see Fig. 1). In such a model, transcriptional factors (TFs) and cytokines precisely conduct HSCs differentiation into mature blood cells (MBCs) (Seita & Weissman, 2010). This classical developmental scheme, while helping understand the HSCs differentiation

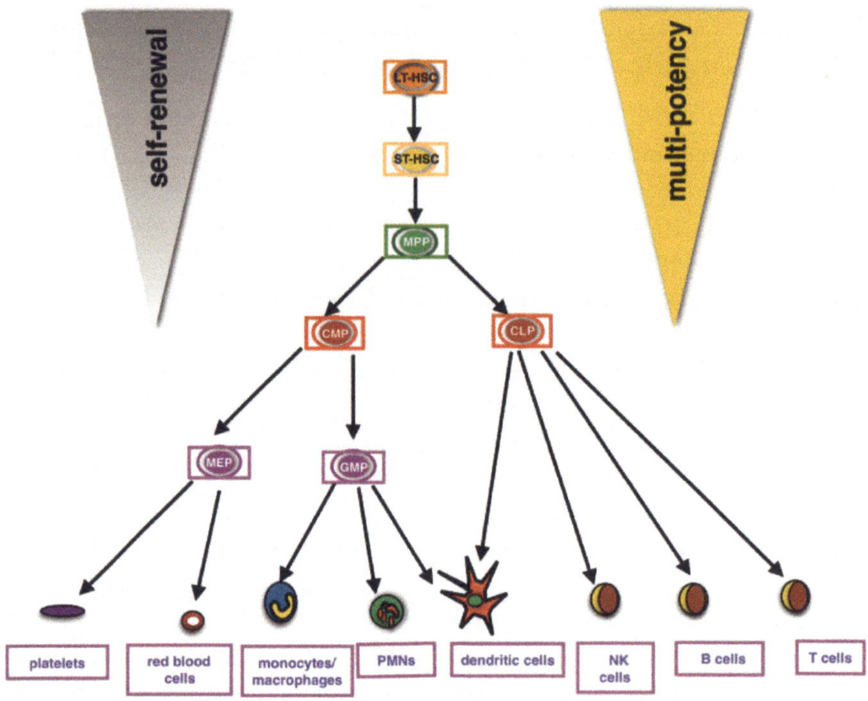

Fig. 1 *Hierarchical model of hematopoiesis.* LT-HSC, long-term hematopoietic stem cell; ST-HSC, short-term hematopoietic stem cell; MPP, multi-potent progenitor; CMP, common myeloid progenitor; CLP, common lymphoid progenitor; GMP, granulocyte-macrophage progenitor; MEP, megacaryocyte-erytroid progenitor; PMNs, polimorphonucleates (granulocytes); NK cells, natural killer cells.

process, oversimplifies the complexity of hematopoietic stem and progenitor cells (also called HSPCs) biology. Thanks to advances in single cells technology and genetic mouse models this hierarchical scheme has been challenged, new types of HSPCs have been identified and new insights into hematopoiesis, especially megakaryopoiesis, gained. A growing body of evidence suggests some cellular heterogeneity within the HSCs pool identifying HSCs subtypes with distinct heritable self-renewal and lineage commitment potentials such as balanced HSCs (Ba), myeloid biased HSCs (My-BI), lymphoid biased (Ly-BI) and platelet biased HSCs, the latter being at the apex of the HSCs hierarchy (Sanjuan-Pla et al., 2013). Similar observations have been also produced for the MPPs pool (Pietras et al., 2015). Indeed, several different and/or revised models have been postulated in recent years and will not be discussed here (for a complete review, the reader is referred to Cheng, Zheng, & Cheng, 2020). As previously mentioned it is widely accepted the age-related immune-hematopoietic decline being primarily an HSCs aging-dependent phenomenon (Geiger et al., 2013). With advancing age the number of phenotypically defined HSCs increases substantially in both humans and mice, the latter showing high variability depending on the mouse strains considered (Beerman et al., 2010; Pang et al., 2011; Rossi et al., 2005). Despite being augmented in frequency, aged HSCs display functional decline (Dykstra, Olthof, Schreuder, Ritsema, & de Haan, 2011). Diminished per cell self-renewal, regenerative potential upon stress, competitive repopulation and engraftment ability, altered homing back to BM and mobilization (Florian et al., 2012; Janzen et al., 2006; Kim, Moon, & Spangrude, 2003; Liang, Van Zant, & Szilvassy, 2005; Xing et al., 2006) were actually observed in aged murine HSCs. Similarly, regenerative capacity diminishes with age in human HSCs, the latter having reduced transplantation success (Kollman et al., 2001). Skewing of lineage potentials from lymphopoiesis to myelopoiesis was also observed with age: aged HSCs exhibit a markedly decreased output of lymphoid and erythroid lineages whereas myeloid lineage is maintained or even increased (Beerman et al., 2010; Pang et al., 2011). In more details CMPs and GMPs are frequently increased at the expense of CLPs and B lymphopoiesis, with early B progenitors number being reduced with advancing age (Henry, Marusyk, Zaberezhnyy, Adane, & DeGregori, 2010; Kuranda et al., 2011; Pang et al., 2011; Rossi et al., 2005). In mice clonal analyses of HSCs highlighted an higher number of myeloid biased HSCs in aged BM with respect to the young counterpart, thus postulating the aged-related myeloid skewing being HSCs pool composition dependent (Beerman et al., 2010; Challen, Boles, Chambers, & Goodell, 2010; Muller-Sieburg & Sieburg, 2006). Composition shift toward myeloid and platelet biased HSCs *via* clonal expansion and/or selection (Beerman et al., 2010; Cho, Sieburg, & Muller-Sieburg, 2008; Muller-Sieburg & Sieburg, 2006), clonal evolution of lymphoid biased into platelet biased or myeloid biased HSCs *via* cell intrinsic changes (Waterstrat & Van Zant, 2009), altered MPPs differentiation into CLPs (Busch et al., 2015) as well as disturbance in proliferation and survival of progenitors were thought to contribute to the age-induced myeloid lineage skewing (Dolence, Gwin, Shapiro, & Medina, 2014). While cell intrinsic factors contribution to HSCs aged-related functional decay has been largely established (Latchney & Calvi, 2017), the role of extrinsic factors (niche derived and/or systemic) is less defined with, however, growing and exciting evidence sustaining their part in hematopoietic aging (Morrison & Scadden, 2014). It has became more and more evident both these factors may participate to a complex network able to promote HSCs age-related decline (Geiger et al., 2013; Geiger, Koehler, & Gunzer, 2007; Latchney & Calvi, 2017; Vas, Senger, Dorr, Niebel, & Geiger, 2012). Altered transcriptional landscapes with upregulation of myeloid genes, such as runt-related transcriptional factor 1 *(runx1),* stress genes and/or genes involved in the nuclear factor κB (NF-κB) pathway (Chambers et al., 2007; Rossi et al., 2005; Sun et al., 2014), downregulation of lymphoid genes, cell cycle and DNA repair genes as well as epigenetic modifiers genes, such as sirtuins (Chambers et al., 2007; Sun et al., 2014) have been recorded by aged HSCs. In addition, impaired epigenetic signatures (at both DNA and histones levels) are currently emerging as important mechanisms involved in HSCs aging (Beerman et al., 2013; Florian et al., 2012; Rando & Chang, 2012). DNA damage and telomere attrition, mitochondrial DNA (mtDNA) mutations, increased reactive oxygen species (ROS) and oxidative damage, altered polarity and clonal selection were also reported to intrinsically contribute to HSCs aging (Latchney & Calvi, 2017). In a similar way, chemokines (i.e., RANTES, CXCL12), hormones and/or growth factors (GH, IGF-1) levels from the niche microenvironment or from the circulation were shown to have a crucial part in the establishment and maintenance of several phenotypes described for aged HSCs (Ergen, Boles, & Goodell, 2012; French et al., 2002; Young, Eudy, Loberg, Bell, & Trowbridge, 2019; Young, Loberg, Bell, & Trowbridge, 2017; Young, Loberg, Eudy, Bell, & Trowbridge, 2018). Furthermore external inflammatory cues may also act as a driving force for HSCs aging (Mann et al., 2018; Montecino-Rodriguez et al., 2019; Pietras et al., 2016; Yamashita & Passegue, 2019). The increased outputs of myeloid cells observed with age are paralleled by decreased outputs of T cells precursors and immature B cells pool, the latter contributing along with downstream phenomena (such as thymus involution), to the adaptive immune response decay observed with age (Linton & Dorshkind, 2004; Miller & Allman, 2005). Erythropoiesis also declines in the elderly, frequently causing anemia (Berliner, 2013).

3. Innate immunosenescence

Though the myeloid dominant skewed hematopoiesis observed with aging is well established and mostly conserved in humans and mice (Wang et al., 2011), senescence of the innate immune system takes shape of a more complex and puzzled phenomenon (Gomez, Nomellini, Faunce, & Kovacs, 2008). Despite cumulative findings highlighting the innate immune cells becoming dysfunctional with age (Solana et al., 2006) with impairment of cell number and functions, along with large amounts of data showing a decrease in many innate immune responses and functions (such as phagocytosis, chemotaxis, oxidative burst, TLR induced cytokines production in monocytes/macrophages and neutrophils) (Brubaker, Rendon, Ramirez, Choudhry, & Kovacs, 2013; Butcher et al., 2001; Linehan et al., 2014; Mariani et al., 2002; Montgomery & Shaw, 2015) several evidence of preservation and/or hyperactivation of the innate immunity with age (Butcher, Chahel, & Lord, 2000; El Mezayen, El Gazzar, Myer, & High, 2009; Lutz, Moore, Bradley, Shelton, & Lutgendorf, 2005; Mariani et al., 2002; Tan et al., 2012) has been emerging. Experimental settings (*in vivo vs in vitro*), sources of cells (bone marrow *vs* spleen, human *vs* mice and/or rats), mouse strains, stimuli used, activation state of the cells and different separation procedures may account for the partial heterogeneity of the data from the literature (for more details, the reader is referred to Kovacs et al., 2009). Besides, different factors (nutrients, hormones, chronic viral infection, diabetes and other metabolic syndromes) may influence several innate immune cells functions independently from the aging process (Montgomery & Shaw, 2015). DCs, the most potent antigen presenting cells (APCs), are subdivided in classical (mouse)/myeloid (human) DCs (cDCs, mDCs), plasmocytoid DCs (pDCs) all arising from DCs precursors in the BM (Geissmann et al., 2010; Mildner & Jung, 2014) and monocytes-derived DCs (MoDCs) originating from monocytes infiltrating lymphoid and nonlymphoid organs upon inflammation or infection (Mildner, Yona, & Jung, 2013). DCs represent sentinels bridging innate and adaptive immunity (Mellman & Steinman, 2001) as they sense pathogens in periphery and present them to T cells in secondary lymphoid organs, thus contributing to initiation of näive T cell response (Sallusto & Lanzavecchia, 2002) and to effective restimulation of memory T cells (Mehlhop-Williams & Bevan, 2014). In such a picture, age-related DCs dysfunction may have a strong impact on the immune status in the elderly, in particular on vaccine response, reactivity to self, ability to fight infections, all features being altered in aged subjects (Dorrington & Bowdish, 2013; Watad et al., 2017). Major data on DCs biology in old age come from studies in humans (Agrawal, Agrawal, & Gupta, 2017; Agrawal & Gupta, 2011). In the elderly the number of circulating subtypes of DCs is either conserved (MoDCs, mDCs) or reduced (pDCs) (Agrawal, Agrawal, & Gupta, 2007; Jing et al., 2009). Cytokines production is also altered with subtypes specific features (Agrawal et al., 2007, 2017;

Agrawal, Tay, Ton, Agrawal, & Gupta, 2009; Qian et al., 2011; Sridharan et al., 2011) and a general deficiency in priming naïve T cells was observed in human aged DCs (Briceno et al., 2016; Sridharan et al., 2011; Wong, Magnusson, & Ho, 2010). It is generally assumed that DCs in elderly individuals display enhanced basal level of activation (preactivation), low and chronic production of proinflammatory cytokines (inflamm-aging) and reduction of tolerance in steady state as well as altered pro/anti-inflammatory cytokines production (IL-10, IL-6, TNF-α) and decreased ability to fight infection upon activation (Agrawal et al., 2017). A more puzzled picture comes from studies in mice with DCs total number and composition in lymphoid tissues varying, being unaltered, increased or decreased (Pereira, de Souza, Borges, & Bonorino, 2011; Tan et al., 2012; Wong et al., 2010). Similarly, cytokines production, co-stimulatory molecules upregulation, antigen uptake, processing and presentation and more in general T cells priming were shown to be reduced, preserved or even increased (cytokines production) with aging (Chougnet et al., 2015; El Mezayen et al., 2009; Li, Smithey, Rudd, & Nikolich-Zugich, 2012; Pereira et al., 2011; Shen, Tesar, Du, & Goldstein, 2009; Stout-Delgado, Yang, Walker, Tesar, & Goldstein, 2008; Tesar et al., 2006). In addition, aged murine DCs failed to upregulate C-C Chemokine Receptor 7 (CCR7) and thus to accumulate in draining lymph nodes upon immunization (Grolleau-Julius, Harning, Abernathy, & Yung, 2008). Though DCs senescence appears to be very complex and a common and unifying biological mechanism is still to be identified, a major consensus on a reduced pDCs mediated immunity in old beings is present, whereas impairment in cDCs (myeloid DCs) functions and its direction is a little bit less clear, with however a general agreement on an overall reduced functionality of cDCs, with *in vitro* studies gaining major importance when sustained by *in vivo* approach (Shaw, Goldstein, & Montgomery, 2013; Wong & Goldstein, 2013).

4. Hormones and immunosenescence

4.1 GH/IGF-1 axis (somatotropic axis)

HSCs biology, immune cells development and function are largely controlled by several hormones, cytokines and hematopoietic growth factors (Alexander, 1998; Bateman, Singh, Kral, & Solomon, 1989; Broudy, 1997; Kelley, Arkins, Minshall, Liu, & Dantzer, 1996). Along with age complex alterations in the hormonal networks occur (Lamberts, van den Beld, & van der Lely, 1997) contributing to the aging process of the immune system.

Growth hormone (GH) is a peptide hormone mainly produced by the anterior pituitary gland that promotes protein synthesis (Kostyo, 1968), cell division, regeneration and growth (Wells & de Vos, 1993). In particular this pleiotropic hormone exerts anabolic effects by regulating many growth-related and metabolic physiological processes (Brooks & Waters, 2010). It is now widely accepted GH production by extra pituitary tissues such as mammary glands (Mol et al., 1995) and various cell types and organs in the immune system (Varma, Sabharwal, Sheridan, & Malarkey, 1993). Though at the beginning and for a long time GH has been believed to mainly exert its physiological effects through "endocrine" liver-produced insulin-like growth factor 1 (IGF-1) (Salmon & Daughaday, 1957), subsequently several findings challenged the "somatomedin hypothesis" (Daughaday, 1989). It is nowadays largely accepted GH and IGF-1 having independent physiological functions and effects *in vivo* (Pasarica, Zachwieja, Dejonge, Redman, & Smith, 2007; Powell-Braxton et al., 1993), though they collaborate with distinct relevance and functional specificity to several biologic processes, such as global and longitudinal bone growth (Liu & LeRoith, 1999; Lupu, Terwilliger, Lee, Segre, & Efstratiadis, 2001). Besides being systemic factors, they were shown to locally act in autocrine and paracrine manners (Florini, Ewton, & Coolican, 1996; Sabharwal & Varma, 1996). IGF-1 belongs to the IGF system, a very complex system with many actors in play comprising three ligands (IGF-1, IGF-2, insulin), their receptors (IGF-1 receptor (IGF-1R), mannose 6-phosphate/IGF-2 receptor (M6P/IGF-2R), insulin receptor (IR) and hybrid IR/IGF-1R), at least six IGF-binding proteins (IGFBP1–6), acid labile subunit (ALS) and IGF-1 binding proteins proteases. Binding of the ligands to their receptors (with low or high affinity) causes potent proliferative, differentiative, anti-apoptotic and metabolic effects by activating common intracellular pathways, the insulin and insulin-like growth factor signaling (IIS) pathways (For more details the reader is referred to Annunziata, Granata, & Ghigo, 2011).

4.1.1 Somatotropic axis and immune system

Huge amounts of *in vivo* and *in vitro* data have uncovered for the GH/IGF-1 axis important roles in immune system development, maintenance, repair and functions (Clark, 1997). Within the immune system GH biological effects are primarily mediated by IGF-1 (Sabharwal & Varma, 1996). Pioneering studies in rats by Smith and colleagues (Smith, 1930) showed thymus atrophy following hypophysectomies. Besides, GH administration was shown to rescue and prevent a thymus-dependent immunodeficiency in hypophysectomized rodents as well as to increase T activity in aged rodents (Baroni, Fabris, & Bertoli, 1967; Murphy, Durum, & Longo, 1992; Nagy, Berczi, & Friesen, 1983; Taub & Longo, 2005). Indeed, a strictly connection between age-related GH/IGF-1 axis decline and thymus atrophy has been largely documented (Patel & Taub, 2009). Furthermore GH or IGF-1 administration to old mice and dogs (Clark, 1997) and GH infusion in HIV infected patients (Napolitano et al., 2002) were shown to reverse thymic involution (for a deeper description of GH effects on thymus, the reader is referred to Savino, Mendes-da-Cruz, Lepletier, & Dardenne, 2016). GH and/or IGF-1 contribution to lymphopoiesis (B and T) was largely shown to occur even earlier, in the BM, influencing proliferation, differentiation and survival of progenitors (multi and lineage restricted) as well as T lymphocytes precursors migration and their engraftment into the thymus (Clark, Strasser, McCabe, Robbins, & Jardieu, 1993; Dardenne, Smaniotto, de Mello-Coelho, Villa-Verde, & Savino, 2009; Hanley, Napolitano, & McCune, 2005; Landreth, Narayanan, & Dorshkind, 1992; Murphy et al., 1992; Welniak, Sun, & Murphy, 2002). Besides playing important parts in lymphopoiesis, GH and IGF-1 have been identified as important mediators of erytropoiesis, granulopoiesis, trombopoiesis and myelopoiesis (Chen et al., 2018; Li et al., 1997; Merchav, Tatarsky, & Hochberg, 1988; Welniak et al., 2004). Recently, IIS was shown to be essential for early lineage commitment of MPPs. *Insr* (insulin receptor) genetic disruption was actually shown to cause MPPs to differentiate into myeloid cells (Xia et al., 2015), postulating that multiple members of the IGF-system (insulin and/or IGF-1) may have a role in dictating balanced hematopoiesis and immune system. This postulation has been recently supported by studies by Young et al. (2019, 2017, 2018) identifying IGF-1 as a crucial factor able to reverse many hallmarks of aged HSCs (see later). In vertebrates activation of IIS in HSCs by Akt activation leads to loss of quiescence and aberrant cell cycling (Kharas et al., 2010; Perry et al., 2011). To conclude, GH and IGF-1 were shown to promote BM and splenic hematopoietic growth *in vitro* and *in vivo* in both humans and mice (Kawa et al., 2015; Tsarfaty, Longo, & Murphy, 1994) to control thymus, spleen and lymph nodes size by increasing lymphocytes and Thy + cells number and their function (Clark, 1997; Clark et al., 1993). GH and/or IGF-1 administration are usually used to promote hematopoiesis and immune system reconstitution after irradiation and transplantation in mice, primates and humans (Alpdogan et al., 2003; Chen et al., 2010; Tian et al., 1998).

4.1.2 Somatotropic axis and aging

GH/IGF-1 axis decline is one of the most robust and well characterized endocrine alterations occurring with normal aging, with progressive withdrawal of these hormones being responsible for several biological and physiological modifications occurring with aging in humans and rodents (Le Roith, Bondy, Yakar, Liu, & Butler, 2001). In some tissues, amelioration of the age-related decline was found to be achieved by growth hormone-releasing hormone (GHRH), GH and IGF-1 therapy (Baker et al., 2012; Blackman et al., 2002; Falomir-Lockhart, Dolcetti, Garcia-Segura, Herenu, & Bellini, 2019; Liu et al., 2007; Rudman et al., 1990; Schwerdt et al., 2018; Thornton, Ingram, & Sonntag, 2000). Similarly, in rodents, *in vivo* GH administration and *in vitro* IGF-1 treatment have been shown to reverse the age-induced loss of hematopoiesis (French et al., 2002) as well as the myeloid skewed outputs of aged HSCs (Young et al., 2018). IIS pathways are the most conserved aging-controlling pathways in evolution as they were already present in invertebrates such *S. cerevisae, C. elegans* and *D. melanogaster* with mutants disrupting this signaling showing lifespan extension (Carter, Ramsey, & Sonntag, 2002). Accordingly, polymorphisms and mutations attenuating mammals IIS cascade have been linked to longevity in both humans and model organisms, further sustaining a role of nutrients and bioenergetics in determining lifespan (Fontana, Klein, & Holloszy, 2010). In mice circulating IGF-1 levels negatively relate to median lifespan in female in a study enrolling 31 mouse strains (Yuan et al., 2009), female *IGF-1r*(+/−) mice live 33% longer than WT females (Holzenberger et al., 2003) and late targeting of IGF-1R was recently shown to increase health span and lifespan in female mice (Mao et al., 2018). In humans partial loss of function mutations in the *IGF-1r* gene are over-presented in centenarians, whereas plasma IGF-1 levels are either increased or reduced according to the study considered (Vitale, Pellegrino, Vollery, & Hofland, 2019). Thus, paradoxically, reduced GH and IGF-1 plasma levels and/or IIS pathways have been observed either in normal aging or in rodent genetic models of both premature aging (such as ATM deficient mice) (Peretz, Jensen, Baserga, & Glazer, 2001) and prolonged lifespan (Ortega-Molina et al., 2012) as well as in response to caloric restriction (CR), the only nonpharmacological intervention known to ameliorate health span and increase lifespan across species (Lopez-Lluch & Navas, 2016). It is likely that in higher eukaryotes IIS may play a multifaceted influence on lifespan control, maybe secondary to a more complex IGF system (comprising its up-stream GH-mediated regulation).

Indeed, physiological and/or pathological decreases of the GH/IGF-1 axis/IIS pathways are thought to be an adaptive survival response to extend lifespan upon internal and external stress cues (nutrients, hormones, DNA damage, proinflammatory cytokines) (Garinis, van der Horst, Vijg, & Hoeijmakers, 2008; Niedernhofer et al., 2006; O'Connor et al., 2008). Nevertheless, the truly role of the GH/IGF-1 axis in aging and longevity has been and still is at the center of a scientific controversy.

Among IGF-1 signaling molecules, Sir2 (silent information regulator 2), a repressor with NAD^+ dependent deacetylase activity, was shown to mediate CR longevity effects in yeast and over-expression of yeast *SIR2* gene and its homologs in worms and flies was largely demonstrated to increase lifespan (Kaeberlein, McVey, & Guarente, 1999; Lin, Defossez, & Guarente, 2000). Similarly, multiple mammal members of sirtuins family were uncovered to modulate IIS pathways in nutrient enough and caloric restriction conditions (Zhou, Tang, & Chen, 2018). In mice deletion of members of the sirtuins family such as SIRT6 and SIRT7, is associated with an aging-like phenotype (Mostoslavsky et al., 2006; Vazquez et al., 2016) while SIRT6 over-expression causes lifespan extension in male mice (Kanfi et al., 2012). In these models IGF-1 systemic levels are reduced and they similarly do in case of SIRT1 and SIRT6 neuronal ablation (in the latter cases in parallel with low GH plasma levels) on normal and CR diet (Cohen, Supinski, Bonkowski, Donmez, & Guarente, 2009; Guarente, 2013; Schwer et al., 2010). SIRT1 was actually suggested to control the somatotropic axis by positively regulating pituitary GH and maybe hepatic IGF-1 production (Chen et al., 2008; Cohen et al., 2009). However, other works reported SIRT1 being a negative regulator of GH/IGF-1 expression and synthesis in pituitary cells, liver and other cell types (Monteserin-Garcia et al., 2013; Sansone et al., 2013; Yamamoto et al., 2013). As we will see later a similar role in modulating somatotropic axis may be postulated for SIRT6 and considering the SIRT6 resembling mouse phenotype it will be not hazardous to hypothesize for SIRT7 a similar part. During evolution, along with a more complex organization of the IIS, sirtuins may have acquired additive functions (i.e., somatotropic axis modulation) to fine control mammalian lifespan in an overarching way. While these data identify sirtuins as key and potent mediators interconnecting somatotropic axis and organismal aging, they seem to overcomplicate the understanding of an already complex phenomenon. Many other age-dependent biological processes are under control of sirtuins (see later), as well as multiple CR anti-aging effects, making them attractive targets to develop therapeutics for age-related diseases, such as type II diabetes (Milne et al., 2007). In accordance, many studies have been performed aiming at better understanding sirtuins biology in connection with the aging phenomenon. However, the knowledge of their part in immunosenescence is still in its infancy. In this scenario, I will describe the principal characteristics of sirtuins family with a particular accent on SIRT6, its contribution to the aging process and its role in immunosenescence.

5. Sirtuins

5.1 General characteristics

Sirtuins (Sir2-like proteins) comprise an universally conserved and ancient family of NAD^+ dependent enzymes, playing crucial roles in many biological processes such as metabolism, genome integrity, epigenetic and transcriptional regulation, aging, lifespan extension and cancer. The reader is referred to several excellent reviews summarizing these important functions (Bosch-Presegue & Vaquero, 2015; Choi & Mostoslavsky, 2014; Houtkooper, Pirinen, & Auwerx, 2012; Saunders & Verdin, 2007). Sir2, the yeast founding member, was initially discovered in *S. cerevisae* as a factor necessary for the epigenetic silencing of *HML HMR* at the mating type loci (Shore, Squire, & Nasmyth, 1984). This early discovery paved the way for the identification of Sir2 homologs in almost all organisms (Frye, 1999; Greiss & Gartner, 2009). Procaryotes have one or two sirtuins, whereas most eukaryotes have multiple sirtuins, with mammals harboring seven members (SIRT1 through 7), the latter showing, as an evolutionary adaption to increased complexity, different cellular localization, tissue distribution, substrate specificity and some emerging biochemical diversities (Bheda, Jing, Wolberger, & Lin, 2016). X-ray crystallography studies have contributed to establish sirtuins tridimensional structures alone or in various liganded forms (for a detailed list of X-ray structures the reader is referred to Bheda et al., 2016, supplementary table 1), which in turn have been extensively described (for a detailed review we refer the reader to Sanders, Jackson, & Marmorstein, 2010). From a biochemical point of view sirtuins are categorized as Class III histone deacetylases (Class III HDACs) (Seto & Yoshida, 2014). However, in contrast to the other classes of HDACs, sirtuins have the unique property to use nicotinamide adenine dinucleotide (NAD^+) as a co-substrate to perform their enzymatic activity: they were shown to couple NAD^+ cleavage to the removal of an acetyl moiety from the ε amino-group of lysine residues within protein targets (Imai, Armstrong, Kaeberlein, & Guarente, 2000; Landry et al., 2000) (see Fig. 2) to yield the deacetylated substrate, nicotinamide (NAM)

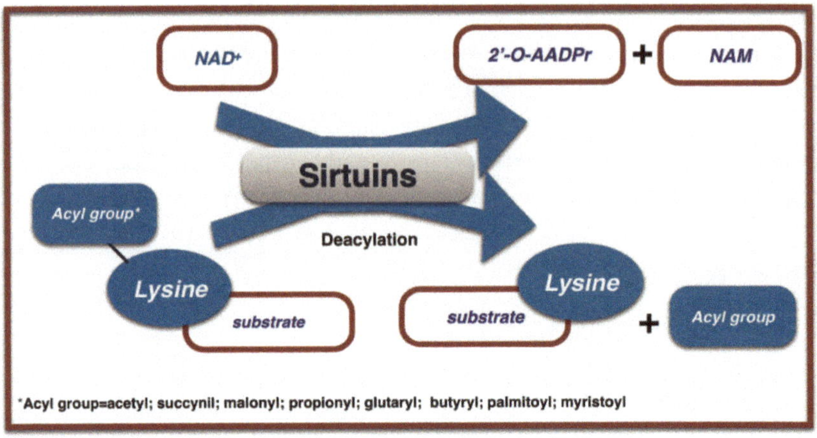

Fig. 2 *General mechanism of sirtuins enzimatic activity.* Sirtuin proteins couple substrate deacylation to NAD^+ cleavage, with production of 2′-O-AADPr, NAM and deacylated substrate. NAD^+, Nicotinamide adenine dinucleotide; 2′-O-AADPr, 2-O′-acetyl-ADP-ribose; NAM, nicotinamide.

and 2′-*O*-acetyl-ADP-ribose (2′-O-AADPr) (Jackson & Denu, 2002; Sauve et al., 2001). Thanks to their NAD^+ dependency these enzymes actually act as nutrients and energy sensors able to mediate cell responses for adaption to nutrient stress, such as CR. CR induces an increase of the intracellular NAD^+/NADH ratio, which activates sirtuins (Cantó & Auwerx, 2009). Though deacetylation is largely considered the main enzymatic activity of sirtuins, recent studies have uncovered for multiple members of this family alternative/additional activities, such as desuccinylation, demalonylation, ADP-ribosylation and others (deeply reviewed in Bheda et al., 2016). Since their discovery the number of sirtuins substrates identified has enormously increased ranging from histones (H3K16ac, H3K56ac, H3K9ac, H1ac) to nonhistone targets (p53, NFkB, GDH, PARP1) (Bosch-Presegue & Vaquero, 2014; Martinez-Redondo & Vaquero, 2013).

5.2 Sirtuins as master regulators of aging

Aging is widely considered a multifactorial phenomenon and many theories have been postulated so far to connect this multifaceted physiological process to the cellular and molecular hallmarks largely shown to contribute to aging (Jin, 2010). Genome instability, telomere attrition, mitochondria dysfunction, epigenetic alterations, proteostasis and dysregulated intercellular communication (i.e., cytokines and hormones) have been largely demonstrated to take part inter-connectedly to aging (Lopez-Otin, Blasco, Partridge, Serrano, & Kroemer, 2013). A great boost to the interest in sirtuins biology came from early studies in lower eukaryotes showing for yeast Sir2, flies and worms homologs a remarkable longevity activity (Guarente, 2011). Subsequently these enzymes have been extensively studied as anti-aging factors uncovering for many members important roles in genome stability (Sir2, SIRT1–4,6,7), DNA damage response, repair and prevention (Sir2, SIRT1 through SIRT7), telomere maintenance (Sir2, SIRT1, SIRT6), oxidative and genotoxic stress protection (SIRT1 through SIRT7), facultative heterochromatin (FH) formation, transcriptional silencing and epigenetic pattern maintenance (Sir2, SIRT1–SIRT3, SIRT6–SIRT7), mitochondria homeostasis (SIRT1 through SIRT7), protein turnover and autophagy (SIRT1, SIRT3, SIRT6, SIRT7), as nutrients sensors (Sir2, SIRT1 through SIRT7) and mediators at different levels of the IIS/IGF system/somatotropic axis (SIRT1–SIRT3, SIRT6, SIRT7) (Bosch-Presegue & Vaquero, 2014; Chang & Guarente, 2014; Choi & Mostoslavsky, 2014; Elkhwanky & Hakkola, 2018; Fang, 2019; Jing & Lin, 2015; Lee, 2019; Sun et al., 2019; Tang et al., 2019; Vakhrusheva et al., 2008; Westerheide, Anckar, Stevens, Sistonen, & Morimoto, 2009; Xu et al., 2014, 2019; Zhang, Head, & Yu, 2016; Zhou et al., 2018). In mice several sirtuins are implicated in the amelioration and delay of aging *in vivo,* with total body and hypothalamus specific over-expression of SIRT1, the mammal closest homolog to yeast Sir2, improving health during aging and extending lifespan (Herranz et al., 2010; Satoh et al., 2013). More compelling evidence for sirtuins-mediated longevity effects comes from SIRT6, whose deficiency in mouse is connected with a severe progeroid syndrome whereas its over-expression in male mice has been shown to increase lifespan (see Section 4.1.2). Similarly, SIRT7 null mice exhibits an aging-like phenotype, whereas no SIRT7 over-expressing mice has been, to my acknowledge, described so far. The beneficial effects of sirtuins on health and lifespan may rely on their multiple and integrative functions at different levels in key processes connected with genome instability prevention, metabolic, mitochondria and protein homeostasis maintenance, oxidative and genotoxic stress control, "young" transcriptome and epigenome maintenance as well as regulation of local and systemic inflammation and other types of (neuro)endocrine and paracrine intercellular communication (i.e., IGF system/somatotropic axis). Taking part to these processes sirtuins may contribute to prevent cell senescence, tissue and organismal aging (Bosch-Presegue & Vaquero, 2015; Lopez-Otin et al., 2013).

5.3 SIRT6 mouse models and *in vitro* studies: Insights into SIRT6 biological functions

SIRT6 is a widely expressed nuclear NAD^+ dependent deacetylase, with H3K9ac and H356ac as main targets (Kugel & Mostoslavsky, 2014; Michishita, Park, Burneskis, Barrett, & Horikawa, 2005) and recently described cytoplasmatic functions (Simeoni et al., 2013). Among sirtuins KO models the most extreme phenotype is reported for SIRT6 null mice (Mostoslavsky et al., 2006), the latter showing severely retarded growth rates and lethality. These mice are under-sized and develop a severe progeroid syndrome, with a number of abnormalities including lymphopenia, osteopenia, lordokyphosis, low plasma IGF-1 and glucose levels, the latter leading to death by 4 weeks by hypoglycemia. At a cellular level SIRT6 depletion is associated with genome instability and defective base excision repair (BER) (Mostoslavsky et al., 2006). Increased apoptosis rate by thymocytes as well as altered proliferation ability by mouse embryonal fibroblasts (MEFs) were also observed in absence of SIRT6. This initial analysis paved the way for following *in vivo* and *in vitro* studies aiming at deeply defining SIRT6 functions, at both cellular and organismal level. SIRT6 mediated H3K56ac deacetylation was shown to contribute to genome stability *in vivo* (Yang, Zwaans, Eckersdorff, & Lombard, 2009). SIRT6 was also reported to have a part in telomere stability and function in both humans (H3K9) and mice (H3K56) (Michishita et al., 2008, 2009). Besides, in humans, maintenance of telomere position effect (TPE) is strictly SIRT6 dependent (Tennen, Bua, Wright, & Chua, 2011). Under oxidative stress SIRT6 participates to single-strand and double-strand breaks repair (SSBR, DSBR) in a Poly ADP-ribose polymerase 1 (PARP1) dependent manner (Mao et al., 2011; Xu et al., 2015). In addition, in response to DSBs SIRT6-mediated H3K9ac deacetylation was proven to play a crucial role in the positioning of DNA dependent protein kinase (DNA-PK), SNF2H (sucrose non fermenting 2 homolog) and other factors on DNA lesions (McCord et al., 2009; Toiber et al., 2013). SIRT6 was shown to further exert his protective role under oxidative conditions by modulating the anti-oxidative NRF2-HO-1 axis in human mesenchymal stem cells (hMSCs) (Pan et al., 2016), the latter showing, in absence of SIRT6, increased ROS and accelerated functional decay. Similar observations were obtained by Zhai et al. (2016) with SIRT6 protecting hMSCs from premature senescence. Additional studies highlighted for SIRT6 a role as co-repressor of NF-κB, hypoxia-inducible factor-1α (HIF-1α), c-Myc and c-Jun mediated pathways in a tissue- and context-dependent manner (Kawahara et al., 2009; Sebastian et al., 2012; Sundaresan et al., 2012; Zhong & Mostoslavsky, 2010) confirming and uncovering for this sirtuin key roles in glicolysis (HIF-1α), inflammation and aging (NF-κB and c-Jun) as well as in ribosome biosynthesis (c-Myc). SIRT6 role in metabolism goes beyond its contribution to glycolysis, with this protein being crucial for gluconeogenesis, lipid metabolism, insulin secretion, adipolysis and adipogenesis (for a deeper review, the reader is referred to Kuang et al., 2018) as well as for IIS, IGF system and somatotropic axis. The complex phenotype of SIRT6 null mouse model recapitulates the multitasking essence of this enzyme and its multifaceted contribution to different biological phenomena, such as aging, metabolism, growth and development, straighten the concept of an integrated and coordinated *modus operandi* of sirtuin enzymes. In absence of SIRT6 altered glycolysis occurs and a metabolic syndrome develops, genome instability is observed as well as altered body growth, thymus and spleen development and maintenance, reduced number of $CD4^+ CD8^+$ thymocytes and splenic lymphocytes, the latter characteristics being likely in line with an attenuation of GH/IGF-1axis/IIS pathways (Taub & Longo, 2005). SIRT6 link to somatotropic axis has been further documented by studies in mice overexpressing SIRT6 and with neural ablation of SIRT6 (see previous paragraphs). Besides, SIRT6 was also shown to interact *in vitro* with c-Jun and co-repress IGF-1 signaling related genes and their downstream targets (including *Igf2, Igf2r, Foxo1, Igf1r, Mapk3, Pten, Gsk3b, Irs2, Akt3* and *Ins2,* and others) contributing to IIS down-modulation (Sundaresan et al., 2012). Interestingly in this work *ins1* and *Igf1* have been indicated to be positively regulated by SIRT6. In fact *Igf1* expression was strictly reduced in both SIRT6 null cardiomyocytes and skeletal muscle cells (Sundaresan et al., 2012). These latter data are in line with previous observations made by Kawahara et al. (2011), who indicated *Igf1, ghr, Igf1bp2/5, Igf2bp1, Insr* genes as possible targets of SIRT6 deacetylase activity. Besides, IIS attenuation was shown to be essential for SIRT6 anti-aging promotion of autopaghy (Lu et al., 2016). SIRT6 may also modulate IGF-1 signaling through the above described DNA damage protection mechanisms. Attenuation of IGF-1 signaling by DNA damage is a largely described phenomenon (Niedernhofer et al., 2006). On the whole these *in vivo and in vitro* data let us postulate a closely relationship between SIRT6, somatotropic axis/IIS pathways and aging.

5.4 SIRT6 and immunosenescence

Thanks to a collaboration between our group and Mostonslansky's lab, we had the great chance to work on SIRT6 null mice and carry on a study aiming at deeply characterizing the impact of SIRT6 absence on the innate immune system. At that time the SIRT6 "era" was at the very beginning and few information on its role within the immune system were available, being limited to its pro-inflammatory part in TNF-α production (Van Gool et al., 2009). In an effort to study the consequences of SIRT6 depletion on pro-inflammatory cytokines production and secretion, we generated bone marrow-derived dendritic cells (BMDCs) and stimulated them with different toll like receptors (TLRs) ligands (LPS/TLR4, CpG/TLR9, POLY:IC/TLR3, R848/TLR7/8) (see Fig. 3 for acronymous). The secretion profile recorded in absence of SIRT6 was variable, being both stimulus and cytokines dependent. Similarly to what observed in a previous study (Van Gool et al., 2009) SIRT6 deficiency caused a significant reduction of TNF-α secretion upon CpG stimulation whereas treatment with LPS was shown to induce significantly higher amounts of TNF-α when SIRT6 was lacking (see Fig. 3E). TLR9 endosomal localization seemed not to be relevant for the "defective" secretion observed, as stimulation of endosomal-localized TLR7 and TLR3, induced the same pattern of secretion as LPS did. A similar secretion profile was also observed for IL-6 when SIRT6 was lacking (not shown). In contrast, secretion of IL-12 was recorded to be significantly reduced in absence of SIRT6 (Fig. 3F) upon all the stimuli analyzed and a similar trend was also observed for IL-2 (Lasigliè D, unpublished data) and for Cxcl1 and Cxcl2 (Bauer et al., 2012). Thus, SIRT6 contribution to cytokines expression was shown to be more complex then uncovered before being either anti-inflammatory or pro-inflammatory and depending on cytokines considered and stimuli encountered. These data were further confirmed by kinetic studies, FACS analysis (Fig. 3A–C) and qPCR (not shown) and subsequently additive important parts in cDCs maturation have been unrevealed for this sirtuin. Spontaneous *in vitro* maturation was demonstrated to be defective upon SIRT6 depletion, as proved by a reduced percentage of mature and immature DCs (Fig. 4). Altered TLRs stimulation dependent maturation was also displayed by SIRT6 null BMDCs. Consistent with these findings, SIRT6 −/− BMDCs exhibited increased endocytic activity and impaired allostimulatory capacity, the latter being hallmarks of cDCs immaturity (Lasiglie et al., 2016). Similarly, pharmacological inhibition with S6, a SIRT6 inhibitor, strongly impaired human MoDCs ability to stimulate T cells proliferation in mixed leukocytes reaction (MLR) (Lasiglie et al., 2016). In addition, *in vitro* migration to CCL21 was altered when mouse SIRT6 was loss, in line with a lower degree of CCR7 expression in both steady state and upon TLRs stimulation (Lasiglie et al., 2016). These data contributed to shed light for the first time on a crucial part

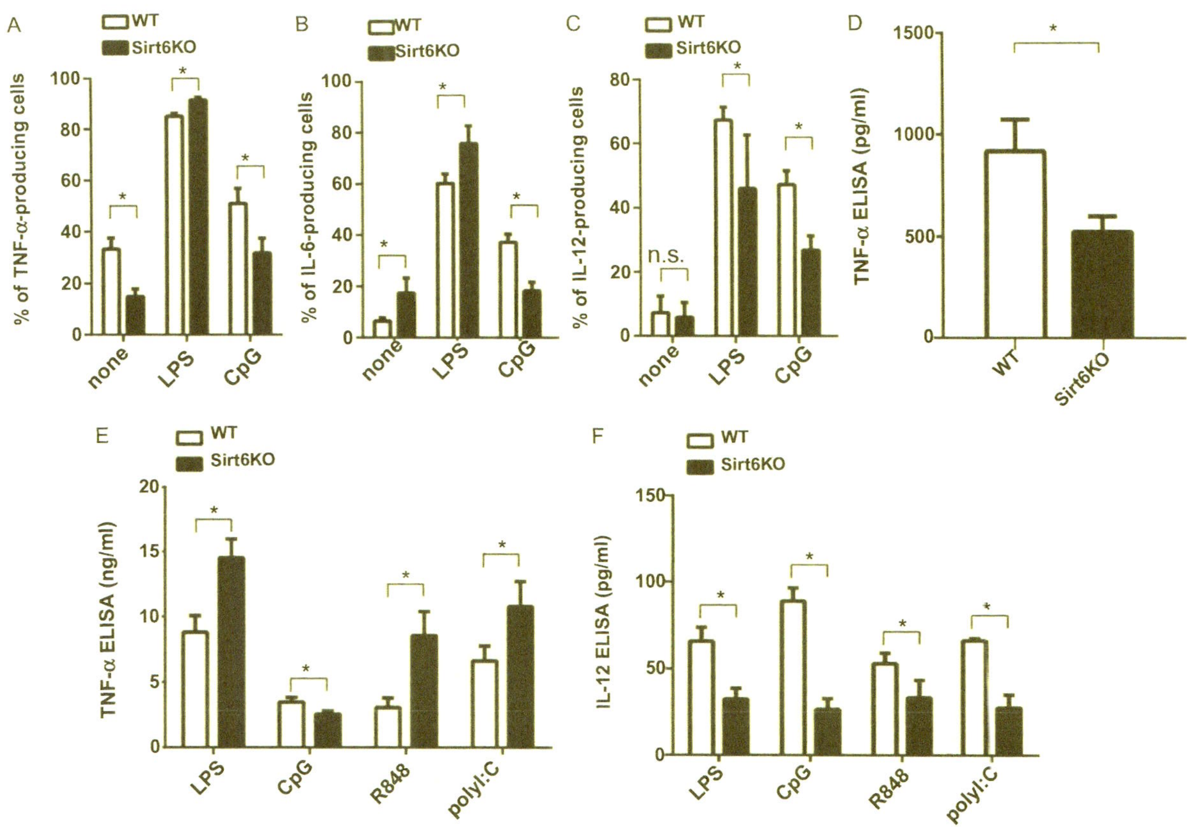

Fig. 3 *Cytokines profile in absence of SIRT6*. Production and secretion of TNF-α (A, D, E), IL-12 (C and F) IL-6 (B) by BMDCs in steady state and upon TLRs stimulation (LPS/TLR4, CpG/TLR9, R848/TLR7/8; POLY I:C/TLR3) as evaluated by FACS and ELISA. BMDCs, bone marrow-derived dendritic cells; TLR, Toll like receptor; LPS: lipopolysaccharide; CpG, short synthetic ss DNA containing unmethylated CpG motif; R848, resiquimod (synthetic short ssRNA); Poly I:C, polyinosinic:polycytidynic acid (synthetic dsRNA); TNF-α, tumor necrosis factor alpha; IL-12, interleukin 12; IL-6, interleukin 6; ss, single strand; ds, double strand. FACS, Fluorescence activated cell sorter; ELISA, enzyme-linked immunosorbent assay.

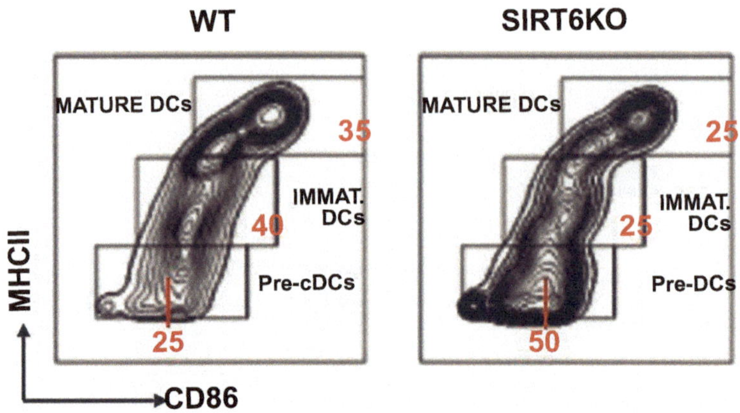

Fig. 4 *In vitro cDCs spontaneous maturation in absence of SIRT6.* 6 days *in vitro* cultured BMDCs: Percentage of mature, immature and Pre-cDCs has been evaluated by FACS. Analyses were performed by evaluating MHCII and CD86 expression on gated CD11c^{+} population. MHCII, Major histocompatibility complex II; CD86, cluster of differentiation 86; CD11c, cluster of differentiation 11c; BMDCs, bone marrow-derived dendritic cells; FACS, fluorescence activated cell sorter; Pre-cDCs, pre-mature cDCs.

of SIRT6 in DCs biology in both humans and mice. Pharmacological inhibition of sirtuins with cambinol and sirnitol was already demonstrated to alter cytokines production by DCs, macrophages and splenocytes as well as to impact lung DCs maturation and migration as well as a DCs driven Th2 response in a mouse model of airway inflammation, fueling these important observations. Furthermore, SIRT1 was reported to play an essential role in DCs in maturation and activation in both humans and mice (Alvarez et al., 2012; Legutko et al., 2011; Liu et al., 2015; Owczarczyk et al., 2015; Woo et al., 2016). The contribution of the remaining sirtuins to DCs biology is less documented and clear and remains at the moment mostly elusive, with evidence of a role of SIRT3 in DCs cytokines production (Ciarlo et al., 2017) as well as of SIRT2 in mediating DCs immune response to *Salmonella* (Gogoi et al., 2018). As altered cytokines production, maturation and migration are typical features of aged DCs (Wong & Goldstein, 2013) SIRT6 appears to play a major role in DCs aging process. Epigenetic and transcriptional regulation as well as the establishment of a DCs regulatory TFs circuitry were shown to be crucial for DCs lineage commitment and function as well as HDACs pharmacological inhibition was reported to negative influence myeloid and DCs commitment as well as DCs biology (Bartels, Geest, Bierings, Buitenhuis, & Coffer, 2010; Chauvistré et al., 2014; Frikeche et al., 2012; Lin et al., 2015; Song et al., 2011; Tian, Meng, & Zhang, 2017; Zenke & Hieronymus, 2006). Besides, growth factors and cytokines signaling (IL-6, IGF-1, TNF-α, GM-CSF, Flt3l, TGF-β1) both in steady state and inflammatory conditions may additionally contribute to DCs commitment, differentiation and function (Ding et al., 2011; Felker et al., 2010; Inaba et al., 1992; Lehner et al., 2012; Naik et al., 2005; Park et al., 2004; Sere et al., 2012). It is likely the age-related SIRT6 reduction (Grabowska, Sikora, & Bielak-Zmijewska, 2017) may alter the chromatin remodeling and the transcriptional program (comprising transcriptional factors and cytokines) crucial for a proper DCs ontogeny. In our work supplementation of high amounts of TNF-α was shown to partially rescue the *in vitro* defective spontaneous DCs maturation observed in absence of SIRT6. In particular TNF-α was demonstrated to be crucial for the immature-mature transition, whereas no effects was recorded for the Pre-mature (Pre) DCs-immature DCs passage (Lasiglie et al., 2016). Similarly, others characteristics, such as endocytosis, cannot be recovered by TNF-α supplementation (Lasigliè D, unpublished data), highlighting contribution of additional SIRT6 dependent mechanisms in dictating DCs function. In their elegant work Kawahara et al. (2011) identified in MEFs and by chromatin immunoprecipitation (ChIP) analysis, in both steady state and upon TNF-α treatment, thousands potential targets of SIRT6. Promoters bound by SIRT6 were proven to be enriched in genes involved in myeloid differentiation, DCs development, DCs subsets specification and function (*cebpa, gil1, gata2, tal1, runx1, nfil3, id2, flt3, flt4, irf2, relb, notch2, Stat5, fos, jun, tcf4, ets1, ly6c, nfkb1, nfkb2, nfatc1, nfatc2, il12, il6, il2, il27, il4, tgfb2, igf1*), immune system development and maintenance (*ghr, igf1*), cytoskeleton organization, intracellular trafficking, cell cycle progression (*cdc42, Rhoc, rab4b, rad9b*) and many others. These intriguing observations led us to explore the possibility that SIRT6 may have a role in DCs commitment, myeloid differentiation and more in general in hematopoiesis.

5.5 SIRT6, hematopoiesis and DCs commitment

First indications of a role of SIRT6 in DCs commitment came from observations of a reduced frequency of $CD11c^+$ cells generated when culturing total SIRT6 null BM cells in presence of GM-CSF (see Fig. 5 for acronymous). Time course analyses showed this defect being present early, as 3 day BM cultures already presented an altered number of $CD11c^+$ cells in absence of SIRT6 (Fig. 5A). BM cultures supplementation with physiological concentrations of TNF-α only partially rescued $CD11c^+$ frequency (Lasiglie et al., 2016) while higher concentrations of TNF-α even reduced the recovery of $CD11c^+$ SIRT6 null cells causing a switch toward the macrophage lineage

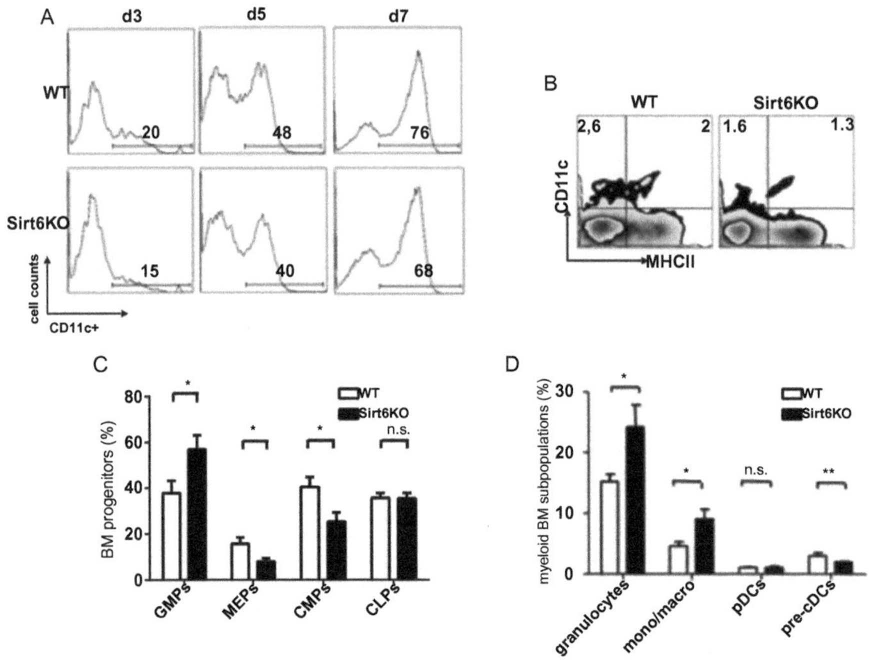

Fig. 5 *cDCs commitment and hematopoiesis in absence of SIRT6.* (A) Time course of CD11c$^+$ expression by total BM cells cultured *in vitro* in presence of GM-CSF. Percentage of CD11c$^+$ was determined by FACS. (B) *Ex vivo* FACS analysis of Pre-cDCs. (C) and (D) Frequency of BM progenitors and myeloid BM subpopulations as determined by FACS. GMP, granulocyte-macrophage progenitor; MEP, megacaryocyte-erythroid progenitor; CMP, common myeloid progenitor; CLP, common lymphoid progenitor; pDC, plasmocytoid dendritic cells; Pre-cDCs, pre-mature cDCs; mono/macro, monocytes/macrophages; GM-CSF, granulocyte-macrophage colony stimulating factor; FACS, fluorescence activated cell sorter.

(Lasigliè D, unpublished data). In like manner to what described for cDCs maturation, TNF-α failed to completely recovery the defective cDCs differentiation observed in absence of SIRT6, postulating supplementary SIRT6 dependent processes governing this process (Lasiglie et al., 2016). In humans SIRT6 appears to be crucial for the proper differentiation of monocytes to MoDCs, as pharmacological inhibition with S6 strongly blocked this process (Lasiglie et al., 2016). *Ex vivo* immune-phenotypical analyses of BM subpopulations and progenitors highlighted the reduced CD11c$^+$ frequency observed *in vitro* being already present *in vivo*. Indeed a significant decrease of Pre-cDCs frequency was observed in absence of SIRT6 in parallel to a robust expansion of granulocytes and monocytes/macrophages (see Fig. 5B and D). The number of pDCs seemed not to be altered and will need further studies. Consistent with these findings progenitors enumeration pointed out an aberrant myeloid lineage commitment with a robust reduction of CMPs and MEPs in favor of GMPs (Fig. 5C). As the latter gives rise to granulocytes, monocyte/macrophage and Pre-DCs, it is likely that the absence of SIRT6 may impact different steps of the myeloid commitment determining the expansion of some myeloid subpopulations (granulocytes, mono/macrophage) and the reduction of others (Pre-cDCs) (for more details on DCs commitment, the reader is referred to Mildner & Jung, 2014). Besides, in absence of SIRT6 a significant decrease in the frequency of B cell lineage committed $B220^{+}CD19^{+}$ cells was observed (Lasigliè D. personal communication). In harmony with reduced MEPs, platelet count and mean corpuscular hemoglobin concentration were significantly reduced in peripheral blood (Lasiglie et al., 2016) and a trend toward a reduced erythroid lineage commitment ($Tr119^{+}$ cells) was also observed when SIRT6 was lacking (Lasigliè D, personal communication). On the whole SIRT6 deficiency was proven to contribute to a switch toward myelopoiesis, a largely described age-related phenomenon (Rossi et al., 2005). Similarly, increased number of immune-phenotypically defined HSCs, another hallmark of aged HSCs, was observed in absence of SIRT6 (Lasigliè D personal communication) underlining a key role of SIRT6 in early stages of immunosenescence. A similar BM phenotype has been described when conditionally ablating SIRT6 in HSCs, with loss of quiescence in HSCs and increased frequency of GMPs (Wang et al., 2016). Other members of the sirtuins family have been described to take part to HSCs homeostasis and function as well as to balanced myeloid *vs* lymphoid lineage specification, such as SIRT1, SIRT2, SIRT3 and SIRT7 (Brown et al., 2013; Luo et al., 2019; Mohrin et al., 2015; Rimmele et al., 2014) underscoring for these sirtuins a conserved part in such processes. Though the hallmarks of aged HSCs are well defined the supremacy of intrinsic factors *vs* environmental cues (systemic or from the niche) in mediating HSCs aging is still under debate. Systemic or local factors such as GH and IGF-1 have been extensively indicated to play important parts in hematopoiesis, immune cells function regulation, lymphoid organs growth and maintenance (Clark, 1997) while SIRT6 is now emerging as a paramount factor for the regulation of the GH/IGF-1 axis/IGF system/IIS by acting at different levels and in different ways (see before). This is particular evident when analyzing SIRT6 null mice, the latter showing thymus atrophy, reduced spleen size, BM cellularity and osteopenia. The overall skeletal pool of IGF-1 captured in the bone matrix derives from paracrine and systemic IGF-1. During bone resorption IGF-1 is released from matrix storage by osteoclasts thus inducing osteoblasts differentiation from MSCs. MSCs themselves, pre-osteoblasts and osteoblasts in turn produce IGF-1 together with pre-osteoclasts, osteoclasts and osteocytes in a vitious cycle connecting bone resorption and formation. In fact systemic and niche-derived (autocrine and paracrine) IGF-1 pools are crucial for osteoclasts and osteoblasts proliferation, survival and differentiation, as well as for bone matrix deposition and resorption (Canalis, 1993; Hill, Reynolds, & Meikle, 1995; Wang, Bikle, & Chang, 2013; Xian et al., 2012; Zhang et al., 2002). Osteoblasts express SIRT6 and a role for SIRT6 in bone metabolism has been unrevealed (Sugatani, Agapova, Malluche, & Hruska, 2015). Though SIRT6 was proved to participate to bone remodeling by repressing *Runx2*, *Osx* and *dkk1* gene, it is not uncorrect to postulate it may contribute to bone homeostasis by positively regulating IGF-1 production (local and systemic). IGF-1 reduction from the BM niche has been recently identified as a top candidate regulator for hematopoietic stem and progenitor cells aging (Young et al., 2017). Transient and direct restoration of "young" IGF-1 levels in the BM niche was actually shown to cause a dose-dependent suppression of myeloid cells overproduction and skewing, restore B cells frequency and rescue other canonical markers of aged HSCs, thus phenotypically and functionally rebalancing hematopoiesis (Young et al., 2019, 2018). Different cell types take part to the BM niche, which regulates number, retention, quiescence, migration and differentiation of stem cells and progenitors cells (Orkin & Zon, 2008). Among BM niche components, osteoblasts were indicated to have a crucial role in supporting HSCs/progenitors biology (Geiger et al., 2013). Thus reduced systemic/local IGF-1 levels (i.e., from the osteoblastic niche) may impact HSCs/progenitors function, contributing to the BM myeloid skewed phenotype and to other hallmarks of aged HSCs (Caselli et al., 2013) observed when SIRT6 is lacking. While I was writing down this chapter, Young et al. (2019) identified stromal MSCs as the main source of IGF-1 in young BM and the MSCs-related IGF-1 decline occurring with age advancement was elegantly shown to be responsible for HSCs senescence. While several cell types may contribute to local IGF-1 production and thus it may be hazardous to simply ascribe it to a single cell type, IGF-1 signaling relevance seems out of question. Though IGF-1 is likely the factor which influences more HSCs aging, other systemic and/or niche-derived factors may cooperate (Geiger et al., 2013). IL-6 is a pleiotropic cytokine with largely described roles in inflammation, bone metabolism, osteoblasts and osteoclasts differentiation

(Ishimi et al., 1990; Kaneshiro et al., 2014; Peruzzi et al., 2012; Udagawa et al., 1995). In a context of systemic inflammation and infection, IL-6 is produced by circulating and organ resident mature myeloid and lymphoid cells. Besides osteoblasts, MSCs and other stromal cells as well as ST-HSCs and a subset of MPPs in the BM niche were also reported to produce IL-6, supporting MPPs proliferation and dictating myelopoiesis at the expense of lymphopoiesis (Maeda et al., 2009; Mirantes, Passegue, & Pietras, 2014; Tie et al., 2019; Zhao et al., 2014). In addition, *in vitro* studies demonstrated over-expression of IL-6 (and TNF-α) may impact IGF-1 activity (Lazarus, Moldawer, & Lowry, 1993). In this direction senescent murine SIRT6 null MSCs were proven to produce huge amounts of IL-6 (Lasiglié D personal communication). Thus SIRT6 may also contribute to hematopoietic aging by regulating IL-6 expression. As previously mentioned IL-6 production is part to a wider picture of low and chronic systemic inflammation (inflammaging) observed with age in both human and mice (Franceschi et al., 2000). The importance of environmental cues, in particular from the inflammatory milieu, for HSCs aging (Montecino-Rodriguez et al., 2019) is becoming more and more evident without undermining the relevance of intrinsic factors. Though DNA damage, ROS accumulation and erosion of telomeres are conserved and well known senescence mechanisms, new research techniques have uncovered a growing significance of epigenetic and transcriptional landscapes changes in determining HSCs aging, with histone modifiers such as *hdac1*, *hdac2*, *hdac6* and *sirt2*, *sirt3*, *sirt7* expression becoming reduced in aged HSCs (Chambers et al., 2007). In this context SIRT3 and SIRT7 over-expression was shown to rejuvenate aged HSCs (Latchney & Calvi, 2017). In such a scenario, SIRT6 appears as an intriguing factor to be targeted to reverse the age-dependent HSCs defects. The multitasking essence of SIRT6 is already remarkable when considering HSCs senescence, with this sirtuin able to prevent such a phenomenon at different level, starting from telomere and genome integrity maintenance, epigenetic signature and transcriptional control and finishing by intercellular paracrine and/or systemic communication (GH, IGF-1, IL-6, TNF-α) regulation. Thus by controlling intrinsic and extrinsic factors involved in HSCs aging in an integrated fashion, SIRT6 may be considered a crucial regulator of HSCs biology. While the increased proliferation and the following HSCs expansion observed in absence of SIRT6 may be connected to an aberrant activation of the Wnt pathway occurring simultaneously to an altered (IGF-1/insulin) Akt signaling (Kharas et al., 2010; Perry et al., 2011; Sundaresan et al., 2012; Wang et al., 2016), subsequent reduced self-renewal and functional exhaustion may be attributable to telomere shortening, altered

DNA repair, increased ROS and the following DNA damage accumulation (Rossi et al., 2005). SIRT6 dependent IGF-1downmodulaton (direct or cytokines and/or DNA damage-dependent) (Niedernhofer et al., 2006; O'Connor et al., 2008) may then contribute to the BM myeloid skewed phenotype and the altered B lymphopoiesis observed when ablating SIRT6, concomitantly to a SIRT6-related myeloid and inflammation biased transcriptional profile and maybe to a putative systemic and local IL-6 and TNF-α overproduction. SIRT6 contribution to DCs biology may be less defined, with however evidence of a mainly epigenetic part in DCs differentiation: indeed several TFs and cytokines (such as IL-6 and maybe TGFβ2) (Thomas et al., 2001) crucial for DCs development and functional specification may be under SIRT6 transcriptional control (Kawahara et al., 2011). The overall picture is surely more complex than previously described, taking into account the various SIRT6 functions and the growing numbers of substrates continuously discovered for SIRT6.

6. Future directions and perspectives

SIRT6 null mice has represented a powerful tool to evaluate SIRT6 contribution to cytokines production as well as to unravel its crucial part in immunosenescence. SIRT6 was shown to have a prominent role in DCs aging, with cell number and several phenotypical and functional properties of DCs being altered when SIRT6 was ablated. These discoveries paved the way to wider considerations on the role of SIRT6 in DCs commitment and hematopoiesis, contributing to shed light on an important role of this sirtuin in the HSCs aging process. Several sirtuins were already proved to contribute to HSCs aging (see before), thus becoming attractive targets for dampening and/or reversing the age-associated HSCs defects. In such a view SIRT6, being at the crossroad between intrinsic and extrinsic avenues leading to HSCs aging, and thus to immunosenescence, may be targetable for the development of reprogramming and rejuvenating HSCs agents, with the aim to improve immune response, reduce vaccine failure as well as to decrease the risk for age-dependent myeloid malignancies. Thus, a more detailed analysis and understanding of the contribution of SIRT6 to HSCs aging and immunosenescence (i.e., considering other immune system populations and/or districts and over-expressing SIRT6 in aged HSCs) will be crucial for future discoveries of treatments aiming at ameliorating immune function in the elderly and combating age-related hematopoietic diseases.

Aknowledgments

I apologize with all the colleagues I was unable to cite due to space restriction. Sirtuins literature is enormous and some aspects of sirtuins biology, such as metabolism and their part in CR have been deliberately unaddressed in this chapter for the sake of brevity and clarity, despite the largely documented interconnection with the aging process. Similarly, NFkB–mediated SIRT6 effects on aging and lifespan were consciously untouched. A warm thank to my former boss Dr. A. Nencioni, University of Genoa, for having helped me and supported during my work on SIRT6. All the considerations made in this chapter on the role of SIRT6 in dendritic cells and HSCs cell biology stemmed from my previous works in his lab. Besides I would like to thank Dr. G. Litwark for the great chance he gave to me. The possibility to write down a chapter during a sort of sabbatical leave was truly important to me as let me keep in touch to what has been and still is one of the strongest passion of my life. A special word of thanks to my friend and colleague Dr. Giulia Amico for helping me during the preparation of this chapter.

References

Agrawal, A., Agrawal, S., Cao, J. N., Su, H., Osann, K., & Gupta, S. (2007). Altered innate immune functioning of dendritic cells in elderly humans: A role of phosphoinositide 3-kinase-signaling pathway. *Journal of Immunology*, *178*(11), 6912–6922. https://doi.org/10.4049/jimmunol.178.11.6912.

Agrawal, A., Agrawal, S., & Gupta, S. (2007). Dendritic cells in human aging. *Experimental Gerontology*, *42*(5), 421–426. https://doi.org/10.1016/j.exger.2006.11.007.

Agrawal, A., Agrawal, S., & Gupta, S. (2017). Role of dendritic cells in inflammation and loss of tolerance in the elderly. *Frontiers in Immunology*, *8*, 896. https://doi.org/10.3389/fimmu.2017.00896.

Agrawal, A., & Gupta, S. (2011). Impact of aging on dendritic cell functions in humans. *Ageing Research Reviews*, *10*(3), 336–345. https://doi.org/10.1016/j.arr.2010.06.004.

Agrawal, A., Tay, J., Ton, S., Agrawal, S., & Gupta, S. (2009). Increased reactivity of dendritic cells from aged subjects to self-antigen, the human DNA. *Journal of Immunology (Baltimore, Md.: 1950)*, *182*(2), 1138–1145. https://doi.org/10.4049/jimmunol.182.2.1138.

Akashi, K., Traver, D., Miyamoto, T., & Weissman, I. L. (2000). A clonogenic common myeloid progenitor that gives rise to all myeloid lineages. *Nature*, *404*(6774), 193–197. https://doi.org/10.1038/35004599.

Alexander, W. S. (1998). Cytokines in hematopoiesis. *International Reviews of Immunology*, *16*(5–6), 651–682. https://doi.org/10.3109/08830189809043013.

Alpdogan, O., Muriglan, S. J., Kappel, B. J., Doubrovina, E., Schmaltz, C., Schiro, R., et al. (2003). Insulin-like growth factor-I enhances lymphoid and myeloid reconstitution after allogeneic bone marrow transplantation. *Transplantation*, *75*(12), 1977–1983. https://doi.org/10.1097/01.TP.0000070167.81584.A2.

Alvarez, Y., Rodriguez, M., Municio, C., Hugo, E., Alonso, S., Ibarrola, N., et al. (2012). Sirtuin 1 is a key regulator of the interleukin-12 p70/interleukin-23 balance in human dendritic cells. *The Journal of Biological Chemistry*, *287*(42), 35689–35701. https://doi.org/10.1074/jbc.M112.391839.

Annunziata, M., Granata, R., & Ghigo, E. (2011). The IGF system. *Acta Diabetologica*, *48*(1), 1–9. https://doi.org/10.1007/s00592-010-0227-z.

Baker, L. D., Barsness, S. M., Borson, S., Merriam, G. R., Friedman, S. D., Craft, S., et al. (2012). Effects of growth hormone-releasing hormone on cognitive function in adults with mild cognitive impairment and healthy older adults: Results of a controlled trial. *Archives of Neurology*, *69*(11), 1420–1429. https://doi.org/10.1001/archneurol.2012.1970.

Baroni, C., Fabris, N., & Bertoli, G. (1967). Age dependency of the primary immune response in the hereditary pituitary dwarf and normal Snell-Bagg mouse. *Experientia, 23*(12), 1059–1060. https://doi.org/10.1007/bf02136450.

Bartels, M., Geest, C. R., Bierings, M., Buitenhuis, M., & Coffer, P. J. (2010). Histone deacetylase inhibition modulates cell fate decisions during myeloid differentiation. *Haematologica, 95*(7), 1052–1060. https://doi.org/10.3324/haematol.2009.008870.

Bateman, A., Singh, A., Kral, T., & Solomon, S. (1989). The immune-hypothalamic-pituitary-adrenal axis. *Endocrine Reviews, 10*(1), 92–112. https://doi.org/10.1210/edrv-10-1-92.

Bauer, I., Grozio, A., Lasiglie, D., Basile, G., Sturla, L., Magnone, M., et al. (2012). The NAD+-dependent histone deacetylase SIRT6 promotes cytokine production and migration in pancreatic cancer cells by regulating Ca2+ responses. *The Journal of Biological Chemistry, 287*(49), 40924–40937. https://doi.org/10.1074/jbc.M112.405837.

Beerman, I., Bhattacharya, D., Zandi, S., Sigvardsson, M., Weissman, I. L., Bryder, D., et al. (2010). Functionally distinct hematopoietic stem cells modulate hematopoietic lineage potential during aging by a mechanism of clonal expansion. *Proceedings of the National Academy of Sciences of the United States of America, 107*(12), 5465–5470. https://doi.org/10.1073/pnas.1000834107.

Beerman, I., Bock, C., Garrison, B. S., Smith, Z. D., Gu, H., Meissner, A., et al. (2013). Proliferation-dependent alterations of the DNA methylation landscape underlie hematopoietic stem cell aging. *Cell Stem Cell, 12*(4), 413–425. https://doi.org/10.1016/j.stem.2013.01.017.

Berliner, N. (2013). Anemia in the elderly. *Transactions of the American Clinical and Climatological Association, 124*, 230–237.

Bheda, P., Jing, H., Wolberger, C., & Lin, H. (2016). The substrate specificity of sirtuins. *Annual Review of Biochemistry, 85*, 405–429. https://doi.org/10.1146/annurev-biochem-060815-014537.

Blackman, M. R., Sorkin, J. D., Munzer, T., Bellantoni, M. F., Busby-Whitehead, J., Stevens, T. E., et al. (2002). Growth hormone and sex steroid administration in healthy aged women and men: A randomized controlled trial. *JAMA, 288*(18), 2282–2292. https://doi.org/10.1001/jama.288.18.2282.

Bosch-Presegue, L., & Vaquero, A. (2014). Sirtuins in stress response: Guardians of the genome. *Oncogene, 33*(29), 3764–3775. https://doi.org/10.1038/onc.2013.344.

Bosch-Presegue, L., & Vaquero, A. (2015). Sirtuin-dependent epigenetic regulation in the maintenance of genome integrity. *The FEBS Journal, 282*(9), 1745–1767. https://doi.org/10.1111/febs.13053.

Briceno, O., Lissina, A., Wanke, K., Afonso, G., von Braun, A., Ragon, K., et al. (2016). Reduced naive CD8(+) T-cell priming efficacy in elderly adults. *Aging Cell, 15*(1), 14–21. https://doi.org/10.1111/acel.12384.

Brooks, A. J., & Waters, M. J. (2010). The growth hormone receptor: Mechanism of activation and clinical implications. *Nature Reviews. Endocrinology, 6*(9), 515–525. https://doi.org/10.1038/nrendo.2010.123.

Broudy, V. C. (1997). Stem cell factor and hematopoiesis. *Blood, 90*(4), 1345–1364.

Brown, K., Xie, S., Qiu, X., Mohrin, M., Shin, J., Liu, Y., et al. (2013). SIRT3 reverses aging-associated degeneration. *Cell Reports, 3*(2), 319–327. https://doi.org/10.1016/j.celrep.2013.01.005.

Brubaker, A. L., Rendon, J. L., Ramirez, L., Choudhry, M. A., & Kovacs, E. J. (2013). Reduced neutrophil chemotaxis and infiltration contributes to delayed resolution of cutaneous wound infection with advanced age. *Journal of Immunology, 190*(4), 1746–1757. https://doi.org/10.4049/jimmunol.1201213.

Busch, K., Klapproth, K., Barile, M., Flossdorf, M., Holland-Letz, T., Schlenner, S. M., et al. (2015). Fundamental properties of unperturbed haematopoiesis from stem cells in vivo. *Nature, 518*(7540), 542–546. https://doi.org/10.1038/nature14242.

Butcher, S. K., Chahal, H., Nayak, L., Sinclair, A., Henriquez, N. V., Sapey, E., et al. (2001). Senescence in innate immune responses: Reduced neutrophil phagocytic capacity and CD16 expression in elderly humans. *Journal of Leukocyte Biology*, *70*(6), 881–886.

Butcher, S., Chahel, H., & Lord, J. M. (2000). Review article: Ageing and the neutrophil: No appetite for killing? *Immunology*, *100*(4), 411–416. https://doi.org/10.1046/j.1365-2567.2000.00079.x.

Canalis, E. (1993). Insulin like growth factors and the local regulation of bone formation. *Bone*, *14*(3), 273–276. https://doi.org/10.1016/8756-3282(93)90151-y.

Cantó, C., & Auwerx, J. (2009). Caloric restriction, SIRT1 and longevity. *Trends in Endocrinology and Metabolism: TEM*, *20*(7), 325–331. https://doi.org/10.1016/j.tem.2009.03.008.

Carter, C. S., Ramsey, M. M., & Sonntag, W. E. (2002). A critical analysis of the role of growth hormone and IGF-1 in aging and lifespan. *Trends in Genetics*, *18*(6), 295–301. https://doi.org/10.1016/S0168-9525(02)02696-3.

Caselli, A., Olson, T. S., Otsuru, S., Chen, X., Hofmann, T. J., Nah, H.-D., et al. (2013). IGF-1-mediated osteoblastic niche expansion enhances long-term hematopoietic stem cell engraftment after murine bone marrow transplantation. *Stem Cells*, *31*(10), 2193–2204. https://doi.org/10.1002/stem.1463.

Challen, G. A., Boles, N. C., Chambers, S. M., & Goodell, M. A. (2010). Distinct hematopoietic stem cell subtypes are differentially regulated by TGF-beta1. *Cell Stem Cell*, *6*(3), 265–278. https://doi.org/10.1016/j.stem.2010.02.002.

Chambers, S. M., Shaw, C. A., Gatza, C., Fisk, C. J., Donehower, L. A., & Goodell, M. A. (2007). Aging hematopoietic stem cells decline in function and exhibit epigenetic dysregulation. *PLoS Biology*, *5*(8), e201. https://doi.org/10.1371/journal.pbio.0050201.

Chang, H. C., & Guarente, L. (2014). SIRT1 and other sirtuins in metabolism. *Trends in Endocrinology and Metabolism*, *25*(3), 138–145. https://doi.org/10.1016/j.tem.2013.12.001.

Chauvistré, H., Küstermann, C., Rehage, N., Klisch, T., Mitzka, S., Felker, P., et al. (2014). Dendritic cell development requires histone deacetylase activity. *European Journal of Immunology*, *44*(8), 2478–2488. https://doi.org/10.1002/eji.201344150.

Chen, D., Bruno, J., Easlon, E., Lin, S. J., Cheng, H. L., Alt, F. W., et al. (2008). Tissue-specific regulation of SIRT1 by calorie restriction. *Genes & Development*, *22*(13), 1753–1757. https://doi.org/10.1101/gad.1650608.

Chen, B. J., Deoliveira, D., Spasojevic, I., Sempowski, G. D., Jiang, C., Owzar, K., et al. (2010). Growth hormone mitigates against lethal irradiation and enhances hematologic and immune recovery in mice and nonhuman primates. *PLoS One*, *5*(6), e11056. https://doi.org/10.1371/journal.pone.0011056.

Chen, S., Hu, M., Shen, M., Wang, S., Wang, C., Chen, F., et al. (2018). IGF-1 facilitates thrombopoiesis primarily through Akt activation. *Blood*, *132*(2), 210–222. https://doi.org/10.1182/blood-2018-01-825927.

Cheng, H., Zheng, Z., & Cheng, T. (2020). New paradigms on hematopoietic stem cell differentiation. *Protein & Cell*, *11*(1), 34–44. https://doi.org/10.1007/s13238-019-0633-0.

Cho, R. H., Sieburg, H. B., & Muller-Sieburg, C. E. (2008). A new mechanism for the aging of hematopoietic stem cells: Aging changes the clonal composition of the stem cell compartment but not individual stem cells. *Blood*, *111*(12), 5553–5561. https://doi.org/10.1182/blood-2007-11-123547.

Choi, J. E., & Mostoslavsky, R. (2014). Sirtuins, metabolism, and DNA repair. *Current Opinion in Genetics & Development*, *26*, 24–32. https://doi.org/10.1016/j.gde.2014.05.005.

Chougnet, C. A., Thacker, R. I., Shehata, H. M., Hennies, C. M., Lehn, M. A., Lages, C. S., et al. (2015). Loss of phagocytic and antigen cross-presenting capacity in aging dendritic cells is associated with mitochondrial dysfunction. *Journal of Immunology*, *195*(6), 2624–2632. https://doi.org/10.4049/jimmunol.1501006.

Ciarlo, E., Heinonen, T., Lugrin, J., Acha-Orbea, H., Le Roy, D., Auwerx, J., et al. (2017). Sirtuin 3 deficiency does not alter host defenses against bacterial and fungal infections. *Scientific Reports*, 7(1), 3853. https://doi.org/10.1038/s41598-017-04263-x.

Clark, R. (1997). The somatogenic hormones and insulin-like growth factor-1: Stimulators of lymphopoiesis and immune function. *Endocrine Reviews*, *18*(2), 157–179. https://doi.org/10.1210/edrv.18.2.0296.

Clark, R., Strasser, J., McCabe, S., Robbins, K., & Jardieu, P. (1993). Insulin-like growth factor-1 stimulation of lymphopoiesis. *The Journal of Clinical Investigation*, *92*(2), 540–548. https://doi.org/10.1172/JCI116621.

Cohen, D. E., Supinski, A. M., Bonkowski, M. S., Donmez, G., & Guarente, L. P. (2009). Neuronal SIRT1 regulates endocrine and behavioral responses to calorie restriction. *Genes & Development*, *23*(24), 2812–2817. https://doi.org/10.1101/gad.1839209.

Dardenne, M., Smaniotto, S., de Mello-Coelho, V., Villa-Verde, D. M., & Savino, W. (2009). Growth hormone modulates migration of developing T cells. *Annals of the New York Academy of Sciences*, *1153*, 1–5. https://doi.org/10.1111/j.1749-6632.2008.03977.x.

Daughaday, W. H. (1989). Fetal lung fibroblasts secrete and respond to insulin-like growth factors. *American Journal of Respiratory Cell and Molecular Biology*, *1*(1), 11–12. https://doi.org/10.1165/ajrcmb/1.1.11.

Ding, X., Yang, W., Shi, X., Du, P., Su, L., Qin, Z., et al. (2011). TNF receptor 1 mediates dendritic cell maturation and CD8 T cell response through two distinct mechanisms. *Journal of Immunology*, *187*(3), 1184–1191. https://doi.org/10.4049/jimmunol.1002902.

Dolence, J. J., Gwin, K. A., Shapiro, M. B., & Medina, K. L. (2014). Flt3 signaling regulates the proliferation, survival, and maintenance of multipotent hematopoietic progenitors that generate B cell precursors. *Experimental Hematology*, *42*(5), 380–393.e383. https://doi.org/10.1016/j.exphem.2014.01.001.

Dorrington, M. G., & Bowdish, D. M. (2013). Immunosenescence and novel vaccination strategies for the elderly. *Frontiers in Immunology*, *4*, 171. https://doi.org/10.3389/fimmu.2013.00171.

Dorshkind, K., Montecino-Rodriguez, E., & Signer, R. A. (2009). The ageing immune system: Is it ever too old to become young again? *Nature Reviews. Immunology*, *9*(1), 57–62. https://doi.org/10.1038/nri2471.

Dykstra, B., Olthof, S., Schreuder, J., Ritsema, M., & de Haan, G. (2011). Clonal analysis reveals multiple functional defects of aged murine hematopoietic stem cells. *The Journal of Experimental Medicine*, *208*(13), 2691–2703. https://doi.org/10.1084/jem.20111490.

El Mezayen, R., El Gazzar, M., Myer, R., & High, K. P. (2009). Aging-dependent upregulation of IL-23p19 gene expression in dendritic cells is associated with differential transcription factor binding and histone modifications. *Aging Cell*, *8*(5), 553–565. https://doi.org/10.1111/j.1474-9726.2009.00502.x.

Elkhwanky, M. S., & Hakkola, J. (2018). Extranuclear sirtuins and metabolic stress. *Antioxidants & Redox Signaling*, *28*(8), 662–676. https://doi.org/10.1089/ars.2017.7270.

Ergen, A. V., Boles, N. C., & Goodell, M. A. (2012). Rantes/Ccl5 influences hematopoietic stem cell subtypes and causes myeloid skewing. *Blood*, *119*(11), 2500–2509. https://doi.org/10.1182/blood-2011-11-391730.

Falomir-Lockhart, E., Dolcetti, F. J. C., Garcia-Segura, L. M., Herenu, C. B., & Bellini, M. J. (2019). IGF1 gene therapy modifies microglia in the striatum of senile rats. *Frontiers in Aging Neuroscience*, *11*, 48. https://doi.org/10.3389/fnagi.2019.00048.

Fang, E. F. (2019). Mitophagy and NAD(+) inhibit Alzheimer disease. *Autophagy*, *15*(6), 1112–1114. https://doi.org/10.1080/15548627.2019.1596497.

Felker, P., Sere, K., Lin, Q., Becker, C., Hristov, M., Hieronymus, T., et al. (2010). TGF-beta1 accelerates dendritic cell differentiation from common dendritic cell progenitors and directs subset specification toward conventional dendritic cells. *Journal of Immunology*, *185*(9), 5326–5335. https://doi.org/10.4049/jimmunol.0903950.

Florian, M. C., Dorr, K., Niebel, A., Daria, D., Schrezenmeier, H., Rojewski, M., et al. (2012). Cdc42 activity regulates hematopoietic stem cell aging and rejuvenation. *Cell Stem Cell*, *10*(5), 520–530. https://doi.org/10.1016/j.stem.2012.04.007.

Florini, J. R., Ewton, D. Z., & Coolican, S. A. (1996). Growth hormone and the insulin-like growth factor system in myogenesis. *Endocrine Reviews*, *17*(5), 481–517. https://doi.org/10.1210/edrv-17-5-481.

Fontana, L., Klein, S., & Holloszy, J. O. (2010). Effects of long-term calorie restriction and endurance exercise on glucose tolerance, insulin action, and adipokine production. *Age (Dordrecht, Netherlands)*, *32*(1), 97–108. https://doi.org/10.1007/s11357-009-9118-z.

Franceschi, C., Bonafe, M., Valensin, S., Olivieri, F., De Luca, M., Ottaviani, E., et al. (2000). Inflamm-aging. An evolutionary perspective on immunosenescence. *Annals of the New York Academy of Sciences*, *908*, 244–254. https://doi.org/10.1111/j.1749-6632.2000.tb06651.x.

French, R. A., Broussard, S. R., Meier, W. A., Minshall, C., Arkins, S., Zachary, J. F., et al. (2002). Age-associated loss of bone marrow hematopoietic cells is reversed by GH and accompanies thymic reconstitution. *Endocrinology*, *143*(2), 690–699. https://doi.org/10.1210/endo.143.2.8612.

Frikeche, J., Peric, Z., Brissot, E., Gregoire, M., Gaugler, B., & Mohty, M. (2012). Impact of HDAC inhibitors on dendritic cell functions. *Experimental Hematology*, *40*(10), 783–791. https://doi.org/10.1016/j.exphem.2012.06.008.

Frye, R. A. (1999). Characterization of five human cDNAs with homology to the yeast SIR2 gene: Sir2-like proteins (sirtuins) metabolize NAD and may have protein ADP-ribosyltransferase activity. *Biochemical and Biophysical Research Communications*, *260*(1), 273–279. https://doi.org/10.1006/bbrc.1999.0897.

Garinis, G. A., van der Horst, G. T., Vijg, J., & Hoeijmakers, J. H. (2008). DNA damage and ageing: New-age ideas for an age-old problem. *Nature Cell Biology*, *10*(11), 1241–1247. https://doi.org/10.1038/ncb1108-1241.

Geiger, H., de Haan, G., & Florian, M. C. (2013). The ageing haematopoietic stem cell compartment. *Nature Reviews. Immunology*, *13*(5), 376–389. https://doi.org/10.1038/nri3433.

Geiger, H., Koehler, A., & Gunzer, M. (2007). Stem cells, aging, niche, adhesion and Cdc42: A model for changes in cell-cell interactions and hematopoietic stem cell aging. *Cell Cycle*, *6*(8), 884–887. https://doi.org/10.4161/cc.6.8.4131.

Geissmann, F., Manz, M. G., Jung, S., Sieweke, M. H., Merad, M., & Ley, K. (2010). Development of monocytes, macrophages, and dendritic cells. *Science*, *327*(5966), 656–661. https://doi.org/10.1126/science.1178331.

Gogoi, M., Chandra, K., Sarikhani, M., Ramani, R., Sundaresan, N. R., & Chakravortty, D. (2018). Salmonella escapes adaptive immune response via SIRT2 mediated modulation of innate immune response in dendritic cells. *PLoS Pathogens*, *14*(11), e1007437. https://doi.org/10.1371/journal.ppat.1007437.

Gomez, C. R., Nomellini, V., Faunce, D. E., & Kovacs, E. J. (2008). Innate immunity and aging. *Experimental Gerontology*, *43*(8), 718–728. https://doi.org/10.1016/j.exger.2008.05.016.

Goronzy, J. J., & Weyand, C. M. (2013). Understanding immunosenescence to improve responses to vaccines. *Nature Immunology*, *14*(5), 428–436. https://doi.org/10.1038/ni.2588.

Grabowska, W., Sikora, E., & Bielak-Zmijewska, A. (2017). Sirtuins, a promising target in slowing down the ageing process. *Biogerontology*, *18*(4), 447–476. https://doi.org/10.1007/s10522-017-9685-9.

Greiss, S., & Gartner, A. (2009). Sirtuin/Sir2 phylogeny, evolutionary considerations and structural conservation. *Molecules and Cells*, *28*(5), 407–415. https://doi.org/10.1007/s10059-009-0169-x.

Grolleau-Julius, A., Harning, E. K., Abernathy, L. M., & Yung, R. L. (2008). Impaired dendritic cell function in aging leads to defective antitumor immunity. *Cancer Research*, *68*(15), 6341–6349. https://doi.org/10.1158/0008-5472.CAN-07-5769.

Guarente, L. (2011). Sirtuins, aging, and metabolism. *Cold Spring Harbor Symposia on Quantitative Biology*, *76*, 81–90. https://doi.org/10.1101/sqb.2011.76.010629.

Guarente, L. (2013). Calorie restriction and sirtuins revisited. *Genes & Development*, *27*(19), 2072–2085. https://doi.org/10.1101/gad.227439.113.

Hanley, M. B., Napolitano, L. A., & McCune, J. M. (2005). Growth hormone-induced stimulation of multilineage human hematopoiesis. *Stem Cells*, *23*(8), 1170–1179. https://doi.org/10.1634/stemcells.2004-0322.

Henry, C. J., Marusyk, A., Zaberezhnyy, V., Adane, B., & DeGregori, J. (2010). Declining lymphoid progenitor fitness promotes aging-associated leukemogenesis. *Proceedings of the National Academy of Sciences of the United States of America*, *107*(50), 21713–21718. https://doi.org/10.1073/pnas.1005486107.

Herranz, D., Munoz-Martin, M., Canamero, M., Mulero, F., Martinez-Pastor, B., Fernandez-Capetillo, O., et al. (2010). Sirt1 improves healthy ageing and protects from metabolic syndrome-associated cancer. *Nature Communications*, *1*, 3. https://doi.org/10.1038/ncomms1001.

Hill, P. A., Reynolds, J. J., & Meikle, M. C. (1995). Osteoblasts mediate insulin-like growth factor-I and -II stimulation of osteoclast formation and function. *Endocrinology*, *136*(1), 124–131. https://doi.org/10.1210/endo.136.1.7828521.

Holzenberger, M., Dupont, J., Ducos, B., Leneuve, P., Géloën, A., Even, P. C., et al. (2003). IGF-1 receptor regulates lifespan and resistance to oxidative stress in mice. *Nature*, *421*(6919), 182–187. https://doi.org/10.1038/nature01298.

Houtkooper, R. H., Pirinen, E., & Auwerx, J. (2012). Sirtuins as regulators of metabolism and healthspan. *Nature Reviews. Molecular Cell Biology*, *13*(4), 225–238. https://doi.org/10.1038/nrm3293.

Imai, S., Armstrong, C. M., Kaeberlein, M., & Guarente, L. (2000). Transcriptional silencing and longevity protein Sir2 is an NAD-dependent histone deacetylase. *Nature*, *403*(6771), 795–800. https://doi.org/10.1038/35001622.

Inaba, K., Inaba, M., Romani, N., Aya, H., Deguchi, M., Ikehara, S., et al. (1992). Generation of large numbers of dendritic cells from mouse bone marrow cultures supplemented with granulocyte/macrophage colony-stimulating factor. *The Journal of Experimental Medicine*, *176*(6), 1693–1702. https://doi.org/10.1084/jem.176.6.1693.

Ishimi, Y., Miyaura, C., Jin, C. H., Akatsu, T., Abe, E., Nakamura, Y., et al. (1990). IL-6 is produced by osteoblasts and induces bone resorption. *Journal of Immunology*, *145*(10), 3297–3303.

Jackson, M. D., & Denu, J. M. (2002). Structural identification of 2′- and 3'-O-acetyl-ADP-ribose as novel metabolites derived from the Sir2 family of beta -NAD+-dependent histone/protein deacetylases. *The Journal of Biological Chemistry*, *277*(21), 18535–18544. https://doi.org/10.1074/jbc.M200671200.

Janzen, V., Forkert, R., Fleming, H. E., Saito, Y., Waring, M. T., Dombkowski, D. M., et al. (2006). Stem-cell ageing modified by the cyclin-dependent kinase inhibitor p16INK4a. *Nature*, *443*(7110), 421–426. https://doi.org/10.1038/nature05159.

Jin, K. (2010). Modern biological theories of aging. *Aging and Disease*, *1*(2), 72–74.

Jing, H., & Lin, H. (2015). Sirtuins in epigenetic regulation. *Chemical Reviews*, *115*(6), 2350–2375. https://doi.org/10.1021/cr500457h.

Jing, Y., Shaheen, E., Drake, R. R., Chen, N., Gravenstein, S., & Deng, Y. (2009). Aging is associated with a numerical and functional decline in plasmacytoid dendritic cells, whereas myeloid dendritic cells are relatively unaltered in human peripheral blood. *Human Immunology*, *70*(10), 777–784. https://doi.org/10.1016/j.humimm.2009.07.005.

Kaeberlein, M., McVey, M., & Guarente, L. (1999). The SIR2/3/4 complex and SIR2 alone promote longevity in *Saccharomyces cerevisiae* by two different mechanisms. *Genes & Development*, *13*(19), 2570–2580. https://doi.org/10.1101/gad.13.19.2570.

Kaneshiro, S., Ebina, K., Shi, K., Higuchi, C., Hirao, M., Okamoto, M., et al. (2014). IL-6 negatively regulates osteoblast differentiation through the SHP2/MEK2 and SHP2/Akt2 pathways in vitro. *Journal of Bone and Mineral Metabolism*, *32*(4), 378–392. https://doi.org/10.1007/s00774-013-0514-1.

Kanfi, Y., Naiman, S., Amir, G., Peshti, V., Zinman, G., Nahum, L., et al. (2012). The sirtuin SIRT6 regulates lifespan in male mice. *Nature*, *483*(7388), 218–221. https://doi.org/10.1038/nature10815.

Katz, J. M., Plowden, J., Renshaw-Hoelscher, M., Lu, X., Tumpey, T. M., & Sambhara, S. (2004). Immunity to influenza: The challenges of protecting an aging population. *Immunologic Research*, *29*(1–3), 113–124. https://doi.org/10.1385/IR:29:1-3:113.

Kaushansky, K. (2006). Hematopoietic growth factors, signaling and the chronic myeloproliferative disorders. *Cytokine & Growth Factor Reviews*, *17*(6), 423–430. https://doi.org/10.1016/j.cytogfr.2006.09.005.

Kawa, M. P., Stecewicz, I., Piecyk, K., Pius-Sadowska, E., Paczkowska, E., Roginska, D., et al. (2015). Effects of growth hormone therapeutic supplementation on hematopoietic stem/progenitor cells in children with growth hormone deficiency: Focus on proliferation and differentiation capabilities. *Endocrine*, *50*(1), 162–175. https://doi.org/10.1007/s12020-015-0591-0.

Kawahara, T. L., Michishita, E., Adler, A. S., Damian, M., Berber, E., Lin, M., et al. (2009). SIRT6 links histone H3 lysine 9 deacetylation to NF-kappaB-dependent gene expression and organismal life span. *Cell*, *136*(1), 62–74. https://doi.org/10.1016/j.cell.2008.10.052.

Kawahara, T. L., Rapicavoli, N. A., Wu, A. R., Qu, K., Quake, S. R., & Chang, H. Y. (2011). Dynamic chromatin localization of Sirt6 shapes stress- and aging-related transcriptional networks. *PLoS Genetics*, 7(6), e1002153. https://doi.org/10.1371/journal.pgen.1002153.

Kelley, K. W., Arkins, S., Minshall, C., Liu, Q., & Dantzer, R. (1996). Growth hormone, growth factors and hematopoiesis. *Hormone Research*, *45*(1–2), 38–45. https://doi.org/10.1159/000184757.

Kharas, M. G., Okabe, R., Ganis, J. J., Gozo, M., Khandan, T., Paktinat, M., et al. (2010). Constitutively active AKT depletes hematopoietic stem cells and induces leukemia in mice. *Blood*, *115*(7), 1406–1415. https://doi.org/10.1182/blood-2009-06-229443.

Kim, M., Moon, H. B., & Spangrude, G. J. (2003). Major age-related changes of mouse hematopoietic stem/progenitor cells. *Annals of the New York Academy of Sciences*, *996*, 195–208. https://doi.org/10.1111/j.1749-6632.2003.tb03247.x.

Kollman, C., Howe, C. W., Anasetti, C., Antin, J. H., Davies, S. M., Filipovich, A. H., et al. (2001). Donor characteristics as risk factors in recipients after transplantation of bone marrow from unrelated donors: The effect of donor age. *Blood*, *98*(7), 2043–2051. https://doi.org/10.1182/blood.v98.7.2043.

Kondo, M., Weissman, I. L., & Akashi, K. (1997). Identification of clonogenic common lymphoid progenitors in mouse bone marrow. *Cell*, *91*(5), 661–672. https://doi.org/10.1016/s0092-8674(00)80453-5.

Kostyo, J. L. (1968). Rapid effects of growth hormone on amino acid transport and protein synthesis. *Annals of the New York Academy of Sciences*, *148*(2), 389–407. https://doi.org/10.1111/j.1749-6632.1968.tb20365.x.

Kovacs, E. J., Palmer, J. L., Fortin, C. F., Fulop, T., Jr., Goldstein, D. R., & Linton, P. J. (2009). Aging and innate immunity in the mouse: Impact of intrinsic and extrinsic factors. *Trends in Immunology*, *30*(7), 319–324. https://doi.org/10.1016/j.it.2009.03.012.

Kuang, J., Chen, L., Tang, Q., Zhang, J., Li, Y., & He, J. (2018). The role of Sirt6 in obesity and diabetes. *Frontiers in Physiology*, *9*, 135. https://doi.org/10.3389/fphys.2018.00135.

Kugel, S., & Mostoslavsky, R. (2014). Chromatin and beyond: The multitasking roles for SIRT6. *Trends in Biochemical Sciences*, *39*(2), 72–81. https://doi.org/10.1016/j.tibs.2013.12.002.

Kuranda, K., Vargaftig, J., de la Rochere, P., Dosquet, C., Charron, D., Bardin, F., et al. (2011). Age-related changes in human hematopoietic stem/progenitor cells. *Aging Cell*, *10*(3), 542–546. https://doi.org/10.1111/j.1474-9726.2011.00675.x.

Lamberts, S. W., van den Beld, A. W., & van der Lely, A. J. (1997). The endocrinology of aging. *Science*, *278*(5337), 419–424. https://doi.org/10.1126/science.278.5337.419.

Landreth, K. S., Narayanan, R., & Dorshkind, K. (1992). Insulin-like growth factor-I regulates pro-B cell differentiation. *Blood*, *80*(5), 1207–1212.

Landry, J., Sutton, A., Tafrov, S. T., Heller, R. C., Stebbins, J., Pillus, L., et al. (2000). The silencing protein SIR2 and its homologs are NAD-dependent protein deacetylases. *Proceedings of the National Academy of Sciences of the United States of America*, *97*(11), 5807–5811. https://doi.org/10.1073/pnas.110148297.

Lasiglie, D., Boero, S., Bauer, I., Morando, S., Damonte, P., Cea, M., et al. (2016). Sirt6 regulates dendritic cell differentiation, maturation, and function. *Aging (Albany NY)*, *8*(1), 34–49. https://doi.org/10.18632/aging.100870.

Latchney, S. E., & Calvi, L. M. (2017). The aging hematopoietic stem cell niche: Phenotypic and functional changes and mechanisms that contribute to hematopoietic aging. *Seminars in Hematology*, *54*(1), 25–32. https://doi.org/10.1053/j.seminhematol.2016.10.001.

Lazarus, D. D., Moldawer, L. L., & Lowry, S. F. (1993). Insulin-like growth factor-1 activity is inhibited by interleukin-1 alpha, tumor necrosis factor-alpha, and interleukin-6. *Lymphokine and Cytokine Research*, *12*(4), 219–223.

Le Roith, D., Bondy, C., Yakar, S., Liu, J. L., & Butler, A. (2001). The somatomedin hypothesis: 2001. *Endocrine Reviews*, *22*(1), 53–74. https://doi.org/10.1210/edrv.22.1.0419.

Lee, I. H. (2019). Mechanisms and disease implications of sirtuin-mediated autophagic regulation. *Experimental & Molecular Medicine*, *51*(9), 102. https://doi.org/10.1038/s12276-019-0302-7.

Legutko, A., Marichal, T., Fievez, L., Bedoret, D., Mayer, A., de Vries, H., et al. (2011). Sirtuin 1 promotes Th2 responses and airway allergy by repressing peroxisome proliferator-activated receptor-gamma activity in dendritic cells. *Journal of Immunology*, *187*(9), 4517–4529. https://doi.org/10.4049/jimmunol.1101493.

Lehner, M., Kellert, B., Proff, J., Schmid, M. A., Diessenbacher, P., Ensser, A., et al. (2012). Autocrine TNF is critical for the survival of human dendritic cells by regulating BAK, BCL-2, and FLIPL. *Journal of Immunology*, *188*(10), 4810–4818. https://doi.org/10.4049/jimmunol.1101610.

Li, Y. M., Schacher, D. H., Liu, Q., Arkins, S., Rebeiz, N., McCusker, R. H., Jr., et al. (1997). Regulation of myeloid growth and differentiation by the insulin-like growth factor I receptor. *Endocrinology*, *138*(1), 362–368. https://doi.org/10.1210/endo.138.1.4847.

Li, G., Smithey, M. J., Rudd, B. D., & Nikolich-Zugich, J. (2012). Age-associated alterations in CD8alpha + dendritic cells impair CD8 T-cell expansion in response to an intracellular bacterium. *Aging Cell*, *11*(6), 968–977. https://doi.org/10.1111/j.1474-9726.2012.00867.x.

Liang, Y., Van Zant, G., & Szilvassy, S. J. (2005). Effects of aging on the homing and engraftment of murine hematopoietic stem and progenitor cells. *Blood*, *106*(4), 1479–1487. https://doi.org/10.1182/blood-2004-11-4282.

Lin, Q., Chauvistre, H., Costa, I. G., Gusmao, E. G., Mitzka, S., Hanzelmann, S., et al. (2015). Epigenetic program and transcription factor circuitry of dendritic cell development. *Nucleic Acids Research*, *43*(20), 9680–9693. https://doi.org/10.1093/nar/gkv1056.

Lin, S. J., Defossez, P. A., & Guarente, L. (2000). Requirement of NAD and SIR2 for life-span extension by calorie restriction in *Saccharomyces cerevisiae*. *Science*, *289*(5487), 2126–2128. https://doi.org/10.1126/science.289.5487.2126.

Linehan, E., Dombrowski, Y., Snoddy, R., Fallon, P. G., Kissenpfennig, A., & Fitzgerald, D. C. (2014). Aging impairs peritoneal but not bone marrow-derived macrophage phagocytosis. *Aging Cell*, *13*(4), 699–708. https://doi.org/10.1111/acel.12223.

Linton, P. J., & Dorshkind, K. (2004). Age-related changes in lymphocyte development and function. *Nature Immunology*, *5*(2), 133–139. https://doi.org/10.1038/ni1033.

Liu, G., Bi, Y., Xue, L., Zhang, Y., Yang, H., Chen, X., et al. (2015). Dendritic cell SIRT1-HIF1alpha axis programs the differentiation of CD4+ T cells through IL-12 and TGF-beta1. *Proceedings of the National Academy of Sciences of the United States of America*, *112*(9), E957–E965. https://doi.org/10.1073/pnas.1420419112.

Liu, H., Bravata, D. M., Olkin, I., Nayak, S., Roberts, B., Garber, A. M., et al. (2007). Systematic review: The safety and efficacy of growth hormone in the healthy elderly. *Annals of Internal Medicine*, *146*(2), 104–115. https://doi.org/10.7326/0003-4819-146-2-200701160-00005.

Liu, J. L., & LeRoith, D. (1999). Insulin-like growth factor I is essential for postnatal growth in response to growth hormone. *Endocrinology*, *140*(11), 5178–5184. https://doi.org/10.1210/endo.140.11.7151.

Lopez-Lluch, G., & Navas, P. (2016). Calorie restriction as an intervention in ageing. *The Journal of Physiology*, *594*(8), 2043–2060. https://doi.org/10.1113/JP270543.

Lopez-Otin, C., Blasco, M. A., Partridge, L., Serrano, M., & Kroemer, G. (2013). The hallmarks of aging. *Cell*, *153*(6), 1194–1217. https://doi.org/10.1016/j.cell.2013.05.039.

Lu, J., Sun, D., Liu, Z., Li, M., Hong, H., Liu, C., et al. (2016). SIRT6 suppresses isoproterenol-induced cardiac hypertrophy through activation of autophagy. *Translational Research*, *172*, 96–112.e116. https://doi.org/10.1016/j.trsl.2016.03.002.

Luo, H., Mu, W. C., Karki, R., Chiang, H. H., Mohrin, M., Shin, J. J., et al. (2019). Mitochondrial stress-initiated aberrant activation of the NLRP3 inflammasome regulates the functional deterioration of hematopoietic stem cell aging. *Cell Reports*, *26*(4), 945–954.e944. https://doi.org/10.1016/j.celrep.2018.12.101.

Lupu, F., Terwilliger, J. D., Lee, K., Segre, G. V., & Efstratiadis, A. (2001). Roles of growth hormone and insulin-like growth factor 1 in mouse postnatal growth. *Developmental Biology*, *229*(1), 141–162. https://doi.org/10.1006/dbio.2000.9975.

Lutz, C. T., Moore, M. B., Bradley, S., Shelton, B. J., & Lutgendorf, S. K. (2005). Reciprocal age related change in natural killer cell receptors for MHC class *I. Mechanisms of Ageing and Development*, *126*(6–7), 722–731. https://doi.org/10.1016/j.mad.2005.01.004.

Maeda, K., Malykhin, A., Teague-Weber, B. N., Sun, X. H., Farris, A. D., & Coggeshall, K. M. (2009). Interleukin-6 aborts lymphopoiesis and elevates production of myeloid cells in systemic lupus erythematosus-prone B6.Sle1.Yaa animals. *Blood*, *113*(19), 4534–4540. https://doi.org/10.1182/blood-2008-12-192559.

Mann, M., Mehta, A., de Boer, C. G., Kowalczyk, M. S., Lee, K., Haldeman, P., et al. (2018). Heterogeneous responses of hematopoietic stem cells to inflammatory stimuli are altered with age. *Cell Reports*, *25*(11), 2992–3005.e2995. https://doi.org/10.1016/j.celrep.2018.11.056.

Manz, M. G., Miyamoto, T., Akashi, K., & Weissman, I. L. (2002). Prospective isolation of human clonogenic common myeloid progenitors. *Proceedings of the National Academy of Sciences of the United States of America*, *99*(18), 11872–11877. https://doi.org/10.1073/pnas.172384399.

Mao, Z., Hine, C., Tian, X., Van Meter, M., Au, M., Vaidya, A., et al. (2011). SIRT6 promotes DNA repair under stress by activating PARP1. *Science*, *332*(6036), 1443–1446. https://doi.org/10.1126/science.1202723.

Mao, K., Quipildor, G. F., Tabrizian, T., Novaj, A., Guan, F., Walters, R. O., et al. (2018). Late-life targeting of the IGF-1 receptor improves healthspan and lifespan in female mice. *Nature Communications*, *9*(1), 2394. https://doi.org/10.1038/s41467-018-04805-5.

Mariani, E., Meneghetti, A., Neri, S., Ravaglia, G., Forti, P., Cattini, L., et al. (2002). Chemokine production by natural killer cells from nonagenarians. *European Journal of Immunology*, *32*(6), 1524–1529. https://doi.org/10.1002/1521-4141(200206)32:6<1524::AID-IMMU1524>3.0.CO;2-E.

Martinez-Redondo, P., & Vaquero, A. (2013). The diversity of histone versus nonhistone sirtuin substrates. *Genes & Cancer*, *4*(3–4), 148–163. https://doi.org/10.1177/1947601913483767.

McCord, R. A., Michishita, E., Hong, T., Berber, E., Boxer, L. D., Kusumoto, R., et al. (2009). SIRT6 stabilizes DNA-dependent protein kinase at chromatin for DNA double-strand break repair. *Aging (Albany NY)*, *1*(1), 109–121. https://doi.org/10.18632/aging.100011.

Mehlhop-Williams, E. R., & Bevan, M. J. (2014). Memory CD8+ T cells exhibit increased antigen threshold requirements for recall proliferation. *The Journal of Experimental Medicine*, *211*(2), 345–356. https://doi.org/10.1084/jem.20131271.

Mellman, I., & Steinman, R. M. (2001). Dendritic cells: Specialized and regulated antigen processing machines. *Cell*, *106*(3), 255–258. https://doi.org/10.1016/s0092-8674(01)00449-4.

Méndez-Ferrer, S., Lucas, D., Battista, M., & Frenette, P. S. (2008). Haematopoietic stem cell release is regulated by circadian oscillations. *Nature*, *452*(7186), 442–447. https://doi.org/10.1038/nature06685.

Merchav, S., Tatarsky, I., & Hochberg, Z. (1988). Enhancement of human granulopoiesis in vitro by biosynthetic insulin-like growth factor I/somatomedin C and human growth hormone. *The Journal of Clinical Investigation*, *81*(3), 791–797. https://doi.org/10.1172/JCI113385.

Michishita, E., McCord, R. A., Berber, E., Kioi, M., Padilla-Nash, H., Damian, M., et al. (2008). SIRT6 is a histone H3 lysine 9 deacetylase that modulates telomeric chromatin. *Nature*, *452*(7186), 492–496. https://doi.org/10.1038/nature06736.

Michishita, E., McCord, R. A., Boxer, L. D., Barber, M. F., Hong, T., Gozani, O., et al. (2009). Cell cycle-dependent deacetylation of telomeric histone H3 lysine K56 by human SIRT6. *Cell Cycle*, *8*(16), 2664–2666. https://doi.org/10.4161/cc.8.16.9367.

Michishita, E., Park, J. Y., Burneskis, J. M., Barrett, J. C., & Horikawa, I. (2005). Evolutionarily conserved and nonconserved cellular localizations and functions of human SIRT proteins. *Molecular Biology of the Cell*, *16*(10), 4623–4635. https://doi.org/10.1091/mbc.e05-01-0033.

Mildner, A., & Jung, S. (2014). Development and function of dendritic cell subsets. *Immunity*, *40*(5), 642–656. https://doi.org/10.1016/j.immuni.2014.04.016.

Mildner, A., Yona, S., & Jung, S. (2013). A close encounter of the third kind: Monocyte-derived cells. *Advances in Immunology*, *120*, 69–103. https://doi.org/10.1016/B978-0-12-417028-5.00003-X.

Miller, J. P., & Allman, D. (2005). Linking age-related defects in B lymphopoiesis to the aging of hematopoietic stem cells. *Seminars in Immunology*, *17*(5), 321–329. https://doi.org/10.1016/j.smim.2005.05.003.

Milne, J. C., Lambert, P. D., Schenk, S., Carney, D. P., Smith, J. J., Gagne, D. J., et al. (2007). Small molecule activators of SIRT1 as therapeutics for the treatment of type 2 diabetes. *Nature*, *450*(7170), 712–716. https://doi.org/10.1038/nature06261.

Mirantes, C., Passegue, E., & Pietras, E. M. (2014). Pro-inflammatory cytokines: Emerging players regulating HSC function in normal and diseased hematopoiesis. *Experimental Cell Research*, *329*(2), 248–254. https://doi.org/10.1016/j.yexcr.2014.08.017.

Mohrin, M., Shin, J., Liu, Y., Brown, K., Luo, H., Xi, Y., et al. (2015). Stem cell aging. A mitochondrial UPR-mediated metabolic checkpoint regulates hematopoietic stem cell aging. *Science, 347*(6228), 1374–1377. https://doi.org/10.1126/science.aaa2361.

Mol, J. A., van Garderen, E., Selman, P. J., Wolfswinkel, J., Rijinberk, A., & Rutteman, G. R. (1995). Growth hormone mRNA in mammary gland tumors of dogs and cats. *The Journal of Clinical Investigation, 95*(5), 2028–2034. https://doi.org/10.1172/JCI117888.

Montecino-Rodriguez, E., Kong, Y., Casero, D., Rouault, A., Dorshkind, K., & Pioli, P. D. (2019). Lymphoid-biased hematopoietic stem cells are maintained with age and efficiently generate lymphoid progeny. *Stem Cell Reports, 12*(3), 584–596. https://doi.org/10.1016/j.stemcr.2019.01.016.

Monteserin-Garcia, J., Al-Massadi, O., Seoane, L. M., Alvarez, C. V., Shan, B., Stalla, J., et al. (2013). Sirt1 inhibits the transcription factor CREB to regulate pituitary growth hormone synthesis. *The FASEB Journal, 27*(4), 1561–1571. https://doi.org/10.1096/fj.12-220129.

Montgomery, R. R., & Shaw, A. C. (2015). Paradoxical changes in innate immunity in aging: Recent progress and new directions. *Journal of Leukocyte Biology, 98*(6), 937–943. https://doi.org/10.1189/jlb.5MR0315-104R.

Morrison, S. J., & Scadden, D. T. (2014). The bone marrow niche for haematopoietic stem cells. *Nature, 505*(7483), 327–334. https://doi.org/10.1038/nature12984.

Mostoslavsky, R., Chua, K. F., Lombard, D. B., Pang, W. W., Fischer, M. R., Gellon, L., et al. (2006). Genomic instability and aging-like phenotype in the absence of mammalian SIRT6. *Cell, 124*(2), 315–329. https://doi.org/10.1016/j.cell.2005.11.044.

Muller-Sieburg, C. E., & Sieburg, H. B. (2006). Clonal diversity of the stem cell compartment. *Current Opinion in Hematology, 13*(4), 243–248. https://doi.org/10.1097/01.moh.0000231421.00407.65.

Murphy, W. J., Durum, S. K., & Longo, D. L. (1992). Role of neuroendocrine hormones in murine T cell development. Growth hormone exerts thymopoietic effects in vivo. *Journal of Immunology, 149*(12), 3851–3857.

Nagy, E., Berczi, I., & Friesen, H. G. (1983). Regulation of immunity in rats by lactogenic and growth hormones. *Acta Endocrinologica, 102*(3), 351–357. https://doi.org/10.1530/acta.0.1020351.

Naik, S. H., Proietto, A. I., Wilson, N. S., Dakic, A., Schnorrer, P., Fuchsberger, M., et al. (2005). Cutting edge: Generation of splenic CD8+ and CD8- dendritic cell equivalents in Fms-like tyrosine kinase 3 ligand bone marrow cultures. *Journal of Immunology, 174*(11), 6592–6597. https://doi.org/10.4049/jimmunol.174.11.6592.

Napolitano, L. A., Lo, J. C., Gotway, M. B., Mulligan, K., Barbour, J. D., Schmidt, D., et al. (2002). Increased thymic mass and circulating naive CD4 T cells in HIV-1-infected adults treated with growth hormone. *AIDS, 16*(8), 1103–1111. https://doi.org/10.1097/00002030-200205240-00003.

Niedernhofer, L. J., Garinis, G. A., Raams, A., Lalai, A. S., Robinson, A. R., Appeldoorn, E., et al. (2006). A new progeroid syndrome reveals that genotoxic stress suppresses the somatotroph axis. *Nature, 444*(7122), 1038–1043. https://doi.org/10.1038/nature05456.

O'Connor, J. C., McCusker, R. H., Strle, K., Johnson, R. W., Dantzer, R., & Kelley, K. W. (2008). Regulation of IGF-I function by proinflammatory cytokines: At the interface of immunology and endocrinology. *Cellular Immunology, 252*(1–2), 91–110. https://doi.org/10.1016/j.cellimm.2007.09.010.

Orkin, S. H., & Zon, L. I. (2008). Hematopoiesis: An evolving paradigm for stem cell biology. *Cell, 132*(4), 631–644. https://doi.org/10.1016/j.cell.2008.01.025.

Ortega-Molina, A., Efeyan, A., Lopez-Guadamillas, E., Munoz-Martin, M., Gomez-Lopez, G., Canamero, M., et al. (2012). Pten positively regulates brown adipose function, energy expenditure, and longevity. *Cell Metabolism, 15*(3), 382–394. https://doi.org/10.1016/j.cmet.2012.02.001.

Owczarczyk, A. B., Schaller, M. A., Reed, M., Rasky, A. J., Lombard, D. B., & Lukacs, N. W. (2015). Sirtuin 1 regulates dendritic cell activation and autophagy during respiratory syncytial virus-induced immune responses. *Journal of Immunology, 195*(4), 1637–1646. https://doi.org/10.4049/jimmunol.1500326.

Pan, H., Guan, D., Liu, X., Li, J., Wang, L., Wu, J., et al. (2016). SIRT6 safeguards human mesenchymal stem cells from oxidative stress by coactivating NRF2. *Cell Research, 26*(2), 190–205. https://doi.org/10.1038/cr.2016.4.

Pang, W. W., Price, E. A., Sahoo, D., Beerman, I., Maloney, W. J., Rossi, D. J., et al. (2011). Human bone marrow hematopoietic stem cells are increased in frequency and myeloid-biased with age. *Proceedings of the National Academy of Sciences of the United States of America, 108*(50), 20012–20017. https://doi.org/10.1073/pnas.1116110108.

Park, S. J., Nakagawa, T., Kitamura, H., Atsumi, T., Kamon, H., Sawa, S., et al. (2004). IL-6 regulates in vivo dendritic cell differentiation through STAT3 activation. *Journal of Immunology, 173*(6), 3844–3854. https://doi.org/10.4049/jimmunol.173.6.3844.

Pasarica, M., Zachwieja, J. J., Dejonge, L., Redman, S., & Smith, S. R. (2007). Effect of growth hormone on body composition and visceral adiposity in middle-aged men with visceral obesity. *The Journal of Clinical Endocrinology and Metabolism, 92*(11), 4265–4270. https://doi.org/10.1210/jc.2007-0786.

Patel, K., & Taub, D. D. (2009). Role of neuropeptides, hormones, and growth factors in regulating thymopoiesis in middle to old age. *F1000 Biology Reports, 1*, 42. https://doi.org/10.3410/B1-42.

Pereira, L. F., de Souza, A. P., Borges, T. J., & Bonorino, C. (2011). Impaired in vivo CD4+ T cell expansion and differentiation in aged mice is not solely due to T cell defects: Decreased stimulation by aged dendritic cells. *Mechanisms of Ageing and Development, 132*(4), 187–194. https://doi.org/10.1016/j.mad.2011.03.005.

Peretz, S., Jensen, R., Baserga, R., & Glazer, P. M. (2001). ATM-dependent expression of the insulin-like growth factor-I receptor in a pathway regulating radiation response. *Proceedings of the National Academy of Sciences of the United States of America, 98*(4), 1676–1681. https://doi.org/10.1073/pnas.041416598.

Perry, J. M., He, X. C., Sugimura, R., Grindley, J. C., Haug, J. S., Ding, S., et al. (2011). Cooperation between both Wnt/{beta}-catenin and PTEN/PI3K/Akt signaling promotes primitive hematopoietic stem cell self-renewal and expansion. *Genes & Development, 25*(18), 1928–1942. https://doi.org/10.1101/gad.17421911.

Peruzzi, B., Cappariello, A., Del Fattore, A., Rucci, N., De Benedetti, F., & Teti, A. (2012). c-Src and IL-6 inhibit osteoblast differentiation and integrate IGFBP5 signalling. *Nature Communications, 3*, 630. https://doi.org/10.1038/ncomms1651.

Pietras, E. M., Mirantes-Barbeito, C., Fong, S., Loeffler, D., Kovtonyuk, L. V., Zhang, S. Y., et al. (2016). Chronic interleukin-1 exposure drives haematopoietic stem cells towards precocious myeloid differentiation at the expense of self-renewal. *Nature Cell Biology, 18*(6), 607–618. https://doi.org/10.1038/ncb3346.

Pietras, E. M., Reynaud, D., Kang, Y. A., Carlin, D., Calero-Nieto, F. J., Leavitt, A. D., et al. (2015). Functionally distinct subsets of lineage-biased multipotent progenitors control blood production in normal and regenerative conditions. *Cell Stem Cell, 17*(1), 35–46. https://doi.org/10.1016/j.stem.2015.05.003.

Powell-Braxton, L., Hollingshead, P., Warburton, C., Dowd, M., Pitts-Meek, S., Dalton, D., et al. (1993). IGF-I is required for normal embryonic growth in mice. *Genes & Development*, 7(12B), 2609–2617. https://doi.org/10.1101/gad.7.12b.2609.

Qian, F., Wang, X., Zhang, L., Lin, A., Zhao, H., Fikrig, E., et al. (2011). Impaired interferon signaling in dendritic cells from older donors infected in vitro with West Nile virus. *The Journal of Infectious Diseases, 203*(10), 1415–1424. https://doi.org/10.1093/infdis/jir048.

Rando, T. A., & Chang, H. Y. (2012). Aging, rejuvenation, and epigenetic reprogramming: Resetting the aging clock. *Cell, 148*(1–2), 46–57. https://doi.org/10.1016/j.cell.2012.01.003.

Rimmele, P., Bigarella, C. L., Liang, R., Izac, B., Dieguez-Gonzalez, R., Barbet, G., et al. (2014). Aging-like phenotype and defective lineage specification in SIRT1-deleted hematopoietic stem and progenitor cells. *Stem Cell Reports, 3*(1), 44–59. https://doi.org/10.1016/j.stemcr.2014.04.015.

Rossi, D. J., Bryder, D., Zahn, J. M., Ahlenius, H., Sonu, R., Wagers, A. J., et al. (2005). Cell intrinsic alterations underlie hematopoietic stem cell aging. *Proceedings of the National Academy of Sciences of the United States of America, 102*(26), 9194–9199. https://doi.org/10.1073/pnas.0503280102.

Rossi, D. J., Jamieson, C. H., & Weissman, I. L. (2008). Stems cells and the pathways to aging and cancer. *Cell, 132*(4), 681–696. https://doi.org/10.1016/j.cell.2008.01.036.

Rudman, D., Feller, A. G., Nagraj, H. S., Gergans, G. A., Lalitha, P. Y., Goldberg, A. F., et al. (1990). Effects of human growth hormone in men over 60 years old. *The New England Journal of Medicine, 323*(1), 1–6. https://doi.org/10.1056/NEJM199007053230101.

Sabharwal, P., & Varma, S. (1996). Growth hormone synthesized and secreted by human thymocytes acts via insulin-like growth factor I as an autocrine and paracrine growth factor. *The Journal of Clinical Endocrinology and Metabolism, 81*(7), 2663–2669. https://doi.org/10.1210/jcem.81.7.8675594.

Sallusto, F., & Lanzavecchia, A. (2002). The instructive role of dendritic cells on T-cell responses. *Arthritis Research, 4*(Suppl 3), S127–S132. https://doi.org/10.1186/ar567.

Salmon, W. D., Jr., & Daughaday, W. H. (1957). A hormonally controlled serum factor which stimulates sulfate incorporation by cartilage in vitro. *The Journal of Laboratory and Clinical Medicine, 49*(6), 825–836.

Sanders, B. D., Jackson, B., & Marmorstein, R. (2010). Structural basis for sirtuin function: What we know and what we don't. *Biochimica et Biophysica Acta, 1804*(8), 1604–1616. https://doi.org/10.1016/j.bbapap.2009.09.009.

Sanjuan-Pla, A., Macaulay, I. C., Jensen, C. T., Woll, P. S., Luis, T. C., Mead, A., et al. (2013). Platelet-biased stem cells reside at the apex of the haematopoietic stem-cell hierarchy. *Nature, 502*(7470), 232–236. https://doi.org/10.1038/nature12495.

Sansone, L., Reali, V., Pellegrini, L., Villanova, L., Aventaggiato, M., Marfe, G., et al. (2013). SIRT1 silencing confers neuroprotection through IGF-1 pathway activation. *Journal of Cellular Physiology, 228*(8), 1754–1761. https://doi.org/10.1002/jcp.24334.

Satoh, A., Brace, C. S., Rensing, N., Cliften, P., Wozniak, D. F., Herzog, E. D., et al. (2013). Sirt1 extends life span and delays aging in mice through the regulation of Nk2 homeobox 1 in the DMH and LH. *Cell Metabolism, 18*(3), 416–430. https://doi.org/10.1016/j.cmet.2013.07.013.

Saunders, L. R., & Verdin, E. (2007). Sirtuins: Critical regulators at the crossroads between cancer and aging. *Oncogene, 26*(37), 5489–5504. https://doi.org/10.1038/sj.onc.1210616.

Sauve, A. A., Celic, I., Avalos, J., Deng, H., Boeke, J. D., & Schramm, V. L. (2001). Chemistry of gene silencing: The mechanism of NAD+-dependent deacetylation reactions. *Biochemistry, 40*(51), 15456–15463. https://doi.org/10.1021/bi011858j.

Savino, W., Mendes-da-Cruz, D. A., Lepletier, A., & Dardenne, M. (2016). Hormonal control of T-cell development in health and disease. *Nature Reviews. Endocrinology, 12*(2), 77–89. https://doi.org/10.1038/nrendo.2015.168.

Schwer, B., Schumacher, B., Lombard, D. B., Xiao, C., Kurtev, M. V., Gao, J., et al. (2010). Neural sirtuin 6 (Sirt6) ablation attenuates somatic growth and causes obesity. *Proceedings of the National Academy of Sciences of the United States of America, 107*(50), 21790–21794. https://doi.org/10.1073/pnas.1016306107.

Schwerdt, J. I., Lopez-Leon, M., Console, G. M., Brown, O. A., Morel, G. R., Spinedi, E., et al. (2018). Rejuvenating effect of long-term insulin-like growth factor-I gene therapy in the hypothalamus of aged rats with dopaminergic dysfunction. *Rejuvenation Research*, *21*(2), 102–108. https://doi.org/10.1089/rej.2017.1935.

Sebastian, C., Zwaans, B. M., Silberman, D. M., Gymrek, M., Goren, A., Zhong, L., et al. (2012). The histone deacetylase SIRT6 is a tumor suppressor that controls cancer metabolism. *Cell*, *151*(6), 1185–1199. https://doi.org/10.1016/j.cell.2012.10.047.

Seita, J., & Weissman, I. L. (2010). Hematopoietic stem cell: Self-renewal versus differentiation. *Wiley Interdisciplinary Reviews. Systems Biology and Medicine*, *2*(6), 640–653. https://doi.org/10.1002/wsbm.86.

Sere, K. M., Lin, Q., Felker, P., Rehage, N., Klisch, T., Ortseifer, I., et al. (2012). Dendritic cell lineage commitment is instructed by distinct cytokine signals. *European Journal of Cell Biology*, *91*(6–7), 515–523. https://doi.org/10.1016/j.ejcb.2011.09.007.

Seto, E., & Yoshida, M. (2014). Erasers of histone acetylation: The histone deacetylase enzymes. *Cold Spring Harbor Perspectives in Biology*, *6*(4), a018713. https://doi.org/10.1101/cshperspect.a018713.

Shaw, A. C., Goldstein, D. R., & Montgomery, R. R. (2013). Age-dependent dysregulation of innate immunity. *Nature Reviews. Immunology*, *13*(12), 875–887. https://doi.org/10.1038/nri3547.

Shen, H., Tesar, B. M., Du, W., & Goldstein, D. R. (2009). Aging impairs recipient T cell intrinsic and extrinsic factors in response to transplantation. *PLoS One*, *4*(1), e4097. https://doi.org/10.1371/journal.pone.0004097.

Shore, D., Squire, M., & Nasmyth, K. A. (1984). Characterization of two genes required for the position-effect control of yeast mating-type genes. *The EMBO Journal*, *3*(12), 2817–2823.

Simeoni, F., Tasselli, L., Tanaka, S., Villanova, L., Hayashi, M., Kubota, K., et al. (2013). Proteomic analysis of the SIRT6 interactome: Novel links to genome maintenance and cellular stress signaling. *Scientific Reports*, *3*, 3085. https://doi.org/10.1038/srep03085.

Smith, P. E. (1930). The effect of hypophysectomy upon the involution of the thymus in the rat. *The Anatomical Record*, *47*(1), 119–129. https://doi.org/10.1002/ar.1090470110.

Solana, R., Pawelec, G., & Tarazona, R. (2006). Aging and innate immunity. *Immunity*, *24*(5), 491–494. https://doi.org/10.1016/j.immuni.2006.05.003.

Song, W., Tai, Y. T., Tian, Z., Hideshima, T., Chauhan, D., Nanjappa, P., et al. (2011). HDAC inhibition by LBH589 affects the phenotype and function of human myeloid dendritic cells. *Leukemia*, *25*(1), 161–168. https://doi.org/10.1038/leu.2010.244.

Sridharan, A., Esposo, M., Kaushal, K., Tay, J., Osann, K., Agrawal, S., et al. (2011). Age-associated impaired plasmacytoid dendritic cell functions lead to decreased CD4 and CD8 T cell immunity. *Age (Dordrecht, Netherlands)*, *33*(3), 363–376. https://doi.org/10.1007/s11357-010-9191-3.

Stout-Delgado, H. W., Yang, X., Walker, W. E., Tesar, B. M., & Goldstein, D. R. (2008). Aging impairs IFN regulatory factor 7 up-regulation in plasmacytoid dendritic cells during TLR9 activation. *Journal of Immunology*, *181*(10), 6747–6756. https://doi.org/10.4049/jimmunol.181.10.6747.

Sugatani, T., Agapova, O., Malluche, H. H., & Hruska, K. A. (2015). SIRT6 deficiency culminates in low-turnover osteopenia. *Bone*, *81*, 168–177. https://doi.org/10.1016/j.bone.2015.07.018.

Sun, D., Luo, M., Jeong, M., Rodriguez, B., Xia, Z., Hannah, R., et al. (2014). Epigenomic profiling of young and aged HSCs reveals concerted changes during aging that reinforce self-renewal. *Cell Stem Cell*, *14*(5), 673–688. https://doi.org/10.1016/j.stem.2014.03.002.

Sun, X., Wang, S., Gai, J., Guan, J., Li, J., Li, Y., et al. (2019). SIRT5 promotes cisplatin resistance in ovarian cancer by suppressing DNA damage in a ROS-dependent manner via regulation of the Nrf2/HO-1 pathway. *Frontiers in Oncology*, *9*, 754. https://doi.org/10.3389/fonc.2019.00754.

Sundaresan, N. R., Vasudevan, P., Zhong, L., Kim, G., Samant, S., Parekh, V., et al. (2012). The sirtuin SIRT6 blocks IGF-Akt signaling and development of cardiac hypertrophy by targeting c-Jun. *Nature Medicine*, *18*(11), 1643–1650. https://doi.org/10.1038/nm.2961.

Tan, S. Y., Cavanagh, L. L., d'Advigor, W., Shackel, N., Fazekas de St Groth, B., & Weninger, W. (2012). Phenotype and functions of conventional dendritic cells are not compromised in aged mice. *Immunology and Cell Biology*, *90*(7), 722–732. https://doi.org/10.1038/icb.2011.104.

Tang, M., Li, Z., Zhang, C., Lu, X., Tu, B., Cao, Z., et al. (2019). SIRT7-mediated ATM deacetylation is essential for its deactivation and DNA damage repair. *Science Advances*, *5*(3), eaav1118. https://doi.org/10.1126/sciadv.aav1118.

Taub, D. D., & Longo, D. L. (2005). Insights into thymic aging and regeneration. *Immunological Reviews*, *205*, 72–93. https://doi.org/10.1111/j.0105-2896.2005.00275.x.

Tennen, R. I., Bua, D. J., Wright, W. E., & Chua, K. F. (2011). SIRT6 is required for maintenance of telomere position effect in human cells. *Nature Communications*, *2*, 433. https://doi.org/10.1038/ncomms1443.

Tesar, B. M., Walker, W. E., Unternaehrer, J., Joshi, N. S., Chandele, A., Haynes, L., et al. (2006). Murine [corrected] myeloid dendritic cell-dependent toll-like receptor immunity is preserved with aging. *Aging Cell*, *5*(6), 473–486. https://doi.org/10.1111/j.1474-9726.2006.00245.x.

Thomas, R. M., Belsito, D. V., Huang, C., Chen Lz, L. Z., Ormsby, I., Simmons, W. J., et al. (2001). Appearance of Langerhans cells in the epidermis of Tgfb1(−/−) SCID mice: Paracrine and autocrine effects of transforming growth factor-beta 1 and -beta 2(1). *The Journal of Investigative Dermatology*, *117*(6), 1574–1580. https://doi.org/10.1046/j.0022-202x.2001.01550.x.

Thompson, H. L., Smithey, M. J., Surh, C. D., & Nikolich-Zugich, J. (2017). Functional and homeostatic impact of age-related changes in lymph node stroma. *Frontiers in Immunology*, *8*, 706. https://doi.org/10.3389/fimmu.2017.00706.

Thornton, P. L., Ingram, R. L., & Sonntag, W. E. (2000). Chronic [D-Ala2]-growth hormone-releasing hormone administration attenuates age-related deficits in spatial memory. *The Journals of Gerontology. Series A, Biological Sciences and Medical Sciences*, *55*(2), B106–B112. https://doi.org/10.1093/gerona/55.2.b106.

Tian, Y., Meng, L., & Zhang, Y. (2017). Epigenetic regulation of dendritic cell development and function. *Cancer Journal*, *23*(5), 302–307. https://doi.org/10.1097/PPO.0000000000000280.

Tian, Z. G., Woody, M. A., Sun, R., Welniak, L. A., Raziuddin, A., Funakoshi, S., et al. (1998). Recombinant human growth hormone promotes hematopoietic reconstitution after syngeneic bone marrow transplantation in mice. *Stem Cells*, *16*(3), 193–199. https://doi.org/10.1002/stem.160193.

Tie, R., Li, H., Cai, S., Liang, Z., Shan, W., Wang, B., et al. (2019). Interleukin-6 signaling regulates hematopoietic stem cell emergence. *Experimental & Molecular Medicine*, *51*(10), 124. https://doi.org/10.1038/s12276-019-0320-5.

Toiber, D., Erdel, F., Bouazoune, K., Silberman, D. M., Zhong, L., Mulligan, P., et al. (2013). SIRT6 recruits SNF2H to DNA break sites, preventing genomic instability through chromatin remodeling. *Molecular Cell*, *51*(4), 454–468. https://doi.org/10.1016/j.molcel.2013.06.018.

Tsarfaty, G., Longo, D. L., & Murphy, W. J. (1994). Human insulin-like growth factor I exerts hematopoietic growth-promoting effects after in vivo administration. *Experimental Hematology*, *22*(13), 1273–1277.

Udagawa, N., Takahashi, N., Katagiri, T., Tamura, T., Wada, S., Findlay, D. M., et al. (1995). Interleukin (IL)-6 induction of osteoclast differentiation depends on IL-6 receptors expressed on osteoblastic cells but not on osteoclast progenitors. *The Journal of Experimental Medicine*, *182*(5), 1461–1468. https://doi.org/10.1084/jem.182.5.1461.

Vakhrusheva, O., Smolka, C., Gajawada, P., Kostin, S., Boettger, T., Kubin, T., et al. (2008). Sirt7 increases stress resistance of cardiomyocytes and prevents apoptosis and inflammatory cardiomyopathy in mice. *Circulation Research*, *102*(6), 703–710. https://doi.org/10.1161/CIRCRESAHA.107.164558.

Van Gool, F., Galli, M., Gueydan, C., Kruys, V., Prevot, P. P., Bedalov, A., et al. (2009). Intracellular NAD levels regulate tumor necrosis factor protein synthesis in a sirtuin-dependent manner. *Nature Medicine*, *15*(2), 206–210. https://doi.org/10.1038/nm.1906.

Varma, S., Sabharwal, P., Sheridan, J. F., & Malarkey, W. B. (1993). Growth hormone secretion by human peripheral blood mononuclear cells detected by an enzyme-linked immunoplaque assay. *The Journal of Clinical Endocrinology and Metabolism*, *76*(1), 49–53. https://doi.org/10.1210/jcem.76.1.8421102.

Vas, V., Senger, K., Dorr, K., Niebel, A., & Geiger, H. (2012). Aging of the microenvironment influences clonality in hematopoiesis. *PLoS One*, *7*(8), e42080. https://doi.org/10.1371/journal.pone.0042080.

Vazquez, B. N., Thackray, J. K., Simonet, N. G., Kane-Goldsmith, N., Martinez-Redondo, P., Nguyen, T., et al. (2016). SIRT7 promotes genome integrity and modulates non-homologous end joining DNA repair. *The EMBO Journal*, *35*(14), 1488–1503. https://doi.org/10.15252/embj.201593499.

Vitale, G., Pellegrino, G., Vollery, M., & Hofland, L. J. (2019). ROLE of IGF-1 system in the modulation of longevity: Controversies and new insights from a centenarians' perspective. *Frontiers in Endocrinology*, *10*, 27. https://doi.org/10.3389/fendo.2019.00027.

Wagner, W., Horn, P., Bork, S., & Ho, A. D. (2008). Aging of hematopoietic stem cells is regulated by the stem cell niche. *Experimental Gerontology*, *43*(11), 974–980. https://doi.org/10.1016/j.exger.2008.04.007.

Wang, Y., Bikle, D. D., & Chang, W. (2013). Autocrine and paracrine actions of IGF-I signaling in skeletal development. *Bone Research*, *1*(3), 249–259. https://doi.org/10.4248/BR201303003.

Wang, H., Diao, D., Shi, Z., Zhu, X., Gao, Y., Gao, S., et al. (2016). SIRT6 controls hematopoietic stem cell homeostasis through epigenetic regulation of Wnt signaling. *Cell Stem Cell*, *18*(4), 495–507. https://doi.org/10.1016/j.stem.2016.03.005.

Wang, J., Geiger, H., & Rudolph, K. L. (2011). Immunoaging induced by hematopoietic stem cell aging. *Current Opinion in Immunology*, *23*(4), 532–536. https://doi.org/10.1016/j.coi.2011.05.004.

Watad, A., Bragazzi, N. L., Adawi, M., Amital, H., Toubi, E., Porat, B. S., et al. (2017). Autoimmunity in the elderly: Insights from basic science and clinics—A mini-review. *Gerontology*, *63*(6), 515–523. https://doi.org/10.1159/000478012.

Waterstrat, A., & Van Zant, G. (2009). Effects of aging on hematopoietic stem and progenitor cells. *Current Opinion in Immunology*, *21*(4), 408–413. https://doi.org/10.1016/j.coi.2009.05.002.

Weissman, I. L., & Shizuru, J. A. (2008). The origins of the identification and isolation of hematopoietic stem cells, and their capability to induce donor-specific transplantation tolerance and treat autoimmune diseases. *Blood*, *112*(9), 3543–3553. https://doi.org/10.1182/blood-2008-08-078220.

Wells, J. A., & de Vos, A. M. (1993). Structure and function of human growth hormone: Implications for the hematopoietins. *Annual Review of Biophysics and Biomolecular Structure*, *22*, 329–351. https://doi.org/10.1146/annurev.bb.22.060193.001553.

Welniak, L. A., Karas, M., Yakar, S., Anver, M. R., Murphy, W. J., & LeRoith, D. (2004). Effects of organ-specific loss of insulin-like growth factor-I production on murine hematopoiesis. *Biology of Blood and Marrow Transplantation*, *10*(1), 32–39. https://doi.org/10.1016/j.bbmt.2003.09.008.

Welniak, L. A., Sun, R., & Murphy, W. J. (2002). The role of growth hormone in T-cell development and reconstitution. *Journal of Leukocyte Biology, 71*(3), 381–387.

Weng, N. P. (2006). Aging of the immune system: How much can the adaptive immune system adapt? *Immunity, 24*(5), 495–499. https://doi.org/10.1016/j.immuni.2006.05.001.

Westerheide, S. D., Anckar, J., Stevens, S. M., Jr., Sistonen, L., & Morimoto, R. I. (2009). Stress-inducible regulation of heat shock factor 1 by the deacetylase SIRT1. *Science, 323*(5917), 1063–1066. https://doi.org/10.1126/science.1165946.

Wilkinson, A. C., & Yamazaki, S. (2018). The hematopoietic stem cell diet. *International Journal of Hematology, 107*(6), 634–641. https://doi.org/10.1007/s12185-018-2451-1.

Wong, C., & Goldstein, D. R. (2013). Impact of aging on antigen presentation cell function of dendritic cells. *Current Opinion in Immunology, 25*(4), 535–541. https://doi.org/10.1016/j.coi.2013.05.016.

Wong, C. P., Magnusson, K. R., & Ho, E. (2010). Aging is associated with altered dendritic cells subset distribution and impaired proinflammatory cytokine production. *Experimental Gerontology, 45*(2), 163–169. https://doi.org/10.1016/j.exger.2009.11.005.

Woo, S. J., Lee, S. M., Lim, H. S., Hah, Y. S., Jung, I. D., Park, Y. M., et al. (2016). Myeloid deletion of SIRT1 suppresses collagen-induced arthritis in mice by modulating dendritic cell maturation. *Experimental & Molecular Medicine, 48*, e221. https://doi.org/10.1038/emm.2015.124.

Xia, P., Wang, S., Du, Y., Huang, G., Satoh, T., Akira, S., et al. (2015). Insulin-InsR signaling drives multipotent progenitor differentiation toward lymphoid lineages. *The Journal of Experimental Medicine, 212*(13), 2305–2321. https://doi.org/10.1084/jem.20150618.

Xian, L., Wu, X., Pang, L., Lou, M., Rosen, C. J., Qiu, T., et al. (2012). Matrix IGF-1 maintains bone mass by activation of mTOR in mesenchymal stem cells. *Nature Medicine, 18*(7), 1095–1101. https://doi.org/10.1038/nm.2793.

Xing, Z., Ryan, M. A., Daria, D., Nattamai, K. J., Van Zant, G., Wang, L., et al. (2006). Increased hematopoietic stem cell mobilization in aged mice. *Blood, 108*(7), 2190–2197. https://doi.org/10.1182/blood-2005-12-010272.

Xu, Y., Li, F., Lv, L., Li, T., Zhou, X., Deng, C. X., et al. (2014). Oxidative stress activates SIRT2 to deacetylate and stimulate phosphoglycerate mutase. *Cancer Research, 74*(13), 3630–3642. https://doi.org/10.1158/0008-5472.CAN-13-3615.

Xu, D., Wu, L., Jiang, X., Yang, L., Cheng, J., Chen, H., et al. (2019). SIRT2 inhibition results in meiotic arrest, mitochondrial dysfunction, and disturbance of redox homeostasis during bovine oocyte maturation. *International Journal of Molecular Sciences, 20*(6), 1365. https://doi.org/10.3390/ijms20061365.

Xu, Z., Zhang, L., Zhang, W., Meng, D., Zhang, H., Jiang, Y., et al. (2015). SIRT6 rescues the age related decline in base excision repair in a PARP1-dependent manner. *Cell Cycle, 14*(2), 269–276. https://doi.org/10.4161/15384101.2014.980641.

Yamamoto, M., Iguchi, G., Fukuoka, H., Suda, K., Bando, H., Takahashi, M., et al. (2013). SIRT1 regulates adaptive response of the growth hormone—Insulin-like growth factor-I axis under fasting conditions in liver. *Proceedings of the National Academy of Sciences of the United States of America, 110*(37), 14948–14953. https://doi.org/10.1073/pnas.1220606110.

Yamashita, M., & Passegue, E. (2019). TNF-alpha coordinates hematopoietic stem cell survival and myeloid regeneration. *Cell Stem Cell, 25*(3), 357–372.e357. https://doi.org/10.1016/j.stem.2019.05.019.

Yang, B., Zwaans, B. M., Eckersdorff, M., & Lombard, D. B. (2009). The sirtuin SIRT6 deacetylates H3 K56Ac in vivo to promote genomic stability. *Cell Cycle, 8*(16), 2662–2663. https://doi.org/10.4161/cc.8.16.9329.

Young, K., Eudy, E., Loberg, M., Bell, R., & Trowbridge, J. (2019). Decline in insulin-like growth factor-1 (IGF1) from aged mesenchymal stromal cells is a targetable mechanism to rescue hematopoietic stem cell aging. *Blood, 134*(Suppl. 1), 526. https://doi.org/10.1182/blood-2019-124534.

Young, K., Loberg, M., Bell, R., & Trowbridge, J. J. (2017). Decreased IGF-1 expression in the bone marrow microenvironment drives hematopoietic aging. *Blood, 130*(Suppl. 1), 2440. https://doi.org/10.1182/blood.V130.Suppl_1.2440.2440.

Young, K., Loberg, M., Eudy, E., Bell, R., & Trowbridge, J. J. (2018). The middle-aged bone marrow microenvironment drives hematopoietic stem cell aging through altered production of insulin-like growth factor-1 (IGF1). *Blood, 132*(Suppl. 1), 5091. https://doi.org/10.1182/blood-2018-99-111055.

Yuan, R., Tsaih, S. W., Petkova, S. B., Marin de Evsikova, C., Xing, S., Marion, M. A., et al. (2009). Aging in inbred strains of mice: Study design and interim report on median lifespans and circulating IGF1 levels. *Aging Cell, 8*(3), 277–287. https://doi.org/10.1111/j.1474-9726.2009.00478.x.

Zenke, M., & Hieronymus, T. (2006). Towards an understanding of the transcription factor network of dendritic cell development. *Trends in Immunology, 27*(3), 140–145. https://doi.org/10.1016/j.it.2005.12.007.

Zhai, X. Y., Yan, P., Zhang, J., Song, H. F., Yin, W. J., Gong, H., et al. (2016). Knockdown of SIRT6 enables human bone marrow mesenchymal stem cell senescence. *Rejuvenation Research, 19*(5), 373–384. https://doi.org/10.1089/rej.2015.1770.

Zhang, H., Head, P. S. E., & Yu, D. S. (2016). SIRT2 orchestrates the DNA damage response. *Cell Cycle (Georgetown, Texas), 15*(16), 2089–2090. https://doi.org/10.1080/15384101.2016.1184517.

Zhang, M., Xuan, S., Bouxsein, M. L., von Stechow, D., Akeno, N., Faugere, M. C., et al. (2002). Osteoblast-specific knockout of the insulin-like growth factor (IGF) receptor gene reveals an essential role of IGF signaling in bone matrix mineralization. *The Journal of Biological Chemistry, 277*(46), 44005–44012. https://doi.org/10.1074/jbc.M208265200.

Zhao, J. L., Ma, C., O'Connell, R. M., Mehta, A., DiLoreto, R., Heath, J. R., et al. (2014). Conversion of danger signals into cytokine signals by hematopoietic stem and progenitor cells for regulation of stress-induced hematopoiesis. *Cell Stem Cell, 14*(4), 445–459. https://doi.org/10.1016/j.stem.2014.01.007.

Zhong, L., & Mostoslavsky, R. (2010). SIRT6: A master epigenetic gatekeeper of glucose metabolism. *Transcription, 1*(1), 17–21. https://doi.org/10.4161/trns.1.1.12143.

Zhou, S., Tang, X., & Chen, H. Z. (2018). Sirtuins and insulin resistance. *Frontiers in Endocrinology, 9*, 748. https://doi.org/10.3389/fendo.2018.00748.

CHAPTER ELEVEN

Alterations in cyclic nucleotide signaling are implicated in healthy aging and age-related pathologies of the brain

Nicole Gorny and Michy P. Kelly*
Department of Anatomy & Neurobiology, University of Maryland School of Medicine, Baltimore, MD, United States
*Corresponding author: e-mail address: michy.kelly@uscmed.sc.edu

Contents

Abstract

It is not only important to consider how hormones may change with age, but also how downstream signaling pathways that couple to hormone receptors may change. Among these hormone-coupled signaling pathways are the 3′,5′-cyclic guanosine monophosphate (cGMP) and 3′,5′-cyclic adenosine monophosphate (cAMP) intracellular second messenger cascades. Here, we test the hypothesis that dysfunction of cAMP and/or cGMP synthesis, execution, and/or degradation occurs in the brain during healthy and pathological diseases such as Alzheimer's disease, Parkinson's disease, and Huntington's disease. Although most studies report lower cyclic nucleotide signaling in the aged brain, with further reductions noted in the context of age-related diseases, there are select examples where cAMP signaling may be elevated in select tissues. Thus, therapeutics would need to target cAMP/cGMP in a tissue-specific manner if efficacy for select symptoms is to be achieved without worsening others.

Vitamins and Hormones, Volume 115
ISSN 0083-6729
https://doi.org/10.1016/bs.vh.2020.12.012

1. Introduction

It is not only important to consider how hormones may change with age, but also how downstream signaling pathways that couple to hormone receptors may change. Among these hormone-coupled signaling pathways are the 3′,5′-cyclic guanosine monophosphate (cGMP) and 3′,5′-cyclic adenosine monophosphate (cAMP) second messenger cascades. Specifically, natriuretic peptide hormones (i.e., ANP, BNP and CNP) bind transmembrane guanylyl cyclases to stimulate the formation of cGMP, while antidiuretic hormone, calcitonin, epinephrine, follicle stimulating hormone, glucagon, luteinizing hormone, norepinephrine, and thyroid-stimulating hormone bind to G protein-coupled receptors that activate adenylyl cyclase to catalyze the formation of cAMP. 3′,5′-cyclic nucleotides are responsible for regulating numerous functions within the central nervous system (CNS), including development of neural circuitry, neurogenesis, apoptosis, neural plasticity, sensorimotor filtering, sleep, mood stability, memory, and other aspects of cognitive performance (Averaimo & Nicol, 2014; Favilla, Abel, & Kelly, 2008; Fiscus, 2002; Giralt, Saavedra, Alberch, & Perez-Navarro, 2012; Heckman, Blokland, Ramaekers, & Prickaerts, 2015; Kanes et al., 2007; Kaplan, Kent, Charron, & Fournier, 2014; Kelly et al., 2007, 2009; Luo, Phan, Yang, Garelick, & Storm, 2013; Merz, Herold, & Lie, 2011; Nicol & Gaspar, 2014). Healthy aging, alongside age-related diseases such as Alzheimer's disease (AD), Parkinson's disease (PD), and Huntington's disease (HD), are characterized by the deterioration of many of the above processes (e.g., Apple, Solano-Fonseca, & Kokovay, 2017; Dean, Weiss, Morris, & Chasens, 2017; Ffytche et al., 2017; Kane et al., 2017); thus, it stands to reason cyclic nucleotide signaling in these patient populations may be impaired.

Numerous molecules are involved in the synthesis, execution, and breakdown of cAMP and cGMP signaling cascades (Fig. 1). As described above, particulate guanylyl cyclases (pGCs) synthesize cGMP when activated by natriuretic peptides, but so do soluble guanylyl cyclases (sGCs) when activated by nitric oxide (NO) (Kobialka & Gorczyca, 2000). Also, as noted above, cAMP is synthesized by transmembrane adenylyl cyclases (ACs) when activated by the G-protein Gαs (Neves, Ram, & Iyengar, 2002). Additionally, cAMP can be generated by soluble ACs, which can be stimulated by calcium and bicarbonate and are localized to mammalian glia cells and neurons (Chen, Martinez, Milner, Buck, & Levin, 2013). cAMP activates a number

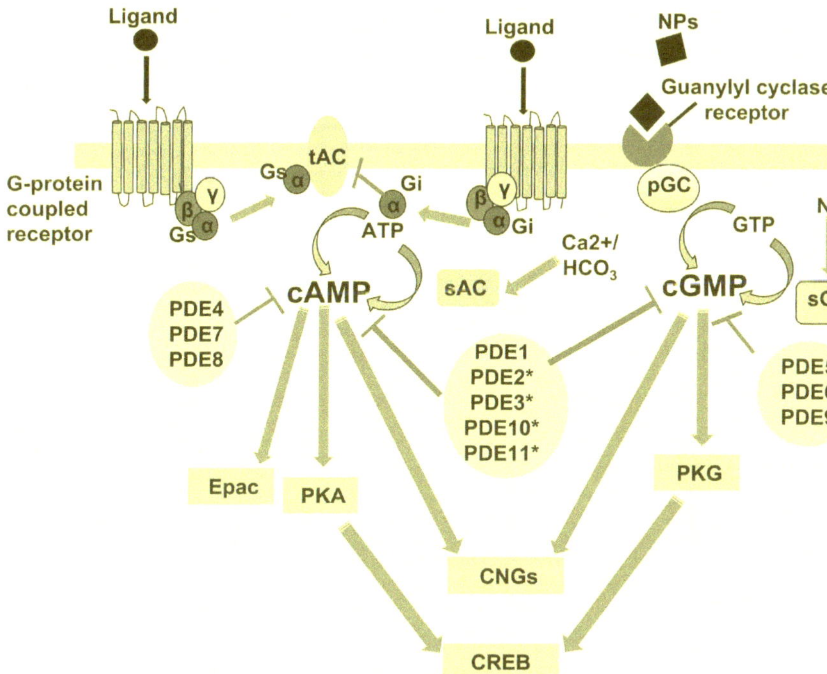

Fig. 1 Signals involved in cAMP and cGMP synthesis, execution, and break down. Upon stimulation, G protein-coupled receptors (GPCRs) facilitate the displacement of GDP for GTP, activating heterotrimeric G-proteins comprised of a β, γ, and either a stimulatory (Gs) or an inhibitory (Gi) α subunit that stimulates transmembrane adenylyl cyclase (tAC). Simultaneously, cytosolic bicarbonate (HCO3) and calcium stimulate soluble AC (sAC). ACs synthesize the formation of cAMP, 3′5′-cyclic adenosine monophosphate, from ATP. cAMP can then amplify the initial signal by acting on exchange protein activated by cAMP (Epac), cyclic nucleotide gated channels (CNGs), and protein kinase A (PKA). In contrast, particulate guanylyl cyclase (pGC)-coupled receptors are activated by natriuretic peptides (NPs), while soluble GC (sGC) is stimulated by membrane-permeable nitric oxide (NO). GCs convert GTP to 3′,5′-cyclic guanosine monophosphate (cGMP), which activates downstream targets such as CNGs and protein kinase G (PKG). cAMP or cGMP activation of PKA and/or PKG, respectively, subsequently results in phosphorylation and activation of the transcription factor cAMP response element binding protein (CREB). 3′,5′-cyclic nucleotide phosphodiesterases (PDEs), of which 11 families exist, hydrolyze cGMP and/or cAMP. Select PDEs (as indicated by *) are also regulated allosterically by cGMP or cAMP. *Figure from Kelly, M. P. (2018). Cyclic nucleotide signaling changes associated with normal aging and age-related diseases of the brain.* Cellular Signalling, 42*, 281–291. doi: 10.1016/j.cellsig.2017.11.004 with permission from Elsevier.*

of downstream molecules, such as protein kinase A (PKA), cyclic nucleotide gated ion channels, exchange protein activated by cAMP (Epac), and popeye-domain containing (POPDC) proteins (Baillie, Tejeda, & Kelly, 2019). Likewise, cGMP activates protein kinase G (PKG) in addition to cyclic nucleotide-gated channels. Initiation of either the cGMP or cAMP signaling cascade results in the downstream activation of cAMP response element binding protein (CREB), which promotes CRE-dependent gene transcription. Phosphodiesterases (PDEs), of which 11 families exist, are responsible for the inactivation of cAMP and cGMP, and several PDE families are themselves allosterically modulated by the cyclic nucleotides (Francis, Blount, & Corbin, 2011). In the following review, we assess the relevant literature to test the hypothesis that dysfunction of cAMP/cGMP synthesis, execution, and/or degradation occurs in the brain during healthy and pathological aging.

2. Healthy aging

Studies investigating the causal relationship between age and changes in cAMP synthesis show mixed results. Increasing age is correlated with reductions in basal and Gαs-stimulated AC activity in human brain samples (region not specified, Cowburn et al., 1992). On the contrary, animal studies testing age-related changes in AC activity report inconsistent conclusions. Approximately half of these animal studies, regardless of brain region (hippocampus, striatum, cerebellum, and cortex), showed age-linked reductions in AC activity, while the other half showed no detectable change in AC activity correlated with age (Araki, Kato, Fujiwara, & Itoyama, 1995; Hara, Onodera, Kato, & Kogure, 1992; Makman et al., 1978, 1980; Puri & Volicer, 1981; Sugawa & May, 1993, 1994).

Studies reporting alterations in cAMP with age in human tissue are scarce, while those in rodent brains are inconsistent in their results. Despite this, some overall trends emerge (Fig. 2B). White blood cell basal cAMP levels appear to decrease with age in comparison to young counterparts in both humans and rodents (Birkenfeld & Ben-Zvi, 1984; Fulop, Kekessy, & Foris, 1990; Kostromin, Berdyshev, Demidov, & Kuibeda, 1984; Tam & Walford, 1980). Serum also saw a reduction in cAMP levels in aged vs. young adult rodents (Hu et al., 2007). In contrast, there was no change in cAMP levels from aged vs. young human cerebral microvessels (Grammas, Roher, & Ball, 1994; Martinez, Hernandez, & Hernanz, 2001). Rodent hippocampi do not appear to experience a change in basal cAMP levels with age (Hara et al., 1992; Puri & Volicer, 1981; Schmidt & Thornberry, 1978; Titus, Furones, Kang, & Atkins, 2013), unless previously exposed to traumatic brain injury (TBI). In such a case, cAMP levels are significantly reduced

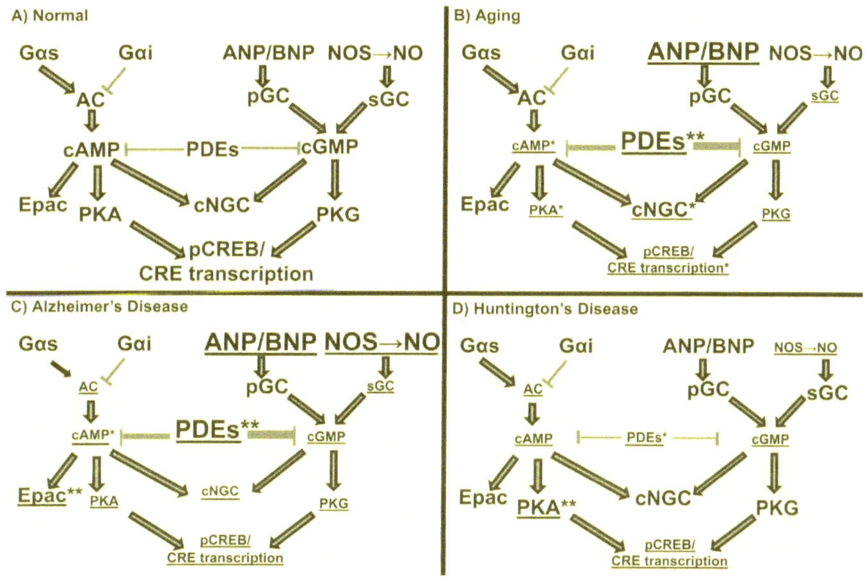

Fig. 2 Reductions in cGMP and cAMP signaling are found in the brain during normal aging as well as age-related diseases. (A) Canonical signaling present in young adult brain. Targets that are underscored are those that appear to change in response to disease or age relative to normal baseline, and direction of these changes are indicated by a change in relative font size (e.g., larger font = increases noted, smaller font = decreases noted). (B) Studies across multiple brain regions suggest aging is correlated with reductions in either basal or stimulated cAMP levels, cGMP levels, PKA activity, PKG activity, soluble guanylyl cyclase (sGC) activity, CREB activity (measured either by CRE-dependent transcription rates or by phosphorylation), and upregulated phosphodiesterase (PDE) activity. (C) cAMP/cGMP signaling is further reduced in Alzheimer's disease (AD) patients/models relative to age-matched healthy controls. AD is characterized by greater reductions in cyclic nucleotide gated channels (cNGC) and adenylyl cyclase (AC) activity, in addition to the targets mentioned above. (D) Patients/models of Huntington's disease also demonstrate greater reductions in AC activity, cNGC activity, cyclic nucleotide signaling relative to age-matched controls. Notably, most HD studies report a decrease in PDE expression/activity rather than the increase characteristic of AD. *select tissues demonstrate opposite effect (e.g., amplified cAMP signaling in HD hippocampus or aging prefrontal cortex or HD hippocampus)—see main text. **changes are correlated with aging/disease but directionality of change (increase or decrease) is isoform and brain region specific—see main text. *Figure from Kelly, M. P. (2018). Cyclic nucleotide signaling changes associated with normal aging and age-related diseases of the brain.* Cellular Signalling, 42, *281-291. doi: 10.1016/j.cellsig.2017.11.004 with permission from Elsevier.*

in old rodent hippocampus when compared to young rodents (Titus et al., 2013). Although basal cAMP levels do not change with age in the adult rodent cerebellum (Hara et al., 1992; Puri & Volicer, 1981) (but see Austin, Connole, Kett, & Collins, 1978), cAMP levels stimulated by norepinephrine and kainite do diminish with age (Schmidt & Thornberry, 1978). Conversely, adult rodent cortex, thalamus, and hypothalamus all exhibited decreases in basal cAMP levels with age (Berg & Zimmerman, 1975; Hara et al., 1992; Puri & Volicer, 1981; Zimmerman & Berg, 1974) (but see Puri & Volicer, 1981; Schmidt & Thornberry, 1978; Titus et al., 2013). Thus, it is interesting that prefrontal cortex injection of a cAMP analogue appears to mimic, while a cAMP blocker rescues, age-related deficits in working memory (as opposed to blocking and mimicking, respectively) (Ramos et al., 2003; Vandesquille et al., 2013; Wang et al., 2011). The reversal of working memory deficits associated with cAMP blockers also appears contradictory in the face of age-related changes in PKA activity. Aged rodents exhibit a decrease in PKA activity when compared to young adult rodents in prefrontal cortex (Karege et al., 2001), hippocampus (Karege, Lambercy, Schwald, Steimer, & Cisse, 2001; Karege, Schwald, et al., 2001), and serum (Hu et al., 2007). This age-related loss in PKA activity may be conserved as a similar change in activity has been noted in aged fly brain (Laviada, Galve-Roperh, Malpartida, & Haro, 1997). However, a blanket reduction in PKA activity across all rodent brain tissues is not characteristic of aging, as PKA activity notably increases with age in rodent cerebral microvessels (Cashman & Grammas, 1995). This may suggest that the relationship between aging and subsequent effects on cAMP expression is brain-tissue specific.

Age-related changes in expression and/or activity of cAMP-PDE vary greatly, depending on isoform and tissue type (Table 1). For instance, in aged rat serum cAMP-PDE expression appears unchanged (Hu et al., 2007), but in aged rodent cortex and hippocampus there is an increase in high Km cAMP-PDE activity (Chalimoniuk & Strosznajder, 1998; Stancheva & Alova, 1991). It is unclear which isoform(s) are responsible for this upregulation in cAMP-PDE activity in cortex, as PDE4 and PDE7A expression/activity are downregulated in cortex and PDE8 is unchanged (Carlyle et al., 2014; Kelly, 2014; Kelly et al., 2014; Tohda et al., 1996) (Table 1). Reduced expression of PDE7A mRNA in aged rat cortex is noteworthy, given three replication cohorts and a joint analysis suggest a link between a single nucleotide polymorphism (SNP) in this isoform and age-related cognitive deficits (De Jager et al., 2012). PDE4 expression and activity also appear to decrease in both rat and monkey striatum and cortex (Carlyle et al., 2014; Kelly, 2014; Kelly et al., 2014; Tohda et al., 1996), dorsolateral prefrontal cortex included (Carlyle et al., 2014). A similar effect has been noted in rat cerebellum (Kato et al., 1998; Tohda et al., 1996), although results did not replicate in monkeys (Harada et al., 2002). The age-related changes observed in PDE4 are bolstered by the fact that rolipram, a PDE4 inhibitor, mimics age-related impairments

Table 1 Reported differences in phosphodiesterases (PDEs) seen in aged vs. young adults.

	Hippocampus	Cortex	Striatum	Cerebellum	Other
PDE1	Increased protein expression of select 1C isoforms (Kelly et al., 2014)	No change 1A, 1B, or 1C mRNA (Kelly et al., 2014)	No change 1A, 1B, or 1C mRNA (Kelly et al., 2014)	No change 1A, 1B, or 1C mRNA (Kelly et al., 2014)	
PDE2	No change 2A mRNA (Kelly et al., 2014)	No change 2A mRNA (Kelly et al., 2014)	Decreased 2A mRNA (Kelly et al., 2014)	No change 2A mRNA (Kelly et al., 2014)	
PDE3	No change 3A or 3B mRNA (Kelly et al., 2014)	No change 3A or 3B mRNA (Kelly et al., 2014)	No change 3A or 3B mRNA (Kelly et al., 2014)	No change 3A or 3B mRNA (Kelly et al., 2014)	
PDE4	No change 4A, 4B or 4D mRNA; decreased PDE4 binding in situ but not in cytosolic fractions (Harada et al., 2002; Kato, Araki, Chen, Itoyama, & Kogure, 1998; Kelly et al., 2014; Tohda, Murayama, Nogiri, & Nomura, 1996)	Decreased 4D but not 4A or 4B mRNA in rat; decreased PDE4A5 protein in monkey in dlPFC; decreased PDE4 binding in rat and monkey; decreased PDE4 activity rat (Carlyle et al., 2014; Harada et al., 2002; Kato et al., 1998; Kelly et al., 2014; Tohda et al., 1996)	Decreased 4A mRNA but no change 4B or 4D mRNA; decreased PDE4 binding in rat and monkey; decreased PDE4 activity rat (Harada et al., 2002; Kato et al., 1998; Kelly et al., 2014; Tohda et al., 1996)	Decreased 4D mRNA but no change 4A or 4B mRNA; decreased PDE4 binding in rat but no change monkey; decreased PDE4 activity rat (Harada et al., 2002; Kato et al., 1998; Kelly et al., 2014; Tohda et al., 1996)	Decreased PDE4 binding in thalamus, SN, & BS; increased PDE4 activity in basal forebrain (Asanuma et al., 1996; Kato et al., 1998)
PDE5	No change 5A mRNA (Kelly et al., 2014)	No change 5A mRNA (Kelly et al., 2014)	No change 5A mRNA (Kelly et al., 2014)	Increased 5A mRNA (Kelly et al., 2014)	
PDE7	No change 7A mRNA (Kelly et al., 2014)	Decreased 7A mRNA (Kelly et al., 2014)	No change 7A mRNA (Kelly et al., 2014)	No change 7A mRNA (Kelly et al., 2014)	SNP linked to age-related cognitive (De Jager et al., 2012)

Continued

Table 1 Reported differences in phosphodiesterases (PDEs) seen in aged vs. young adults.—cont'd

	Hippocampus	Cortex	Striatum	Cerebellum	Other
PDE8	Increased 8A and 8B mRNA and protein expression, with possible change in compartmentalization (Kelly et al., 2014)	No change 8A or 8B mRNA (Kelly et al., 2014)	Increased 8A mRNA, no change 8B mRNA (Kelly et al., 2014)	No change 8A or 8B mRNA (Kelly et al., 2014)	
PDE9	No change 9A mRNA (Kelly et al., 2014)	No change 9A mRNA (Kelly et al., 2014)	No change 9A mRNA (Kelly et al., 2014)	No change 9A mRNA (Kelly et al., 2014)	
PDE10	No change 10A mRNA (Kelly et al., 2014)	No change 10A mRNA (Kelly et al., 2014)	No change 10A mRNA (Kelly et al., 2014)	Increased 10A mRNA (Kelly et al., 2014)	
PDE11	Increased 11A mRNA and protein expression with possible change in compartmentalization (Kelly et al., 2014)	No change 11A mRNA (Kelly et al., 2014)	No change 11A mRNA (Kelly et al., 2014)	No change 11A mRNA (Kelly et al., 2014)	
PDE activity	Increased high (but not low) Km cAMP-PDE activity; large increase in cGMP-PDE activity (Chalimoniuk & Strosznajder, 1998; Stancheva & Alova, 1991)	Increased high (but not low) Km cAMP-PDE activity; small increase in cGMP-PDE activity (Chalimoniuk & Strosznajder, 1998; Stancheva & Alova, 1991), but see (Zimmerman & Berg, 1975)	Findings conflict: 1 study no change in high or low Km cAMP-PDE activity; 1 study reduced cAMP-PDE activity (Puri & Volicer, 1977; Stancheva & Alova, 1991)	Large increase in cGMP-PDE activity (Chalimoniuk & Strosznajder, 1998)	No change cAMP-PDE or cGMP-PDE activity in superior cervical ganglia or serum (Giorgi, Squitti, Bonsi, Paggi, & Toschi, 1994; Hu et al., 2007)

dlPFC, dorsolateral prefrontal cortex; SN, substantia nigra; BS, brainstem; SNP, single nucleotide polymorphism.

Table from Kelly, M. P. (2018). Cyclic nucleotide signaling changes associated with normal aging and age-related diseases of the brain. *Cellular Signalling*, *42*, 281–291. doi: 10.1016/j.cellsig.2017.11.004 with permission from Elsevier.

in working memory in animals (Ramos et al., 2003), but are contradicted by studies in elderly humans reporting a PDE4 inhibitor improves working memory (Blokland et al., 2019). Perhaps the pro-cognitive effects of the PDE4 inhibitor are related to the fact that PDE4 activity reportedly increases with age in the basal forebrain (Asanuma et al., 1996). Despite downregulation of hippocampal PDE4 expression (Harada et al., 2002; Kato et al., 1998) (but see Tohda et al., 1996), expression of PDE1C, PDE8A, and PDE11A appears to increase with age in rodent hippocampus (Kelly et al., 2014). Age-related increases in these PDE isoforms may also be responsible for apparent increases in high Km cAMP-PDE described above (Stancheva & Alova, 1991) in addition to upregulation of cGMP-PDE activity in aged hippocampus (Chalimoniuk & Strosznajder, 1998). Thus, regional variability in the expression and activity of PDEs suggest it is necessary to localize therapeutic targets in a region-specific manner, such that effectiveness is maximized and side effects are avoided.

Reinforcing claims that increased cAMP-PDE activity plays a role in the neuropathology of age-linked cognitive deficits, numerous cAMP-PDE inhibitors have proven therapeutically efficacious in clinical and preclinical studies. For instance, caffeine inhibits cAMP-PDE activity in addition to adenosine receptors (Boswell-Smith, Spina, & Page, 2006). Reports focused on high caffeine consumption in elderly women suggest a protective effect against cognitive decline (Driscoll et al., 2016) and a decrease in the rate of cognitive impairment (Vercambre, Berr, Ritchie, & Kang, 2013). These results are mirrored in rodent models, such that chronic caffeine treatment averts cognitive decline associated with age (Costa, Botton, Mioranzza, Souza, & Porciuncula, 2008; Leite, Wilhelm, Jesse, Brandao, & Nogueira, 2011; Vila-Luna et al., 2012) (but see Shukitt-Hale, Miller, Chu, Lyle, & Joseph, 2013). Consistent with the above, cAMP-specific and dual-specific PDE inhibitors are reported to enhance cognition in aged subjects. A drug containing the PDE1 inhibitor Vinpocetin improved memory performance in elderly participants (Richter, Herzog, Eyal, & Cohen, 2011), a result also produced by two novel PDE4 inhibitors from Tetra Discovery Partners and Dart Neuroscience (c.f. Baillie et al., 2019). Animal models in which PDE2 (Domek-Lopacinska & Strosznajder, 2008; Nakashima, Suzuki, Shiraishi, & Iwashita, 2019), PDE3 (Yanai & Endo, 2019; Yanai, Toyohara, Ishiwata, Ito, & Endo, 2017), PDE4 (Bach et al., 1999; de Lima et al., 2008; Drott, Desire, Drouin, Pando, & Haun, 2010; Kumar & Singh, 2017b), and PDE8B (Tsai et al., 2012) are blocked by pharmaceutical compounds or "knocked out" via gene deletion have shown resistance to brain aging pathologies. Results from these studies indicate cAMP signaling cascades may prove a viable therapeutic target in treating age-related cognitive decline when administered in a region-specific manner.

Studies examining the role of the NOS-sGC pathway in age-related decline are few. Findings with regard to NOS are contradictory. Blum-Degen and colleagues saw no change in NOS activity in aged human parietal cortex, cerebellum, or nucleus accumbens when compared to young counterparts (Blum-Degen et al., 1999). However, the animal literature equally reports increased, decreased, or stable NOS activity in aged hippocampus, cortex, and cerebellum without apparent discrepancies in methodology able to explain the inconsistencies (Chalimoniuk & Strosznajder, 1998; Jesko, Chalimoniuk, & Strosznajder, 2003; Law, O'Donnell, Gauthier, & Quirion, 2002; Liu, Smith, Appleton, Darlington, & Bilkey, 2003, 2004, 2005; Mollace, Rodino, Massoud, Rotiroti, & Nistico, 1995; Strosznajder, Jesko, Zambrzycka, Eckert, & Chalimoniuk, 2004; Vallebuona & Raiteri, 1995; Yu, Juang, Lee, Liu, & Cheng, 2000). Changes in GC expression and/or activity are more consistent across species. For example, sGC expression is shown to decrease between ages 60 and 90 in human nucleus accumbens as well as parietal, frontal, prefrontal, and orbital frontal cortex (Ibarra, Nedvetsky, Gerlach, Riederer, & Schmidt, 2001). Likewise, aged rodent hippocampus and cerebellum show an age-related decrease in GC activity (Chalimoniuk & Strosznajder, 1998; Vallebuona & Raiteri, 1995). Consistent with the apparent age-related disruption of sGC activity, stimulation of sGC with NO donors rescued memory deficits in both rats (Paul, Reddy, & Ekambaram, 2005) and crickets (Matsumoto, Matsumoto, Takahashi, & Mizunami, 2016). Together, these studies suggest cGMP synthesis largely decreases with age in the brain (Fig. 2B).

Despite the above results implying cGMP synthesis via sGC is reduced, other studies suggest cGMP synthesis via pGCs may be increased. Reported changes in natriuretic peptides, which serve as pGC receptor agonists, have thus far been focused on circulating concentrations. In humans, ANP concentration is increased fourfold in healthy elderly subjects when compared to young adults, and this increase in ANP levels remains correlated with aging between 70 and 102 years of age (Davis, Fish, Minaker, & Elahi, 1996). Although BNP levels remained consistent between ages 30 and 72 (Marksteiner et al., 2014), an increase in circulating BNP levels was observed after the age of 75 among individuals without dementia (Daniels et al., 2011; Hiltunen et al., 2013; Nilsson, Gustafson, & Hultberg, 2008). Upregulation of BNP in humans over 60 is correlated with increased risk of developing a cognitive impairment (Hiltunen et al., 2013), poorer performance, and more rapid decline on tests measuring cognition, such as the mini mental state exam (MMSE), logical memory tests, and Trails B test (Daniels et al., 2011; Feinkohl et al., 2012; Gunstad et al., 2006; van Vliet et al., 2014). It is necessary to acknowledge that many of the above studies reported a relationship between high BNP and increased incidence of cardiovascular disease (CVD), presenting yet another risk factor for development of a cognitive disorder (Daniels et al., 2011; Nilsson et al., 2008; van Vliet et al., 2014). However, two such studies report that after controlling for CVD factors, elevation of BNP remained independently correlated with poor cognitive performance (Daniels et al., 2011; Feinkohl et al., 2012); a 3rd study examining a group of older adults that all had CVD reinforced these results, as elevated BNP was still associated with lower MMSE scores (Gunstad et al., 2006). Consequently, it appears that in elderly humans elevated BNP is a factor leading to deficits in cognitive performance regardless of the presence of CVD pathologies. The effect of aging on ANP and/or BNP expression and activity in the brain is not yet clear.

Age-related regulation of cGMP itself appears to vary between the circulation and the brain. Serum collected from aged vs. young mice (Fulop et al., 1990; Tam & Walford, 1980) and white blood cells taken from aged vs. young adult human subjects (Hu et al., 2007) suggest cGMP levels, in addition to circulating ANP and BNP concentrations, increase with age. Conversely, cGMP localized to platelets appear downregulated in aged vs. young adult humans (Kawamoto et al., 2005). The effects of aging on cGMP levels in the human brain have not been studied extensively, although rodent models show an age-related decrease in cGMP levels in cerebellum (Chalimoniuk & Strosznajder, 1998; Puri & Volicer, 1981; Scavone et al., 2005; Schmidt & Thornberry, 1978) (but see Vallebuona & Raiteri, 1995) and hippocampus (Chalimoniuk & Strosznajder, 1998; Vallebuona & Raiteri, 1995) (but see Puri & Volicer, 1981), but not cortex (Chalimoniuk & Strosznajder, 1998) when compared to age-matched young adult rodents. Consistent with reported decreases in cGMP levels, PKG activity also decreases with age in rat cerebellum (Scavone et al., 2005). Furthermore, aging appears to severely impair glutamate-activated cGMP accumulation in the hippocampus and cerebellum of aged vs. young adult rodents (Chalimoniuk & Strosznajder, 1998; Piedrafita, Cauli, Montoliu, & Felipo, 2007; Vallebuona & Raiteri, 1995). These age-related decreases in cGMP signaling may be driven somewhat by increases in cGMP-PDE activity in aged vs. young rodent cerebellum and hippocampus (Chalimoniuk & Strosznajder, 1998), likely attributed to overexpression of PDE5A and PDE10A in the cerebellum and PDE11A in the hippocampus (Kelly et al., 2014). Inhibitors of PDE2 (Domek-Lopacinska & Strosznajder, 2008; Nakashima et al., 2019), PDE3 (Yanai et al., 2017), and PDE5 reversed cognitive deficits characteristic of aging in LTP, pCREB, and/or memory in animal models (Devan et al., 2014; Orejana, Barros-Minones, Jordan, Puerta, & Aguirre, 2012; Palmeri,Privitera, Giunta, Loreto, & Puzzo, 2013; VerPlank, Tyrkalska, Fleming, Rubinsztein, & Goldberg, 2020) (but see Domek-Lopacinska & Strosznajder, 2008), suggesting that stimulation of cGMP signaling may prove effective in treating age-associated memory impairments.

Taken altogether, the majority of evidence argues for reduced cyclic nucleotide signaling in the aging brain, which is paralleled by a reduction in CREB function. Studies focused on CREB mainly address hippocampus-localized signaling, although some delve into CREB signaling in specific subregions of cortex as well. In humans, age is correlated with decreased levels of CREB in the hippocampus (Yamamoto-Sasaki, Ozawa, Saito, Rosler, & Riederer, 1999). Likewise, numerous studies show a reduction in basal and/or training-induced hippocampal CREB phosphorylation (pCREB) in aged vs. young adult rodents (Dagnas, Guillou, Prevot, & Mons, 2013; Hattiangady, Rao, Shetty, & Shetty, 2005; Kudo, Wati, Qiao, Arita, & Kanba, 2005; Marchese et al., 2017; Porte, Buhot, & Mons, 2008; Titus et al., 2013; Xu et al., 2010), with only a single study suggesting age-linked pCREB deficiency is circumscribed by an existing cognitive disorder (Xu et al., 2010). Coinciding with these results, age-related neurophysiological and memory deficits are effectively rescued in rodents by overexpression of CREB in the hippocampus (Ma et al., 2019; Marchese et al., 2017; Yu, Curlik, Oh, Yin, & Disterhoft, 2017). pCREB shows variable patterns in adult rodent cortex, with a notable age-related decrease in parietal regions (Titus et al., 2013) but an age-related upsurge in prefrontal cortical regions (Ramos et al., 2003; Vandesquille et al., 2013). Once more, age-related changes in cyclic nucleotide signaling prove to be brain region-specific, with decreased activity in hippocampus and some cortical regions, but upregulation in prefrontal cortex.

3. Alzheimer's disease, dementia, and mild cognitive impairment (MCI)

Human, rodent, and in vitro models examining dementia-related pathologies suggest that alterations in cAMP signaling pathways are contributing factors. Notably, AC activity stimulated by Gαs is reduced in the frontal cortex (Cowburn, O'Neill, Ravid, Alafuzoff, et al., 1992), temporal cortex (Bonkale et al., 1996; Cowburn, O'Neill, Ravid, Alafuzoff, et al., 1992), occipital cortex (Cowburn, O'Neill, Ravid, Alafuzoff, et al., 1992), hippocampus (Ohm, Bohl, & Lemmer, 1989, 1991; O'Neill et al., 1994; Schnecko, Witte, Bohl, Ohm, & Lemmer, 1994), and cerebellum (Cowburn, O'Neill, Ravid, Alafuzoff, et al., 1992; Schnecko et al., 1994) (but see Bonkale et al., 1996) of patients with AD or dementia compared to age-matched controls. The relationship between cognitive decline and both basal and forskolin-activated AC is difficult to determine given conflicting results across various studies, although some general trends do emerge. The majority of studies found that AC activity in the cerebellum (Bonkale et al., 1996; Cowburn, O'Neill, Ravid, Alafuzoff, et al., 1992) (but see Schnecko et al., 1994) and occipital cortex (Cowburn, O'Neill, Ravid, Alafuzoff, et al., 1992) is not altered in AD models, while temporal cortical (Cowburn, O'Neill, Ravid, Alafuzoff, et al., 1992; Cowburn, O'Neill, Ravid, Winblad, & Fowler, 1992) (but see Bonkale et al., 1996) and hippocampal (both basal and/or forskolin stimulated) AC activity is reduced (Kumar & Singh, 2017c; Ohm et al., 1989, 1991; Ohm, Schmitt, Bohl, & Lemmer, 1997; Schnecko et al., 1994) (but see Kumar & Singh, 2017c; O'Neill et al., 1994). Reported AC activity in the frontal cortex is too inconsistent to draw even general conclusions as to the effects AD has on this region (Cowburn, O'Neill, Ravid, Alafuzoff, et al., 1992; Cowburn, O'Neill, Ravid, Winblad, & Fowler, 1992). Decreased temporal cortical and hippocampal AC activity seen in AD patients appears to occur independently from upregulations in Gαi activity/expression (Brinton, Yamazaki, Gonzalez, O'Neill, & Schreiber, 1998; Cowburn, O'Neill, Ravid, Winblad, & Fowler, 1992; Schnecko et al., 1994), downregulated Gαs expression (O'Neill et al., 1994), or global reductions in AC expression noted across the brain (Dewar, Horsburgh, Graham, Brooks, & McCulloch, 1990); however, isolated reductions in the expression of AC1 and AC2 (not AC5–AC8) may drive the hippocampal and cortical reduction of AC activity (Yamamoto et al., 1997, 2000).

Following evidence that cAMP synthesis in dampened in AD patients, both cAMP levels and cAMP effector molecules also appear to be reduced in select tissues. For instance, polymorphonuclear granulocytes harvested from AD patients had lower cAMP levels compared to age-matched controls (Fulop et al., 1990). According to in vitro and rodent studies, exposure to Aβ1–42 or an upregulation of BACE1 (β-site amyloid precursor protein-cleaving enzyme 1) produced a corresponding reduction in cAMP levels measured in the hippocampus (Ettcheto et al., 2018; Zhang et al., 2014) and brain lysates (Chen et al., 2012), respectively. Furthermore, independent studies found that AD patients with severe amyloid aggregates saw a notable reduction in cAMP binding only in the cytosolic fraction of entorhinal cortex and subiculum, while the particulate fraction showed normal levels of cAMP binding (Bonkale, Cowburn, Ohm, Bogdanovic, & Fastbom, 1999). Notably, reduction of PKA expression and/or activity is also reported in human, rodent, and in vitro AD models (Chen et al., 2012; Du et al., 2014; Kim, Nairn, Cairns, & Lubec, 2001; Liang, Liu, Grundke-Iqbal, Iqbal, & Gong, 2007; Shi et al., 2011; Vitolo et al., 2002) (but see Bonkale et al., 1996). The role of cAMP expression in AD pathology is supported by the fact that hippocampal deficits can be reversed by increased cAMP levels, through either forskolin-stimulated AC activity (Wang, Rowan, & Anwyl, 2009) or a reduction in cAMP-degrading PDE expression/activity (Cheng et al., 2010; Gong et al., 2004; Perez-Gonzalez et al., 2013; Sierksma et al., 2013, 2014; Vitolo et al., 2002; Wang et al., 2012, 2016; Zhang et al., 2014). Additionally, treatments indirectly targeting cAMP/cGMP signaling appear to attenuate cognitive impairments in AD mice (Chen et al., 2019; Kumar & Singh, 2017a; Tang et al., 2019; Wang et al., 2019; Xia et al., 2019; Zhang, Shen, Wu, Zhang, & Xing, 2020; Zhu, Yang, Xie, & Wan, 2018). Importantly, PKA appears to be a necessary element in reversing and/or preventing Aβ1–42-induced hippocampal deficits via cAMP-elevating agents (Vitolo et al., 2002; Wang et al., 2009; Wang, Chen, et al., 2016; Zhang et al., 2020). Thus, evidence strongly supports the notion that a reduction in PKA-mediated cAMP signaling is characteristic of AD (Fig. 2C).

The role of HCN channels and Epac in AD pathology is less clear than that of PKA. AD patients show reduced HCN1 expression in the temporal lobe, and both murine and N2A-cell models imply a loss of HCN1 function is sufficient to heighten Aβ accumulation (Saito et al., 2012). On the contrary, in vivo rat models in which Aβ1–42 is injected into CA1 resulted in increased HCN1 mRNA expression; this indicates HCN1's role in AD pathology is either species-specific or part of a complex negative feedback loop with Aβ1–42 (Eslamizade et al., 2015). A feedback loop hypothesis is supported by in vivo studies in which administration of Aβ1–42 into rat CA1 decreased HCN2 mRNA expression (Eslamizade et al., 2015); related in vitro studies suggest Aβ secretion should also be reduced (Frykman et al., 2017). Unlike HCN channels, which appear to be related to AD in a species-specific or co-regulatory way, Epacs are associated with AD in an isoform-specific manner. Specifically, Epac2 expression is reduced in the frontal cortex of patients with AD, while Epac1 expression is increased in the same region (McPhee et al., 2005). The interplay between the simultaneous up- and downregulation of these isoforms is not clearly defined, but Epac1 activity is known to regulate secretion of sAPPα, the protective soluble form of APP (role of Epac2 not reported, Maillet et al., 2003; Zaldua, Gastineau, Hoshino, Lezoualc'h, & Zugaza, 2007). Further studies are needed to fully understand how HCN and Epac function impact AD-related deficits.

AD pathology may also be driven by localized cAMP signaling upregulation. For instance, cerebral vessels of AD patients show an increase in cAMP levels compared to healthy controls (Grammas et al., 1994; Martinez et al., 2001). Results depicting cAMP levels in CSF of AD patients varied, with two studies finding levels consistent with age-matched controls, and a third study noting elevated cAMP in patients with AD (Martinez et al., 1999). It should be noted that the latter study, in which CSF cAMP was elevated in AD patients, also reported a corresponding increase in CSF tau protein expression (Martinez et al., 1999). Interestingly, PKA is strongly related to the formation of Tau neurofibrillary tangles (NFTs) and the hyperphosphorylation of Tau, specifically at residues S198, S199, S214, and S409 (Jicha et al., 1999; Robertson et al., 1993; Wang et al., 2018; Ye et al., 2020; Zhang, Li, Wang, Liu, & Wang, 2006). In fact, Tau-pS214 and Tau-pS409 are exclusively identified in AD tissue (Jicha et al., 1999). The specificity of Tau phosphorylation potentially explains why low doses, but not high doses, of drugs blocking PKA activity reverse AD symptomatology (Amini et al., 2015; Eftekharzadeh et al., 2012; Ueda, Yagami, Kageyama, & Kawasaki, 1996). Recent reports suggest a PKA regulatory subunit, PKA RII, complexes with tau and other kinase molecules in AD to alter cAMP/PKA signaling (Ko et al., 2019). Furthermore, increased Tau expression subsequently reduces PKA activity and downstream CREB activation in a negative feedback loop mediated by nuclear PKA RII upregulation (Ye et al., 2020). This is in stark contrast to the global reduction of PKA activity noted in both AD patients and animal models (as described above, Chen et al., 2012; Du et al., 2014; Kim et al., 2001; Liang et al., 2007; Shi et al., 2011; Vitolo et al., 2002), as well as apparent attenuation and/or prevention of AD-correlated deficits following cAMP/PKA/CREB upregulation (Bate & Williams, 2015; Cameron, Whiteley, Day, Parachikova, & Baillie, 2017; Cheng et al., 2010; Du et al., 2014; Gong et al., 2004; Maki et al., 2014; Perez-Torres & Mengod, 2003; Pugazhenthi, Wang, Pham, Sze, & Eckman, 2011; Sierksma et al., 2013, 2014; Smith, Pozueta, Gong, Arancio, & Shelanski, 2009; Taguchi et al., 2013; Tai, Chien, Chang, & Yang, 2017; Vitolo et al., 2002; Wang et al., 2009, 2012; Wang, Chen, et al., 2016; Yiu, Rashid, & Josselyn, 2011; Zhang et al., 2014).

However, phosphorylation of Tau via PKA does not always stimulate the production of neurofibrillary tangles. For instance, PKA-mediated phosphorylation of Tau-S214 following Thr2122 phosphorylation by glycogen synthase kinase-3β (GSK-3β) results in the formation of paired helical filaments (PHFs); whereas, PKA-mediated phosphorylation of S214 prior to Thr212 phosphorylation by GSK-3β does not result in PHFs (Zheng-Fischhofer et al., 1998). Phosphorylation of Tau via PKA reduces its affinity for microtubules and the formation of PHFs (Schneider, Biernat, von Bergen, Mandelkow, & Mandelkow, 1999), possibly due to steric blocking of Tau sites by PKA-added phosphate groups, preventing other kinases access to Tau (Raghunandan & Ingram, 1995; Shea & Cressman, 1999). Interestingly, a recent preprint identified two rare missense mutations in the dual-specific PDE11A in patients with early onset AD and found the mutations increased hyperphosphorylation of Tau in a PKA-dependent manner (https://doi.org/10.21203/rs.3.rs-36875/v1).

Additional evidence implicates cAMP-specific or dual specific PDE isoforms in AD pathology. PDE3 expression, for instance, is increased in AD cerebrovessels (Maki et al., 2014). In cultured microglia, amyloid-β 1–42, in both soluble and fibrous form, lead to increases in PDE4B expression and subsequent neuroinflammation (Sebastiani et al., 2006). Early stages of AD are characterized by increased PDE4B and PDE4A mRNA expression in entorhinal cortex, but late-stage AD patients see a reduction in PDE4A mRNA expression in frontal cortex and CA2 (Perez-Torres & Mengod, 2003). PDE4D isoforms show a range of expression in the hippocampus of AD patients, with PDE4D1 mRNA doubled, PDE4D2 and PDE4D3 unchanged, and PDE4D5–9 markedly reduced (McLachlan et al., 2007). PDE4D expression remains unchanged in temporal cortex (Ugarte et al., 2015), but is increased in the putamen of AD patients (Perez-Torres & Mengod, 2003). Continuing the isoform-specific changes in PDE associated with AD, PDE7A mRNA expression is decreased in dentate gyrus (DG) of hippocampus, while expression of PDE8B mRNA is increased in CA2 and DG of hippocampus (Perez-Torres et al., 2003). The dual-specific PDE10A in temporal cortex (Ugarte et al., 2015) and the dual-specific PDE2A in cortex, cerebellum, hippocampus, and striatum of AD patients showed no detectable change in expression (Reyes-Irisarri, Markerink-Van Ittersum, Mengod, & de Vente, 2007). Clearly, changes in PDE expression/activity associated with AD are both isoform and brain-region specific.

Despite conflicting evidence surrounding the upregulation of cAMP-PDE expression and/or activity in AD pathology, PDE inhibitors have proven effective in both human and animal studies (Gulisano, Tropea, Arancio, Palmeri, & Puzzo, 2018; Jankowska, Wesolowska, Pawlowski, & Chlon-Rzepa, 2019; Nabavi et al., 2019). Women over 65 reporting high consumption of caffeine, a naturally occurring PDE inhibitor, saw a significantly reduced incidence of dementia (Driscoll et al., 2016), and AD mouse models report caffeine reversed cognitive deficits, Tau hyperphosphorylation, and Aβ accumulation (Arendash et al., 2006, 2009; Laurent et al., 2014). Vinpocetine, a PDE1 inhibitor, had mixed effects on patients with MCI or AD (Szatmari & Whitehouse, 2003; Thal, Salmon, Lasker, Bower, & Klauber, 1989; Valikovics, Csanyi, & Nemeth, 2012), but the PDE3 inhibitor cilostazol and several PDE4 inhibitors appear to mitigate symptoms associated with MCI or dementia in rodents and humans (Arai & Takahashi, 2009; Feng et al., 2019; Guo et al., 2017; Ihara et al., 2014; Prickaerts, Heckman, & Blokland, 2017; Ricciarelli et al., 2017; Sakurai et al., 2013; Saletu et al., 1992; Taguchi et al., 2013; Tai, Chen, Chien, & Yang, 2017; Tai, Chien, et al., 2017; Tibbo, Tejeda, & Baillie, 2019; Treves & Korczyn, 1999; Xu et al., 2018; Yanai & Endo, 2019). Along with PDE3 and PDE4 inhibitors, PDE2 and PDE7 inhibitors also aid in reversal or prevention of AD-related cognitive deficits in rodent models (Bartolome et al., 2018; Cheng et al., 2010; Costa et al., 2007; Gong et al., 2004; Maki et al., 2014; Park et al., 2011; Perez-Gonzalez et al., 2013; Perez-Torres & Mengod, 2003; Ruan et al., 2019; Sierksma et al., 2013; Sierksma et al., 2014; Smith et al., 2009; Wang et al., 2012, 2017; Wang, Chen, et al., 2016) and Aβ cytotoxicity in vitro models (Bate & Williams, 2015; Cameron et al., 2017; Du et al., 2014; Perez-Gonzalez et al., 2013; Vitolo et al., 2002). Thus, the results of these studies together imply inhibiting cAMP-PDE activity may be an appropriate therapeutic strategy to attenuate of AD and dementia-related symptoms.

Regarding cGMP synthesis, numerous studies reported elevated BNP levels in blood and/or plasma are associated with incidence, severity, or risk of developing mild cognitive impairment (MCI), MCI conversion to AD, AD, and/or vascular dementia (VaD) (Hilal et al., 2015; Hu et al., 2012; Kara et al., 2017; Kondziella, Gothlin, Fu, Zetterberg, & Wallin, 2009; Marksteiner et al., 2014; Mirza et al., 2016; Naito, Naka, & Watanabe, 2009; Soares et al., 2012; Tynkkynen et al., 2015, 2017) (but see Hiltunen et al., 2013). Patient populations in many of these studies were notably older than comparative healthy controls, thus introducing age as a confound (Hilal et al., 2015; Kara et al., 2017; Tynkkynen et al., 2015, 2017). That said, two studies reported BNP levels increased in patients with AD and/or MCI when compared to healthy, age-matched controls (Marksteiner et al., 2014; Soares et al., 2012), suggesting age-related and disease-related BNP elevations may occur independently. BNP and ANP are also well known CVD markers, and CVD alone increases risk of developing dementia. In light of this, it is interesting that select studies reported a correlation between elevated BNP/ANP levels and dementia even after controlling for CVD risk factors (Kara et al., 2017; Mirza et al., 2016; Naito et al., 2009; Tynkkynen et al., 2015, 2017). In fact, Tykkynen and colleagues suggested BNP elevation may be a particularly promising circulating marker of neuronal damage, possibly reflecting disease-causing mechanisms in the brain (Tynkkynen et al., 2017). Thus, it is interesting that Hu and colleagues found plasma BNP levels both elevated and correlated with CSF Aβ1–42 levels in patients with MCI and AD (Hu et al., 2012). In contrast, Llano and colleagues reported no significant change in BNP in MCI/AD patients vs. controls, although they did identify BNP as one of four components of a protein signature differentiating diagnosis (Llano, Devanarayan, Simon, & Alzheimer's Disease Neuroimaging, 2013).

The association between dementia and ANP is not as well researched. ANP levels are not specifically correlated with measured cognitive function of patients with dementia (Naito et al., 2009), but increased ANP levels were associated with incidence of dementia (Buerger et al., 2009; Holm et al., 2017) or the transition from MCI to more severe AD/dementia symptomology (Schneider et al., 2011). Unfortunately, studies analyzing ANP regulation face the challenge noted above of finding age-matched healthy controls; thus, increased age in the patient population is a common confound (Buerger et al., 2009; Holm et al., 2017; Schneider et al., 2011). ANP's role in vasoconstriction may also contribute to the rate of conversion from MCI to AD, given Schneider and colleagues findings suggesting anti-hypertensive treatment in patients with higher baseline ANP (but not low baseline ANP) see a slower transition to AD (Schneider et al., 2011). This is particularly true in patients under 75 years of age, implicating ANP biomarker as a potential selection criterion for use of anti-hypertensive therapies to treat AD and/or dementia-related deficits (Schneider et al., 2011).

In addition to apparent upregulation of signals upstream of pGCs, AD is also correlated with amplified signaling upstream of sGCs. AD patients show reduced CSF expression of an intrinsic NOS inhibitor (Abe, Tohgi, Murata, Isobe, & Sato, 2001), as well as increased platelet NOS activity (Kawamoto et al., 2005). NOS expression also appears to be upregulated in hippocampal tissue harvested from AD patients (Yew, Wong, Li, Lai, & Yu, 1999) and Aβ1–40-treated microglia and monocytes (Combs, Karlo, Kao, & Landreth, 2001). Increased NOS activity leads to subsequent increases in NO levels, which stimulate sGC activity and resultant cGMP production. One proposed mechanism suggests NOS signaling may increase in response to the downregulation of sGC activity/expression in AD. In the superior temporal cortex of AD patients, a decrease in sGC activity- but not pGC- was identified (Bonkale, Winblad, Ravid, & Cowburn, 1995). Reactive astrocytes of AD patients (Baltrons, Pifarre, Ferrer, Carot, & Garcia, 2004) and Aβ1–40- or Aβ25–35-treated cultured astrocytes saw a similar decrease in sGC activity and/or expression (Baltrons, Pedraza, Heneka, & Garcia, 2002). Bolstering the claim that AD corresponds to a loss of sGC signaling, CSF of AD patients had markedly reduced cGMP levels compared to matched controls, and this decrease correlated with poorer performance on the mini mental state (MMSE) (Hesse et al., 2017; Ugarte et al., 2015) (but see Martinez et al., 1999). Platelets harvested from AD patients also show a reduction in cGMP levels (Kawamoto et al., 2005). In AD mouse models, hippocampal cGMP levels do not appear to increase even after NMDA receptor activation (Duszczyk, Kuszczyk, Guridi, Lazarewicz, & Sadowski, 2012); furthermore, hippocampal slices treated with Aβ1–42 failed to show a corresponding LTP-induced increase in cGMP, suggesting a disruption in cGMP signaling (Puzzo et al., 2005). That said, one study suggests Aβ may be needed for cGMP-induced LTP and memory (Palmeri et al., 2017). Although Aβ1–42-induced LTP deficits were rescued by elevating cGMP (Puzzo et al., 2005), it is important to note that elevating cGMP can upregulate Aβ peptide production and APP aggregation independent of PKG (Calcagno et al., 2017). Increasing cGMP also rescues LTP deficits found in AD synaptosomes (Prieto et al., 2017), microglial inflammation (Paris et al., 1999), cytotoxicity (Wirtz-Brugger & Giovanni, 2000), and synaptosomal glutamate and glucose transport deficits (Mattson, Guo, & Geiger, 1999). Attenuation of these AD-related deficits via elevation of cGMP (through NO donors, sGC stimulators, cGMP analogues, PDE inhibitors) appears to be dependent upon PKG (Puzzo et al., 2005; Wirtz-Brugger & Giovanni, 2000; Zhang et al., 2013). Thus, it is difficult to reconcile the positive effects methylene blue treatment has on AD, given it is known to inhibit NOS and sGC, dampening cGMP signaling

(Oz, Lorke, Hasan, & Petroianu, 2011). It should be noted, however, that methylene blue targets a number of molecular and cellular components aside from those related to cGMP signaling (Oz et al., 2011). Collectively, these studies support impaired sGC/cGMP/PKG signaling in AD pathology (Fig. 2C).

Consistent with AD being associated with a reduction in cGMP synthesis, PDE-mediated cGMP degradation appears to increase. In fact, cGMP-specific PDE5A expression increased 5-fold in the temporal cortex of AD patients compared to healthy controls (Ugarte et al., 2015), despite consistent measured expression of cGMP-specific PDE9A or dual-specific PDE2A and PDE10A (Reyes-Irisarri et al., 2007; Ugarte et al., 2015). Supporting a proposed disease-related upregulation of PDE5A, inhibitors of the isoform reduce cytotoxicity, Tau hyperphosphorylation, memory deficits, synaptic dysfunction, and/or Aβ burden while also improving LTP deficits in synaptosomes from AD mouse models (Cuadrado-Tejedor et al., 2011, 2017; El-Bakly et al., 2019; Fiorito et al., 2013, 2017; Garcia-Barroso et al., 2013; Huang et al., 2020; Liu et al., 2018; Mao et al., 2018; Ni et al., 2018; Puzzo et al., 2009, 2014; Rabal et al., 2018, 2019a; Yin et al., 2018; Zhang et al., 2013; Zhang, Seo, Li, Nam, & Yang, 2018; Zhu et al., 2019; Zuccarello et al., 2020) and patients (Prieto et al., 2017). At least some protective effects of PDE5A inhibitors appear to be mediated by PKG (Zhang et al., 2013). An increase in PDE5A combined with a decrease in sGC would be anticipated to specifically diminish cGMP localized to the cytosol, with no effect expected on membrane/particulate cGMP levels. Despite this, protective sAPPα actually seems to increase cGMP by stimulating pGC (Barger & Mattson, 1995). This may be attributed to isoform-specific regulation of distinct cGMP pools, with PDE5A regulating cGMP downstream of sGC and PDE9A regulating cGMP downstream of pGCs (Lee et al., 2015). In fact, PDE9A inhibitors have been reported to improve plasticity, cytotoxicity, and/or disrupted memory in murine AD models (Cuadrado-Tejedor et al., 2019; Kroker et al., 2014; Li et al., 2016; Rabal et al., 2019b; Yu et al., 2017), but do not appear to recover cognition or behavior in AD patients (Schwam et al., 2014), likely due to PDE9A inhibitors' inability to target the AD-compromised cytosolic pool of cGMP (Harms, Menniti, & Schmidt, 2019).

In addition to the reduction of cyclic nucleotide signaling measure in AD, CRE-mediated transcription (Arvanitis et al., 2007; Echeverria, Ducatenzeiler, Chen, & Cuello, 2005; Giambarella, Murayama, Ikezu, Fujita, & Nishimoto, 1997; Ikezu et al., 1996; Pugazhenthi et al., 2011;

Satoh, Tabunoki, & Arima, 2009; Saura, 2012) and pCREB levels also appear to be reduced in human patients as well as rodent and in vitro models (Chen et al., 2012; Ettcheto et al., 2018; Sierksma et al., 2014; Tong, Thornton, Balazs, & Cotman, 2001; Wang et al., 2012; Yamamoto-Sasaki et al., 1999; Yiu et al., 2011; Zhang et al., 2014) (but see Muller, Cardenas, Mei, Cheung, & Foskett, 2011). Interestingly, overexpression of CREB appears to rescue AD-related hippocampal deficits (Pugazhenthi et al., 2011; Yiu et al., 2011). In fact, restoration of pCREB to normal levels corresponds with the reversal/prevention of Aβ1–42-induced hippocampal deficits via therapies targeting cAMP elevation (Cheng et al., 2010; Du et al., 2014; Gong et al., 2020; Perez-Gonzalez et al., 2013; Wang, Chen, et al., 2016; Wang et al., 2009) (but see, Sierksma et al., 2014). Notably, normalization of CREB activity via therapies that indirectly target the transcription factor (i.e., not through cAMP/cGMP cascades), also attenuate AD pathologies in rodent models (Al Rihani, Lan, & Kaddoumi, 2019; Du, Zhu, & Si, 2020; Guo et al., 2017; Liu et al., 2017; Singh et al., 2017; Wang et al., 2020). Thus, a loss of function in CREB is implicated in AD-related cognitive decline.

4. Parkinson's disease

Parkinson's disease is triggered by degradation of dopaminergic neurons localized to the substantia nigra pars compacta (SNpc) innervating the striatum. It is widely accepted that PD is defined by loss of motor control (i.e., bradykinesia and resting tremor) as well as cognitive decline in later stages of the disease (Ffytche et al., 2017). Both 6-hydroxy-dopamine (6-OHDA) and 1-methyl-4-phenyl-1,2,3,6-tetrahydropyridine (MPTP) toxins are commonly employed in PD animal models to specifically damage SNpc dopaminergic neurons (Schober, 2004). Precise and localized injection of 6-OHDA into the SNpc allows uptake by neuronal dopamine transporter (DAT) and accumulation of the toxin within mitochondria (Schober, 2004). MPTP, however, is able to cross the blood brain barrier and can therefore be administered systemically. Once it crosses the blood brain barrier, MPTP is engulfed by astrocytes and converted to its active form MPP+. DAT on neurons then allow MPP+ to pass into the cells, where it concentrates in mitochondria and synapses (Schober, 2004). Of the two commonly used toxins, MPTP is regarded as the stronger given evidence that it spontaneously induces Parkinsonism in humans (Schober, 2004).

Reports of cAMP signaling changes in PD are highly variable. With regard to CSF cAMP, one study found cAMP levels were stable across clinical and control groups (Oeckl et al., 2012) while another found a reduction of CSF cAMP in PD patients (Volicer, Beal, Direnfeld, Marquis, & Albert, 1986). In vitro, PD-associated mutations in LRRK2 interfere with PKA signaling while PD-associated mutations in alpha synuclein do the same for CREB signaling (Greggio, Bubacco, & Russo, 2017; Kim, Park, Moon, & Choi, 2014; Muda et al., 2014). Indeed, recent reports suggest LRRK2 may alter PKA activity in a PDE4-dependent manner (Russo, 2019; Russo et al., 2018). That said, this relationship between LRRK2, PKA and PDE4 signaling has yet to be fully understood in the context of patient tissue since PD patients exhibit a marked decrease in PDE4 expression in the caudate, frontal cortices, and thalamic nuclei that is correlated with poorer cognitive performance (Niccolini et al., 2017). PDE10A is also downregulated in the brain of patients with PD (Niccolini et al., 2015). It remains to be determined if the reductions in PDE4 and PDE10, which would be expected to increase cAMP and cGMP, are a driver of PD-related cognitive deficits or an attempt to compensate for lost function, as seen in Huntington's disease (Section 5).

Increases in cyclic nucleotide signaling, specifically that of cGMP, have been associated with therapeutic approaches using electrophysiological and pharmacological treatments in humans. According to a meta-analysis analyzing 13 separate studies, high consumption of caffeine appears to protect against PD risk, possibly due to caffeine's ability to inhibit PDEs and subsequently increase cAMP and cGMP (Qi & Li, 2014). Indeed, a recent in vitro study found a correlation between increased PDE7 expression and progression of PD pathology in a human dopaminergic cell line, an effect that was replicated in rodent models (Morales-Garcia, Alonso-Gil, Santos, & Perez-Castillo, 2020). That said, numerous studies reveal Parkinsonism is characterized by a decrease in PDE4D (Kaut et al., 2017), PDE8B (Azuma et al., 2015), and PDE10A (Niccolini et al., 2015; Pagano et al., 2019) expression, with the magnitude of PDE10A loss in striatum of PD animal models correlating with severity of disease burden (Pagano et al., 2019). An emerging treatment for PD employs deep brain stimulation (DBS) of the subthalamic nucleus (STN) to temporarily upregulate striatal cGMP signaling (Galati et al., 2006; Stefani et al., 2005, 2006). A dopamine restoring drug, L-DOPA, increases serum (Brodacki et al., 2011; Chalimoniuk & Stepien, 2004) (but not plasma Navarro et al., 1998) and CSF (Navarro et al., 1998) cGMP levels. The fact that effective treatments for PD upregulate cGMP implies that disease pathology is marked by a reduction in cGMP signaling. Unfortunately, direct evidence supporting this relationship is weak. One study examining striatum of PD patients found NOS expression decreases, which could lower sGC synthesis of cGMP (Bockelmann, Wolf, Ransmayr, & Riederer, 1994); however, several studies report untreated PD patients show no basal difference in CSF cGMP levels when compared to healthy controls (Ikeda, Sato, Yuasa, Miyatake, & Murota, 1995; Oeckl et al., 2012; Volicer et al., 1986). Clearly further work is needed to examine cGMP signaling is a tissue-specific manner.

In general, rodent models employing 6-OHDA point to a decrease in cGMP and increase in cAMP being related to PD. Administration of 6-OHDA to the substantia nigra appears to reduce cGMP levels in the striatum, globus pallidus (GP), and cortex and increase striatal cAMP (Giorgi et al., 2008; Sancesario et al., 2004, 2014). Further, 6-OHDA lesioning leads to amplified PKA activity downstream of D1 receptors in striatal projection neurons, apparently due to a loss of PDE activity (Mariani, Longueville, Girault, Herve, & Gervasi, 2019). Downregulation of PDE10A in particular (Pagano et al., 2019; Sancesario et al., 2014)—which is mirrored in PD patients (Niccolini et al., 2015)—may account for the increase in cAMP and indirectly contribute to the downregulation of cGMP since PDE10 inhibition both increases cAMP and suppresses NO signaling in PD mouse models (Lee et al., 2019). Indeed, reductions in NO stimulation of sGC have been noted in this model (Barthwal, Srivastava, & Dikshit, 2001; Sancesario et al., 2004). In contrast, PDE1B expression is increased, which would also contribute to reductions in striatal cGMP (Sancesario et al., 2004, 2014). Interestingly, other mouse studies have similarly shown an opposing regulation of PDE1B vs. PDE10A, suggesting these two families are regulated in tandem (Hufgard et al., 2017; Oliva et al., 2012). It is possible that PD-associated reductions in PDE10A signaling reflect an attempt to compensate for lost signaling, because simultaneous treatment with L-DOPA and a PDE10A inhibitor reduced L-DOPA-induced dyskinesias in 6-OHDA lesioned rats (Arakawa, Yuge, & Maehara, 2020). Further, stimulation of effector molecules downstream of cAMP (PKA, CREB, etc.) attenuates PD pathologies (Baluchnejadmojarad et al., 2017; Bergantin, 2020; Ham et al., 2017) (but see Wu et al., 2018). It will be of interest to future studies to specifically localize these decreases in cGMP and increases in cAMP to determine if they are occurring in different pathways (e.g., striatopallidal vs. striatonigral) and/or cell populations (e.g., neurons vs. glia).

Unlike 6-OHDA, MPTP does not appear to produce consistent changes in cyclic nucleotide signaling across species. In vivo administration of MPTP increases GCβ1 expression, GC activity, NOS expression and activity, and cGMP levels in murine striatum and midbrain (Chalimoniuk & Langfort, 2007; Chalimoniuk, Langfort, Lukacova, & Marsala, 2004). An MPTP-induced increase in striatal cGMP activity drastically deviates from results seen in 6-OHDA models, in which the toxin was associated with a decrease in cGMP signaling. In contrast, rats treated with MPTP did exhibit a decrease in striatal cGMP, but also cAMP (Sharma & Deshmukh, 2015), as opposed to the bimodal regulation of the two nucleotides seen in the 6-OHDA model. Even though reports using the MPTP model conflict with regard to how the toxin impacts cyclic nucleotide signaling, they do support the concept that upregulating cGMP and/or cAMP signaling cascades may prove therapeutically beneficial. Administration of the PDE1 inhibitor vinpocetine reversed both the motoric and cAMP/cGMP deficits caused by MPTP treatment in rats (Sharma & Deshmukh, 2015). Vasonatrin, a synthetic natriuretic peptide, reversed cytotoxicity of in vitro mouse striatal neurons by way of increasing cGMP signaling (Zhao et al., 2013). Further, a low dose of rolipram (Yang, Calingasan, Lorenzo, & Beal, 2008) or moderate dose of roflupram (Zhong et al., 2020)—both cAMP-specific PDE4 inhibitors—also attenuate MPTP-induced neurotoxicity. Not only do cAMP/cGMP elevating agents show promise in reversing the primary pathology associated with the lesion, a PDE10 inhibitor was able to reverse L-DOPA-induced dyskinesias in MPTP-treated primates (Beck, Maehara, Chang, & Papa, 2018). This study, coupled with the one above describing similar effects in 6-OHDA lesioned rats (Arakawa et al., 2020), suggest cyclic nucleotide therapeutic targets may hold promise not only as primary treatment options, but also as an adjunctive treatments for PD (Nthenge-Ngumbau & Mohanakumar, 2018).

5. Huntington's disease

Huntington's disease is genetically characterized by an expansion mutation in the *huntingtin* gene (mHTT), resulting in the repetition of a CAG trinucleotide. Uncontrolled choreiform movements are common to HD, and research suggests this symptom may be the result of medium spiny neuron degeneration within the striatum's indirect pathway. It is important to note, however, that many patients experience depression and cognitive decline well before motor dysfunction due to disrupted hippocampal and cortical signaling. Both types of symptoms, along with the neurodegenerative pathology, are replicated in HD mouse models (Giralt, Saavedra, et al., 2012).

Altered cAMP signaling cascades within the hippocampus, striatum, and cortex are implicated in both human and animal models of Huntington's disease (Fig. 2D). Patients diagnosed with HD show a marked decrease in cAMP levels within the parietal cortex and lymphoblastoid cells (Gines et al., 2003). HD mouse models also report a decrease in cAMP localized to the striatum (Gines et al., 2003; Niewiadomska-Cimicka et al., 2017) and cortex (Gines et al., 2003), with changes in the striatal pool attributed to either downregulation of synthesis by AC5 and/or upregu lation of degradation by PDE4 (Tanaka et al., 2017). Indeed, forskolin-induced activation of AC rescued physical and cognitive deficits in HD rat models, while also increasing striatal, cortical, and hippocampal ATP levels (Mehan, Parveen, & Kalra, 2017). Interestingly, although HD mouse models show an increase in PDE4B expression and PDE4 activity in both striatum and cortex (Tanaka et al., 2017), expression of PDE4A mRNA remains stable (Hebb, Robertson, & Denovan-Wright, 2004). This increased PDE4 activity is likely caused by mHTT sequestering of DISC1, thereby disinhibiting PDE4B (Tanaka et al., 2017). Mechanistic studies imply that depression-like symptoms, but not motor deficits, seen in HD mouse models are driven by this increase in striatal PDE4 activity (Tanaka et al., 2017). It follows that an increase in PDE4 function would coincide with a decrease in cAMP levels and a consequent reduction of PKA activity. mHTT appears to aggregate this reduction in PKA activity by impairing proteosomal degradation of PKA regulatory subunits, mHTT in the striatum of an HD mouse model (Lin et al., 2013). In stark contrast, PKA activity in the hippocampus is increases in HD mice, possibly explained by impaired PDE4 signaling (Giralt et al., 2011, 2013). Although levels of PDE4A1 and PDE4A5 remain normal, PDE4AX and both PDE4D1 and PDE4D3 are downregulated in the hippocampus of an HD mouse model, resulting in an increase in PKA activity in the cytosolic fraction only (Giralt et al., 2011, 2013). Furthermore, human embryonic stem cell-induced neurons expressing mHTT show disruption of PKA inhibition, reinforcing the argument that HD is characterized by increased PKA activity in select neuronal tissue (Feyeux et al., 2012).

cGMP levels also appear to be affected by HD pathology. HD mouse models have shown reduced levels of cGMP in the hippocampus, potentially due to lower sGC activity stimulated by NO (Saavedra, Giralt, Arumi, Alberch, & Perez-Navarro, 2013). PDE10A and PDE1B (Ahmad et al., 2014; Beaumont et al., 2016; Haggkvist et al., 2017; Hebb et al., 2004; Hu, McCaw, Hebb, Gomez, & Denovan-Wright, 2004; Leuti et al., 2013; Miller et al., 2014; Niewiadomska-Cimicka et al., 2017; Russell et al., 2014;

Wilson et al., 2016), but not PDE5A or PDE9A (Saavedra et al., 2013), expression appears to be downregulated in HD. In fact, a recent study suggests PDE10A is the earliest imaging biomarker capable of detecting neural changes associated with HD, though these changes may be secondary to cAMP-mediated dopaminergic signaling (Fazio et al., 2020). This reduction in PDE10A expression is thought to be an attempt to compensate for loss elsewhere in the signaling cascade, because PDE10A inhibitors actually rescue cognitive and physiological deficits in HD animal models (Beaumont et al., 2016; Giampa et al., 2009, 2010; Giralt et al., 2013; Harada, Suzuki, & Kimura, 2017). Although PDE9A expression does not appear to change in HD (Saavedra et al., 2013), inhibition of PDE9A also facilitates cGMP signaling in a manner that alleviates striatal pathology and motor/cognitive deficits in HD rat models (Chakroborty et al., 2020). Mechanistically, PDE10A may improve HD pathologies by reversing disease-related reductions in CRE-mediated transcription (Giralt et al., 2012), given the ability of PDE10 inhibitors to increase pCREB in murine cortex, striatum, and hippocampus (Beaumont et al., 2016; Giampa et al., 2009, 2010; Giralt et al., 2013). Indeed, restoration of pCREB levels is a common theme of many experimental therapeutics that ameliorated both the physical and cognitive pathologies characteristic of HD, including compounds targeting neuroinflammatory pathways (Abd-Elrahman & Ferguson, 2019; El-Abhar, Abd El Fattah, Wadie, & El-Tanbouly, 2018; Lee et al., 2019; Paldino, Giampa, Montagna, Angeloni, & Fusco, 2019). The mechanism by which PDE10A inhibitors and other experimental therapeutics might increase hippocampal pCREB has yet to be clearly defined, but it is likely related to restoration of PKG activity, since increased PKA activity is already observed in the diseased hippocampus (Giralt et al., 2011, 2013).

6. Conclusions and future directions

In conclusion, aging-related brain diseases and age-related cognitive decline appear to be associated with changes in cAMP and cGMP signaling cascades in a region and/or subcellular compartment-specific manner. For instance, decreased hippocampal cAMP signaling correlated with both aging and AD despite reports of increased cAMP signaling in prefrontal cortex and cerebral vessels. HD, on the other hand, is characterized by an increase in hippocampal cAMP signaling and a decrease in both striatal and cortical cAMP signaling. Thus, it is clear cyclic nucleotide signaling is regulated in a region-specific manner, and proposed therapies must target localized mechanisms specifically in order to prove effective. In addition to regional specificity, attention must be paid to the fact that many cyclic nucleotide deficits occur in a subcellular compartment-specific manner, as exemplified by reports showing a change in cytosolic but not membrane compartments in AD.

By taking advantage of regional and compartment-specific differences in PDE isoform expression, it may very well be possible to target cyclic nucleotide cascades in a region- and compartment-specific manner. PDE11A is a promising therapeutic target, given it is selectively expressed in the hippocampus (Kelly et al., 2014) and particularly enriched in the cytosolic- rather than membrane or nuclear- compartment (Kelly, 2018). Furthermore, PDE11A's cytosolic enrichment is amplified by a disruption in homodimerization (Pathak et al., 2017), allowing localized cAMP and cGMP modification within the cytosol of the hippocampus, specifically in the CA1 and subiculum sub-regions. In accordance with the role this particular isoform plays in social interaction behaviors, social memory formation, and mood stabilization (Hegde et al., 2016, 2016; Kelly et al., 2010; Pathak et al., 2017), PDE11A may be an appropriate biological target for modulating cyclic nucleotide-mediated diseases such as AD. In contrast to cytosolic hippocampal cAMP/cGMP disturbances in AD, HD patients show striatal deficits in cAMP. As such, specifically targeting PDE10A—which is enriched in the striatum and diminished in the hippocampus—may be an ideal approach for treating HD patients without exacerbating the already heightened hippocampal cAMP signaling. Fortunately, PDE inhibitors and other drugs that upregulate cyclic nucleotide signaling have demonstrated promising results in clinical and animal models, suggesting targeting this particular signaling cascade is a viable treatment for age-related diseases of the brain, particular those for which attenuation of neuroinflammation is a goal (Datta et al., 2020; Wruck & Adjaye, 2020).

Acknowledgments

This work was supported by NIMH 1R01MH101130, NIA 1R01AG054839, NIA 1R01061200, and start-up funds from the University of South Carolina. The authors would like to thank Neema Patel for assistance with Fig. 1.

References

Abd-Elrahman, K. S., & Ferguson, S. S. G. (2019). Modulation of mTOR and CREB pathways following mGluR5 blockade contribute to improved Huntington's pathology in zQ175 mice. *Molecular Brain*, *12*(1), 35. https://doi.org/10.1186/s13041-019-0456-1.

Abe, T., Tohgi, H., Murata, T., Isobe, C., & Sato, C. (2001). Reduction in asymmetrical dimethylarginine, an endogenous nitric oxide synthase inhibitor, in the cerebrospinal fluid during aging and in patients with Alzheimer's disease. *Neuroscience Letters*, *312*(3), 177–179.

Ahmad, R., Bourgeois, S., Postnov, A., Schmidt, M. E., Bormans, G., Van Laere, K., et al. (2014). PET imaging shows loss of striatal PDE10A in patients with Huntington disease. *Neurology*, *82*(3), 279–281. https://doi.org/10.1212/WNL.0000000000000037.

Al Rihani, S. B., Lan, R. S., & Kaddoumi, A. (2019). Granisetron alleviates Alzheimer's disease pathology in TgSwDI mice through calmodulin-dependent protein kinase II/cAMP-response element binding protein pathway. *Journal of Alzheimer's Disease*, *72*(4), 1097–1117. https://doi.org/10.3233/JAD-190849.

Amini, E., Nassireslami, E., Payandemehr, B., Khodagholi, F., Foolad, F., Khalaj, S., et al. (2015). Paradoxical role of PKA inhibitor on amyloid beta-induced memory deficit. *Physiology & Behavior*, *149*, 76–85. https://doi.org/10.1016/j.physbeh.2015.05.029.

Apple, D. M., Solano-Fonseca, R., & Kokovay, E. (2017). Neurogenesis in the aging brain. *Biochemical Pharmacology*, *141*, 77–85. https://doi.org/10.1016/j.bcp.2017.06.116.

Arai, H., & Takahashi, T. (2009). A combination therapy of donepezil and cilostazol for patients with moderate Alzheimer disease: Pilot follow-up study. *The American Journal of Geriatric Psychiatry*, *17*(4), 353–354. https://doi.org/10.1097/JGP.0b013e31819431ea.

Arakawa, K., Yuge, N., & Maehara, S. (2020). Ameliorative effects of a phosphodiesterase 10A inhibitor, MR1916 on L-DOPA-induced dyskinesia in parkinsonian rats. *Pharmacological Reports*, *72*(2), 443–448. https://doi.org/10.1007/s43440-020-00060-y.

Araki, T., Kato, H., Fujiwara, T., & Itoyama, Y. (1995). Age-related changes in bindings of second messengers in the rat brain. *Brain Research*, *704*(2), 227–232.

Arendash, G. W., Mori, T., Cao, C., Mamcarz, M., Runfeldt, M., Dickson, A., et al. (2009). Caffeine reverses cognitive impairment and decreases brain amyloid-beta levels in aged Alzheimer's disease mice. *Journal of Alzheimer's Disease*, *17*(3), 661–680. https://doi.org/10.3233/JAD-2009-1087.

Arendash, G. W., Schleif, W., Rezai-Zadeh, K., Jackson, E. K., Zacharia, L. C., Cracchiolo, J. R., et al. (2006). Caffeine protects Alzheimer's mice against cognitive impairment and reduces brain beta-amyloid production. *Neuroscience*, *142*(4), 941–952. https://doi.org/10.1016/j.neuroscience.2006.07.021.

Arvanitis, D. N., Ducatenzeiler, A., Ou, J. N., Grodstein, E., Andrews, S. D., Tendulkar, S. R., et al. (2007). High intracellular concentrations of amyloid-beta block nuclear translocation of phosphorylated CREB. *Journal of Neurochemistry*, *103*(1), 216–228. https://doi.org/10.1111/j.1471-4159.2007.04704.x.

Asanuma, M., Nishibayashi, S., Iwata, E., Kondo, Y., Nakanishi, T., Vargas, M. G., et al. (1996). Alterations of cAMP response element-binding activity in the aged rat brain in response to administration of rolipram, a cAMP-specific phosphodiesterase inhibitor. *Brain Research. Molecular Brain Research*, *41*(1–2), 210–215.

Austin, J. H., Connole, E., Kett, D., & Collins, J. (1978). Studies in aging of the brain. V. Reduced norepinephrine, dopamine, and cyclic AMP in rat brain with advancing age. *AGE*, *1*(4), 121–124.

Averaimo, S., & Nicol, X. (2014). Intermingled cAMP, cGMP and calcium spatiotemporal dynamics in developing neuronal circuits. *Frontiers in Cellular Neuroscience*, *8*, 376. https://doi.org/10.3389/fncel.2014.00376.

Azuma, R., Ishikawa, K., Hirata, K., Hashimoto, Y., Takahashi, M., Ishii, K., et al. (2015). A novel mutation of PDE8B Gene in a Japanese family with autosomal-dominant striatal degeneration. *Movement Disorders*, *30*(14), 1964–1967. https://doi.org/10.1002/mds.26345.

Bach, M. E., Barad, M., Son, H., Zhuo, M., Lu, Y. F., Shih, R., et al. (1999). Age-related defects in spatial memory are correlated with defects in the late phase of hippocampal long-term potentiation in vitro and are attenuated by drugs that enhance the cAMP signaling pathway. *Proceedings of the National Academy of Sciences of the United States of America*, *96*(9), 5280–5285.

Baillie, G. S., Tejeda, G. S., & Kelly, M. P. (2019). Therapeutic targeting of 3',5'-cyclic nucleotide phosphodiesterases: Inhibition and beyond. *Nature Reviews. Drug Discovery, 18*(10), 770–796. https://doi.org/10.1038/s41573-019-0033-4.

Baltrons, M. A., Pedraza, C. E., Heneka, M. T., & Garcia, A. (2002). Beta-amyloid peptides decrease soluble guanylyl cyclase expression in astroglial cells. *Neurobiology of Disease, 10*(2), 139–149.

Baltrons, M. A., Pifarre, P., Ferrer, I., Carot, J. M., & Garcia, A. (2004). Reduced expression of NO-sensitive guanylyl cyclase in reactive astrocytes of Alzheimer disease, Creutzfeldt-Jakob disease, and multiple sclerosis brains. *Neurobiology of Disease, 17*(3), 462–472. https://doi.org/10.1016/j.nbd.2004.07.022.

Baluchnejadmojarad, T., Eftekhari, S. M., Jamali-Raeufy, N., Haghani, S., Zeinali, H., & Roghani, M. (2017). The anti-aging protein klotho alleviates injury of nigrostriatal dopaminergic pathway in 6-hydroxydopamine rat model of Parkinson's disease: Involvement of PKA/CaMKII/CREB signaling. *Experimental Gerontology, 100*, 70–76. https://doi.org/10.1016/j.exger.2017.10.023.

Barger, S. W., & Mattson, M. P. (1995). The secreted form of the Alzheimer's beta-amyloid precursor protein stimulates a membrane-associated guanylate cyclase. *The Biochemical Journal, 311*(Pt. 1), 45–47.

Barthwal, M. K., Srivastava, N., & Dikshit, M. (2001). Role of nitric oxide in a progressive neurodegeneration model of Parkinson's disease in the rat. *Redox Report, 6*(5), 297–302. https://doi.org/10.1179/135100001101536436.

Bartolome, F., de la Cueva, M., Pascual, C., Antequera, D., Fernandez, T., Gil, C., et al. (2018). Amyloid beta-induced impairments on mitochondrial dynamics, hippocampal neurogenesis, and memory are restored by phosphodiesterase 7 inhibition. *Alzheimer's Research & Therapy, 10*(1), 24. https://doi.org/10.1186/s13195-018-0352-4.

Bate, C., & Williams, A. (2015). cAMP-inhibits cytoplasmic phospholipase A(2) and protects neurons against amyloid-beta-induced synapse damage. *Biology (Basel), 4*(3), 591–606. https://doi.org/10.3390/biology4030591.

Beaumont, V., Zhong, S., Lin, H., Xu, W., Bradaia, A., Steidl, E., et al. (2016). Phosphodiesterase 10A inhibition improves cortico-basal ganglia function in Huntington's disease models. *Neuron, 92*(6), 1220–1237. https://doi.org/10.1016/j.neuron.2016.10.064.

Beck, G., Maehara, S., Chang, P. L., & Papa, S. M. (2018). A selective phosphodiesterase 10A inhibitor reduces L-Dopa-induced dyskinesias in parkinsonian monkeys. *Movement Disorders, 33*(5), 805–814. https://doi.org/10.1002/mds.27341.

Berg, A., & Zimmerman, I. D. (1975). Effects of electrical stimulation and norepinephrine on cyclic-AMP levels in the cerebral cortex of the aging rat. *Mechanisms of Ageing and Development, 4*(5–6), 377–383.

Bergantin, L. B. (2020). The interplay between depression and Parkinson's disease: Learning the link through Ca^{2+}/cAMP signalling. *Current Protein & Peptide Science, 21*(12), 1223–1228. https://doi.org/10.2174/1389203721666200206161912.

Birkenfeld, A., & Ben-Zvi, A. (1984). Age associated changes in intracellular cyclic adenosine monophosphate. *Clinical and Experimental Immunology, 55*(3), 651–654.

Blokland, A., Van Duinen, M. A., Sambeth, A., Heckman, P. R. A., Tsai, M., Lahu, G., et al. (2019). Acute treatment with the PDE4 inhibitor roflumilast improves verbal word memory in healthy old individuals: A double-blind placebo-controlled study. *Neurobiology of Aging, 77*, 37–43. https://doi.org/10.1016/j.neurobiolaging.2019.01.014.

Blum-Degen, D., Heinemann, T., Lan, J., Pedersen, V., Leblhuber, F., Paulus, W., et al. (1999). Characterization and regional distribution of nitric oxide synthase in the human brain during normal ageing. *Brain Research, 834*(1–2), 128–135.

Bockelmann, R., Wolf, G., Ransmayr, G., & Riederer, P. (1994). NADPH-diaphorase/nitric oxide synthase containing neurons in normal and Parkinson's disease putamen. *Journal of Neural Transmission. Parkinson's Disease and Dementia Section, 7*(2), 115–121.

Bonkale, W. L., Cowburn, R. F., Ohm, T. G., Bogdanovic, N., & Fastbom, J. (1999). A quantitative autoradiographic study of [3H]cAMP binding to cytosolic and particulate protein kinase A in post-mortem brain staged for Alzheimer's disease neurofibrillary changes and amyloid deposits. *Brain Research*, *818*(2), 383–396.

Bonkale, W. L., Fastbom, J., Wiehager, B., Ravid, R., Winblad, B., & Cowburn, R. F. (1996). Impaired G-protein-stimulated adenylyl cyclase activity in Alzheimer's disease brain is not accompanied by reduced cyclic-AMP-dependent protein kinase A activity. *Brain Research*, *737*(1–2), 155–161.

Bonkale, W. L., Winblad, B., Ravid, R., & Cowburn, R. F. (1995). Reduced nitric oxide responsive soluble guanylyl cyclase activity in the superior temporal cortex of patients with Alzheimer's disease. *Neuroscience Letters*, *187*(1), 5–8.

Boswell-Smith, V., Spina, D., & Page, C. P. (2006). Phosphodiesterase inhibitors. *British Journal of Pharmacology*, *147*(Suppl. 1), S252–S257. https://doi.org/10.1038/sj.bjp.0706495.

Brinton, R. D., Yamazaki, R., Gonzalez, C. M., O'Neill, K., & Schreiber, S. S. (1998). Vasopressin-induction of the immediate early gene, NGFI-A, in cultured hippocampal glial cells. *Brain Research. Molecular Brain Research*, *57*(1), 73–85.

Brodacki, B., Chalimoniuk, M., Wesolowska, J., Staszewski, J., Chrapusta, S. J., Stepien, A., et al. (2011). cGMP level in idiopathic Parkinson's disease patients with and without cardiovascular disease—A pilot study. *Parkinsonism & Related Disorders*, *17*(9), 689–692. https://doi.org/10.1016/j.parkreldis.2011.07.003.

Buerger, K., Ernst, A., Ewers, M., Uspenskaya, O., Omerovic, M., Morgenthaler, N. G., et al. (2009). Blood-based microcirculation markers in Alzheimer's disease-diagnostic value of midregional pro-atrial natriuretic peptide/C-terminal endothelin-1 precursor fragment ratio. *Biological Psychiatry*, *65*(11), 979–984. https://doi.org/10.1016/j.biopsych.2009.01.032.

Calcagno, E., Caudano, F., Passalacqua, M., Pronzato, M. A., Fedele, E., & Ricciarelli, R. (2017). Investigating the amyloid-beta enhancing effect of cGMP in neuro2a cells. *Mechanisms of Ageing and Development*, *166*, 1–5. https://doi.org/10.1016/j.mad.2017.08.001.

Cameron, R. T., Whiteley, E., Day, J. P., Parachikova, A. I., & Baillie, G. S. (2017). Selective inhibition of phosphodiesterases 4, 5 and 9 induces HSP20 phosphorylation and attenuates amyloid beta 1-42-mediated cytotoxicity. *FEBS Open Bio*, 7(1), 64–73. https://doi.org/10.1002/2211-5463.12156.

Carlyle, B. C., Nairn, A. C., Wang, M., Yang, Y., Jin, L. E., Simen, A. A., et al. (2014). cAMP-PKA phosphorylation of tau confers risk for degeneration in aging association cortex. *Proceedings of the National Academy of Sciences of the United States of America*, *111*(13), 5036–5041. https://doi.org/10.1073/pnas.1322360111.

Cashman, R. E., & Grammas, P. (1995). cAMP-dependent protein kinase in cerebral microvessels in aging and Alzheimer disease. *Molecular and Chemical Neuropathology*, *26*(3), 247–258.

Chakroborty, S., Manfredsson, F. P., Dec, A. M., Campbell, P. W., Stutzmann, G. E., Beaumont, V., et al. (2020). Phosphodiesterase 9A inhibition facilitates corticostriatal transmission in wild-type and transgenic rats that model Huntington's disease. *Frontiers in Neuroscience*, *14*, 466. https://doi.org/10.3389/fnins.2020.00466.

Chalimoniuk, M., & Langfort, J. (2007). The effect of subchronic, intermittent L-DOPA treatment on neuronal nitric oxide synthase and soluble guanylyl cyclase expression and activity in the striatum and midbrain of normal and MPTP-treated mice. *Neurochemistry International*, *50*(6), 821–833. https://doi.org/10.1016/j.neuint.2007.02.002.

Chalimoniuk, M., Langfort, J., Lukacova, N., & Marsala, J. (2004). Upregulation of guanylyl cyclase expression and activity in striatum of MPTP-induced Parkinsonism in mice. *Biochemical and Biophysical Research Communications*, *324*(1), 118–126. https://doi.org/10.1016/j.bbrc.2004.09.028.

Chalimoniuk, M., & Stepien, A. (2004). Influence of the therapy with pergolide mesylate plus L-DOPA and with L-DOPA alone on serum cGMP level in PD patients. *Polish Journal of Pharmacology*, *56*(5), 647–650.

Chalimoniuk, M., & Strosznajder, J. B. (1998). Aging modulates nitric oxide synthesis and cGMP levels in hippocampus and cerebellum. Effects of amyloid beta peptide. *Molecular and Chemical Neuropathology*, *35*(1–3), 77–95.

Chen, Y., Huang, X., Zhang, Y. W., Rockenstein, E., Bu, G., Golde, T. E., et al. (2012). Alzheimer's beta-secretase (BACE1) regulates the cAMP/PKA/CREB pathway independently of beta-amyloid. *The Journal of Neuroscience*, *32*(33), 11390–11395. https://doi.org/10.1523/JNEUROSCI.0757-12.2012.

Chen, Y. R., Li, Y. H., Hsieh, T. C., Wang, C. M., Cheng, K. C., Wang, L., et al. (2019). Aging-induced Akt activation involves in aging-related pathologies and Abeta-induced toxicity. *Aging Cell*, *18*(4), e12989. https://doi.org/10.1111/acel.12989.

Chen, J., Martinez, J., Milner, T. A., Buck, J., & Levin, L. R. (2013). Neuronal expression of soluble adenylyl cyclase in the mammalian brain. *Brain Research*, *1518*, 1–8. https://doi.org/10.1016/j.brainres.2013.04.027.

Cheng, Y. F., Wang, C., Lin, H. B., Li, Y. F., Huang, Y., Xu, J. P., et al. (2010). Inhibition of phosphodiesterase-4 reverses memory deficits produced by Abeta25-35 or Abeta1-40 peptide in rats. *Psychopharmacology*, *212*(2), 181–191. https://doi.org/10.1007/s00213-010-1943-3.

Combs, C. K., Karlo, J. C., Kao, S. C., & Landreth, G. E. (2001). Beta-amyloid stimulation of microglia and monocytes results in TNFalpha-dependent expression of inducible nitric oxide synthase and neuronal apoptosis. *The Journal of Neuroscience*, *21*(4), 1179–1188.

Costa, M. S., Botton, P. H., Mioranzza, S., Souza, D. O., & Porciuncula, L. O. (2008). Caffeine prevents age-associated recognition memory decline and changes brain-derived neurotrophic factor and tirosine kinase receptor (TrkB) content in mice. *Neuroscience*, *153*(4), 1071–1078. https://doi.org/10.1016/j.neuroscience.2008.03.038.

Costa, D. A., Cracchiolo, J. R., Bachstetter, A. D., Hughes, T. F., Bales, K. R., Paul, S. M., et al. (2007). Enrichment improves cognition in AD mice by amyloid-related and unrelated mechanisms. *Neurobiology of Aging*, *28*(6), 831–844. https://doi.org/10.1016/j.neurobiolaging.2006.04.009.

Cowburn, R. F., O'Neill, C., Ravid, R., Alafuzoff, I., Winblad, B., & Fowler, C. J. (1992). Adenylyl cyclase activity in postmortem human brain: Evidence of altered G protein mediation in Alzheimer's disease. *Journal of Neurochemistry*, *58*(4), 1409–1419.

Cowburn, R. F., O'Neill, C., Ravid, R., Winblad, B., & Fowler, C. J. (1992). Preservation of Gi-protein inhibited adenylyl cyclase activity in the brains of patients with Alzheimer's disease. *Neuroscience Letters*, *141*(1), 16–20.

Cuadrado-Tejedor, M., Garcia-Barroso, C., Sanchez-Arias, J. A., Rabal, O., Perez-Gonzalez, M., Mederos, S., et al. (2017). A First-In-Class small-molecule that acts as a dual inhibitor of HDAC and PDE5 and that rescues hippocampal synaptic impairment in Alzheimer's disease mice. *Neuropsychopharmacology*, *42*(2), 524–539. https://doi.org/10.1038/npp.2016.163.

Cuadrado-Tejedor, M., Hervias, I., Ricobaraza, A., Puerta, E., Perez-Roldan, J. M., Garcia-Barroso, C., et al. (2011). Sildenafil restores cognitive function without affecting beta-amyloid burden in a mouse model of Alzheimer's disease. *British Journal of Pharmacology*, *164*(8), 2029–2041. https://doi.org/10.1111/j.1476-5381.2011.01517.x.

Cuadrado-Tejedor, M., Perez-Gonzalez, M., Garcia-Munoz, C., Muruzabal, D., Garcia-Barroso, C., Rabal, O., et al. (2019). Taking advantage of the selectivity of histone deacetylases and phosphodiesterase inhibitors to design better therapeutic strategies to treat Alzheimer's disease. *Frontiers in Aging Neuroscience*, *11*, 149. https://doi.org/10.3389/fnagi.2019.00149.

Dagnas, M., Guillou, J. L., Prevot, T., & Mons, N. (2013). HDAC inhibition facilitates the switch between memory systems in young but not aged mice. *The Journal of Neuroscience, 33*(5), 1954–1963. https://doi.org/10.1523/JNEUROSCI.3453-12.2013.

Daniels, L. B., Laughlin, G. A., Kritz-Silverstein, D., Clopton, P., Chen, W. C., Maisel, A. S., et al. (2011). Elevated natriuretic peptide levels and cognitive function in community-dwelling older adults. *The American Journal of Medicine, 124*(7), 670 e671–678. https://doi.org/10.1016/j.amjmed.2011.02.027.

Datta, D., Leslie, S. N., Morozov, Y. M., Duque, A., Rakic, P., van Dyck, C. H., et al. (2020). Classical complement cascade initiating C1q protein within neurons in the aged rhesus macaque dorsolateral prefrontal cortex. *Journal of Neuroinflammation, 17*(1), 8. https://doi.org/10.1186/s12974-019-1683-1.

Davis, K. M., Fish, L. C., Minaker, K. L., & Elahi, D. (1996). Atrial natriuretic peptide levels in the elderly: Differentiating normal aging changes from disease. *The Journals of Gerontology. Series A, Biological Sciences and Medical Sciences, 51*(3), M95–101.

De Jager, P. L., Shulman, J. M., Chibnik, L. B., Keenan, B. T., Raj, T., Wilson, R. S., et al. (2012). A genome-wide scan for common variants affecting the rate of age-related cognitive decline. *Neurobiology of Aging, 33*(5), 1017 e1011–1015. https://doi.org/10.1016/j.neurobiolaging.2011.09.033.

de Lima, M. N., Presti-Torres, J., Garcia, V. A., Guimaraes, M. R., Scalco, F. S., Roesler, R., et al. (2008). Amelioration of recognition memory impairment associated with iron loading or aging by the type 4-specific phosphodiesterase inhibitor rolipram in rats. *Neuropharmacology, 55*(5), 788–792. https://doi.org/10.1016/j.neuropharm.2008.06.025.

Dean, G. E., Weiss, C., Morris, J. L., & Chasens, E. R. (2017). Impaired sleep: A multifaceted geriatric syndrome. *The Nursing Clinics of North America, 52*(3), 387–404. https://doi.org/10.1016/j.cnur.2017.04.009.

Devan, B. D., Pistell, P. J., Duffy, K. B., Kelley-Bell, B., Spangler, E. L., & Ingram, D. K. (2014). Phosphodiesterase inhibition facilitates cognitive restoration in rodent models of age-related memory decline. *NeuroRehabilitation, 34*(1), 101–111. https://doi.org/10.3233/NRE-131022.

Dewar, D., Horsburgh, K., Graham, D. I., Brooks, D. N., & McCulloch, J. (1990). Selective alterations of high affinity [3H]forskolin binding sites in Alzheimer's disease: A quantitative autoradiographic study. *Brain Research, 511*(2), 241–248.

Domek-Lopacinska, K., & Strosznajder, J. B. (2008). The effect of selective inhibition of cyclic GMP hydrolyzing phosphodiesterases 2 and 5 on learning and memory processes and nitric oxide synthase activity in brain during aging. *Brain Research, 1216*, 68–77. https://doi.org/10.1016/j.brainres.2008.02.108.

Driscoll, I., Shumaker, S. A., Snively, B. M., Margolis, K. L., Manson, J. E., Vitolins, M. Z., et al. (2016). Relationships between caffeine intake and risk for probable dementia or global cognitive impairment: The women's health initiative memory study. *The Journals of Gerontology. Series A, Biological Sciences and Medical Sciences, 71*(12), 1596–1602. https://doi.org/10.1093/gerona/glw078.

Drott, J., Desire, L., Drouin, D., Pando, M., & Haun, F. (2010). Etazolate improves performance in a foraging and homing task in aged rats. *European Journal of Pharmacology, 634*(1–3), 95–100. https://doi.org/10.1016/j.ejphar.2010.02.036.

Du, H., Guo, L., Wu, X., Sosunov, A. A., McKhann, G. M., Chen, J. X., et al. (2014). Cyclophilin D deficiency rescues Abeta-impaired PKA/CREB signaling and alleviates synaptic degeneration. *Biochimica et Biophysica Acta, 1842*(12 Pt. A), 2517–2527. https://doi.org/10.1016/j.bbadis.2013.03.004.

Du, Q., Zhu, X., & Si, J. (2020). Angelica polysaccharide ameliorates memory impairment in Alzheimer's disease rat through activating BDNF/TrkB/CREB pathway. *Experimental Biology and Medicine (Maywood, N.J.), 245*(1), 1–10. https://doi.org/10.1177/1535370219894558.

Duszczyk, M., Kuszczyk, M., Guridi, M., Lazarewicz, J. W., & Sadowski, M. J. (2012). In vivo hippocampal microdialysis reveals impairment of NMDA receptor-cGMP signaling in APP(SW) and APP(SW)/PS1(L166P) Alzheimer's transgenic mice. *Neurochemistry International, 61*(7), 976–980. https://doi.org/10.1016/j.neuint.2012.07.017.

Echeverria, V., Ducatenzeiler, A., Chen, C. H., & Cuello, A. C. (2005). Endogenous beta-amyloid peptide synthesis modulates cAMP response element-regulated gene expression in PC12 cells. *Neuroscience, 135*(4), 1193–1202. https://doi.org/10.1016/j.neuroscience.2005.06.057.

Eftekharzadeh, B., Ramin, M., Khodagholi, F., Moradi, S., Tabrizian, K., Sharif, R., et al. (2012). Inhibition of PKA attenuates memory deficits induced by beta-amyloid (1–42), and decreases oxidative stress and NF-kappaB transcription factors. *Behavioural Brain Research, 226*(1), 301–308. https://doi.org/10.1016/j.bbr.2011.08.015.

El-Abhar, H., Abd El Fattah, M. A., Wadie, W., & El-Tanbouly, D. M. (2018). Cilostazol disrupts TLR-4, Akt/GSK-3beta/CREB, and IL-6/JAK-2/STAT-3/SOCS-3 crosstalk in a rat model of Huntington's disease. *PLoS One, 13*(9), e0203837. https://doi.org/10.1371/journal.pone.0203837.

El-Bakly, W., Wagdy, O., Sobhy, A., Abo Elenain, O., Riad, M. S., El Sayed, M., et al. (2019). The efficacy and underlying mechanism of phosphodiesterase- 5 inhibitors in preventing cognitive impairment and Alzheimer pathology: A systematic review of animal studies. *Behavioural Brain Research, 372*, 112004. https://doi.org/10.1016/j.bbr.2019.112004.

Eslamizade, M. J., Saffarzadeh, F., Mousavi, S. M., Meftahi, G. H., Hosseinmardi, N., Mehdizadeh, M., et al. (2015). Alterations in CA1 pyramidal neuronal intrinsic excitability mediated by Ih channel currents in a rat model of amyloid beta pathology. *Neuroscience, 305*, 279–292. https://doi.org/10.1016/j.neuroscience.2015.07.087.

Ettcheto, M., Abad, S., Petrov, D., Pedros, I., Busquets, O., Sanchez-Lopez, E., et al. (2018). Early preclinical changes in hippocampal CREB-binding protein expression in a mouse model of familial Alzheimer's disease. *Molecular Neurobiology, 55*(6), 4885–4895. https://doi.org/10.1007/s12035-017-0690-4.

Favilla, C., Abel, T., & Kelly, M. P. (2008). Chronic Galphas signaling in the striatum increases anxiety-related behaviors independent of developmental effects. *The Journal of Neuroscience, 28*(51), 13952–13956. https://doi.org/10.1523/JNEUROSCI.4986-08.2008.

Fazio, P., Fitzer-Attas, C. J., Mrzljak, L., Bronzova, J., Nag, S., Warner, J. H., et al. (2020). PET molecular imaging of phosphodiesterase 10A: An early biomarker of Huntington's disease progression. *Movement Disorders, 35*(4), 606–615. https://doi.org/10.1002/mds.27963.

Feinkohl, I., Sattar, N., Welsh, P., Reynolds, R. M., Deary, I. J., Strachan, M. W., et al. (2012). Association of N-terminal pro-brain natriuretic peptide with cognitive function and depression in elderly people with type 2 diabetes. *PLoS One*, 7(9), e44569. https://doi.org/10.1371/journal.pone.0044569.

Feng, H., Wang, C., He, W., Wu, X., Li, S., Zeng, Z., et al. (2019). Roflumilast ameliorates cognitive impairment in APP/PS1 mice via cAMP/CREB/BDNF signaling and anti-neuroinflammatory effects. *Metabolic Brain Disease, 34*(2), 583–591. https://doi.org/10.1007/s11011-018-0374-4.

Feyeux, M., Bourgois-Rocha, F., Redfern, A., Giles, P., Lefort, N., Aubert, S., et al. (2012). Early transcriptional changes linked to naturally occurring Huntington's disease mutations in neural derivatives of human embryonic stem cells. *Human Molecular Genetics, 21*(17), 3883–3895. https://doi.org/10.1093/hmg/dds216.

Ffytche, D. H., Creese, B., Politis, M., Chaudhuri, K. R., Weintraub, D., Ballard, C., et al. (2017). The psychosis spectrum in Parkinson disease. *Nature Reviews. Neurology, 13*(2), 81–95. https://doi.org/10.1038/nrneurol.2016.200.

Fiorito, J., Saeed, F., Zhang, H., Staniszewski, A., Feng, Y., Francis, Y. I., et al. (2013). Synthesis of quinoline derivatives: Discovery of a potent and selective phosphodiesterase 5 inhibitor for the treatment of Alzheimer's disease. *European Journal of Medicinal Chemistry*, *60*, 285–294. https://doi.org/10.1016/j.ejmech.2012.12.009.

Fiorito, J., Vendome, J., Saeed, F., Staniszewski, A., Zhang, H., Yan, S., et al. (2017). Identification of a novel 1,2,3,4-tetrahydrobenzo[b][1,6]naphthyridine analogue as a potent phosphodiesterase 5 inhibitor with improved aqueous solubility for the treatment of Alzheimer's disease. *Journal of Medicinal Chemistry*, *60*(21), 8858–8875. https://doi.org/10.1021/acs.jmedchem.7b00979.

Fiscus, R. R. (2002). Involvement of cyclic GMP and protein kinase G in the regulation of apoptosis and survival in neural cells. *Neurosignals*, *11*(4), 175–190.

Francis, S. H., Blount, M. A., & Corbin, J. D. (2011). Mammalian cyclic nucleotide phosphodiesterases: Molecular mechanisms and physiological functions. *Physiological Reviews*, *91*(2), 651–690. https://doi.org/10.1152/physrev.00030.2010.

Frykman, S., Inoue, M., Ikeda, A., Teranishi, Y., Kihara, T., Lundgren, J. L., et al. (2017). Maturation and processing of the amyloid precursor protein is regulated by the potassium/sodium hyperpolarization-activated cyclic nucleotide-gated ion channel 2 (HCN2). *Biochemical and Biophysical Research Communications*, *483*(1), 352–358. https://doi.org/10.1016/j.bbrc.2016.12.140.

Fulop, T., Jr., Kekessy, D., & Foris, G. (1990). Altered post-receptorial signal transduction mechanism under various stimulation in polymorphonuclear granulocytes of Alzheimer's disease. *Mechanisms of Ageing and Development*, *52*(2–3), 277–285.

Galati, S., Mazzone, P., Fedele, E., Pisani, A., Peppe, A., Pierantozzi, M., et al. (2006). Biochemical and electrophysiological changes of substantia nigra pars reticulata driven by subthalamic stimulation in patients with Parkinson's disease. *The European Journal of Neuroscience*, *23*(11), 2923–2928. https://doi.org/10.1111/j.1460-9568.2006.04816.x.

Garcia-Barroso, C., Ricobaraza, A., Pascual-Lucas, M., Unceta, N., Rico, A. J., Goicolea, M. A., et al. (2013). Tadalafil crosses the blood-brain barrier and reverses cognitive dysfunction in a mouse model of AD. *Neuropharmacology*, *64*, 114–123. https://doi.org/10.1016/j.neuropharm.2012.06.052.

Giambarella, U., Murayama, Y., Ikezu, T., Fujita, T., & Nishimoto, I. (1997). Potential CRE suppression by familial Alzheimer's mutants of APP independent of adenylyl cyclase regulation. *FEBS Letters*, *412*(1), 97–101.

Giampa, C., Laurenti, D., Anzilotti, S., Bernardi, G., Menniti, F. S., & Fusco, F. R. (2010). Inhibition of the striatal specific phosphodiesterase PDE10A ameliorates striatal and cortical pathology in R6/2 mouse model of Huntington's disease. *PLoS One*, *5*(10), e13417. https://doi.org/10.1371/journal.pone.0013417.

Giampa, C., Patassini, S., Borreca, A., Laurenti, D., Marullo, F., Bernardi, G., et al. (2009). Phosphodiesterase 10 inhibition reduces striatal excitotoxicity in the quinolinic acid model of Huntington's disease. *Neurobiology of Disease*, *34*(3), 450–456. https://doi.org/10.1016/j.nbd.2009.02.014.

Gines, S., Seong, I. S., Fossale, E., Ivanova, E., Trettel, F., Gusella, J. F., et al. (2003). Specific progressive cAMP reduction implicates energy deficit in presymptomatic Huntington's disease knock-in mice. *Human Molecular Genetics*, *12*(5), 497–508.

Giorgi, M., D'Angelo, V., Esposito, Z., Nuccetelli, V., Sorge, R., Martorana, A., et al. (2008). Lowered cAMP and cGMP signalling in the brain during levodopa-induced dyskinesias in hemiparkinsonian rats: New aspects in the pathogenetic mechanisms. *The European Journal of Neuroscience*, *28*(5), 941–950. https://doi.org/10.1111/j.1460-9568.2008.06387.x.

Giorgi, M., Squitti, R., Bonsi, P., Paggi, P., & Toschi, G. (1994). Activities of 3':5' cyclic nucleotide phosphodiesterases in the superior cervical ganglion of rat: Characterization, compartmentalization and observations in young and old animals. *Neurochemistry International*, *25*(5), 493–500.

Giralt, A., Puigdellivol, M., Carreton, O., Paoletti, P., Valero, J., Parra-Damas, A., et al. (2012). Long-term memory deficits in Huntington's disease are associated with reduced CBP histone acetylase activity. *Human Molecular Genetics*, *21*(6), 1203–1216. https://doi.org/10.1093/hmg/ddr552.

Giralt, A., Saavedra, A., Alberch, J., & Perez-Navarro, E. (2012). Cognitive dysfunction in Huntington's disease: Humans, mouse models and molecular mechanisms. *Journal of Huntington's Disease*, *1*(2), 155–173. https://doi.org/10.3233/JHD-120023.

Giralt, A., Saavedra, A., Carreton, O., Arumi, H., Tyebji, S., Alberch, J., et al. (2013). PDE10 inhibition increases GluA1 and CREB phosphorylation and improves spatial and recognition memories in a Huntington's disease mouse model. *Hippocampus*, *23*(8), 684–695. https://doi.org/10.1002/hipo.22128.

Giralt, A., Saavedra, A., Carreton, O., Xifro, X., Alberch, J., & Perez-Navarro, E. (2011). Increased PKA signaling disrupts recognition memory and spatial memory: Role in Huntington's disease. *Human Molecular Genetics*, *20*(21), 4232–4247. https://doi.org/10.1093/hmg/ddr351.

Gong, Y., Chen, J., Jin, Y., Wang, C., Zheng, M., & He, L. (2020). GW9508 ameliorates cognitive impairment via the cAMP-CREB and JNK pathways in APPswe/PS1dE9 mouse model of Alzheimer's disease. *Neuropharmacology*, *164*, 107899. https://doi.org/10.1016/j.neuropharm.2019.107899.

Gong, B., Vitolo, O. V., Trinchese, F., Liu, S., Shelanski, M., & Arancio, O. (2004). Persistent improvement in synaptic and cognitive functions in an Alzheimer mouse model after rolipram treatment. *The Journal of Clinical Investigation*, *114*(11), 1624–1634. https://doi.org/10.1172/JCI22831.

Grammas, P., Roher, A. E., & Ball, M. J. (1994). Increased accumulation of cAMP in cerebral microvessels in Alzheimer's disease. *Neurobiology of Aging*, *15*(1), 113–116.

Greggio, E., Bubacco, L., & Russo, I. (2017). Cross-talk between LRRK2 and PKA: Implication for Parkinson's disease? *Biochemical Society Transactions*, *45*(1), 261–267. https://doi.org/10.1042/BST20160396.

Gulisano, W., Tropea, M. R., Arancio, O., Palmeri, A., & Puzzo, D. (2018). Sub-efficacious doses of phosphodiesterase 4 and 5 inhibitors improve memory in a mouse model of Alzheimer's disease. *Neuropharmacology*, *138*, 151–159. https://doi.org/10.1016/j.neuropharm.2018.06.002.

Gunstad, J., Poppas, A., Smeal, S., Paul, R. H., Tate, D. F., Jefferson, A. L., et al. (2006). Relation of brain natriuretic peptide levels to cognitive dysfunction in adults $>$55 years of age with cardiovascular disease. *The American Journal of Cardiology*, *98*(4), 538–540. https://doi.org/10.1016/j.amjcard.2006.02.062.

Guo, H., Cheng, Y., Wang, C., Wu, J., Zou, Z., Niu, B., et al. (2017). FFPM, a PDE4 inhibitor, reverses learning and memory deficits in APP/PS1 transgenic mice via cAMP/PKA/CREB signaling and anti-inflammatory effects. *Neuropharmacology*, *116*, 260–269. https://doi.org/10.1016/j.neuropharm.2017.01.004.

Haggkvist, J., Toth, M., Tari, L., Varnas, K., Svedberg, M., Forsberg, A., et al. (2017). Longitudinal small-animal PET imaging of the zQ175 mouse model of Huntington Disease shows in vivo changes of molecular targets in the striatum and cerebral cortex. *Journal of Nuclear Medicine*, *58*(4), 617–622. https://doi.org/10.2967/jnumed.116.180497.

Ham, S., Lee, Y. I., Jo, M., Kim, H., Kang, H., Jo, A., et al. (2017). Hydrocortisone-induced parkin prevents dopaminergic cell death via CREB pathway in Parkinson's disease model. *Scientific Reports*, 7(1), 525. https://doi.org/10.1038/s41598-017-00614-w.

Hara, H., Onodera, H., Kato, H., & Kogure, K. (1992). Effects of aging on signal transmission and transduction systems in the gerbil brain: Morphological and autoradiographic study. *Neuroscience*, *46*(2), 475–488.

Harada, N., Nishiyama, S., Ohba, H., Sato, K., Kakiuchi, T., & Tsukada, H. (2002). Age differences in phosphodiesterase type-IV and its functional response to dopamine D1 receptor modulation in the living brain: A PET study in conscious monkeys. *Synapse, 44*(3), 139–145. https://doi.org/10.1002/syn.10067.

Harada, A., Suzuki, K., & Kimura, H. (2017). TAK-063, a novel phosphodiesterase 10A inhibitor, protects from striatal neurodegeneration and ameliorates behavioral deficits in the R6/2 mouse model of Huntington's disease. *The Journal of Pharmacology and Experimental Therapeutics, 360*(1), 75–83. https://doi.org/10.1124/jpet.116.237388.

Harms, J. F., Menniti, F. S., & Schmidt, C. J. (2019). Phosphodiesterase 9A in brain regulates cGMP signaling independent of nitric-oxide. *Frontiers in Neuroscience, 13*, 837. https://doi.org/10.3389/fnins.2019.00837.

Hattiangady, B., Rao, M. S., Shetty, G. A., & Shetty, A. K. (2005). Brain-derived neurotrophic factor, phosphorylated cyclic AMP response element binding protein and neuropeptide Y decline as early as middle age in the dentate gyrus and CA1 and CA3 subfields of the hippocampus. *Experimental Neurology, 195*(2), 353–371. https://doi.org/10.1016/j.expneurol.2005.05.014.

Hebb, A. L., Robertson, H. A., & Denovan-Wright, E. M. (2004). Striatal phosphodiesterase mRNA and protein levels are reduced in Huntington's disease transgenic mice prior to the onset of motor symptoms. *Neuroscience, 123*(4), 967–981.

Heckman, P. R., Blokland, A., Ramaekers, J., & Prickaerts, J. (2015). PDE and cognitive processing: Beyond the memory domain. *Neurobiology of Learning and Memory, 119*, 108–122. https://doi.org/10.1016/j.nlm.2014.10.011.

Hegde, S., Capell, W. R., Ibrahim, B. A., Klett, J., Patel, N. S., Sougiannis, A. T., et al. (2016). Phosphodiesterase 11A (PDE11A), enriched in ventral hippocampus neurons, is required for consolidation of social but not nonsocial memories in mice. *Neuropsychopharmacology, 41*(12), 2920–2931. https://doi.org/10.1038/npp.2016.106.

Hegde, S., Ji, H., Oliver, D., Patel, N. S., Poupore, N., Shtutman, M., et al. (2016). PDE11A regulates social behaviors and is a key mechanism by which social experience sculpts the brain. *Neuroscience, 335*, 151–169. https://doi.org/10.1016/j.neuroscience.2016.08.019.

Hesse, R., Lausser, L., Gummert, P., Schmid, F., Wahler, A., Schnack, C., et al. (2017). Reduced cGMP levels in CSF of AD patients correlate with severity of dementia and current depression. *Alzheimer's Research & Therapy, 9*(1), 17. https://doi.org/10.1186/s13195-017-0245-y.

Hilal, S., Chai, Y. L., Ikram, M. K., Elangovan, S., Yeow, T. B., Xin, X., et al. (2015). Markers of cardiac dysfunction in cognitive impairment and dementia. *Medicine (Baltimore), 94*(1), e297. https://doi.org/10.1097/MD.0000000000000297.

Hiltunen, M., Kerola, T., Kettunen, R., Hartikainen, S., Sulkava, R., Vuolteenaho, O., et al. (2013). The prognostic capacity of B-type natriuretic peptide on cognitive disorder varies by age. *Annals of Medicine, 45*(1), 74–78. https://doi.org/10.3109/07853890.2012.663929.

Holm, H., Nagga, K., Nilsson, E. D., Ricci, F., Melander, O., Hansson, O., et al. (2017). Biomarkers of microvascular endothelial dysfunction predict incident dementia: A population-based prospective study. *Journal of Internal Medicine, 282*(1), 94–101. https://doi.org/10.1111/joim.12621.

Hu, W. T., Holtzman, D. M., Fagan, A. M., Shaw, L. M., Perrin, R., Arnold, S. E., et al. (2012). Plasma multianalyte profiling in mild cognitive impairment and Alzheimer disease. *Neurology, 79*(9), 897–905. https://doi.org/10.1212/WNL.0b013e318266fa70.

Hu, H., McCaw, E. A., Hebb, A. L., Gomez, G. T., & Denovan-Wright, E. M. (2004). Mutant huntingtin affects the rate of transcription of striatum-specific isoforms of phosphodiesterase 10A. *The European Journal of Neuroscience, 20*(12), 3351–3363. https://doi.org/10.1111/j.1460-9568.2004.03796.x.

Hu, R., Yuan, B., Wei, X., Zhao, L., Tang, J., & Chen, D. (2007). Enhanced cAMP/PKA pathway by seabuckthorn fatty acids in aged rats. *Journal of Ethnopharmacology, 111*(2), 248–254. https://doi.org/10.1016/j.jep.2006.11.016.

Huang, X. F., Dong, Y. H., Wang, J. H., Ke, H. M., Song, G. Q., & Xu, D. F. (2020). Novel PDE5 inhibitors derived from rutaecarpine for the treatment of Alzheimer's disease. *Bioorganic & Medicinal Chemistry Letters*, *30*(9), 127097. https://doi.org/10.1016/j.bmcl.2020.127097.

Hufgard, J. R., Williams, M. T., Skelton, M. R., Grubisha, O., Ferreira, F. M., Sanger, H., et al. (2017). Phosphodiesterase-1b (Pde1b) knockout mice are resistant to forced swim and tail suspension induced immobility and show upregulation of Pde10a. *Psychopharmacology*, *234*(12), 1803–1813. https://doi.org/10.1007/s00213-017-4587-8.

Ibarra, C., Nedvetsky, P. I., Gerlach, M., Riederer, P., & Schmidt, H. H. (2001). Regional and age-dependent expression of the nitric oxide receptor, soluble guanylyl cyclase, in the human brain. *Brain Research*, *907*(1–2), 54–60.

Ihara, M., Nishino, M., Taguchi, A., Yamamoto, Y., Hattori, Y., Saito, S., et al. (2014). Cilostazol add-on therapy in patients with mild dementia receiving donepezil: A retrospective study. *PLoS One*, *9*(2), e89516. https://doi.org/10.1371/journal.pone.0089516.

Ikeda, M., Sato, I., Yuasa, T., Miyatake, T., & Murota, S. (1995). Nitrite, nitrate and cGMP in the cerebrospinal fluid in degenerative neurologic diseases. *Journal of Neural Transmission. General Section*, *100*(3), 263–267.

Ikezu, T., Okamoto, T., Komatsuzaki, K., Matsui, T., Martyn, J. A., & Nishimoto, I. (1996). Negative transactivation of cAMP response element by familial Alzheimer's mutants of APP. *The EMBO Journal*, *15*(10), 2468–2475.

Jankowska, A., Wesolowska, A., Pawlowski, M., & Chlon-Rzepa, G. (2019). Multifunctional ligands targeting phosphodiesterase as the future strategy for the symptomatic and disease-modifying treatment of Alzheimer's disease. *Current Medicinal Chemistry*, *27*(32), 5351–5373. https://doi.org/10.2174/0929867326666190620095623.

Jesko, H., Chalimoniuk, M., & Strosznajder, J. B. (2003). Activation of constitutive nitric oxide synthase(s) and absence of inducible isoform in aged rat brain. *Neurochemistry International*, *42*(4), 315–322.

Jicha, G. A., Weaver, C., Lane, E., Vianna, C., Kress, Y., Rockwood, J., et al. (1999). cAMP-dependent protein kinase phosphorylations on tau in Alzheimer's disease. *The Journal of Neuroscience*, *19*(17), 7486–7494.

Kane, R. L., Butler, M., Fink, H. A., Brasure, M., Davila, H., Desai, P., et al. (2017). *Interventions to prevent age-related cognitive decline, mild cognitive impairment, and clinical Alzheimer's-type dementia*. Rockville, MD: Agency for Healthcare Research and Quality.

Kanes, S. J., Tokarczyk, J., Siegel, S. J., Bilker, W., Abel, T., & Kelly, M. P. (2007). Rolipram: A specific phosphodiesterase 4 inhibitor with potential antipsychotic activity. *Neuroscience*, *144*(1), 239–246. https://doi.org/10.1016/j.neuroscience.2006.09.026.

Kaplan, A., Kent, C. B., Charron, F., & Fournier, A. E. (2014). Switching responses: Spatial and temporal regulators of axon guidance. *Molecular Neurobiology*, *49*(2), 1077–1086. https://doi.org/10.1007/s12035-013-8582-8.

Kara, K., Mahabadi, A. A., Weimar, C., Winkler, A., Neumann, T., Kalsch, H., et al. (2017). N-terminal Pro-B Type natriuretic peptide is associated with mild cognitive impairment in the general population. *Journal of Alzheimer's Disease*, *55*(1), 359–369. https://doi.org/10.3233/JAD-160635.

Karege, F., Lambercy, C., Schwald, M., Steimer, T., & Cisse, M. (2001). Differential changes of cAMP-dependent protein kinase activity and 3H-cAMP binding sites in rat hippocampus during maturation and aging. *Neuroscience Letters*, *315*(1–2), 89–92.

Karege, F., Schwald, M., Lambercy, C., Murama, J. J., Cisse, M., & Malafosse, A. (2001). A non-radioactive assay for the cAMP-dependent protein kinase activity in rat brain homogenates and age-related changes in hippocampus and cortex. *Brain Research*, *903*(1–2), 86–93.

Kato, H., Araki, T., Chen, T., Itoyama, Y., & Kogure, K. (1998). Effect of rolipram on age-related changes in cyclic AMP-selective phosphodiesterase in the rat brain: An autoradiographic study. *Methods and Findings in Experimental and Clinical Pharmacology*, *20*(5), 403–408.

Kaut, O., Schmitt, I., Tost, J., Busato, F., Liu, Y., Hofmann, P., et al. (2017). Epigenome-wide DNA methylation analysis in siblings and monozygotic twins discordant for sporadic Parkinson's disease revealed different epigenetic patterns in peripheral blood mononuclear cells. *Neurogenetics*, *18*(1), 7–22. https://doi.org/10.1007/s10048-016-0497-x.

Kawamoto, E. M., Munhoz, C. D., Glezer, I., Bahia, V. S., Caramelli, P., Nitrini, R., et al. (2005). Oxidative state in platelets and erythrocytes in aging and Alzheimer's disease. *Neurobiology of Aging*, *26*(6), 857–864. https://doi.org/10.1016/j.neurobiolaging.2004.08.011.

Kelly, M. P. (2014). Putting together the pieces of phosphodiesterase distribution patterns in the brain: A jigsaw puzzle of cyclic nucleotide regulation. In N. J. Brandon, & A. R. West (Eds.), *Cyclic nucleotide phosphodiesterases in the central nervous system: From biology to disease* (pp. 47–58). New Jersey: John Wiley & Sons, Inc.

Kelly, M. P. (2018). PDE11A. In S. Choi (Ed.), *Encyclopedia of signaling molecules* (pp. 3804–3826). New York, NY: Springer New York.

Kelly, M. P., Adamowicz, W., Bove, S., Hartman, A. J., Mariga, A., Pathak, G., et al. (2014). Select 3',5'-cyclic nucleotide phosphodiesterases exhibit altered expression in the aged rodent brain. *Cellular Signalling*, *26*(2), 383–397. https://doi.org/10.1016/j.cellsig.2013.10.007.

Kelly, M. P., Isiegas, C., Cheung, Y. F., Tokarczyk, J., Yang, X., Esposito, M. F., et al. (2007). Constitutive activation of Galphas within forebrain neurons causes deficits in sensorimotor gating because of PKA-dependent decreases in cAMP. *Neuropsychopharmacology*, *32*(3), 577–588. https://doi.org/10.1038/sj.npp.1301099.

Kelly, M. P., Logue, S. F., Brennan, J., Day, J. P., Lakkaraju, S., Jiang, L., et al. (2010). Phosphodiesterase 11A in brain is enriched in ventral hippocampus and deletion causes psychiatric disease-related phenotypes. *Proceedings of the National Academy of Sciences of the United States of America*, *107*(18), 8457–8462. https://doi.org/10.1073/pnas.1000730107.

Kelly, M. P., Stein, J. M., Vecsey, C. G., Favilla, C., Yang, X., Bizily, S. F., et al. (2009). Developmental etiology for neuroanatomical and cognitive deficits in mice overexpressing Galphas, a G-protein subunit genetically linked to schizophrenia. *Molecular Psychiatry*, *14*(4), 398–415. 347 https://doi.org/10.1038/mp.2008.124.

Kim, S. H., Nairn, A. C., Cairns, N., & Lubec, G. (2001). Decreased levels of ARPP-19 and PKA in brains of Down syndrome and Alzheimer's disease. *Journal of Neural Transmission. Supplementum*, (61), 263–272.

Kim, S., Park, J. M., Moon, J., & Choi, H. J. (2014). Alpha-synuclein interferes with cAMP/PKA-dependent upregulation of dopamine beta-hydroxylase and is associated with abnormal adaptive responses to immobilization stress. *Experimental Neurology*, *252*, 63–74. https://doi.org/10.1016/j.expneurol.2013.11.009.

Ko, H. J., Chiou, S. J., Wong, Y. H., Wang, Y. H., Lai, Y., Chou, C. H., et al. (2019). GSKIP-mediated anchoring increases phosphorylation of Tau by PKA but not by GSK3beta via cAMP/PKA/GSKIP/GSK3/Tau axis signaling in cerebrospinal fluid and iPS cells in Alzheimer disease. *Journal of Clinical Medicine*, *8*(10), 1751. https://doi.org/10.3390/jcm8101751.

Kobialka, M., & Gorczyca, W. A. (2000). Particulate guanylyl cyclases: Multiple mechanisms of activation. *Acta Biochimica Polonica*, *47*(3), 517–528.

Kondziella, D., Gothlin, M., Fu, M., Zetterberg, H., & Wallin, A. (2009). B-type natriuretic peptide plasma levels are elevated in subcortical vascular dementia. *Neuroreport*, *20*(9), 825–827. https://doi.org/10.1097/WNR.0b013e328326f82f.

Kostromin, A. P., Berdyshev, G. D., Demidov, S. V., & Kuibeda, V. V. (1984). Age-associated changes in content of cyclic adenosine-3', 5'-monophosphate (cAMP), cyclic guanosine-3',5'-monophosphate (cGMP) and in activity of phosphodiesterase-cAM (pDE-cAM) in spleen T-lymphocytes from C3HA mice. *Zeitschrift für Alternsforschung*, *39*(6), 351–355.

Kroker, K. S., Mathis, C., Marti, A., Cassel, J. C., Rosenbrock, H., & Dorner-Ciossek, C. (2014). PDE9A inhibition rescues amyloid beta-induced deficits in synaptic plasticity and cognition. *Neurobiology of Aging*, *35*(9), 2072–2078. https://doi.org/10.1016/j.neurobiolaging.2014.03.023.

Kudo, K., Wati, H., Qiao, C., Arita, J., & Kanba, S. (2005). Age-related disturbance of memory and CREB phosphorylation in CA1 area of hippocampus of rats. *Brain Research*, *1054*(1), 30–37. https://doi.org/10.1016/j.brainres.2005.06.045.

Kumar, A., & Singh, N. (2017a). Calcineurin inhibitors improve memory loss and neuropathological changes in mouse model of dementia. *Pharmacology, Biochemistry, and Behavior*, *153*, 147–159. https://doi.org/10.1016/j.pbb.2016.12.018.

Kumar, A., & Singh, N. (2017b). Inhibitor of phosphodiestearse-4 improves memory deficits, oxidative stress, neuroinflammation and neuropathological alterations in mouse models of dementia of Alzheimer's Type. *Biomedicine & Pharmacotherapy*, *88*, 698–707. https://doi.org/10.1016/j.biopha.2017.01.059.

Kumar, A., & Singh, N. (2017c). Pharmacological activation of protein kinase A improves memory loss and neuropathological changes in a mouse model of dementia of Alzheimer's type. *Behavioural Pharmacology*, *28*(2 and 3-Spec Issue), 187–198. https://doi.org/10.1097/FBP.0000000000000294.

Laurent, C., Eddarkaoui, S., Derisbourg, M., Leboucher, A., Demeyer, D., Carrier, S., et al. (2014). Beneficial effects of caffeine in a transgenic model of Alzheimer's disease-like tau pathology. *Neurobiology of Aging*, *35*(9), 2079–2090. https://doi.org/10.1016/j.neurobiolaging.2014.03.027.

Laviada, I. D., Galve-Roperh, I., Malpartida, J. M., & Haro, A. (1997). cAMP signalling mechanisms with aging in the Ceratitis capitata brain. *Mechanisms of Ageing and Development*, *97*(1), 45–53.

Law, A., O'Donnell, J., Gauthier, S., & Quirion, R. (2002). Neuronal and inducible nitric oxide synthase expressions and activities in the hippocampi and cortices of young adult, aged cognitively unimpaired, and impaired Long-Evans rats. *Neuroscience*, *112*(2), 267–275.

Lee, Y. Y., Park, J. S., Leem, Y. H., Park, J. E., Kim, D. Y., Choi, Y. H., et al. (2019). The phosphodiesterase 10 inhibitor papaverine exerts anti-inflammatory and neuroprotective effects via the PKA signaling pathway in neuroinflammation and Parkinson's disease mouse models. *Journal of Neuroinflammation*, *16*(1), 246. https://doi.org/10.1186/s12974-019-1649-3.

Lee, D. I., Zhu, G., Sasaki, T., Cho, G. S., Hamdani, N., Holewinski, R., et al. (2015). Phosphodiesterase 9A controls nitric-oxide-independent cGMP and hypertrophic heart disease. *Nature*, *519*(7544), 472–476. https://doi.org/10.1038/nature14332.

Leite, M. R., Wilhelm, E. A., Jesse, C. R., Brandao, R., & Nogueira, C. W. (2011). Protective effect of caffeine and a selective A2A receptor antagonist on impairment of memory and oxidative stress of aged rats. *Experimental Gerontology*, *46*(4), 309–315. https://doi.org/10.1016/j.exger.2010.11.034.

Leuti, A., Laurenti, D., Giampa, C., Montagna, E., Dato, C., Anzilotti, S., et al. (2013). Phosphodiesterase 10A (PDE10A) localization in the R6/2 mouse model of Huntington's disease. *Neurobiology of Disease*, *52*, 104–116. https://doi.org/10.1016/j.nbd.2012.11.016.

Li, J., Liu, C. N., Wei, N., Li, X. D., Liu, Y. Y., Yang, R., et al. (2016). Protective effects of BAY 73-6691, a selective inhibitor of phosphodiesterase 9, on amyloid-beta peptides-induced oxidative stress in in-vivo and in-vitro models of Alzheimer's disease. *Brain Research*, *1642*, 327–335. https://doi.org/10.1016/j.brainres.2016.04.011.

Liang, Z., Liu, F., Grundke-Iqbal, I., Iqbal, K., & Gong, C. X. (2007). Down-regulation of cAMP-dependent protein kinase by over-activated calpain in Alzheimer disease brain. *Journal of Neurochemistry*, *103*(6), 2462–2470. https://doi.org/10.1111/j.1471-4159.2007.04942.x.

Lin, J. T., Chang, W. C., Chen, H. M., Lai, H. L., Chen, C. Y., Tao, M. H., et al. (2013). Regulation of feedback between protein kinase A and the proteasome system worsens Huntington's disease. *Molecular and Cellular Biology*, *33*(5), 1073–1084. https://doi.org/10.1128/MCB.01434-12.

Liu, S., Li, X., Gao, J., Liu, Y., Shi, J., & Gong, Q. (2018). Icariside II, a phosphodiesterase-5 inhibitor, attenuates beta-amyloid-induced cognitive deficits via BDNF/TrkB/CREB signaling. *Cellular Physiology and Biochemistry*, *49*(3), 985. https://doi.org/10.1159/000493232.

Liu, Y., Ni, C., Li, Z., Yang, N., Zhou, Y., Rong, X., et al. (2017). Prophylactic melatonin attenuates isoflurane-induced cognitive impairment in aged rats through hippocampal melatonin receptor 2—cAMP response element binding signalling. *Basic & Clinical Pharmacology & Toxicology*, *120*(3), 219–226. https://doi.org/10.1111/bcpt.12652.

Liu, P., Smith, P. F., Appleton, I., Darlington, C. L., & Bilkey, D. K. (2003). Regional variations and age-related changes in nitric oxide synthase and arginase in the sub-regions of the hippocampus. *Neuroscience*, *119*(3), 679–687.

Liu, P., Smith, P. F., Appleton, I., Darlington, C. L., & Bilkey, D. K. (2004). Age-related changes in nitric oxide synthase and arginase in the rat prefrontal cortex. *Neurobiology of Aging*, *25*(4), 547–552. https://doi.org/10.1016/j.neurobiolaging.2003.07.003.

Liu, P., Smith, P. F., Appleton, I., Darlington, C. L., & Bilkey, D. K. (2005). Hippocampal nitric oxide synthase and arginase and age-associated behavioral deficits. *Hippocampus*, *15*(5), 642–655. https://doi.org/10.1002/hipo.20085.

Llano, D. A., Devanarayan, V., Simon, A. J., & Alzheimer's Disease Neuroimaging Initiative. (2013). Evaluation of plasma proteomic data for Alzheimer disease state classification and for the prediction of progression from mild cognitive impairment to Alzheimer disease. *Alzheimer Disease and Associated Disorders*, *27*(3), 233–243. https://doi.org/10.1097/WAD.0b013e31826d597a.

Luo, J., Phan, T. X., Yang, Y., Garelick, M. G., & Storm, D. R. (2013). Increases in cAMP, MAPK activity, and CREB phosphorylation during REM sleep: Implications for REM sleep and memory consolidation. *The Journal of Neuroscience*, *33*(15), 6460–6468. https://doi.org/10.1523/JNEUROSCI.5018-12.2013.

Ma, Q., Geng, Y., Wang, H. L., Han, B., Wang, Y. Y., Li, X. L., et al. (2019). High frequency repetitive transcranial magnetic stimulation alleviates cognitive impairment and modulates hippocampal synaptic structural plasticity in aged mice. *Frontiers in Aging Neuroscience*, *11*, 235. https://doi.org/10.3389/fnagi.2019.00235.

Maillet, M., Robert, S. J., Cacquevel, M., Gastineau, M., Vivien, D., Bertoglio, J., et al. (2003). Crosstalk between Rap1 and Rac regulates secretion of sAPPalpha. *Nature Cell Biology*, *5*(7), 633–639. https://doi.org/10.1038/ncb1007.

Maki, T., Okamoto, Y., Carare, R. O., Hase, Y., Hattori, Y., Hawkes, C. A., et al. (2014). Phosphodiesterase III inhibitor promotes drainage of cerebrovascular beta-amyloid. *Annals of Clinical Translational Neurology*, *1*(8), 519–533. https://doi.org/10.1002/acn3.79.

Makman, M. H., Ahn, H. S., Thal, L. J., Dvorkin, B., Horowitz, S. G., Sharpless, N. S., et al. (1978). Biogenic amine-stimulated adenylate cyclase and spiroperidol-binding sites in rabbit brain: Evidence for selective loss of receptors with aging. *Advances in Experimental Medicine and Biology*, *113*, 211–230.

Makman, M. H., Ahn, H. S., Thal, L. J., Sharpless, N. S., Dvorkin, B., Horowitz, S. G., et al. (1980). Evidence for selective loss of brain dopamine- and histamine-stimulated adenylate cyclase activities in rabbits with aging. *Brain Research*, *192*(1), 177–183.

Mao, F., Wang, H., Ni, W., Zheng, X., Wang, M., Bao, K., et al. (2018). Design, synthesis, and biological evaluation of orally available first-generation dual-target selective inhibitors of acetylcholinesterase (AChE) and phosphodiesterase 5 (PDE5) for the treatment of Alzheimer's disease. *ACS Chemical Neuroscience*, *9*(2), 328–345. https://doi.org/10.1021/acschemneuro.7b00345.

Marchese, E., Di Maria, V., Samengo, D., Pani, G., Michetti, F., & Geloso, M. C. (2017). Post-natal deletion of neuronal cAMP responsive-element binding (CREB)-1 promotes pro-inflammatory changes in the mouse hippocampus. *Neurochemical Research*, *42*(8), 2230–2245. https://doi.org/10.1007/s11064-017-2233-9.

Mariani, L. L., Longueville, S., Girault, J. A., Herve, D., & Gervasi, N. (2019). Differential enhancement of ERK, PKA and Ca(2+) signaling in direct and indirect striatal neurons of Parkinsonian mice. *Neurobiology of Disease*, *130*, 104506. https://doi.org/10.1016/j.nbd.2019.104506.

Marksteiner, J., Imarhiagbe, D., Defrancesco, M., Deisenhammer, E. A., Kemmler, G., & Humpel, C. (2014). Analysis of 27 vascular-related proteins reveals that NT-proBNP is a potential biomarker for Alzheimer's disease and mild cognitive impairment: A pilot-study. *Experimental Gerontology*, *50*, 114–121. https://doi.org/10.1016/j.exger.2013.12.001.

Martinez, M., Fernandez, E., Frank, A., Guaza, C., de la Fuente, M., & Hernanz, A. (1999). Increased cerebrospinal fluid cAMP levels in Alzheimer's disease. *Brain Research*, *846*(2), 265–267.

Martinez, M., Hernandez, A. I., & Hernanz, A. (2001). Increased cAMP immunostaining in cerebral vessels in Alzheimer's disease. *Brain Research*, *922*(1), 148–152.

Matsumoto, Y., Matsumoto, C. S., Takahashi, T., & Mizunami, M. (2016). Activation of NO-cGMP signaling rescues age-related memory impairment in crickets. *Frontiers in Behavioral Neuroscience*, *10*, 166. https://doi.org/10.3389/fnbeh.2016.00166.

Mattson, M. P., Guo, Z. H., & Geiger, J. D. (1999). Secreted form of amyloid precursor protein enhances basal glucose and glutamate transport and protects against oxidative impairment of glucose and glutamate transport in synaptosomes by a cyclic GMP-mediated mechanism. *Journal of Neurochemistry*, *73*(2), 532–537.

McLachlan, C. S., Chen, M. L., Lynex, C. N., Goh, D. L., Brenner, S., & Tay, S. K. (2007). Changes in PDE4D isoforms in the hippocampus of a patient with advanced Alzheimer disease. *Archives of Neurology*, *64*(3), 456–457. https://doi.org/10.1001/archneur.64.3.456.

McPhee, I., Gibson, L. C., Kewney, J., Darroch, C., Stevens, P. A., Spinks, D., et al. (2005). Cyclic nucleotide signalling: A molecular approach to drug discovery for Alzheimer's disease. *Biochemical Society Transactions*, *33*(Pt. 6), 1330–1332.

Mehan, S., Parveen, S., & Kalra, S. (2017). Adenyl cyclase activator forskolin protects against Huntington's disease-like neurodegenerative disorders. *Neural Regeneration Research*, *12*(2), 290–300. https://doi.org/10.4103/1673-5374.200812.

Merz, K., Herold, S., & Lie, D. C. (2011). CREB in adult neurogenesis—Master and partner in the development of adult-born neurons? *The European Journal of Neuroscience*, *33*(6), 1078–1086. https://doi.org/10.1111/j.1460-9568.2011.07606.x.

Miller, S., Hill Della Puppa, G., Reidling, J., Marcora, E., Thompson, L. M., & Treanor, J. (2014). Comparison of phosphodiesterase 10A, dopamine receptors D1 and D2 and dopamine transporter ligand binding in the striatum of the R6/2 and BACHD mouse models of Huntington's disease. *Journal of Huntington's Disease*, *3*(4), 333–341. https://doi.org/10.3233/JHD-140129.

Mirza, S. S., de Bruijn, R. F., Koudstaal, P. J., van den Meiracker, A. H., Franco, O. H., Hofman, A., et al. (2016). The N-terminal pro B-type natriuretic peptide, and risk of dementia and cognitive decline: A 10-year follow-up study in the general population. *Journal of Neurology, Neurosurgery, and Psychiatry*, *87*(4), 356–362. https://doi.org/10.1136/jnnp-2014-309968.

Mollace, V., Rodino, P., Massoud, R., Rotiroti, D., & Nistico, G. (1995). Age-dependent changes of NO synthase activity in the rat brain. *Biochemical and Biophysical Research Communications*, *215*(3), 822–827. https://doi.org/10.1006/bbrc.1995.2537.

Morales-Garcia, J. A., Alonso-Gil, S., Santos, A., & Perez-Castillo, A. (2020). Phosphodiesterase 7 regulation in cellular and rodent models of Parkinson's disease. *Molecular Neurobiology*, *57*(2), 806–822. https://doi.org/10.1007/s12035-019-01745-z.

Muda, K., Bertinetti, D., Gesellchen, F., Hermann, J. S., von Zweydorf, F., Geerlof, A., et al. (2014). Parkinson-related LRRK2 mutation R1441C/G/H impairs PKA phosphorylation of LRRK2 and disrupts its interaction with 14-3-3. *Proceedings of the National Academy of Sciences of the United States of America*, *111*(1), E34–E43. https://doi.org/10.1073/pnas.1312701111.

Muller, M., Cardenas, C., Mei, L., Cheung, K. H., & Foskett, J. K. (2011). Constitutive cAMP response element binding protein (CREB) activation by Alzheimer's disease presenilin-driven inositol trisphosphate receptor (InsP3R) Ca2+ signaling. *Proceedings of the National Academy of Sciences of the United States of America*, *108*(32), 13293–13298. https://doi.org/10.1073/pnas.1109297108.

Nabavi, S. M., Talarek, S., Listos, J., Nabavi, S. F., Devi, K. P., Roberto de Oliveira, M., et al. (2019). Phosphodiesterase inhibitors say NO to Alzheimer's disease. *Food and Chemical Toxicology*, *134*, 110822. https://doi.org/10.1016/j.fct.2019.110822.

Naito, J., Naka, Y., & Watanabe, H. (2009). Clinical impression of brain natriuretic peptide levels in demented patients without cardiovascular disease. *Geriatrics & Gerontology International*, *9*(3), 242–245. https://doi.org/10.1111/j.1447-0594.2009.00526.x.

Nakashima, M., Suzuki, N., Shiraishi, E., & Iwashita, H. (2019). TAK-915, a phosphodiesterase 2A inhibitor, ameliorates the cognitive impairment associated with aging in rodent models. *Behavioural Brain Research*, *376*, 112192. https://doi.org/10.1016/j.bbr.2019.112192.

Navarro, J. A., Jimenez-Jimenez, F. J., Molina, J. A., Benito-Leon, J., Cisneros, E., Gasalla, T., et al. (1998). Cerebrospinal fluid cyclic guanosine 3'5' monophosphate levels in Parkinson's disease. *Journal of the Neurological Sciences*, *155*(1), 92–94.

Neves, S. R., Ram, P. T., & Iyengar, R. (2002). G protein pathways. *Science*, *296*(5573), 1636–1639. https://doi.org/10.1126/science.1071550.

Ni, W., Wang, H., Li, X., Zheng, X., Wang, M., Zhang, J., et al. (2018). Novel tadalafil derivatives ameliorates scopolamine-induced cognitive impairment in mice via inhibition of acetylcholinesterase (AChE) and phosphodiesterase 5 (PDE5). *ACS Chemical Neuroscience*, *9*(7), 1625–1636. https://doi.org/10.1021/acschemneuro.8b00014.

Niccolini, F., Foltynie, T., Reis Marques, T., Muhlert, N., Tziortzi, A. C., Searle, G. E., et al. (2015). Loss of phosphodiesterase 10A expression is associated with progression and severity in Parkinson's disease. *Brain*, *138*(Pt. 10), 3003–3015. https://doi.org/10.1093/brain/awv219.

Niccolini, F., Wilson, H., Pagano, G., Coello, C., Mehta, M. A., Searle, G. E., et al. (2017). Loss of phosphodiesterase 4 in Parkinson disease: Relevance to cognitive deficits. *Neurology*, *89*(6), 586–593. https://doi.org/10.1212/WNL.0000000000004201.

Nicol, X., & Gaspar, P. (2014). Routes to cAMP: Shaping neuronal connectivity with distinct adenylate cyclases. *The European Journal of Neuroscience*, *39*(11), 1742–1751. https://doi.org/10.1111/ejn.12543.

Niewiadomska-Cimicka, A., Krzyzosiak, A., Ye, T., Podlesny-Drabiniok, A., Dembele, D., Dolle, P., et al. (2017). Genome-wide analysis of RARbeta transcriptional targets in mouse striatum links retinoic acid signaling with Huntington's disease and other neurodegenerative disorders. *Molecular Neurobiology*, *54*(5), 3859–3878. https://doi.org/10.1007/s12035-016-0010-4.

Nilsson, K., Gustafson, L., & Hultberg, B. (2008). Homocysteine, cystatin C and N-terminal-pro brain natriuretic peptide. Vascular risk markers in elderly patients with mental illness. *Dementia and Geriatric Cognitive Disorders*, *25*(1), 88–96. https://doi.org/10.1159/000111994.

Nthenge-Ngumbau, D. N., & Mohanakumar, K. P. (2018). Can cyclic nucleotide phosphodiesterase inhibitors be drugs for Parkinson's disease? *Molecular Neurobiology*, *55*(1), 822–834. https://doi.org/10.1007/s12035-016-0355-8.

Oeckl, P., Steinacker, P., Lehnert, S., Jesse, S., Kretzschmar, H. A., Ludolph, A. C., et al. (2012). CSF concentrations of cAMP and cGMP are lower in patients with Creutzfeldt-Jakob disease but not Parkinson's disease and amyotrophic lateral sclerosis. *PLoS One*, 7(3), e32664. https://doi.org/10.1371/journal.pone.0032664.

Ohm, T. G., Bohl, J., & Lemmer, B. (1989). Reduced cAMP-signal transduction in postmortem hippocampus of demented old people. *Progress in Clinical and Biological Research*, *317*, 501–509.

Ohm, T. G., Bohl, J., & Lemmer, B. (1991). Reduced basal and stimulated (isoprenaline, Gpp(NH)p, forskolin) adenylate cyclase activity in Alzheimer's disease correlated with histopathological changes. *Brain Research*, *540*(1–2), 229–236.

Ohm, T. G., Schmitt, M., Bohl, J., & Lemmer, B. (1997). Decrease in adenylate cyclase activity antecedes neurofibrillary tangle formation. *Neurobiology of Aging*, *18*(3), 275–279.

Oliva, A. A., Jr., Kang, Y., Furones, C., Alonso, O. F., Bruno, O., Dietrich, W. D., et al. (2012). Phosphodiesterase isoform-specific expression induced by traumatic brain injury. *Journal of Neurochemistry*, *123*(6), 1019–1029. https://doi.org/10.1111/jnc.12049.

O'Neill, C., Wiehager, B., Fowler, C. J., Ravid, R., Winblad, B., & Cowburn, R. F. (1994). Regionally selective alterations in G protein subunit levels in the Alzheimer's disease brain. *Brain Research*, *636*(2), 193–201.

Orejana, L., Barros-Minones, L., Jordan, J., Puerta, E., & Aguirre, N. (2012). Sildenafil ameliorates cognitive deficits and tau pathology in a senescence-accelerated mouse model. *Neurobiology of Aging*, *33*(3), 625 e611–620. https://doi.org/10.1016/j.neurobiolaging.2011.03.018.

Oz, M., Lorke, D. E., Hasan, M., & Petroianu, G. A. (2011). Cellular and molecular actions of Methylene Blue in the nervous system. *Medicinal Research Reviews*, *31*(1), 93–117. https://doi.org/10.1002/med.20177.

Pagano, G., Niccolini, F., Wilson, H., Yousaf, T., Khan, N. L., Martino, D., et al. (2019). Comparison of phosphodiesterase 10A and dopamine transporter levels as markers of disease burden in early Parkinson's disease. *Movement Disorders*, *34*(10), 1505–1515. https://doi.org/10.1002/mds.27733.

Paldino, E., Giampa, C., Montagna, E., Angeloni, C., & Fusco, F. R. (2019). Modulation of phospho-CREB by systemically administered recombinant BDNF in the hippocampus of the R6/2 mouse model of Huntington's disease. *Neuroscience Journal*, *2019*, 8363274. https://doi.org/10.1155/2019/8363274.

Palmeri, A., Privitera, L., Giunta, S., Loreto, C., & Puzzo, D. (2013). Inhibition of phosphodiesterase-5 rescues age-related impairment of synaptic plasticity and memory. *Behavioural Brain Research*, *240*, 11–20. https://doi.org/10.1016/j.bbr.2012.10.060.

Palmeri, A., Ricciarelli, R., Gulisano, W., Rivera, D., Rebosio, C., Calcagno, E., et al. (2017). Amyloid-beta peptide is needed for cGMP-induced long-term potentiation and memory. *The Journal of Neuroscience*, *37*(29), 6926–6937. https://doi.org/10.1523/JNEUROSCI.3607-16.2017.

Paris, D., Town, T., Parker, T. A., Tan, J., Humphrey, J., Crawford, F., et al. (1999). Inhibition of Alzheimer's beta-amyloid induced vasoactivity and proinflammatory response in microglia by a cGMP-dependent mechanism. *Experimental Neurology*, *157*(1), 211–221. https://doi.org/10.1006/exnr.1999.7055.

Park, S. H., Kim, J. H., Bae, S. S., Hong, K. W., Lee, D. S., Leem, J. Y., et al. (2011). Protective effect of the phosphodiesterase III inhibitor cilostazol on amyloid beta-induced cognitive deficits associated with decreased amyloid beta accumulation. *Biochemical and Biophysical Research Communications*, *408*(4), 602–608. https://doi.org/10.1016/j.bbrc.2011.04.068.

Pathak, G., Agostino, M. J., Bishara, K., Capell, W. R., Fisher, J. L., Hegde, S., et al. (2017). PDE11A negatively regulates lithium responsivity. *Molecular Psychiatry*, *22*(12), 1714–1724. https://doi.org/10.1038/mp.2016.155.

Paul, V., Reddy, L., & Ekambaram, P. (2005). A reversal by L-arginine and sodium nitroprusside of ageing-induced memory impairment in rats by increasing nitric oxide concentration in the hippocampus. *Indian Journal of Physiology and Pharmacology*, *49*(2), 179–186.

Perez-Gonzalez, R., Pascual, C., Antequera, D., Bolos, M., Redondo, M., Perez, D. I., et al. (2013). Phosphodiesterase 7 inhibitor reduced cognitive impairment and pathological hallmarks in a mouse model of Alzheimer's disease. *Neurobiology of Aging*, *34*(9), 2133–2145. https://doi.org/10.1016/j.neurobiolaging.2013.03.011.

Perez-Torres, S., Cortes, R., Tolnay, M., Probst, A., Palacios, J. M., & Mengod, G. (2003). Alterations on phosphodiesterase type 7 and 8 isozyme mRNA expression in Alzheimer's disease brains examined by in situ hybridization. *Experimental Neurology*, *182*(2), 322–334.

Perez-Torres, S., & Mengod, G. (2003). cAMP-specific phosphodiesterases expression in Alzheimer's disease brains. *International Congress Series*, *1251*, 127–138.

Piedrafita, B., Cauli, O., Montoliu, C., & Felipo, V. (2007). The function of the glutamate-nitric oxide-cGMP pathway in brain in vivo and learning ability decrease in parallel in mature compared with young rats. *Learning & Memory*, *14*(4), 254–258. https://doi.org/10.1101/lm.541307.

Porte, Y., Buhot, M. C., & Mons, N. (2008). Alteration of CREB phosphorylation and spatial memory deficits in aged 129T2/Sv mice. *Neurobiology of Aging*, *29*(10), 1533–1546. https://doi.org/10.1016/j.neurobiolaging.2007.03.023.

Prickaerts, J., Heckman, P. R. A., & Blokland, A. (2017). Investigational phosphodiesterase inhibitors in phase I and phase II clinical trials for Alzheimer's disease. *Expert Opinion on Investigational Drugs*, *26*(9), 1033–1048. https://doi.org/10.1080/13543784.2017.1364360.

Prieto, G. A., Trieu, B. H., Dang, C. T., Bilousova, T., Gylys, K. H., Berchtold, N. C., et al. (2017). Pharmacological rescue of long-term potentiation in Alzheimer diseased synapses. *The Journal of Neuroscience*, *37*(5), 1197–1212. https://doi.org/10.1523/JNEUROSCI.2774-16.2016.

Pugazhenthi, S., Wang, M., Pham, S., Sze, C. I., & Eckman, C. B. (2011). Downregulation of CREB expression in Alzheimer's brain and in Abeta-treated rat hippocampal neurons. *Molecular Neurodegeneration*, *6*, 60. https://doi.org/10.1186/1750-1326-6-60.

Puri, S. K., & Volicer, L. (1977). Effect of aging on cyclic AMP levels and adenylate cyclase and phosphodiesterase activities in the rat corpus striatum. *Mechanisms of Ageing and Development*, *6*(1), 53–58.

Puri, S. K., & Volicer, L. (1981). Age-related changes of cyclic nucleotide levels in rat brain regions. *Mechanisms of Ageing and Development*, *15*(3), 239–242.

Puzzo, D., Loreto, C., Giunta, S., Musumeci, G., Frasca, G., Podda, M. V., et al. (2014). Effect of phosphodiesterase-5 inhibition on apoptosis and beta amyloid load in aged mice. *Neurobiology of Aging*, *35*(3), 520–531. https://doi.org/10.1016/j.neurobiolaging.2013.09.002.

Puzzo, D., Staniszewski, A., Deng, S. X., Privitera, L., Leznik, E., Liu, S., et al. (2009). Phosphodiesterase 5 inhibition improves synaptic function, memory, and amyloid-beta load in an Alzheimer's disease mouse model. *The Journal of Neuroscience*, *29*(25), 8075–8086. https://doi.org/10.1523/JNEUROSCI.0864-09.2009.

Puzzo, D., Vitolo, O., Trinchese, F., Jacob, J. P., Palmeri, A., & Arancio, O. (2005). Amyloid-beta peptide inhibits activation of the nitric oxide/cGMP/cAMP-responsive element-binding protein pathway during hippocampal synaptic plasticity. *The Journal of Neuroscience*, *25*(29), 6887–6897. https://doi.org/10.1523/JNEUROSCI.5291-04.2005.

Qi, H., & Li, S. (2014). Dose-response meta-analysis on coffee, tea and caffeine consumption with risk of Parkinson's disease. *Geriatrics & Gerontology International, 14*(2), 430–439. https://doi.org/10.1111/ggi.12123.

Rabal, O., Sanchez-Arias, J. A., Cuadrado-Tejedor, M., de Miguel, I., Perez-Gonzalez, M., Garcia-Barroso, C., et al. (2018). Design, synthesis, biological evaluation and in vivo testing of dual phosphodiesterase 5 (PDE5) and histone deacetylase 6 (HDAC6)-selective inhibitors for the treatment of Alzheimer's disease. *European Journal of Medicinal Chemistry, 150*, 506–524. https://doi.org/10.1016/j.ejmech.2018.03.005.

Rabal, O., Sanchez-Arias, J. A., Cuadrado-Tejedor, M., de Miguel, I., Perez-Gonzalez, M., Garcia-Barroso, C., et al. (2019a). Discovery of in vivo chemical probes for treating Alzheimer's disease: Dual phosphodiesterase 5 (PDE5) and class I histone deacetylase selective inhibitors. *ACS Chemical Neuroscience, 10*(3), 1765–1782. https://doi.org/10.1021/acschemneuro.8b00648.

Rabal, O., Sanchez-Arias, J. A., Cuadrado-Tejedor, M., de Miguel, I., Perez-Gonzalez, M., Garcia-Barroso, C., et al. (2019b). Multitarget approach for the treatment of Alzheimer's disease: Inhibition of phosphodiesterase 9 (PDE9) and histone deacetylases (HDACs) covering diverse selectivity profiles. *ACS Chemical Neuroscience, 10*(9), 4076–4101. https://doi.org/10.1021/acschemneuro.9b00303.

Raghunandan, R., & Ingram, V. M. (1995). Hyperphosphorylation of the cytoskeletal protein Tau by the MAP-kinase PK40erk2: Regulation by prior phosphorylation with cAMP-dependent protein kinase A. *Biochemical and Biophysical Research Communications, 215*(3), 1056–1066. https://doi.org/10.1006/bbrc.1995.2571.

Ramos, B. P., Birnbaum, S. G., Lindenmayer, I., Newton, S. S., Duman, R. S., & Arnsten, A. F. (2003). Dysregulation of protein kinase a signaling in the aged prefrontal cortex: New strategy for treating age-related cognitive decline.[see comment]. *Neuron, 40*(4), 835–845.

Reyes-Irisarri, E., Markerink-Van Ittersum, M., Mengod, G., & de Vente, J. (2007). Expression of the cGMP-specific phosphodiesterases 2 and 9 in normal and Alzheimer's disease human brains. *The European Journal of Neuroscience, 25*(11), 3332–3338. https://doi.org/10.1111/j.1460-9568.2007.05589.x.

Ricciarelli, R., Brullo, C., Prickaerts, J., Arancio, O., Villa, C., Rebosio, C., et al. (2017). Memory-enhancing effects of GEBR-32a, a new PDE4D inhibitor holding promise for the treatment of Alzheimer's disease. *Scientific Reports*, 7, 46320. https://doi.org/10.1038/srep46320.

Richter, Y., Herzog, Y., Eyal, I., & Cohen, T. (2011). Cognitex supplementation in elderly adults with memory complaints: An uncontrolled open label trial. *Journal of Dietary Supplements, 8*(2), 158–168. https://doi.org/10.3109/19390211.2011.569514.

Robertson, J., Loviny, T. L., Goedert, M., Jakes, R., Murray, K. J., Anderton, B. H., et al. (1993). Phosphorylation of tau by cyclic-AMP-dependent protein kinase. *Dementia, 4*(5), 256–263.

Ruan, L., Du, K., Tao, M., Shan, C., Ye, R., Tang, Y., et al. (2019). Phosphodiesterase-2 inhibitor Bay 60-7550 ameliorates Abeta-Induced cognitive and memory impairment via regulation of the HPA axis. *Frontiers in Cellular Neuroscience, 13*, 432. https://doi.org/10.3389/fncel.2019.00432.

Russell, D. S., Barret, O., Jennings, D. L., Friedman, J. H., Tamagnan, G. D., Thomae, D., et al. (2014). The phosphodiesterase 10 positron emission tomography tracer, [18F] MNI-659, as a novel biomarker for early Huntington disease. *JAMA Neurology, 71*(12), 1520–1528. https://doi.org/10.1001/jamaneurol.2014.1954.

Russo, I. (2019). The role of LRRK2 on PKA-NFkappaB pathway in microglia cells: Implications for Parkinson's disease. *Neural Regeneration Research, 14*(10), 1713–1714. https://doi.org/10.4103/1673-5374.257523.

Russo, I., Di Benedetto, G., Kaganovich, A., Ding, J., Mercatelli, D., Morari, M., et al. (2018). Leucine-rich repeat kinase 2 controls protein kinase A activation state through phosphodiesterase 4. *Journal of Neuroinflammation*, *15*(1), 297. https://doi.org/10.1186/s12974-018-1337-8.

Saavedra, A., Giralt, A., Arumi, H., Alberch, J., & Perez-Navarro, E. (2013). Regulation of hippocampal cGMP levels as a candidate to treat cognitive deficits in Huntington's disease. *PLoS One*, *8*(9), e73664. https://doi.org/10.1371/journal.pone.0073664.

Saito, Y., Inoue, T., Zhu, G., Kimura, N., Okada, M., Nishimura, M., et al. (2012). Hyperpolarization-activated cyclic nucleotide gated channels: A potential molecular link between epileptic seizures and Abeta generation in Alzheimer's disease. *Molecular Neurodegeneration*, 7, 50. https://doi.org/10.1186/1750-1326-7-50.

Sakurai, H., Hanyu, H., Sato, T., Kume, K., Hirao, K., Kanetaka, H., et al. (2013). Effects of cilostazol on cognition and regional cerebral blood flow in patients with Alzheimer's disease and cerebrovascular disease: A pilot study. *Geriatrics & Gerontology International*, *13*(1), 90–97. https://doi.org/10.1111/j.1447-0594.2012.00866.x.

Saletu, B., Anderer, P., Fischhof, P. K., Lorenz, H., Barousch, R., & Bohmer, F. (1992). EEG mapping and psychopharmacological studies with denbufylline in SDAT and MID. *Biological Psychiatry*, *32*(8), 668–681.

Sancesario, G., Giorgi, M., D'Angelo, V., Modica, A., Martorana, A., Morello, M., et al. (2004). Down-regulation of nitrergic transmission in the rat striatum after chronic nigrostriatal deafferentation. *The European Journal of Neuroscience*, *20*(4), 989–1000. https://doi.org/10.1111/j.1460-9568.2004.03566.x.

Sancesario, G., Morrone, L. A., D'Angelo, V., Castelli, V., Ferrazzoli, D., Sica, F., et al. (2014). Levodopa-induced dyskinesias are associated with transient down-regulation of cAMP and cGMP in the caudate-putamen of hemiparkinsonian rats: Reduced synthesis or increased catabolism? *Neurochemistry International*, *79*, 44–56. https://doi.org/10.1016/j.neuint.2014.10.004.

Satoh, J., Tabunoki, H., & Arima, K. (2009). Molecular network analysis suggests aberrant CREB-mediated gene regulation in the Alzheimer disease hippocampus. *Disease Markers*, *27*(5), 239–252. https://doi.org/10.3233/DMA-2009-0670.

Saura, C. A. (2012). CREB-regulated transcription coactivator 1-dependent transcription in Alzheimer's disease mice. *Neurodegenerative Diseases*, *10*(1–4), 250–252. https://doi.org/10.1159/000333341.

Scavone, C., Munhoz, C. D., Kawamoto, E. M., Glezer, I., de Sa Lima, L., Marcourakis, T., et al. (2005). Age-related changes in cyclic GMP and PKG-stimulated cerebellar Na, K-ATPase activity. *Neurobiology of Aging*, *26*(6), 907–916. https://doi.org/10.1016/j.neurobiolaging.2004.08.013.

Schmidt, M. J., & Thornberry, J. F. (1978). Cyclic AMP and cyclic GMP accumulation in vitro in brain regions of young, old and aged rats. *Brain Research*, *139*(1), 169–177.

Schnecko, A., Witte, K., Bohl, J., Ohm, T., & Lemmer, B. (1994). Adenylyl cyclase activity in Alzheimer's disease brain: Stimulatory and inhibitory signal transduction pathways are differently affected. *Brain Research*, *644*(2), 291–296.

Schneider, A., Biernat, J., von Bergen, M., Mandelkow, E., & Mandelkow, E. M. (1999). Phosphorylation that detaches tau protein from microtubules (Ser262, Ser214) also protects it against aggregation into Alzheimer paired helical filaments. *Biochemistry*, *38*(12), 3549–3558. https://doi.org/10.1021/bi981874p.

Schneider, P., Buerger, K., Teipel, S., Uspenskaya, O., Hartmann, O., Hansson, O., et al. (2011). Antihypertensive therapy is associated with reduced rate of conversion to Alzheimer's disease in midregional proatrial natriuretic peptide stratified subjects with mild cognitive impairment. *Biological Psychiatry*, *70*(2), 145–151. https://doi.org/10.1016/j.biopsych.2011.01.036.

Schober, A. (2004). Classic toxin-induced animal models of Parkinson's disease: 6-OHDA and MPTP. *Cell and Tissue Research, 318*(1), 215–224. https://doi.org/10.1007/s00441-004-0938-y.

Schwam, E. M., Nicholas, T., Chew, R., Billing, C. B., Davidson, W., Ambrose, D., et al. (2014). A multicenter, double-blind, placebo-controlled trial of the PDE9A inhibitor, PF-04447943, in Alzheimer's disease. *Current Alzheimer Research, 11*(5), 413–421.

Sebastiani, G., Morissette, C., Lagace, C., Boule, M., Ouellette, M. J., McLaughlin, R. W., et al. (2006). The cAMP-specific phosphodiesterase 4B mediates Abeta-induced microglial activation. *Neurobiology of Aging, 27*(5), 691–701. https://doi.org/10.1016/j.neurobiolaging.2005.03.024.

Sharma, S., & Deshmukh, R. (2015). Vinpocetine attenuates MPTP-induced motor deficit and biochemical abnormalities in Wistar rats. *Neuroscience, 286*, 393–403. https://doi.org/10.1016/j.neuroscience.2014.12.008.

Shea, T. B., & Cressman, C. M. (1999). The order of exposure of tau to signal transduction kinases alters the generation of "AD-like" phosphoepitopes. *Cellular and Molecular Neurobiology, 19*(2), 223–233.

Shi, J., Qian, W., Yin, X., Iqbal, K., Grundke-Iqbal, I., Gu, X., et al. (2011). Cyclic AMP-dependent protein kinase regulates the alternative splicing of tau exon 10: A mechanism involved in tau pathology of Alzheimer disease. *The Journal of Biological Chemistry, 286*(16), 14639–14648. https://doi.org/10.1074/jbc.M110.204453.

Shukitt-Hale, B., Miller, M. G., Chu, Y. F., Lyle, B. J., & Joseph, J. A. (2013). Coffee, but not caffeine, has positive effects on cognition and psychomotor behavior in aging. *Age (Dordrecht, Netherlands), 35*(6), 2183–2192. https://doi.org/10.1007/s11357-012-9509-4.

Sierksma, A. S., Rutten, K., Sydlik, S., Rostamian, S., Steinbusch, H. W., van den Hove, D. L., et al. (2013). Chronic phosphodiesterase type 2 inhibition improves memory in the APPswe/PS1dE9 mouse model of Alzheimer's disease. *Neuropharmacology, 64*, 124–136. https://doi.org/10.1016/j.neuropharm.2012.06.048.

Sierksma, A. S., van den Hove, D. L., Pfau, F., Philippens, M., Bruno, O., Fedele, E., et al. (2014). Improvement of spatial memory function in APPswe/PS1dE9 mice after chronic inhibition of phosphodiesterase type 4D. *Neuropharmacology*, 77, 120–130. https://doi.org/10.1016/j.neuropharm.2013.09.015.

Singh, A. K., Kashyap, M. P., Tripathi, V. K., Singh, S., Garg, G., & Rizvi, S. I. (2017). Neuroprotection through rapamycin-induced activation of autophagy and PI3K/Akt1/mTOR/CREB signaling against amyloid-beta-induced oxidative stress, synaptic/neurotransmission dysfunction, and neurodegeneration in adult rats. *Molecular Neurobiology, 54*(8), 5815–5828. https://doi.org/10.1007/s12035-016-0129-3.

Smith, D. L., Pozueta, J., Gong, B., Arancio, O., & Shelanski, M. (2009). Reversal of long-term dendritic spine alterations in Alzheimer disease models. *Proceedings of the National Academy of Sciences of the United States of America, 106*(39), 16877–16882. https://doi.org/10.1073/pnas.0908706106.

Soares, H. D., Potter, W. Z., Pickering, E., Kuhn, M., Immermann, F. W., Shera, D. M., et al. (2012). Plasma biomarkers associated with the apolipoprotein E genotype and Alzheimer disease. *Archives of Neurology, 69*(10), 1310–1317. https://doi.org/10.1001/archneurol.2012.1070.

Stancheva, S. L., & Alova, L. G. (1991). Age-related changes of cyclic AMP phosphodiesterase activity in rat brain regions and a new phosphodiesterase inhibitor—Nootropic agent adafenoxate. *General Pharmacology, 22*(5), 955–958.

Stefani, A., Fedele, E., Galati, S., Pepicelli, O., Frasca, S., Pierantozzi, M., et al. (2005). Subthalamic stimulation activates internal pallidus: Evidence from cGMP microdialysis in PD patients. *Annals of Neurology, 57*(3), 448–452. https://doi.org/10.1002/ana.20402.

Stefani, A., Fedele, E., Galati, S., Raiteri, M., Pepicelli, O., Brusa, L., et al. (2006). Deep brain stimulation in Parkinson's disease patients: Biochemical evidence. *Journal of Neural Transmission. Supplementum*, (70), 401–408.

Strosznajder, J. B., Jesko, H., Zambrzycka, A., Eckert, A., & Chalimoniuk, M. (2004). Age-related alteration of activity and gene expression of endothelial nitric oxide synthase in different parts of the brain in rats. *Neuroscience Letters*, *370*(2–3), 175–179. https://doi.org/10.1016/j.neulet.2004.08.013.

Sugawa, M., & May, T. (1993). Age-related alteration in signal transduction: Involvement of the cAMP cascade. *Brain Research*, *618*(1), 57–62.

Sugawa, M., & May, T. (1994). Signal transduction in aging. *Archives of Gerontology and Geriatrics*, *19*(Suppl. 1), 235–246.

Szatmari, S. Z., & Whitehouse, P. J. (2003). Vinpocetine for cognitive impairment and dementia. *Cochrane Database of Systematic Reviews*, (1), CD003119. https://doi.org/10.1002/14651858.CD003119.

Taguchi, A., Takata, Y., Ihara, M., Kasahara, Y., Tsuji, M., Nishino, M., et al. (2013). Cilostazol improves cognitive function in patients with mild cognitive impairment: A retrospective analysis. *Psychogeriatrics*, *13*(3), 164–169. https://doi.org/10.1111/psyg.12021.

Tai, S. Y., Chen, C. H., Chien, C. Y., & Yang, Y. H. (2017). Cilostazol as an add-on therapy for patients with Alzheimer's disease in Taiwan: A case control study. *BMC Neurology*, *17*(1), 40. https://doi.org/10.1186/s12883-017-0800-y.

Tai, S. Y., Chien, C. Y., Chang, Y. H., & Yang, Y. H. (2017). Cilostazol use is associated with reduced risk of dementia: A nationwide cohort study. *Neurotherapeutics*, *14*(3), 784–791. https://doi.org/10.1007/s13311-017-0512-4.

Tam, C. F., & Walford, R. L. (1980). Alterations in cyclic nucleotides and cyclase-specific activities in T lymphocytes of aging normal humans and patients with Down's syndrome. *Journal of Immunology*, *125*(4), 1665–1670.

Tanaka, M., Ishizuka, K., Nekooki-Machida, Y., Endo, R., Takashima, N., Sasaki, H., et al. (2017). Aggregation of scaffolding protein DISC1 dysregulates phosphodiesterase 4 in Huntington's disease. *The Journal of Clinical Investigation*, *127*(4), 1438–1450. https://doi.org/10.1172/JCI85594.

Tang, H., Ma, M., Wu, Y., Deng, M. F., Hu, F., Almansoub, H., et al. (2019). Activation of MT2 receptor ameliorates dendritic abnormalities in Alzheimer's disease via C/EBPalpha/miR-125b pathway. *Aging Cell*, *18*(2), e12902. https://doi.org/10.1111/acel.12902.

Thal, L. J., Salmon, D. P., Lasker, B., Bower, D., & Klauber, M. R. (1989). The safety and lack of efficacy of vinpocetine in Alzheimer's disease. *Journal of the American Geriatrics Society*, *37*(6), 515–520.

Tibbo, A. J., Tejeda, G. S., & Baillie, G. S. (2019). Understanding PDE4's function in Alzheimer's disease; a target for novel therapeutic approaches. *Biochemical Society Transactions*, *47*(5), 1557–1565. https://doi.org/10.1042/BST20190763.

Titus, D. J., Furones, C., Kang, Y., & Atkins, C. M. (2013). Age-dependent alterations in cAMP signaling contribute to synaptic plasticity deficits following traumatic brain injury. *Neuroscience*, *231*, 182–194. https://doi.org/10.1016/j.neuroscience.2012.12.002.

Tohda, M., Murayama, T., Nogiri, S., & Nomura, Y. (1996). Influence of aging on rolipram-sensitive phosphodiesterase activity and [3H]rolipram binding in the rat brain. *Biological & Pharmaceutical Bulletin*, *19*(2), 300–302.

Tong, L., Thornton, P. L., Balazs, R., & Cotman, C. W. (2001). Beta-amyloid-(1-42) impairs activity-dependent cAMP-response element-binding protein signaling in neurons at concentrations in which cell survival is not compromised. *The Journal of Biological Chemistry*, *276*(20), 17301–17306. https://doi.org/10.1074/jbc.M010450200.

Treves, T. A., & Korczyn, A. D. (1999). Denbufylline in dementia: A double-blind controlled study. *Dementia and Geriatric Cognitive Disorders*, *10*(6), 505–510. doi: 17197.

Tsai, L. C., Chan, G. C., Nangle, S. N., Shimizu-Albergine, M., Jones, G. L., Storm, D. R., et al. (2012). Inactivation of Pde8b enhances memory, motor performance, and protects against age-induced motor coordination decay. *Genes, Brain, and Behavior, 11*(7), 837–847. https://doi.org/10.1111/j.1601-183X.2012.00836.x.

Tynkkynen, J., Hernesniemi, J. A., Laatikainen, T., Havulinna, A. S., Salo, P., Blankenberg, S., et al. (2017). High-sensitivity cardiac troponin I and NT-proBNP as predictors of incident dementia and Alzheimer's disease: The FINRISK study. *Journal of Neurology, 264*(3), 503–511. https://doi.org/10.1007/s00415-016-8378-7.

Tynkkynen, J., Laatikainen, T., Salomaa, V., Havulinna, A. S., Blankenberg, S., Zeller, T., et al. (2015). NT-proBNP and the risk of dementia: A prospective cohort study with 14 years of follow-up. *Journal of Alzheimer's Disease, 44*(3), 1007–1013. https://doi.org/10.3233/JAD-141809.

Ueda, K., Yagami, T., Kageyama, H., & Kawasaki, K. (1996). Protein kinase inhibitor attenuates apoptotic cell death induced by amyloid beta protein in culture of the rat cerebral cortex. *Neuroscience Letters, 203*(3), 175–178.

Ugarte, A., Gil-Bea, F., Garcia-Barroso, C., Cedazo-Minguez, A., Ramirez, M. J., Franco, R., et al. (2015). Decreased levels of guanosine 3', 5'-monophosphate (cGMP) in cerebrospinal fluid (CSF) are associated with cognitive decline and amyloid pathology in Alzheimer's disease. *Neuropathology and Applied Neurobiology, 41*(4), 471–482. https://doi.org/10.1111/nan.12203.

Valikovics, A., Csanyi, A., & Nemeth, L. (2012). Study of the effects of vinpocetin on cognitive functions. *Ideggyógyászati Szemle, 65*(3–4), 115–120.

Vallebuona, F., & Raiteri, M. (1995). Age-related changes in the NMDA receptor/nitric oxide/cGMP pathway in the hippocampus and cerebellum of freely moving rats subjected to transcerebral microdialysis. *The European Journal of Neuroscience,* 7(4), 694–701.

van Vliet, P., Sabayan, B., Wijsman, L. W., Poortvliet, R. K., Mooijaart, S. P., de Ruijter, W., et al. (2014). NT-proBNP, blood pressure, and cognitive decline in the oldest old: The Leiden 85-plus study. *Neurology, 83*(13), 1192–1199. https://doi.org/10.1212/WNL.0000000000000820.

Vandesquille, M., Baudonnat, M., Decorte, L., Louis, C., Lestage, P., & Beracochea, D. (2013). Working memory deficits and related disinhibition of the cAMP/PKA/CREB are alleviated by prefrontal alpha4beta2*-nAChRs stimulation in aged mice. *Neurobiology of Aging, 34*(6), 1599–1609. https://doi.org/10.1016/j.neurobiolaging.2012.10.006.

Vercambre, M. N., Berr, C., Ritchie, K., & Kang, J. H. (2013). Caffeine and cognitive decline in elderly women at high vascular risk. *Journal of Alzheimer's Disease, 35*(2), 413–421. https://doi.org/10.3233/JAD-122371.

VerPlank, J. J. S., Tyrkalska, S. D., Fleming, A., Rubinsztein, D. C., & Goldberg, A. L. (2020). cGMP via PKG activates 26S proteasomes and enhances degradation of proteins, including ones that cause neurodegenerative diseases. *Proceedings of the National Academy of Sciences of the United States of America, 117*(25), 14220–14230. https://doi.org/10.1073/pnas.2003277117.

Vila-Luna, S., Cabrera-Isidoro, S., Vila-Luna, L., Juarez-Diaz, I., Bata-Garcia, J. L., Alvarez-Cervera, F. J., et al. (2012). Chronic caffeine consumption prevents cognitive decline from young to middle age in rats, and is associated with increased length, branching, and spine density of basal dendrites in CA1 hippocampal neurons. *Neuroscience, 202*, 384–395. https://doi.org/10.1016/j.neuroscience.2011.11.053.

Vitolo, O. V., Sant'Angelo, A., Costanzo, V., Battaglia, F., Arancio, O., & Shelanski, M. (2002). Amyloid beta-peptide inhibition of the PKA/CREB pathway and long-term potentiation: Reversibility by drugs that enhance cAMP signaling. *Proceedings of the National Academy of Sciences of the United States of America, 99*(20), 13217–13221. https://doi.org/10.1073/pnas.172504199.

Volicer, L., Beal, M. F., Direnfeld, L. K., Marquis, J. K., & Albert, M. L. (1986). CSF cyclic nucleotides and somatostatin in Parkinson's disease. *Neurology*, *36*(1), 89–92.
Wang, Z., Cao, M., Xiang, H., Wang, W., Feng, X., & Yang, X. (2019). WBQ5187, a multitarget directed agent, ameliorates cognitive impairment in a transgenic mouse model of Alzheimer's disease and modulates cerebral beta-amyloid, gliosis, cAMP levels, and neurodegeneration. *ACS Chemical Neuroscience*, *10*(12), 4787–4799. https://doi.org/10.1021/acschemneuro.9b00409.
Wang, G., Chen, L., Pan, X., Chen, J., Wang, L., Wang, W., et al. (2016). The effect of resveratrol on beta amyloid-induced memory impairment involves inhibition of phosphodiesterase-4 related signaling. *Oncotarget*, 7(14), 17380–17392. https://doi.org/10.18632/oncotarget.8041.
Wang, X. L., Deng, Y. X., Gao, Y. M., Dong, Y. T., Wang, F., Guan, Z. Z., et al. (2020). Activation of alpha7 nAChR by PNU-282987 improves synaptic and cognitive functions through restoring the expression of synaptic-associated proteins and the CaM-CaMKII-CREB signaling pathway. *Aging (Albany NY)*, *12*(1), 543–570. https://doi.org/10.18632/aging.102640.
Wang, M., Gamo, N. J., Yang, Y., Jin, L. E., Wang, X. J., Laubach, M., et al. (2011). Neuronal basis of age-related working memory decline. *Nature*, *476*(7359), 210–213. https://doi.org/10.1038/nature10243.
Wang, H. H., Li, Y., Li, A., Yan, F., Li, Z. L., Liu, Z. Y., et al. (2018). Forskolin induces hyperphosphorylation of Tau accompanied by cell cycle reactivation in primary hippocampal neurons. *Molecular Neurobiology*, *55*, 696–706. https://doi.org/10.1007/s12035-016-0348-7.
Wang, Q. W., Rowan, M. J., & Anwyl, R. (2009). Inhibition of LTP by beta-amyloid is prevented by activation of beta2 adrenoceptors and stimulation of the cAMP/PKA signalling pathway. *Neurobiology of Aging*, *30*(10), 1608–1613. https://doi.org/10.1016/j.neurobiolaging.2007.12.004.
Wang, L., Xiaokaiti, Y., Wang, G., Xu, X., Chen, L., Huang, X., et al. (2017). Inhibition of PDE2 reverses beta amyloid induced memory impairment through regulation of PKA/PKG-dependent neuro-inflammatory and apoptotic pathways. *Scientific Reports*, 7(1), 12044. https://doi.org/10.1038/s41598-017-08070-2.
Wang, C., Yang, X. M., Zhuo, Y. Y., Zhou, H., Lin, H. B., Cheng, Y. F., et al. (2012). The phosphodiesterase-4 inhibitor rolipram reverses Abeta-induced cognitive impairment and neuroinflammatory and apoptotic responses in rats. *The International Journal of Neuropsychopharmacology*, *15*(6), 749–766. https://doi.org/10.1017/S1461145711000836.
Wilson, H., Niccolini, F., Haider, S., Marques, T. R., Pagano, G., Coello, C., et al. (2016). Loss of extra-striatal phosphodiesterase 10A expression in early premanifest Huntington's disease gene carriers. *Journal of the Neurological Sciences*, *368*, 243–248. https://doi.org/10.1016/j.jns.2016.07.033.
Wirtz-Brugger, F., & Giovanni, A. (2000). Guanosine 3',5'-cyclic monophosphate mediated inhibition of cell death induced by nerve growth factor withdrawal and beta-amyloid: Protective effects of propentofylline. *Neuroscience*, *99*(4), 737–750.
Wruck, W., & Adjaye, J. (2020). Meta-analysis of human prefrontal cortex reveals activation of GFAP and decline of synaptic transmission in the aging brain. *Acta Neuropathologica Communications*, *8*(1), 26. https://doi.org/10.1186/s40478-020-00907-8.
Wu, D. M., Wang, S., Wen, X., Han, X. R., Wang, Y. J., Shen, M., et al. (2018). Inhibition of microRNA-200a upregulates the expression of striatal dopamine receptor D2 to repress apoptosis of striatum via the cAMP/PKA signaling pathway in rats with Parkinson's disease. *Cellular Physiology and Biochemistry*, *51*(4), 1600–1615. https://doi.org/10.1159/000495649.
Xia, L., Zhu, X., Zhao, Y., Yang, G., Zuo, X., Xie, P., et al. (2019). Genome-wide RNA sequencing analysis reveals that IGF-2 attenuates memory decline, oxidative stress and

amyloid plaques in an Alzheimer's disease mouse model (AD) by activating the PI3K/AKT/CREB signaling pathway. *International Psychogeriatrics*, *31*, 1–13. https://doi.org/10.1017/S1041610219000383.

Xu, J., Rong, S., Xie, B., Sun, Z., Deng, Q., Wu, H., et al. (2010). Memory impairment in cognitively impaired aged rats associated with decreased hippocampal CREB phosphorylation: Reversal by procyanidins extracted from the lotus seedpod. *The Journals of Gerontology. Series A, Biological Sciences and Medical Sciences*, *65*(9), 933–940. https://doi.org/10.1093/gerona/glq094.

Xu, Y., Zhu, N., Xu, W., Ye, H., Liu, K., Wu, F., et al. (2018). Inhibition of phosphodiesterase-4 reverses abeta-induced memory impairment by regulation of HPA axis related cAMP signaling. *Frontiers in Aging Neuroscience*, *10*, 204. https://doi.org/10.3389/fnagi.2018.00204.

Yamamoto, M., Gotz, M. E., Ozawa, H., Luckhaus, C., Saito, T., Rosler, M., et al. (2000). Hippocampal level of neural specific adenylyl cyclase type I is decreased in Alzheimer's disease. *Biochimica et Biophysica Acta*, *1535*(1), 60–68.

Yamamoto, M., Ozawa, H., Saito, T., Hatta, S., Riederer, P., & Takahata, N. (1997). Ca2+/CaM-sensitive adenylyl cyclase activity is decreased in the Alzheimer's brain: Possible relation to type I adenylyl cyclase. *Journal of Neural Transmission (Vienna)*, *104*(6–7), 721–732. https://doi.org/10.1007/BF01291889.

Yamamoto-Sasaki, M., Ozawa, H., Saito, T., Rosler, M., & Riederer, P. (1999). Impaired phosphorylation of cyclic AMP response element binding protein in the hippocampus of dementia of the Alzheimer type. *Brain Research*, *824*(2), 300–303.

Yanai, S., & Endo, S. (2019). PDE3 inhibitors repurposed as treatments for age-related cognitive impairment. *Molecular Neurobiology*, *56*(6), 4306–4316. https://doi.org/10.1007/s12035-018-1374-4.

Yanai, S., Toyohara, J., Ishiwata, K., Ito, H., & Endo, S. (2017). Long-term cilostazol administration ameliorates memory decline in senescence-accelerated mouse prone 8 (SAMP8) through a dual effect on cAMP and blood-brain barrier. *Neuropharmacology*, *116*, 247–259. https://doi.org/10.1016/j.neuropharm.2016.12.006.

Yang, L., Calingasan, N. Y., Lorenzo, B. J., & Beal, M. F. (2008). Attenuation of MPTP neurotoxicity by rolipram, a specific inhibitor of phosphodiesterase IV. *Experimental Neurology*, *211*(1), 311–314. https://doi.org/10.1016/j.expneurol.2007.02.010.

Ye, J., Yin, Y., Liu, H., Fang, L., Tao, X., Wei, L., et al. (2020). Tau inhibits PKA by nuclear proteasome-dependent PKAR2alpha elevation with suppressed CREB/GluA1 phosphorylation. *Aging Cell*, *19*(1), e13055. https://doi.org/10.1111/acel.13055.

Yew, D. T., Wong, H. W., Li, W. P., Lai, H. W., & Yu, W. H. (1999). Nitric oxide synthase neurons in different areas of normal aged and Alzheimer's brains. *Neuroscience*, *89*(3), 675–686.

Yin, C., Deng, Y., Liu, Y., Gao, J., Yan, L., & Gong, Q. (2018). Icariside II ameliorates cognitive impairments induced by chronic cerebral hypoperfusion by inhibiting the amyloidogenic pathway: Involvement of BDNF/TrkB/CREB signaling and up-regulation of PPARalpha and PPARgamma in rats. *Frontiers in Pharmacology*, *9*, 1211. https://doi.org/10.3389/fphar.2018.01211.

Yiu, A. P., Rashid, A. J., & Josselyn, S. A. (2011). Increasing CREB function in the CA1 region of dorsal hippocampus rescues the spatial memory deficits in a mouse model of Alzheimer's disease. *Neuropsychopharmacology*, *36*(11), 2169–2186. https://doi.org/10.1038/npp.2011.107.

Yu, X. W., Curlik, D. M., Oh, M. M., Yin, J. C., & Disterhoft, J. F. (2017). CREB overexpression in dorsal CA1 ameliorates long-term memory deficits in aged rats. *Elife*, *6*, e19358. https://doi.org/10.7554/eLife.19358.

Yu, Y. F., Huang, Y. D., Zhang, C., Wu, X. N., Zhou, Q., Wu, D., et al. (2017). Discovery of novel pyrazolopyrimidinone derivatives as phosphodiesterase 9A inhibitors capable of

inhibiting butyrylcholinesterase for treatment of Alzheimer's disease. *ACS Chemical Neuroscience*, *8*(11), 2522–2534. https://doi.org/10.1021/acschemneuro.7b00268.
Yu, W., Juang, S., Lee, J., Liu, T., & Cheng, J. (2000). Decrease of neuronal nitric oxide synthase in the cerebellum of aged rats. *Neuroscience Letters*, *291*(1), 37–40.
Zaldua, N., Gastineau, M., Hoshino, M., Lezoualc'h, F., & Zugaza, J. L. (2007). Epac signaling pathway involves STEF, a guanine nucleotide exchange factor for Rac, to regulate APP processing. *FEBS Letters*, *581*(30), 5814–5818.
Zhang, C., Cheng, Y., Wang, H., Wang, C., Wilson, S. P., Xu, J., et al. (2014). RNA interference-mediated knockdown of long-form phosphodiesterase-4D (PDE4D) enzyme reverses amyloid-beta42-induced memory deficits in mice. *Journal of Alzheimer's Disease*, *38*(2), 269–280. https://doi.org/10.3233/JAD-122236.
Zhang, J., Guo, J., Zhao, X., Chen, Z., Wang, G., Liu, A., et al. (2013). Phosphodiesterase-5 inhibitor sildenafil prevents neuroinflammation, lowers beta-amyloid levels and improves cognitive performance in APP/PS1 transgenic mice. *Behavioural Brain Research*, *250*, 230–237. https://doi.org/10.1016/j.bbr.2013.05.017.
Zhang, Y., Li, H. L., Wang, D. L., Liu, S. J., & Wang, J. Z. (2006). A transitory activation of protein kinase-A induces a sustained tau hyperphosphorylation at multiple sites in N2a cells-imply a new mechanism in Alzheimer pathology. *Journal of Neural Transmission (Vienna)*, *113*(10), 1487–1497. https://doi.org/10.1007/s00702-005-0421-2.
Zhang, L., Seo, J. H., Li, H., Nam, G., & Yang, H. O. (2018). The phosphodiesterase 5 inhibitor, KJH-1002, reverses a mouse model of amnesia by activating a cGMP/cAMP response element binding protein pathway and decreasing oxidative damage. *British Journal of Pharmacology*, *175*(16), 3347–3360. https://doi.org/10.1111/bph.14377.
Zhang, Z., Shen, Q., Wu, X., Zhang, D., & Xing, D. (2020). Activation of PKA/SIRT1 signaling pathway by photobiomodulation therapy reduces Abeta levels in Alzheimer's disease models. *Aging Cell*, *19*(1), e13054. https://doi.org/10.1111/acel.13054.
Zhao, H. K., Chen, B. Y., Chang, R., Wang, J. B., Ni, F., Yang, L., et al. (2013). Vasonatrin peptide, a novel protector of dopaminergic neurons against the injuries induced by n-methyl-4-phenylpyridiniums. *Peptides*, *49*, 117–122. https://doi.org/10.1016/j.peptides.2013.09.005.
Zheng-Fischhofer, Q., Biernat, J., Mandelkow, E. M., Illenberger, S., Godemann, R., & Mandelkow, E. (1998). Sequential phosphorylation of Tau by glycogen synthase kinase-3beta and protein kinase A at Thr212 and Ser214 generates the Alzheimer-specific epitope of antibody AT100 and requires a paired-helical-filament-like conformation. *European Journal of Biochemistry*, *252*(3), 542–552.
Zhong, J., Dong, W., Qin, Y., Xie, J., Xiao, J., Xu, J., et al. (2020). Roflupram exerts neuroprotection via activation of CREB/PGC-1alpha signalling in experimental models of Parkinson's disease. *British Journal of Pharmacology*, *177*(10), 2333–2350. https://doi.org/10.1111/bph.14983.
Zhu, G., Yang, S., Xie, Z., & Wan, X. (2018). Synaptic modification by L-theanine, a natural constituent in green tea, rescues the impairment of hippocampal long-term potentiation and memory in AD mice. *Neuropharmacology*, *138*, 331–340. https://doi.org/10.1016/j.neuropharm.2018.06.030.
Zhu, L., Zhang, Z., Hou, X. J., Wang, Y. F., Yang, J. Y., & Wu, C. F. (2019). Inhibition of PDE5 attenuates streptozotocin-induced neuroinflammation and tau hyperphosphorylation in a streptozotocin-treated rat model. *Brain Research*, *1722*, 146344. https://doi.org/10.1016/j.brainres.2019.146344.
Zimmerman, I., & Berg, A. (1974). Levels of adenosine 3',5'-cyclic monophosphate in the cerebral cortex of aging rats. *Mechanisms of Ageing and Development*, *3*(1), 33–36.
Zimmerman, I. D., & Berg, A. P. (1975). Phosphodiesterase and adenyl-cyclase activities in the cerebral cotex of the aging rat. *Mechanisms of Ageing and Development*, *4*(2), 89–96.
Zuccarello, E., Acquarone, E., Calcagno, E., Argyrousi, E. K., Deng, S. X., Landry, D. W., et al. (2020). Development of novel phosphodiesterase 5 inhibitors for the therapy of Alzheimer's disease. *Biochemical Pharmacology*, *176*, 113818. https://doi.org/10.1016/j.bcp.2020.113818.

CHAPTER TWELVE

Aging and FGF23-klotho system

Makoto Kuro-o*
Division of Anti-Aging Medicine, Center for Molecular Medicine, Jichi Medical University, Shimotsuke, Tochigi, Japan
*Corresponding author: e-mail address: mkuroo@jichi.ac.jp

Contents

Abstract

During the evolution of skeletons, vertebrates acquired the bone made of calcium phosphate. By keeping the extracellular fluid in a supersaturated condition regarding calcium and phosphate, vertebrates create the bone when and where they want simply by providing a cue for precipitation. To secure this strategy, a new endocrine system has evolved that strictly controls the extracellular phosphate concentration. In response to phosphate intake, fibroblast growth factor-23 (FGF23) is secreted from the bone and acts on the kidney through binding to its receptor Klotho to increase urinary phosphate excretion and maintain phosphate homeostasis. The FGF23-Klotho endocrine system, when disrupted, results in hyperphosphatemia and ectopic precipitation of calcium phosphate in mice and humans. In addition to disturbed phosphate homeostasis, mice lacking Klotho suffer from premature aging. They exhibit multiple organ atrophy, arteriosclerosis characterized by vascular calcification, cardiac hypertrophy, sarcopenia, cognition impairment, frailty, and a shortened life span associated with chronic non-infectious inflammation. Restoration of the phosphate balance by placing Klotho- or FGF23-deficient mice on low phosphate diet rescued them from the aging-like phenotypes, indicating that phosphate was responsible for the accelerated aging. The similar pathophysiology is universally observed in patients with chronic kidney disease (CKD), rendering advanced CKD a clinical model of accelerated aging. CKD patients bear colloidal nanoparticles containing calcium phosphate in the blood, which are termed calciprotein particles (CPPs). CPPs have the ability to induce cell damage and inflammation, potentially contributing

Vitamins and Hormones, Volume 115
ISSN 0083-6729
https://doi.org/10.1016/bs.vh.2020.12.013

to accelerated aging. Terrestrial vertebrates with the bone made of calcium phosphate may be destined to age due to ectopic calcium phosphate.

1. Discovery of the *klotho* gene

The *klotho* gene was identified as a gene disrupted in a mouse strain that inherited a syndrome resembling human aging in an autosomal recessive manner (Kuro-o, Matsumura, Aizawa, et al., 1997). The founder of this strain was a transgenic mouse carrying a transgene that was supposed to express sodium-proton exchanger type I and develop salt-sensitive hypertension (Kuro-o, Hanaoka, Hiroi, et al., 1995). In this particular strain, however, the transgene was not expressed due to extensive methylation that occurred around the transgene integration site (Azuma, Koyama, Kikuchi, et al., 2012). In addition, insertion of the transgene generated an ~8 kb deletion within the 5′-flanking region of the downstream endogenous gene. As a result, expression of this gene was severely suppressed. When homozygous for the transgene, mice developed complex aging-like phenotypes, including atrophy of multiple organs (gonads, thymus, skin, fat, liver), vascular calcification, cardiac hypertrophy, sarcopenia, osteopenia, cognition impairment, hearing disturbance, frailty, and premature death around 2–3 months of age (Faul, Amaral, Oskouei, et al., 2011; Kamemori, Ohyama, Kurabayashi, et al., 2002; Kawaguchi, Manabe, Miyaura, et al., 1999; Kuro-o et al., 1997; Nagai, Yamada, Kim, et al., 2003). We speculated that the gene disrupted by insertion of the transgene might be an "aging-suppressor gene" that accelerated aging when its expression was lost. The putative aging-suppressor gene was named *klotho* after a Greek goddess who spins the thread of life.

The *klotho* gene was identified in 1997 by positional cloning using the transgene as a marker (Kuro-o et al., 1997). It was a new gene encoding a type-I single-pass transmembrane protein that was expressed predominantly in the kidney (distal convoluted tubules), brain (choroid plexus), and parathyroid glands (Ben-Dov, Galitzer, Lavi-Moshayoff, et al., 2007). The fact that *klotho* expression was limited to specific cell types and that mice lacking *klotho* expression developed disorders in multiple organs where endogenous *klotho* expression was undetectable suggested involvement of a humoral factor(s) in Klotho protein function. Although Klotho protein function had not been clear, transgenic mice that overexpressed Klotho was reported to live longer than wild-type mice (Kurosu, Yamamoto, Clark, et al., 2005), indicating that the *klotho* gene was a *de novo* aging-suppressor gene that shortened life span when disrupted and extended life span when overexpressed.

2. Klotho protein function

Although the *klotho* gene belonged to the family 1 glycosidases, recombinant Klotho protein barely had enzymatic activity, because two catalytic glutamates critical for the hydrolytic activity and thus conserved in all family 1 glycosidases were replaced with other amino acids in Klotho protein (Chen, Liu, Goetz, et al., 2018). Therefore, Klotho protein function remained unclear for nearly a decade after the discovery of the *klotho* gene.

A clue to the understanding of Klotho protein function was a paper on fibroblast growth factor-23 (FGF23) knockout mice published in 2004 (Shimada, Kakitani, Yamazaki, et al., 2004). The *FGF23* gene had been identified in 2000 as a causative gene of a rare hereditary disorder called autosomal dominant hypophosphatemic rickets (ADHR) (White, Evans, O'Rlordan, et al., 2000), patients with which developed disturbed bone mineralization associated with low serum phosphate levels due to phosphate wasting into urine. Shortly thereafter, FGF23 was reported to function as a phosphaturic hormone that increased phosphate excretion into urine by suppressing phosphate reabsorption in renal proximal tubules (Shimada, Hasegawa, Yamazaki, et al., 2004). ADHR patients had missense mutations in the *FGF23* gene that conferred resistance to proteolytic inactivation on the mutant FGF23 proteins (White, Carn, Lorenz-Depiereux, et al., 2001), resulting in elevated circulating FGF23 levels and excess urinary phosphate excretion. In addition to phosphaturic activity, FGF23 lowers circulating levels of 1,25-dihydroxyvitamin D_3 by suppressing expression of the *Cyp27b1* gene encoding 1α-hydroxylase necessary for synthesis of 1,25-dihydroxyvitamin D_3 and by inducing expression of the *Cyp24* gene encoding 24-hydroxylase that inactivates 1,25-dihydroxyvitamin D_3 (Shimada, Hasegawa, et al., 2004). In fact, serum 1,25-dihydroxyvitamin D_3 levels were decreased in ADHR patients (Winters, Graham, Williams, et al., 1958).

Because FGF23 is one of the fibroblast growth factor (FGF) family members, it was thought to bind to FGF receptors (FGFRs). FGFRs are receptor tyrosine kinases encoded by 4 distinct genes (FGFR1 ~ 4) with multiple splice variants. However, the affinity of FGF23 to any FGFR isoforms (K_D = 220–730 nM) turned out to be too low to bind to FGFRs effectively under the physiological concentration of FGF23 (~1 pM) (Yu, Ibrahimi, Goetz, et al., 2005). Thus, identity of FGF23 receptor had been unclear.

Mice lacking FGF23 (*Fgf23*$^{-/-}$ mice) showed high serum phosphate and 1,25-dihydroxyvitamin D_3 levels (Shimada, Kakitani, et al., 2004). These phenotypes were predictable, because FGF23 functions as a counterregulatory phosphaturic hormone for vitamin D. However, it was unexpected that *Fgf23*$^{-/-}$ mice exhibited growth retardation, sarcopenia, vascular calcification, and shortened life span (Shimada, Kakitani, et al., 2004), which were reminiscent of mice lacking Klotho. On the other hand, mice lacking Klotho showed high serum phosphate and 1,25-dihydroxyvitamin D_3 levels (Tsujikawa, Kurotaki, Fujimori, et al., 2003). The striking similarity in the phenotypes between mice lacking FGF23 and Klotho led us to hypothesize that Klotho might function as the FGF23 receptor.

In 2006, we reported that Klotho protein functioned as the obligate co-receptor for FGF23. Klotho protein forms constitutive binary complexes with specific FGFR isoforms (FGFR1c, FGFR3c, and FGFR4), to which FGF23 binds with much higher affinity than to FGFRs or Klotho alone (Kurosu, Ogawa, Miyoshi, et al., 2006). This finding was confirmed later by an independent group (Urakawa, Yamazaki, Shimada, et al., 2006). Namely, the physiological receptor for FGF23 was not FGFRs or Klotho but the Klotho-FGFR complexes. Although various organs express FGFR1c, FGFR3c, and/or FGFR4, Klotho expression is limited primarily to renal tubules in the kidney, choroid plexus in the brain, and parathyroid glands. Thus, these organs that express Klotho endogenously are regarded as potential target organs of FGF23.

3. Structure of Klotho protein

In 2018, the atomic structure of the Klotho-FGFR1c-FGF23 ternary complex was solved (Fig. 1) (Chen et al., 2018; Kuro-o, 2018). Klotho protein, like its namesake, sent out a long "thread" composed of a 49-amino acid stretch. This thread was termed the receptor binding arm (RBA), through which Klotho protein captured FGFR1c. Although RBA appeared to have an intrinsically disordered structure, once bound to FGFR1c, RBA took a solid structure to form a groove between Klotho and FGFR1c, into

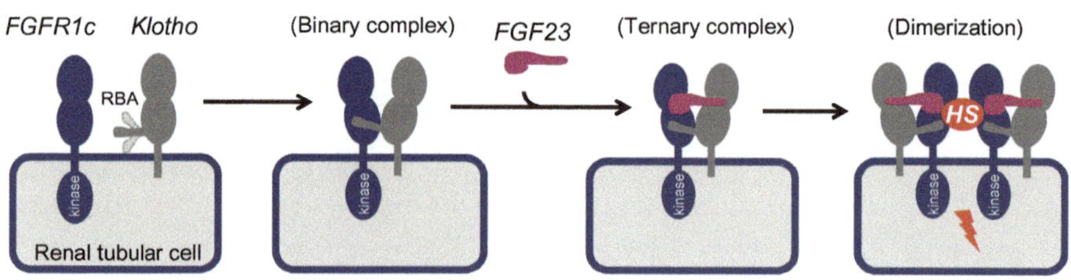

Fig. 1 Structure of Klotho protein. Dimerization of the FGFR-Klotho-FGF23 complex is mediated by heparan sulfate (HS) and required for activation of FGFR tyrosine kinase.

which FGF23 fitted. Furthermore, dimerization of the Klotho-FGFR-FGF23 ternary complexes was necessary to activate the FGFR tyrosine kinase and the canonical FGF signaling pathway. Dimerization was dependent on binding of the ternary complexes to heparan sulfate, a cofactor also required for FGFRs to bind to paracrine FGFs (e.g. FGF2), and required to activate the FGFR tyrosine kinase (Chen et al., 2018).

4. FGF23 and kidney

In response to phosphate intake, FGF23 is secreted from the bone, circulates in the blood, and binds to the Klotho-FGFR complexes expressed in the kidney to promote urinary phosphate excretion. The FGF23-Klotho endocrine system is indispensable for maintaining phosphate homeostasis, because loss-of-function mutations in either FGF23 or Klotho cause phosphate retention phenotypes including ectopic calcification and hyperphosphatemia, whereas gain-of-function mutations of FGF23 lead to phosphate wasting phenotypes including disturbed bone mineralization and hypophosphatemia in rodents and humans (Kuro-o, 2019). Although function of the FGF23-Klotho endocrine system has become clear, there are a few issues that need to be addressed.

One of the unsolved questions is the discrepancy between the facts that Klotho is expressed predominantly in distal convoluted tubules (Kuro-o et al., 1997) and that FGF23 regulates functions of proximal tubules, including reabsorption of phosphate and synthesis of 1,25-dihydroxyvitamin D_3. One possible explanation is that FGF23 may act on distal convoluted tubules where Klotho is most abundantly expressed to let them secrete a paracrine factor that controls the functions of neighboring proximal tubules (Kuro-o, 2012). A candidate of such putative paracrine factor is soluble Klotho protein. Klotho protein is a substrate of membrane-anchored secretases and subject to ectodomain shedding, resulting in secretion of the entire extracellular domain, or soluble Klotho, into the extracellular space (Bloch, Sineshchekova, Reichenbach, et al., 2009; Chen, Podvin, Gillespie, et al., 2007). Soluble Klotho protein is detectable in systemic circulation and urine, and reported to exert multiple functions independently of FGF23, including down-regulation of sodium-phosphate co-transporter type 2a (Npt2a) expressed on the apical membrane of proximal tubules (Hu, Shi, Zhang, et al., 2010). NaPi-IIa functions as a major entry gate of phosphate reabsorption and is down-regulated (i.e., decrease in expression and plasma membrane localization) by administration of FGF23 (Shimada, Hasegawa, et al., 2004), which is the basis of phosphatuic activity of FGF23. Several lines of evidence support the hypothesis that soluble Klotho may mediate phosphaturic activity of FGF23. First, injection of FGF23 increased soluble Klotho protein in the blood and urine within 3 h (Takenaka, Watanabe, Inoue, et al., 2013). Second, soluble Klotho protein was detectable in the proximal tubular fluid (Hu et al., 2010). Third, cultured proximal tubular cells transported soluble Klotho protein from the basolateral side to the apical side through transcellular transcytosis (Hu et al., 2010). Lastly, injection of recombinant soluble Klotho induced phosphaturia not only in wild-type mice but also in *Fgf23*$^{-/-}$ mice (Hu et al., 2010). Recombinant soluble Klotho protein was reported to remove glycans of Npt2a through its putative glycosidase activity in cultured renal tubular cells and in brush border membrane vesicles, which promoted internalization and degradation of Npt2a (Hu et al., 2010). These findings suggest that soluble Klotho may be a phosphaturic factor and that FGF23 may function as a hormone that stimulates release of soluble Klotho protein from distal convoluted tubules to induce phosphaturia. This hypothesis remains to be proven.

Another unsolved question on the FGF23-Klotho endocrine axis was the mechanism by which osteoblasts/osteocytes sense phosphate intake to induce secretion and expression of FGF23. A possible hypothesis is that these cells may express a putative "phosphate-sensing receptor" that detects postprandial increase in the blood phosphate levels and transmits signals culminating in secretion and expression of FGF23. However, such "phosphate-sensing receptor" has not been identified yet. In addition, some studies provided data that contradicted this hypothesis. First, not only serum phosphate but also serum calcium levels were correlated with serum FGF23 levels (Quinn, Thomsen, Pang, et al., 2013). Second, increase in serum phosphate levels failed to increase serum FGF23 in the presence of hypocalcemia (Quinn et al., 2013; Rodriguez-Ortiz, Lopez, Munoz-Castaneda, et al., 2012). Conversely, increase in serum calcium levels failed to increase FGF23 in the presence of hypophosphatemia (Quinn et al., 2013). These findings suggest that both phosphate and calcium may participate in the regulation of FGF23 secretion and expression cooperatively. Akiyama et al. recently reported that it was not phosphate or calcium but calciprotein particles that induced FGF23 expression in cultured osteoblasts (Akiyama, Miura, Hayashi, et al., 2019). Calciprotein particles (CPPs) are defined as colloidal nanoparticles composed of solid-phase calcium-phosphate and serum protein fetuin-A (Heiss, DuChesne, Denecke, et al., 2003). Because blood is a supersaturated solution in terms of calcium and phosphate ions, even a modest and transient postprandial increase in phosphate concentration in the blood can trigger precipitation of amorphous calcium phosphate. These precipitates are quickly adsorbed by fetuin-A, which has enormous capacity to bind to calcium phosphate, and prevented from growing to large crystals. As a result, nascent CPPs are generated consisting of a single fetuin-A molecule laden with amorphous calcium-phosphate precipitates, which are called calciprotein monomers (CPMs). CPMs spontaneously undergo self-aggregation and transition of calcium phosphate from the amorphous phase to the crystalline phase over time to become primary and secondary CPPs, respectively (Fig. 2) (Kuro-o, 2019). Among these CPPs with different physical properties, CPMs were identified as the most potent inducer of FGF23 expression (Akiyama et al., 2019). On the other hand, secondary CPPs containing crystalline calcium phosphate were less potent in inducing FGF23 expression in cultured osteoblasts, but reported to have the ability to induce calcification in cultured vascular smooth muscle cells (Aghagolzadeh, Bachtler, Bijarnia, et al., 2016) and trigger innate immune responses in cultured macrophages (Smith, Hanssen, McMahon, et al., 2013). Moreover, recent clinical studies indicated that blood CPP levels were correlated with parameters of vascular calcification (coronary artery calcification score), vascular stiffness (pulse wave velocity), and inflammation (hs-CRP) (Hamano, Matsui, Mikami, et al., 2010; Smith, Ford, Tomlinson, et al., 2012). Considering the activity of CPPs observed in cultured cells, one may speculate that these correlations identified in the clinical studies may be actually a causation. Namely, CPPs may be a causative agent of arteriosclerosis and chronic non-infectious inflammation. In this regard, the ability of CPMs to induce FGF23 expression may be of physiological significance in the regulation of circulating FGF23 levels. FGF23 can be viewed as a negative regulator of CPP formation, because

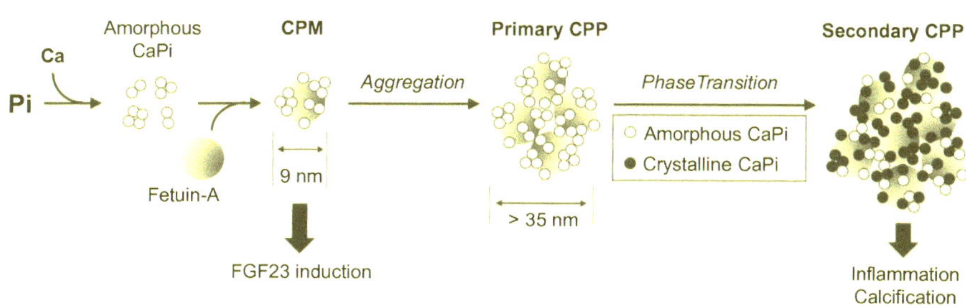

Fig. 2 Formation and maturation of calciprotein particles (CPPs). The diameter of calciprotein monomer (CPM) is around 9 nm. Primary CPPs are composed of amorphous calcium phosphate (CaPi) with a diameter of 35 nm or over. Secondary CPPs contain crystalline CaPi and larger than primary CPPs.

FGF23 reduces blood phosphate levels by its phosphaturic activity and suppresses calcium absorption in the digestive tract by its counterregulatory activity for 1,25-dihydroxyvitamin D_3. Thus, the ability of CPMs to induce FGF23 expression likely contributes to a negative feedback loop that prevents excess formation of CPMs, eventually reducing the risk for formation of secondary CPPs that may accelerate aging as discussed later. However, the mechanism by which osteoblasts sense CPMs and a signaling pathway(s) that induces FGF23 expression remain to be determined.

5. FGF23 and parathyroid

Besides kidney, parathyroid glands express Klotho endogenously and thus identified as a target organ of FGF23 (Ben-Dov et al., 2007). In parathyroid, FGF23 suppresses production and secretion of parathyroid hormone (PTH) (Ben-Dov et al., 2007), whereas PTH reciprocally induces FGF23 expression (Lavi-Moshayoff, Wasserman, Meir, et al., 2010). In addition, FGF23 lowers circulating 1,25-dihydroxyvitamin D_3 levels, whereas 1,25-dihydroxyvitamin D_3 directly transactivates FGF23 gene expression (Masuyama, Stockmans, Torrekens, et al., 2006). It has been known for decades that PTH stimulates 1,25-dihydroxyvitamin D_3 synthesis in the kidney and that 1,25-dihydroxyvitamin D_3 suppresses PTH secretion (Pazianas & Miller, 2020). Thus, the three hormones that control mineral metabolism, FGF23, PTH, and 1,25-dihydroxyvitamin D_3, regulate each other through negative feedback loops (Fig. 3). FGF23 was shown to suppress secretion and expression of PTH when applied to parathyroid glands *ex vivo* and when injected to rodents (Ben-Dov et al., 2007). However, 1,25-dihydroxyvitamin D_3 and calcium appear predominate over FGF23 as negative regulators of PTH, because mice lacking FGF23 showed low PTH, high calcium, and high 1,25-dihydroxyvitamin D_3 (Stubbs, Liu, Tang, et al., 2007).

Like FGF23, PTH exerts phosphaturic activity by suppressing phosphate reabsorption at renal proximal tubules. In fact, serum levels of PTH, but

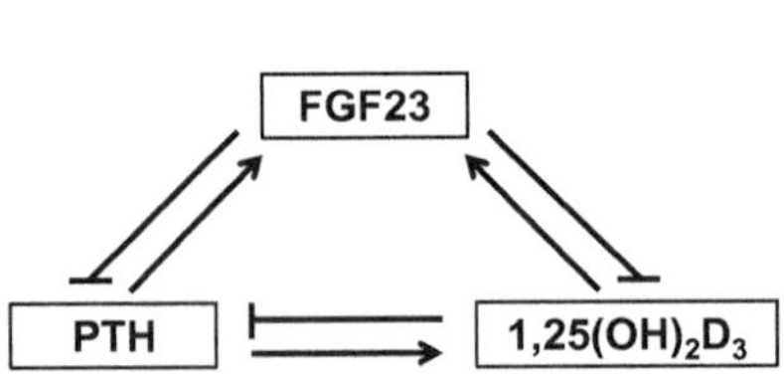

Fig. 3 Negative feedback loops composed of FGF23, parathyroid hormone (PTH), and active vitamin D (1,25(OH)$_2$D$_3$).

barely FGF23, were increased transiently during the course of postprandial increase in serum phosphate levels being restored to the base line within a few hours (Akiyama et al., 2019). Of note, the crystal structure of calcium-sensing receptor (CaSR) revealed that the extracellular domain of CaSR had multiple binding sites not only for calcium but also for phosphate. Furthermore, calcium ions stabilize the active state of CaSR, whereas phosphate ions alleviate the inactive conformation (Geng, Mosyak, Kurinov, et al., 2016). These findings suggest that CaSR may function as a "phosphate-sensing receptor" in parathyroid glands to induce PTH secretion in response to postprandial increase in blood phosphate levels. In contrast, FGF23 was significantly increased when mice were placed on high phosphate diet for a few days or longer, which was associated with increase in blood CPP levels (Akiyama et al., 2019). It appears that PTH responds to a short-term (within hours) fluctuation of the blood phosphate concentration, whereas FGF23 is required for a long-term (within days) maintenance of phosphate homeostasis.

6. FGF23 and chronic kidney disease

Chronic kidney disease (CKD) is defined as a state of impaired renal function, regardless of its cause, that persists 3 months or longer. Since CKD occurs not only as a result of kidney diseases such as glomerulonephritis and polycystic kidney disease, but also as a complication of diabetes and hypertension in many cases, CKD is very prevalent in the aging society, affecting more than 10% of the total population (Hill, Fatoba, Oke, et al., 2016). Because CKD is associated with arteriosclerosis, increased cardiovascular events and mortality, CKD is listed as one of the major health problems worldwide. CKD progression can be viewed as a process of decrease in the functional nephron number that occurs during the natural cause of aging and is accelerated by disorders that affect kidney (Kuro-o, 2019). The severity of CKD is classified from stage 1 through the severest stage 5, based on the estimated glomerular filtration rate (eGFR). As the CKD stage advances with decrease in eGFR, disturbed mineral metabolism becomes evident, which is characterized by high FGF23, low 1,25-dihydroxyvitamin D_3, high PTH, and hyperphosphatemia occurring in this order during CKD progression (Fig. 4) (Isakova, Wahl, Vargas, et al., 2011). Thus, hyperphosphatemia is a terminal symptom of CKD observed in stage 4 or 5. Why then are blood phosphate levels maintained within normal range in the earlier stage CKD patients whose nephron number is supposed to be decreasing? It is likely

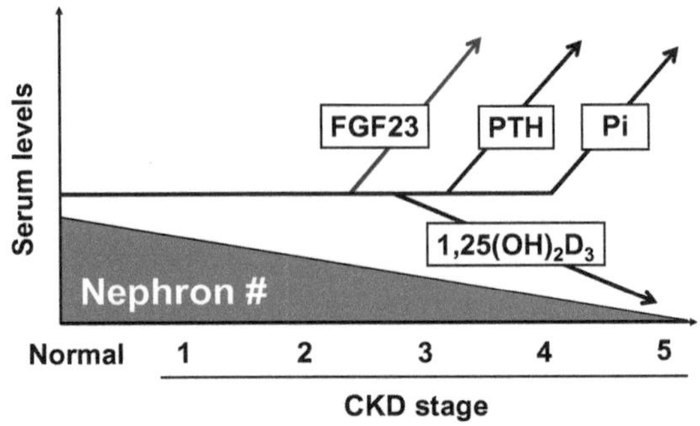

Fig. 4 Changes in circulating levels of FGF23, parathyroid hormone (PTH), active vitamin D ($1,25(OH)_2D_3$), and phosphate (Pi) during the course of CKD progression.

because increase in FGF23, a phosphaturic hormone that increases phosphate excretion per nephron, compensates for the decrease in the nephron number and maintains phosphate homeostasis. In fact, a clinical study indicated that increased phosphaturia is independently associated with decline of eGFR in stage 2–3 CKD patients with normal blood phosphate levels (Santamaria, Diaz-Tocados, Pendon-Ruiz de Mier, et al., 2018). However, the adaptive increase in FGF23 causes decrease in 1,25-dihydroxyvitamin D_3, which increases PTH, resulting in the state of CKD-MBD (mineral-bone disorder). Despite the progressive increase in the two phosphaturic hormones (FGF23 and PTH), hyperphosphatemia ensues in the end-stage CKD when the residual nephron number becomes too low to excrete ingested phosphate into urine. Hyperphosphatemia is a major risk for arteriosclerosis, cardiovascular events, and poor mortality. Thus, phosphate plays a universal role in pathophysiology of CKD and its complications, regardless of the underlying disorders (Kuro-o, 2013a, 2013b).

7. Evolution of Klotho

Life is thought created from the sea about 3.5 billion years ago (Ohno, 1997). Consistent with this notion, composition of sea water is similar to that of human body. In fact, nine out of the 10 most abundant elements (H, O, Na, Cl, S, K, Ca, C, N) are identical between sea water and human body (Bowen, 1979). The exception is phosphorus. Phosphorus is ranked among the top 10 in human body but not in sea water, indicating that organisms have accumulated phosphorus selectively at some time point during evolution. Accumulation of phosphorus occurred ~400 million years ago in the early Devonian era, when bony fishes acquired the skeleton made of calcium phosphate (Hu, Shiizaki, Kuro-o, et al., 2013). Calcium hydroxyapatite is the major bone mineral in vertebrates. Crystalline calcium hydroxyapatite is the hardest biomaterial in vertebrates and much harder than calcium carbonate (Swapna, 2011), which is the major inorganic constituent in the skeleton of invertebrates. Acquisition of the strong bone made of calcium phosphate might be a prerequisite for evolution of terrestrial vertebrates that need to support their body weight on land without the help of buoyancy.

In vertebrates, the extracellular fluid is maintained in a super-saturated condition in terms of calcium and phosphate (Lenton, Nylander, Teixeira, et al., 2015). This metastable state enables vertebrates to determine where to make the bone simply by controlling where to provide a "cue" for nucleation of calcium phosphate. In fact, bone mineralization starts when osteoblasts secrete bone matrix (e.g. collagen and ground substance) into the extracellular space, to which calcium phosphate salts are deposited that eventually grow into crystalline calcium hydroxyapatite (John, 2015). To make this strategy possible, vertebrates must maintain extracellular phosphate concentration within a narrow range to avoid undesired nucleation to occur in the extraosseous tissues. This demand is met by evolving the FGF23-Klotho endocrine system.

8. Phosphate and aging

Mice lacking FGF23 or Klotho fail to maintain phosphate homeostasis due to inability to excrete ingested phosphate into urine, causing hyperphosphatemia and calcification in extraosseous tissues such as arteries (vascular calcification). Besides, they develop complex phenotypes resembling aging as mentioned earlier (Hu et al., 2013). The aging-like phenotypes mostly disappear when phosphate balance is restored by placing these mice on low phosphate diet, leading us to the notion that phosphate accelerates aging (Kuro-o, 2010). Several lines of circumstantial evidence support this notion. For example, an inverse correlation is observed between serum phosphate levels and average life span in different species in mammals (Kuro-o, 2010). In humans, CKD patients in advanced stages develop hyperphosphatemia, vascular calcification, and aging-like phenotypes as seen in mice lacking Klotho (Hruska, Mathew, Lund, et al., 2008). CKD patients exhibit low Klotho expression and are treated effectively by prescription of low phosphate diet and phosphate binders that prevent dietary phosphate from being adsorbed from intestine. Thus, CKD patients and mice lacking Klotho are comparable not only in their signs and symptoms but also in their pathophysiology and response to the treatment. These findings have culminated in the concept that CKD may be viewed as a clinical model for premature aging (Stenvinkel & Larsson, 2013).

Although the mechanism by which phosphate accelerates aging remains elusive, it is noteworthy that high extracellular phosphate is toxic at the cellular level, because the phosphate toxicity may contribute to accelerated aging at the organismal level. Vascular endothelial cells undergo apoptosis when cultured in high phosphate medium (1.4–2.8 mM) (Di Marco, Hausberg, Hillebrand, et al., 2008). Vascular smooth muscle cells undergo phenotypic transition to osteoblastic cells in the high phosphate medium, followed by deposition of calcium phosphate around them (Jono, McKee, Murry, et al., 2000). Of note, increase in phosphate by a few mM can be a cue for precipitation of calcium phosphate in regular tissue culture medium containing ~2 mM calcium and ~1 mM phosphate, raising the possibility that phosphate toxicity might be not due to phosphate *per se* but due to calcium phosphate precipitates. This was indeed the case, because high phosphate medium failed to induce phosphate toxicity in the presence of inhibitors for precipitation of calcium phosphate, such as pyrophosphate or phosphonoformic acid (Sage, Lu, Tintut, et al., 2011; Villa-Bellosta & Sorribas, 2009). Conversely, addition of insoluble calcium phosphate to the medium recapitulated phosphate toxicity without increasing phosphate concentration in the medium (Ewence, Bootman, Roderick, et al., 2008). *In vivo*, however, multiple endogenous inhibitors for precipitation and/or growth of calcium phosphate (e.g. pyrophosphate, matrix gla protein, fetuin-A) are present in tissues, blood, and urine. Despite that, tiny calcium phosphate precipitates indeed appear in the blood under some circumstances (Hamano et al., 2010; Smith et al., 2012), which bind to serum protein fetuin-A and form calciprotein particles (CPPs). Considering that the true culprit of phosphate toxicity is calcium phosphate precipitates, it is intriguing to speculate that correlation found in the clinical studies might actually be causality, and if so, chemicals that inhibit precipitation of calcium phosphate, such as bisphosphonate, citrate, and magnesium, might prevent CPP formation and be useful for the treatment of CKD complications (Kuro-o, 2013b).

9. Concluding remarks

Another FGF-Klotho endocrine system has been identified that has profound impact on aging process. Fibroblast growth factor-21 (FGF21) is a liver-derived hormone that binds to FGFR1c complexed with βKlotho (Kurosu, Choi, Ogawa, et al., 2007; Ogawa, Kurosu, Yamamoto, et al., 2007), which shares ~40% amino acid identity with Klotho (Ito, Kinoshita, Shiraishi, et al., 2000). FGF21 induces responses to stress, including activation of the sympathetic nervous system and the hypothalamus-pituitary--adrenal endocrine axis, through acting on the suprachiasmatic nucleus in hypothalamus where βKlotho is expressed (Bookout, de Groot, Owen, et al., 2013). Importantly, mice overexpressing FGF21 have longer life span than wild-type mice (Zhang, Xie, Berglund, et al., 2012), suggesting that FGF21 may function as an anti-aging hormone. Recent clinical studies indicated that circulating FGF21 levels were increased progressively in CKD patients since early stages (Kohara, Masuda, Shiizaki, et al., 2017; Lin, Zhou, Liu, et al., 2011). Although the mechanism by which CKD increases FGF21 remains to be determined, FGF21 appears to be required to survive CKD, because mice lacking FGF21 showed poorer prognosis than wild-type mice when introduced with CKD (Nakano, Shiizaki, Miura, et al., 2019). However, the high serum FGF21 levels induced adverse effects, including blood pressure dysregulation. CKD patients universally exhibit enhanced sympathetic nerve activity and augmented blood pressure elevating response during physical activity, which are also observed in CKD mice but not in CKD mice lacking FGF21 (Nakano et al., 2019). Thus, increased FGF21 in CKD is a survival response but causes blood pressure dysregulation. Likewise, increased FGF23 in CKD is required to maintain phosphate homeostasis but results in mineral bone disorder. Pathophysiology of CKD may be viewed as a trade-off between benefits (securing survival and maintaining phosphate homeostasis) and risks (blood pressure dysregulation and mineral bone disorder) mediated by FGF-Klotho endocrine axes.

References

Aghagolzadeh, P., Bachtler, M., Bijarnia, R., et al. (2016). Calcification of vascular smooth muscle cells is induced by secondary calciprotein particles and enhanced by tumor necrosis factor-alpha. *Atherosclerosis*, *251*, 404–414.

Akiyama, K., Miura, Y., Hayashi, H., et al. (2019). Calciprotein particles regulate fibroblast growth factor-23 expression in osteoblasts. *Kidney International*, *97*, 702–712.

Azuma, M., Koyama, D., Kikuchi, J., et al. (2012). Promoter methylation confers kidney-specific expression of the Klotho gene. *The FASEB Journal*, *26*, 4264–4274.

Ben-Dov, I. Z., Galitzer, H., Lavi-Moshayoff, V., et al. (2007). The parathyroid is a target organ for FGF23 in rats. *The Journal of Clinical Investigation*, *117*, 4003–4008.

Bloch, L., Sineshchekova, O., Reichenbach, D., et al. (2009). Klotho is a substrate for alpha-, beta- and gamma-secretase. *FEBS Letters*, *583*, 3221–3224.

Bookout, A. L., de Groot, M. H., Owen, B. M., et al. (2013). FGF21 regulates metabolism and circadian behavior by acting on the nervous system. *Nature Medicine*, *19*, 1147–1152.

Bowen, H. J. M. (1979). *Environmental chemistry of the elements*. London: Achademic Press.

Chen, G., Liu, Y., Goetz, R., et al. (2018). Alpha-Klotho is a non-enzymatic molecular scaffold for FGF23 hormone signalling. *Nature, 553*, 461–466.

Chen, C. D., Podvin, S., Gillespie, E., et al. (2007). Insulin stimulates the cleavage and release of the extracellular domain of Klotho by ADAM10 and ADAM17. *Proceedings of the National Academy of Sciences of the United States of America, 104*, 19796–19801.

Di Marco, G. S., Hausberg, M., Hillebrand, U., et al. (2008). Increased inorganic phosphate induces human endothelial cell apoptosis in vitro. *American Journal of Physiology. Renal Physiology, 294*, F1381–F1387.

Ewence, A. E., Bootman, M., Roderick, H. L., et al. (2008). Calcium phosphate crystals induce cell death in human vascular smooth muscle cells: A potential mechanism in atherosclerotic plaque destabilization. *Circulation Research, 103*, e28–e34.

Faul, C., Amaral, A. P., Oskouei, B., et al. (2011). FGF23 induces left ventricular hypertrophy. *The Journal of Clinical Investigation, 121*, 4393–4408.

Geng, Y., Mosyak, L., Kurinov, I., et al. (2016). Structural mechanism of ligand activation in human calcium-sensing receptor. *eLife, 5*, e13662.

Hamano, T., Matsui, I., Mikami, S., et al. (2010). Fetuin-mineral complex reflects extraosseous calcification stress in CKD. *Journal of the American Society of Nephrology, 21*, 1998–2007.

Heiss, A., DuChesne, A., Denecke, B., et al. (2003). Structural basis of calcification inhibition by alpha 2-HS glycoprotein/fetuin-A. Formation of colloidal calciprotein particles. *The Journal of Biological Chemistry, 278*, 13333–13341.

Hill, N. R., Fatoba, S. T., Oke, J. L., et al. (2016). Global prevalence of chronic kidney disease—A systematic review and meta-analysis. *PLoS One, 11*, e0158765.

Hruska, K. A., Mathew, S., Lund, R., et al. (2008). Hyperphosphatemia of chronic kidney disease. *Kidney International, 74*, 148–157.

Hu, M. C., Shi, M., Zhang, J., et al. (2010). Klotho: A novel phosphaturic substance acting as an autocrine enzyme in the renal proximal tubule. *The FASEB Journal, 24*, 3438–3450.

Hu, M. C., Shiizaki, K., Kuro-o, M., et al. (2013). Fibroblast growth factor 23 and klotho: Physiology and pathophysiology of an endocrine network of mineral metabolism. *Annual Review of Physiology, 75*, 503–533.

Isakova, T., Wahl, P., Vargas, G. S., et al. (2011). Fibroblast growth factor 23 is elevated before parathyroid hormone and phosphate in chronic kidney disease. *Kidney International, 79*, 1370–1378.

Ito, S., Kinoshita, S., Shiraishi, N., et al. (2000). Molecular cloning and expression analyses of mouse betaklotho, which encodes a novel Klotho family protein. *Mechanisms of Development, 98*, 115–119.

John, H. (2015). Parathyroid hormone, calcitonin, calcium and phosphate metabolism, vitamin D, bone, and teeth. In *Guyton and hall textbook of medical physiology* (13th ed.). Saunders.

Jono, S., McKee, M. D., Murry, C. E., et al. (2000). Phosphate regulation of vascular smooth muscle cell calcification. *Circulation Research, 87*, E10–E17.

Kamemori, M., Ohyama, Y., Kurabayashi, M., et al. (2002). Expression of Klotho protein in the inner ear. *Hearing Research, 171*, 103–110.

Kawaguchi, H., Manabe, N., Miyaura, C., et al. (1999). Independent impairment of osteoblast and osteoclast differentiation in klotho mouse exhibiting low-turnover osteopenia. *The Journal of Clinical Investigation, 104*, 229–237.

Kohara, M., Masuda, T., Shiizaki, K., et al. (2017). Association between circulating fibroblast growth factor 21 and mortality in end-stage renal disease. *PLoS One, 12*, e0178971.

Kuro-o, M. (2010). A potential link between phosphate and aging—Lessons from Klotho-deficient mice. *Mechanisms of Ageing and Development, 131*, 270–275.

Kuro-o, M. (2012). Klotho in health and disease. *Current Opinion in Nephrology and Hypertension, 21*, 362–368.

Kuro-o, M. (2013a). A phosphate-centric paradigm for pathophysiology and therapy of chronic kidney disease. *Kidney International. Supplement*, *3*, 420–426.
Kuro-o, M. (2013b). Klotho, phosphate and FGF-23 in ageing and disturbed mineral metabolism. *Nature Reviews. Nephrology*, *9*, 650–660.
Kuro-o, M. (2018). Ageing-related receptors resolved. *Nature*, *553*, 409–410.
Kuro-o, M. (2019). The Klotho proteins in health and disease. *Nature Reviews. Nephrology*, *15*, 27–44.
Kuro-o, M., Hanaoka, K., Hiroi, Y., et al. (1995). Salt-sensitive hypertension in transgenic mice overexpressing Na(+)-proton exchanger. *Circulation Research*, *76*, 148–153.
Kuro-o, M., Matsumura, Y., Aizawa, H., et al. (1997). Mutation of the mouse klotho gene leads to a syndrome resembling ageing. *Nature*, *390*, 45–51.
Kurosu, H., Choi, M., Ogawa, Y., et al. (2007). Tissue-specific expression of betaKlotho and fibroblast growth factor (FGF) receptor isoforms determines metabolic activity of FGF19 and FGF21. *The Journal of Biological Chemistry*, *282*, 26687–26695.
Kurosu, H., Ogawa, Y., Miyoshi, M., et al. (2006). Regulation of fibroblast growth factor-23 signaling by klotho. *The Journal of Biological Chemistry*, *281*, 6120–6123.
Kurosu, H., Yamamoto, M., Clark, J. D., et al. (2005). Suppression of aging in mice by the hormone Klotho. *Science*, *309*, 1829–1833.
Lavi-Moshayoff, V., Wasserman, G., Meir, T., et al. (2010). PTH increases FGF23 gene expression and mediates the high-FGF23 levels of experimental kidney failure: A bone parathyroid feedback loop. *American Journal of Physiology. Renal Physiology*, *299*, F882–F889.
Lenton, S., Nylander, T., Teixeira, S. C., et al. (2015). A review of the biology of calcium phosphate sequestration with special reference to milk. *Dairy Science & Technology*, *95*, 3–14.
Lin, Z., Zhou, Z., Liu, Y., et al. (2011). Circulating FGF21 levels are progressively increased from the early to end stages of chronic kidney diseases and are associated with renal function in Chinese. *PLoS One*, *6*, e18398.
Masuyama, R., Stockmans, I., Torrekens, S., et al. (2006). Vitamin D receptor in chondrocytes promotes osteoclastogenesis and regulates FGF23 production in osteoblasts. *The Journal of Clinical Investigation*, *116*, 3150–3159.
Nagai, T., Yamada, K., Kim, H. C., et al. (2003). Cognition impairment in the genetic model of aging klotho gene mutant mice: A role of oxidative stress. *The FASEB Journal*, *17*, 50–52.
Nakano, T., Shiizaki, K., Miura, Y., et al. (2019). Increased fibroblast growth factor-21 in chronic kidney disease is a trade-off between survival benefit and blood pressure dysregulation. *Scientific Reports*, *9*, 19247.
Ogawa, Y., Kurosu, H., Yamamoto, M., et al. (2007). βKlotho is required for metabolic activity of fibroblast growth factor 21. *Proceedings of the National Academy of Sciences of the United States of America*, *104*, 7432–7437.
Ohno, S. (1997). The reason for as well as the consequence of the Cambrian explosion in animal evolution. *Journal of Molecular Evolution*, *44*(Suppl. 1), S23–S27.
Pazianas, M., & Miller, P. D. (2020). Current understanding of mineral and bone disorders of chronic kidney disease and the scientific grounds on the use of exogenous parathyroid hormone in its management. *Journal of Bone Metabolism*, *27*, 1–13.
Quinn, S. J., Thomsen, A. R., Pang, J. L., et al. (2013). Interactions between calcium and phosphorus in the regulation of the production of fibroblast growth factor 23 in vivo. *American Journal of Physiology. Endocrinology and Metabolism*, *304*, E310–E320.
Rodriguez-Ortiz, M. E., Lopez, I., Munoz-Castaneda, J. R., et al. (2012). Calcium deficiency reduces circulating levels of FGF23. *Journal of the American Society of Nephrology*, *23*, 1190–1197.
Sage, A. P., Lu, J., Tintut, Y., et al. (2011). Hyperphosphatemia-induced nanocrystals upregulate the expression of bone morphogenetic protein-2 and osteopontin genes in mouse smooth muscle cells in vitro. *Kidney International*, *79*, 414–422.

Santamaria, R., Diaz-Tocados, J. M., Pendon-Ruiz de Mier, M. V., et al. (2018). Increased Phosphaturia accelerates the decline in renal function: A search for mechanisms. *Scientific Reports*, *8*, 13701.

Shimada, T., Hasegawa, H., Yamazaki, Y., et al. (2004). FGF-23 is a potent regulator of vitamin D metabolism and phosphate homeostasis. *Journal of Bone and Mineral Research*, *19*, 429–435.

Shimada, T., Kakitani, M., Yamazaki, Y., et al. (2004). Targeted ablation of Fgf23 demonstrates an essential physiological role of FGF23 in phosphate and vitamin D metabolism. *The Journal of Clinical Investigation*, *113*, 561–568.

Smith, E. R., Ford, M. L., Tomlinson, L. A., et al. (2012). Phosphorylated fetuin-A-containing calciprotein particles are associated with aortic stiffness and a procalcific milieu in patients with pre-dialysis CKD. *Nephrology, Dialysis, Transplantation*, *27*, 1957–1966.

Smith, E. R., Hanssen, E., McMahon, L. P., et al. (2013). Fetuin-A-containing calciprotein particles reduce mineral stress in the macrophage. *PLoS One*, *8*, e60904.

Stenvinkel, P., & Larsson, T. E. (2013). Chronic kidney disease: A clinical model of premature aging. *American Journal of Kidney Diseases*, *62*, 339–351.

Stubbs, J. R., Liu, S., Tang, W., et al. (2007). Role of hyperphosphatemia and 1,25-dihydroxyvitamin D in vascular calcification and mortality in fibroblastic growth factor 23 null mice. *Journal of the American Society of Nephrology*, *18*, 2116–2124.

Swapna, M. (2011). *Applied mineralogy: Applications in industry and environment*. Netherlands: Springer.

Takenaka, T., Watanabe, Y., Inoue, T., et al. (2013). Fibroblast growth factor 23 enhances renal klotho abundance. *Pflügers Archiv*, *465*, 935–943.

Tsujikawa, H., Kurotaki, Y., Fujimori, T., et al. (2003). Klotho, a gene related to a syndrome resembling human premature aging, functions in a negative regulatory circuit of vitamin D endocrine system. *Molecular Endocrinology*, *17*, 2393–2403.

Urakawa, I., Yamazaki, Y., Shimada, T., et al. (2006). Klotho converts canonical FGF receptor into a specific receptor for FGF23. *Nature*, *444*, 770–774.

Villa-Bellosta, R., & Sorribas, V. (2009). Phosphonoformic acid prevents vascular smooth muscle cell calcification by inhibiting calcium-phosphate deposition. *Arteriosclerosis, Thrombosis, and Vascular Biology*, *29*, 761–766.

White, K. E., Carn, G., Lorenz-Depiereux, B., et al. (2001). Autosomal-dominant hypophosphatemic rickets (ADHR) mutations stabilize FGF-23. *Kidney International*, *60*, 2079–2086.

White, K. E., Evans, W. E., O'Rlordan, J. L. H., et al. (2000). Autosomal dominant hypophosphataemic rickets is associated with mutations in FGF23. *Nature Genetics*, *26*, 345–348.

Winters, R. W., Graham, J. B., Williams, T. F., et al. (1958). A genetic study of familial hypophosphatemia and vitamin D resistant rickets with a review of the literature. *Medicine (Baltimore)*, *37*, 97–142.

Yu, X., Ibrahimi, O. A., Goetz, R., et al. (2005). Analysis of the biochemical mechanisms for the endocrine actions of fibroblast growth factor-23. *Endocrinology*, *146*, 4647–4656.

Zhang, Y., Xie, Y., Berglund, E. D., et al. (2012). The starvation hormone, fibroblast growth factor-21, extends lifespan in mice. *eLife*, *1*, e00065.

CHAPTER THIRTEEN

Aging and sex hormones in males

Maria Chiara Decaroli[a,b], Sara De Vincentis[a,b], and Vincenzo Rochira[a,b,*]

[a]Unit of Endocrinology, Department of Biomedical, Metabolic and Neural Sciences, University of Modena and Reggio Emilia, Modena, Italy

[b]Department of Medical Specialties, Azienda Ospedaliero-Universitaria of Modena, Ospedale Civile di Baggiovara, Modena, Italy

*Corresponding author: e-mail address: vincenzo.rochira@unimore.it

Contents

Abstract

Over the past two decades several large cohort studies have been performed to disclose the changes of sex hormone in elderly and their clinical significance. Beyond the decline of total testosterone, aging is accompanied by a sex hormone binding globulin (SHBG) increase, a steeper free testosterone decline, while gonadotropins may be increased or inappropriately normal, with important contribution of comorbidities (e.g., obesity) to these changes. Actually, it has become firm the concept that the biochemical finding of testosterone deficiency alone is not sufficient for diagnosing hypogonadism in older men. The definition of late-onset hypogonadism (LOH) includes low serum testosterone

Vitamins and Hormones, Volume 115
ISSN 0083-6729
https://doi.org/10.1016/bs.vh.2020.12.014

levels coupled with signs and symptoms related to hypogonadism. Indeed, the combination of multiple factors all contributing to the testosterone decline, with other concurrent comorbidities further overlapping, makes the clinical correlates of LOH highly heterogeneous. For all these reasons both the diagnosis and the therapeutic management of LOH, especially the decision about starting testosterone replacement treatment, remain challenging.

1. Introduction

Aging and hormones are strictly related. The endocrine system undergoes different age-related changes that have been partly characterized. For example, at central level the thyrotropin hormone (TSH) shows an increasing trend upon aging, whereas the secretion tends to progressively decrease during lifespan for other pituitary hormones such as growth hormone (GH) and its mediator insulin-like growth factor-1 (IGF-1) (van den Beld et al., 2018; Veldhuis, 2013). Aging has a substantial impact even on the hypothalamic-pituitary-gonadal axis resulting in the final loss of function of gonads in a gender-specific manner. The menopause represents the moment of the loss of gonadal function in females, characterized by both physiological, drastic drop of estradiol and significant increase in gonadotropins levels (Roeca, Al-Safi, & Santoro, 2000; Veldhuis, 2013). Unlike for females, the male reproductive function gradually worsens with advancing age as a consequence of the progressive reduction of testicular activity in concomitance with a moderate increase of gonadotropins levels (van den Beld et al., 2018; Veldhuis, 2013). However, the accumulation of comorbidities with aging may be a trigger for the development also of hypogonadotropic hypogonadism in men through a change in the hypothalamic set-point (Araujo et al., 2004; Rochira, 2017; Tajar et al., 2010). Across lifespan the sex hormones production is completely different comparing men to women also before aging, starting from puberty (Decaroli & Rochira, 2017). In adulthood, serum testosterone is about 10-fold greater in men than in women and it starts to progressively decline after the fourth decade of life (Decaroli & Rochira, 2017; Kaufman, Lapauw, Mahmoud, T'Sjoen, & Huhtaniemi, 2019). On the other hand, serum estradiol widely varies according to the different phases of menstrual cycle in women during fertile period (Roeca et al., 2000), always being more than 10-fold higher than in men (Decaroli & Rochira, 2017). After 50s, serum estradiol levels decline in both sexes, with a sudden reduction in females at menopause (Roeca et al., 2000) and a progressive reduction in males (Decaroli & Rochira, 2017).

The extent, clinical significance, and treatment options of age-related deterioration of the hypothalamic–pituitary–testicular axis are currently a topical and controversial issue. This chapter presents an overview of the current knowledge on reproductive hormones changes in aging men, the factors driving and modulating these changes and their clinical consequences.

2. Influence of aging on testosterone and its metabolites: Findings from the literature

The serum testosterone mean levels in male population decline with aging, starting from the third and fourth decades of life and going on gradually across lifespan (Kaufman et al., 2019). The reduction of circulating testosterone is the result of the combination of multiple concurrent factors: beyond the aging process per se, there are several lifestyle changes and comorbidities that may contribute in reducing testosterone levels. The countless possible different combinations of all these factors may also explain the high inter-individual heterogeneity of changes in serum testosterone with advancing age.

Changes of serum testosterone in aging men have been investigated since '70s. First data were obtained from retrospective and cross-sectional studies, mainly based on inpatients with a poor health status (Davidson et al., 1983; Stearns et al., 1974), thus not representative for the general aging population (Feldman et al., 2002; Gray, Berlin, McKinlay, & Longcope, 1991; Gray, Feldman, McKinlay, & Longcope, 1991; Seftel, 2006). These older studies reported controversial data, some of them finding no significant age-related decline in circulating sex steroids upon aging (Davidson et al., 1983; Kaufman & Vermeulen, 2005; Stearns et al., 1974). Afterward, a substantial amount of new data has become available from different large cohorts and the decline of sex steroids was proved thanks to properly-designed and prospective studies. All these recent studies differed each other for several aspects mainly related to the study design (e.g., cohort selection, duration of the study, primary endpoints) and the methodological approach (e.g., laboratory assay used for steroids assessment), making the comparison of findings not always easy. Of them, the European Male Aging Study (EMAS) is one of the most important and large cohort studies, involving eight European centers. The EMAS estimated a cross-sectional testosterone decline around 0.40% per year in men aged between 40 and 79 years (Wu et al., 2008). This estimation raised to 0.54% when the decline was longitudinally assessed during a 4.3-year follow-up (Camacho et al., 2013). Overall, notwithstanding the abovementioned methodological differences between large prospective studies, the majority of them found a gradual decline of total testosterone levels upon aging. Considering longitudinal studies, it has been estimated that serum total testosterone declines with a rate of about 0.3%–1.8% per year (Camacho et al., 2013; Harman, Metter, Tobin, Pearson, & Blackman, 2001; Lapauw et al., 2008; Liu et al., 2007; Mohr, Bhasin, Link, O'Donnell, & McKinlay, 2006; Morley et al., 1997; Shi, Araujo, Martin, O'Loughlin, & Wittert, 2013; Travison, Araujo, O'Donnell, Kupelian, & McKinlay, 2007; Zmuda et al., 1997), corresponding approximately to a reduction of 3.2–3.5 ng/dL (0.110–0.121 nmol/L) per year (Decaroli & Rochira, 2017; Kaufman et al., 2019). In line with this observation, even the prevalence of biochemical hypogonadism seems to increase with advancing age, being estimated around 20%–30% in men over 60 years with an increasing trend through decades (Araujo et al., 2004; Harman et al., 2001; Mulligan, Frick, Zuraw, Stemhagen, & McWhirter, 2006; Schneider et al., 2009). However, the definition of the prevalence of male hypogonadism in aging men remains tricky, facing with several unresolved controversies such as different cutoffs suggested for total testosterone (for details see Section 5).

Findings on changes of free testosterone upon aging are more uniform than what reported for total testosterone. The age-related decline of free testosterone is unequivocally proved, generally of the order of 1.5%–2% per year (Gray, Feldman, et al., 1991; Wu et al., 2008; Yeap et al., 2007). This reduction is also due to the constant finding of the age-related increase of sex hormone-binding globulin (SHBG) by around 3% per year (Antonio et al., 2016) as discussed below.

Besides testosterone, age-related changes also occur in circulating levels of its two active metabolites, estradiol and dihydrotestosterone (DHT). Importantly, however, data on age-related changes in estradiol and DHT are inconsistent mainly due to poor accuracy of the immunoassays in older studies (Jasuja et al., 2013) since only the results of mass spectrometry-based measurements can be considered reliable (Ohlsson et al., 2013; Swerdloff, Dudley, Page, Wang, & Salameh, 2017). Indeed, although several longitudinal studies have shown that serum estradiol declines in parallel with serum testosterone in men (Feldman et al., 2002; Huhtaniemi et al., 2012; Wu et al., 2010), other studies have reported the concentrations of estradiol to be relatively stable (Handelsman et al., 2015; Yeap et al., 2018) or even increased in aging men (Hsu et al., 2015). Probably a higher conversion rate of testosterone to estradiol in adipose tissue, together with increased SHBG levels, may explain in turn an increase in estradiol levels in many but not all patients (Rochira & Carani, 2017). Actually, the hypothesis of age-dependent decrease of estradiol, as well as testosterone, seems to be the most accurate, supported also by the decrease of free estradiol in aging men (Decaroli & Rochira, 2017; Orwoll et al., 2006; Rochira & Carani, 2017). DHT circulating levels are 7- to 10-fold lower than those of testosterone, but they are a poor indicator of the action of this hormone that it is mainly determined by in situ conversion of testosterone in DHT (Swerdloff et al., 2017). Available data suggest that DHT levels change minimally before the age of 65 years but they decrease thereafter by 0.5% per year (Handelsman et al., 2015).

It should be considered that a high degree of heterogeneity of sex steroids changes is present accounting for a different scenario when we move from population studies to the single patient. In particular, individual differences involving factors such as concomitant diseases, adiposity, enzyme expression and activity (e.g., aromatase), androgen receptor activity and expression (e.g., length of CAG repeats) play a crucial role in the determination of the amount of circulating sex steroids and binding globulin.

3. Mechanisms underlying age-related testosterone decline

According to the site where the hypothalamus–pituitary–testis axis is impaired, two principal forms of hypogonadism can be distinguished (Table 1): (a) primary hypogonadism (or hypergonadotropic hypogonadism), in which testis function is disrupted, testosterone production is reduced and, as a consequence, gonadotropins levels increase and (b) secondary hypogonadism (or hypogonadotropic hypogonadism) (Tajar et al., 2010), in which hypothalamic and/or pituitary failure leads to low gonadotropins levels, resulting in low testicular stimulation and testosterone synthesis (Bhasin et al., 2018; Petak, Nankin, Spark, Swerdloff, & Rodriguez-Rigau, 2002; Rochira, 2017).

Aging per se seems to have only minor effects on function at the central level of the axis. In aged men it has been observed an increased luteinizing hormone (LH) pulse and frequency, probably as a result of the reduced hypothalamic gonadotropin-releasing hormone (GnRH) secretion, with reduced gonadal feedback and consequent increase in LH secretion (Veldhuis et al., 2009). Thus, the gradual loss of testicular capacity to produce testosterone is initially counteracted by increased gonadotropin

Table 1 Classification of male hypogonadism.

Classification of hypogonadism	Site of impaired function	Hormonal pattern
Primary (or hypergonadotropic) hypogonadism	Testis	↓ Testosterone, ↑ LH
Compensated hypogonadism	Testis	↔ Testosterone, ↑ LH
Secondary (or hypogonadotropic) hypogonadism	Hypothalamic–pituitary unit	↓ Testosterone, ↓ LH
Mixed forms of hypogonadism	Testis and hypothalamic–pituitary unit	↓ Testosterone, ↔ LH

T, testosterone; LH, luteinizing hormone; ↓, reduced; ↑, increased; ↔, in the normal range.

secretion, defining a condition of "compensated hypogonadism" (high LH, normal testosterone) (Table 1). This is a relatively common condition with an estimated prevalence of 9.4% in the EMAS population (Tajar et al., 2010; Wu et al., 2010). When the stimulated testicular reserve capacity is exhausted, the men develop the typical age-related primary hypogonadism (high LH, low testosterone) (Table 1).

3.1 Changes of SHBG levels

Circulating levels of SHBG are affected by several factors and/or comorbidities that can interplay in opposite manners (Table 2). For example, SHBG levels tend to increase with aging of about 2.7% per year (Harman et al., 2001; Lapauw et al., 2008; Travison, Araujo, Kupelian, O'Donnell, & McKinlay, 2007). Mechanisms leading to the increase of SHBG levels with age have not been fully elucidated yet. The long-held concept that insulin negatively regulates hepatic SHBG production through an unknown mechanism has been overcome. More recent in vitro research supports that the aging-associated increase in levels of the anti-inflammatory cytokine adiponectin may stimulate the SHBG production (Lanfranco, Zitzmann, Simoni, & Nieschlag, 2004; Simó, Sáez-López, Barbosa-Desongles, Hernández, & Selva, 2015). Furthermore, a decreased inhibitory effect of the IGF-1 action has been proposed as another cause of increased SHBG with advancing age (Vermeulen, Kaufman, & Giagulli, 1996). The accumulation of comorbidities during aging represents, however, the most important factor influencing circulating levels of SHBG since several diseases significantly impact on their synthesis (Rochira, 2017). In summary, whatever the underlying mechanism, the overall increase of SHBG levels with age might modify in

Table 2 Mechanisms and most common causes of male hypogonadism related to aging.

Districts and factors involved	Mechanism	Common causes
Testes aging	Reduced Leydig cell function and number, LH resistance (uncertain)	Aging
Reduced pituitary LH secretion	Change in the hypothalamic-pituitary set point as an adaptive condition aiming at sparing energy	Poor health status, frailty, comorbidity burden (accumulation of chronic diseases)
Increased aromatase activity/ expression	Increased production of E2 (increased E2 to T ratio) and E2-related inhibition of Gns production	Obesity, fat redistribution
Altered SHBG synthesis	Changes in the rate of circulating free T	Concomitant metabolic diseases (HIV, liver diseases, etc.); age-related decrease of IGF-1; increase of cytokines production
Drugs	Increase of PRL (e.g., psychotropic drugs), inhibition of Gns (e.g., opioids) and (blockade of LH receptor (e.g., therapy for prostate cancer); changes in SHBG, decrease of T synthesis, etc.	Several drugs, the use of whom increases with age

E2, estradiol; T, testosterone; Gns, gonadotropins; PRL, prolactin.

percentage the amount of different testosterone fractions with a consequent increase of the SHBG-bound serum testosterone which is not bioavailable (Vermeulen et al., 1996).

3.2 Impairment of testicular function

Most of aged patients, but not all, present a reduced testicular function. In this case, testosterone production decreases in face of increasing gonadotropins levels, indicating primary testicular failure. Several factors have been identified as possible responsible for the aging-related primary impairment of testicular function (Table 1). There is in vivo evidence of an aging-related decline of Leydig cell steroidogenesis. A few studies have investigated the endogenous steroid profile in testis tissue of young and elderly men suggesting that aging reduces the activity of almost all steroidogenic enzymes with no specific age-sensitive steps (Kaufman et al., 2019). Probably both a decreased number of functional LH receptors and an impaired responsiveness of aged Leydig cells to LH stimulation can contribute to a reduced testicular steroidogenesis (Kaufman et al., 2019) (Table 2). Furthermore, different endogenous (e.g., reduced Leydig cell antioxidant capacity) and extrinsic factors (e.g., environmental factors, drug exposure) may lead to an overall increased oxidative stress, contributing to the age-related decline of Leydig cell function (Table 2). Finally, the reduced steroidogenic capacity of Leydig cell can be also caused by a reduction of the total number of Leydig cells (Neaves, Johnson, Porter, Parker, & Petty, 1984) and by a decreased testicular perfusion due to atherosclerotic alterations (Table 2). However, available data on this issue are inconclusive and mostly obtained from heterogeneous samples.

3.3 Gonadotropin secretion

As part of the hypothalamic–pituitary–testicular axis, even the gonadotropin secretion changes as the male ages. This is due on one hand to chronological aging, but mostly to aging-associated comorbidities (Corona et al., 2016; Decaroli & Rochira, 2017; Rochira, 2017) (Table 2). Several comorbidities in fact, especially obesity and overweight, can take part in this sequence of events, causing predominantly secondary hypogonadism where multiple mechanisms impair the hypothalamic–pituitary function (Corona, Vignozzi, Sforza, Mannucci, & Maggi, 2015; Rigon, Ronsoni, Hohl, & Sande-Lee, 2019). In general, secondary hypogonadism occurs as an adaptive attempt for sparing energy in presence of poor health conditions (Corona et al., 2020, 2016; Rochira, 2017) (Table 2, Fig. 1). In many cases, both the central and testicular impairment overlap making the classification of this forms less evident from a clinical standpoint (Kaufman et al., 2019) (Table 1) (for further details see below paragraphs).

3.4 Aromatase conversion of testosterone to estradiol

The aromatase enzyme determines conversion of serum testosterone in estradiol and, in general, of androgens in estrogens (Rochira & Carani, 2017; Rochira et al., 2000). Aromatase is widely expressed in a large number of tissues and, in particular, in the adipose tissue, that represents the second source of estrogens in men other than the testis (Rochira & Carani, 2017; Rochira, Kara, & Carani, 2015; Rochira et al., 2000). Estrogens strongly

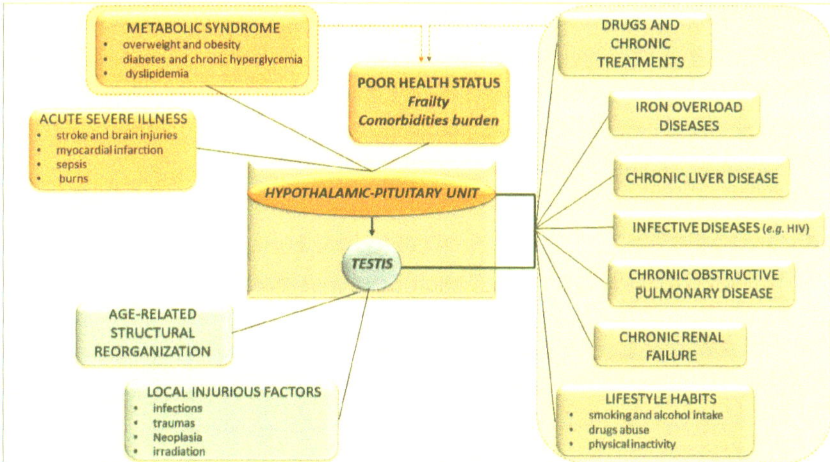

Fig. 1 Principal causes of hypogonadism on-set in aging men. Factors impairing mainly the hypothalamus-pituitary function are in orange, factors impairing mainly the gonadal function are in light blue, while factors acting at both levels are in gray. Chronic comorbidities, drugs and bad lifestyle habits all concur in determining a poor health status, that can lower the hypothalamic set-point of endogenous testosterone levels.

inhibit the secretion of gonadotropins (Rochira et al., 2000, 2006), acting at both hypothalamus and pituitary levels, thus an over-expression of aromatase can lead to hypogonadotropic hypogonadism through the increased production of estrogens (Rochira & Carani, 2017; Rochira et al., 2006) (Table 2). Therefore, the excess of body fat (and especially visceral fat) that characterizes obesity and/or the alterations of body fat distribution such as in obesity (Cohen, 2001; Kelly & Jones, 2015; Russell & Grossmann, 2019) and in the Human Immunodeficiency Virus (HIV)-related lipodystrophy (Rochira & Guaraldi, 2014) is associated to over-expression of aromatase and increased production of estrogens, this latter leading to inhibition of gonadotropins incretion (Rochira & Guaraldi, 2014; Rochira et al., 2000, 2006).

3.5 Drugs

Many drugs can determine the onset of hypogonadism through both central and/or testicular impairment (Table 2). Considering drugs acting directly on the hypothalamus-pituitary unit, the GnRH agonists provide constant serum levels of GnRH that overcome the pulsatility of endogenous GnRH and consequently downregulate GnRH receptors on the anterior pituitary gland, leading to decreased gonadotropin production; on the contrary, GnRH antagonists exert direct inhibition of pituitary GnRH receptors (Van Poppel & Abrahamsson, 2020). Gonadotropin synthesis can be also suppressed indirectly by drugs inducing hyperprolactinemia, such as neuroleptic and prokinetic agents (Melmed et al., 2011). Several drugs may inhibit the secretion of testosterone such as in case of inhibitors of steroidogenetic enzymes or may impact on SHBG synthesis.

4. Hormonal assays: Tips, tricks and pitfalls

The total amount of circulating serum testosterone consists of the sum of testosterone bound to serum proteins and unbound free testosterone, the latter representing the circulating fraction promptly available for binding the androgen receptor in target tissues (Handelsman, 2000). The fraction of testosterone bound to serum proteins includes testosterone bound to SHBG (54%), testosterone bound to albumin (44%), and only a minimal part of testosterone bound to other proteins such as cortisol-binding protein. Due to the weak strength of binding to albumin, the albumin-bound testosterone together with unbound free testosterone represents the bioavailable testosterone (Handelsman, 2000). As explained above, SHBG tends to physiologically increase with age accounting for an increase of the SHBG-bound serum testosterone and a greater decline of serum bioavailable testosterone. In addition, several other factors may influence serum SHBG in older men (see the paragraph above for further details). Hence, information on the different fractions of testosterone, and not only total testosterone, are of relevance in the evaluation of gonadal status during aging. The hormonal parameters needed to evaluate the male gonadal status are serum total testosterone, free testosterone, gonadotropins, and SHBG, especially when chronic systemic diseases are present.

For the diagnosis of hypogonadism, serum testosterone should be consistently and unequivocally low as measured on two separate occasions in blood samples in the morning and in a fasting condition (Bhasin et al., 2018), the same apply to the subgroup of older men (Corona et al., 2020; Wang et al., 2008). There is no consensus among guidelines on the lower total testosterone limit to be used as cut-off for testosterone deficiency, actually ranging from 230 to 320 ng/dL (Bhasin et al., 2018; Nieschlag, 2019; Wang et al., 2008). Moreover, reference ranges vary between laboratories due to methodological issues (Dittadi, Matteucci, Meneghetti, & Ndreu, 2018; Rosner, Auchus, Azziz, Sluss, & Raff, 2007; Simoni, Fanelli, Roli, & Pagotto, 2012; Wang, Gay, Botelho, Caudill, & Vesper, 2014).

Several methods are available to measure the different fractions of circulating testosterone. When the currently available immunoassays for SHBG and gonadotropins appear adequate, a fair amount of uncertainty and controversy still surrounds the approaches on how to monitor circulating testosterone levels, both total and free.

The diagnostic and therapeutic management of male gonadal function is mostly based on serum total testosterone. In clinical practice this is commonly assayed with immunoassays (Simoni et al., 2012; Wang et al., 2008), although mass spectrometry, particularly liquid chromatography-tandem mass spectrometry (LC-MS/MS), is considered the gold standard for the measurement of steroid hormones (Bhasin et al., 2018; Simoni et al., 2012; Wang et al., 2014). At present, the use of mass spectrometry is mainly restricted for research purposes because its availability is still limited by cost, need of expensive equipment, technical demands, and limited commercial applications (Bhasin et al., 2018; Wang et al., 2014).

The poor quality of testosterone assays has been a long-term target of complaints and it is still a controversial issue. Recent studies on large sample numbers have shown that testosterone measurements by immunoassays offered good sensitivity and specificity in eugonadal men, and lower, but still acceptable, accuracy in hypogonadal men compared to mass spectrometry (Dittadi et al., 2018; Fanelli et al., 2011; Haring et al., 2013; Huhtaniemi et al., 2012). In general, for clinical purposes it is important to validate the assay and to perform continuous quality control, independently from the methodology used (Rosner et al., 2007; Rosner & Vesper, 2010; Simoni et al., 2012).

This issue becomes even more complicated focusing on the measurement of serum free testosterone (Goldman et al., 2017). Indeed, the measurement of free serum testosterone might be performed using equilibrium dialysis, which is considered the reference method, ultrafiltration methods, or direct free testosterone assays. Unfortunately, all these methods are technically demanding, sensitive to errors, and seldom available for clinical diagnostics (Simoni et al., 2012). In particular, the direct measurement of serum free testosterone with commercially available kits is poorly accurate and its use must be avoided in the clinic, as remarked on the basis of strong evidence by all clinical guidelines on hypogonadism (Bhasin et al., 2018; Corona et al., 2020; Morales et al., 2015; Wang et al., 2008) and expert opinion papers (Corona, Rastrelli, & Maggi, 2013; Huhtaniemi, 2014; Huhtaniemi & Forti, 2011; Rochira, 2017). Alternatively, different computational algorithms have been elaborated for the estimation of free testosterone starting from total testosterone, SHBG, and albumin concentrations (Simoni et al., 2012; Vermeulen, Verdonck, & Kaufman, 1999). Even though it is well known that calculated free testosterone overestimates the value obtained by equilibrium dialysis (Ly et al., 2010), it remains largely employed both for research and clinical purposes.

Finally, techniques for the measurement of bioavailable serum testosterone are based on previous precipitation and separation of SHBG-bound testosterone. For this reason, these methods are not simple and are not routinely employed in clinical laboratories. Similarly to free testosterone, even bioavailable testosterone can be calculated from serum albumin, SHBG, and total testosterone (Simoni et al., 2012). Importantly, whereas it is not clear the advantage of measuring bioavailable testosterone rather than free testosterone, proof for the usefulness of free testosterone assessment in the diagnosis of hypogonadism of elderly men was provided by large cohort studies. In particular, the EMAS data have demonstrated that the use of total testosterone alone is likely to result in overdiagnosis of hypogonadism, especially in men who are obese and when the total testosterone levels are borderline close to the lowest end of the normal range (Antonio et al., 2016; Rastrelli et al., 2015). There is thus agreement among current guidelines in recommending free testosterone assessment only as a complementary tool in symptomatic men whose total testosterone is near the lower limit of the normal range and in conditions that affect SHBG concentrations and render total testosterone a less reliable index of gonadal function (Bhasin et al., 2018; Corona et al., 2020; Lunenfeld et al., 2015; Morales et al., 2015; Wang et al., 2008).

Serum estradiol assessment is even more challenging. Although current clinical guidelines do not recommend the measurement of serum estrogens within the clinical study of male hypogonadism, it could be actually helpful to establish if a hypogonadal patient presents also relative hypoestrogenism or, on the contrary, to identify an over-treatment with T therapy through the detection of hyperestrogenism (Rochira & Carani, 2017). However, commercially-available immunoassay-based techniques lack of specificity especially at lower serum estradiol concentration, resulting thus unreliable in men (Ohlsson et al., 2013; Rochira & Carani, 2017). Mass spectrometry assays represent nowadays the gold standard methodology also for the assessment of serum estradiol levels, anyhow they are used only for research purposes because of the same above-mentioned critical issues concerning T assessment (Ohlsson et al., 2013; Rochira & Carani, 2017). In addition, standardized reference ranges of estradiol assessed by mass spectrometry for males are still lacking (Rochira & Carani, 2017).

The increasing availability of LC-MS/MS tools also in clinical laboratories allows to use this methodology for clinical purposes and will represent a standard for both androgens and estrogens assay in the future, thus improving the diagnostic cut-off for the diagnosis of male hypogonadism in adult and older men (Decaroli & Rochira, 2017; Rochira, 2017; Rochira & Carani, 2017; Simoni et al., 2012).

5. The clinical diagnosis of hypogonadism in aging men

The amount of data arose from the studies mentioned in the paragraph above has led to deep changes in the theoretical and clinical approach to testosterone deficiency in aging men. First, almost all scientific societies and clinical guidelines agree that the finding of biochemical testosterone deficiency should not be necessarily considered per se significant, but it should be contextualized together with concomitant signs and symptoms suggestive for hypogonadism (Bhasin et al., 2018; Corona et al., 2020; Morales et al., 2015; Wang et al., 2008), as strongly suggested also by several expert opinion papers (Corona et al., 2016; Corona, Rastrelli, Vignozzi, Mannucci, & Maggi, 2012; Huhtaniemi, 2014). In this scenario, it has been adopted the term "late-onset hypogonadism" (LOH) to identify a condition characterized by three sexual symptoms alongside with low serum testosterone (Wu et al., 2010), but this term still needs an unequivocal definition to be reached (Nieschlag, 2019). As described below, some aspects of the evaluation and management of hypogonadism do not differ depending on patient's age (e.g., classification of hypogonadism), whereas for other aspects such as clinical correlates or etiology, the patient's age does matter.

5.1 Classification of LOH

In the context of reduced testosterone levels, LOH can be diagnosed only after having excluded organic and/or genetic diseases of the hypothalamic–pituitary–gonadal axis. Moreover, in the clinical setting of aging men quite frequently occurs a condition of compensated hypogonadism, characterized by preserved normal testosterone secretion arising from increased gonadotropins levels; it may be considered as a form of subclinical primary hypogonadism that could progress to overt hypogonadism overtime (Swee & Gan, 2019; Tajar et al., 2010). In fact, as well as in other body tissues, aging determines morphological and functional modifications in the testis that progressively lead to its impairment, and this can explain why primary hypogonadism and its "subclinical" form of compensated hypogonadism strongly correlate with age (Huhtaniemi & Forti, 2011; Tajar et al., 2010).

On the contrary, secondary hypogonadism seems to be linked more strictly with general health status and comorbidities rather than with age (Grossmann & Matsumoto, 2017; Rochira, 2017; Tajar et al., 2010).

However, while in young people the classification of hypogonadism is usually well definite, with a clear cause (typically organic and/or genetic) for its onset that can be detected in the majority of cases, among aging men multiple factors contribute to its development, with primary and secondary forms that can finally overlap (Corona et al., 2020).

5.2 Causes of LOH

As mentioned before, many different factors, some of which are still unknown, are involved in LOH onset, with final overlap between primary and secondary forms of hypogonadism. Fig. 1 summarizes the main causes of LOH, which will be discussed below.

Aging implies reorganization of testis structure, with degeneration of germinal epithelium, reduction in number of Sertoli and Leydig cells and increasing of connective tissue. Such changes justify the decrease of testosterone secretion with time, thus hypergonadotropic and compensated hypogonadism onset can be considered almost in part a para-physiological phenomenon (Swee & Gan, 2019; Tajar et al., 2010) (Table 2). Moreover, other local factors whose incidence increase with aging such as testicular infections, traumas, neoplasia and pelvic irradiation can be elapsing and concur to final gonadal failure (Bhasin et al., 2018).

Comorbidities as obvious tend to increase in number, to become chronic and worsening with aging and they play a major role in determining low testosterone levels, mostly leading to hypogonadotropic or mixed forms of hypogonadism (Corona et al., 2014; Decaroli & Rochira, 2017; Rochira, 2017). Pathologic conditions and chronic diseases mainly involved in impairing testosterone production can be divided in three main groups as follows.

5.2.1 Metabolic disorders

Males affected by metabolic diseases and metabolic syndrome have high risk to develop testosterone deficiency and, at the same time, hypogonadism contributes in worsening metabolic dysfunction. With this in view, a bidirectional link seems to exist between low testosterone levels and metabolic disorders that can bring to a kind of vicious circle (Braga et al., 2020). Overweight and obesity are the component of metabolic syndrome mainly involved in hypogonadism onset (Tajar et al., 2010). In fact, weight excess determines hypogonadotropic hypogonadism likely through an inflammatory mechanism at hypothalamus-pituitary level that results in decrease of GnRH release. In addition, the peripheral increased conversion of testosterone to estradiol due to higher expression of aromatase in the adipose tissue brings to a negative feedback at hypothalamus–pituitary level, decreasing gonadotropin incretion (Rochira & Carani, 2017; Rochira et al., 2000, 2006) (Table 2). At the same time low testosterone levels concur to the reduction of lean body mass and the increase of fat mass, in particular visceral adiposity, with consequent further deterioration of metabolic status (Braga et al., 2020; Corona et al., 2015; Kelly & Jones, 2015), and the establishment of a vicious circle (Cohen, 2001; Rochira, 2017). Lifestyle interventions targeting weight loss have demonstrated to increase testosterone levels in hypogonadal men in proportional relation with the amount of weight lost, up to completely restoring gonadal function (Corona et al., 2013). Similarly, the administration of exogenous testosterone is able to partially revert obesity (Saad, Yassin, Doros, & Haider, 2016). Considering other features of metabolic syndrome, chronic hyperglycemia of not well-controlled diabetes mellitus and insulin resistance can lead to hypogonadism by reducing GnRH pulsatile release (Braga et al., 2020; Dandona, Dhindsa, Chaudhuri, Bhatia, & Topiwala, 2008; Dhindsa et al., 2004) and dyslipidemia also has been linked somehow to secondary hypogonadism. On the contrary, hypertension does not seem to be involved in gonadal function regulation and impairment (Corona et al., 2013).

5.2.2 Chronic diseases

Advanced chronic liver disease determines impairment of both the hypothalamus–pituitary unit function and the local testicular testosterone synthesis, the latter especially in alcoholic liver disease because of direct effects of ethanol on Leydig cells. Moreover, altered hepatic metabolism of steroid hormones brings to increased levels of estrogens as in severe liver insufficiency (e.g., cirrhosis) (Burra, 2013; Kalyani, Gavini, & Dobs, 2007); high serum estradiol contributes, in turn, to a negative feedback on gonadotropins at the hypothalamus–pituitary level (Rochira et al., 2000, 2006) (Table 2). Another mechanism involves alteration in the synthesis of SHBG by the liver, which is reduced or increased in case of viral liver diseases, depending on the severity of liver insufficiency (decreased in case of liver cirrhosis) (Chaudhury et al., 2019; Gursoy et al., 2004; Rochira, 2017) (Table 2). Similarly, chronic renal failure affects testosterone production acting both at testicular levels, with reduction of Leydig cells sensitivity to LH stimulation, and at central levels, with reduction of LH release (Kalyani et al., 2007; Rochira, 2017). Chronic Obstructive Pulmonary Disease (COPD) is also linked to hypogonadism, whose onset in this setting seems to be attributable to the factors involved in COPD development (e.g., smoking), its clinical peculiarities (hypoxemia and hypercapnia) and treatment (glucocorticoids) (Balasubramanian & Naing, 2012; Kalyani et al., 2007). Among chronic infectious diseases, the association between HIV-infection and gonadal impairment is nowadays well known, with testosterone decline that occurs earlier and to a greater extent in comparison with HIV-uninfected men (Rochira & Guaraldi, 2014; Rochira et al., 2011). Furthermore, in iron overload diseases such as beta-thalassemia and hemochromatosis both pituitary gland and testis are preferential target of iron deposition, with consequent oxidative damage and function disruption (Pelusi, Gasparini, Bianchi, & Pasquali, 2016; Rund & Rachmilewitz, 2005).

5.2.3 Acute severe illnesses

Acute pathological conditions such as stroke, myocardial infarction, traumatic brain injury, sepsis and burns, determine hypothalamic–pituitary gonadal axis suppression, whose degree is related to the severity of the pathological condition and tends to be greater in aging men (Rochira, 2017). Onset of hypogonadism in acute illnesses probably represents a defense mechanism aiming to spare energy when health status is poor (Table 2, Fig. 1), in analogy with the low triiodothyronine (T3) syndrome often described in critical hospitalized patients. In general, testosterone production is restored once the acute illness is resolved (Kalyani et al., 2007; Rochira, 2017).

Aging people often undergoes polytherapy and hypogonadism can occur as a related side effect. Among chronic treatments determining hypogonadism onset in older people the main ones are:

- Opioids and glucocorticoids: they lead to central suppression of hypothalamus–pituitary–gonadal axis;
- Drugs increasing serum prolactin (e.g., psychotropic drugs);
- Chemotherapy drugs: they determine cytotoxic testicular and systemic effects;
- GnRH analogues and antiandrogen agents: they target to switch off testosterone synthesis and action, respectively and are diffusely used in the treatment of prostate cancer (Kalyani et al., 2007; Mohler et al., 2019).

Finally, bad lifestyle habits such as smoking, alcohol intake, substance and drug abuse, junk food excessive consumption and physical inactivity contribute in impairing testosterone production, both directly and by increasing the risk of developing comorbidities (Corona et al., 2020; Duca, Aversa, Condorelli, Calogero, & La Vignera, 2019).

Aiming to have an overview of health status in elderly people, in the last decades the notion of frailty has been developed in the geriatric field to characterize poor health status considering not only physical, but also psychological and social deterioration. According with the concept of frailty, the higher a subject is frail, the more has a reduced ability to deal with stressing events, with consequent increased risk of disability and death (Table 2, Fig. 1). There is evidence showing a strict correlation between frail status and hypogonadism in aging men, confirming indirectly the role of comorbidities on hypogonadism onset, but also highlighting the impact of poor general health status on impairment of gonadal function (Hyde et al., 2010). Accordingly, the accumulation of morbidity and the measurement of frailty represent two different approaches to measure health status (Corona et al., 2014; Rochira, 2017). With this in view, hypogonadism in aging men could represent a kind of protective mechanism, where the hypothalamic–pituitary unit reacts to poor health status by lowering the set-point of endogenous testosterone levels, aiming in this way to reduce physical and sexual activity, to spare energy and avoid fertility in unhealthy conditions (Corona et al., 2014; Decaroli & Rochira, 2017). Similarly, clinical condition characterized by accentuated aging in adult, young men (Guaraldi, Silva, & Stentarelli, 2014) are associated to premature occurrence of hypogonadism such as in HIV-infected men (Rochira & Guaraldi, 2014; Rochira et al., 2011).

5.3 Diagnosis of LOH

5.3.1 Signs and symptoms

Clinical identification of LOH can be very difficult, since signs and symptoms of hypogonadism in elderly men are most of time non-specific, vague and often overlapping with those of comorbidities and aging per se (Corona et al., 2020, 2013; Decaroli & Rochira, 2017; Travison et al., 2008; Wu et al., 2010). Anyhow, clinical features of hypogonadism can be divided in the three following groups. (a) *Sexual area:* signs and symptoms belonging to sexual domain mostly correlate with low testosterone and hypogonadism should always be suspected when low libido, erectile dysfunction and poor morning erection are present (Granata, Rochira, Lerchl, Marrama, & Carani, 1997; Maggi, Filippi, Vignozzi, & Rastrelli, 2020; Wu et al., 2010). In addition, reduction of testicular volume is often observed, because of atrophy of testis structures and tissues. Oligoposia together with oligospermia or azoospermia are common, resulting in impaired fertility and difficulty to fatherhood (Bhasin et al., 2018; Corona et al., 2016). (b) *Physical area:* low testosterone levels determine an increase in visceral adiposity and decrease in muscle mass, with consequent reduced strength, impaired physical activity, and limitation in walking ability. Moreover, the inversion of estradiol to testosterone ratio can determine gynecomastia and breast discomfort, especially in over-weight men (Sansone, Romanelli, Sansone, Lenzi, & Di Luigi, 2017). Reduction of body hair mostly in androgen sensitive areas (e.g., pubic and axillary hair, beard) can be found, with reduction in shaving. Anemia can occur as a consequence of lack of testosterone contribute on stimulating erythropoiesis (Rochira, Zirilli, Madeo, Maffei, & Carani, 2009). Furthermore, low testosterone and consequent low estradiol levels impact on bone health, leading to reduced bone mineral density, osteoporosis and osteoporotic fractures in severe cases (Rochira, 2020). Finally, if hypogonadism onset is acute, hot-flush and sweats can occur, similarly to menopausal women (Corona et al., 2016; Wu et al., 2010). (c) *Psychological area:* depressed mood, decreased vitality with loss of energy and fatigue can be observed in hypogonadal men, although they are very non-specific symptoms (Corona et al., 2016; Wu et al., 2010). Fig. 2 summarizes signs and symptoms of LOH in relation to their specificity.

5.3.2 Laboratory and radiological analysis

A precise, definitive cut-off of total testosterone levels for the biochemical diagnosis of hypogonadism in general male population has not been established so far (Nieschlag, 2019); however, there is agreement in considering total testosterone as deficient when it is below a value between the range of 230–320 ng/dL (Corona et al., 2020; Travison et al., 2017). In addition to biochemical evaluation, specific clinical symptoms of hypogonadism have always to be present in order to establish a diagnosis of LOH (Corona et al., 2020, 2013; Wu et al., 2010).

Since testosterone presents a circadian rhythm with peak in the morning and its incretion is suppressed by food, two measurements in different days of testosterone in the morning fasting should be taken in order to confirm the presence of hypogonadism (Bhasin et al., 2018; Corona et al., 2020; Morales et al., 2015; Wang et al., 2008).

LC-MS/MS is nowadays the gold-standard in the measurement of sexual steroids, presenting higher sensitivity and specificity than commercially available assays especially in the lower range of total testosterone values; anyway, where LC-MS/MS is not available, well-standardized automated

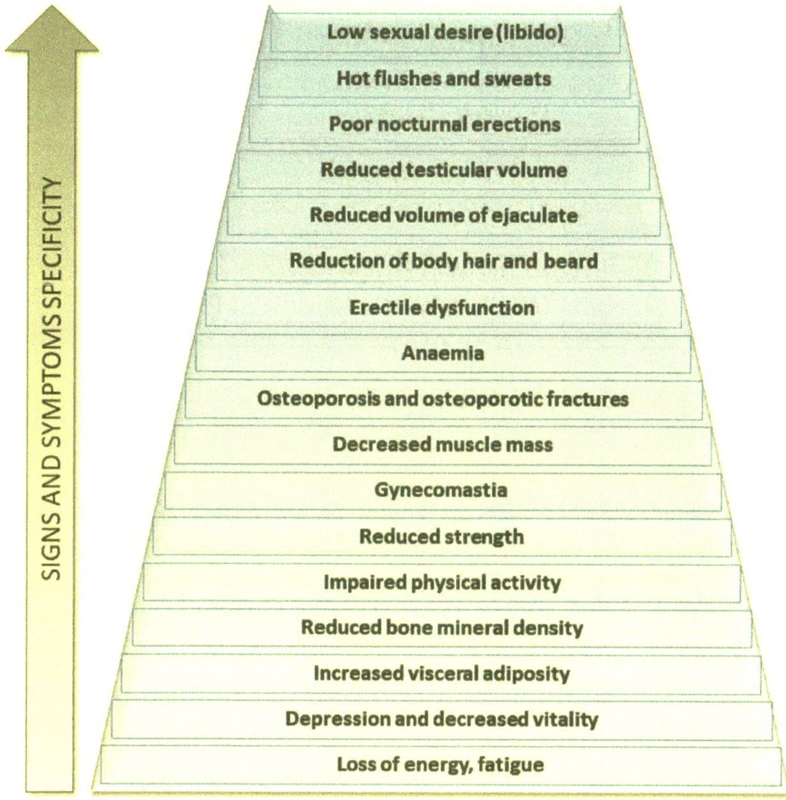

Fig. 2 Signs and symptoms of late onset hypogonadism in relation to their specificity.

platform testosterone immunoassays can offer quite reliable results (Bhasin et al., 2018; Fanelli et al., 2011; Trost & Mulhall, 2016).

Circulating serum testosterone is mainly bound to SHBG and albumin, but only the small fraction of non-protein bound or free testosterone determines physiological effects; thus, when total testosterone is around the lower cut-off of normality range, increased levels of SHBG determine a reduction of the quote of biologically active free testosterone, leading to hypogonadism. Since aging per se brings to increased levels of SHBG and comorbidities also can determine abnormalities (mainly increase) in its production, evaluation of SHBG is essential in the setting of aging people for detecting low free testosterone levels. Laboratory assays for direct determination of free testosterone lack of precision and must be avoided. The calculation of free testosterone according to validated formulas, based on total testosterone, SHBG and albumin, is cheap and reliable, thus should be preferred (Bhasin et al., 2018).

Free testosterone cut-off for the diagnosis of hypogonadism has been estimated to be 64 pg/mL (Bhasin et al., 2018) (Free testosterone calculator available online: http://www.issam.ch/freetesto.htm).

The evaluation of LH and follicle-stimulating hormone (FSH) is helpful to classify the type of hypogonadism, especially when very low testosterone levels are detected and/or an organic cause of hypothalamus-pituitary-gonadal axis impairment is suspected (Bhasin et al., 2018; Corona et al., 2020; Petak et al., 2002). It should be emphasized, in fact, that the diagnosis of LOH can be made only after having excluded genetic and/or organic causes of hypogonadism.

When low serum testosterone is associated to very high gonadotropins levels and clinical features evocate a possible Klinefelter Syndrome still undiagnosed in adulthood (e.g., small testis, gynecomastia, eunuchoid habitus, and infertility), karyotype analysis should be performed (Bhasin et al., 2018; Bonomi et al., 2017). On the other side, when overt hypogonadotropic hypogonadism is found serum prolactin and other pituitary hormones should be assessed. In presence of marked not drug-related hyperprolactinemia, hypopituitarism and/or clinical evidence of possible pituitary mass effect (e.g., headache, visual field deficit) a pituitary magnetic resonance imaging (MRI) is recommended (Bhasin et al., 2018; Corona et al., 2020).

When hypogonadism is confirmed, bone mineral densitometry at femoral and lumbar site should be taken to evaluate bone status; moreover, if densitometric osteoporosis is diagnosed and bone pain or a reduction in height are observed, the dorso-lumbosacral spine X-ray coupled with morphometry become mandatory in order to confirm or detect symptomatic and asymptomatic vertebral fractures, respectively (Rochira, 2020; Rochira, Antonio, & Vanderschueren, 2018).

6. Management of LOH

6.1 Lifestyle modification

In general, a correct lifestyle can prevent age-related decline of testosterone levels, whereas in hypogonadal men lifestyle modification can improve up to restore normal testosterone levels (Grossmann & Matsumoto, 2017).

It has been demonstrated that weight loss obtained through diet and/or physical activity can restore the hypothalamus–pituitary–gonadal axis function, leading to an increase of testosterone serum levels that is proportional to the amount of weight lost. Moreover, an improvement of erectile function can be achieved thanks to weight loss (Kim et al., 2017; Rigon et al., 2019).

Physical exercise should be encouraged not only to drop in weight, but also for the proven direct effectiveness on metabolic disorders and general health status (Di Luigi, Romanelli, Sgrò, & Lenzi, 2012).

It is important to improve chronic comorbidities management when possible; in particular, if diabetes mellitus is present, optimization of glycemic control is desirable, as well as the normalization of cholesterol and triglycerides levels in case of dyslipidemia (Rochira, 2017).

Finally, withdrawing and/or dose reduction of drugs interfering with testosterone secretion should be always performed when possible (Corona et al., 2020; Grossmann & Matsumoto, 2017; Huhtaniemi & Forti, 2011).

6.2 Testosterone replacement therapy (TRT)

Testosterone replacement therapy (TRT) has to be considered only in men with documented low testosterone levels complaining of symptoms (in particular those involving sexual function) attributable to hypogonadism, in absence of clinical contraindications. Testosterone cannot be administered in presence of:

- Active prostate or breast cancer;
- Unevaluated prostate nodule or induration;
- Unevaluated Prostate Specific Antigen (PSA) >4 ng/mL (or >3 ng/mL in men with anamnestic high risk of prostate cancer);
- Basal hematocrit value >48%;
- Severe lower urinary tract symptoms;
- Uncontrolled or poorly controlled congestive heart failure (NYHA 3 and 4);
- Thrombophilia;
- Myocardial infarction or stroke within 6 months (Bhasin et al., 2018; Morales et al., 2015; Rochira, 2017).

With this in view, an accurate anamnesis together with digital rectal examination, PSA and hematocrit measurement should always be performed before starting TRT (Bhasin et al., 2018; Corona et al., 2020; Wang et al., 2008). In patients with history of low grade, organ-limited prostate cancer who have undergone radical prostatectomy and have no detectable signs of residual disease (in particular undetectable PSA levels) after 2 years or more from surgery TRT can be evaluated when overt, symptomatic hypogonadism is present (Bhasin et al., 2018; Corona et al., 2020; Wang et al., 2008).

According to the parameters listed above, clinician should identity men candidate to testosterone therapy and discuss with them the opportunity to start TRT, illustrating possible risks and benefits and informing about the lack of data on long-term effects (Corona et al., 2020; Huhtaniemi & Forti, 2011). To prove effectiveness and safety, a short (e.g., 3–6 months) trial of TRT can be initially proposed, aiming to achieve serum total testosterone levels in the mid-normal range for young healthy men (Bhasin et al., 2018; Wang et al., 2008).

Although several testosterone formulations (oral, transdermal, parenteral and nasal preparations together with buccal and pellet implants) have been developed, in aging men with LOH short-acting transdermal testosterone gel or oral formulations are preferable. In fact, its short action reduces risk of overdose and consents rapid discontinuation in case of adverse effects; in addition, the dosage of these formulations can be easily personalized, consenting to obtain target testosterone levels (Corona et al., 2020; Wang et al., 2008). The oral formulations are mainly affected by the first liver passage that results in the metabolization and inactivation of most of the testosterone absorbed through the gut. However, novel promising formulations able to bypass the liver and to be absorbed directly into the lymphatic system will probably change this paradigm in the next future (Swerdloff et al., 2020; Swerdloff & Dudley, 2020).

Effects of TRT are multiple and present a different impact depending on the clinical feature considered.

The main benefits of TRT can be seen in sexual function, with improvement of sexual desire, sexual satisfaction and, even if to a lesser extent, of erectile function. Sexual advantages seem to be proportional with the increase of testosterone levels in serum (Bhasin et al., 2018; Corona et al., 2020; Snyder et al., 2016, 2018).

TRT consistently improves body composition, with reduction of intra-abdominal and intra-muscular fat mass and increase of lean mass and muscle strength, however enhancement of global physical function and performance in aging men seems to be scanty (Bhasin et al., 2018; Corona et al., 2020; Huhtaniemi & Forti, 2011). Similarly, improvement of metabolic parameters such as fasting glycaemia, triglycerides levels and blood pressure does not seem to be clinically significant with TRT ongoing (Haider et al., 2020). With this in view, it is important to emphasize that TRT must not be started with the only purpose to ameliorate body composition and metabolic status if clinical and biochemical diagnostic criteria of hypogonadism are not present (Bhasin et al., 2018; Corona et al., 2020; Rochira, 2017).

TRT can improve mood, reducing depressive symptoms and increasing vitality together with self-perception of energy (Bhasin et al., 2018; Corona et al., 2020; Snyder et al., 2016, 2018), whereas no effects on memory and cognitive functions have been demonstrated (Snyder et al., 2016, 2018).

Considering bone health, TRT has shown to increase bone mineral density (BMD) and bone strength both at vertebral and femoral sites, however no data are available about a possible efficacy in preventing osteoporotic fractures (Bhasin et al., 2018; Snyder et al., 2017), thus TRT in the context of LOH may improve BMD, but is not recommended alone for patients with a high risk of fracture; for these patients bone active therapy is mandatory (Rochira, 2020; Rochira et al., 2018).

Finally, good results can be expected from TRT in mild forms of anemia, where normalization of hemoglobin levels in the majority of cases is achieved (Snyder et al., 2018).

To evaluate clinical benefits and the possible onset of adverse effects, a regular and careful clinical and bio-humoral follow-up, including at least PSA, hematocrit and total testosterone levels measurement together with morphological evaluation of the prostate, should be performed after the initiation of TRT (Bhasin et al., 2018; Corona et al., 2013).

Among adverse effects, erythrocytosis is the most common occurring in older patients and TRT must be interrupted when hematocrit is above 54%, because of the increased risk of neuro-occlusive or cardiovascular events (Bhasin et al., 2018; Corona et al., 2020). In addition, TRT seems to correlate with an increased risk of venous thromboembolism, especially within the first 3 months of therapy, even if the risk seems to be lower in aging men in comparison with younger men (Walker et al., 2019). Induction or worsening of obstructive sleep apnea is an uncommon adverse effect to keep in mind. Less serious potential adverse effects are gynecomastia, acneic skin and male pattern balding (Bhasin et al., 2018). Moreover, there can be side effects strictly linked to the formulation of T used and, in case of transdermal gel preparation, skin reactions and irritations can occur, as well as the transmission of T to other people in strict contact when the application site gets not covered (Bhasin et al., 2018; Corona et al., 2020).

There is no clear evidence of increased prostate cancer risk in men on TRT, since no appropriate, randomized prospective trials have been realized so far. The apparent increased detection of prostate cancer among people on TRT is probably due mainly to the active clinical surveillance, which is established after starting therapy (Bhasin et al., 2018; Huhtaniemi & Forti, 2011).

Although TRT has been associated with an increase in non-calcified coronary artery plaque volume, there is not consistent evidence of increased cardiovascular and coronary heart disease risk in hypogonadal aging men with treatment ongoing (Bhasin et al., 2018; Snyder et al., 2018). The risk of major cardiovascular events remains to be fully elucidated and seems to be confined to older men with associated very poor health status (Basaria et al., 2015).

Finally, TRT suppresses at least in part spermatogenesis, thus it should be avoided in men who desire fertility, eventually in favor of other pharmacological treatments (Bhasin et al., 2018; Corona et al., 2013).

6.3 Other pharmacological options

When fertility is required and only if a secondary form of LOH is present, thus with testicular function potentially intact, gonadotropin therapy can be proposed. Treatment of choice is human chorionic gonadotropin (hCG), with induction of fertility (defined as the appearance of at least one spermatozoon in seminal fluid) that can be obtained in about 75% of patients. Association of FSH to hCG can lead to even better results (Corona et al., 2020; Rastrelli, Corona, Mannucci, & Maggi, 2014). Of course, the desire of fatherhood is not so frequent in older men, but it should be taken into account.

Selective estrogen receptor modulators (SERMs) explicate an antagonist action on the estrogen receptors located at hypothalamus-pituitary level, in this way they inhibit the estrogens-mediated negative feedback determining an increase of LH and FSH secretion and, consequently, testicular stimulation. With this in view, SERMs could represent theoretically an alternative to testosterone or gonadotropin treatment in men with secondary functional LOH associated especially in those with high estrogen levels (e.g., obese patients). Nevertheless, the use of SERMs in the clinical setting of male hypogonadism is *off-label* and data about efficacy on the restoration of normal serum testosterone are limited to short period of therapy and are still conflicting and inconsistent (Awouters, Vanderschueren, & Antonio, 2019; Corona et al., 2020, 2013; Wheeler, Sharma, Kavoussi, Smith, & Costabile, 2019).

Selective androgen receptor modulators (SARMs) are being developed with the aim to decrease unwanted possible side effects of TRT (such as gynecomastia) maintaining anabolic and sexual effects of androgens, thus represent a possible interesting therapeutic option in future perspective (Corona et al., 2013; Narayanan, Coss, & Dalton, 2018).

Clinical management of LOH—The practical corner

Measurement of serum testosterone suffers from several limitations and should be performed with some precautions.

Practical issues:

- ➢ Patients should undergo blood collection for testosterone assay in the morning fasting
- ➢ Two measurements in different days are required to confirm the presence of hypogonadism
- ➢ In presence of comorbidities or other conditions affecting SHBG levels, the calculation of free testosterone is more reliable than total testosterone for the evaluation of gonadal status

The mere detection of testosterone deficiency is not sufficient to diagnose LOH. The biochemical data should be contextualized with signs and/or symptoms related to hypogonadism.

Practical issues:

- ➢ Symptoms and signs related to hypogonadism should be all accurately checked (e.g., distribution of body and facial hair, gynecomastia, muscle mass, reduced bone density).
- ➢ The interview should include sexual issues: sexual desire impairment and information about erectile function.
- ➢ The most specific symptoms related to hypogonadism are: low libido, impaired nocturnal erections, reduced volume of the ejaculate, hot flushes.

Lifestyle habits and comorbidities, especially overweight and obesity, can worsen the reduction of testosterone levels.

Practical issues:

- ➢ A detailed medical history should be performed in order to rule out any comorbidity that could affect the functionality of the hypothalamic-pituitary-gonadal axis.
- ➢ Physical examination of patients should include anthropometric measurements (height, weight, waist and hip circumferences) in order to calculate the body mass index (BMI) and waist-to-hip ratio.
- ➢ Lifestyle modifications can prevent the worsening of testosterone decline and the occurrence of comorbidities.

A pharmacological treatment should be considered in case of confirmed LOH, avoiding it when unnecessary and selecting patients deserving TRT.

Practical issues:

- ➢ Contraindications to TRT should be accurately checked: breast or prostate cancer, hematocrit >48%, uncontrolled congestive heart failure, thrombophilia, myocardial infarction or stroke within 6 months, severe lower urinary tract symptoms.
- ➢ TRT should be personalized in order to choose the therapeutic option (in terms of dosage and formulation) that best fits for each patient.

An adequate follow-up to evaluate benefits and side effects should be planned for those patients with ongoing TRT.

7. Conclusions

Changes in sex hormones in aging men and their potential clinical consequences are of increasing clinical relevancy because of an increasing proportion of elderly men in the population. The decline of sex steroids with advancing age depends on multiple endogenous (i.e., aging process per se) and extrinsic (e.g., lifestyle habits, comorbidities) factors, leading to high inter-individual differences. An accurate clinical evaluation, including medical history and detection of signs and symptoms suggestive for hypogonadism, is required in order to assure each patient the best management. At the same time, this may help the clinician in addressing the prescription of replacement treatment, avoiding it when unnecessary or contraindicated. However, the diagnostic and therapeutic work-out of LOH still presents several pitfalls, mainly due to the low specificity of clinical correlates and the analytical limits of testosterone measurement with the gold standard assay rarely available in clinical practice.

References

Antonio, L., Wu, F. C., O'Neill, T. W., Pye, S. R., Ahern, T. B., Laurent, M. R., et al. (2016). Low free testosterone is associated with hypogonadal signs and symptoms in men with normal total testosterone. *The Journal of Clinical Endocrinology and Metabolism*, *101*(7), 2647–2657. https://doi.org/10.1210/jc.2015-4106.

Araujo, A. B., O'Donnell, A. B., Brambilla, D. J., Simpson, W. B., Longcope, C., Matsumoto, A. M., et al. (2004). Prevalence and incidence of androgen deficiency in middle-aged and older men: Estimates from the Massachusetts Male Aging Study. *The Journal of Clinical Endocrinology and Metabolism*, *89*(12), 5920–5926. https://doi.org/10.1210/jc.2003-031719.

Awouters, M., Vanderschueren, D., & Antonio, L. (2019). Aromatase inhibitors and selective estrogen receptor modulators: Unconventional therapies for functional hypogonadism? *Andrology*, *8*, 1590–1597. https://doi.org/10.1111/andr.12725.

Balasubramanian, V., & Naing, S. (2012). Hypogonadism in chronic obstructive pulmonary disease: Incidence and effects. *Current Opinion in Pulmonary Medicine*, *18*(2), 112–117. https://doi.org/10.1097/MCP.0b013e32834feb37.

Basaria, S., Travison, T. G., Alford, D., Knapp, P. E., Teeter, K., Cahalan, C., et al. (2015). Effects of testosterone replacement in men with opioid-induced androgen deficiency: A randomized controlled trial. *Pain*, *156*(2), 280–288. https://doi.org/10.1097/01.j.pain.0000460308.86819.aa.

Bhasin, S., Brito, J. P., Cunningham, G. R., Hayes, F. J., Hodis, H. N., Matsumoto, A. M., et al. (2018). Testosterone therapy in men with hypogonadism: An endocrine society clinical practice guideline. *The Journal of Clinical Endocrinology and Metabolism*, *103*(5), 1715–1744. https://doi.org/10.1210/jc.2018-00229.

Bonomi, M., Rochira, V., Pasquali, D., Balercia, G., Jannini, E. A., & Ferlin, A. (2017). Klinefelter syndrome (KS): Genetics, clinical phenotype and hypogonadism. *Journal of Endocrinological Investigation*, *40*(2), 123–134. https://doi.org/10.1007/s40618-016-0541-6.

Braga, P. C., Pereira, S. C., Ribeiro, J. C., Sousa, M., Monteiro, M. P., Oliveira, P. F., et al. (2020). Late-onset hypogonadism and lifestyle-related metabolic disorders. *Andrology*, *8*, 1530–1538. https://doi.org/10.1111/andr.12765.

Burra, P. (2013). Liver abnormalities and endocrine diseases. *Best Practice & Research. Clinical Gastroenterology*, *27*(4), 553–563. https://doi.org/10.1016/j.bpg.2013.06.014.

Camacho, E. M., Huhtaniemi, I. T., O'Neill, T. W., Finn, J. D., Pye, S. R., Lee, D. M., et al. (2013). Age-associated changes in hypothalamic-pituitary-testicular function in middle-aged and older men are modified by weight change and lifestyle factors: Longitudinal results from the European Male Ageing Study. *European Journal of Endocrinology*, *168*(3), 445–455. https://doi.org/10.1530/eje-12-0890.

Chaudhury, C. S., Mee, T., Chairez, C., McLaughlin, M., Silk, R., Gross, C., et al. (2019). Testosterone in men with chronic hepatitis C infection and after hepatitis C viral clearance. *Clinical Infectious Diseases*, *69*(4), 571–576. https://doi.org/10.1093/cid/ciy965.

Cohen, P. G. (2001). Aromatase, adiposity, aging and disease. The hypogonadal-metabolic-atherogenic-disease and aging connection. *Medical Hypotheses*, *56*(6), 702–708. https://doi.org/10.1054/mehy.2000.1169.

Corona, G., Goulis, D. G., Huhtaniemi, I., Zitzmann, M., Toppari, J., Forti, G., et al. (2020). European Academy of Andrology (EAA) guidelines* on investigation, treatment and monitoring of functional hypogonadism in males. *Andrology*, *8*, 970–987. https://doi.org/10.1111/andr.12770.

Corona, G., Maseroli, E., Rastrelli, G., Francomano, D., Aversa, A., Hackett, G. I., et al. (2016). Is late-onset hypogonadotropic hypogonadism a specific age-dependent disease, or merely an epiphenomenon caused by accumulating disease-burden? *Minerva Endocrinologica*, *41*(2), 196–210.

Corona, G., Rastrelli, G., & Maggi, M. (2013). Diagnosis and treatment of late-onset hypogonadism: systematic review and meta-analysis of TRT outcomes. *Best Practice & Research. Clinical Endocrinology & Metabolism*, *27*(4), 557–579. https://doi.org/10.1016/j.beem.2013.05.002.

Corona, G., Rastrelli, G., Maseroli, E., Fralassi, N., Sforza, A., Forti, G., et al. (2014). Low testosterone syndrome protects subjects with high cardiovascular risk burden from major adverse cardiovascular events. *Andrology*, *2*(5), 741–747. https://doi.org/10.1111/j.2047-2927.2014.00241.x.

Corona, G., Rastrelli, G., Vignozzi, L., Mannucci, E., & Maggi, M. (2012). How to recognize late-onset hypogonadism in men with sexual dysfunction. *Asian Journal of Andrology*, *14*(2), 251–259. https://doi.org/10.1038/aja.2011.138.

Corona, G., Vignozzi, L., Sforza, A., Mannucci, E., & Maggi, M. (2015). Obesity and late-onset hypogonadism. *Molecular and Cellular Endocrinology*, *418*(Pt. 2), 120–133. https://doi.org/10.1016/j.mce.2015.06.031.

Dandona, P., Dhindsa, S., Chaudhuri, A., Bhatia, V., & Topiwala, S. (2008). Hypogonadotrophic hypogonadism in type 2 diabetes. *The Aging Male*, *11*(3), 107–117. https://doi.org/10.1080/13685530802317934.

Davidson, J. M., Chen, J. J., Crapo, L., Gray, G. D., Greenleaf, W. J., & Catania, J. A. (1983). Hormonal changes and sexual function in aging men. *The Journal of Clinical Endocrinology and Metabolism*, *57*(1), 71–77. https://doi.org/10.1210/jcem-57-1-71.

Decaroli, M. C., & Rochira, V. (2017). Aging and sex hormones in males. *Virulence*, *8*(5), 545–570. https://doi.org/10.1080/21505594.2016.1259053.

Dhindsa, S., Prabhakar, S., Sethi, M., Bandyopadhyay, A., Chaudhuri, A., & Dandona, P. (2004). Frequent occurrence of hypogonadotropic hypogonadism in type 2 diabetes. *The Journal of Clinical Endocrinology and Metabolism*, *89*(11), 5462–5468. https://doi.org/10.1210/jc.2004-0804.

Di Luigi, L., Romanelli, F., Sgrò, P., & Lenzi, A. (2012). Andrological aspects of physical exercise and sport medicine. *Endocrine*, *42*(2), 278–284. https://doi.org/10.1007/s12020-012-9655-6.

Dittadi, R., Matteucci, M., Meneghetti, E., & Ndreu, R. (2018). Reassessment of the access testosterone chemiluminescence assay and comparison with LC-MS method. *Journal of Clinical Laboratory Analysis*, *32*(3), e22286. https://doi.org/10.1002/jcla.22286.

Duca, Y., Aversa, A., Condorelli, R. A., Calogero, A. E., & La Vignera, S. (2019). Substance abuse and male hypogonadism. *Journal of Clinical Medicine*, *8*(5), 732. https://doi.org/10.3390/jcm8050732.

Fanelli, F., Belluomo, I., Di Lallo, V. D., Cuomo, G., De Iasio, R., Baccini, M., et al. (2011). Serum steroid profiling by isotopic dilution-liquid chromatography-mass spectrometry: Comparison with current immunoassays and reference intervals in healthy adults. *Steroids*, *76*(3), 244–253. https://doi.org/10.1016/j.steroids.2010.11.005.

Feldman, H. A., Longcope, C., Derby, C. A., Johannes, C. B., Araujo, A. B., Coviello, A. D., et al. (2002). Age trends in the level of serum testosterone and other hormones in middle-aged men: Longitudinal results from the Massachusetts male aging study. *The Journal of Clinical Endocrinology and Metabolism*, *87*(2), 589–598. https://doi.org/10.1210/jcem.87.2.8201.

Goldman, A. L., Bhasin, S., Wu, F. C. W., Krishna, M., Matsumoto, A. M., & Jasuja, R. (2017). A reappraisal of testosterone's binding in circulation: Physiological and clinical implications. *Endocrine Reviews*, *38*(4), 302–324. https://doi.org/10.1210/er.2017-00025.

Granata, A. R., Rochira, V., Lerchl, A., Marrama, P., & Carani, C. (1997). Relationship between sleep-related erections and testosterone levels in men. *Journal of Andrology*, *18*(5), 522–527.

Gray, A., Berlin, J. A., McKinlay, J. B., & Longcope, C. (1991). An examination of research design effects on the association of testosterone and male aging: Results of a meta-analysis. *Journal of Clinical Epidemiology*, *44*(7), 671–684. https://doi.org/10.1016/0895-4356(91)90028-8.

Gray, A., Feldman, H. A., McKinlay, J. B., & Longcope, C. (1991). Age, disease, and changing sex hormone levels in middle-aged men: Results of the Massachusetts Male Aging Study. *The Journal of Clinical Endocrinology and Metabolism*, *73*(5), 1016–1025. https://doi.org/10.1210/jcem-73-5-1016.

Grossmann, M., & Matsumoto, A. M. (2017). A perspective on middle-aged and older men with functional hypogonadism: Focus on holistic management. *The Journal of Clinical Endocrinology and Metabolism*, *102*(3), 1067–1075. https://doi.org/10.1210/jc.2016-3580.

Guaraldi, G., Silva, A. R., & Stentarelli, C. (2014). Multimorbidity and functional status assessment. *Current Opinion in HIV and AIDS*, *9*(4), 386–397. https://doi.org/10.1097/coh.0000000000000079.

Gursoy, S., Baskol, M., Ozbakir, O., Guven, K., Kelestimur, F., & Yucesoy, M. (2004). Hypothalamo-pituitary gonadal axis in men with chronic hepatitis. *Hepato-Gastroenterology*, *51*(57), 787–790.

Haider, K. S., Haider, A., Saad, F., Doros, G., Hanefeld, M., Dhindsa, S., et al. (2020). Remission of type 2 diabetes following long-term treatment with injectable testosterone undecanoate in patients with hypogonadism and type 2 diabetes: 11-year data from a real-world registry study. *Diabetes, Obesity & Metabolism*, *22*, 2055–2068. https://doi.org/10.1111/dom.14122.

Handelsman, D. J. (2000). Androgen physiology, pharmacology and abuse. In K. R. Feingold, B. Anawalt, A. Boyce, G. Chrousos, K. Dungan, A. Grossman, & D. P. Wilson (Eds.), *Endotext* MDText.com, Inc.

Handelsman, D. J., Yeap, B., Flicker, L., Martin, S., Wittert, G. A., & Ly, L. P. (2015). Age-specific population centiles for androgen status in men. *European Journal of Endocrinology*, *173*(6), 809–817. https://doi.org/10.1530/eje-15-0380.

Haring, R., Baumeister, S. E., Nauck, M., Völzke, H., Keevil, B. G., Brabant, G., et al. (2013). Testosterone and cardiometabolic risk in the general Population—The impact of measurement method on risk associations: A comparative study between immunoassay and mass spectrometry. *European Journal of Endocrinology, 169*(4), 463–470. https://doi.org/10.1530/eje-13-0222.

Harman, S. M., Metter, E. J., Tobin, J. D., Pearson, J., & Blackman, M. R. (2001). Longitudinal effects of aging on serum total and free testosterone levels in healthy men. Baltimore Longitudinal Study of Aging. *The Journal of Clinical Endocrinology and Metabolism, 86*(2), 724–731. https://doi.org/10.1210/jcem.86.2.7219.

Hsu, B., Cumming, R. G., Waite, L. M., Blyth, F. M., Naganathan, V., Le Couteur, D. G., et al. (2015). Longitudinal relationships between reproductive hormones and cognitive decline in older men: The concord health and ageing in men project. *The Journal of Clinical Endocrinology and Metabolism, 100*(6), 2223–2230. https://doi.org/10.1210/jc.2015-1016.

Huhtaniemi, I. (2014). Late-onset hypogonadism: Current concepts and controversies of pathogenesis, diagnosis and treatment. *Asian Journal of Andrology, 16*(2), 192–202. https://doi.org/10.4103/1008-682x.122336.

Huhtaniemi, I., & Forti, G. (2011). Male late-onset hypogonadism: Pathogenesis, diagnosis and treatment. *Nature Reviews. Urology, 8*(6), 335–344. https://doi.org/10.1038/nrurol.2011.47.

Huhtaniemi, I. T., Tajar, A., Lee, D. M., O'Neill, T. W., Finn, J. D., Bartfai, G., et al. (2012). Comparison of serum testosterone and estradiol measurements in 3174 European men using platform immunoassay and mass spectrometry; relevance for the diagnostics in aging men. *European Journal of Endocrinology, 166*(6), 983–991. https://doi.org/10.1530/eje-11-1051.

Hyde, Z., Flicker, L., Almeida, O. P., Hankey, G. J., McCaul, K. A., Chubb, S. A., et al. (2010). Low free testosterone predicts frailty in older men: The health in men study. *The Journal of Clinical Endocrinology and Metabolism, 95*(7), 3165–3172. https://doi.org/10.1210/jc.2009-2754.

Jasuja, G. K., Travison, T. G., Davda, M., Murabito, J. M., Basaria, S., Zhang, A., et al. (2013). Age trends in estradiol and estrone levels measured using liquid chromatography tandem mass spectrometry in community-dwelling men of the Framingham Heart Study. *The Journals of Gerontology. Series A, Biological Sciences and Medical Sciences, 68*(6), 733–740. https://doi.org/10.1093/gerona/gls216.

Kalyani, R. R., Gavini, S., & Dobs, A. S. (2007). Male hypogonadism in systemic disease. *Endocrinology and Metabolism Clinics of North America, 36*(2), 333–348. https://doi.org/10.1016/j.ecl.2007.03.014.

Kaufman, J. M., Lapauw, B., Mahmoud, A., T'Sjoen, G., & Huhtaniemi, I. T. (2019). Aging and the male reproductive system. *Endocrine Reviews, 40*(4), 906–972. https://doi.org/10.1210/er.2018-00178.

Kaufman, J. M., & Vermeulen, A. (2005). The decline of androgen levels in elderly men and its clinical and therapeutic implications. *Endocrine Reviews, 26*(6), 833–876. https://doi.org/10.1210/er.2004-0013.

Kelly, D. M., & Jones, T. H. (2015). Testosterone and obesity. *Obesity Reviews, 16*(7), 581–606. https://doi.org/10.1111/obr.12282.

Kim, C., Dabelea, D., Kalyani, R. R., Christophi, C. A., Bray, G. A., Pi-Sunyer, X., et al. (2017). Changes in visceral adiposity, subcutaneous adiposity, and sex hormones in the diabetes prevention program. *The Journal of Clinical Endocrinology and Metabolism, 102*(9), 3381–3389. https://doi.org/10.1210/jc.2017-00967.

Lanfranco, F., Zitzmann, M., Simoni, M., & Nieschlag, E. (2004). Serum adiponectin levels in hypogonadal males: Influence of testosterone replacement therapy. *Clinical Endocrinology, 60*(4), 500–507. https://doi.org/10.1111/j.1365-2265.2004.02007.x.

Lapauw, B., Goemaere, S., Zmierczak, H., Van Pottelbergh, I., Mahmoud, A., Taes, Y., et al. (2008). The decline of serum testosterone levels in community-dwelling men over

70 years of age: Descriptive data and predictors of longitudinal changes. *European Journal of Endocrinology, 159*(4), 459–468. https://doi.org/10.1530/eje-07-0873.

Liu, P. Y., Beilin, J., Meier, C., Nguyen, T. V., Center, J. R., Leedman, P. J., et al. (2007). Age-related changes in serum testosterone and sex hormone binding globulin in Australian men: Longitudinal analyses of two geographically separate regional cohorts. *The Journal of Clinical Endocrinology and Metabolism, 92*(9), 3599–3603. https://doi.org/10.1210/jc.2007-0862.

Lunenfeld, B., Mskhalaya, G., Zitzmann, M., Arver, S., Kalinchenko, S., Tishova, Y., et al. (2015). Recommendations on the diagnosis, treatment and monitoring of hypogonadism in men. *The Aging Male, 18*(1), 5–15. https://doi.org/10.3109/13685538.2015.1004049.

Ly, L. P., Sartorius, G., Hull, L., Leung, A., Swerdloff, R. S., Wang, C., et al. (2010). Accuracy of calculated free testosterone formulae in men. *Clinical Endocrinology, 73*(3), 382–388. https://doi.org/10.1111/j.1365-2265.2010.03804.x.

Maggi, M., Filippi, S., Vignozzi, L., & Rastrelli, G. (2020). Controversial aspects of testosterone in the regulation of sexual function in late-onset hypogonadism. *Andrology, 8*, 1580–1589. https://doi.org/10.1111/andr.12794.

Melmed, S., Casanueva, F. F., Hoffman, A. R., Kleinberg, D. L., Montori, V. M., Schlechte, J. A., et al. (2011). Diagnosis and treatment of hyperprolactinemia: An Endocrine Society clinical practice guideline. *The Journal of Clinical Endocrinology and Metabolism, 96*(2), 273–288. https://doi.org/10.1210/jc.2010-1692.

Mohler, J. L., Antonarakis, E. S., Armstrong, A. J., D'Amico, A. V., Davis, B. J., Dorff, T., et al. (2019). Prostate cancer, version 2.2019, NCCN Clinical Practice Guidelines in Oncology. *Journal of the National Comprehensive Cancer Network, 17*(5), 479–505. https://doi.org/10.6004/jnccn.2019.0023.

Mohr, B. A., Bhasin, S., Link, C. L., O'Donnell, A. B., & McKinlay, J. B. (2006). The effect of changes in adiposity on testosterone levels in older men: longitudinal results from the Massachusetts Male Aging Study. *European Journal of Endocrinology, 155*(3), 443–452. https://doi.org/10.1530/eje.1.02241.

Morales, A., Bebb, R. A., Manjoo, P., Assimakopoulos, P., Axler, J., Collier, C., et al. (2015). Diagnosis and management of testosterone deficiency syndrome in men: clinical practice guideline. *Canadian Medical Association Journal, 187*(18), 1369–1377. https://doi.org/10.1503/cmaj.150033.

Morley, J. E., Kaiser, F. E., Perry, H. M., 3rd, Patrick, P., Morley, P. M., Stauber, P. M., et al. (1997). Longitudinal changes in testosterone, luteinizing hormone, and follicle-stimulating hormone in healthy older men. *Metabolism, 46*(4), 410–413. https://doi.org/10.1016/s0026-0495(97)90057-3.

Mulligan, T., Frick, M. F., Zuraw, Q. C., Stemhagen, A., & McWhirter, C. (2006). Prevalence of hypogonadism in males aged at least 45 years: The HIM study. *International Journal of Clinical Practice, 60*(7), 762–769. https://doi.org/10.1111/j.1742-1241.2006.00992.x.

Narayanan, R., Coss, C. C., & Dalton, J. T. (2018). Development of selective androgen receptor modulators (SARMs). *Molecular and Cellular Endocrinology, 465*, 134–142. https://doi.org/10.1016/j.mce.2017.06.013.

Neaves, W. B., Johnson, L., Porter, J. C., Parker, C. R., Jr., & Petty, C. S. (1984). Leydig cell numbers, daily sperm production, and serum gonadotropin levels in aging men. *The Journal of Clinical Endocrinology and Metabolism, 59*(4), 756–763. https://doi.org/10.1210/jcem-59-4-756.

Nieschlag, E. (2019). Late-onset hypogonadism: A concept comes of age. *Andrology, 8*, 1506–1511. https://doi.org/10.1111/andr.12719.

Ohlsson, C., Nilsson, M. E., Tivesten, A., Ryberg, H., Mellstrom, D., Karlsson, M. K., et al. (2013). Comparisons of immunoassay and mass spectrometry measurements of serum estradiol levels and their influence on clinical association studies in men. *The Journal of Clinical Endocrinology and Metabolism, 98*(6), E1097–E1102. https://doi.org/10.1210/jc.2012-3861.

Orwoll, E., Lambert, L. C., Marshall, L. M., Phipps, K., Blank, J., Barrett-Connor, E., et al. (2006). Testosterone and estradiol among older men. *The Journal of Clinical Endocrinology and Metabolism, 91*(4), 1336–1344. https://doi.org/10.1210/jc.2005-1830.

Pelusi, C., Gasparini, D. I., Bianchi, N., & Pasquali, R. (2016). Endocrine dysfunction in hereditary hemochromatosis. *Journal of Endocrinological Investigation, 39*(8), 837–847. https://doi.org/10.1007/s40618-016-0451-7.

Petak, S. M., Nankin, H. R., Spark, R. F., Swerdloff, R. S., & Rodriguez-Rigau, L. J. (2002). American Association of Clinical Endocrinologists Medical Guidelines for clinical practice for the evaluation and treatment of hypogonadism in adult male patients--2002 update. *Endocrine Practice, 8*(6), 440–456.

Rastrelli, G., Carter, E. L., Ahern, T., Finn, J. D., Antonio, L., O'Neill, T. W., et al. (2015). Development of and recovery from secondary hypogonadism in aging men: Prospective results from the EMAS. *The Journal of Clinical Endocrinology and Metabolism, 100*(8), 3172–3182. https://doi.org/10.1210/jc.2015-1571.

Rastrelli, G., Corona, G., Mannucci, E., & Maggi, M. (2014). Factors affecting spermatogenesis upon gonadotropin-replacement therapy: A meta-analytic study. *Andrology, 2*(6), 794–808. https://doi.org/10.1111/andr.262.

Rigon, F. A., Ronsoni, M. F., Hohl, A., & van de Sande-Lee, S. (2019). Effects of bariatric surgery in male obesity-associated hypogonadism. *Obesity Surgery, 29*(7), 2115–2125. https://doi.org/10.1007/s11695-019-03829-0.

Rochira, V. (2017). Hypogonadism in systemic diseases. In M. Simoni, & I. Huhtaniemi (Eds.), *Endocrinology of the Testis and Male Reproduction* (pp. 1–51). Springer International Publishing. https://doi.org/10.1007/978-3-319-29456-8_28-1.

Rochira, V. (2020). Late-onset hypogonadism: Bone health. *Andrology, 8*, 1539–1550. https://doi.org/10.1111/andr.12827.

Rochira, V., Antonio, L., & Vanderschueren, D. (2018). EAA clinical guideline on management of bone health in the andrological outpatient clinic. *Andrology, 6*(2), 272–285. https://doi.org/10.1111/andr.12470.

Rochira, V., & Carani, C. (2017). Estrogen deficiency in men. In M. Simoni, & I. T. Huhtaniemi (Eds.), *Endocrinology of the testis and male reproduction* (pp. 797–828). Springer International Publishing. https://doi.org/10.1007/978-3-319-44441-3_27.

Rochira, V., & Guaraldi, G. (2014). Hypogonadism in the HIV-infected man. *Endocrinology and Metabolism Clinics of North America, 43*(3), 709–730. https://doi.org/10.1016/j.ecl.2014.06.005.

Rochira, V., Kara, E., & Carani, C. (2015). The endocrine role of estrogens on human male skeleton. *International Journal of Endocrinology, 2015*, 165215. https://doi.org/10.1155/2015/165215.

Rochira, V., Madeo, B., Diazzi, C., Zirilli, L., Daniele, S., & Carani, C. (2000). Estrogens and male reproduction. In K. R. Feingold, B. Anawalt, A. Boyce, G. Chrousos, K. Dungan, A. Grossman, & D. P. Wilson (Eds.), *Endotext* MDText.com, Inc. Copyright © 2000–2020, MDText.com, Inc.

Rochira, V., Zirilli, L., Genazzani, A. D., Balestrieri, A., Aranda, C., Fabre, B., et al. (2006). Hypothalamic-pituitary-gonadal axis in two men with aromatase deficiency: Evidence that circulating estrogens are required at the hypothalamic level for the integrity of gonadotropin negative feedback. *European Journal of Endocrinology, 155*(4), 513–522. https://doi.org/10.1530/eje.1.02254.

Rochira, V., Zirilli, L., Madeo, B., Maffei, L., & Carani, C. (2009). Testosterone action on erythropoiesis does not require its aromatization to estrogen: Insights from the testosterone and estrogen treatment of two aromatase-deficient men. *The Journal of Steroid Biochemistry and Molecular Biology, 113*(3–5), 189–194. https://doi.org/10.1016/j.jsbmb.2008.12.007.

Rochira, V., Zirilli, L., Orlando, G., Santi, D., Brigante, G., Diazzi, C., et al. (2011). Premature decline of serum total testosterone in HIV-infected men in the HAART-era. *PLoS One, 6*(12), e28512. https://doi.org/10.1371/journal.pone.0028512.

Roeca, C., Al-Safi, Z., & Santoro, N. (2000). The postmenopausal women. In K. R. Feingold, B. Anawalt, A. Boyce, G. Chrousos, K. Dungan, A. Grossman, & D. P. Wilson (Eds.), *Endotext* MDText.com, Inc.

Rosner, W., Auchus, R. J., Azziz, R., Sluss, P. M., & Raff, H. (2007). Position statement: Utility, limitations, and pitfalls in measuring testosterone: An Endocrine Society position statement. *The Journal of Clinical Endocrinology and Metabolism, 92*(2), 405–413. https://doi.org/10.1210/jc.2006-1864.

Rosner, W., & Vesper, H. (2010). Toward excellence in testosterone testing: A consensus statement. *The Journal of Clinical Endocrinology and Metabolism, 95*(10), 4542–4548. https://doi.org/10.1210/jc.2010-1314.

Rund, D., & Rachmilewitz, E. (2005). Beta-thalassemia. *The New England Journal of Medicine, 353*(11), 1135–1146. https://doi.org/10.1056/NEJMra050436.

Russell, N., & Grossmann, M. (2019). MECHANISMS IN ENDOCRINOLOGY: Estradiol as a male hormone. *European Journal of Endocrinology, 181*(1), R23–R43. https://doi.org/10.1530/eje-18-1000.

Saad, F., Yassin, A., Doros, G., & Haider, A. (2016). Effects of long-term treatment with testosterone on weight and waist size in 411 hypogonadal men with obesity classes I-III: Observational data from two registry studies. *International Journal of Obesity, 40*(1), 162–170. https://doi.org/10.1038/ijo.2015.139.

Sansone, A., Romanelli, F., Sansone, M., Lenzi, A., & Di Luigi, L. (2017). Gynecomastia and hormones. *Endocrine, 55*(1), 37–44. https://doi.org/10.1007/s12020-016-0975-9.

Schneider, H. J., Sievers, C., Klotsche, J., Bohler, S., Pittrow, D., Lehnert, H., et al. (2009). Prevalence of low male testosterone levels in primary care in Germany: Cross-sectional results from the DETECT study. *Clinical Endocrinology, 70*(3), 446–454. https://doi.org/10.1111/j.1365-2265.2008.03370.x.

Seftel, A. D. (2006). Male hypogonadism. Part I: Epidemiology of hypogonadism. *International Journal of Impotence Research, 18*(2), 115–120. https://doi.org/10.1038/sj.ijir.3901397.

Shi, Z., Araujo, A. B., Martin, S., O'Loughlin, P., & Wittert, G. A. (2013). Longitudinal changes in testosterone over five years in community-dwelling men. *The Journal of Clinical Endocrinology and Metabolism, 98*(8), 3289–3297. https://doi.org/10.1210/jc.2012-3842.

Simó, R., Sáez-López, C., Barbosa-Desongles, A., Hernández, C., & Selva, D. M. (2015). Novel insights in SHBG regulation and clinical implications. *Trends in Endocrinology and Metabolism, 26*(7), 376–383. https://doi.org/10.1016/j.tem.2015.05.001.

Simoni, M., Fanelli, F., Roli, L., & Pagotto, U. (2012). Methodology for measuring testosterone, dihydrotestosterone and sex hormone-binding globulin in a clinical setting. In E. Nieschlag, & H. M. Behre (Eds.), *Testosterone action, deficiency, substitution* (pp. 60–86). Cambridge University Press.

Snyder, P. J., Bhasin, S., Cunningham, G. R., Matsumoto, A. M., Stephens-Shields, A. J., Cauley, J. A., et al. (2016). Effects of testosterone treatment in older men. *The New England Journal of Medicine, 374*(7), 611–624. https://doi.org/10.1056/NEJMoa1506119.

Snyder, P. J., Bhasin, S., Cunningham, G. R., Matsumoto, A. M., Stephens-Shields, A. J., Cauley, J. A., et al. (2018). Lessons from the testosterone trials. *Endocrine Reviews, 39*(3), 369–386. https://doi.org/10.1210/er.2017-00234.

Snyder, P. J., Kopperdahl, D. L., Stephens-Shields, A. J., Ellenberg, S. S., Cauley, J. A., Ensrud, K. E., et al. (2017). Effect of testosterone treatment on volumetric bone density and strength in older men with low testosterone: a controlled clinical trial. *JAMA Internal Medicine*, *177*(4), 471–479. https://doi.org/10.1001/jamainternmed.2016.9539.

Stearns, E. L., MacDonnell, J. A., Kaufman, B. J., Padua, R., Lucman, T. S., Winter, J. S., et al. (1974). Declining testicular function with age. Hormonal and clinical correlates. *The American Journal of Medicine*, *57*(5), 761–766. https://doi.org/10.1016/0002-9343(74)90850-x.

Swee, D. S., & Gan, E. H. (2019). Late-onset hypogonadism as primary testicular failure. *Frontiers in Endocrinology*, *10*, 372. https://doi.org/10.3389/fendo.2019.00372.

Swerdloff, R. S., & Dudley, R. E. (2020). A new oral testosterone undecanoate therapy comes of age for the treatment of hypogonadal men. *Therapeutic Advances in Urology*, *12*. 1756287220937232. https://doi.org/10.1177/1756287220937232.

Swerdloff, R. S., Dudley, R. E., Page, S. T., Wang, C., & Salameh, W. A. (2017). Dihydrotestosterone: Biochemistry, physiology, and clinical implications of elevated blood levels. *Endocrine Reviews*, *38*(3), 220–254. https://doi.org/10.1210/er.2016-1067.

Swerdloff, R. S., Wang, C., White, W. B., Kaminetsky, J., Gittelman, M. C., Longstreth, J. A., et al. (2020). A new oral testosterone undecanoate formulation restores testosterone to normal concentrations in hypogonadal men. *The Journal of Clinical Endocrinology and Metabolism*, *105*(8), 2515–2531. https://doi.org/10.1210/clinem/dgaa238.

Tajar, A., Forti, G., O'Neill, T. W., Lee, D. M., Silman, A. J., Finn, J. D., et al. (2010). Characteristics of secondary, primary, and compensated hypogonadism in aging men: Evidence from the European Male Ageing Study. *The Journal of Clinical Endocrinology and Metabolism*, *95*(4), 1810–1818. https://doi.org/10.1210/jc.2009-1796.

Travison, T. G., Araujo, A. B., Kupelian, V., O'Donnell, A. B., & McKinlay, J. B. (2007). The relative contributions of aging, health, and lifestyle factors to serum testosterone decline in men. *The Journal of Clinical Endocrinology and Metabolism*, *92*(2), 549–555. https://doi.org/10.1210/jc.2006-1859.

Travison, T. G., Araujo, A. B., O'Donnell, A. B., Kupelian, V., & McKinlay, J. B. (2007). A population-level decline in serum testosterone levels in American men. *The Journal of Clinical Endocrinology and Metabolism*, *92*(1), 196–202. https://doi.org/10.1210/jc.2006-1375.

Travison, T. G., Shackelton, R., Araujo, A. B., Hall, S. A., Williams, R. E., Clark, R. V., et al. (2008). The natural history of symptomatic androgen deficiency in men: Onset, progression, and spontaneous remission. *Journal of the American Geriatrics Society*, *56*(5), 831–839. https://doi.org/10.1111/j.1532-5415.2008.01679.x.

Travison, T. G., Vesper, H. W., Orwoll, E., Wu, F., Kaufman, J. M., Wang, Y., et al. (2017). Harmonized reference ranges for circulating testosterone levels in men of four Cohort Studies in the United States and Europe. *The Journal of Clinical Endocrinology and Metabolism*, *102*(4), 1161–1173. https://doi.org/10.1210/jc.2016-2935.

Trost, L. W., & Mulhall, J. P. (2016). Challenges in testosterone measurement, data interpretation, and methodological appraisal of interventional trials. *The Journal of Sexual Medicine*, *13*(7), 1029–1046. https://doi.org/10.1016/j.jsxm.2016.04.068.

van den Beld, A. W., Kaufman, J. M., Zillikens, M. C., Lamberts, S. W. J., Egan, J. M., & van der Lely, A. J. (2018). The physiology of endocrine systems with ageing. *The Lancet Diabetes and Endocrinology*, *6*(8), 647–658. https://doi.org/10.1016/s2213-8587(18)30026-3.

Van Poppel, H., & Abrahamsson, P. A. (2020). Considerations for the use of gonadotropin-releasing hormone agonists and antagonists in patients with prostate cancer. *International Journal of Urology*, *27*, 830–837. https://doi.org/10.1111/iju.14303.

Veldhuis, J. D. (2013). Changes in pituitary function with ageing and implications for patient care. *Nature Reviews. Endocrinology, 9*(4), 205–215. https://doi.org/10.1038/nrendo.2013.38.

Veldhuis, J. D., Keenan, D. M., Liu, P. Y., Iranmanesh, A., Takahashi, P. Y., & Nehra, A. X. (2009). The aging male hypothalamic-pituitary-gonadal axis: Pulsatility and feedback. *Molecular and Cellular Endocrinology, 299*(1), 14–22. https://doi.org/10.1016/j.mce.2008.09.005.

Vermeulen, A., Kaufman, J. M., & Giagulli, V. A. (1996). Influence of some biological indexes on sex hormone-binding globulin and androgen levels in aging or obese males. *The Journal of Clinical Endocrinology and Metabolism, 81*(5), 1821–1826. https://doi.org/10.1210/jcem.81.5.8626841.

Vermeulen, A., Verdonck, L., & Kaufman, J. M. (1999). A critical evaluation of simple methods for the estimation of free testosterone in serum. *The Journal of Clinical Endocrinology and Metabolism, 84*(10), 3666–3672. https://doi.org/10.1210/jcem.84.10.6079.

Walker, R. F., Zakai, N. A., MacLehose, R. F., Cowan, L. T., Adam, T. J., Alonso, A., et al. (2019). Association of testosterone therapy with risk of venous thromboembolism among men with and without hypogonadism. *JAMA Internal Medicine, 180*(2), 190–197. https://doi.org/10.1001/jamainternmed.2019.5135.

Wang, Y., Gay, G. D., Botelho, J. C., Caudill, S. P., & Vesper, H. W. (2014). Total testosterone quantitative measurement in serum by LC-MS/MS. *Clinica Chimica Acta, 436*, 263–267. https://doi.org/10.1016/j.cca.2014.06.009.

Wang, C., Nieschlag, E., Swerdloff, R., Behre, H. M., Hellstrom, W. J., Gooren, L. J., et al. (2008). Investigation, treatment and monitoring of late-onset hypogonadism in males: ISA, ISSAM, EAU, EAA and ASA recommendations. *European Journal of Endocrinology, 159*(5), 507–514. https://doi.org/10.1530/eje-08-0601.

Wheeler, K. M., Sharma, D., Kavoussi, P. K., Smith, R. P., & Costabile, R. (2019). Clomiphene citrate for the treatment of hypogonadism. *Sexual Medicine Reviews*, 7(2), 272–276. https://doi.org/10.1016/j.sxmr.2018.10.001.

Wu, F. C., Tajar, A., Beynon, J. M., Pye, S. R., Silman, A. J., Finn, J. D., et al. (2010). Identification of late-onset hypogonadism in middle-aged and elderly men. *The New England Journal of Medicine, 363*(2), 123–135. https://doi.org/10.1056/NEJMoa0911101.

Wu, F. C., Tajar, A., Pye, S. R., Silman, A. J., Finn, J. D., O'Neill, T. W., et al. (2008). Hypothalamic-pituitary-testicular axis disruptions in older men are differentially linked to age and modifiable risk factors: The European Male Aging Study. *The Journal of Clinical Endocrinology and Metabolism, 93*(7), 2737–2745. https://doi.org/10.1210/jc.2007-1972.

Yeap, B. B., Almeida, O. P., Hyde, Z., Norman, P. E., Chubb, S. A., Jamrozik, K., et al. (2007). In men older than 70 years, total testosterone remains stable while free testosterone declines with age. The Health in Men Study. *European Journal of Endocrinology, 156*(5), 585–594. https://doi.org/10.1530/eje-06-0714.

Yeap, B. B., Manning, L., Chubb, S. A. P., Handelsman, D. J., Almeida, O. P., Hankey, G. J., et al. (2018). Progressive impairment of testicular endocrine function in ageing men: Testosterone and dihydrotestosterone decrease, and luteinizing hormone increases, in men transitioning from the 8th to 9th decades of life. *Clinical Endocrinology, 88*(1), 88–95. https://doi.org/10.1111/cen.13484.

Zmuda, J. M., Cauley, J. A., Kriska, A., Glynn, N. W., Gutai, J. P., & Kuller, L. H. (1997). Longitudinal relation between endogenous testosterone and cardiovascular disease risk factors in middle-aged men. A 13-year follow-up of former Multiple Risk Factor Intervention Trial participants. *American Journal of Epidemiology, 146*(8), 609–617. https://doi.org/10.1093/oxfordjournals.aje.a009326.

CHAPTER FOURTEEN

Sexual dimorphism, aging and immunity

Veena Taneja*
Department of Immunology and Rheumatology, Mayo Clinic, Rochester, MN, United States
*Corresponding author: e-mail address: taneja.veena@mayo.edu

Contents

Abstract

Health and lifespan disparities between sexes are dependent on the immune responses. Men and women have different life styles which determine the environment, nutritional requirements and their interactions with the sex hormones. Sexual dimorphism in innate and adaptive immunity determines responses to infections and other environmental factors regulating health and diseases. Sex hormones regulate immune responses through the expression of receptors which differ for female and male hormones. Estrogen receptors are expressed in brain, lymphoid tissue cells and many immune cells while androgen receptors are limited in expression. Genetic, epigenetic factors and X chromosome linked immune function genes are important in enhanced adaptive immunity in females, leading to production of higher levels of antibodies compared to males. Different nutritional requirements and hormonal control of the mucosal microbiome and its function regulate mucosal immunity. Hormonal changes during various aspects of life and during aging control immune senescence. Evolutionarily,

Vitamins and Hormones, Volume 115
ISSN 0083-6729
https://doi.org/10.1016/bs.vh.2020.12.015

females have an advantage during young age when they are protected from infections by heightened immune reactivity though during aging that can lead to pathologies. Considering the sexual dimorphism in immunity, guidelines need to be established for sex-based treatments for optimal response.

1. Introduction

Sexual dimorphism in diseases has led scientists to investigate the sex-biased differences in the immune system that direct the final outcome. Aging is associated with changes in the immune cell populations, thereby altering innate and adaptive immune system. Aging immune system is characterized by low grade inflammation, also called inflammaging. Some studies have suggested that innate immune system changes lead to chronic inflammation in females while adaptive immune system changes are prominent in males. The immune dysregulation during aging leads to various chronic conditions like autoimmunity and cancers and have sex-specific prevalence for various conditions (Taneja, 2018).

Autoimmunity occurs as a result of abnormal immune response to self-proteins which generally become severe during aging. Autoimmune diseases affect around 5–10% of population in western world. Why autoimmunity onset happens in a subset of individuals is unknown though the studies till date suggest an interaction of environmental and host genetic factors play a role in the causation of most of the autoimmune diseases. According to the Genome-wide association studies (GWAS), the most significant association between autoimmune diseases and genetic factors has been the presence of certain genes in the Major histocompatibility complex class (MHC), located on the short arm of chromosome 6. Within the MHC complex, HLA class II genes are the most polymorphic genes in the human genome. All autoimmune diseases have been found to be associated with the presence of certain class II alleles (Mangalam, Rajagopalan, Taneja, & David, 2008). Although the mechanism by which HLA alleles predispose to various diseases is unknown, antigen presentation by the class II alleles has been suggested as one of the potential mechanisms (Behrens et al., 2010; Taneja, 2015). However, of the many alleles and sub alleles of HLA-DR, only 3 HLA DR alleles, DR2, DR3 and DR4 and their linked HLA-DQ alleles have been associated with autoimmune diseases. HLA molecules are involved in presenting antigens and generating an immune response to pathogens for clearing infections. However, not all class II alleles generate similar responses. Certain alleles like DR4 generate robust response and clear infection;

however, it is associated with aging-associated rheumatoid arthritis (RA). Antigen presentation in the thymus is involved in selection of T cells thus based on the host's class II allele, selection of T cell repertoire could vary (Moller, Bohme, Valugerdi, Ridderstad, & Olerup, 1990). Accumulating evidence that the autoimmune associated class II alleles are commonly represented in the population suggests they have been positively selected during evolution. Women generate robust cellular and humoral response to infections though during aging they are much more susceptible to dysregulated immune response-associated autoimmunity, suggesting immune cell functions are sex-specific. However, not all individuals carrying these genes develop autoimmunity suggesting the involvement of other factors.

Sociological differences between men and women as well as occupation, smoking habits, alcohol consumption and health differences during aging have been implicated in the health and autoimmunity (Borchers & Gershwin, 2012). Differential exposure to environmental factors due to the type of work such as sunlight, have been suggested to impact the susceptibility to developing an autoimmune disease. Multiples sclerosis has been inversely associated with exposure to sunlight (D'hooghe, Haentjens, Nagels, Garmyn, & De Keyser, 2012). Recent studies have shed light on the sex-specific impact of smoking on immunity (Koo et al., 2020). Women show a significant increase in expression of certain genes like autophagy.

There is evidence that genetics and epigenetics, sex chromosomes, sex-linked gene regulation of immune responses and their interactions with environmental factors may have the potential for augmenting autoimmunity in women.

2. Epigenetics in sex-bias response

Besides genetic factors, the sex-bias of diseases may be related to parental chromosomal differences and epigenetics. Epigenetics is a phenomenon which can change gene expression affecting gene function without changing the underlying DNA sequences. Epigenetics occurs in cellular processes in normal physiology and either silences or activates gene expression and can be one of the reasons for differential expression of genes in individuals. Recently, a role of epigenetics has received attention in pathogenesis of various sex-biased autoimmune diseases. Epigenetic changes in autoimmunity can result from external factors, nutrition and environmental, or they could be methyl marks imprinted due to the sex of a parent. Parental inheritance of sex-biased autoimmunity has been studied extensively.

Although not consistent, a significant maternal inheritance of multiples sclerosis has been reported (Chao et al., 2010; Herrera et al., 2008; Hoppenbrouwers et al., 2008). For rheumatoid arthritis, inheritance was based on the susceptible genetic factor from affected or unaffected parent suggesting genetic susceptibility has a critical role in the onset of disease (Taneja, Mehra, Anand, & Malaviya, 1993).

Environmental factors may lead to epigenetic changes which can be heritable. Within cells, epigenetics leads to the expression of required genes so that each cell performs its intended functions. Epigenetic silencing of X chromosome genes in females is an important phenomenon to ensure that double dosage of genes is not expressed. It is hypothesized that an alteration of this phenomenon can lead to double dosage of certain genes on X chromosomes in various diseases, for example a higher expression of TLR7 in lupus patients suggest that TLR7 gene on inactivated X chromosome may have become active enhancing the expression (Shen et al., 2010). TLR7 can bind endogenous single stranded RNA to elicit immune response to pathogens (Hemmi et al., 2002). Thus epigenetics can regulate gene expression and function. Three interrelated mechanisms interact to regulate gene expression, Histone modifications, DNA methylation and microRNA-associated silencing.

2.1 Methylation

DNA methylation adds a methyl group where cytosine is present next to guanine, also called CPG islands, within the gene sequence resulting in transcriptional silencing. Methylation requires one of the 3 enzymes called DNA methyl transferases (DNMTs). Using whole blood, a difference in DNA methylation with 3% of CPG sites being methylated differentially between males and females were observed (Pelikan et al., 2018). Genome-wide studies in immune cells showed that cell-specific methylated regions determine a role in sexual dimorphic immune response (Mamrut et al., 2015; Singmann et al., 2015). Estrogen induces expression of miR148a which suppresses DNMT1 leading to hypomethylation of genes that may have a role in pathophysiology of diseases (Weeding & Sawalha, 2018). Estrogen can downregulate DNMT 1 expression thereby enhancing global DNA hypomethylation in CD4 T cells, as observed in patients with lupus (Wu et al., 2014). In patients with rheumatoid arthritis, fibroblasts are the key mediators of inflammation. DNA hypomethylation in IL-6 and CXCL12 genes in fibroblasts leads to activation and retaining of activated cells and unrestricted production of IL-6 causing osteoclast differentiation and bone resorption (Firestein & McInnes, 2017). Distinctive methylation signature of RA synoviocytes includes genes regulating cell signaling, trafficking, and extracellular matrix pathways (Arend & Firestein, 2012). Most autoimmune diseases show sex-bias affecting women predominantly and are strongly associated with class II genes of the Major histocompatibility complex (MHC). Expression of HLA class II locus has been shown to be regulated by histone tail methylation, increasing the risk of autoimmunity (Pelikan et al., 2018). A reduction in methylation in T cells of HLA-DRB1*1501 was reported to result in increased susceptibility to multiple sclerosis (Kular et al., 2018). Methylation of estrogen receptor alpha promoter in the brain has also been shown to contribute to sex differences by imprinting methyl marks on the DNA (Edelmann & Auger, 2011). Female brains have higher methylation as compared to males (Nugent et al., 2015). A comparison of sex-differential methylation in brains in patients with autism spectrum disorder (ASD) showed that genes associated with the disorder were hypermethylated in females leading to low expression and lower likelihood of developing disorder (Xia et al., 2019). Indeed, males are 3–4 times likely to ASD.

2.2 Histone modifications

Histone modifications can occur via methylation, addition of a methyl group, and acetylation, addition of an acetyl group, to lysine amino acids in the histones. Histones are primary components of chromatin and their modifications can alter chromatin structure, which based on the modification, determines whether DNA can be transcribed. Histone acetylation leads to gene expression while deacetylation is associated with gene silencing. Similar to DNMTs, histone modifications require enzymes like histone acetylates (HAT) and deacetylases (HDAC). Inactivation of one of the parental X chromosome is random and occurs via modification of histone (Dixon-McDougall & Brown, 2016). Genetic imprinting via epigenetic modifications determines gene copies to be expressed by the cells thus controlling the differential expression in cells. Sex determining region Y (SRY), a gene on the Y chromosome, through the high-mobility group box interacts with the Kruppel-associated box domain to recruit histone-modifying enzymes leading to chromatin remodeling (Oh, Li, & Lau, 2005). Histone modifications can also change functions in various cells. During brain differentiation, HDAC2 and HDAC4 bind to promoters of Esr1 (estrogen receptor1) and Cyp19 (cytochrome P450 family 19 subfamily A member 1) leading to higher gene expression impacting steroidogenesis. This as well as H3 methylation in the cortical region is much more prevalent in males compared to females (Tsai, Grant, & Rissman, 2009).

In T regulatory cells, acetylation of histone H4 at the FoxP3 locus when exposed to androgen leads to an increase in expression of FoXP3 (Walecki et al., 2015). T regulatory cells (Tregs) can suppress inflammatory response. That could be one of the reasons males have higher numbers of FoxP3 expressing Tregs (Afshan, Afzal, & Qureshi, 2012). A study with transgenic mice suggested that percentage of Tregs may also be dependent on the HLA class II polymorphism suggesting that epigenetics and MHC determine the selection of T cells in the thymus (Behrens et al., 2010; Luckey et al., 2014; Taneja & David, 2010). Autoimmune regulator (AIRE), a gene expressed in medullary thymic epithelial cells (mTEC), is a transcription regulator. It regulates central tolerance by inducing naturally occurring Tregs through controlling the expression of tissue specific antigens which is a key factor in negative selection of autoreactive T cells and hence autoimmunity (Hinterberger et al., 2010). Using mouse models, estrogen was shown to downregulate the expression of AIRE in mTEC via methylation of multiple sites which increased susceptibility to autoimmunity (Dragin et al., 2016). On the other hand, dihydrotestosterone (DHT) increases the expression of AIRE (Hinterberger et al., 2010). In humans, reduced expression of AIRE has been observed in patients with autoimmune diseases suffering from thymoma. Also, reduced expression of intrathymic AIRE was present in individuals with hypothyroidism and Down syndrome (Gimenez-Barcons et al., 2014). However, a genetic defect causing deficiency of AIRE led to similar autoimmune disease, type 1 polyglandular syndrome, in both sexes suggesting AIRE and sex hormones interaction plays a role in sexual dimorphism (Kahaly & Frommer, 2018). Estrogen acts via ligand-dependent or ligand independent mediated pathways to regulate gene expression and signaling pathways to control various cellular functions. 17β Estradiol (E2) can bind estrogen receptors, ERα or ERβ. E2 binds the target gene via motifs estrogen response element (ERE). ERα molecules can recruit chromatin remodeling enzymes, such as HATs and HDAC, via their association with cofactors such as GATA-3, a transcription factor controlling inflammatory and humoral responses, and Forkhead Box A1(FOXA1), a transcription factor involved in embryonic development and gene expression in tissues (Lee, Park, Oh, Kwack, & Park, 2017). These enzymes can further participate in immune response and thereby control pathologies. A role of HDAC inhibitors in suppressing arthritis in animal models elicit important role of epigenetics in diseases (Cantley et al., 2015).

2.3 MicroRNA

Epigenetics in parental chromosome also called genomic imprinting, and in particular, microRNA (miRNA) imprinting, can also influence the sex-bias observed in immunity thereby contributing to health or disease. miRNA are small non-protein coding RNA, encoded within the genome, that control post transcriptional gene expression. Expressed ubiquitously, miRNA constitute 3% of the human genome and control functions of most of the human genes. Evolutionarily, they are conserved across species. The expression of various miRNAs is controlled by other epigenetic mechanisms like methylation and acetylation. miRNA inhibit translation of mRNA to protein by binding to a specific 3′ untranslated region of an mRNA. During immune cell development, immune homeostasis and tolerance require maintaining a control of the expression and function of miRNA. Evidence for a role of miRNA in differentiation of T and B cells has been shown in mouse models (Dai et al., 2013). Besides differences in the expression of miRNA on X and Y chromosomes, sex differences in gonadal and non-gonadal tissues have also been documented. At least 40 miRNA been shown to have sex-specific expression in prefrontal cortex which target genes involved in Wnt signaling, that controls cell fate and organogenesis, and TGFβ pathways, implicated in neurological diseases (Zaits & Rennert, 2014).

A significant role of miRNA has been documented in immunity and autoimmunity. Females generate better innate and adaptive immune response compared to males. The major reason for this difference has been attributed to the genetic differences in sex chromosome, sex hormones and epigenetic differences in immune related genes. The X chromosome linked immune genes FoxP3, a transcription marker for T regulatory cells, and CD40L, co-stimulatory molecule, regulate T cell and B cell responses. Genes for cytokine receptors, IL-1, IL-9R, IL-2Rγ chain, IL-13RA1 present on the X chromosome are all involved in adaptive immunity. Also, genes for TLR7 and TLR8, involved in innate immunity, are present on the X chromosome. Expression of many of these genes is controlled by miRNA as discussed in another section.

Dysregulated miRNA expression has been reported in many diseases. Both lupus and RA occur predominantly in females. Sexually dimorphic expression of many miRNA, miR31, miR127, miR148a, miR155, miR182 and miR379, have been observed in spleens of the animal model of lupus in NZB/W mice (Dai et al., 2013). miR31 has been suggested to regulate Tregs differentiation and function (Rouas et al., 2009) and hence have a role in various autoimmune diseases. Role of miR155 has been extensively studied in various diseases including RA and lupus. Stimulation of macrophages through TLRs or increased levels of IFNs leads to activation. miR55 regulates polarization of macrophages and causes increased production of TNFα by inhibiting phosphatidylinositol-3,4,5-trisphosphate 5-phosphatase 1 (SHIP-1). This can cause activation of inflammatory cells and in combination with activation of macrophages leads to Th17 response. This concept is supported by a high miR155 expression in Th17 cells and low expression in Tregs (Alivernini et al., 2018). A central role of miR155 has been shown in the differentiation of T cells in to Th1, Th2, Th17, follicular T helper cells (Tfh) and Tregs (Alivernini et al., 2018). In RA joints, increased expression of miR155 is associated with differentiation of osteoclasts and is correlated with disease severity. Also, in activated B cells, an increased expression of miR155 can augment antibody production and isotype switch. Since RA is characterized by the presence of preclinical antibodies, miR155 has a significant role in RA as well as other antibody mediated diseases like lupus. Circulating B cells of seropositive RA patients have higher expression of miR155 as compared to psoriatic arthritis patients (Alivernini et al., 2016) and are a prognostic marker for treatment response. Further, miR155 suppresses the TGFβ signaling molecule SMAD5, leading to proliferation of germinal center B cells which can amplify inflammation as besides producing antibodies, B cells also present antigen. miR155 targets CD1d in B cells and impairs lipid antigen presentation to natural killer (NK) cells (Liu et al., 2015). Women have higher numbers of B cells and are more likely to be affected by an increased miR155 expression. miR155 also regulates function of dendritic cells (DCs) and downregulates DC-SIGN by inhibiting the transcription factor PU.1 while increasing CD11a expression which interacts with intercellular adhesion molecule 1 (ICAM-1), facilitating migration of monocytes to inflamed site (Syed, Amin, & Rabquer, 2015).

Epigenetic mechanisms can be affected by hormones as well as environmental factors such as nutrition and smoking. Cigarette smoking in conjunction with genetic factors has been associated with increased diseases severity (Klareskog, Padyukov, Lorentzen, & Alfredsson, 2006; Vassallo et al., 2014; Wang et al., 2019). Smoking is associated with epigenetic changes and posttranslational modifications like citrullination. Long term smoking can cause epigenetic changes which can predict mortality in diseases

(Corley et al., 2019). Smoke-induced DNA methylation showed sex-specific differences in genes associated with many immunological functions in the lung tissue with higher effects in women (Koo et al., 2020). Nutrition prenatally and during early life determines the epigenetic changes via enzymes DNMT, HATs, and HDAC (Lillycrop, Hoile, Grenfell, & Burdge, 2014). For example, folate metabolism and resveratrol are linked to DNA methylation (Choi & Friso, 2010). High fat diet has been associated with obesity through histone deacetylation in gene expression of pancreatic islets causing dysregulation of fatty acid signaling (Nammo et al., 2018). One year follow up of individuals, above the age of 65 years, on Mediterranean diet showed significant changes in methylation patterns in immunosenescence markers associated with aging (Gensous et al., 2020). The differences were much more noticeable in women in one of the populations studied, suggesting sex-specific diet-based epigenetic changes may determine the immunity and aging process.

3. Regulation of epigenetics by sex hormones

Sex hormones, estrogen and androgens, can also regulate the expression of miRNA. Androgen receptors bind DNA sequences called androgen response element (ARE) in the regulatory region of the target gene to activate downstream signaling which is essential for spermatogenesis. Testosterone has been shown to regulate miR-471, which targets Sertoli cells markers, Forkhead/winged-helix transcription factor (FoxD1) and the desmosomal cadherin (Dsc1) genes (Panneerdoss et al., 2012). Sequencing has revealed that there are around 86 miRNAs that are differentially expressed between androgen dependent and non-dependent prostate cancer (Xu et al., 2010). Androgen has been shown to promote many miRs including miR19a, miR133, which regulate male reproductive function. Androgen receptor binds the promoter of miR125b and upregulates its expression reducing the expression of BAK1 (Bcl2 homologous antagonist/killer), involved in mitochondrial apoptotic process (Shi et al., 2007). Unlike estrogen receptors, there are no reports that ARs are expressed on immune cells. Generally androgens suppress immune response and thus may protect from autoimmune diseases like RA and lupus. However, there is limited information on the role of androgens in miRNA control of immune response.

Estrogen receptors, on the other hand, are expressed on immune cells and hence E2 can control cellular immune functions (Yang & Wang, 2011). While many miRNAs regulate estrogen function, there is a feedback loop by which estrogen can regulate miRs functions and vice versa. Estrogen can control miRNA via 2 methods, miRNA biogenesis or via direct binding of the receptor to the estrogen binding site of the promoter of miRNA. Transcription of a gene can be controlled indirectly by estrogen via miRNA (Bjornstrom & Sjoberg, 2005; Dai & Ahmed, 2014). Estrogen can induce miR21, miR23a and miR221 which target ERs.

Another way estrogen can control miRNA expression is by inducing estrogen-regulated transcription factors, c-Myc and E2F, which further induces miR17 thereby forming a negative feedback loop to regulate estrogen responses. Estrogen-induced miR21, miR98 and let-7 play an essential role in refining estrogen-mediated responses via ERα (Castellano et al., 2009). Many estrogen regulated miRNA like miR146 and miR148 have been implicated in autoimmunity eliciting the mechanism by which estrogen contributes to autoimmune diseases (Dai et al., 2008). miR 221/222 is increased by TNFα and regulates proliferation and survival of cells.

4. Sex chromosome and dimorphic immune response

X chromosome harbors many genes known to regulate immune response. Although the common paradigm is that one of the two X chromosomes in women with XX is transcriptionally inactivated, there is evidence that certain genes can occur in double doses. In females, one of the 2 copies of X chromosome is randomly transcriptionally inactivated in each cell. Different cell populations may have maternal or paternal X chromosome inactivated resulting in variable responses. Males have XY chromosomes and lack dosage dependent inactivation. This makes males susceptible to diseases with X-linked genetic mutations. Evidence for the role of sex hormones comes from the significant differences observed in males and females in all the vertebrates including humans.

The increased expression of certain genes can regulate innate and adaptive immunity. The genes linked with immune response on X chromosome include IL-2R γ chain, IL-3Rα chain, IL-13α chain, IL-1R associated kinase 1 (IRAK1) TLR7, TLR8, GATA1, FOXP3 and CD40L. TLR 7 and TLR8 can sense viral RNA or DNA and generate robust response to a virus and clear the infection. Women produce higher IFNα in response to stimulation via TLR7, 9 while men produce high IL-10 (Bouman, Heineman, & Faas, 2005). On the other hand, CD40L is involved in B cell activation. Women are known to have greater numbers of B cells and consequently produce higher levels of antibodies than men (Klein & Flanagan, 2016). Lupus patients have increased expression of CD40L as well as TLR7 suggesting X-linked genes in causation of autoimmunity (Lu et al., 2007; Shen et al., 2010). Demethylation of CD40L in CD4+ T cells leads to increased expression of CD40L which can enhance immune reactivity. Thus gene dosage may explain enhanced immune reactivity in women. In contrast, a mutation in FoxP3 gene leads to polyendocrinopathy in males (Fish, 2008). It is theorized that 10–15% of the X-linked genes may be activated due to alterations in inactivation or under some conditions or via hormone-dependent epigenetic modification (Dai & Ahmed, 2014). Indeed X chromosome has around 113 micro RNAs (miR) which constitutes 7% of the miRNA in human genome while Y chromosome has only 2 miRNAs (Hewagama et al., 2013). miRNA are involved in epigenetic and transcriptional modification of gene expression and X-linked miRNA have shown to enhance immune response (Pinheiro, Dejager, & Libert, 2011). For example, both miR 98 and miR221 are X linked and have been shown to be increased in lupus patients (Hewagama et al., 2013). miR 223, another X-linked gene, is expressed in myeloid cells and reduced expression of miR223 is associated with cancer (Johnnidis et al., 2008). Further evidence that X-linked genes are involved in immune response comes from studies showing that Turner syndrome with 45 X rarely develop lupus while Klinefelter syndrome with 47XXY develop lupus with similar incidence as women with 46XX (Scofield et al., 2008). Besides, inactivation of paternal or maternal X chromosome can be cell-specific in females which can change function of the cell.

However, if estrogen dependent events and the X-linked genes were the only factors in the causation of sex-biased autoimmunity, all women would likely have autoimmune diseases. Thus other factors also are involved in autoimmune diseases, including genetic susceptibility and environmental factors.

5. Sex hormones and immunity

While autoimmune diseases occur predominantly in women, certain non-reproductive cancers are observed more often in men. Even though women suffer from autoimmunity, they have a longer life span than men. During aging, immune cell profile changes in both sexes though the variability in immune cells can contribute to sex-specific differences in immunity. Most autoimmune diseases have strong genetic disposition, which suggests that other factors that are required for the causation of autoimmunity have significant sex-specific impact on the immune system. With the present knowledge about the role of sex hormones, OMICS and immune system, it has become clear that interactions between these factors and genetic factors of the host determine the final outcome of any infection and abnormality of immune system.

There are significant differences in the innate and adaptive immune system of men and women which might depend on the immune cell differences. Males have much higher frequency of NK cells while antigen presenting capabilities as well as phagocytic activity of macrophages in females is much higher than males (Abdullah et al., 2012). Also, reduced numbers of type 2 innate lymphoid cells (ILCs) in females can cause dysregulation in innate immunity. Many of these differences are due to the interactions of environment and genetic factors with sex hormones. The numbers of total T cells are lower in men, however, women have higher CD4:CD8 ratio while men have higher CD8+ T cell (Uppal, Verma, & Dhot, 2003). Proliferation studies have pointed to a higher numbers of proliferating T cells with higher levels of Th1 cytokines in females than males (Klein & Flanagan, 2016). Females have higher number of B cells and produce higher levels of immunoglobulins than males (Klein & Flanagan, 2016). During aging, a change in immune cell profile with a decline in CD4+ and CD8+ cells occurs in both sexes. However lesser memory T regulatory cells and NK cells were associated with aged women as compared to aged men (Hirokawa et al., 2013; Lee et al., 1996). Since T cells and B cells as well NK cells produce cytokines, these differences explain the sexual dimorphism in immune response and overall immunity in health and diseased conditions.

Estrogen can control functions of many immune cells because of the presence of estrogen receptors. Many immune cells including T cells and B cells, macrophages, DCs, epithelial cells and endothelial cells express estrogen receptors. Estrogen is expressed as a dimer constituted of ERα and ERβ. Expression of ERα is more common in immune cells while ERβ shows much restricted expression. Most immune cells express ER receptors though expression levels may differ. While CD4 T cells express more of ERα, B cells preferentially express ERβ. This differential expression of ERα and ERβ regulates and maintains immune response. Estrogen can trigger signaling by binding to ER via estrogen response element (ERE) binding sites as well as membrane-associated G-protein coupled receptors (GPCRs). Estrogen has been shown to influence T cell development and lymphopoiesis as a reduction in thymic cellularity was observed in the presence of high doses of exogenous estrogen, leading to thymic atrophy (Brown & Su, 2019). Estrogen also exerts a strong influence on B cells in maturation of B cells leading to immunoglobulin production and isotype switching (Paavonen, Andersson, & Adlercreutz, 1981). Estradiol treatment increases the secretion of IL-6 and IL-10 thereby inducing polyclonal B cell activation and upregulating Bcl-2 expression leading to resistance to apoptosis (Medina, Strasser, & Kincade, 2000) which can enhance the survival of autoreactive B cells (Bynoe, Grimaldi, & Diamond, 2000) and exacerbate autoimmunity. A role of estrogen is also indicated in the alteration of tolerance induction of naive immature B cells. However, estrogen's impact on cells is dependent on the E2 levels. High levels of E2 inhibit CD8+ T cell induced suppression of B cells, augmenting antibody response (Clerici, Bergamasco, Ferrario, & Villa, 1991). B cells have been shown to present HLA-mediated antigen presentation and generate differential cellular response in both sexes. Studies in humanized mouse model of arthritis suggested that while B cells produce DRB1*0401-restricted peptides efficiently in females, DQ8-restricted peptides were presented better in male DRB1*0401/DQ8 mice (Behrens et al., 2010; Behrens, Luckey, Luthra, David, & Taneja, 2015; Taneja et al., 2007, 2008). Male mice implanted with 17Eβ showed an increase in B220 numbers which might be due to an increased survival of B cells. This suggested that APCs work differently in both sexes. While males produced high IL-10 levels, females produced higher IL-13 in arthritic and naïve mice. IL-13 is increased in patients with RA (Hitchon et al., 2004) and it can induce MHC expression leading to B cells activation and antibody production. Similar sex-difference in IL-13 has been shown in an animal model for multiple sclerosis (Sinha et al., 2008). Interestingly, HLA-DR4 expression is higher in female patients with RA (Behrens et al., 2010) and it is associated with seropositivity. Since B cells are critical to the onset of RA, studies were conducted with B cell depletion in mice. Arthritic humanized mice treated with anti-CD20 antibodies to deplete B cells showed that treatment efficacy is different in males and females. B cell depletion was optimal for males if started before plasma cell pool is established (Behrens et al., 2010). In patients with RA, anti-CD20 treatment has shown variable efficacy (Edwards et al., 2004). However, B cell depletion treatment is not based on the sex of the patient. Similarly, treatment for cancers has shown sex-specific outcomes. In patients with melanoma, treatment with immune checkpoint inhibitor programmed cell death ligand 1 (PDL1) showed better efficacy in females than males (Lin et al., 2010).

There is also an evidence of estrogen-mediated activation of DCs which can lead to the activation of autoreactive T cells and B cells and production of TNF-α, IL-1, IL-6 and IL-10 by monocytes (Bouman, Moes, Heineman, de Leij, & Faas, 2001). E2, in the presence of granulocyte-macrophage colony-stimulating factor (GMCSF), promotes differentiation of monocytes in to DCs that express TLR7 and TLR9 and produce IFN-γ, forming the basis of dimorphic response in sexes. E2 regulates DC differentiation through ERα. On the other hand, progesterone has been suggested to suppress DC function (Nalbandian & Kovats, 2005). Testosterone potentiates Th1, IL-1 and IL-12, response from monocytes (Posma, Moes, Heineman, & Faas, 2004).

Impact of estrogen is based on its levels in the host. During pregnancy, very high levels of estrogen are present which lead to differentiation of Tregs and generation of Th2 response resulting in the remission of symptoms. Moreover there is suppression of new B cells so antibodies are not increased. Mice treated with high doses of E2 show decreased IL-17 production (Wang et al., 2009). Action of estrogen is dependent on the ratio of estrogen and testosterone. When estrogen is low, in the presence of prolactin, there is a higher expression of Th1. Since CD4 cells express ERα, estrogen can directly act on them. E2, at lower concentrations, can augment CD4+ T cell proliferation and IFN-γ production as well as the IFN-γ mediated effects including production of nitric oxide (NO), inducible nitric oxide synthase (iNOS), and (cyclooxygenase) Cox2 (Moulton, 2018). This can activate macrophages and produce inflammatory cytokines. E2 enhances expression of CCR5 on T cells (Mo et al., 2005) thereby contributing to immune cell trafficking and inflammatory macrophage inflammatory protein (MIP)-1b and stromal cell-derived factor (SDF)-1b. Dendritic cells, CD11b + CD103-, express Esr1 through which E2 can regulate interferon regulatory factor-4 (IRF-4) production. These functions of E2, based on levels, in regulating immune cells, result in sex-specific immunity in health and disease.

6. Sex-hormones in autoimmunity

Among the many autoimmune diseases affecting various organs, neurological, gastrointestinal, rheumatologic and systemic, women have a higher prevalence for most of the autoimmune conditions with 80% patients being women though both sexes are equally susceptible to some conditions (Ngo, Steyn, & McCombe, 2014). However, among rheumatologic diseases, most conditions are prevalent in women except Spondyloarthritis (Spa) which occurs more often in men. While rheumatologic diseases like rheumatoid arthritis (RA) and Spa have a strong genetic disposition, sexual dimorphic nature of these conditions suggest a role of sex-hormones. Moreover RA undergoes remission during pregnancy suggesting a role of sex hormones in disease severity. This sexual-dimorphism in rheumatologic diseases provides an opportunity to understand the immunologic differences involved in sex-specific immunity.

The basis of sex-bias in autoimmunity is multiplex and involves genetics, epigenetics, sex chromosomes and sex-hormone regulation of immunity. Men and women have different requirements in life due to various life events including nutritional as well as hormonal. Sex hormones impact many aspects of human life. During pubertal maturation, pregnancy, and menopause, change in hormonal levels can alter immunity and impact immunity and susceptibility immune related conditions and autoimmunity. After puberty, with the change in hormonal status, females generate much higher adaptive immune responses while males show increased innate immunity (Lamason et al., 2006). Higher levels of androgens post pubertal suppress immune responses via reduction of innate immune cell responses including NK cells and macrophages while it leads to increasing numbers of T regulatory cells and levels of IL-10 and TGFβ. Androgen deficiency is associated with higher levels of inflammatory cytokines and antibody titers as well as CD4/CD8 ratio than with androgen sufficient men (Bobjer, Katrinaki, Tsatsanis, Giwercman, & Giwercman, 2013; Liva & Voskuhl, 2001; Pergola et al., 2008).

During pregnancy, estrogen levels are high which has been associated with improved symptoms of autoimmune disease. However, based on the changes in immunity due to sex hormones, autoimmune disease like lupus increases after puberty and decreases after menopause and becomes worse during pregnancy. Thus ratio of estrogen to androgen determines the outcome of certain diseases. Lupus patients have a high estrogen/androgen ratio. The differential effects of female hormones (estrogen, progesterone, and prolactin) and male hormone (androgens) on various immune cells, through direct or indirect effects, may also explain sex-bias in specific diseases. Men with abnormal ratio of androgens/E2 are much more likely to develop autoimmunity as shown for RA patients (Tengstrand, Carlstrom, Fellander-Tsai, & Hafstrom, 2003). Many autoimmune diseases manifest during aging or when there is a hormonal disturbance. During aging, a change in immune cell profiles occurs in both sexes, though women have much lower levels of memory T regulatory cells which can bias the immune response toward proinflammatory.

7. Sex-bias in mucosal immunity

There is some evidence of infectious etiology of sex-bias autoimmune diseases like rheumatoid arthritis and multiple sclerosis (Balakrishnan & Taneja, 2018; Legato, 2017). The concept that resident commensal bacteria can be causative first came into focus when microbial DNA, likely to be of gut origin, were sequenced from synovial fluid of RA patients (Holoshitz, Koning, Coligan, Debruyn, & Strober, 1989; Warden, 1909) as well as recurrence of periodontitis by a resident bacteria, *Porphyromonas gingivalis* (Mikuls et al., 2009; Parahitiyawa et al., 2010). There is a plethora of evidence that the gut microbiome has a strong impact on immune response locally and systemically (Balakrishnan & Taneja, 2018). The human intestine is colonized by trillions of bacteria that support various functions. The gut microbiota protects the intestine from environmental agents to maintain a healthy gut. Gut colonization with commensals begins at birth and forms a mutualistic relationship with the host stabilizing around 3 years. Each individual has a unique intestinal microbiota which is dependent on the genotype, hormones, aging and environmental factors including smoking, infections and diet. Diet is the strongest influencing factor for the gut microbiota. Men and women have different dietary needs which can influence the microbial composition. In humans, men showed much more noticeable diet-based effects on the microbiome than women suggesting diet-based effect on colonic ecosystem can impact sex-bias immunity (Bolnick et al., 2014; Dominianni et al., 2015). However, dietary and sex effect on the gut microbiome were restricted to a commonly present taxa, Fusobacteriaceae. Microbiota can also impact via producing metabolites as well as byproducts of dietary nutrients. Sex was correlated to the functional aspects of the microbes, such as aspartate biosynthesis in males and inosine monophosphate biosynthesis in females (Arumugam et al., 2011).

Gut microbes metabolize sex steroid hormones as well as interact with steroids, influencing their function as well as regulating the intestinal steroid balance and metabolic activity (Adlercreutz, Pulkkinen, Hamalainen, & Korpela, 1984; Hughes & Sperandio, 2008; Lombardi, Goldin, Boutin, & Gorbach, 1978). Certain microbes produce metabolites which have structural similarities with E2. This can skew the hormonal balance in the gut. Dietary fibers, metabolized by the gut microbes, can also increase systemic estrogen levels thereby impacting the gut microbiota (Dominianni et al., 2015; Gaskins et al., 2009). Intestinal commensal *Slackia* sp., is involved in inter-conversion of B-estradiol and estrone (Tamura, Dudley, Nei, & Kumar, 2007). Gut microbiota can change the intestinal metabolic landscape by 17β reduction of androgen leading to changes in the systemic testosterone levels. For example *Clostridium scindens* encodes hydroxysteroid dehydrogenases which can convert glucocorticoids in to androgens (Ridlon et al., 2013). Support for the role of gut microbiota in sex-specific immunity was shown in an experimental model of diabetes where germ free mice do not exhibit sex-bias in diabetes. In addition, testosterone-driven protection from diabetes was observed in females when microbiota from male mice was transferred (Markle et al., 2013; Yurkovetskiy et al., 2013). This suggests that steroids derived from commensals can impact host physiology and immunity. Also, host's hormones determine the expansion and colonization of the microbes that either influence autoimmunity or tolerance. Indeed, estrogen has a strong influence on the composition of gut microbiota as epithelial cells as well as immune cells including B cells and T cells express estrogen receptors. These receptors can bind E2 and lead to signaling of downstream events in the gut. Thus hormone levels may define the colonic ecosystem in a host. Hormonal changes during puberty are associated with different alpha diversity between males and females mice though major differences occurred in male mice while females did not show significant differences suggesting a strong influence of androgens on microbiota (Yurkovetskiy et al., 2013).

However in humans, studies on the sex as the driver of gut microbial composition are inconsistent. Certain studies have shown a difference in the presence of certain taxa based on the sex of an individual. According to one study, Bacteroidetes were present in higher abundance in men compared to women (Dominianni et al., 2015) while another study showed Prevotella group was abundant in men (Mueller et al., 2006). In an ethnically different population, *Bacteroides thetaiotaomicron* was observed to be higher in men than women in one study while another study reported increased *Rumnicoccus* abundance in females in Chinese population (Gao et al., 2018; Li et al., 2008). In contrast, in another study no sex differences in microbial composition were reported (Lay et al., 2005). It is possible that ethnicity dependent microbial composition will dictate the sex-specific differences. However, niche based differences in microbial composition based on sex have also been reported. A study from Italy reported that sex-differences are dependent on the niche; while lumen-associated microbiota showed no differences, mucosa associated microbiota showed sex differences with females with abundance of *Bifidobacterium Adolescentis* while males had

Gemmiger formicils (Borgo et al., 2018). Human microbiome project (HMP) Consortium showed sex-dependent differences with males having fewer Bacteroides and higher Prevotella (Ding & Schloss, 2014). Females generally had a higher diversity. In some populations, sex-differences in microbial composition were associated with body mass index (Haro et al., 2016). In a large cohort comprising of 1135 individuals, a study from Netherlands showed that microbial and metabolic functions were associated with sex; however, sex explained only 0.5% of variation (Sinha et al., 2019) suggesting other factors including genetic and environmental factors play a role in determining the microbial composition. It is also likely that lifestyle and culture-related role of sexes and related factors rather than the role of hormones may be reflected in these small differences. However, in humans, due to numerous factors, it is difficult to interpret the factors that influence overall immunity and their role in causality versus consequence of disease. Thus mouse models have been critical in defining the role of gut microbiota in health and disease.

Genetic factors, HLA class II alleles and hormones impact gut microbiome suggesting a possibility of their role in autoimmunity. HLA alleles determine immunity via presentation of antigens and production of cytokines (Taneja, 2015). It is conceivable that immune response defines which commensals can survive the milieu and take advantage of the ecosystem. Thus autoimmune-associated HLA genes might be able to control the microbial composition indirectly. This idea is supported by a study in mice expressing RA-associated HLA-DRB1*0401 mice and non-RA associated DRB1*0402 mice. These transgenic mice do not carry endogenous class II molecules and generate DR-restricted responses to antigens similar to humans (Taneja et al., 2007). The naïve transgenic mice, DRB1*0401 and *0402, produce varied responses to antigen-specific as well as bacterial products such as lipopolysaccharide (LPS) and harbor a different gut microbial composition (Behrens et al., 2010; Gomez et al., 2012; Luckey et al., 2014; Luckey, Gomez, Murray, White, & Taneja, 2013). Only arthritis-resistant mice showed gender-driven gut microbiome (Gomez, Luckey, & Taneja, 2015). In DRB1*0401 mice increased gut permeability caused by the resident bacteria as they aged could lead to systemic immune activation. Expansion of *Clostriadiales* in *0401 mice was associated with proinflammatory cytokine milieu with a decrease in immune regulatory cytokines. Even though arthritis-susceptible mice did not show any sex-bias in microbial composition overall, males had higher abundance of *Bifidobacterium* and females had increased *Clostriadiales* associated inflammation. Since HLA-DR4 determines the sex-bias for arthritis, it is possible that interaction between the host genotype and microbiota influences sex-bias in autoimmunity. *Bifidobacteria*, present in arthritis-resistant mice and male mice are considered beneficial taxa (Vaughan, Heilig, Ben-Amor, & de Vos, 2005). These observations do suggest that HLA-mediated control of dysbiosis and hormonal effect on the gut microbiome are responsible for the sexual dimorphism of disease. A role of gut microbiome in the causation and protection of arthritis has been suggested (Chen et al., 2016; Marietta et al., 2016; Taneja, 2017). One study suggested that genotype has stronger influence on the gut microbiota than sex in animals (Kovacs et al., 2011). Using 2 strains of mice, sex-specific microbial lineages as well as colonic immune functions were reported based on the strains (Elderman et al., 2018). A study using 89 inbred strains of mice showed sex-specific microbiota based on the genotype. In general, males showed increased abundance of *Allobaculum, Anaeroplasma and Erwinia* while females showed increased abundance of species of families, *Clostridiaceae* and *Lachnosperaceae* (Org et al., 2016).

These observations suggest a role of gut microbiome in health as well as gender-bias autoimmunity and further indicate that for defining mechanism of these pathologies, the role of dietary factors and immune-based differences may be more influential for disease progression. Advanced OMICS technology and tremendous efforts of scientists have been able to define the unculturable gut bacteria as well as microbiota of other mucosal surfaces. How these microbes and their metabolites shape immune response in various pathologies still remains unknown. Whether the colonic ecosystem is driven by the sex hormones and genotype of the host or other factors contribute substantially to different gender-based pathologies is unclear. Thus defining specific disease and sex-associated microbial clades as biomarkers will allow defining the mechanism as well as how to modulate the ecosystem.

8. Infections, aging and sex-bias

Most sex-bias diseases including autoimmune diseases occur during aging or are associated with immune senescence. There is evidence that women have stronger immune system, when young, and can fight infections by generating higher T cell and antibody responses. Generally women maintain a high immune reactivity even after viral infections have cleared.

In response to vaccinations, women generate higher protective antibody responses as compared to men. While this provides efficient protection from infections, side effects, if any, can be worse in women than men. One of the main reasons for women having immune advantage to clear infections efficiently is due to the immune function genes and TLRs on the X chromosome which are under the control of miRNAs. TLRs initiate innate immune response via ligation with the pathogen-associated molecular patterns (PAMPs) or damage-associated molecular patterns (DAMPs). While PAMPs are pathogen associated and allow TLRs microbial sensing, DAMPs are endogenous molecules released from cells that are damaged or necrosis.

Many viral infections show sexually dimorphic response. Androgens protect from Herpes simplex virus 1 (HSV1)-mediated inflammatory response though females generate higher HSV-specific IgG and IgM compared to males (Han et al., 2001). Coxsackie virus (CVB3) infection leading to myocarditis, an autoimmune condition, affects men more often than women (McNamara et al., 2011). E2 suppresses IL-17 and enhances Tregs and Th2 response while testosterone induces TH17 cell differentiation making men more susceptible. All strains of influenza also show sexual dimorphism and affect more women than men, which results in higher mortality in women (Klein, Passaretti, Anker, Olukoya, & Pekosz, 2010). Lower levels of estrogen and higher androgen suppress immune response to influenza in men limiting pathology (Hoffmann et al., 2015). Studies in patients infected with human immunodeficiency virus (HIV) have shown that HIV infected women carry 40% less viral load than men (Klein & Flanagan, 2016) confirming that women have stronger immunity and can clear viral infections better at young age. The most recent infectious disease is COVID-19 caused by severe acute respiratory syndrome coronavirus 2 (SARS CoV2). It was observed to have sex-specific pathogenicity in a study of 1099 COVID 19 patients from China (Chen et al., 2020) with increased severity in males as compared to women. While cigarette smoking was suggested to be the main cause of these differences, other genetic and environmental factors still need to be identified. One of the mechanisms, though not proven yet, is the epigenetic control of immunity resulting in a robust T cell response to the virus as well as antibodies production in women that leads to lower risk of mortality. Although aging was associated with increased mortality in both, men still showed higher mortality than women (Scully, Haverfield, Ursin, Tannenbaum, & Klein, 2020). In aged women, comorbidities like autoimmune diseases could increase COVID-19 related complications. Thus sex-specific immunity presents advantage for women though the price is the increased autoimmunity.

9. Conclusions

A strong evidence of sexual dimorphism in immunity is present and women have stronger immune system. This augmented immune response can lead to enhanced production of pro-inflammatory cytokines causing bystander effects. If immune response to a pathogen shares antigen mimicry with a self-protein, it can perpetuate inflammation precipitating an autoimmune response. This sexual dimorphism is a nature's way of protecting future progeny. Women during their childbirth age are protected from infectious diseases however most autoimmune diseases become severe during aging. This concept is supported by studies in an animal model where arthritis-susceptible HLA-DRB1*0401 mice can clear H1N1 influenza strains and when vaccinated, generate response and clear H3N2 when challenged (Luckey, Weaver, Osborne, Billadeau, & Taneja, 2019). However, DRB1*0401 mice are susceptible to develop arthritis (Taneja et al., 2007). On the other hand, arthritis-resistant DRB1*0402 mice, cleared H1N1 but did not develop cross antibodies to H3N2 and succumbed to infection. The study did not study sex-differences. Females show increased cellular and humoral response, differences in the numbers of various immune cells, responses generated by cells, proinflammatory cytokine production, and efficient antigen presentation, all of which can influence predisposition to diseases (Fig. 1). Increased T cell and B cell responses in females are associated with effective resistance to infections but consequently increased susceptibility to autoimmune diseases. Males, on the other hand, have lower proinflammatory cytokines, TNFα, while increased levels of IL-10 are produced which can dampen immune response. Women have been suggested to have longer life span and that could also be related to sex hormones and epigenetic changes. The studies point to sex as a driver of immunity in health and disease. Few studies in humans have determined sex-specific response to treatments. However, there are no guidelines to use patient's sex in to consideration for treatment. Taking a sex-informed approach to understanding immunopathology and treatments options will help provide optimal outcomes for all.

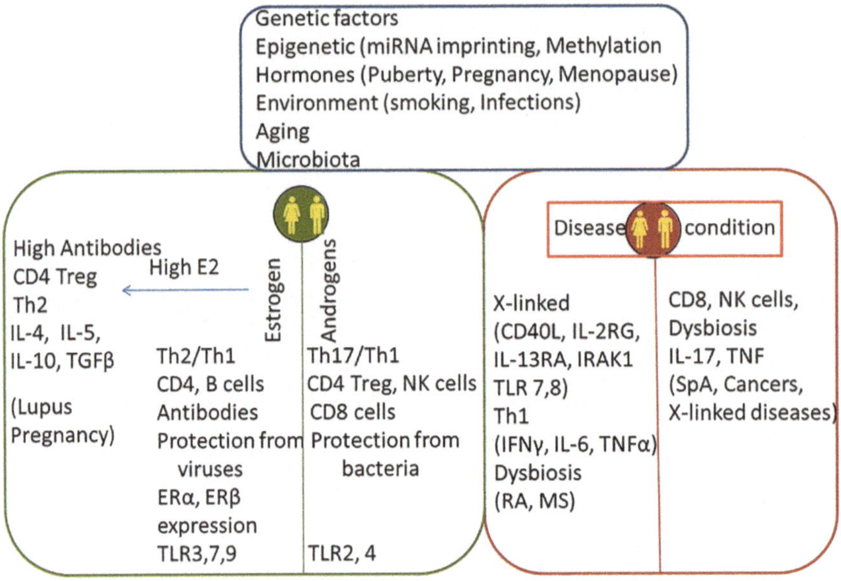

Fig. 1 Sexual dimorphism in health and disease. Factors affecting sexual dimorphism in immunity include the differences in immune cell profile of women and men. Women have more B cells and CD4+ cells and produce higher antibodies and Th1/Th2 responses. Estrogen receptors (ERs) ERα and ERβ are expressed on immune cells and thus estrogen can regulate function directly via binding to receptors and via epigenetics. Men have higher CD8 cells, CD4 T regulatory (Treg) cells and NK cells and generate Th1/Th17 responses. Women express higher toll like receptors (TLRs) 3, 7, 9, ligands for viruses while men express TLR2, 4, ligands for bacteria thus providing respective protection in sexes. The immune response is influenced by the genetics and epigenetic imprinting as well as hormones, environment, and microbiota and changes associated with the immune system during aging. High estrogen levels, such as in pregnancy, change immune cell profile and immune response leading to remission of Th1/Th17 diseases while causing flares for antibody-dependent diseases like systemic lupus erythematosus (lupus). In diseased conditions, X chromosome linked immune genes and TLR7, 8, low estrogen accompanied with microbial dysbiosis and epigenetic changes caused by miRNA on X chromosome can lead to increased proinflammatory conditions in women causing autoimmunity. In men, low androgens reduce suppression of T cell proliferation and dysbiosis accompanied with NK cells producing IL-17 and TNF-α, can result in spondyloarthropathies (SpA) or non-reproductive cancers and X-linked pathologies.

Acknowledgments

Veena Taneja's work has been supported by the Department of Defense, National Institute of Health, Arthritis Foundation and Mayo Clinic.

References

Abdullah, M., Chai, P. S., Chong, M. Y., Tohit, E. R. M., Ramasamy, R., Pei, C. P., et al. (2012). Gender effect on in vitro lymphocyte subset levels of healthy individuals. *Cellular Immunology*, *272*(2), 214–219. Retrieved from <Go to ISI>://WOS:000299302400015. https://doi.org/10.1016/j.cellimm.2011.10.009.

Adlercreutz, H., Pulkkinen, M. O., Hamalainen, E. K., & Korpela, J. T. (1984). Studies on the role of intestinal bacteria in metabolism of synthetic and natural steroid hormones. *Journal of Steroid Biochemistry*, *20*(1), 217–229. Retrieved from http://www.ncbi.nlm.nih.gov/pubmed/6231418.

Afshan, G., Afzal, N., & Qureshi, S. (2012). CD4(+)CD25(hi) regulatory T cells in healthy males and females mediate gender difference in the prevalence of autoimmune diseases. *Clinical Laboratory*, *58*(5–6), 567–571. Retrieved from <Go to ISI>://WOS:000305711100026.

Alivernini, S., Gremese, E., McSharry, C., Tolusso, B., Ferraccioli, G., McInnes, I. B., et al. (2018). MicroRNA-155-at the critical interface of innate and adaptive immunity in arthritis. *Frontiers in Immunology*, *8*, 1932. Retrieved from <Go to ISI>://WOS:000485400700002. doi:ARTN 1932. https://doi.org/10.3389/fimmu.2017.01932.

Alivernini, S., Kurowska-Stolarska, M., Tolusso, B., Benvenuto, R., Elmesmari, A., Canestri, S., et al. (2016). MicroRNA-155 influences B-cell function through PU.1 in rheumatoid arthritis. *Nature Communications*, 7, 12970. Retrieved from <Go to ISI>://WOS:000385445600001. doi:ARTN 12970. https://doi.org/10.1038/ncomms12970.

Arend, W. P., & Firestein, G. S. (2012). Pre-rheumatoid arthritis: predisposition and transition to clinical synovitis. *Nature Reviews Rheumatology*, *8*(10), 573–586. Retrieved from <Go to ISI>://WOS:000309514600007. https://doi.org/10.1038/nrrheum.2012.134.

Arumugam, M., Raes, J., Pelletier, E., Le Paslier, D., Yamada, T., Mende, D. R., et al. (2011). Enterotypes of the human gut microbiome. *Nature*, *473*(7346), 174–180. Retrieved from http://www.ncbi.nlm.nih.gov/pubmed/21508958. https://doi.org/10.1038/nature09944.

Balakrishnan, B., & Taneja, V. (2018). Microbial modulation of the gut microbiome for treating autoimmune diseases. Expert Review of Gastroenterology & Hepatology. *12*(10), 985–996. Retrieved from <Go to ISI>://WOS:000447195700006. https://doi.org/10.1080/17474124.2018.1517044.

Behrens, M., Luckey, D., Luthra, H., David, C., & Taneja, V. (2015). B cells influence sex specificity of arthritis via myeloid suppressors and chemokines in humanized mice. *Clinical Immunology*, *178*, 10–19. Retrieved from http://www.ncbi.nlm.nih.gov/pubmed/26057130. https://doi.org/10.1016/j.clim.2015.05.015.

Behrens, M., Trejo, T., Luthra, H., Griffiths, M., David, C. S., & Taneja, V. (2010). Mechanism by which HLA-DR4 regulates sex-bias of arthritis in humanized mice. *Journal of Autoimmunity*, *35*(1), 1–9. Retrieved from <Go to ISI>://WOS:000279200700001. https://doi.org/10.1016/j.jaut.2009.12.007.

Bjornstrom, L., & Sjoberg, M. (2005). Mechanisms of estrogen receptor signaling: Convergence of genomic and nongenomic actions on target genes. *Molecular Endocrinology*, *19*(4), 833–842. Retrieved from <Go to ISI>://WOS:000227826200001. https://doi.org/10.1210/me.2004-0486.

Bobjer, J., Katrinaki, M., Tsatsanis, C., Giwercman, Y. L., & Giwercman, A. (2013). Negative association between testosterone concentration and Inflammatory markers in young men: A Nested Cross-Sectional Study. *PLoS One*, *8*(4), e61466. https://doi.org/10.1371/journal.pone.0061466. Retrieved from < Go to ISI >://WOS:000317908700038. doi:ARTN e61466.

Bolnick, D. I., Snowberg, L. K., Hirsch, P. E., Lauber, C. L., Org, E., Parks, B., et al. (2014). Individual diet has sex-dependent effects on vertebrate gut microbiota. *Nature*

Communications, *5*, 4500. Retrieved from http://www.ncbi.nlm.nih.gov/pubmed/25072318. https://doi.org/10.1038/ncomms5500.

Borchers, A. T., & Gershwin, M. E. (2012). Sociological differences between women and men: Implications for autoimmunity. *Autoimmunity Reviews*, *11*(6–7), A413–A421. Retrieved from <Go to ISI>://WOS:000302973000005. https://doi.org/10.1016/j.autrev.2011.11.016.

Borgo, F., Garbossa, S., Riva, A., Severgnini, M., Luigiano, C., Benetti, A., et al. (2018). Body mass index and sex affect diverse Microbial niches within the gut. *Frontiers in Microbiology*, *9*, 213. Retrieved from <Go to ISI>://WOS:000425130800001. doi: ARTN 213. https://doi.org/10.3389/fmicb.2018.00213.

Bouman, A., Heineman, M. J., & Faas, M. M. (2005). Sex hormones and the immune response in humans. *Human Reproduction Update*, *11*(4), 411–423. Retrieved from <Go to ISI>://WOS:000230309600007. https://doi.org/10.1093/humupd/dmi008.

Bouman, A., Moes, H., Heineman, M. J., de Leij, L. F. M. H., & Faas, M. M. (2001). The immune response during the luteal phase of the ovarian cycle: increasing sensitivity of human monocytes to endotoxin. *Fertility and Sterility*, *76*(3), 555–559. Retrieved from <Go to ISI>://WOS:000170863300022. https://doi.org/10.1016/S0015-0282(01)01971-9.

Brown, M. A., & Su, M. A. (2019). An inconvenient variable: Sex hormones and their impact on T cell responses. *Journal of Immunology*, *202*(7), 1927–1933. Retrieved from <Go to ISI>://WOS:000461705000003. https://doi.org/10.4049/jimmunol.1801403.

Bynoe, M. S., Grimaldi, C. M., & Diamond, B. (2000). Estrogen up-regulates Bcl-2 and blocks tolerance induction of naive B cells. *Proceedings of the National Academy of Sciences of the United States of America*, *97*(6), 2703–2708. Retrieved from <Go to ISI>://WOS:000085941400055. https://doi.org/10.1073/pnas.040577497.

Cantley, M. D., Fairlie, D. P., Bartold, P. M., Marino, V., Gupta, P. K., & Haynes, D. R. (2015). Inhibiting histone deacetylase 1 suppresses both inflammation and bone loss in arthritis. *Rheumatology*, *54*(9), 1713–1723. Retrieved from <Go to ISI>://WOS:000362848500025. https://doi.org/10.1093/rheumatology/kev022.

Castellano, L., Giamas, G., Jacob, J., Coombes, R. C., Lucchesi, W., Thiruchelvam, P., et al. (2009). The estrogen receptor-alpha-induced microRNA signature regulates itself and its transcriptional response. *Proceedings of the National Academy of Sciences of the United States of America*, *106*(37), 15732–15737. Retrieved from <Go to ISI>://WOS:000269806600042. https://doi.org/10.1073/pnas.0906947106.

Chao, M. J., Herrera, B. M., Ramagopalan, S. V., Deluca, G., Handunetthi, L., Orton, S. M., et al. (2010). Parent-of-origin effects at the major histocompatibility complex in multiple sclerosis. *Human Molecular Genetics*, *19*(18), 3679–3689. Retrieved from https://doi.org/10.1093/hmg/ddq282. https://doi.org/10.1093/hmg/ddq282.

Chen, J., Wright, K., Davis, J. M., Jeraldo, P., Marietta, E. V., Murray, J., et al. (2016). An expansion of rare lineage intestinal microbes characterizes rheumatoid arthritis. *Genome Medicine*, *8*(1), 43. Retrieved from http://www.ncbi.nlm.nih.gov/pubmed/27102666. https://doi.org/10.1186/s13073-016-0299-7.

Chen, N. S., Zhou, M., Dong, X., Qu, J. M., Gong, F. Y., Han, Y., et al. (2020). Epidemiological and clinical characteristics of 99 cases of 2019 novel coronavirus pneumonia in Wuhan, China: a descriptive study. *Lancet*, *395*(10223), 507–513. Retrieved from <Go to ISI>://WOS:000514576900033. https://doi.org/10.1016/S0140-6736(20)30211-7.

Choi, S. W., & Friso, S. (2010). Epigenetics: A new bridge between nutrition and health. *Advances in Nutrition*, *1*(1), 8–16. Retrieved from <Go to ISI>://WOS:000208589000004. https://doi.org/10.3945/an.110.1004.

Clerici, E., Bergamasco, E., Ferrario, E., & Villa, M. L. (1991). Influence of sex steroids on the antigen-specific primary antibody response in vitro. *Journal of Clinical & Laboratory Immunology, 34*(2), 71–78. Retrieved from https://www.ncbi.nlm.nih.gov/pubmed/1667422.

Corley, J., Cox, S. R., Harris, S. E., Hernandez, M. V., Maniega, S. M., Bastin, M. E., et al. (2019). Epigenetic signatures of smoking associate with cognitive function, brain structure, and mental and physical health outcomes in the Lothian Birth Cohort 1936. *Translational Psychiatry, 9*, 248. Retrieved from <Go to ISI>://WOS:000489970700001. https://doi.org/10.1038/s41398-019-0576-5.

Dai, R. J., & Ahmed, S. A. (2014). Sexual dimorphism of miRNA expression: A new perspective in understanding the sex bias of autoimmune diseases. *Therapeutics and Clinical Risk Management, 10*, 151–163. Retrieved from <Go to ISI>://WOS:000332212900001. https://doi.org/10.2147/Tcrm.S33517.

Dai, R., McReynolds, S., Leroith, T., Heid, B., Liang, Z., & Ahmed, S. A. (2013). Sex differences in the expression of lupus-associated miRNAs in splenocytes from lupus-prone NZB/WF1 mice. *Biology of Sex Differences, 4*(1), 19. Retrieved from https://www.ncbi.nlm.nih.gov/pubmed/24175965. https://doi.org/10.1186/2042-6410-4-19.

Dai, R. J., Phillips, R. A., Zhang, Y., Khan, D., Crasta, O., & Ahmed, S. A. (2008). Suppression of LPS-induced Interferon-gamma and nitric oxide in splenic lymphocytes by select estrogen-regulated microRNAs: a novel mechanism of immune modulation. *Blood, 112*(12), 4591–4597. Retrieved from <Go to ISI>://WOS:000261217000036. https://doi.org/10.1182/blood-2008-04-152488.

D'hooghe, M. B., Haentjens, P., Nagels, G., Garmyn, M., & De Keyser, J. (2012). Sunlight exposure and sun sensitivity associated with disability progression in multiple sclerosis. *Multiple Sclerosis Journal, 18*(4), 451–459. https://doi.org/10.1177/1352458511423778. Retrieved from <Go to ISI>://WOS:000302289900012.

Ding, T., & Schloss, P. D. (2014). Dynamics and associations of microbial community types across the human body. *Nature, 509*(7500), 357–360. Retrieved from <Go to ISI>://WOS:000336121200038. https://doi.org/10.1038/nature13178.

Dixon-McDougall, T., & Brown, C. (2016). The making of a Barr body: the mosaic of factors that eXIST on the mammalian inactive X chromosome. *Biochemistry and Cell Biology, 94*(1), 56–70. Retrieved from <Go to ISI>://WOS:000370037600007. https://doi.org/10.1139/bcb-2015-0016.

Dominianni, C., Sinha, R., Goedert, J. J., Pei, Z. H., Yang, L. Y., Hayes, R. B., et al. (2015). Sex, body mass index, and dietary fiber intake influence the human gut microbiome. *PLoS One, 10*(4), e0124599. Retrieved from <Go to ISI>://WOS:000353015800147. https://doi.org/10.1371/journal.pone.0124599.

Dragin, N., Bismuth, J., Cizeron-Clairac, G., Biferi, M. G., Berthault, C., Serraf, A., et al. (2016). Estrogen-mediated downregulation of AIRE influences sexual dimorphism in autoimmune diseases. *Journal of Clinical Investigation, 126*(4), 1525–1537. Retrieved from <Go to ISI>://WOS:000373522300039. https://doi.org/10.1172/Jci81894.

Edelmann, M. N., & Auger, A. P. (2011). Epigenetic impact of simulated maternal grooming on estrogen receptor alpha within the developing amygdala. *Brain, Behavior, and Immunity, 25*(7), 1299–1304. Retrieved from <Go to ISI>://WOS:000295554300004. https://doi.org/10.1016/j.bbi.2011.02.009.

Edwards, J. C. W., Szczepanski, L., Szechinski, J., Filipowicz-Sosnowska, A., Emery, P., Close, D. R., et al. (2004). Efficacy of B-cell-targeted therapy with rituximab in patients with rheumatoid arthritis. *New England Journal of Medicine, 350*(25), 2572–2581. Retrieved from <Go to ISI>://WOS:000222054700008. https://doi.org/10.1056/NEJMoa032534.

Elderman, M., Hugenholtz, F., Belzer, C., Boekschoten, M., van Beek, A., de Haan, B., et al. (2018). Sex and strain dependent differences in mucosal immunology and microbiota composition in mice. *Biology of Sex Differences*, *9*, 26. Retrieved from <Go to ISI>://WOS:000435611000001. https://doi.org/10.1186/s13293-018-0186-6.

Firestein, G. S., & McInnes, I. B. (2017). Immunopathogenesis of rheumatoid arthritis. *Immunity*, *46*(2), 183–196. Retrieved from <Go to ISI>://WOS:000396410900008. https://doi.org/10.1016/j.immuni.2017.02.006.

Fish, E. N. (2008). The X-files in immunity: Sex-based differences predispose immune responses. *Nature Reviews. Immunology*, *8*(9), 737–744. Retrieved from http://www.ncbi.nlm.nih.gov/pubmed/18728636. https://doi.org/10.1038/nri2394.

Gao, X. F., Zhang, M. R., Xue, J. M., Huang, J. D., Zhuang, R. H., Zhou, X. L., et al. (2018). Body mass index differences in the gut microbiota are gender specific. *Frontiers in Microbiology*, *9*, 1250. Retrieved from <Go to ISI>://WOS:000436098900001. https://doi.org/10.3389/fmicb.2018.01250.

Gaskins, A. J., Mumford, S. L., Zhang, C. L., Wactawski-Wende, J., Hovey, K. M., Whitcomb, B. W., et al. (2009). Effect of daily fiber intake on reproductive function: the BioCycle Study. *American Journal of Clinical Nutrition*, *90*(4), 1061–1069. Retrieved from <Go to ISI>://WOS:000269956700023. https://doi.org/10.3945/ajcn.2009.27990.

Gensous, N., Garagnani, P., Santoro, A., Giuliani, C., Ostan, R., Fabbri, C., et al. (2020). One-year Mediterranean diet promotes epigenetic rejuvenation with country- and sex-specific effects: A pilot study from the NU-AGE project. *Geroscience*, *42*(2), 687–701. Retrieved from <Go to ISI>://WOS:000509149200001. https://doi.org/10.1007/s11357-019-00149-0.

Gimenez-Barcons, M., Casteras, A., Armengol, M., Porta, E., Correa, P. A., Marin, A., et al. (2014). Autoimmune Predisposition in Down Syndrome May Result from a Partial Central Tolerance Failure due to Insufficient Intrathymic Expression of AIRE and Peripheral Antigens. *Journal of Immunology*, *193*(8), 3872–3879. Retrieved from <Go to ISI>://WOS:000344078200006. https://doi.org/10.4049/jimmunol.1400223.

Gomez, A., Luckey, D., & Taneja, V. (2015). The gut microbiome in autoimmunity: Sex matters. *Clinical Immunology*, *159*(2), 154–162. Retrieved from http://www.ncbi.nlm.nih.gov/pubmed/25956531. https://doi.org/10.1016/j.clim.2015.04.016.

Gomez, A., Luckey, D., Yeoman, C. J., Marietta, E. V., Berg Miller, M. E., Murray, J. A., et al. (2012). Loss of sex and age driven differences in the gut microbiome characterize arthritis-susceptible 0401 mice but not arthritis-resistant 0402 mice. *PLoS One*, 7(4), e36095. Retrieved from http://www.ncbi.nlm.nih.gov/pubmed/22553482. https://doi.org/10.1371/journal.pone.0036095.

Han, X., Lundberg, P., Tanamachi, B., Openshaw, H., Longmate, J., & Cantin, E. (2001). Gender influences herpes simplex virus type 1 infection in normal and gamma interferon-mutant mice. *Journal of Virology*, *75*, 3048. 2001. 75(16), 7789–7789. Retrieved from <Go to ISI>://WOS:000170093000063.

Haro, C., Rangel-Zuniga, O. A., Alcala-Diaz, J. F., Gomez-Delgado, F., Perez-Martinez, P., Delgado-Lista, J., et al. (2016). Intestinal microbiota is influenced by gender and body mass index. *PLoS One*, *11*(5), e0154090. Retrieved from <Go to ISI>://WOS:000376882500003. https://doi.org/10.1371/journal.pone.0154090.

Hemmi, H., Kaisho, T., Takeuchi, O., Sato, S., Sanjo, H., Hoshino, K., et al. (2002). Small anti-viral compounds activate immune cells via the TLR7 MyD88-dependent signaling pathway. *Nature Immunology*, *3*(2), 196–200. Retrieved from <Go to ISI>://WOS:000173562400021. https://doi.org/10.1038/ni758.

Herrera, B. M., Ramagopalan, S. V., Lincoln, M. R., Orton, S. M., Chao, M. J., Sadovnick, A. D., et al. (2008). Parent-of-origin effects in MS: Observations from avuncular pairs. *Neurology*, *71*(11), 799–803. Retrieved from <Go to ISI>://WOS:000259039000005. https://doi.org/10.1212/01.wnl.0000312377.50395.00.

Hewagama, A., Gorelik, G., Patel, D., Liyanarachchi, P., McCune, W. J., Somers, E., et al. (2013). Overexpression of X-Linked genes in T cells from women with lupus. *Journal of Autoimmunity*, *41*, 60–71. Retrieved from <Go to ISI>://WOS:000318331800008. https://doi.org/10.1016/j.jaut.2012.12.006.

Hinterberger, M., Aichinger, M., da Costa, O. P., Voehringer, D., Hoffmann, R., & Klein, L. (2010). Autonomous role of medullary thymic epithelial cells in central CD4(+) T cell tolerance. *Nature Immunology*, *11*(6), 512–519. Retrieved from <Go to ISI>://WOS:000277821300012. https://doi.org/10.1038/ni.1874.

Hirokawa, K., Utsuyama, M., Hayashi, Y., Kitagawa, M., Makinodan, T., & Fulop, T. (2013). Slower immune system aging in women versus men in the Japanese population. *Immunity & Ageing*, *10*(1), 19. Retrieved from http://www.ncbi.nlm.nih.gov/pubmed/23675689. https://doi.org/10.1186/1742-4933-10-19.

Hitchon, C. A., Alex, P., Erdile, L. B., Frank, M. B., Dozmorov, I., Tang, Y. H., et al. (2004). A distinct multicytokine profile is associated with anti-cyclical citrullinated peptide antibodies in patients with early untreated inflammatory arthritis. *Journal of Rheumatology*, *31*(12), 2336–2346. Retrieved from <Go to ISI>://WOS:000225577800007.

Hoffmann, J., Otte, A., Thiele, S., Lotter, H., Shu, Y. L., & Gabriel, G. (2015). Sex differences in H7N9 influenza A virus pathogenesis. *Vaccine*, *33*(49), 6949–6954. Retrieved from <Go to ISI>://WOS:000366775900002. https://doi.org/10.1016/j.vaccine.2015.08.044.

Holoshitz, J., Koning, F., Coligan, J. E., Debruyn, J., & Strober, S. (1989). Isolation of Cd4- Cd8- mycobacteria-reactive lymphocyte-T clones from rheumatoid-arthritis synovial-fluid. *Nature*, *339*(6221), 226–229. https://doi.org/10.1038/339226a0.

Hoppenbrouwers, I. A., Liu, F., Aulchenko, Y. S., Ebers, G. C., Oostra, B. A., van Duijn, C. M., et al. (2008). Maternal transmission of multiple sclerosis in a Dutch population. *Archives of Neurology*, *65*(3), 345–348. Retrieved from <Go to ISI>://WOS:000253881900008. https://doi.org/10.1001/archneurol.2007.63.

Hughes, D. T., & Sperandio, V. (2008). Inter-kingdom signalling: Communication between bacteria and their hosts. *Nature Reviews. Microbiology*, *6*(2), 111–120. Retrieved from http://www.ncbi.nlm.nih.gov/pubmed/18197168. https://doi.org/10.1038/nrmicro1836.

Johnnidis, J. B., Harris, M. H., Wheeler, R. T., Stehling-Sun, S., Lam, M. H., Kirak, O., et al. (2008). Regulation of progenitor cell proliferation and granulocyte function by microRNA-223. *Nature*, *451*(7182), 1125–1129. https://doi.org/10.1038/nature06607. Retrieved from <Go to ISI>://WOS:000253453600046.

Kahaly, G. J., & Frommer, L. (2018). Polyglandular autoimmune syndromes. *Journal of Endocrinological Investigation*, *41*(1), 91–98. Retrieved from https://www.ncbi.nlm.nih.gov/pubmed/28819917. https://doi.org/10.1007/s40618-017-0740-9.

Klareskog, L., Padyukov, L., Lorentzen, J., & Alfredsson, L. (2006). Mechanisms of disease: Genetic susceptibility and environmental triggers in the development of rheumatoid arthritis. *Nature Clinical Practice. Rheumatology*, *2*(8), 425–433. Retrieved from http://www.ncbi.nlm.nih.gov/entrez/query.fcgi?cmd=Retrieve&db=PubMed&dopt=Citation&list_uids=16932734.

Klein, S. L., & Flanagan, K. L. (2016). Sex differences in immune responses. *Nature Reviews. Immunology*, *16*(10), 626–638. Retrieved from https://www.ncbi.nlm.nih.gov/pubmed/27546235. https://doi.org/10.1038/nri.2016.90.

Klein, S. L., Passaretti, C., Anker, M., Olukoya, P., & Pekosz, A. (2010). The impact of sex, gender and pregnancy on 2009 H1N1 disease. *Biology of Sex Differences*, *1*(1), 5. Retrieved from https://www.ncbi.nlm.nih.gov/pubmed/21208468. https://doi.org/10.1186/2042-6410-1-5.

Koo, H. K., Morrow, J., Kachroo, P., Tantisira, K., Weiss, S. T., Hersh, C. P., et al. (2020). Sex-specific associations with DNA methylation in lung tissue demonstrate smoking interactions. *Epigenetics*, 1–12. Retrieved from https://www.ncbi.nlm.nih.gov/pubmed/32962511. https://doi.org/10.1080/15592294.2020.1819662.

Kovacs, A., Ben-Jacob, N., Tayem, H., Halperin, E., Iraqi, F. A., & Gophna, U. (2011). Genotype is a stronger determinant than sex of the mouse gut microbiota. *Microbial Ecology*, *61*(2), 423–428. Retrieved from http://www.ncbi.nlm.nih.gov/pubmed/21181142. https://doi.org/10.1007/s00248-010-9787-2.

Kular, L., Liu, Y., Ruhrmann, S., Zheleznyakova, G., Marabita, F., Gomez-Cabrero, D., et al. (2018). DNA methylation as a mediator of HLA-DRB1(star)15:01 and a protective variant in multiple sclerosis. *Nature Communications*, *9*, 2397. Retrieved from <Go to ISI>://WOS:000435538500015. https://doi.org/10.1038/s41467-018-04732-5.

Lamason, R., Zhao, P., Rawat, R., Davis, A., Hall, J. C., Chae, J. J., et al. (2006). Sexual dimorphism in immune response genes as a function of puberty. *BMC Immunology*, *7*, 2. Retrieved from <Go to ISI>://WOS:000236025000001. https://doi.org/10.1186/1471-2172-7-2.

Lay, C., Rigottier-Gois, L., Holmstrom, K., Rajilic, M., Vaughan, E. E., de Vos, W. M., et al. (2005). Colonic microbiota signatures across five northern European countries. *Applied and Environmental Microbiology*, *71*(7), 4153–4155. Retrieved from <Go to ISI>://WOS:000230445700099. https://doi.org/10.1128/Aem.71.7.4153-4155.2005.

Lee, J. Y., Park, Y. J., Oh, N., Kwack, K. B., & Park, K. S. (2017). A transcriptional complex composed of ER(alpha), GATA3, FOXA1 and ELL3 regulates IL-20 expression in breast cancer cells. *Oncotarget*, *8*(26), 42752–42760. Retrieved from <Go to ISI>://WOS:000405493400082. 10.18632/oncotarget.17459.

Lee, B. W., Yap, H. K., Chew, F. T., Quah, T. C., Prabhakaran, K., Chan, G. S., et al. (1996). Age- and sex-related changes in lymphocyte subpopulations of healthy Asian subjects: From birth to adulthood. *Cytometry*, *26*(1), 8–15. http://www.ncbi.nlm.nih.gov/pubmed/8809475. Retrieved from doi:10.1002/(SICI)1097-0320(19960315)26:1<8::AID-CYTO2>3.0.CO;2-E.

Legato, M. J. (2017). *Principles of gender-specific medicine : gender in the genomic era* (3rd ed.). London; San Diego, CA: Elsevier/Academic Press.

Li, M., Wang, B. H., Zhang, M. H., Rantalainen, M., Wang, S. Y., Zhou, H. K., et al. (2008). Symbiotic gut microbes modulate human metabolic phenotypes. *Proceedings of the National Academy of Sciences of the United States of America*, *105*(6), 2117–2122. Retrieved from <Go to ISI>://WOS:000253261900062. https://doi.org/10.1073/pnas.0712038105.

Lillycrop, K. A., Hoile, S. P., Grenfell, L., & Burdge, G. C. (2014). DNA methylation, ageing and the influence of early life nutrition. *Proceedings of the Nutrition Society*, *73*(3), 413–421. Retrieved from <Go to ISI>://WOS:000340177700008. https://doi.org/10.1017/S0029665114000081.

Lin, P. Y., Sun, L., Thibodeaux, S. R., Ludwig, S. M., Vadlamudi, R. K., Hurez, V. J., et al. (2010). B7-H1-dependent sex-related differences in tumor immunity and immunotherapy responses. *Journal of Immunology*, *185*(5), 2747–2753. Retrieved from https://www.ncbi.nlm.nih.gov/pubmed/20686128. https://doi.org/10.4049/jimmunol.1000496.

Liu, F., Fan, H. Y., Ren, D. S., Dong, G. J., Hu, E. L., Ji, J. J., et al. (2015). TLR9-induced miR-155 and Ets-1 decrease expression of CD1d on B cells in SLE. *European Journal of Immunology*, *45*(7), 1934–1945. Retrieved from <Go to ISI>://WOS:000357687700008. https://doi.org/10.1002/eji.201445286.

Liva, S. M., & Voskuhl, R. R. (2001). Testosterone acts directly on CD4(+) T lymphocytes to increase IL-10 production. Journal of Immunology *167*(4), 2060–2067. Retrieved from <Go to ISI>://WOS:000170949600030. https://doi.org/10.4049/jimmunol.167.4.2060.

Lombardi, P., Goldin, B., Boutin, E., & Gorbach, S. L. (1978). Metabolism of androgens and estrogens by human fecal microorganisms. *Journal of Steroid Biochemistry*, *9*(8), 795–801. Retrieved from https://www.ncbi.nlm.nih.gov/pubmed/713557.

Lu, Q., Wu, A., Tesmer, L., Ray, D., Yousif, N., & Richardson, B. (2007). Demethylation of CD40LG on the inactive X in T cells from women with lupus. *Journal of Immunology*, *179*(9), 6352–6358. Retrieved from <Go to ISI>://WOS:000250388000084. https://doi.org/10.4049/jimmunol.179.9.6352.

Luckey, D., Behrens, M., Smart, M., Luthra, H., David, C. S., & Taneja, V. (2014). DRB1*0402 may influence arthritis by promoting naive CD4+ T-cell differentiation in to regulatory T cells. *European Journal of Immunology*, *44*(11), 3429–3438. Retrieved from http://www.ncbi.nlm.nih.gov/pubmed/25103892. https://doi.org/10.1002/eji.201344424.

Luckey, D., Gomez, A., Murray, J., White, B., & Taneja, V. (2013). Bugs & us: The role of the gut in autoimmunity. *The Indian Journal of Medical Research*, *138*(5), 732–743. Retrieved from https://www.ncbi.nlm.nih.gov/pubmed/24434325.

Luckey, D., Weaver, E. A., Osborne, D. G., Billadeau, D. D., & Taneja, V. (2019). Immunity to influenza is dependent on MHC II polymorphism: Study with 2 HLA transgenic strains. *Scientific Reports*, *9*(1), 19061. https://doi.org/10.1038/s41598-019-55503-1. Retrieved from https://www.ncbi.nlm.nih.gov/pubmed/31836763.

Mamrut, S., Avidan, N., Staun-Ram, E., Ginzburg, E., Truffault, F., Berrih-Aknin, S., et al. (2015). Integrative analysis of methylome and transcriptome in human blood identifies extensive sex- and immune cell-specific differentially methylated regions. *Epigenetics*, *10*(10), 943–957. Retrieved from <Go to ISI>://WOS:000362556600006. https://doi.org/10.1080/15592294.2015.1084462.

Mangalam, A. K., Rajagopalan, G., Taneja, V., & David, C. S. (2008). HLA class II transgenic mice mimic human inflammatory diseases. *Advances in Immunology*, *97*, 65–147. Retrieved from http://www.ncbi.nlm.nih.gov/pubmed/18501769. https://doi.org/10.1016/S0065-2776(08)00002-3.

Marietta, E. V., Murray, J. A., Luckey, D. H., Jeraldo, P. R., Lamba, A., Patel, R., et al. (2016). Suppression of inflammatory arthritis by human gut-derived Prevotella histicola in humanized mice. *Arthritis & Rheumatology*, *68*(12), 2878–2888. Retrieved from https://www.ncbi.nlm.nih.gov/pubmed/27337150. https://doi.org/10.1002/art.39785.

Markle, J. G., Frank, D. N., Mortin-Toth, S., Robertson, C. E., Feazel, L. M., Rolle-Kampczyk, U., et al. (2013). Sex differences in the gut microbiome drive hormone-dependent regulation of autoimmunity. *Science*, *339*(6123), 1084–1088. Retrieved from http://www.ncbi.nlm.nih.gov/pubmed/23328391. https://doi.org/10.1126/science.1233521.

McNamara, D. M., Starling, R. C., Cooper, L. T., Boehmer, J. P., Mather, P. J., Janosko, K. M., et al. (2011). Clinical and demographic predictors of outcomes in recent onset dilated cardiomyopathy results of the IMAC (Intervention in Myocarditis and Acute Cardiomyopathy)-2 Study. *Journal of the American College of Cardiology*, *58*(11), 1112–1118. Retrieved from <Go to ISI>://WOS:000294449200005. https://doi.org/10.1016/j.jacc.2011.05.033.

Medina, K. L., Strasser, A., & Kincade, P. W. (2000). Estrogen influences the differentiation, proliferation, and survival of early B-lineage precursors. *Blood*, *95*(6), 2059–2067. Retrieved from <Go to ISI>://WOS:000085837100025. https://doi.org/10.1182/blood.V95.6.2059.

Mikuls, T. R., Payne, J. B., Reinhardt, R. A., Thiele, G. M., Maziarz, E., Cannella, A. C., et al. (2009). Antibody responses to *Porphyromonas gingivalis* (*P. gingivalis*) in subjects with rheumatoid arthritis and periodontitis. *International Immunopharmacology*, *9*(1), 38–42. https://doi.org/10.1016/j.intimp.2008.09.008.

Mo, R., Chen, J., Grolleau-Julius, A., Murphy, H. S., Richardson, B. C., & Yung, R. L. (2005). Estrogen regulates CCR gene expression and function in T lymphocytes. *Journal of Immunology*, *174*(10), 6023–6029. Retrieved from https://www.ncbi.nlm.nih.gov/pubmed/15879095.

Moller, E., Bohme, J., Valugerdi, M. A., Ridderstad, A., & Olerup, O. (1990). Speculations on mechanisms of HLA associations with autoimmune diseases and the specificity of "autoreactive" T lymphocytes. *Immunological Reviews*, *118*, 5–19. Retrieved from http://www.ncbi.nlm.nih.gov/entrez/query.fcgi?cmd=Retrieve&db=PubMed&dopt=Citation&list_uids=1706682.

Moulton, V. R. (2018). Sex hormones in acquired immunity and autoimmune disease. *Frontiers in Immunology*, *9*, 2279. Retrieved from <Go to ISI>://WOS:000446346600001. https://doi.org/10.3389/fimmu.2018.02279.

Mueller, S., Saunier, K., Hanisch, C., Norin, E., Alm, L., Midtvedt, T., et al. (2006). Differences in fecal microbiota in different European study populations in relation to age, gender, and country: A cross-sectional study. *Applied and Environmental Microbiology*, *72*(2), 1027–1033. Retrieved from http://www.ncbi.nlm.nih.gov/entrez/query.fcgi?cmd=Retrieve&db=PubMed&dopt=Citation&list_uids=16461645.

Nalbandian, G., & Kovats, S. (2005). Understanding sex biases in immunity—Effects of estrogen on the differentiation and function of antigen-presenting cells. *Immunologic Research*, *31*(2), 91–106. Retrieved from <Go to ISI>://WOS:000228982100002. https://doi.org/10.1385/Ir:31:2:091.

Nammo, T., Udagawa, H., Funahashi, N., Kawaguchi, M., Uebanso, T., Hiramoto, M., et al. (2018). Genome-wide profiling of histone H3K27 acetylation featured fatty acid signalling in pancreatic beta cells in diet-induced obesity in mice. *Diabetologia*, *61*(12), 2608–2620. Retrieved from <Go to ISI>://WOS:000449290200015. https://doi.org/10.1007/s00125-018-4735-7.

Ngo, S. T., Steyn, F. J., & McCombe, P. A. (2014). Gender differences in autoimmune disease. *Frontiers in Neuroendocrinology*, *35*(3), 347–369. Retrieved from https://www.ncbi.nlm.nih.gov/pubmed/24793874. https://doi.org/10.1016/j.yfrne.2014.04.004.

Nugent, B. M., Wright, C. L., Shetty, A. C., Hodes, G. E., Lenz, K. M., Mahurkar, A., et al. (2015). Brain feminization requires active repression of masculinization via DNA methylation. *Nature Neuroscience*, *18*(5), 690–697. Retrieved from <Go to ISI>://WOS:000353636900016. https://doi.org/10.1038/nn.3988.

Oh, H. J., Li, Y., & Lau, Y. F. (2005). Sry associates with the heterochromatin protein 1 complex by interacting with a KRAB domain protein. *Biology of Reproduction*, *72*(2), 407–415. Retrieved from https://www.ncbi.nlm.nih.gov/pubmed/15469996. https://doi.org/10.1095/biolreprod.104.034447.

Org, E., Mehrabian, M., Parks, B. W., Shipkova, P., Liu, X. Q., Drake, T. A., et al. (2016). Sex differences and hormonal effects on gut microbiota composition in mice. *Gut Microbes*, 7(4), 313–322. Retrieved from <Go to ISI>://WOS:000425290600004. https://doi.org/10.1080/19490976.2016.1203502.

Paavonen, T., Andersson, L. C., & Adlercreutz, H. (1981). Sex-hormone regulation of invitro immune-response—estradiol enhances human B-cell maturation via inhibition of suppressor T-cells in pokeweed mitogen-stimulated cultures. *Journal of Experimental Medicine*, *154*(6), 1935–1945. Retrieved from <Go to ISI>://WOS:A1981MU61700017. https://doi.org/10.1084/jem.154.6.1935.

Panneerdoss, S., Chang, Y. F., Buddavarapu, K. C., Chen, H. I., Shetty, G., Wang, H., et al. (2012). Androgen-responsive microRNAs in mouse Sertoli cells. *PLoS One*, 7(7), e41146. Retrieved from https://www.ncbi.nlm.nih.gov/pubmed/22911753. https://doi.org/10.1371/journal.pone.0041146.

Parahitiyawa, N. B., Scully, C., Leung, W. K., Yam, W. C., Jin, L. J., & Samaranayake, L. P. (2010). Exploring the oral bacterial flora: Current status and future directions. *Oral Diseases*, *16*(2), 136–145. https://doi.org/10.1111/j.1601-0825.2009.01607.x.

Pelikan, R. C., Kelly, J. A., Fu, Y., Lareau, C. A., Tessneer, K. L., Wiley, G. B., et al. (2018). Enhancer histone-QTLs are enriched on autoimmune risk haplotypes and influence gene expression within chromatin networks. *Nature Communications*, *9*, 2905.

Retrieved from <go to ISI>://WOS:000439687600003. https://doi.org/10.1038/s41467-018-05328-9.

Pergola, C., Dodt, G., Rossi, A., Neunhoeffer, E., Lawrenz, B., Northoff, H., et al. (2008). ERK-mediated regulation of leukotriene biosynthesis by androgens: A molecular basis for gender differences in inflammation and asthma. *Proceedings of the National Academy of Sciences of the United States of America*, *105*(50), 19881–19886. Retrieved from <Go to ISI>://WOS:000261802300056. https://doi.org/10.1073/pnas.0809120105.

Pinheiro, I., Dejager, L., & Libert, C. (2011). X-chromosome-located microRNAs in immunity: might they explain male/female differences? The X chromosome-genomic context may affect X-located miRNAs and downstream signaling, thereby contributing to the enhanced immune response of females. *BioEssays: News and Reviews in Molecular, Cellular and Developmental Biology*, *33*(11), 791–802. Retrieved from <Go to ISI>://MEDLINE:21953569.

Posma, E., Moes, H., Heineman, M. J., & Faas, M. M. (2004). The effect of testosterone on cytokine production in the specific and non-specific immune response. *American Journal of Reproductive Immunology*, *52*(4), 237–243. Retrieved from <Go to ISI>://WOS:000224562200002. https://doi.org/10.1111/j.1600-0897.2004.00216.x.

Ridlon, J. M., Ikegawa, S., Alves, J. M. P., Zhou, B., Kobayashi, A., Iida, T., et al. (2013). Clostridium scindens: A human gut microbe with a high potential to convert glucocorticoids into androgens. *Journal of Lipid Research*, *54*(9), 2437–2449. Retrieved from <Go to ISI>://MEDLINE:23772041.

Rouas, R., Fawad-Kazan, H., El Zein, N., Lewalle, P., Rothe, F., Simion, A., et al. (2009). Human natural Treg microRNA signature: Role of microRNA-31 and microRNA-21 in FOXP3 expression. *European Journal of Immunology*, *39*(6), 1608–1618. Retrieved from <Go to ISI>://WOS:000267481400018. https://doi.org/10.1002/eji.200838509.

Scofield, R. H., Bruner, G. R., Namjou, B., Kimberly, R. P., Ramsey-Goldman, R., Petri, M., et al. (2008). Klinefelter's syndrome (47,XXY) in male systemic lupus erythematosus patients: Support for the notion of a gene-dose effect from the X chromosome. *Arthritis and Rheumatism*, *58*(8), 2511–2517. Retrieved from https://www.ncbi.nlm.nih.gov/pubmed/18668569. https://doi.org/10.1002/art.23701.

Scully, E. P., Haverfield, J., Ursin, R. L., Tannenbaum, C., & Klein, S. L. (2020). Considering how biological sex impacts immune responses and COVID-19 outcomes. *Nature Reviews Immunology*, *20*(7), 442–447. Retrieved from <Go to ISI>://WOS:000565303700014. https://doi.org/10.1038/s41577-020-0348-8.

Shen, N., Fu, Q. O., Deng, Y., Qian, X. X., Zhao, J. A., Kaufman, K. M., et al. (2010). Sex-specific association of X-linked Toll-like receptor 7 (TLR7) with male systemic lupus erythematosus. *Proceedings of the National Academy of Sciences of the United States of America*, *107*(36), 15838–15843. Retrieved from <Go to ISI>://WOS:000281637800040. https://doi.org/10.1073/pnas.1001337107.

Shi, X. B., Xue, L., Yang, J., Ma, A. H., Zhao, J., Xu, M., et al. (2007). An androgen-regulated miRNA suppresses Bak1 expression and induces androgen-independent growth of prostate cancer cells. *Proceedings of the National Academy of Sciences of the United States of America*, *104*(50), 19983–19988. Retrieved from <Go to ISI>://WOS:000251752200061. https://doi.org/10.1073/pnas.0706641104.

Singmann, P., Shem-Tov, D., Wahl, S., Grallert, H., Fiorito, G., Shin, S. Y., et al. (2015). Characterization of whole-genome autosomal differences of DNA methylation between men and women. *Epigenetics & Chromatin*, *8*, 43. Retrieved from <Go to ISI>://WOS:000363085300001. https://doi.org/10.1186/s13072-015-0035-3.

Sinha, S., Kaler, L. J., Proctor, T. M., Teuscher, C., Vandenbark, A. A., & Offner, H. (2008). IL-13-Mediated gender difference in susceptibility to autoimmune encephalomyelitis. *Journal of Immunology*, *180*(4), 2679–2685. Retrieved from <Go to ISI>://WOS:000253005600084. https://doi.org/10.4049/jimmunol.180.4.2679.

Sinha, T., Vila, A. V., Garmaeva, S., Jankipersadsing, S. A., Imhann, F., Collij, V., et al. (2019). Analysis of 1135 gut metagenomes identifies sex-specific resistome profiles. *Gut Microbes, 10*(3), 358–366. Retrieved from <Go to ISI>://WOS:000468733100007. https://doi.org/10.1080/19490976.2018.1528822.

Syed, S., Amin, M., & Rabquer, B. (2015). miR155 expression is increased by inflammation and modulates the expression of CD11a in monocytes. *FASEB Journal, 29*, 634.6. Retrieved from < Go to ISI >://WOS:000361470505369.

Tamura, K., Dudley, J., Nei, M., & Kumar, S. (2007). MEGA4: Molecular evolutionary genetics analysis (MEGA) software version 4.0. *Molecular Biology and Evolution, 24*(8), 1596–1599. Retrieved from <Go to ISI>://WOS:000248848400005. https://doi.org/10.1093/molbev/msm092.

Taneja, V. (2015). Cytokines pre-determined by genetic factors are involved in pathogenesis of rheumatoid arthritis. *Cytokine, 75*(2), 216–221. Retrieved from https://www.ncbi.nlm.nih.gov/pubmed/25541434. https://doi.org/10.1016/j.cyto.2014.11.028.

Taneja, V. (2017). Microbiome in 2016: T follicular helper cells and the gut microbiome in arthritis. *Nature Reviews Rheumatology, 13*(2), 72–74. Retrieved from https://www.ncbi.nlm.nih.gov/pubmed/28077867. https://doi.org/10.1038/nrrheum.2016.222.

Taneja, V. (2018). Sex hormones determine immune response. *Frontiers in Immunology, 9*, 1931. Retrieved from https://www.ncbi.nlm.nih.gov/pubmed/30210492. https://doi.org/10.3389/fimmu.2018.01931.

Taneja, V., Behrens, M., Basal, E., Sparks, J., Griffiths, M. M., Luthra, H., et al. (2008). Delineating the role of the HLA-DR4 "shared epitope" in susceptibility versus resistance to develop arthritis. *Journal of Immunology, 181*(4), 2869–2877. Retrieved from <Go to ISI>://WOS:000258345300067.

Taneja, V., Behrens, M., Mangalam, A., Griffiths, M. M., Luthra, H. S., & David, C. S. (2007). New humanized HLA-DR4-transgenic mice that mimic the sex bias of rheumatoid arthritis. *Arthritis and Rheumatism, 56*(1), 69–78. Retrieved from <Go to ISI>://WOS:000243395900010. https://doi.org/10.1002/art.22213.

Taneja, V., & David, C. S. (2010). Role of HLA class II genes in susceptibility/resistance to inflammatory arthritis: Studies with humanized mice. *Immunological Reviews, 233*(1), 62–78. Retrieved from http://www.ncbi.nlm.nih.gov/pubmed/20192993. https://doi.org/10.1111/j.0105-2896.2009.00858.x.

Taneja, V., Mehra, N. K., Anand, C., & Malaviya, A. N. (1993). HLA-linked susceptibility to rheumatoid arthritis. A study of forty-one multicase families from northern India. *Arthritis and Rheumatism, 36*(10), 1380–1386. Retrieved from https://www.ncbi.nlm.nih.gov/pubmed/8216398. https://doi.org/10.1002/art.1780361009.

Tengstrand, B., Carlstrom, K., Fellander-Tsai, L., & Hafstrom, I. (2003). Abnormal levels of serum dehydroepiandrosterone, estrone, and estradiol in men with rheumatoid arthritis: High correlation between serum estradiol and current degree of inflammation. *The Journal of Rheumatology, 30*(11), 2338–2343. Retrieved from http://www.ncbi.nlm.nih.gov/pubmed/14677174.

Tsai, H. W., Grant, P. A., & Rissman, E. F. (2009). Sex differences in histone modifications in the neonatal mouse brain. *Epigenetics, 4*(1), 47–53. Retrieved from <Go to ISI>://WOS:000263517400010. https://doi.org/10.4161/epi.4.1.7288.

Uppal, S. S., Verma, S., & Dhot, P. S. (2003). Normal values of CD4 and CD8 lymphocyte subsets in healthy Indian adults and the effects of sex, age, ethnicity, and smoking. *Cytometry Part B-Clinical Cytometry, 52B*(1), 32–36. Retrieved from <Go to ISI>://WOS:000182426300004. https://doi.org/10.1002/cyto.b.10011.

Vassallo, R., Luckey, D., Behrens, M., Madden, B., Luthra, H., David, C., et al. (2014). Cellular and humoral immunity in arthritis are profoundly influenced by the interaction between cigarette smoke effects and host HLA-DR and DQ genes. *Clinical Immunology,*

152(1–2), 25–35. Retrieved from http://www.ncbi.nlm.nih.gov/pubmed/24631425. https://doi.org/10.1016/j.clim.2014.02.002.

Vaughan, E. E., Heilig, H. G. H. J., Ben-Amor, K., & de Vos, W. M. (2005). Diversity, vitality and activities of intestinal lactic acid bacteria and bifidobacteria assessed by molecular approaches. *FEMS Microbiology Reviews*, *29*(3), 477–490. Retrieved from <Go to ISI>://WOS:000230861700006. doi:10.1016/j.femsre.2005.04.009.

Walecki, M., Eisel, F., Klug, J., Baal, N., Paradowska-Dogan, A., Wahle, E., et al. (2015). Androgen receptor modulates Foxp3 expression in CD4(+)CD25(+)Foxp3(+) regulatory T-cells. *Molecular Biology of the Cell*, *26*(15), 2845–2857. Retrieved from <Go to ISI>://WOS:000359012500011. https://doi.org/10.1091/mbc.E14-08-1323.

Wang, C., Dehghani, B., Li, Y., Kaler, L. J., Vandenbark, A. A., & Offner, H. (2009). Oestrogen modulates experimental autoimmune encephalomyelitis and interleukin-17 production via programmed death 1. *Immunology*, *126*(3), 329–335. Retrieved from <Go to ISI>://WOS:000263038700004. https://doi.org/10.1111/j.1365-2567.2008.03051.x.

Wang, D., Zhang, J., Lau, J., Wang, S. H., Taneja, V., Matteson, E. L., et al. (2019). Mechanisms of lung disease development in rheumatoid arthritis. *Nature Reviews Rheumatology*, *15*(10), 581–596. Retrieved from <Go to ISI>://WOS:000487334300010. https://doi.org/10.1038/s41584-019-0275-x.

Warden, C. C. (1909). The toxemic factor in rheumatoid arthritis. *California State Journal of Medicine*, 7(8), 299–301. Retrieved from http://www.ncbi.nlm.nih.gov/pubmed/18734746.

Weeding, E., & Sawalha, A. H. (2018). Deoxyribonucleic acid methylation in systemic lupus erythematosus: implications for future clinical practice. *Frontiers in Immunology*, *9*, 875. Retrieved from <Go to ISI>://WOS:000430745300001. https://doi.org/10.3389/fimmu.2018.00875.

Wu, Z., Sun, Y., Mei, X., Zhang, C., Pan, W., & Shi, W. (2014). 17 beta-oestradiol enhances global DNA hypomethylation in CD4-positive T cells from female patients with lupus, through overexpression of oestrogen receptor-a-mediated downregulation of DNMT1. *Clinical and Experimental Dermatology*, *39*(4), 525–532. Retrieved from <Go to ISI>://WOS:000336212500020. https://doi.org/10.1111/ced.12346.

Xia, Y., Dai, R., Wang, K., Jiao, C., Zhang, C., Xu, Y., et al. (2019). Sex-differential DNA methylation and associated regulation networks in human brain implicated in the sex-biased risks of psychiatric disorders. *Molecular Psychiatry*. https://doi.org/10.1038/s41380-019-0416-2. Retrieved from https://www.ncbi.nlm.nih.gov/pubmed/30976086.

Xu, G., Wu, J. Y., Zhou, L. L., Chen, B. H., Sun, Z. S., Zhao, F. Q., et al. (2010). Characterization of the small RNA transcriptomes of androgen dependent and independent prostate cancer cell line by deep sequencing. *PLoS One*, *5*(11), e15519. Retrieved from <Go to ISI>://WOS:000284755100090. https://doi.org/10.1371/journal.pone.0015519.

Yang, Z. H., & Wang, L. (2011). Regulation of microRNA expression and function by nuclear receptor signaling. *Cell & Bioscience*, *1*, 31. Retrieved from <Go to ISI>://WOS:000307058100001. https://doi.org/10.1186/2045-3701-1-31.

Yurkovetskiy, L., Burrows, M., Khan, A. A., Graham, L., Volchkov, P., Becker, L., et al. (2013). Gender bias in autoimmunity is influenced by microbiota. *Immunity*, *39*(2), 400–412. Retrieved from http://www.ncbi.nlm.nih.gov/pubmed/23973225. https://doi.org/10.1016/j.immuni.2013.08.013.

Zaits, M. N., & Rennert, O. M. (2014). Identification of differentially expressed microRNAs across the developing human brain. *Molecular Psychiatry*, *19*(7), 848–852. Retrieved from <Go to ISI>://WOS:000338185300014. https://doi.org/10.1038/mp.2013.93.

CHAPTER FIFTEEN

Hormones and bone loss across the menopause transition

Arun S. Karlamangla*, Albert Shieh, and Gail A. Greendale
Division of Geriatrics, Department of Medicine, David Geffen School of Medicine at UCLA, Los Angeles, CA, United States
*Corresponding author: e-mail address: akarlamangla@mednet.ucla.edu

Contents

Abstract

The menopause transition is a critical period for bone health in women, with rapid losses in bone mass and strength occurring over an approximately 3-year window bracketing the date of the final menstrual period. The onset of the rapid bone loss phase is preceded by large changes in sex steroid hormones, measurements of which may be clinically useful in predicting the onset of the rapid loss phase and identifying the women who will lose the most bone mass during this rapid bone loss phase. Here we summarize recent and new findings related to the ability of sex hormone levels to (1) determine if a woman in her 5th decade of life is about to enter or has already entered the rapid phase of bone loss, and (2) if she will lose more than the average amount of bone mass over the menopause transition.

1. Osteoporosis and the menopause transition

Osteoporosis and low bone mass affect over 50 million U.S. adults, and disproportionately affect older women (Wright et al., 2014). One in two women experiences an osteoporotic fracture in her lifetime (US Preventive Services Task Force, 2011), and many women have multiple fractures with major morbidity and even loss of independence and reduced survival (Forsen, Søgaard, Meyer, Edna, & Kopjar, 1999;

Vitamins and Hormones, Volume 115
ISSN 0083-6729
https://doi.org/10.1016/bs.vh.2020.12.016

Gehlbach, Burge, Puleo, & Klar, 2003). In the U.S., osteoporotic fractures affect nearly 1.5 million older women every year (Van Staa, Dennison, Leufkens, & Cooper, 2001) and cost the U.S. health system nearly $20 billion dollars annually (Blume & Curtis, 2011).

Appropriately, the focus of fracture prevention has been in older women. However, women start losing bone strength in midlife, and interventions to reduce the risk of osteoporosis and fractures in older ages may need to start in midlife (National Osteoporosis Foundation, 2020; Riis, Hansen, Jensen, Overgaard, & Christiansen, 1996). The midlife menopause transition (MT) in women is a critical period of change for bone health, with many, but not all, women losing a substantial amount of bone mass and quality. Over a 3-year period bracketing the date of the final menstrual period (FMP), rapid decline in bone mineral density (BMD), cortical thickness, trabecular connectivity (as indexed by the trabecular bone score), and composite indices of bone strength, is the norm (Greendale et al., 2012, 2020; Guthrie, Dennerstein, Taffe, Lehert, & Burger, 2004; Ishii et al., 2013; Nagaraj et al., 2019; Pedrazzoni et al., 2014; Recker, Lappe, Davies, & Heaney, 2000; Seifert-Klauss et al., 2012). The magnitude of these declines is striking; the average White American woman loses 2.5% of her lumbar spine BMD and 1.8% of her femoral neck BMD every year during this MT-related rapid loss phase, and continues to lose bone mass post-MT, albeit at a slightly slower rate—Fig. 1. As a result, the cumulative decline in BMD over this rapid loss phase of the MT and the subsequent 3 postmenopausal years is around 10% (Greendale et al., 2012), which translates to a 1-unit decline in T score. The larger the decline, the greater the fracture risk. Indeed, each 0.5% per year faster decline in lumbar spine BMD during this period (MT and following 3 years) is associated with 22% greater hazard for incident fracture, independent of the baseline pre-MT BMD (Shieh, Huang, Han, Karlamangla, & Greendale, 2020). This has led us and other researchers to suggest that midlife, specifically early in the MT, may be the opportune time to intervene to stave off osteoporosis in women in later years (Blumenfeld & Lavie, 2020; Greendale & Eriksen, 2017; Karlamangla, Burnett-Bowie, & Crandall, 2018; Zaidi et al., 2009).

An effective and efficient midlife intervention would be *timed* before substantial bone loss has occurred and would *target* women who are likely to lose the most bone mass over the MT. Appropriately timing the intervention is predicated on the ability to predict or detect the onset of the rapid phase of MT-related bone loss. The rapid bone loss phase starts *1 year before* the FMP, which is *2 years before* the date is clinically determined by the

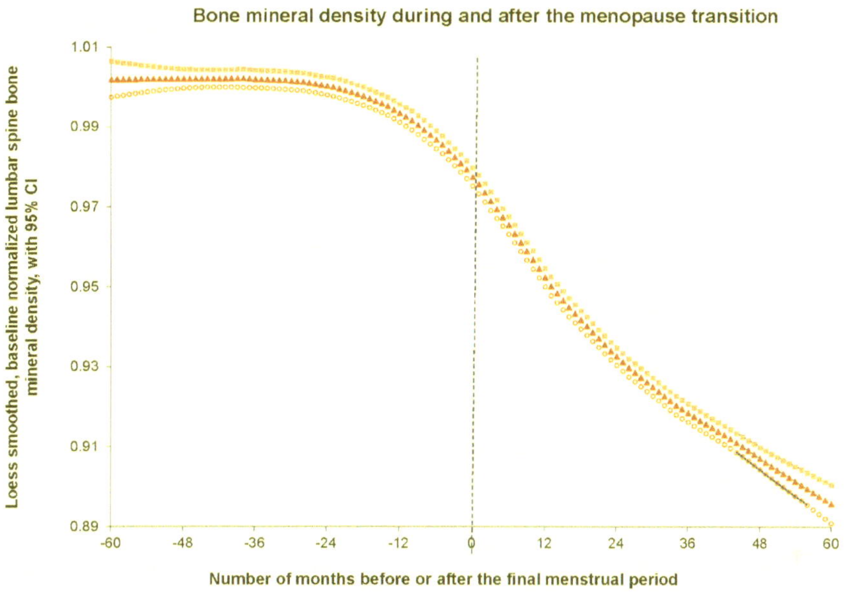

Fig. 1 LOESS-smoothed plot of the trajectory of lumbar spine bone mineral density over the menopause transition, as fraction of starting level. Progression along the menopause transition is indexed by time (in months) before (negative numbers) or after (positive numbers) the final menstrual period.

subsequent occurrence of 12 consecutive months of amenorrhea (Greendale et al., 2012). Changes in menstrual bleeding patterns that characterize the MT before the FMP *cannot* be used to mark *the onset* of the rapid bone loss phase, because of substantial variability in how long different women experience menstrual irregularity before their FMP. Repeated monitoring of BMD by dual x-ray absorptiometry (DXA) to detect loss after it starts is also not an option, because multiple years have to pass before the change in BMD is sufficiently large to exceed the precision error in DXA measurements (Leslie, Majumdar, Morin, & Lix, 2015). Hormonal markers of progression along the MT may, on the other hand, have the potential to predict the onset of the rapid loss phase *before* much bone mass has been lost.

Targeting interventions appropriately is also important, because different women lose different amounts of bone mass over the MT. There are significant variations in the amount of bone mass that is lost, by race/ethnicity (African American women lose a smaller fraction of their bone mass than White women, on average) and by the age at which the woman reaches her FMP (with a later age at FMP associated with less loss) (Greendale et al., 2012). There is also variability by bone size, with wider femoral necks being associated with greater loss of bone mass (Jepsen et al., 2017). The *Study of Women's Health Across the Nation* (SWAN) followed women from before the MT to several years after the MT, with repeat BMD measurements every 18 months on average. In the SWAN cohort, the standard deviation (SD) in the annualized rate of change in BMD over the rapid bone loss phase was 2.1% per year at both the lumbar spine and the femoral neck. This is of the same order as the mean rate of BMD change over this period, −2.4% per year at the lumbar spine and −1.7% per year at the femoral neck (Shieh et al., 2016), demonstrating that not all women lose large amounts of bone mass during this period, although many do. Identifying the women who will lose the most bone mass over the MT's rapid loss phase is key to efficiently delivering early interventions that moderate the loss.

At this point, however, the efficacy of early interventions (before substantial BMD loss has occurred) has not been established (Becker & Cohen, 2019). As a potential short-duration treatment to prevent rapid bone loss during and after the MT, hormone therapy is not a suitable option, because it merely delays bone loss to after discontinuation of sex hormone therapy (Greendale, Espeland, Slone, Marcus, & Barrett-Connor, 2002; Saarelainen et al., 2016). Nor is denosumab a viable option, because its discontinuation leads to rapid BMD loss that can exceed prior gains (Bone et al., 2011; Cummings et al., 2018). Bisphosphonates, however, can prevent BMD decline in perimenopausal women, and zoledronic acid, in particular, is an intriguing option because a single dose of zoledronic acid stabilizes BMD for at least 36 months (Hosking et al., 1998; Reid et al., 2013). To design clinical trials that test the efficacy of such early interventions, including behavioral interventions, in reducing the risk of later-life fractures, we need to be able to tell when to intervene and in whom.

The rest of this chapter is devoted to the twin problems of predicting the onset of MT-related rapid bone loss and identifying the women slated to lose the most bone mass over the MT. Our primary focus here is on sex hormone levels (and changes therein) as predictors of the onset of the rapid loss phase. Although serial measurement of BMD cannot be used to detect ongoing loss before significant loss has occurred, serum or urinary measurements of bone turnover markers might be able to provide information about the magnitude of ongoing and future bone loss. We briefly address the potential use of bone resorption markers to identify the women who will lose the most bone mass over the MT, and provide directions for future work that might combine measurement of sex hormone levels and bone turnover markers.

2. Sex hormones changes over the menopause transition

The MT is a time of large changes in the hormonal milieu; circulating levels of follicular stimulating hormone (FSH) start increasing rapidly and serum estradiol (E2) levels start decreasing sharply around 2 years before the FMP (Randolph et al., 2011; Sowers et al., 2008a, 2008b). These changes, which cause the menstrual irregularity that is the clinical manifestation of the MT, also mark ovarian aging and the march towards the FMP. Within-woman changes in E2 and FSH, when combined with age and menstrual bleeding patterns, can indicate whether or not a perimenopausal woman has reached the "1 year before the FMP" milestone in her progression along the MT. The discrimination ability of the approach for the FMP -1y time-point is high; the area under the receiver operating curve (AUC) is 0.93 (Greendale, Ishii, Huang, & Karlamangla, 2013). This suggests that sex hormones should be able to predict the onset of the rapid phase of MT-related bone loss.

However, just as there is between-women variability in the length of the MT so that menstrual bleeding irregularity by itself cannot determine how far a woman is from her FMP, there is also variability between women in the timing of the onset of MT-related bone loss. On average, the rapid phase of bone loss starts 1 year before the FMP, but it can start earlier in some women and later in others. Despite this variability in when bone loss starts relative to the FMP date, sex hormone levels may still mark the onset of bone loss because perimenopausal change in hormone levels is likely the primary driver of the loss in bone mass during the MT (Cauley, 2015; Greendale, Edelstein, & Barrett-Connor, 1997; Khosla et al., 1998; Prior, Seifert-Klauss, & Hale, 2013).

3. Sex hormones and onset of rapid bone loss

Consistent with a causal role for hormonal changes in MT-related bone loss, substantial decreases in circulating levels of E2 and increases in levels of FSH *precede* the onset of the rapid bone loss phase by about a year (Fig. 2). Several longitudinal studies have shown that levels of sex hormones during the MT are indeed associated with rates of BMD decline over the subsequent months to years (Slemenda, Hui, Longcope, & Johnston, 1987;

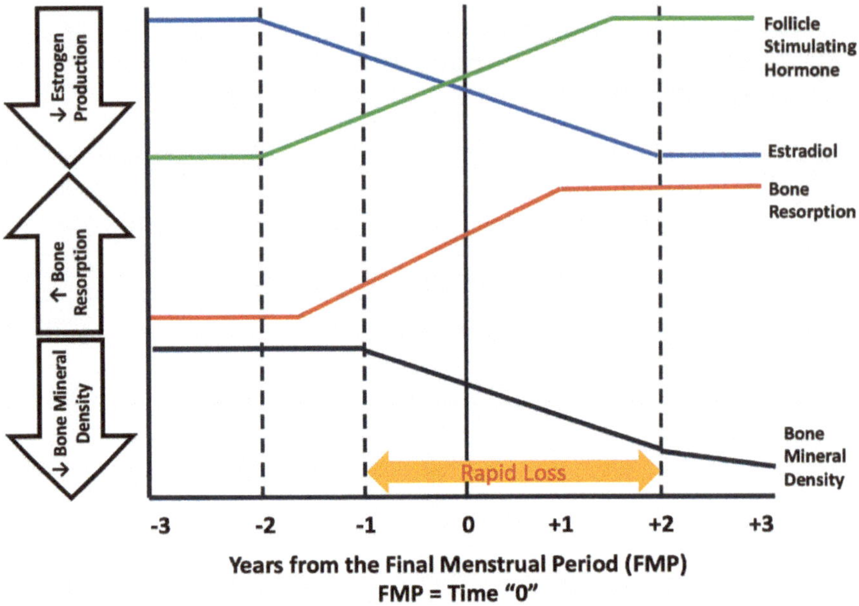

Fig. 2 Legend for online color version. Schematic depiction of trajectories of sex steroid hormone circulating levels (estradiol (blue) and follicular stimulating hormone (green)), bone resorption (red), and bone mineral density (black), over the menopause transition.

Sowers et al., 2006). In the SWAN cohort, every doubling of FSH level was associated with an additional 0.3% decline per year in BMD at both the femoral neck and lumbar spine (Crandall et al., 2013).

These observations spurred us to investigate the ability of E2 and FSH levels (and within-woman changes in levels) to predict or detect the onset of the rapid bone loss phase of the MT in 1559 participants from SWAN. Both E2 and FSH were measured in the follicular phase (between days 2 and 5) of a spontaneous menstrual cycle in women who were menstruating sufficiently regularly to be able to time the follow up study visit to within 90 days of the expected visit date. When a follicular phase blood sample could not be obtained within 90 days of the expected visit date, a random sample was used for the assays. The likelihood of sampling outside the follicular phase window increased as the women progressed through the MT, from 18% in premenopause to 35% in early perimenopause to 98% in late perimenopause. These clinical stages of the MT are defined by menstrual bleeding patterns. SWAN defined premenopause as no change in menstrual regularity in the past year, early perimenopause as decreased regularity in menstrual bleeding but with no gap of 3 or more months, and late perimenopause as irregular menstrual bleeding with gaps of 3–12 months.

We considered the rapid phase of bone loss to be imminent or ongoing if there was significant loss in bone mass by the subsequent SWAN study visit (on average, 18 months after the current visit), relative to an earlier pre- or early perimenopausal baseline. Bone loss was deemed significant if the decline in BMD from the SWAN baseline visit to the SWAN visit subsequent to the E2 or FSH measurement was greater than the *least significant change* (LSC) for the measurement. LSC is the smallest change that is detectable statistically using a two-sided type I error (alpha) of 5%, and equals 2.8 times the coefficient of variation (CV) in the measurement. In the SWAN study, the LSC for lumbar spine BMD was 3.9%; it was 6.2% for femoral neck BMD. We examined the prediction ability of log-transformed E2 and FSH levels, and change from SWAN baseline (when women were either premenopausal or early perimenopausal) to the current visit in log(E2) and log(FSH), separately, in repeated measures, modified Poisson regression models. Models included MT stage and other relevant covariates, and controlled for correlations between repeated observations in the same women.

Adjusted for MT stage (pre- vs. early peri- vs. late perimenopause), age, race/ethnicity (White, Black, Japanese, Chinese), and body mass index (BMI), lower E2 and higher FSH predicted greater probability of significant bone loss in the next 18 months. Each halving (50% reduction) of E2 and each doubling of FSH was associated respectively with 10% and 39% greater likelihood of imminent or ongoing rapid bone loss at the lumbar spine ($P < 0.0001$ for each). Every halving of E2 and each doubling of FSH were also associated respectively with 12% ($P = 0.01$) and 27% ($P < 0.001$) greater likelihood of imminent or ongoing rapid bone loss at the femoral neck (Table 1). FSH was somewhat better than E2 at identifying women who were likely to be losing bone rapidly; the AUC for the lumbar spine site was 0.78 for FSH compared to 0.76 for E2. Tracking within-individual change in either hormone (from an earlier premenopausal or early

Table 1 Associations of estradiol (E2) and follicle stimulating hormone (FSH) with imminent or ongoing rapid bone loss; Study of Women's Health Across the Nation.

	Relative risk (RR) of significant bone loss by the next visit			
	Lumbar spine		**Femoral neck**	
	RR (95% CI)	***AUC***	***RR (95% CI)***	***AUC***
Per 50% reduction of E2	1.10 (1.06, 1.15)	0.756	1.12 (1.02, 1.23)	0.740
Per doubling of FSH	1.39 (1.30, 1.49)	0.782	1.27 (1.11, 1.44)	0.751

AUC = Area under the receiver operating curve.
CI = Confidence interval.

perimenopausal baseline) did not improve the prediction of the onset of rapid bone loss over the use of a single measure of current hormone level; see Shieh et al. (2019a) for detailed results.

These findings lead us to conclude that FSH levels early in the MT can help clinicians prospectively identify women who are about to enter the rapid bone loss phase. This motivated us to refine the prediction models, create a calculator for a woman's probability of imminent or ongoing rapid bone loss using her FSH level, MT stage, and other needed covariates, and examine the sensitivity and specificity of the prediction at various probability thresholds. We used 4381 repeated observations of FSH over the MT from 1376 women in the SWAN cohort. The outcome of interest, imminent or ongoing rapid bone loss, was operationalized as significant loss in bone mass, relative to an earlier pre- or early perimenopausal baseline, in the next 2 years at either (or both) the femoral neck or the lumbar spine. Bone loss was considered significant if the decline in BMD from the SWAN baseline visit to a SWAN follow up visit ~2 years after the FSH measurement (mean 2.2 years, SD 0.3 years) was greater than the site-specific LSC.

In interaction testing, the association of FSH with significant bone loss was different in late perimenopause than in early perimenopause ($P < 0.001$), but not statistically different in premenopause vs. early perimenopause ($P = 0.8$). We therefore created two separate calculators for the prediction/detection of the onset of the rapid phase of MT-related bone loss, one for late perimenopausal women, and one for pre- and early perimenopausal women. There were 607 repeat observations in late perimenopause from 451 women, and 4381 observations in pre- and early perimenopause from 1343 women. FSH predicted imminent/ongoing rapid bone loss in both groups, independent of age, BMI, race/ethnicity, and MT stage. To keep the calculators simple, we pruned the prediction models of covariates in a step-wise fashion, and stopped pruning when the model AUC dropped substantially. The final model for pre- and early perimenopause included FSH, BMI, age, and MT stage (pre- vs. early perimenopause); the final model for late perimenopause included FSH, BMI, and race/ethnicity (East Asian vs. rest). The resulting formulas for the predicted probability of imminent or ongoing rapid bone loss were, by MT stage:

- $(\text{FSH})^{0.229}$ * anti natural-log (0.171*age – 0.018*BMI – 10.26) in premenopause,
- $(\text{FSH})^{0.229}$ * anti-natural-log (0.171*age – 0.018*BMI – 9.65) in early perimenopause,
- $(\text{FSH})^{0.11}$ * anti-natural-log (0.15*East Asian – 0.014*BMI – 0.39) in late perimenopause,

with FSH in mIU/mL, age in years, BMI in kg/m^2, and East Asian referring to Chinese and Japanese participants in SWAN.

The model predicted probability of imminent or ongoing rapid bone loss had a narrow distribution over the 607 observations in late perimenopause, with mean and median of 80% and interquartile range (69%, 90%). As expected, the model-predicted probability of imminent or ongoing bone loss had a wider range in pre- and early perimenopause, with mean of 13.4% (SD 7.9%) in premenopause and 36.7% (SD 22.7%) in early perimenopause. Age also had a big influence on the predicted probability in this group. We therefore examined the performance characteristics of the predictions, stratified further by age. We used the median age specific to MT stage (46 years in premenopause, 48 years in early perimenopause) to create a young vs. old stratum for each of the two MT stages.

Table 2 lists the sensitivity, specificity, positive predictive value (PPV), and negative predictive value (NPV), for various thresholds of the predicted probability. Thresholds were selected around the observed prevalence of imminent or ongoing rapid bone loss in the group (9.3%, 16.4%, 18.0%, 44.2%, and 79.5%, respectively, in young premenopausal, old premenopausal, young early perimenopausal, old early perimenopausal, and late perimenopausal women). We did not divide the late perimenopausal group by age, since age did not improve discrimination (model AUC) in this group. It is apparent from Table 2 that although FSH is a strong independent predictor of imminent or ongoing rapid bone loss during the MT, the onset of MT-related bone loss cannot be predicted with simultaneously high levels of sensitivity and specificity, even after FSH is combined with clinical covariates.

However, bone loss during the MT (and in postmenopause) is the result of increased bone resorption, which raises the possibility that measurement of bone resorption might also assist in the detection of onset of MT-related bone loss. The hormonal changes of the MT lead to an increase in the number of activated bone remodeling units throughout the skeleton and a deficit in bone formation (less formation than resorption) within each remodeling unit (Seeman, 2013). These changes in the skeleton during the MT are reflected in increased levels of circulating markers of both bone resorption and formation (Ebeling et al., 1996), with bone resorption markers increasing proportionally more than markers of bone formation (Gossiel et al., 2018). We next address the potential for markers of bone resorption to predict the onset of MT-related rapid bone loss and to identify the women who will lose the most bone mass.

Table 2 Prediction of imminent or ongoing rapid bone loss.

Predicted probability threshold	Sensitivity (%)	NPV (%)	Specificity (%)	PPV (%)
Premenopausal, under 46 years of age				
>5%	96.4	94.1	5.9	9.5
>10%	7.3	90.3	88.5	6.1
>15%	1.8	90.9	99.8	50.0
Premenopausal, 46 years of age or older				
>10%	96.6	94.0	10.4	17.5
>15%	64.4	88.4	53.0	21.2
>20%	42.3	87.4	78.3	27.6
Early perimenopausal, under 48 years of age				
>10%	96.9	89.6	5.9	18.4
>20%	58.1	87.4	63.4	25.8
>25%	36.0	86.0	86.3	36.5
Early perimenopausal, 48 years of age or older				
>40%	69.0	70.2	57.8	56.5
>45%	58.1	67.0	67.6	58.7
>50%	48.6	65.2	76.6	62.2
Late perimenopausal				
>75%	69.0	33.6	61.4	87.4
>80%	57.7	31.2	74.8	89.9
>85%	43.5	28.6	88.2	93.5

Performance characteristics at various thresholds or predicted probability (based on follicular stimulating hormone level, age, menopause transition stage, body mass index, and race/ethnicity).

4. Bone resorption markers and the rapid phase of bone loss

Whatever the proximal cause of MT-related bone loss, the final pathway is through increase in bone resorption, with insufficiently matched bone formation. Indeed, increases in bone resorption precede the onset of the rapid phase of bone loss in the MT (Fig. 2). Markers of bone resorption, such as urinary N-terminal telopeptide of type I collagen (U-NTX), start rising 2 years before the FMP, peak approximately 1.5 years after the FMP, and plateau thereafter (Sowers et al., 2013). Decreases in levels of circulating E2 and increases in FSH occur on nearly the same time-frame, while decreases in BMD lag by about 6 months, consistent with a causal pathway from hormones to excess bone resorption (relative to bone formation) to BMD decline (Fig. 2). Also consistent with a causal role for disproportionate bone resorption is the observation that a greater level of U-NTX in late perimenopause (which is in the middle of the rapid bone loss phase) is strongly associated with faster decline in BMD during this rapid loss phase (Shieh, Ishii, et al., 2016).

This association of late perimenopausal U-NTX level with the amount of bone lost in the MT's rapid loss phase raises the possibility of using measurements of U-NTX earlier in the MT to identify the women who will lose the most bone mass. Unfortunately, U-NTX levels in pre- and early perimenopause do *not* predict the amount of bone lost over the rapid loss phase of the MT (Shieh, Ishii, et al., 2016), limiting U-NTX's use in early identification of women who are destined to lose the most bone mass. Markers of bone resorption like U-NTX can index bone loss when it is happening (Dresner-Pollak et al., 1996), but cannot signal bone loss in the future; hence the inability to predict the amount of bone mass lost in the MT's rapid phase *before* the loss has commenced. However, resorption markers might be able to signal the onset of the rapid loss phase.

Indeed, U-NTX measured early in the MT, does appear to be associated with imminent (over the next 3–4 years) rapid bone loss (Shieh et al., 2019b). Again, this prediction ability of U-NTX is limited to early perimenopause; it does *not* predict future rapid loss in premenopausal women in their 5th decade of life (average age 46, SD 2.6 years), even with a 3–4 year-long future window (Shieh et al., 2019b).

5. Summary and future directions

Because bone mass declines rapidly during the MT and the decline is accompanied by deleterious and potentially irreversible trabecular micro-architectural changes (Akhter, Lappe, Davies, & Recker, 2007; Greendale et al., 2020), cortical thinning, and increases in cortical porosity (Cooper et al., 2007), the MT may be the optimal time-limited window for early interventions that reduce the risk of debilitating fractures in older ages. To intervene before substantial bone loss has occurred, we need to be able to determine that rapid bone loss is imminent (to time the intervention optimally) and to target the women who are likely to lose the most bone mass during the MT.

Levels of sex hormones and of markers of bone resorption are quantitative indices of physiological changes in the MT proximal to BMD decline. Not surprisingly, they are associated with the rate of BMD decline over the subsequent years. In this chapter, we examined levels of E2, FSH and U-NTX separately as predictors of the onset of MT-related rapid bone loss phase; none were sufficiently strong predictors on their own. There are several potential reasons for this. Circulating levels of E2 and FSH vary over the menstrual cycle, and a single measurement, even if it is in the follicular phase, may not reliably index the average hormone level in a woman with irregular and variable-length menstrual cycles—the hallmark of the MT. Other candidate hormones, like Anti-Mullerian hormone (AMH) whose levels are stable across the menstrual cycle (Hehenkamp et al., 2006) and which also mark progression towards the FMP (Finkelstein et al., 2020), may be better than E2 and FSH in predicting the onset of the MT-related rapid bone loss phase.

Among bone resorption markers, U-NTX has higher within-individual variability than newer markers, such as serum collagen type I C-telopeptide (CTX), which makes CTX a more promising candidate predictor of both the onset and magnitude of rapid bone loss (Szulc, Naylor, Hoyle, Eastell, & Leary, 2017). This possibility is bolstered by the observation that CTX predicts incident fractures better than does U-NTX (Garnero, Sornay-Rendu, Claustrat, & Delmas, 2000). Another intriguing possibility is combining markers of bone resorption with markers of bone formation. Even though resorption increases during the MT, so does bone formation to compensate, and levels of bone formation markers are also associated with the amount of bone lost over the MT (Gutierrez-Buey et al., 2019). With both bone resorption and formation kicking into high gear, the actual amount of bone lost over the MT is determined by the magnitude of imbalance between resorption and formation. Measures of bone resorption on their own do not capture the magnitude of imbalance well enough to accurately predict the rate of ongoing and future bone loss. But, a bone (im)balance index that uses both formation and resorption markers to quantify the excess of resorption over formation does indeed predict the rate of BMD decline over the next 3–4 years in women in their fifth decade of life better than a bone resorption marker on its own (Shieh, Han, et al., 2016).

Combining measurements of AMH with CTX and a bone formation marker such as serum procollagen type I propeptide (P1NP) may provide the best prediction of both onset and magnitude of MT-related bone loss, making possible the optimal timing and efficient targeting of midlife interventions to prevent osteoporotic fractures in older ages.

Acknowledgments

The Study of Women's Health Across the Nation (SWAN) has grant support from the National Institutes of Health (NIH), Department of Health and Human Services, through the National Institute on Aging (NIA), the National Institute of Nursing Research (NINR) and the NIH Office of Research on Women's Health (ORWH) (Grants U01NR004061; U01AG012505, U01AG012535, U01AG012531, U01AG012539, U01AG012546, U01AG012553, U01AG012554, U01AG012495). The content of this review chapter is solely the responsibility of the authors and does not necessarily represent the official views of the NIA, NINR, ORWH or the NIH.

The authors thank Drs. Jane Cauley (University of Pittsburgh), Carrie Karvonen-Gutierrez (University of Michigan), and Carolyn Crandall (University of California, Los Angeles) for their contribution to the original study from which Table 2 is drawn.

References

Akhter, M. P., Lappe, J. M., Davies, K. M., & Recker, R. R. (2007). Transmenopausal changes in the trabecular bone structure. *Bone*, *41*(1), 111–116.

Becker, C. B., & Cohen, A. (2019). Evaluation and treatment of premenopausal osteoporosis. In C. J. Rosen, & J. E. Mulder (Eds.), *UpToDate*. Retrieved on Aug 6, 2020, from https://www.uptodate.com/contents/evaluation-and-treatment-of-premenopausal-osteoporosis.

Blume, S. W., & Curtis, J. R. (2011). Medical costs of osteoporosis in the elderly medicare population. *Osteoporosis International*, *22*(6), 1835–1844.

Blumenfeld, Z., & Lavie, G. (2020). Menopausal transition bone loss: What are the practical implications. *The Journal of Clinical Endocrinology & Metabolism*, *105*(4), e1903–e1905.

Bone, H. G., Bolognese, M. A., Yuen, C. K., Kendler, D. L., Miller, P. D., Yang, Y. C., et al. (2011). Effects of denosumab treatment and discontinuation on bone mineral density and bone turnover markers in postmenopausal women with low bone mass. *The Journal of Clinical Endocrinology & Metabolism*, *96*(4), 972–980.

Cauley, J. A. (2015). Estrogen and bone health in men and women. *Steroids*, *99*, 11–15.

Cooper, D. M., Thomas, C. D. L., Clement, J. G., Turinsky, A. L., Sensen, C. W., & Hallgrímsson, B. (2007). Age-dependent change in the 3D structure of cortical porosity at the human femoral midshaft. *Bone*, *40*(4), 957–965.

Crandall, C. J., Tseng, C. H., Karlamangla, A. S., Finkelstein, J. S., Randolph, J. F., Jr., Thurston, R. C., et al. (2013). Serum sex steroid levels and longitudinal changes in bone density in relation to the final menstrual period. *The Journal of Clinical Endocrinology & Metabolism*, *98*(4), E654–E663.

Cummings, S. R., Ferrari, S., Eastell, R., Gilchrist, N., Jensen, J. E. B., McClung, M., et al. (2018). Vertebral fractures after discontinuation of denosumab: A post hoc analysis of the randomized placebo-controlled FREEDOM trial and its extension. *Journal of Bone and Mineral Research*, *33*(2), 190–198.

Dresner-Pollak, R., Parker, R. A., Poku, M., Thompson, J., Seibel, M. J., & Greenspan, S. L. (1996). Biochemical markers of bone turnover reflect femoral bone loss in elderly women. *Calcified Tissue International, 59*(5), 328–333.

Ebeling, P. R., Atley, L. M., Guthrie, J. R., Burger, H. G., Dennerstein, L., Hopper, J. L., et al. (1996). Bone turnover markers and bone density across the menopausal transition. *The Journal of Clinical Endocrinology & Metabolism, 81*(9), 3366–3371.

Finkelstein, J. S., Lee, H., Karlamangla, A., Neer, R. M., Sluss, P. M., Burnett-Bowie, S. A. M., et al. (2020). Antimullerian hormone and impending menopause in late reproductive age: The study of women's health across the nation. *The Journal of Clinical Endocrinology & Metabolism, 105*(4), e1862–e1871.

Forsen, L., Søgaard, A. J., Meyer, H. E., Edna, T. H., & Kopjar, B. (1999). Survival after hip fracture: Short-and long-term excess mortality according to age and gender. *Osteoporosis International, 10*(1), 73–78.

Garnero, P., Sornay-Rendu, E., Claustrat, B., & Delmas, P. D. (2000). Biochemical markers of bone turnover, endogenous hormones and the risk of fractures in postmenopausal women: The OFELY study. *Journal of Bone and Mineral Research, 15*(8), 1526–1536.

Gehlbach, S. H., Burge, R. T., Puleo, E., & Klar, J. (2003). Hospital care of osteoporosis-related vertebral fractures. *Osteoporosis International, 14*(1), 53–60.

Gossiel, F., Altaher, H., Reid, D. M., Roux, C., Felsenberg, D., Glüer, C. C., et al. (2018). Bone turnover markers after the menopause: T-score approach. *Bone, 111*, 44–48.

Greendale, G. A., Edelstein, S., & Barrett-Connor, E. (1997). Endogenous sex steroids and bone mineral density in older women and men: The Rancho Bernardo Study. *Journal of Bone and Mineral Research, 12*(11), 1833–1843.

Greendale, G., & Eriksen, E. (2017). ASBMR/ECTS Clinical Debate: Anti-resorptive therapy during the menopausal transition prevents bone fractures later in life. In *ASBMR 2017 annual meeting*. Washington, DC: American Society for Bone and Mineral Research. Weblinks: https://www.asbmr.org/education-detail?cid=10bfedc7-c20f-4f0d-ae1e-3130063efd89#.Xuu6y6Z7l8J. http://asbmr.capitalreach.com/console/player/36549?mediaType=audio.

Greendale, G. A., Espeland, M., Slone, S., Marcus, R., & Barrett-Connor, E. (2002). Bone mass response to discontinuation of long-term hormone replacement therapy: Results from the postmenopausal estrogen/progestin interventions (PEPI) safety follow-up study. *Archives of Internal Medicine, 162*(6), 665–672.

Greendale, G. A., Huang, M., Cauley, J. A., Liao, D., Harlow, S., Finkelstein, J. S., et al. (2020). Trabecular bone score declines during the menopause transition: The study of women's health across the nation (SWAN). *The Journal of Clinical Endocrinology & Metabolism, 105*(4), e1872–e1882.

Greendale, G. A., Ishii, S., Huang, M. H., & Karlamangla, A. S. (2013). Predicting the timeline to the final menstrual period: The study of women's health across the nation. *The Journal of Clinical Endocrinology & Metabolism, 98*(4), 1483–1491.

Greendale, G. A., Sowers, M., Han, W., Huang, M. H., Finkelstein, J. S., Crandall, C. J., et al. (2012). Bone mineral density loss in relation to the final menstrual period in a multiethnic cohort: Results from the study of women's health across the nation (SWAN). *Journal of Bone and Mineral Research, 27*(1), 111–118.

Guthrie, J. R., Dennerstein, L., Taffe, J. R., Lehert, P., & Burger, H. G. (2004). The menopausal transition: A 9-year prospective population-based study. The Melbourne women's midlife health project. *Climacteric*, 7(4), 375–389.

Gutierrez-Buey, G., Restituto, P., Botella, S., Monreal, I., Colina, I., Rodríguez-Fraile, M., et al. (2019). Trabecular bone score and bone remodelling markers identify perimenopausal women at high risk of bone loss. *Clinical Endocrinology, 91*(3), 391–399.

Hehenkamp, W. J., Looman, C. W., Themmen, A. P., de Jong, F. H., Te Velde, E. R., & Broekmans, F. J. (2006). Anti-Mullerian hormone levels in the spontaneous menstrual cycle do not show substantial fluctuation. *The Journal of Clinical Endocrinology & Metabolism, 91*(10), 4057–4063.

Hosking, D., Chilvers, C. E., Christiansen, C., Ravn, P., Wasnich, R., Ross, P., et al. (1998). Prevention of bone loss with alendronate in postmenopausal women under 60 years of age. *New England Journal of Medicine, 338*(8), 485–492.

Ishii, S., Cauley, J. A., Greendale, G. A., Crandall, C. J., Huang, M. H., Danielson, M. E., et al. (2013). Trajectories of femoral neck strength in relation to the final menstrual period in a multi-ethnic cohort. *Osteoporosis International, 24*(9), 2471–2481.

Jepsen, K. J., Kozminski, A., Bigelow, E. M., Schlecht, S. H., Goulet, R. W., Harlow, S. D., et al. (2017). Femoral neck external size but not aBMD predicts structural and mass changes for women transitioning through menopause. *Journal of Bone and Mineral Research, 32*(6), 1218–1228.

Karlamangla, A. S., Burnett-Bowie, S. A. M., & Crandall, C. J. (2018). Bone health during the menopause transition and beyond. *Obstetrics and Gynecology Clinics, 45*(4), 695–708.

Khosla, S., Melton, L. J., III, Atkinson, E. J., O'fallon, W. M., Klee, G. G., & Riggs, B. L. (1998). Relationship of serum sex steroid levels and bone turnover markers with bone mineral density in men and women: A key role for bioavailable estrogen. *The Journal of Clinical Endocrinology & Metabolism, 83*(7), 2266–2274.

Leslie, W. D., Majumdar, S. R., Morin, S. N., & Lix, L. M. (2015). Why does rate of bone density loss not predict fracture risk? *The Journal of Clinical Endocrinology & Metabolism, 100*(2), 679–683.

Nagaraj, N., Boudreau, R. M., Danielson, M. E., Greendale, G. A., Karlamangla, A. S., Beck, T. J., et al. (2019). Longitudinal changes in hip geometry in relation to the final menstrual period: Study of women's health across the nation (SWAN). *Bone, 122*, 237–245.

National Osteoporosis Foundation. (2020). *What women need to know*. Retrieved on August 6, 2020, from https://www.nof.org/preventing-fractures/general-facts/what-women-need-to-know/.

Pedrazzoni, M., Casola, A., Verzicco, I., Abbate, B., Vescovini, R., & Sansoni, P. (2014). Longitudinal changes of trabecular bone score after estrogen deprivation: Effect of menopause and aromatase inhibition. *Journal of Endocrinological Investigation, 37*(9), 871–874.

Prior, J. C., Seifert-Klauss, V. R., & Hale, G. (2013). The endocrinology of perimenopause– new definitions and understandings of hormonal and bone changes. In V. Dvornyk (Ed.), *Current topics in menopause* (pp. 54–83). Hong Kong: Benthan Science Publishers.

Randolph, J. F., Jr., Zheng, H., Sowers, M. R., Crandall, C., Crawford, S., Gold, E. B., et al. (2011). Change in follicle-stimulating hormone and estradiol across the menopausal transition: Effect of age at the final menstrual period. *The Journal of Clinical Endocrinology & Metabolism, 96*(3), 746–754.

Recker, R., Lappe, J., Davies, K., & Heaney, R. (2000). Characterization of perimenopausal bone loss: A prospective study. *Journal of Bone and Mineral Research, 15*(10), 1965–1973.

Reid, I. R., Black, D. M., Eastell, R., Bucci-Rechtweg, C., Su, G., Hue, T. F., et al. (2013). HORIZON pivotal fracture trial and HORIZON recurrent fracture trial steering committees. Reduction in the risk of clinical fractures after a single dose of zoledronic acid 5 milligrams. *The Journal of Clinical Endocrinology and Metabolism, 98*(2), 557–563.

Riis, B. J., Hansen, M. A., Jensen, A. M., Overgaard, K., & Christiansen, C. (1996). Low bone mass and fast rate of bone loss at menopause: Equal risk factors for future fracture: A 15-year follow-up study. *Bone, 19*(1), 9–12.

Saarelainen, J., Hassi, S., Honkanen, R., Koivumaa-Honkanen, H., Sirola, J., Kröger, H., et al. (2016). Bone loss and wrist fractures after withdrawal of hormone therapy: The 15-year follow-up of the OSTPRE cohort. *Maturitas, 85*, 49–55.

Seeman, E. (2013). Age-and menopause-related bone loss compromise cortical and trabecular microstructure. *Journals of Gerontology Series A: Biomedical Sciences and Medical Sciences, 68*(10), 1218–1225.

Seifert-Klauss, V., Fillenberg, S., Schneider, H., Luppa, P., Mueller, D., & Kiechle, M. (2012). Bone loss in premenopausal, perimenopausal and postmenopausal women: Results of a prospective observational study over 9 years. *Climacteric, 15*(5), 433–440.

Shieh, A., Greendale, G. A., Cauley, J. A., Karvonen-Gutierrez, C., Crandall, C. J., & Karlamangla, A. S. (2019a). Estradiol and follicle-stimulating hormone as predictors of onset of menopause transition-related Bone loss in pre-and perimenopausal women. *Journal of Bone and Mineral Research, 34*(12), 2246–2253.

Shieh, A., Greendale, G. A., Cauley, J. A., Karvonen-Gutierrez, C., Lo, J., & Karlamangla, A. S. (2019b). Urinary N-telopeptide as predictor of onset of menopause-related Bone loss in pre-and perimenopausal women. *JBMR Plus, 3*(4), e10116.

Shieh, A., Han, W., Ishii, S., Greendale, G. A., Crandall, C. J., & Karlamangla, A. S. (2016). Quantifying the balance between total bone formation and total bone resorption: An index of net bone formation. *The Journal of Clinical Endocrinology & Metabolism, 101*(7), 2802–2809.

Shieh, A., Huang, M. H., Han, W., Karlamangla, A. S., & Greendale, G. A. (2020). The associations of peak bone mineral density and rate of bone mineral density loss during the menopause transition and early postmenopause with subsequent fracture: Results from the study of women's health across the nation (SWAN). In *Oral presentation at the American Society for Bone and Mineral Research 2020 Annual Meeting*. September 2020 virtual event.

Shieh, A., Ishii, S., Greendale, G. A., Cauley, J. A., Lo, J. C., & Karlamangla, A. S. (2016). Urinary N-telopeptide and rate of bone loss over the menopause transition and early postmenopause. *Journal of Bone and Mineral Research, 31*(11), 2057–2064.

Slemenda, C. S. C. C., Hui, S. L., Longcope, C. H., & Johnston, C. C. (1987). Sex steroids and bone mass. A study of changes about the time of menopause. *The Journal of Clinical Investigation, 80*(5), 1261–1269.

Sowers, M. R., Jannausch, M., McConnell, D., Little, R., Greendale, G. A., Finkelstein, J. S., et al. (2006). Hormone predictors of bone mineral density changes during the menopausal transition. *The Journal of Clinical Endocrinology & Metabolism, 91*(4), 1261–1267.

Sowers, M. R., Zheng, H., Greendale, G. A., Neer, R. M., Cauley, J. A., Ellis, J., et al. (2013). Changes in bone resorption across the menopause transition: Effects of reproductive hormones, body size, and ethnicity. *The Journal of Clinical Endocrinology & Metabolism, 98*(7), 2854–2863.

Sowers, M. R., Zheng, H., McConnell, D., Nan, B., Harlow, S. D., & Randolph, J. F., Jr. (2008a). Estradiol rates of change in relation to the final menstrual period in a population-based cohort of women. *The Journal of Clinical Endocrinology & Metabolism, 93*(10), 3847–3852.

Sowers, M. R., Zheng, H., McConnell, D., Nan, B., Harlow, S., & Randolph, J. F., Jr. (2008b). Follicle stimulating hormone and its rate of change in defining menopause transition stages. *The Journal of Clinical Endocrinology & Metabolism, 93*(10), 3958–3964.

Szulc, P., Naylor, K., Hoyle, N. R., Eastell, R., & Leary, E. T. (2017). Use of CTX-I and PINP as bone turnover markers: National Bone Health Alliance recommendations to standardize sample handling and patient preparation to reduce pre-analytical variability. *Osteoporosis International, 28*(9), 2541–2556.

US Preventive Services Task Force. (2011). Screening for osteoporosis: US preventive services task force recommendation statement. *Annals of Internal Medicine*, *154*(5), 356.
Van Staa, T. P., Dennison, E. M., Leufkens, H. A., & Cooper, C. (2001). Epidemiology of fractures in England and Wales. *Bone*, *29*(6), 517–522.
Wright, N. C., Looker, A. C., Saag, K. G., Curtis, J. R., Delzell, E. S., Randall, S., et al. (2014). The recent prevalence of osteoporosis and low bone mass in the United States based on bone mineral density at the femoral neck or lumbar spine. *Journal of Bone and Mineral Research*, *29*(11), 2520–2526.
Zaidi, M., Turner, C. H., Canalis, E., Pacifici, R., Sun, L., Iqbal, J., et al. (2009). Bone loss or lost bone: Rationale and recommendations for the diagnosis and treatment of early postmenopausal bone loss. *Current Osteoporosis Reports*, 7(4), 118.

CHAPTER SIXTEEN

Relaxin abrogates genomic remodeling of the aged heart

Guillermo Romero[a,*] and Guy Salama[b]

[a]Department of Pharmacology & Chemical Biology, University of Pittsburgh, Pittsburgh, PA, United States
[b]Department of Medicine, Heart and Vascular Institute, University of Pittsburgh, Pittsburgh, PA, United States
*Corresponding author: e-mail address: ggr@pitt.edu

Contents

Abstract

"Healthy" aging drives structural and functional changes in the heart including maladaptive electrical remodeling, fibrosis and inflammation, which lower the threshold for cardiovascular diseases such as heart failure (HF) and atrial fibrillation (AF). Despite mixed results in clinical trials, Relaxin-therapy for 2-days reduced mortality by 37% at 180-days post-treatment, in patients with acute decompensated HF.

Relaxin's short lifespan (2 – 3 h) but long-lasting protective actions suggested that relaxin acts at a genomic level to reverse maladaptive remodeling in AF, HF and aging. Our recent studies showed that a 2-week treatment with Relaxin (0.4 mg/kg/day) of aged (24 months old F-344 rats) increases the expression of voltage-gated Na^+ channels (mRNA, Nav1.5 and I_{Na}), connexin-43, abrogates inflammatory and immune responses and reverses myocardial fibrosis and cellular hypertrophy of the aged hearts. Relaxin acts directly at a wide range of cell types in the cardiovascular system that express its cognate GPCR receptor, RXFP1. RNA-seq analysis of young and aged hearts with and without Relaxin treatment revealed that "normal" aging altered the expression of ~10% of genes expressed in the ventricles, including: ion channels, components of fibrosis,

Vitamins and Hormones, Volume 115
ISSN 0083-6729
https://doi.org/10.1016/bs.vh.2020.12.017

hemodynamic biomarkers, immune and inflammatory responses which were reversed by Relaxin. The extensive cardiovascular remodeling caused by Relaxin was mediated through the activation of the Wnt/β-catenin signaling pathway which was otherwise suppressed by in adult cardiomyocytes intracellular by cytosolic Dickkopf1 (Dkk1). Wnt/β-catenin signaling is a mechanism that can explain the pleiotropic actions of Relaxin and the marked reversal of genomic changes that occur in aged hearts.

1. Introduction

It is well-documented that age is the single most important risk factor in the cardiovascular health of humans (Rodgers et al., 2019). By 2030, 20% of the world's population will be aged 65 or older, and cardiovascular disease is the leading cause of death in this age group. In general, people 65 and older are much more likely to suffer a heart attack, experience a stroke, or develop coronary heart disease and heart failure. However, until recently, few studies have addressed the effects of aging on the biochemistry and molecular physiology of the cardiovascular system. This panorama has changed. Recent years have seen a significant increase in the interest of the biomedical and scientific communities in the biology of the aging process in the heart and the vasculature. Several important review articles have been published on the subject (Gude, Broughton, Firouzi, & Sussman, 2018; Liang & Gustafsson, 2020; Meschiari, Ero, Pan, Finkel, & Lindsey, 2017; Rodgers et al., 2019; Ruiz-Meana et al., 2020) and the reader is referred to those sources for additional details on the biochemistry of the aging process. It is increasingly evident that cardiac and vascular aging are intimately linked processes. Aging is associated with changes in vascular resistance, ventricular hypertrophy, endothelial dysfunction, increased blood pressure, and altered lipid/cholesterol metabolism. All these phenomena produce changes in the metabolic function of the heart, which almost invariably result in oxidative damage, inflammation and increased fibrosis. With age, left ventricle contractility and ejection fraction diminish, and the sympathetic regulation of heart rate decreases. These changes are temporarily compensated by increasing cardiac muscle mass, leading to ventricular hypertrophy. This is, however, a short-term solution, as hypertrophy eventually leads to a reduction of cardiac function. In this chapter we will examine some of the more relevant changes in cardiac biochemistry, their relationship to the aging process, and will discuss some of the potential therapeutic interventions that may help ameliorate the deterioration of cardiac function during the aging process.

2. Biochemistry of the aging heart

The heart is a complex organ composed primarily of three distinct cell types: cardiomyocytes (CM), fibroblasts (FB), and endothelium (EC). During the aging process and its accompanying inflammation, the heart is infiltrated with immune cells (IC; primarily macrophages and lymphocytes), and these also play a major role in the physiology of the aging heart. CMs occupy roughly 80% of the volume in the human heart, whereas EC and FB occupy 3.2–5.3% and 1.2–1.9%, respectively (Zhou & Pu, 2016). Given that the mean volume of CMs is roughly 25 times that of ECs and FBs, ECs may outnumber CMs in a healthy heart. ECs, FBs and ICs are not inert cells; on the contrary, they secrete multiple signaling molecules that regulate their own function and that of the surrounding CMs. It is necessary to take into account the biology of all these diverse cells when discussing the biochemistry of the heart. Cardiomyocyte energy metabolism is almost entirely aerobic. The heart is an oxygen-avid organ that extracts roughly 50% of the available arterial oxygen that circulates through the coronary system. Mitochondria occupy over 30% of the volume of a healthy cardiomyocyte (Piquereau et al., 2013) and mitochondrial respiration using fatty acids as main substrates is the primary source of energy for cardiac function. The massive oxygen consumption required for proper cardiac function plays a major role in heart senescence. Oxidative phosphorylation generates free radicals, and these have been linked to cellular senescence since Harman's original proposal in 1956 (Harman, 1956). Traditionally, free radicals, reactive oxygen species (ROS) and oxidative stress have been considered deleterious to cardiac health due to the oxidation of multiple cellular components, such as DNA, membrane components, and macromolecules. These changes accelerate cell death due to apoptosis and/or necrosis and the ensuing inflammatory process further augments the local concentrations of free radicals, which in turn increase apoptosis and necrosis of local tissues. More recently, however, there is mounting evidence that indicates that ROS and oxidative stress may have much subtler modulatory effects in the regulation of cardiac physiology. Importantly, CMs are not the sole source of these compounds: ECs, FBs and ICs produce multiple oxidants that can significantly modify the physiology of the heart.

The effects of ROS in the heart can be classified as acute and chronic. Acute effects are the consequence of the covalent modification of relevant heart proteins, such as ion channels, ion pumps and proteins that regulate contractile activity (Gao, Liu, & Marban, 1996). These responses are varied and depend on the type of ROS species generated. For example, hydroxyl radical (·OH) decreases Ca^{+2} transients and systolic force whereas superoxide radical ($\cdot O_2^-$) decreases systolic force without affecting Ca^{+2} transients (Gao et al., 1996). Chronic effects are those that result from alterations in the DNA or in the transcriptional regulation of important regulatory proteins. Some of these may be a consequence of the transient activation or inactivation of signaling molecules; others may result from chronic local changes in cell composition due to cell differentiation, deposition of extracellular materials, or inflammation. Importantly, acute changes can become chronic due to infiltration with immune cells, limited cell death and local inflammation. All these factors play a major role in age-induced changes in heart biochemistry and physiology.

Mitochondrial metabolism has been identified as the main source of ROS and a key component in the aging process (Balaban, Nemoto, & Finkel, 2005). ROS are generated in the mitochondria as a byproduct of oxidative phosphorylation, a process in which the energy derived from reduced NADH or FADH is used to generate a proton gradient across the mitochondrial inner membrane, this proton gradient serving as the source of energy for the phosphorylation of by the F1/F0 ATPase. It turns out that the electrons donated by NADH and FADH may react directly with oxygen or other electron acceptors to generate free radicals. For a long time, the generation of ROS by the mitochondrial electron transport chain was considered a stochastic "slippage"; however, recent work suggests that the generation of ROS may play an important role in the fine regulation of critical cellular metabolic processes, including energy utilization and stress (Dada et al., 2003; Nemoto, Takeda, Yu, Ferrans, & Finkel, 2000; Werner & Werb, 2002).

Considerable evidence indicates increased ROS production with advancing age both in the heart (Judge, Jang, Smith, Hagen, & Leeuwenburgh, 2005) and in the vasculature (Ungvari et al., 2007). Aged major vessels show increased activity of the NFκB pathway, and this phenomenon is attenuated by hydrogen peroxide scavengers (Ungvari et al., 2007). Mitochondrial dysfunction is associated to aging via several distinct mechanisms. Mitochondrial DNA mutations are substantially more prevalent in aged individuals (Corral-Debrinski et al., 1992). Functionally, aged mammalian tissues show a decreased capacity to produce ATP by oxidative phosphorylation due to the impairment of the activity of complexes I and IV of the mitochondrial respiratory chain (Navarro & Boveris, 2007). Proteomic studies have demonstrated increased age-associated nitration of rat heart mitochondrial proteins; these include Mn-Superoxide dismutase (MnSOD), electron-transfer complex I flavoprotein and F1 ATPase (Kanski, Behring, Pelling, & Schoneich, 2005). The activity of MnSOD, a major antioxidant enzyme, has been found to decrease linearly by up to 67% with age in the brain mitochondria of mice between the ages of 28 and 76 weeks (Kanski et al., 2005). These factors (increased instability of mitochondrial DNA, reduced activity of the respiratory chain enzymes and reduced levels of detoxifying enzymes) result in gradually increasing cell damage leading to cell death and inflammation. These are the landmarks of the aging heart.

Other cell proteins that contribute significantly to the production of ROS are the members of the NADPH oxidase (Nox) family. Nox1, Nox2, Nox4 and Nox5 are expressed in the human heart, although Nox5 is not expressed in rodents (Pandey et al., 2012). Nox2 and Nox4 are the most abundant forms in CMs. Among these, Nox2 has been implicated in cardiovascular aging in a rodent model. Aged F344 rats show an substantial increase in myocardial Nox2 activity accompanied by increased expression of Rac1, a small GTPase required for Nox2 activation (Wang et al., 2010). Sies and Jones recently published an excellent review that contains detailed information on the many potential sources of free radicals; the reader is referred to this article for additional details (Sies & Jones, 2020).

Ion channels are major targets for regulation and modification by REDOX signaling. In particular, the cardiac ryanodine receptor RyR2 is directly regulated by REDOX reactions where thiol oxidation-reduction, open and close the channel resulting in sarcoplasmic reticulum (SR) Ca^{2+} release and Ca^{2+} -reuptake, respectively (Zaidi, Lagenaur, Abramson, Pessah, & Salama, 1989). REDOX regulation of RyR2 was shown to occur at "critical" thiols (low pK_a thiols that are deprotonated at physiological pH), which constitute 2% of the RyR2 cysteine residues and can regulate excitation-contraction coupling in skeletal muscle (Abramson & Salama, 1989; Salama & Abramson, 1984; Salama, Abramson, & Pike, 1992; Zaidi et al., 1989). Nitric oxide and NO donors can preferentially nitrosylate multiple critical cysteine residues to open RyR2 channels to elicit SR Ca^{2+} release (Cheong, Tumbev, Abramson, Salama, & Stoyanovsky, 2005). Many ion channels that play a critical role in cardiomyocyte physiology are direct targets for nitric oxide and other free radicals. For example, oxidation of C143 of the human Cav1.2 channel protein triggers calcium influx, which in turn increases mitochondrial ROS production (Muralidharan, Cserne Szappanos, Ingley, & Hool, 2017); this leads to a signaling cascade that often results in cardiomyocyte hypertrophy. Likewise, the C-terminal PDZ binding motif of the human cardiac sodium channel Nav1.5 interacts directly with the PDZ domain of nNOS, and this interaction mediates nitric oxide (NO)-mediated nitrosylation and regulation of Nav1.5 activity (Ueda et al., 2008). These direct effects of REDOX reactions on the activity of ion channels are usually transient, and the balance is maintained by the presence of multiple antioxidant molecules. Thiol reagents, such a glutathione, play a critical role in the regulation of these transient changes. However, studies in mice show a clear age-dependent decline in the levels of glutathione (GSH), as well as of critical enzymes that regulate REDOX states and GSH levels, such as glutathione peroxidase and glutathione reductase (Hazelton & Lang, 1980, 1985)(PMID:7406884). Thus, aging is expected to have a strong effect on the homeostasis of cardiomyocyte ion channels. These effects have not been studied in detail, but deserve extensive investigation.

Finally, changes in the REDOX balance have a profound influence on the activation of transcriptional regulators, protein kinases, and other signaling molecules. Some examples of these are: Nrf2, NFκB, p53, FOXO and AMPK (Sies & Jones, 2020). Changes in the activity of these factors alters substantially the transcriptome of the cell and lead to additional levels of dysregulation.

3. Structural changes in the aging heart

Spatial and temporal remodeling of the heart is an inevitable consequence of the chronic insult to the myocardium caused by continuous REDOX metabolism. These remodeling phenomena are characterized by wall stiffening, reduced contractility, and overall impaired performance. Fibrosis is frequently observed in the aging heart. This phenomenon is believed to be a consequence of the differentiation of cardiac fibroblasts into myofibroblasts and the subsequent increase in the deposits of connective tissue in the interstitial space.

Cardiac fibrolasts (FB) play an essential role in cardiac physiology. They are responsible for the homeostasis of the extra cellular matrix (ECM), providing a scaffold for cardiomyocytes. ECM structures help distribute mechanical forces through the myocardium and play an important role in efficient electrical conduction (Travers, Kamal, Robbins, Yutzey, & Blaxall, 2016). FB also play a central role in the tissue repair process that follows local tissue injury. Cardiomyocytes have a very limited proliferative capacity; thus, the typical process of injury repair in the heart involves primarily the removal of necrotic tissue followed by its replacement by fibrotic scar tissue whose main role is the preservation of myocardial structure and functional integrity. To perform these functions, FB within the cardiac tissue are activated to myofibroblasts (MF), which secrete ECM proteins and promote a fibrotic environment. This differentiation is initiated and promoted by a multitude of pro-inflammatory factors, such as TGFβ and angiotensin II (Travers et al., 2016).

Although the deposit of ECM proteins alleviates the immediate effects of cellular injury in the heart, excess fibrosis has a detrimental effect. Excessive deposit of ECM proteins leads to distorted organ architecture and impairment of cardiac function (Weber, 2000). Excess fibrosis leads to increased ventricular stiffness and decreased contractility, often accompanied by impaired mechano-electric coupling, arrhythmogenesis, propensity to atrial fibrillation, and increased mortality (de Bakker et al., 1993, 1996; Spach & Boineau, 1997). Furthermore, excess MF can electrically couple to adjacent CMs, reducing conduction velocities and inducing arrythmogenesis (Martin et al., 2019; Rubart et al., 2018). These phenomena are a consequence of the aging process and independent of any additional pathologies. In fact, studies in healthy aged Fischer F344 rats demonstrate: (1) a marked increase in collagen deposits; (2) a substantial decrease in atrial and ventricular conduction velocities accompanied by (3) pronounced arrhythmogenicity (Henry et al., 2016; Martin et al., 2019). Effective interventions must address all of these interrelated phenomena.

4. Relaxin

Relaxin (RLX) was first described by Hisaw (1926). Hisaw originally described RLX as a hormone whose function was the relaxation of tissues involved in parturition—hence its name. Believed to be exclusively a reproductive hormone for many years, RLX has been shown to be a pleiotropic hormone with multiple effects and functions. Initially purified from guinea pig *corpus luteum* extracts in 1930 (Fevold, Hisaw, & Meyer, 1930), RLX is a peptide of the insulin family composed of two peptide chains of 24 and 29 amino acids linked by disulfide bridges (Hossain et al., 2011). The purification and sequencing of the hormone allowed the cloning of the human relaxin gene-1 (*RLN1*) (Hudson et al., 1983). A second relaxin gene, *RLN2*, was identified shortly thereafter (Hudson et al., 1984). It was subsequently shown that humans and primates have *RLN1* and *RLN2*, whereas rodents and other mammals only have *RLN1*. Interestingly, the predominant hormone that circulates during pregnancy in humans is the product of the *RLN2* gene, RLX2. The function of the *RLN1* gene in humans and primates remains to be elucidated.

Further cloning experiments led to the discovery of several additional related peptides. The family now consists of at least seven members and includes, in addition to RLX1 and RLX2, the insulin-like peptides 3, 4, 5 and 6 (INSL3, INSL4, INSL5 and INSL6, respectively), and the more recently discovered relaxin-3 (Bathgate et al., 2013). All members of the family are two-chain polypeptides synthesized as a single precursor cleaved specifically by convertase enzymes. Both peptide chains are linked via disulfide bonds reminiscent of the structure of insulin. A large body of literature regarding the biology of these peptides has been generated in recent years. This chapter deals primarily with RLX2. Readers interested in the biology of other members of the family are directed to the excellent review by Bathgate et al. (2013).

As stated above, RLX was long assumed to be a hormone related to reproduction. In humans, RLX2 blood levels vary during pregnancy and are highest during the first trimester (Szlachter et al., 1982). Interestingly, there is little correlation between circulating RLX levels and pregnancy disorders. The only association reported is a risk of premature delivery in women with very high levels of circulating RLX2 (Goldsmith & Weiss, 2009). However, there is evidence that RLX2 may be synthesized locally in the endometrium, and animal model studies strongly support a role for RLX in early pregnancy maintenance via its local effects on endometrial function. These include angiogenesis, increased numbers of lymphocytes and macrophages, and regulation of the activity of metal metalloproteinases (MMP) (Goldsmith & Weiss, 2009). Importantly, RLX regulates blood flow during pregnancy (Conrad, Jeyabalan, Danielson, Kerchner, & Novak, 2005; Debrah et al., 2006; Debrah, Conrad, Jeyabalan, Danielson, & Shroff, 2005; Vodstrcil et al., 2012).

RLX expression is not limited to reproductive tissues. Messenger RNA (mRNA) data suggests that RLX is locally expressed in the vasculature, leading to the proposal that many of the cardiovascular actions of RLX are a consequence of paracrine/autocrine mechanisms (Novak et al., 2006). Transcriptome analysis has demonstrated the presence of RLX mRNA in multiple tissues, including brain, blood, lungs, liver, kidney, multiple endocrine and cardiovascular tissues (Bathgate et al., 2013; Uhlen et al., 2015). Thus, many actions of RLX may be mediated by local paracrine or autocrine signaling. The study of the functional role of local RLX sources awaits the development of tissue-specific knockouts. Remarkably, the brain has the highest levels of RLX2 mRNA in humans (Uhlen et al., 2015), but the role of RLX2 in the brain remains to be determined.

The phenotype of $RLX^{-/-}$ mice is complex. The homozygous females are fertile and produce normal litters; however, mammary development during and after pregnancy is deficient. Pups are unable to feed and die within 24h unless fostered by a wild-type ($RLX^{+/+}$) female (Zhao et al., 1999). However, the most interesting phenotypes of the knockout animals are not related to their reproductive systems; rather, $RLX^{-/-}$ animals (both male and female) exhibit a number of distinct cardiovascular phenotypes (Leo et al., 2017). Most of the macroscopic vascular actions of RLX appear to be mediated by nitric oxide-dependent vasodilator mechanisms that improve arterial compliance in small resistance-size arteries; however, the actions of RLX on the vasculature are rather heterogeneous, and the responses vary substantially among different vascular beds (Leo et al., 2017). A second phenotype of the RLX knockout mouse was s significant age-related increase in interstitial collagen deposits in multiple organs, such a heart, lung and kidneys (Samuel et al., 2003, 2005). Significant changes in collagen contents in the lungs of knockout animals were detected after 9 months of age, and by 12 months collagen contents was increased by 65% in male $RLX^{-/-}$ mice and by 33% in females (Samuel et al., 2005). Importantly, excess fibrosis was removed by treatment with human recombinant RLX2 (0.5 mg/kg, 14 days) (Samuel et al., 2005).

5. Relaxin signaling

RLX1, RLX2, INSL3-6 and relaxin-3 signal through G protein-coupled receptors of the Relaxin Family Peptide receptor family (RXFP). This is a heterogeneous family that includes 4 distinct receptors, RXFP1-4. Two of these, RXFP1 and RXFP2, are structurally related to the leucine-rich repeat family of receptors (LGR) that also includes the receptors for TSH and LH; before being de-orphaned in 2002, RXFP1 and RXFP2 were known as LGR7 and LGR8 (Hsu et al., 2002). RXFP1 and RXFP2 are characterized by an N-terminal LDLa module followed by a long leucine-rich region that plays a primary role in ligand binding (Bathgate et al., 2013). RXFP3 and RXFP4 are distinctly different and are more closely related to the AT1 Angiotensin II receptor. Interestingly, there is some potential promiscuity in the signaling pathways managed by RLX peptides and their receptors. RLX2 activates both RXFP1 and RXFP2 in vitro (Hsu et al., 2002) although there is no evidence that RXFP2 is a target for RLX2 in vivo. The cognate ligand for RXFP2 is INSL3 (Kumagai et al., 2002). Relaxin-3 appears to interact primarily with RXFP1 (Sudo et al., 2003). However, there is evidence that

Relaxin-3 also interacts with RXFP3 and RXFP4 activating downstream signaling pathways (Liu et al., 2003; Sutton et al., 2005). Thus, the overall panorama of relaxin-family peptide signaling is complex and not fully understood.

It is generally accepted that most of the reported physiological consequences of the administration of recombinant human RLX2 are mediated by RXFP1. Initial evidence based on the expression of RXFP1 in 293T cells suggested that the main signaling pathway downstream of RLX was the activation of cAMP pathways (Hsu et al., 2002). Two distinct mechanisms appear to be involved in this activity: (a) an initial response likely due to the activation of G_s heterotrimeric G proteins, followed by (b) a delayed response sensitive to PI3 kinase and PKCζ inhibitors that involves the activation of G_{i3} (Halls, Bathgate, & Summers, 2006; Nguyen, Yang, Sanborn, & Dessauer, 2003). This biphasic mechanism is probably not universal: published data show that Pertussis toxin (Ptox) treatment increases substantially cAMP responses in cells that normally express endogenous RXFP1, such as HeLa cells, cardiac fibroblasts (Halls & Cooper, 2010), and kidney myofibroblasts (Mookerjee et al., 2009). These observations are inconsistent with the idea that G_{i3} contributes to the production of cAMP in these cells. It has also been proposed that the RXFP1-cAMP pathway may be activated at attomolar concentrations of RLX2 via a signalosome complex that involves AKAP97, PKA, PDE3D3 and β-arrestin 2 (Halls & Cooper, 2010). However, the detection of cAMP generated by sub-picomolar RLX2 depends on the methods used even though these procedures have comparable sensitivities (Halls & Cooper, 2010). Furthermore, cAMP responses to RLX are very small or undetectable in several cell types that express endogenous RXFP1. These include rat renal myofibroblasts (Mookerjee et al., 2009), rat cardiac fibroblasts (Samuel et al., 2004), and human cardiac fibroblasts (Sarwar, Samuel, Bathgate, Stewart, & Summers, 2015). Furthermore, RLX inhibits FSH-dependent cAMP production and CREB activation in Sertoli cells (Nascimento, Macheroni, Lucas, Porto, & Lazari, 2016). Thus, the activation cAMP pathways by RLX/RXFP1 is far from a universal mechanism. Likewise, the literature contains conflicting reports regarding the association of RXFP1 with β-arrestin 2: some reports indicate that this interaction is constitutive (Halls & Cooper, 2010), ligand-induced (Kern & Bryant-Greenwood, 2009), or just does not happen (Bathgate et al., 2013). Furthermore, there is conflicting literature on the internalization of RXFP1 from the plasma membrane: some reports suggest that it does not occur

(Bathgate et al., 2013); whereas others claim that internalization is constitutive but enhanced by the presence of the ligand (Kern & Bryant-Greenwood, 2009).

Independently on the specific mechanisms, RLX2-dependent signaling in cells with endogenous expression of RXFP1 is a complex phenomenon. In a 2015 study, Sarwar et al. investigated cAMP accumulation in human vascular cells of various origins (Sarwar et al., 2015). They found that RXFP1 was expressed at the surface of human umbilical vascular endothelial cells (HUVECs), human umbilical vein smooth muscle cells (HUVSMCs), human umbilical artery smooth muscle cells (HUASMCs) and human cardiac fibroblasts (HCFs). In HUVECs and HUVSMCs, RLX2 increased cAMP, cGMP and pERK1/2 but the RLX dose-response curves were bell-shaped (Sarwar et al., 2015). Similar bell-shaped RLX2 dose-response curves for cGMP and pERK1/2 were observed in HCFs, but RLX2 failed to increase cAMP levels in these cells. Finally, the dose-response curves for cAMP, cGMP and pERK1/2 were sigmoidal in HUASMCs (Sarwar et al., 2015). The simplest explanation for these cell-specific responses is that RXFP1 exists in macromolecular assemblies that contain alternative cellular components depending on the specific proteome of the cell, and that RLX-RXFP1 signaling may be finely regulated by interactions with these components.

Another potential explanation for RLX2-RXFP1 signaling heterogeneity is the fact that the RXFP1 mRNA is subjected to alternative splicing. As many as 29 naturally occurring alternative splicing variants of RXFP1 have been identified (Kern, Hubbard, Amano, & Bryant-Greenwood, 2008; Scott et al., 2006), and some of these have been reported to interfere with RLX2-dependent cAMP signaling, either by reducing RLX binding or by interfering with downstream signaling (Scott et al., 2006). Furthermore, there is substantial evidence that indicates that RXFP1 forms receptor homodimers with itself, and heterodimers with some of the alternative RXFP1 splice products or other GPCRs (Chen et al., 2020; Chow et al., 2014, 2019; Svendsen et al., 2008, 2008). These complexes, in particular the heterodimers, give rise to intricate ligand binding behaviors, including negative cooperativity (Svendsen, Zalesko, et al., 2008).

Many of the physiological actions of RLX2 are mediated by the activation of nitric oxide (NO) signaling. RLX2 is a powerful vasodilator in multiple vascular preparations. There is a large body of evidence indicating that RLX increases the expression and activities of all three types of nitric oxide synthases: nNOS (NOS I), iNOS (NOS II) and eNOS (NOS III) (Bathgate et al., 2013). All three NOS enzymes are regulated by RLX. These mechanisms are complex. Three distinct mechanisms for the activation of NO synthesis by RLX have been proposed: (1) RLX activates eNOS by a mechanism that involves Gβγ-dependent activation of PI3Kinase, followed by activation of the protein kinase Akt, which in turn phosphorylates eNOS at Ser-1179 and activates it (Nistri & Bani, 2003); (2) cAMP-dependent activation of PKA followed by phosphorylation of IκB-α (an inhibitor of NFκB), which results in increased expression of the inducible iNOS (Failli et al., 2002); (3) RLX-promoted activation of ERK kinases activates nNOS, which in turn inhibits TGFβ signaling (Mookerjee et al., 2009).

6. Effects of relaxin in the aged heart

6.1 Inflammation and fibrosis

Fibrosis is a pathological process characterized by persistent low-grade inflammatory injury and increased production of ECM proteins Fibrosis is generally accepted as one of the primary causes for the functional deterioration of the heart during the aging process (Licastro et al., 2005). Indeed, the parallel progression of inflammation and fibrosis has been linked to multiple cardiac pathologies that are over represented in the elderly: diastolic dysfunction, coronary heart disease, atrial fibrillation and heart failure. Inflammation and fibrosis have been linked with myocardial infarction and increased leukocyte infiltration that leads to pathological ventricular remodeling (J. Liu, Wang, & Li, 2016), and RLX has been shown to reduce these anomalies (Perna et al., 2005). Furthermore, the inflammatory cytokines IL-1, IL-6 and TNFα have been linked to multiple cardiac pathologies, such as cardiac hypertrophy, increased susceptibility to arrhythmia, reduced contractility, stroke risk and heart failure (Anker & von Haehling, 2004; Bacchiega et al., 2017; von Haehling, Jankowska, & Anker, 2004).

One of the earliest suggestions for a role of RLX in the aging process derived from the observation that $RLX^{-/-}$ mice develop age-dependent kidney dysfunction characterized by increasing levels of fibrosis in heart, kidneys and other organs (Samuel et al., 2003, 2005). By 9 months of age, $RLX^{-/-}$ males demonstrated a 40–50% increase in heart collagen concentrations when compared with wild-type littermates. This age-dependent change was associated to atrial hypertrophy, increased ventricular stiffness and impaired ventricular diastolic filling (Du et al., 2003). Remarkably, a 14-day treatment with recombinant RLX2 reversed cardiac fibrosis by 40% in the $RLX^{-/-}$ animals and by 58% in a mouse model of fibrotic cardiomyopathy induced by overexpression of β_2-adrenergic receptors in the heart (Samuel et al., 2004).

More recent work has focused on the study of cardiac aging using the Fischer F344xBN rat, a standard aging model used for the past 30 years (Shimokawa et al., 1993). Treatment with recombinant human RLX2 (400 μg/kg/day) significantly decreased the levels of collagen in both atria and ventricles of 24-month-old F344xBN rats (Henry et al., 2016; Martin et al., 2019). This effect was significantly more pronounced in atria (66%) than in ventricles (30%), and it was accompanied by significant reduction of the levels of mRNA for collagen I and collagen III (25–33%) (Henry et al., 2016). Likewise, RLX treatment reduced the expression of TGFβ, a potent pro-fibrotic molecule (Henry et al., 2016). These findings led to the hypothesis that the effects of RLX in the aged heart are mediated primarily by genomic rather than transient hemodynamic effects.

Recently, a full RNAseq transcriptome investigation of the consequences of aging in aged male and female F344 rats showed major changes in inflammatory signaling during the aging process (Martin et al., 2018). Interestingly, gene expression patterns were different in males and females. For example, in females, but not in males, aging increased atrial natriuretic peptide A (NPPA) and Interferon γ (IFNG) mRNA levels by 10 and 6-fold, respectively (Martin et al., 2018). Relaxin treatment (400 g/kg/day recombinant RLX2 for 14 days) had a remarkable effect on multiple pro-inflammatory genes in both males and females. Although these effects were also sex-dependent, similar pro-inflammatory pathways were affected (Martin et al., 2018). In aged females, NPPA, IFNG, CCL19 and CXCL9 were reduced by RLX treatment. In aged males, IL1β and IL6 expression were reduced by RLX. Some of these results were confirmed at the protein level by immunohistochemistry (Martin et al., 2018). These effects are summarized in Table 1. One important result from the RNAseq results was that chemo-attractants levels that were increased in aged females were significantly reduced by RLX treatment. For example, CXCL9, a small cytokine normally secreted by monocytes, endothelial cells and fibroblasts in response to IFNγ, promotes the migration of immune cells such as lymphocytes and macrophages to the site of production (Tokunaga et al., 2018). Reduced CXCL9 expression in treated aged females correlated with reduced numbers of infiltrated macrophages. Males, in contrast, did not show significant changes in the levels of CXCL9 and other chemo-attractants either due to aging or to

Table 1 Inflammatory pathways regulated by aging and RLX in F344xBN rats.

	Females		Males	
Pathway	**Aging**	**RLX**	**Aging**	**RLX**
Hydrogen Peroxide	No effect	Inhibited	No effect	Inhibited
NFκB	Activated	Inhibited	No effect	Inhibited
TLR4	Activated	No effect	No effect	Inhibited
IFNG	Activated	Inhibited	No effect	Inhibited
IL6	Activated	Inhibited	No effect	Inhibited
mTOR	Inhibited	No effect	No effect	Activated
TGFβ1	Activated	Inhibited	No effect	Inhibited
SMAD7	Inhibited	Activated	Activated	No effect

treatment with RLX. Consistently with this, macrophage infiltration in the left ventricles of males was constant, relatively low, and independent of age and treatment (Martin et al., 2018).

7. Tissue remodeling

Age-related changes in the heart include progressive cardiomyocyte hypertrophy, differentiation of cardiac fibroblasts into myofibroblasts, and increased collagen deposition. Some of these changes are compensated in part by opposing phenomena. For example, age-dependent increased collagen production is compensated by increased expression of matrix metalloproteinases, enzymes known to cleave collagen (Meschiari et al., 2017). Many of these age-related changes are reversed by treatment with RLX. RLX decreases fibrosis, differentiation of fibroblasts and cardiomyocyte hypertrophy (Henry et al., 2016; Martin et al., 2019; Mookerjee et al., 2009; Parikh et al., 2013). Perhaps the most striking effect of RLX on cardiac remodeling is the restoration of the structure of cardiac fibers compromised by the aging process (Martin et al., 2019). In young rat (9 month) left ventricles, cardiac fibers are well organized, with clearly defined accumulation of intercalated disc markers (β-catenin, connexin 43) at the contacts between adjacent cardiomyocytes (Fannin et al., 2014; Martin et al., 2019). The hearts of aged F334 rats, in contrast, have disorganized fibers with reduced, diffuse staining for β-catenin and significant lateralization of connexin 43 (Cx43) (see Fig. 1 and (Fannin et al., 2014; Martin et al., 2019)); these maladaptive

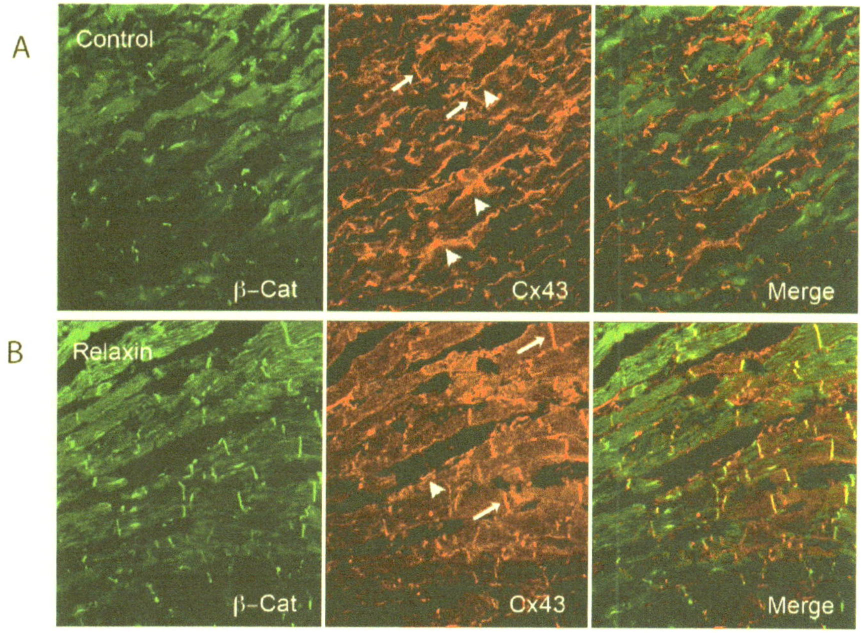

Fig. 1 Effects of RLX on cardiac remodeling. (A) Left ventricle from 24-month-old male F344xBN rats stained with β-catenin (green) and Cx43 (red) antibodies. (B) Left ventricle from a littermate of the animal shown in (A) after 2-week treatment with recombinant human RLX2 (400 μg/kg/day). The *ARROWS* (→) point to the intercalated discs. The *ARROW HEADS* (►) point to regions with lateralized Cx43.

changes are dramatically reduced by recombinant RLX (400 μg/kg/day over 2 weeks; (Martin et al., 2019)). The lateralization of Cx43 is a critical finding. Cx43 is a structural component of the gap junctions that preserve electrical conduction along heart fibers and its localization at the intercalated disc is essential for the anisotropy of the electrical conduction in the heart (van Rijen et al., 2004). Not only is Cx43 an essential component of the gap junction hemichannel; it is also a molecular partner for Nav1.5, the main cardiac sodium channel (Agullo-Pascual et al., 2014). When tested for Nav1.5 expression, both ventricular and atrial cardiomyocytes obtained from aged animals showed reduced levels of Nav1.5 protein and reduced sodium currents; RLX treatment restored Nav1.5 levels and increased I_{Na} to levels comparable to those of young animals (Martin et al., 2019).

The causes of the age-related lateralization of Cx43 and its reversal by RLX are not clear. Cx43 is recruited to the intercalated disc where it binds tightly to other resident proteins. Severe stress such as ischemic heart disease induces Cx43 lateralization (Smith, Green, Peters, Rothery, & Severs, 1991), and it has been suggested that the presence of myofibroblasts may play a significant role in this process (Rubart et al., 2018). Myofibroblasts express Cx43, and recent data have shown that there is limited but measurable electrotonic coupling between cardiomyocytes and surrounding scar tissue in models of heart injury (McArthur, Chilton, Smith, & Nicklin, 2015; Rubart et al., 2018). This coupling has been proposed to result from electrical conductivity across heterocellular junctions. RLX treatment reduces the differentiation of cardiac fibroblasts into myofibroblasts (Martin et al., 2019) and therefore may reduce Cx43 lateralization by this mechanism. Another potential mechanism by which RLX may direct the traffic of Cx43 is via phosphorylation (Parikh et al., 2013); however, Cx43 is phosphorylated in multiple sites and these modifications may have many diverse effects on Cx43 traffic. At this time, the sites of RLX-induced phosphorylation have not been identified.

8. Electrical remodeling

It is a well-documented fact that arrhythmias increase with age (Furberg et al., 1994). The prevalence of atrial fibrillation (AF) increases with age from 2% to 4% before the age of 70 up to 9–15% in 80-year old and older patients (Brembilla-Perrot, 2003). Because of the heterogeneity of the population included in the Furberg study, It is not clear how many of these electrophysiological symptoms are due to direct alterations in atrial electrophysiology due to aging or to underlying non-cardiac problems, such as hypertension, diabetes or kidney disease. Animal studies using the F344xBN rat model, however, showed significant electrophysiological changes associated with age in the absence of any other detected pathologies (Fannin et al., 2014; Henry et al., 2016; Martin et al., 2019; Walker Jr. et al., 2006).

Recent work in animal models has demonstrated marked age-associated arrhythmogenicity in both atria and ventricles (Henry et al., 2016; Martin et al., 2019). Using optical mapping methods in paced, Langendorff-perfused hearts of young and aged F344xBN rats, our lab demonstrated that a single premature pulse sufficed to elicit atrial and ventricular arrhythmias in 24-month old, but not in 9-month old animals (Henry et al., 2016; Martin et al., 2019). Measurement of the electrical conduction velocities along cardiac fibers showed a marked decrease in the aged animals. This was accompanied by a significant reduction in Na^+ currents and impaired expression of the main cardiac Na^+ channel, Nav1.5 (Henry et al., 2016; Martin et al., 2019). Remarkably, RLX treatment reversed these effects, increasing the right atria conduction velocity by almost 2-fold under fast pacing conditions (Henry et al., 2016). Likewise, the sodium current density was increased by 46% in the RLX-treated animals. These changes were accompanied by a significant increase in the expression of Nav1.5 in cardiomyocytes of RLX-treated animals (Henry et al., 2016; Martin et al., 2019). These effects were recapitulated in freshly isolated atrial cardiomyocytes: a 24-h treatment with RLX (10 100nM) increased Nav1.5 expression by 80% (Henry et al., 2016). These studies lead to the conclusion that RLX exerts pleiotropic effects in the aged heart leading to reduced fibrosis and improved electrical compliance.

9. Mechanistic aspects of the effects of RLX in the heart

The effects of RLX can be (1) acute (within minutes) or (2) sustained (over hours or days). Acute effects of RLX on cAMP/PKA signaling in cardiac tissues were initially suggested by the modulation of calcium currents in cardiac pacemaker cells; RLX causes reversible increases in the L-type calcium current, but this effect is eliminated after application of isoproterenol, superfusion with 8-Br-cAMP or administration of a PKA inhibitor (Han, Habuchi, & Giles, 1994). More recent studies have demonstrated that there is significant cross-talk between RLX and β_2-adrenergic signaling in cardiomyocytes, as treatment with RLX induces phosphorylation of the β_2AR leading to the inhibition of β_2 adrenergic responses (Shi et al., 2017). However, the effects of RLX on heart maladaptation belong to the second category; that is: RLX reverses fibrosis and defects in cardiac electrophysiology via mechanisms that involve changes in gene expression over a period of hours or days. Several classical pathways that regulate gene expression downstream of RLX have been described in multiple organs and tissues, including cAMP/CREB, NO via mechanisms dependent or independent of cGMP, NFκB, ERK1/2, tyrosine kinases, PI3Kinase/Akt and, more recently, PPARγ and glucocorticoid receptors (Ahmad, Wang, Nair, & Kapila, 2012; Dschietzig, Bartsch, Stangl, Baumann, & Stangl, 2004; Nascimento et al., 2016; Quattrone, Chiappini, Scapagnini, Bigazzi, & Bani, 2004; Singh, Simpson, & Bennett, 2015). Nevertheless, there is little information regarding the signaling mechanisms that regulate the genomic effects of RLX in the heart.

Recent work has identified the Wnt signaling pathway as an important player in the mediation of RLX genomic signaling. In an investigation of the mechanisms by which RLX reverses maladaptive remodeling of the aged heart, Martin et al. made an unexpected observation: RLX treatment increases the expression of Wnt1 and β-catenin in aged hearts (Martin et al., 2019). This was unexpected because RLX shows anti-fibrotic properties, and the Wnt signaling pathway has been classically considered a promoter of cardiac fibrosis (Akhmetshina et al., 2012; Blyszczuk et al., 2017; Hu, Cao, Wu, Vaziri, & Zhao, 2020; Tao, Yang, Shi, & Li, 2016). Canonic Wnt signaling plays a central role in cardiac embryogenesis, but it was long thought to be quiescent after birth (Dawson, Aflaki, & Nattel, 2013). More recent studies, however, suggest that the Wnt signaling pathway remains active in adult cardiomyocytes. Disruption of canonical Wnt signaling in cardiomyocytes is associated to the development of arrhythmogenic cardiomyopathies (Austin et al., 2019). In cardiac myocytes, β-catenin is concentrated at the intercalated discs where it interacts with N-cadherins, plakoglobin and Cx43 (Swope, Cheng, Gao, Li, & Radice, 2012). Excess cytosolic β-catenin forms a complex with axin and adenomatous polyposis coli (the destruction complex), where it is constitutively phosphorylated by GSK3β and targeted for ubiquitination and destruction by the proteasome. During canonical Wnt signaling, binding of Wnt ligands to Frizzled receptors and the co-receptors LRP5 and LRP6 promotes the binding of Disheveled to the Wnt-receptor complex and the subsequent disruption of the destruction complex. This leads to the accumulation of β-catenin, which enters the nucleus and regulates the activity of the TCF/LEF family of transcriptional regulators (Austin et al., 2019). Disruption of Wnt/β-catenin signaling in cardiomyocytes causes adipogenic and fibrogenic signaling and ventricular arrhythmias (Garcia-Gras et al., 2006; Lorenzon et al., 2017). Wnt signaling plays an important role in the repair mechanisms elicited after injury. In the hearts from heart failure patients, Akt activation and subsequent inhibition of GSK3β protect cardiomyocytes from apoptosis (Lorenzon et al., 2017). Furthermore, Wnt1 is a potent inducer of the expression of Cx43 (Ai, Fischer, Spray, Brown, & Fishman, 2000).

Multiple lines of evidence support the hypothesis that many of the genomic effects of RLX in the aged heart are a consequence of the fine regulation of canonical Wnt signaling in cardiomyocytes. First, RLX treatment induces the expression of Wnt1 in vivo and in cultured cardiomyocytes. Second, RLX reduces the expression of Dickkopf-1 (Dkk1), a naturally occurring inhibitor of canonical Wnt signaling. Third, RLX increases transiently the localization of β-catenin in the nucleus of treated cardiomyocytes.

Fourth, inhibitors of GSK3β mimick the effects of RLX on the induction of Nav1.5. Fifth, the addition of recombinant Dkk1 inhibited the effects of RLX (Martin et al., 2019). Moreover, Wnt signaling also plays an important role in RLX signaling in cardiac fibroblasts. Indeed, RLX inhibited TGFβ-induced differentiation of myofibroblasts by a mechanism that was inhibited by exogenous Dkk1 (Martin et al., 2019). We do not know, at this time the detailed mechanisms by which RLX regulates the expression of Wnt1 and Dkk1, or if these are the only mechanisms in play for the effects of RLX on the remodeling of aged hearts. PKA and adenylyl cyclase inhibitors had no effects on the responses to RLX suggesting a pathway independent of cAMP (Martin et al., 2019). A model summarizing this mechanism is shown in Fig. 2.

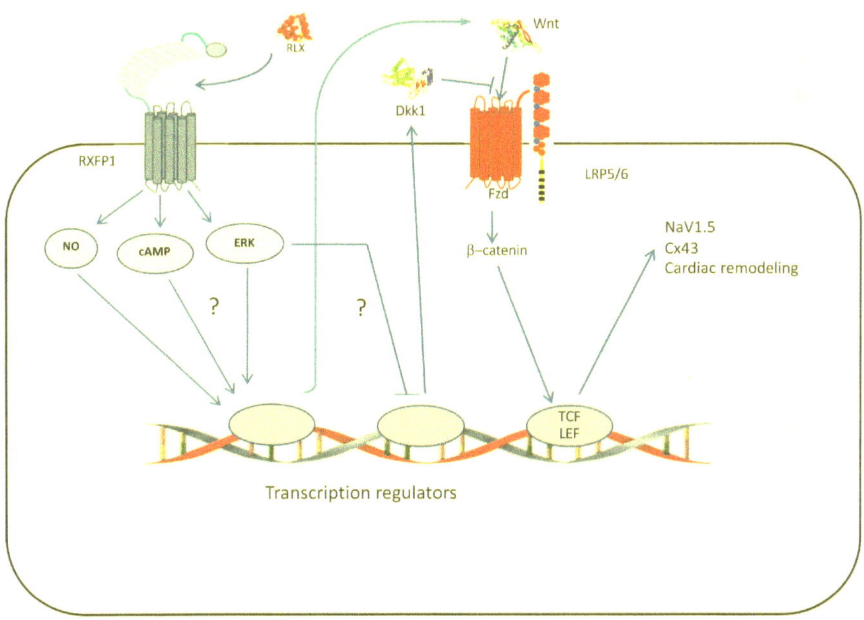

Fig. 2 Relaxin signaling mediated by Wnt/β-catenin pathway. The insulin-like hormone, Relaxin (RLX) acts via its receptor, Rxfp1 to elevate levels of nitric oxide (NO), cyclic adenosine monophosphate (cAMP) and activation of extracellular regulated kinase (ERK). These factors result in transcriptional regulation that suppress Dickkopf 1 and upregulate Wnt1 but details of these pathways are unknown. Wnt1 acting via the Frizzled (Fzd) receptor and its co-receptor LRP5/6 induces the release of β-catenin from a cytoplasmic degradation complex (the scaffold protein Axin2, casein kinase 1, and glycogen synthase kinase-3 (GSK3)). Cytosolic β-catenin traffics with Cx43 to the intercalated disk (see Fig. 1) and non-phosphorylated, active β-catenin enters the nucleus, where it does not bind DNA directly but forms complexes with transcription factors of the Tcf and Lef1 to promote transcription which results in part, in a higher expression of Nav1.5, Cx43 and a cascade of cardiac remodeling.

10. Relaxin in the clinic

The cardiovascular and anti-fibrotic actions of RLX made it a prime candidate for the treatment of multiple diseases. To date, RLX2, also called Serelaxin in the bulk of the recent clinical trials, has been shown to be safe, easily tolerated, and showed some promise for the treatment of multiple cardiac conditions (Unemori, 2017). Among hospitalized patients with acute cardiac failure, 10–15% experience worsening of their condition during hospitalization, and 10–15% die within 90 days after discharge (Gheorghiade, Vaduganathan, Fonarow, & Bonow, 2013; Torre-Amione et al., 2009). In acute heart failure patients, administration of RLX (i.v. infusion) caused significant reductions in systemic vascular resistance, systolic blood pressure and pulmonary artery pressure, with the onset of the effects within 30 min after starting the infusion (Dschietzig et al., 2009; Gheorghiade, Teerlink, & Mebazaa, 2005; Teerlink et al., 2013). Two studies done a few years ago suggested promising results for the treatment of acute HF with RLX. Pre-RELAX-AHF was a double blind, placebo-controlled trial, with 234 patients admitted for acute heart failure. The patients were infused with recombinant RLX2 or placebo for 48 h, and the study suggested beneficial effects of RLX on both dyspnea and post-discharge clinical outcomes (Teerlink et al., 2009). RELAX-AHF was a follow up study, with a much larger patient sample (1161) designed to confirm the effects observed in the previous clinical trial. RELAX-AHF confirmed the beneficial effects of RLX on dyspnea (Teerlink et al., 2013), but also suggested additional beneficial effects of RLX, such as reduced 180-day mortality. Encouraged by these results, a third clinical trial, RELAX-AHF-II, included an even larger number of patients (6545) with the goal of confirming the effects of RLX in the reduction of mortality and lower incidence of worsening heart failure (Metra et al., 2019). Paradoxically, this study failed to confirm the findings of the previous trials. It should be noted, however, that the design of these clinical trials has come under fire recently. A 2017 editorial article published in the American Journal of Cardiology sharply criticized the experimental design of all recent acute heart failure clinical trials. AHF clinical trials over the past 15 years have invariably followed the same short-term treatment protocols (24–72 h infusions of the tested agent). However, the symptoms of AHF develop over a period of weeks and months; thus, the expectations that short treatments can improve AHF outcomes appear unrealistic (McCullough, 2017). In fact, Phase III clinical trials testing the effects of short-term infusions of multiple agents with promising pre-clinical therapeutic profiles (Milrinone, Levosemendan, Tezosentan, Nesitiride, Ularitide and others) have failed to show beneficial effects in AHF (McCullough, 2017). Additional caveats regarding AHF clinical trials have been raised by others, including the heterogeneity of the symptoms in the selected patient population (Teerlink et al., 2013), the timing of therapy, the imprecision of the diagnosis, and the need to align the selected patient cohort with the mechanisms of action of the drugs tested (Mebazaa et al., 2015; Metra et al., 2019).

11. Concluding remarks

Although the failure of the RELAX-AHF-II dampened the enthusiasm initially elicited by previous reports, RLX remains a prime candidate for the treatment of diverse cardiac pathologies due to very strong preclinical findings. It should be noted that most of the animal studies described above used a 14-day continuous administration of the drug rather than the 48-h approach selected for the clinical trials (e.g., Henry et al., 2016; Martin et al., 2018; Martin et al., 2019; Parikh et al., 2013). In fact, it is hard to envision how short-term treatments might elicit the dramatic changes induced by RLX in the heart. Of particular interest are the results on atrial electrical and histological remodeling described by Henry et al. (2016). Atrial fibrillation remains a critical health issue among the elderly with an unmet need for pharmacological interventions. RLX treatment reduces arrhythmogenicity while significantly decreasing heart fibrosis, which is an important contributor to therapeutic resistance in patients with long standing AF (Henry et al., 2016; Martin et al., 2019). Current AF therapies, such as ablation, do not address the massive tissue remodeling that is classically associated with the electrical anomalies. RLX could contribute to a comprehensive therapeutic approach to AF.

Acknowledgment

The authors have no conflicts to declare.

References

Abramson, J. J., & Salama, G. (1989). Critical sulfhydryls regulate calcium release from sarcoplasmic reticulum. *Journal of Bioenergetics and Biomembranes*, *21*(2), 283–294. https://doi.org/10.1007/BF00812073.

Agullo-Pascual, E., Lin, X., Leo-Macias, A., Zhang, M., Liang, F. X., Li, Z., et al. (2014). Super-resolution imaging reveals that loss of the C-terminus of connexin 43 limits microtubule plus-end capture and NaV1.5 localization at the intercalated disc. *Cardiovascular Research*, *104*(2), 371–381. https://doi.org/10.1093/cvr/cvu195.

Ahmad, N., Wang, W., Nair, R., & Kapila, S. (2012). Relaxin induces matrix-metalloproteinases-9 and -13 via RXFP1: Induction of MMP-9 involves the PI3K, ERK, Akt and PKC-zeta pathways. *Molecular and Cellular Endocrinology, 363*(1–2), 46–61. https://doi.org/10.1016/j.mce.2012.07.006.

Ai, Z., Fischer, A., Spray, D. C., Brown, A. M., & Fishman, G. I. (2000). Wnt-1 regulation of connexin43 in cardiac myocytes. *The Journal of Clinical Investigation, 105*(2), 161–171. https://doi.org/10.1172/JCI7798.

Akhmetshina, A., Palumbo, K., Dees, C., Bergmann, C., Venalis, P., Zerr, P., et al. (2012). Activation of canonical Wnt signalling is required for TGF-beta-mediated fibrosis. *Nature Communications, 3*, 735. https://doi.org/10.1038/ncomms1734.

Anker, S. D., & von Haehling, S. (2004). Inflammatory mediators in chronic heart failure: An overview. *Heart, 90*(4), 464–470. https://doi.org/10.1136/hrt.2002.007005.

Austin, K. M., Trembley, M. A., Chandler, S. F., Sanders, S. P., Saffitz, J. E., Abrams, D. J., et al. (2019). Molecular mechanisms of arrhythmogenic cardiomyopathy. *Nature Reviews. Cardiology, 16*(9), 519–537. https://doi.org/10.1038/s41569-019-0200-7.

Bacchiega, B. C., Bacchiega, A. B., Usnayo, M. J., Bedirian, R., Singh, G., & Pinheiro, G. D. (2017). Interleukin 6 inhibition and coronary artery disease in a high-risk population: A prospective community-based clinical study. *Journal of the American Heart Association, 6*(3), e005038. https://doi.org/10.1161/JAHA.116.005038.

Balaban, R. S., Nemoto, S., & Finkel, T. (2005). Mitochondria, oxidants, and aging. *Cell, 120*(4), 483–495. https://doi.org/10.1016/j.cell.2005.02.001.

Bathgate, R. A., Halls, M. L., van der Westhuizen, E. T., Callander, G. E., Kocan, M., & Summers, R. J. (2013). Relaxin family peptides and their receptors. *Physiological Reviews, 93*(1), 405–480. https://doi.org/10.1152/physrev.00001.2012.

Blyszczuk, P., Muller-Edenborn, B., Valenta, T., Osto, E., Stellato, M., Behnke, S., et al. (2017). Transforming growth factor-beta-dependent Wnt secretion controls myofibroblast formation and myocardial fibrosis progression in experimental autoimmune myocarditis. *European Heart Journal, 38*(18), 1413–1425. https://doi.org/10.1093/eurheartj/ehw116.

Brembilla-Perrot, B. (2003). Age-related changes in arrhythmias and electrophysiologic properties. *Cardiac Electrophysiology Review*, 7(1), 88–91. https://doi.org/10.1023/a:1023611727947.

Chen, T. Y., Li, X., Hung, C. H., Bahudhanapati, H., Tan, J., Kass, D. J., et al. (2020). The relaxin family peptide receptor 1 (RXFP1): An emerging player in human health and disease. *Molecular Genetics & Genomic Medicine, 8*(4). https://doi.org/10.1002/mgg3.1194, e1194.

Cheong, E., Tumbev, V., Abramson, J., Salama, G., & Stoyanovsky, D. A. (2005). Nitroxyl triggers Ca2+ release from skeletal and cardiac sarcoplasmic reticulum by oxidizing ryanodine receptors. *Cell Calcium, 37*(1), 87–96. https://doi.org/10.1016/j.ceca.2004.07.001.

Chow, B. S., Kocan, M., Bosnyak, S., Sarwar, M., Wigg, B., Jones, E. S., et al. (2014). Relaxin requires the angiotensin II type 2 receptor to abrogate renal interstitial fibrosis. *Kidney International, 86*(1), 75–85. https://doi.org/10.1038/ki.2013.518.

Chow, B. S. M., Kocan, M., Shen, M., Wang, Y., Han, L., Chew, J. Y., et al. (2019). AT1R-AT2R-RXFP1 functional crosstalk in Myofibroblasts: Impact on the therapeutic targeting of renal and cardiac fibrosis. *Journal of the American Society of Nephrology: JASN, 30*(11), 2191–2207. https://doi.org/10.1681/ASN.2019060597.

Conrad, K. P., Jeyabalan, A., Danielson, L. A., Kerchner, L. J., & Novak, J. (2005). Role of relaxin in maternal renal vasodilation of pregnancy. *Annals of the New York Academy of Sciences, 1041*, 147–154. https://doi.org/10.1196/annals.1282.022.

Corral-Debrinski, M., Horton, T., Lott, M. T., Shoffner, J. M., Beal, M. F., & Wallace, D. C. (1992). Mitochondrial DNA deletions in human brain: Regional variability and increase with advanced age. *Nature Genetics, 2*(4), 324–329. https://doi.org/10.1038/ng1292-324.

Dada, L. A., Chandel, N. S., Ridge, K. M., Pedemonte, C., Bertorello, A. M., & Sznajder, J. I. (2003). Hypoxia-induced endocytosis of Na,K-ATPase in alveolar epithelial cells is mediated by mitochondrial reactive oxygen species and PKC-zeta. *The Journal of Clinical Investigation, 111*(7), 1057–1064. https://doi.org/10.1172/JCI16826.

Dawson, K., Aflaki, M., & Nattel, S. (2013). Role of the Wnt-frizzled system in cardiac pathophysiology: A rapidly developing, poorly understood area with enormous potential. *The Journal of Physiology, 591*(6), 1409–1432. https://doi.org/10.1113/jphysiol.2012.235382.

de Bakker, J. M., van Capelle, F. J., Janse, M. J., Tasseron, S., Vermeulen, J. T., de Jonge, N., et al. (1993). Slow conduction in the infarcted human heart. 'Zigzag' course of activation. *Circulation, 88*(3), 915–926. https://doi.org/10.1161/01.cir.88.3.915.

de Bakker, J. M., van Capelle, F. J., Janse, M. J., Tasseron, S., Vermeulen, J. T., de Jonge, N., et al. (1996). Fractionated electrograms in dilated cardiomyopathy: Origin and relation to abnormal conduction. *Journal of the American College of Cardiology, 27*(5), 1071–1078. https://doi.org/10.1016/0735-1097(95)00612-5.

Debrah, D. O., Conrad, K. P., Jeyabalan, A., Danielson, L. A., & Shroff, S. G. (2005). Relaxin increases cardiac output and reduces systemic arterial load in hypertensive rats. *Hypertension, 46*(4), 745–750. https://doi.org/10.1161/01.HYP.0000184230.52059.33.

Debrah, D. O., Novak, J., Matthews, J. E., Ramirez, R. J., Shroff, S. G., & Conrad, K. P. (2006). Relaxin is essential for systemic vasodilation and increased global arterial compliance during early pregnancy in conscious rats. *Endocrinology, 147*(11), 5126–5131. https://doi.org/10.1210/en.2006-0567.

Dschietzig, T., Bartsch, C., Stangl, V., Baumann, G., & Stangl, K. (2004). Identification of the pregnancy hormone relaxin as glucocorticoid receptor agonist. *The FASEB Journal, 18*(13), 1536–1538. https://doi.org/10.1096/fj.03-1120fje.

Dschietzig, T., Teichman, S., Unemori, E., Wood, S., Boehmer, J., Richter, C., et al. (2009). Intravenous recombinant human relaxin in compensated heart failure: A safety, tolerability, and pharmacodynamic trial. *Journal of Cardiac Failure, 15*(3), 182–190. https://doi.org/10.1016/j.cardfail.2009.01.008.

Du, X. J., Samuel, C. S., Gao, X. M., Zhao, L., Parry, L. J., & Tregear, G. W. (2003). Increased myocardial collagen and ventricular diastolic dysfunction in relaxin deficient mice: A gender-specific phenotype. *Cardiovascular Research, 57*(2), 395–404. https://doi.org/10.1016/s0008-6363(02)00663-6.

Failli, P., Nistri, S., Quattrone, S., Mazzetti, L., Bigazzi, M., Sacchi, T. B., et al. (2002). Relaxin up-regulates inducible nitric oxide synthase expression and nitric oxide generation in rat coronary endothelial cells. *The FASEB Journal, 16*(2), 252–254. https://doi.org/10.1096/fj.01-0569fje.

Fannin, J., Rice, K. M., Thulluri, S., Dornon, L., Arvapalli, R. K., Wehner, P., et al. (2014). Age-associated alterations of cardiac structure and function in the female F344xBN rat heart. *Age (Dordrecht, Netherlands), 36*(4), 9684. https://doi.org/10.1007/s11357-014-9684-6.

Fevold, H. L., Hisaw, F. L., & Meyer, R. K. (1930). The relaxative hormone of the corpus luteum. Its purification and concentration. *Journal of the American Chemical Society, 52*(8), 3340–3348. https://pubs.acs.org/doi/abs/10.1021/ja01371a051.

Furberg, C. D., Psaty, B. M., Manolio, T. A., Gardin, J. M., Smith, V. E., & Rautaharju, P. M. (1994). Prevalence of atrial fibrillation in elderly subjects (the cardiovascular health study). *The American Journal of Cardiology, 74*(3), 236–241. https://doi.org/10.1016/0002-9149(94)90363-8.

Gao, W. D., Liu, Y., & Marban, E. (1996). Selective effects of oxygen free radicals on excitation-contraction coupling in ventricular muscle. Implications for the mechanism of stunned myocardium. *Circulation, 94*(10), 2597–2604. https://doi.org/10.1161/01.cir.94.10.2597.

Garcia-Gras, E., Lombardi, R., Giocondo, M. J., Willerson, J. T., Schneider, M. D., Khoury, D. S., et al. (2006). Suppression of canonical Wnt/beta-catenin signaling by nuclear plakoglobin recapitulates phenotype of arrhythmogenic right ventricular cardiomyopathy. *The Journal of Clinical Investigation*, *116*(7), 2012–2021. https://doi.org/10.1172/JCI27751.

Gheorghiade, M., Teerlink, J. R., & Mebazaa, A. (2005). Pharmacology of new agents for acute heart failure syndromes. *The American Journal of Cardiology*, *96*(6A), 68G–73G. https://doi.org/10.1016/j.amjcard.2005.07.023.

Gheorghiade, M., Vaduganathan, M., Fonarow, G. C., & Bonow, R. O. (2013). Rehospitalization for heart failure: Problems and perspectives. *Journal of the American College of Cardiology*, *61*(4), 391–403. https://doi.org/10.1016/j.jacc.2012.09.038.

Goldsmith, L. T., & Weiss, G. (2009). Relaxin in human pregnancy. *Annals of the New York Academy of Sciences*, *1160*, 130–135. https://doi.org/10.1111/j.1749-6632.2008.03800.x.

Gude, N. A., Broughton, K. M., Firouzi, F., & Sussman, M. A. (2018). Cardiac ageing: Extrinsic and intrinsic factors in cellular renewal and senescence. *Nature Reviews. Cardiology*, *15*(9), 523–542. https://doi.org/10.1038/s41569-018-0061-5.

Halls, M. L., Bathgate, R. A., & Summers, R. J. (2006). Relaxin family peptide receptors RXFP1 and RXFP2 modulate cAMP signaling by distinct mechanisms. *Molecular Pharmacology*, *70*(1), 214–226. https://doi.org/10.1124/mol.105.021691.

Halls, M. L., & Cooper, D. M. (2010). Sub-picomolar relaxin signalling by a pre-assembled RXFP1, AKAP79, AC2, beta-arrestin 2, PDE4D3 complex. *The EMBO Journal*, *29*(16), 2772–2787. https://doi.org/10.1038/emboj.2010.168.

Han, X., Habuchi, Y., & Giles, W. R. (1994). Relaxin increases heart rate by modulating calcium current in cardiac pacemaker cells. *Circulation Research*, *74*(3), 537–541. https://doi.org/10.1161/01.res.74.3.537.

Harman, D. (1956). Aging: A theory based on free radical and radiation chemistry. *Journal of Gerontology*, *11*(3), 298–300. https://doi.org/10.1093/geronj/11.3.298.

Hazelton, G. A., & Lang, C. A. (1980). Glutathione contents of tissues in the aging mouse. *The Biochemical Journal*, *188*(1), 25–30. https://doi.org/10.1042/bj1880025.

Hazelton, G. A., & Lang, C. A. (1985). Glutathione peroxidase and reductase activities in the aging mouse. *Mechanisms of Ageing and Development*, *29*(1), 71–81. https://doi.org/10.1016/0047-6374(85)90048-x.

Henry, B. L., Gabris, B., Li, Q., Martin, B., Giannini, M., Parikh, A., et al. (2016). Relaxin suppresses atrial fibrillation in aged rats by reversing fibrosis and upregulating Na + channels. *Heart Rhythm*, *13*(4), 983–991. https://doi.org/10.1016/j.hrthm.2015.12.030.

Hisaw, F. L. (1926). Experimental relaxation of the pubic ligament of the Guinea pig. *Proceedings of the Society for Experimental Biology and Medicine*, *23*(8), 661–663. https://doi.org/10.3181/00379727-23-3107.

Hossain, M. A., Rosengren, K. J., Samuel, C. S., Shabanpoor, F., Chan, L. J., Bathgate, R. A., et al. (2011). The minimal active structure of human relaxin-2. *The Journal of Biological Chemistry*, *286*(43), 37555–37565. https://doi.org/10.1074/jbc.M111.282194.

Hsu, S. Y., Nakabayashi, K., Nishi, S., Kumagai, J., Kudo, M., Sherwood, O. D., et al. (2002). Activation of orphan receptors by the hormone relaxin. *Science*, *295*(5555), 671–674. https://doi.org/10.1126/science.1065654.

Hu, H. H., Cao, G., Wu, X. Q., Vaziri, N. D., & Zhao, Y. Y. (2020). Wnt signaling pathway in aging-related tissue fibrosis and therapies. *Ageing Research Reviews*, *60*, 101063. https://doi.org/10.1016/j.arr.2020.101063.

Hudson, P., Haley, J., John, M., Cronk, M., Crawford, R., Haralambidis, J., et al. (1983). Structure of a genomic clone encoding biologically active human relaxin. *Nature*, *301*(5901), 628–631. https://doi.org/10.1038/301628a0.

Hudson, P., John, M., Crawford, R., Haralambidis, J., Scanlon, D., Gorman, J., et al. (1984). Relaxin gene expression in human ovaries and the predicted structure of a human preprorelaxin by analysis of cDNA clones. *The EMBO Journal*, *3*(10), 2333–2339. Retrieved from https://www.ncbi.nlm.nih.gov/pubmed/6548702.

Judge, S., Jang, Y. M., Smith, A., Hagen, T., & Leeuwenburgh, C. (2005). Age-associated increases in oxidative stress and antioxidant enzyme activities in cardiac interfibrillar mitochondria: Implications for the mitochondrial theory of aging. *The FASEB Journal*, *19*(3), 419–421. https://doi.org/10.1096/fj.04-2622fje.

Kanski, J., Behring, A., Pelling, J., & Schoneich, C. (2005). Proteomic identification of 3-nitrotyrosine-containing rat cardiac proteins: Effects of biological aging. *American Journal of Physiology. Heart and Circulatory Physiology*, *288*(1), H371–H381. https://doi.org/10.1152/ajpheart.01030.2003.

Kern, A., & Bryant-Greenwood, G. D. (2009). Characterization of relaxin receptor (RXFP1) desensitization and internalization in primary human decidual cells and RXFP1-transfected HEK293 cells. *Endocrinology*, *150*(5), 2419–2428. https://doi.org/10.1210/en.2008-1385.

Kern, A., Hubbard, D., Amano, A., & Bryant-Greenwood, G. D. (2008). Cloning, expression, and functional characterization of relaxin receptor (leucine-rich repeat-containing g protein-coupled receptor 7) splice variants from human fetal membranes. *Endocrinology*, *149*(3), 1277–1294. https://doi.org/10.1210/en.2007-1348.

Kumagai, J., Hsu, S. Y., Matsumi, H., Roh, J. S., Fu, P., Wade, J. D., et al. (2002). INSL3/Leydig insulin-like peptide activates the LGR8 receptor important in testis descent. *The Journal of Biological Chemistry*, *277*(35), 31283–31286. https://doi.org/10.1074/jbc.C200398200.

Leo, C. H., Jelinic, M., Ng, H. H., Marshall, S. A., Novak, J., Tare, M., et al. (2017). Vascular actions of relaxin: Nitric oxide and beyond. *British Journal of Pharmacology*, *174*(10), 1002–1014. https://doi.org/10.1111/bph.13614.

Liang, W. J., & Gustafsson, A. B. (2020). The aging heart: Mitophagy at the Center of Rejuvenation. *Frontiers in Cardiovascular Medicine*, *7*, 18. https://doi.org/10.3389/fcvm.2020.00018.

Licastro, F., Candore, G., Lio, D., Porcellini, E., Colonna-Romano, G., Franceschi, C., et al. (2005). Innate immunity and inflammation in ageing: A key for understanding age-related diseases. *Immunity & Ageing*, *2*, 8. https://doi.org/10.1186/1742-4933-2-8.

Liu, C., Chen, J., Sutton, S., Roland, B., Kuei, C., Farmer, N., et al. (2003). Identification of relaxin-3/INSL7 as a ligand for GPCR142. *The Journal of Biological Chemistry*, *278*(50), 50765–50770. https://doi.org/10.1074/jbc.M308996200.

Liu, J., Wang, H., & Li, J. (2016). Inflammation and inflammatory cells in myocardial infarction and reperfusion injury: A double-edged sword. *Clinical Medicine Insights: Cardiology*, *10*, 79–84. https://doi.org/10.4137/CMC.S33164.

Lorenzon, A., Calore, M., Poloni, G., De Windt, L. J., Braghetta, P., & Rampazzo, A. (2017). Wnt/beta-catenin pathway in arrhythmogenic cardiomyopathy. *Oncotarget*, *8*(36), 60640–60655. https://doi.org/10.18632/oncotarget.17457.

Martin, B., Gabris, B., Barakat, A. F., Henry, B. L., Giannini, M., Reddy, R. P., et al. (2019). Relaxin reverses maladaptive remodeling of the aged heart through Wnt-signaling. *Scientific Reports*, *9*(1), 18545. https://doi.org/10.1038/s41598-019-53867-y.

Martin, B., Gabris-Weber, B. A., Reddy, R., Romero, G., Chattopadhyay, A., & Salama, G. (2018). Relaxin reverses inflammatory and immune signals in aged hearts. *PLoS One*, *13*(1), e0190935. https://doi.org/10.1371/journal.pone.0190935.

McArthur, L., Chilton, L., Smith, G. L., & Nicklin, S. A. (2015). Electrical consequences of cardiac myocyte: Fibroblast coupling. *Biochemical Society Transactions*, *43*(3), 513–518. https://doi.org/10.1042/BST20150035.

McCullough, P. A. (2017). How Trialists and pharmaceutical sponsors have failed us by thinking that acute heart failure is a 48-hour illness. *The American Journal of Cardiology, 120*(3), 505–508. https://doi.org/10.1016/j.amjcard.2017.04.056.

Mebazaa, A., Longrois, D., Metra, M., Mueller, C., Richards, A. M., Roessig, L., et al. (2015). Agents with vasodilator properties in acute heart failure: How to design successful trials. *European Journal of Heart Failure, 17*(7), 652–664. https://doi.org/10.1002/ejhf.294.

Meschiari, C. A., Ero, O. K., Pan, H., Finkel, T., & Lindsey, M. L. (2017). The impact of aging on cardiac extracellular matrix. *Geroscience, 39*(1), 7–18. https://doi.org/10.1007/s11357-017-9959-9.

Metra, M., Teerlink, J. R., Cotter, G., Davison, B. A., Felker, G. M., Filippatos, G., et al. (2019). Effects of Serelaxin in patients with acute heart failure. *The New England Journal of Medicine, 381*(8), 716–726. https://doi.org/10.1056/NEJMoa1801291.

Mookerjee, I., Hewitson, T. D., Halls, M. L., Summers, R. J., Mathai, M. L., Bathgate, R. A., et al. (2009). Relaxin inhibits renal myofibroblast differentiation via RXFP1, the nitric oxide pathway, and Smad2. *The FASEB Journal, 23*(4), 1219–1229. https://doi.org/10.1096/fj.08-120857.

Muralidharan, P., Cserne Szappanos, H., Ingley, E., & Hool, L. C. (2017). The cardiac L-type calcium channel alpha subunit is a target for direct redox modification during oxidative stress-the role of cysteine residues in the alpha interacting domain. *Clinical and Experimental Pharmacology & Physiology, 44*(Suppl 1), 46–54. https://doi.org/10.1111/1440-1681.12750.

Nascimento, A. R., Macheroni, C., Lucas, T. F., Porto, C. S., & Lazari, M. F. (2016). Crosstalk between FSH and relaxin at the end of the proliferative stage of rat Sertoli cells. *Reproduction, 152*(6), 613–628. https://doi.org/10.1530/REP-16-0330.

Navarro, A., & Boveris, A. (2007). The mitochondrial energy transduction system and the aging process. *American Journal of Physiology. Cell Physiology, 292*(2), C670–C686. https://doi.org/10.1152/ajpcell.00213.2006.

Nemoto, S., Takeda, K., Yu, Z. X., Ferrans, V. J., & Finkel, T. (2000). Role for mitochondrial oxidants as regulators of cellular metabolism. *Molecular and Cellular Biology, 20*(19), 7311–7318. https://doi.org/10.1128/mcb.20.19.7311-7318.2000.

Nguyen, B. T., Yang, L., Sanborn, B. M., & Dessauer, C. W. (2003). Phosphoinositide 3-kinase activity is required for biphasic stimulation of cyclic adenosine 3′,5′-monophosphate by relaxin. *Molecular Endocrinology, 17*(6), 1075–1084. https://doi.org/10.1210/me.2002-0284.

Nistri, S., & Bani, D. (2003). Relaxin receptors and nitric oxide synthases: Search for the missing link. *Reproductive Biology and Endocrinology, 1*, 5. https://doi.org/10.1186/1477-7827-1-5.

Novak, J., Parry, L. J., Matthews, J. E., Kerchner, L. J., Indovina, K., Hanley-Yanez, K., et al. (2006). Evidence for local relaxin ligand-receptor expression and function in arteries. *The FASEB Journal, 20*(13), 2352–2362. https://doi.org/10.1096/fj.06-6263com.

Pandey, D., Patel, A., Patel, V., Chen, F., Qian, J., Wang, Y., et al. (2012). Expression and functional significance of NADPH oxidase 5 (Nox5) and its splice variants in human blood vessels. *American Journal of Physiology. Heart and Circulatory Physiology, 302*(10), H1919–H1928. https://doi.org/10.1152/ajpheart.00910.2011.

Parikh, A., Patel, D., McTiernan, C. F., Xiang, W., Haney, J., Yang, L., et al. (2013). Relaxin suppresses atrial fibrillation by reversing fibrosis and myocyte hypertrophy and increasing conduction velocity and sodium current in spontaneously hypertensive rat hearts. *Circulation Research, 113*(3), 313–321. https://doi.org/10.1161/CIRCRESAHA.113.301646.

Perna, A. M., Masini, E., Nistri, S., Briganti, V., Chiappini, L., Stefano, P., et al. (2005). Novel drug development opportunity for relaxin in acute myocardial infarction: Evidences from a swine model. *The FASEB Journal, 19*(11), 1525–1527. https://doi.org/10.1096/fj.04-3664fje.

Piquereau, J., Caffin, F., Novotova, M., Lemaire, C., Veksler, V., Garnier, A., et al. (2013). Mitochondrial dynamics in the adult cardiomyocytes: Which roles for a highly specialized cell? *Frontiers in Physiology*, *4*, 102. https://doi.org/10.3389/fphys.2013.00102.

Quattrone, S., Chiappini, L., Scapagnini, G., Bigazzi, B., & Bani, D. (2004). Relaxin potentiates the expression of inducible nitric oxide synthase by endothelial cells from human umbilical vein in in vitro culture. *Molecular Human Reproduction*, *10*(5), 325–330. https://doi.org/10.1093/molehr/gah040.

Rodgers, J. L., Jones, J., Bolleddu, S. I., Vanthenapalli, S., Rodgers, L. E., Shah, K., et al. (2019). Cardiovascular risks associated with gender and aging. *Journal of Cardiovascular Development and Disease*, *6*(2), 19. https://doi.org/10.3390/jcdd6020019.

Rubart, M., Tao, W., Lu, X. L., Conway, S. J., Reuter, S. P., Lin, S. F., et al. (2018). Electrical coupling between ventricular myocytes and myofibroblasts in the infarcted mouse heart. *Cardiovascular Research*, *114*(3), 389–400. https://doi.org/10.1093/cvr/cvx163.

Ruiz-Meana, M., Bou-Teen, D., Ferdinandy, P., Gyongyosi, M., Pesce, M., Perrino, C., et al. (2020). Cardiomyocyte ageing and cardioprotection: Consensus document from the ESC working groups cell biology of the heart and myocardial function. *Cardiovascular Research*, *116*, 1835–1849. https://doi.org/10.1093/cvr/cvaa132.

Salama, G., & Abramson, J. (1984). Silver ions trigger Ca2+ release by acting at the apparent physiological release site in sarcoplasmic reticulum. *The Journal of Biological Chemistry*, *259*(21), 13363–13369. Retrieved from https://www.ncbi.nlm.nih.gov/pubmed/6208194.

Salama, G., Abramson, J. J., & Pike, G. K. (1992). Sulphydryl reagents trigger Ca2+ release from the sarcoplasmic reticulum of skinned rabbit psoas fibres. *The Journal of Physiology*, *454*, 389–420. https://doi.org/10.1113/jphysiol.1992.sp019270.

Samuel, C. S., Unemori, E. N., Mookerjee, I., Bathgate, R. A., Layfield, S. L., Mak, J., et al. (2004). Relaxin modulates cardiac fibroblast proliferation, differentiation, and collagen production and reverses cardiac fibrosis in vivo. *Endocrinology*, *145*(9), 4125–4133. https://doi.org/10.1210/en.2004-0209.

Samuel, C. S., Zhao, C., Bathgate, R. A., Bond, C. P., Burton, M. D., Parry, L. J., et al. (2003). Relaxin deficiency in mice is associated with an age-related progression of pulmonary fibrosis. *The FASEB Journal*, *17*(1), 121–123. https://doi.org/10.1096/fj.02-0449fje.

Samuel, C. S., Zhao, C., Bathgate, R. A., Du, X. J., Summers, R. J., Amento, E. P., et al. (2005). The relaxin gene-knockout mouse: A model of progressive fibrosis. *Annals of the New York Academy of Sciences*, *1041*, 173–181. https://doi.org/10.1196/annals.1282.025.

Sarwar, M., Samuel, C. S., Bathgate, R. A., Stewart, D. R., & Summers, R. J. (2015). Serelaxin-mediated signal transduction in human vascular cells: Bell-shaped concentration-response curves reflect differential coupling to G proteins. *British Journal of Pharmacology*, *172*(4), 1005–1019. https://doi.org/10.1111/bph.12964.

Scott, D. J., Layfield, S., Yan, Y., Sudo, S., Hsueh, A. J., Tregear, G. W., et al. (2006). Characterization of novel splice variants of LGR7 and LGR8 reveals that receptor signaling is mediated by their unique low density lipoprotein class a modules. *The Journal of Biological Chemistry*, *281*(46), 34942–34954. https://doi.org/10.1074/jbc.M602728200.

Shi, Q., Li, M., Mika, D., Fu, Q., Kim, S., Phan, J., et al. (2017). Heterologous desensitization of cardiac beta-adrenergic signal via hormone-induced betaAR/arrestin/PDE4 complexes. *Cardiovascular Research*, *113*(6), 656–670. https://doi.org/10.1093/cvr/cvx036.

Shimokawa, I., Higami, Y., Hubbard, G. B., McMahan, C. A., Masoro, E. J., & Yu, B. P. (1993). Diet and the suitability of the male Fischer 344 rat as a model for aging research. *Journal of Gerontology*, *48*(1), B27–B32. https://doi.org/10.1093/geronj/48.1.b27.

Sies, H., & Jones, D. P. (2020). Reactive oxygen species (ROS) as pleiotropic physiological signalling agents. *Nature Reviews. Molecular Cell Biology*, *21*, 363–383. https://doi.org/10.1038/s41580-020-0230-3.

Singh, S., Simpson, R. L., & Bennett, R. G. (2015). Relaxin activates peroxisome proliferator-activated receptor gamma (PPARgamma) through a pathway involving PPARgamma coactivator 1alpha (PGC1alpha). *The Journal of Biological Chemistry*, *290*(2), 950–959. https://doi.org/10.1074/jbc.M114.589325.

Smith, J. H., Green, C. R., Peters, N. S., Rothery, S., & Severs, N. J. (1991). Altered patterns of gap junction distribution in ischemic heart disease. An immunohistochemical study of human myocardium using laser scanning confocal microscopy. *The American Journal of Pathology*, *139*(4), 801–821. Retrieved from https://www.ncbi.nlm.nih.gov/pubmed/1656760.

Spach, M. S., & Boineau, J. P. (1997). Microfibrosis produces electrical load variations due to loss of side-to-side cell connections: A major mechanism of structural heart disease arrhythmias. *Pacing and Clinical Electrophysiology*, *20*(2 Pt 2), 397–413. https://doi.org/10.1111/j.1540-8159.1997.tb06199.x.

Sudo, S., Kumagai, J., Nishi, S., Layfield, S., Ferraro, T., Bathgate, R. A., et al. (2003). H3 relaxin is a specific ligand for LGR7 and activates the receptor by interacting with both the ectodomain and the exoloop 2. *The Journal of Biological Chemistry*, *278*(10), 7855–7862. https://doi.org/10.1074/jbc.M212457200.

Sutton, S. W., Bonaventure, P., Kuei, C., Nepomuceno, D., Wu, J., Zhu, J., et al. (2005). G-protein-coupled receptor (GPCR)-142 does not contribute to relaxin-3 binding in the mouse brain: Further support that relaxin-3 is the physiological ligand for GPCR135. *Neuroendocrinology*, *82*(3–4), 139–150. https://doi.org/10.1159/000091267.

Svendsen, A. M., Vrecl, M., Ellis, T. M., Heding, A., Kristensen, J. B., Wade, J. D., et al. (2008). Cooperative binding of insulin-like peptide 3 to a dimeric relaxin family peptide receptor 2. *Endocrinology*, *149*(3), 1113–1120. https://doi.org/10.1210/en.2007-0412.

Svendsen, A. M., Zalesko, A., Konig, J., Vrecl, M., Heding, A., Kristensen, J. B., et al. (2008). Negative cooperativity in H2 relaxin binding to a dimeric relaxin family peptide receptor 1. *Molecular and Cellular Endocrinology*, *296*(1–2), 10–17. https://doi.org/10.1016/j.mce.2008.07.014.

Swope, D., Cheng, L., Gao, E., Li, J., & Radice, G. L. (2012). Loss of cadherin-binding proteins beta-catenin and plakoglobin in the heart leads to gap junction remodeling and arrhythmogenesis. *Molecular and Cellular Biology*, *32*(6), 1056–1067. https://doi.org/10.1128/MCB.06188-11.

Szlachter, B. N., Quagliarello, J., Jewelewicz, R., Osathanondh, R., Spellacy, W. N., & Weiss, G. (1982). Relaxin in normal and pathogenic pregnancies. *Obstetrics and Gynecology*, *59*(2), 167–170. Retrieved from https://www.ncbi.nlm.nih.gov/pubmed/7078861.

Tao, H., Yang, J. J., Shi, K. H., & Li, J. (2016). Wnt signaling pathway in cardiac fibrosis: New insights and directions. *Metabolism*, *65*(2), 30–40. https://doi.org/10.1016/j.metabol.2015.10.013.

Teerlink, J. R., Cotter, G., Davison, B. A., Felker, G. M., Filippatos, G., Greenberg, B. H., et al. (2013). Serelaxin, recombinant human relaxin-2, for treatment of acute heart failure (RELAX-AHF): A randomised, placebo-controlled trial. *Lancet*, *381*(9860), 29–39. https://doi.org/10.1016/S0140-6736(12)61855-8.

Teerlink, J. R., Metra, M., Felker, G. M., Ponikowski, P., Voors, A. A., Weatherley, B. D., et al. (2009). Relaxin for the treatment of patients with acute heart failure (pre-RELAX-AHF): A multicentre, randomised, placebo-controlled, parallel-group, dose-finding phase IIb study. *Lancet*, *373*(9673), 1429–1439. https://doi.org/10.1016/S0140-6736(09)60622-X.

Tokunaga, R., Zhang, W., Naseem, M., Puccini, A., Berger, M. D., Soni, S., et al. (2018). CXCL9, CXCL10, CXCL11/CXCR3 axis for immune activation—A target for novel cancer therapy. *Cancer Treatment Reviews*, *63*, 40–47. https://doi.org/10.1016/j.ctrv.2017.11.007.

Torre-Amione, G., Milo-Cotter, O., Kaluski, E., Perchenet, L., Kobrin, I., Frey, A., et al. (2009). Early worsening heart failure in patients admitted for acute heart failure: Time course, hemodynamic predictors, and outcome. *Journal of Cardiac Failure*, *15*(8), 639–644. https://doi.org/10.1016/j.cardfail.2009.04.001.

Travers, J. G., Kamal, F. A., Robbins, J., Yutzey, K. E., & Blaxall, B. C. (2016). Cardiac fibrosis: The fibroblast awakens. *Circulation Research*, *118*(6), 1021–1040. https://doi.org/10.1161/CIRCRESAHA.115.306565.

Ueda, K., Valdivia, C., Medeiros-Domingo, A., Tester, D. J., Vatta, M., Farrugia, G., et al. (2008). Syntrophin mutation associated with long QT syndrome through activation of the nNOS-SCN5A macromolecular complex. *Proceedings of the National Academy of Sciences of the United States of America*, *105*(27), 9355–9360. https://doi.org/10.1073/pnas.0801294105.

Uhlen, M., Fagerberg, L., Hallstrom, B. M., Lindskog, C., Oksvold, P., Mardinoglu, A., et al. (2015). Proteomics. Tissue-based map of the human proteome. *Science*, *347*(6220), 1260419. https://doi.org/10.1126/science.1260419.

Unemori, E. (2017). Serelaxin in clinical development: Past, present and future. *British Journal of Pharmacology*, *174*(10), 921–932. https://doi.org/10.1111/bph.13695.

Ungvari, Z., Orosz, Z., Rivera, A., Labinskyy, N., Xiangmin, Z., Olson, S., et al. (2007). Resveratrol increases vascular oxidative stress resistance. *American Journal of Physiology. Heart and Circulatory Physiology*, *292*(5), H2417–H2424. https://doi.org/10.1152/ajpheart.01258.2006.

van Rijen, H. V., Eckardt, D., Degen, J., Theis, M., Ott, T., Willecke, K., et al. (2004). Slow conduction and enhanced anisotropy increase the propensity for ventricular tachyarrhythmias in adult mice with induced deletion of connexin43. *Circulation*, *109*(8), 1048–1055. https://doi.org/10.1161/01.CIR.0000117402.70689.75.

Vodstrcil, L. A., Tare, M., Novak, J., Dragomir, N., Ramirez, R. J., Wlodek, M. E., et al. (2012). Relaxin mediates uterine artery compliance during pregnancy and increases uterine blood flow. *The FASEB Journal*, *26*(10), 4035–4044. https://doi.org/10.1096/fj.12-210567.

von Haehling, S., Jankowska, E. A., & Anker, S. D. (2004). Tumour necrosis factor-alpha and the failing heart—Pathophysiology and therapeutic implications. *Basic Research in Cardiology*, *99*(1), 18–28. https://doi.org/10.1007/s00395-003-0433-8.

Walker, E. M., Jr., Nillas, M. S., Mangiarua, E. I., Cansino, S., Morrison, R. G., Perdue, R. R., et al. (2006). Age-associated changes in hearts of male Fischer 344/Brown Norway F1 rats. *Annals of Clinical and Laboratory Science*, *36*(4), 427–438. Retrieved from https://www.ncbi.nlm.nih.gov/pubmed/17127729.

Wang, M., Zhang, J., Walker, S. J., Dworakowski, R., Lakatta, E. G., & Shah, A. M. (2010). Involvement of NADPH oxidase in age-associated cardiac remodeling. *Journal of Molecular and Cellular Cardiology*, *48*(4), 765–772. https://doi.org/10.1016/j.yjmcc.2010.01.006.

Weber, K. T. (2000). Fibrosis and hypertensive heart disease. *Current Opinion in Cardiology*, *15*(4), 264–272. https://doi.org/10.1097/00001573-200007000-00010.

Werner, E., & Werb, Z. (2002). Integrins engage mitochondrial function for signal transduction by a mechanism dependent on rho GTPases. *The Journal of Cell Biology*, *158*(2), 357–368. https://doi.org/10.1083/jcb.200111028.

Zaidi, N. F., Lagenaur, C. F., Abramson, J. J., Pessah, I., & Salama, G. (1989). Reactive disulfides trigger Ca2+ release from sarcoplasmic reticulum via an oxidation reaction. *The Journal of Biological Chemistry*, *264*(36), 21725–21736. Retrieved from https://www.ncbi.nlm.nih.gov/pubmed/2532212.

Zaidi, N. F., Lagenaur, C. F., Hilkert, R. J., Xiong, H., Abramson, J. J., & Salama, G. (1989). Disulfide linkage of biotin identifies a 106-kDa Ca2+ release channel in sarcoplasmic reticulum. *The Journal of Biological Chemistry*, *264*(36), 21737–21747. Retrieved from https://www.ncbi.nlm.nih.gov/pubmed/2480955.

Zhao, L., Roche, P. J., Gunnersen, J. M., Hammond, V. E., Tregear, G. W., Wintour, E. M., et al. (1999). Mice without a functional relaxin gene are unable to deliver milk to their pups. *Endocrinology*, *140*(1), 445–453. https://doi.org/10.1210/endo.140.1.6404.

Zhou, P., & Pu, W. T. (2016). Recounting cardiac cellular composition. *Circulation Research*, *118*(3), 368–370. https://doi.org/10.1161/CIRCRESAHA.116.308139.

CHAPTER SEVENTEEN

Role of FoxO transcription factors in aging-associated cardiovascular diseases

Sneha Mishra, Venkatraman Ravi, and Nagalingam Ravi Sundaresan*
Cardiovascular and Muscle Research Laboratory, Department of Microbiology and Cell Biology, Indian Institute of Science, Bengaluru, India
*Corresponding author: e-mail address: rsundaresan@iisc.ac.in

Contents

Abstract

Aging constitutes a major risk factor toward the development of cardiovascular diseases (CVDs). The aging heart undergoes several changes at the molecular, cellular and physiological levels, which diminishes its contractile function and weakens stress tolerance. Further, old age increases the exposure to risk factors such as hypertension, diabetes and hypercholesterolemia. Notably, research in the past decades have identified FoxO subfamily of the forkhead transcription factors as key players in regulating diverse cellular processes linked to cardiac aging and diseases. In the present chapter, we discuss the important role of FoxO in the development of various aging-associated cardiovascular complications such as cardiac hypertrophy, cardiac fibrosis, heart failure, vascular dysfunction, atherosclerosis, hypertension and myocardial ischemia. Besides, we will also discuss the role of FoxO in cardiometabolic alterations, autophagy and proteasomal degradation, which are implicated in aging-associated cardiac dysfunction.

Vitamins and Hormones, Volume 115
ISSN 0083-6729
https://doi.org/10.1016/bs.vh.2020.12.018

1. Introduction

Aging is characterized by the gradual functional decline of various organ systems in the body with age. The heart, a highly metabolically active organ, is no exception and suffers from various aging-associated detrimental changes. Remarkably, aging constitutes a dominant risk factor for the development of cardiovascular diseases (CVDs), with their prevalence dramatically increasing with advancing age (North & Sinclair, 2012). Cardiac aging is characterized by a progressive decline in structure and function of heart as well as weakened stress-response (Dai, Chen, Johnson, Szeto, & Rabinovitch, 2012; Lee & Kim, 2018). The aging heart undergoes physiological changes such as left ventricular hypertrophy, muscle weakness, interstitial fibrosis, metabolic dysfunction, diastolic dysfunction, valvular degeneration, atrial fibrillation, and reduced maximal exercise capacity (Dai et al., 2012; Paneni, Diaz Canestro, Libby, Luscher, & Camici, 2017). In addition to intrinsic cardiac aging, old age increases the exposure to hypertension, diabetes, hypercholesterolemia and various other risk factors for CVDs (Dai et al., 2012). Collectively, these factors increase the prevalence of CVDs in the older population, thus increasing morbidity and mortality.

Research in the past decades have identified multiple genes and pathways that regulate aging and the development of aging-associated diseases. Among these, the Forkhead box O (FoxO) subfamily of the forkhead box family of transcription factors have gained increasing attention due to their crucial roles in regulating diverse cellular processes implicated in lifespan extension and aging (Martins, Lithgow, & Link, 2016). The Forkhead box (Fox) family of transcription factors consists of 19 different subfamilies designated A-S of which, the subfamily O consists of four isoforms, namely FoxO1, FoxO3a, FoxO4 and FoxO6 in the mammals which share a high degree of homology (Burgering, 2008; Kaestner, Knochel, & Martinez, 2000). Of these, the first three members have been shown to play critical roles in the maintenance of cardiac function. These FoxO transcription factors control a wide range of crucial cellular processes, including stress resistance, metabolism, cell proliferation, DNA-repair processes, apoptosis and autophagy (Wang, Zhou, & Graves, 2014). They majorly act as transcriptional activators by binding to the consensus binding sequence within the promoters of their target genes (Calnan & Brunet, 2008; Furuyama, Nakazawa, Nakano, & Mori, 2000). The FoxO factors are inhibited by the insulin/insulin-like growth factor signaling, through AKT mediated phosphorylation of the FoxO proteins at three conserved residues, which lead to their nuclear exclusion and suppression of their target gene expression (Tzivion, Dobson, & Ramakrishnan, 2011). Besides phosphorylation, the activity of FoxO is also regulated by other post-translational modifications such as acetylation, methylation and ubiquitination (Wang, Yu, & Huang, 2016).

The effects of FoxO can be beneficial or detrimental during the development of cardiovascular diseases depending upon the nature of stress, cell type and expression levels. FoxO acts as a converging point for multiple signaling pathways known to govern aging and aging-associated changes in the heart. As described in this chapter, the complex network of signaling pathways comprising of insulin/IGF-1 signaling, AMPK, JNK as well as Sirtuins regulate the development of aging-associated cardiovascular dysfunction by modulating the activity of FoxO transcription factors. FoxO plays a critical role in cardiac pathologies associated with aging, such as cardiac hypertrophy, cardiac fibrosis, heart failure, vascular dysfunction, atherosclerosis, hypertension and myocardial ischemia. In addition, FoxO is also implicated in aging-associated metabolic and proteostatic derangements in the heart, which leads to compromised cardiac function. The present chapter will discuss the specific role of different FoxO transcription factors in the development of aging-associated cardiovascular complications along with the relevant mechanisms. The various targets of FoxO transcription factors and the associated processes implicated in the pathogenesis of aging-associated cardiovascular diseases are presented in Fig. 1. In addition, changes observed in the heart of various genetic models of FoxO transcription factors are summarized in Table 1.

2. Role of FoxO in cardiac remodeling

Age-related changes in the cellular and molecular processes of the cardiomyocytes, fibroblasts and endothelial cells, triggered by biomechanical stresses, cause morphological and structural changes of the heart, which lead to compromised cardiac function (Chiao & Rabinovitch, 2015). Generally, cardiac aging is characterized by left ventricular hypertrophy and fibrosis that contributes to adverse remodeling of heart, eventually leading to heart failure (Steenman & Lande, 2017). Interestingly, the FoxO transcription factors have been shown to play key roles in many of the cellular changes that contribute to these adverse changes in the heart.

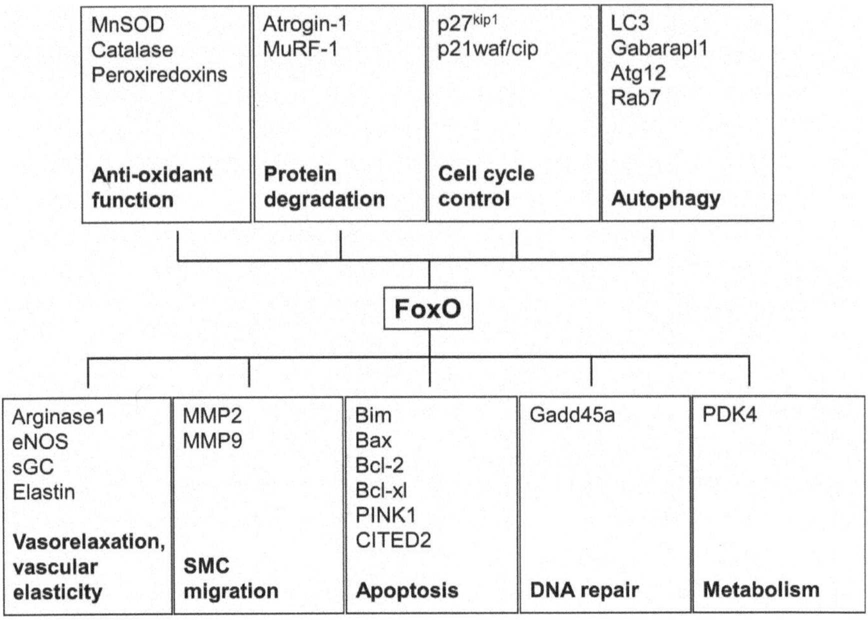

Fig. 1 Diagram indicating the targets of FoxO and the associated processes implicated in the pathogenesis of aging-associated cardiovascular diseases.

2.1 Role of FoxO in the development of cardiac hypertrophy and cardiac dysfunction

Cardiac hypertrophy develops as an adaptive response to biomechanical stress caused by aging associated complications such as hypertension, metabolic diseases, valvular disease and ischemic disease. Although hypertrophy is a compensatory mechanism in the early stages, prolonged stress turns hypertrophy maladaptive and triggers the development of heart failure (Nakamura & Sadoshima, 2018). Studies indicate that members of the FoxO subfamily of transcription factors act as key negative regulators of cardiac hypertrophy. It has been found that, diverse hypertrophic stimuli negatively regulate the members of the group such as the FoxO1, FoxO3a and FoxO4 through AKT-mediated inhibitory phosphorylation (Skurk et al., 2005). Specifically, overexpression of FoxO3a in mouse hearts or isolated cardiomyocytes reduces the cardiomyocyte cell size and antagonizes hypertrophy induced by diverse stimuli, through upregulation of multiple atrophy related genes such as atrogin-1 and MuRF-1 (Skurk et al., 2005). Moreover, expression of constitutively active forms of FoxO1 or FoxO3a in the

Table 1 Changes observed in the heart of various genetic models of FoxO transcription factors.

FoxO genetic models	Affected targets and resultant phenotypes	References
Cardiac specific dFOXO-Tg	• Modest FOXO overexpression in Drosophila hearts protects against aging-associated cardiac dysfunction and stress-induced heart failure	Blice-Baum et al. (2017), Wessells, Fitzgerald, Cypser, Tatar, and Bodmer (2004)
FoxO3a-KO	• Increased MCIP1.4 via calcineurin/NFAT pathway • Increased cardiac hypertrophy	Ni et al. (2006)
	• Reduced MMP-2 expression in vessel • Reduced vessel sprouting, SMC proliferation and vascular remodeling	Diebold, Petry, Burger, Hess, and Gorlach (2011)
	• Increased eNOS activity • Enhanced neovascularisation following ischemia	Potente et al. (2005)
FoxO4-KO	• Reduced MMP-9 expression • Reduced SMC migration	Li, Liang, et al. (2007)
	• Increased atherosclerosis in apoE deficient mice background upon high fat diet (HFD) feeding • Increased macrophages and T-cells recruitment to atherosclerotic lesions in apoE deficient mice background	Zhu, Zhang, Wang, Li, and Liu (2011)
Cardiac specific FoxO1-KO	• Protected from cardiac dysfunction upon HFD feeding • Reduced lipid accumulation and PDK4 induced shift from glycolytic to fatty acid oxidation pathways under HFD-fed conditions	Battiprolu et al. (2012)

Continued

Table 1 Changes observed in the heart of various genetic models of FoxO transcription factors.—cont'd

FoxO genetic models	Affected targets and resultant phenotypes	References
Cardiac specific FoxO1/FoxO3a-KO	• Increased susceptibility to MI and IR injury • Reduced expression of anti-oxidant genes: SOD2 and Catalase • Reduced expression of autophagy genes: Gaparapl1, LC33 • Impaired expression of DNA damage repair gene: Gadd45a	Sengupta, Molkentin, Paik, DePinho, and Yutzey (2011)
	• Protected from cardiac dysfunction upon HFD feeding	Battiprolu et al. (2012)
Smooth muscle cell specific FoxO1-KO	• Increased pulmonary hypertension, vascular remodeling, RV hypertrophy and cardiac dysfunction	Savai et al. (2014)
Endothelial cell specific FoxO1-KO	• Enhanced endothelial proliferation, vascular enlargement	Wilhelm et al. (2016)
Endothelial cell specific FoxO4-KO	• Reduced arginase1 expression • Reduced neutrophil infiltration and cytokine expression post MI	Zhu et al. (2015)
Endothelial cell specific FoxO1/FoxO3a/FoxO4-KO	• Reduced cell adhesion molecules expression in ECs • Reduced ROS levels, inflammation, and apoptosis in ECs • Protection from atherosclerosis in Ldlr-KO mice background upon western type diet	Tsuchiya et al. (2012)

cardiomyocytes blocks the development of agonist-induced hypertrophy by attenuating the pro-hypertrophic calcineurin-NFAT pathway (Ni et al., 2006). In line with this, mice deficient for FoxO3a demonstrate a hypertrophic phenotype accompanied by increased expression of the MCIP1.4, a direct target downstream of the calcineurin/NFAT pathway (Ni et al., 2006). Further, it has been shown that atrogin-1, a downstream target of FoxO transcription factors, directly interacts with calcineurin and promotes its proteasomal degradation (Li et al., 2004). Also, atrogin-1 increases the FoxO activity in a feed-forward loop through non-canonical polyubiquitylation of FoxO1 and FoxO3a, which renders them resistant to AKT mediated inhibitory phosphorylation, thus promoting their nuclear retention (Li, Willis, et al., 2007). Consistently, overexpression of atrogin-1 in the heart increases FoxO transcriptional activity and protects against IGF-1 induced hypertrophy, while atrogin-1 deletion elicits an opposite phenotype (Li, Willis, et al., 2007). Moreover, FoxO3a has been shown to protect against insulin induced hypertrophy through transcriptional activation of catalase, which reduces ROS levels (Tan, Wang, Lv, & Li, 2008). Insulin mediated FoxO3a inactivation reduces catalase expression which leads to increased ROS generation. The increased ROS levels in turn induces myocardin expression which promotes cardiac hypertrophy. However, expression of a constitutively active FoxO3a attenuates insulin-induced myocardin expression and hypertrophy due to ROS detoxification via catalase. In line with this, catalase knockdown results in loss of protection against cardiac hypertrophy even in the presence of constitutively active FoxO3a, indicating a key role for FoxO mediated antioxidant gene expression in regulating hypertrophy (Tan et al., 2008). Further, low to moderate SIRT1 overexpression protects the heart against oxidative stress-induced damage to the myocardium through a FoxO1-dependent increase in the expression of antioxidants such as catalase (Alcendor et al., 2007). While moderate activation of FoxO is beneficial to prevent aging-associated changes in heart, chronic activation causes detrimental effects to the heart. Our recent findings show that TLR2 deficiency leads to accelerated aging-associated cardiac dysfunction with chronic activation of FoxO1, which promotes the expression of atrophy-related ubiquitin ligases and cell death genes. Mechanistically, TLR2 deficiency suppresses AKT signaling which leads to reduced phosphorylation and chronic activation of FoxO1. Consistently, inhibition of FoxO1 improves cardiac function in old TLR2-KO mice, suggesting a key role for FoxO in mediating the effects of TLR2 deficiency on aging-associated cardiac dysfunction (Spurthi et al., 2018).

2.2 Role of FoxO in cardiac fibrosis

Fibrotic remodeling of the heart is a common phenomenon during aging, which is characterized by transdifferentiation of the cardiac fibroblasts into myofibroblasts, which secrete abundant extracellular matrix proteins leading to fibrosis of the heart (Lu et al., 2017). Notably, FoxO factors have been shown to play a key role in mediating the effects of factors like transforming growth factor beta-1 (TGF-β1), which regulates the phenotypic conversion of fibroblasts to myofibroblasts. It has been shown that TGF-β1 treatment promotes the expression, nuclear localization and transcriptional activity of FoxO1 (Vivar et al., 2016). Indeed, depletion of FoxO1 attenuates the TGF-β1-induced fibroblast differentiation as well as the increase in the expression of the cytostatic factor p21waf/cip. On the other hand, overexpression of FoxO1 enhances the TGF-β1-induced expression of fibroblast differentiation markers as well as p21waf/cip, suggesting that FoxO1 is a key mediator of TGF-β1-induced fibroblast differentiation (Vivar et al., 2016). In contrast to FoxO1, TGF-β1 reduces the levels and the activity of FoxO3a by promoting its proteasomal degradation (Vivar et al., 2020). Further, TGF-β1 induced differentiation of fibroblasts is enhanced by FoxO3a depletion. Interestingly, the levels of FoxO3a are regulated in a FoxO1 dependent manner, as FoxO1 depletion abrogates the TGF-β1induced decrease in FoxO3a levels (Vivar et al., 2020). In addition, FoxO3a has been shown to regulate the proliferation of fibroblasts in response to mitogenic stimulation. Notably, serum induction activates the ERK1/2 kinase, which phosphorylates FoxO3a and suppresses its transcriptional activity toward the expression of the cell cycle inhibitor $p27^{kip1}$ (Pramod & Shivakumar, 2014). Consistently, depletion of FoxO3a prevents increased expression of $p27^{kip}$ during ERK 1/2 inhibition (Pramod & Shivakumar, 2014). Furthermore, FoxO has been shown to play a critical role in oxidative stress resistance in fibroblasts. FoxO3a provides resistance to oxidative stress via transcriptional regulation of peroxiredoxin (Prx III), a mitochondrial antioxidant in cardiac fibroblasts (Chiribau et al., 2008). While FoxO3a depletion reduces Prx III expression and protein levels in serum-deprived state, expression of a constitutively active FoxO increases Prx III expression in cardiac fibroblasts (Chiribau et al., 2008). Taken together, FoxO transcription factors play a key role in the development of aging-related cardiac fibrosis, by controlling the proliferation, differentiation and stress resistance in cardiac fibroblasts.

3. Role of FoxO transcription factors in vascular remodeling and dysfunction

Age-related changes in the structure and function of the vasculature is one of the major cause for the increased prevalence of cardiovascular diseases in older population (Ungvari, Kaley, de Cabo, Sonntag, & Csiszar, 2010). Vascular aging is characterized by increased oxidative stress that occurs due to an imbalance in pro-oxidative and anti-oxidative pathways (Li & Fukagawa, 2010). Further, vascular aging is also associated with increased proliferation and migration of vascular smooth muscle cells (VSMCs), as well as increased extracellular matrix generation and inflammatory cytokine levels, resulting in vascular remodeling (Li & Fukagawa, 2010; Orlandi, Bochaton-Piallat, Gabbiani, & Spagnoli, 2006). In addition to VSMC dysfunction, functional alterations in the endothelial cells (ECs) diminish the bioavailability of nitric oxide (NO), leading to reduced endothelium-dependent relaxation and enhanced vasoconstriction (Ungvari et al., 2010). These age-related changes in the VSMCs and ECs together contribute to the development of a pro-atherogenic phenotype, thus increasing the risk for CVDs (Kawashima & Yokoyama, 2004). Notably, studies indicate that, the FoxO transcription factors play key roles in regulating the vascular phenotype by regulating distinct pathways and phenotypes both in the VSMCs and the ECs.

3.1 Role of FoxO transcription factors in smooth muscle cells

VSMCs are phenotypically plastic and can transition between a differentiated myogenic state and a proliferative and migratory phenotype. Tight regulation of this process is essential as excessive proliferation of VSMCs can cause multiple vascular abnormalities, including the development of atherosclerotic lesions and hypertension (Wang et al., 2018). Multiple cytokines and growth factors have been shown to modulate these phenotypic conversions by regulating distinct cellular pathways (Basatemur, Jorgensen, Clarke, Bennett, & Mallat, 2019). It has been shown that IGF-1 promotes a differentiated myogenic phenotype of SMCs by inhibiting the FoxO4 activity through activation of the PI3K/AKT signaling (Liu, Wang, Yanagisawa, & Olson, 2005). Specifically, FoxO4 inhibits the pro-myogenic phenotype of SMCs by directly binding to and inhibiting the activity of myocardin, a key transcriptional coactivator of smooth muscle genes. Consistent with this, depletion of FoxO4 stimulates myocardin activity and promotes SMC differentiation (Liu et al., 2005). Besides, Spry proteins which act as feedback regulators of receptor tyrosine kinase signaling, have been shown to regulates the VSMC phenotype through the AKT/FoxO signaling pathway, albeit with opposing roles for Spry1 and Spry4 (Yang et al., 2013). Spry1 promotes SMC differentiation by activating AKT signaling, which relieves the repressive effect of FoxO3a on the myocardin promoter activity, thus promoting smooth muscle marker expression. On the other hand, Spry4 suppresses SMC differentiation by inhibiting MAPK/ERK as well as AKT activation and reducing myocardin expression (Yang et al., 2013). In addition, physiological mechanical stretch has been shown to increase FoxO3a and decrease FoxO4 in a SIRT1-dependent manner, which helps in maintaining the quiescent contractile differentiated phenotype of VSMCs (Huang et al., 2015). Consistent with this, overexpression of FoxO3a in VSMCs increases the expression of contractile markers, while FoxO4 overexpression has an opposite effect (Huang et al., 2015).

In addition to their role in differentiation, the FoxO factors also impact vascular remodeling by regulating the proliferation and migration of SMCs in response to different stimuli. FoxO4 enhances the migration of SMC through transcriptional upregulation of matrix metalloprotease-9 (MMP9) in response to cytokine signals like TNF-α (Li, Liang, et al., 2007). In line with this, genetic depletion of FoxO4 under in vitro as well as in vivo conditions reduces MMP9 expression leading to suppressed SMC migration, which rescues neointima formation in an in vivo flow cessation injury model. Notably, the migration defect in FoxO4 deficient smooth muscle cells is rescued by exogenous MMP9 addition, suggesting that FoxO4 promotes VSMC migration in response to cytokine signals like TNF-α in MMP9 dependent manner (Li, Liang, et al., 2007). On the other hand, Foxo3a expression exerts an antiproliferative effect on the coronary artery smooth muscle cells (CASMC) through upregulation of the cell cycle inhibitor $p27^{kip1}$ (Abid et al., 2005). In line with this, adenoviral delivery of FoxO3a in a rat carotid arterial injury model prevents the development of neointimal hyperplasia through increased expression of $p27^{kip1}$ and thus prevents progression toward atherosclerosis (Abid et al., 2005). Moreover, constitutively active FoxO3a expression has been shown to prevent the proliferation and migration of rat VSMCs by suppressing the expression of the cysteine-rich angiogenic protein 61 (CYR61), an angiogenic immediate early gene expressed in response to vascular injury and growth factor stimulation (Lee et al., 2007). Consistently, constitutively active FoxO3a expression prevents VSMC proliferation and neointimal formation after balloon injury in the rat carotid artery by suppressing CYR61 expression (Lee et al., 2007). In contrast to these studies, FoxO3a has been shown to mediate the SMC proliferation induced by the vasoactive peptide urotensin-II which is upregulated in several cardiovascular diseases (Diebold et al., 2011). Urotensin-II activates FoxO3a by reducing the interaction between FoxO3a and its inhibitor 14-3-3 through NOX-4 and JNK dependent phosphorylation of the 14-3-3 protein. The activated FoxO3a in turn leads to increased transcription of MMP2, which increases SMC proliferation and vascular remodeling. Consistently, deficiency of FoxO3a or treatment with MMP2 inhibitor abolishes the urotensin-II mediated vascular growth and remodeling (Diebold et al., 2011). While the reasons behind this discrepancy are not clear, it is possible that dose-specific effects might contribute to distinct phenotypes (Diebold et al., 2011).

Studies indicate that FoxO mediated changes in the VSMCs play a key role in regulating vascular elasticity, vascular tone and hypertension. Loss of vascular elasticity is associated with arterial hypertension which later progresses to atherosclerosis. Notably, insulin has been shown to promote the elastic fiber formation in aortic smooth muscle cells by relieving the repressive effect of FoxO1 on elastin gene expression (Shi et al., 2012). In addition, inhibition of FoxO in VSMCs with the FoxO inhibitor AS1842856 reduces the expression of soluble guanylyl cyclase (sGC), which leads to deregulated VSMC relaxation upon nitric oxide (NO) stimulation and vascular remodeling (Galley et al., 2019). In line with this, treatment of mouse thoracic aortas with the FoxO inhibitor AS1842856 reduces sGC expression and attenuates NO-dependent vasorelaxation, suggesting FoxO is beneficial for the maintenance of vascular tone and protects from hypertension via sGC (Galley et al., 2019). Moreover, FoxO1 has been reported to play a vital role in the pathogenesis of pulmonary hypertension (pH) by regulating multiple pathways in pulmonary arteries smooth muscle cell (PASMC) (Savai et al., 2014). Interestingly, FoxO1 is downregulated in pulmonary vessels and PASMCs of PH patients. Further, PASMCs from patients with idiopathic pulmonary arterial hypertension show increased phosphorylation and reduced nuclear retention of FoxO1. In line with this, inhibition or genetic depletion of FoxO1 exhibits features of PH. Notably, PASMCs isolated from arterial smooth muscle cell specific-FoxO1 knockout mice are resistant to apoptosis and exhibit increased proliferation leading to vascular remodeling and cardiac dysfunction. Accordingly, genetic or pharmacological reconstitution of FoxO activity restores the quiescent phenotype of PASMCs in vitro, which is linked to changes in BMP signaling and cell cycle control and also reverses right ventricular hypertrophy and vascular remodeling in animal models of pulmonary hypertension, suggesting a key role for FoxO1 in the development of PH (Savai et al., 2014). Collectively, these findings suggest that the FoxO factors play a crucial role in vascular remodeling and hypertension by controlling various aspects of SMC biology such as proliferation, migration, differentiation and apoptosis.

3.2 Role of FoxO transcription factors in endothelial cells

There is a strong association of aging with defects in angiogenesis. Importantly, angiogenesis is essential for cardiac repair following ischemic injury. Further, defects in angiogenesis could augment the damage post myocardial infarction (Lahteenvuo & Rosenzweig, 2012). Studies demonstrate that FoxO transcription factors have been shown to play an important role in regulating the angiogenesis and vessel forming abilities of ECs by controlling the expression of key genes involved in these processes such as angiopoietin and endothelial nitric oxide synthase (eNOS) (Potente et al., 2005). Notably, FoxO1 and FoxO3a, inhibits the expression of eNOS by binding to its promoter and suppresses endothelial cell migration and tube formation in vitro. Accordingly, aortic explants from FoxO3a deficient mice exhibit enhanced eNOS expression and show increased revascularization and limb perfusion following hind limb ischemia (Potente et al., 2005). Moreover, FoxO1 controls vascular expansion by regulating the switch between quiescent and growth states of ECs by adjusting their metabolic states (Wilhelm et al., 2016). FoxO1 has been shown to maintain endothelial quiescence by reducing the glycolysis and mitochondrial respiration through suppression of the MYC driven anabolism and growth. Consistently, endothelial-specific FoxO1 deletion was shown to increase EC proliferation and lead to vascular enlargement, while FoxO1 overexpression leads to thinning of the vessels and hypobranching (Wilhelm et al., 2016). In addition, FoxO3a has also been shown to exert an antiproliferative effect in the endothelial progenitor cells derived from human umbilical cord blood, through upregulating the expression of cell cycle inhibitor $p27^{kip1}$ and downregulation of pro-proliferative genes such as cyclin-dependent kinase 2 (CDK2), cyclin D1 and proliferating cell nuclear antigen (PCNA) (Sang, Cao, Wang, Liu, & Chen, 2014). In line with this, it has been shown that cytochrome P450-derived epoxyeicosatrienoic acids stimulate endothelial cell proliferation through AKT dependent inactivation of FoxO1 and FoxO3a transcription factors and subsequent downregulation of $p27^{kip1}$ (Potente, Fisslthaler, Busse, & Fleming, 2003).

In addition to its role in angiogenesis, the FoxO transcription factors play a key role in endothelial stress response. Specifically, FoxO3a protects the endothelial cells against oxidative stress by enhancing the expression of mitochondrial antioxidant genes such as MnSOD, catalase, UCP-2, Prx3, Prx5, Trx2 andTrxR2 (Olmos et al., 2009). Importantly, FoxO3a regulates these genes through a cooperative association with the PGC-1α transcription factor, as PGC-1α deficiency abrogates FoxO3a mediated increased expression of antioxidant genes (Olmos et al., 2009). Similarly, upregulation of MnSOD expression by FoxO1 has been shown to mediate the protective effects of SIRT1 activation in hyperglycemia-induced oxidative stress in microvascular ECs (Mortuza, Chen, Feng, Sen, & Chakrabarti, 2013). In addition, FoxO3a has been implicated in the response to hypoxic stress in cardiac microvascular endothelial cells (CMECs) (Zhang et al., 2013). Notably, the hypoxic stress induces apoptosis in CMECs leading to microvascular dysfunction and myocardial ischemia. At the molecular level, hypoxic stress specifically stimulates the expression of FoxO3a but not FoxO1 through HIF-1α, and also reduces AKT activation which leads to increased nuclear FoxO3a. Notably, FoxO3a knockdown rescues hypoxia-induced apoptosis by downregulating pro-apoptotic genes like Bim and Bax and enhancing anti-apoptotic genes like Bcl-2 and Bcl-xl (Zhang et al., 2013). Taken together, the studies suggest that FoxO factors play an important role in endothelial homeostasis by modulating proliferation, migration, apoptosis and angiogenic and vessel forming abilities of the endothelial cells.

3.3 Role of FoxO transcription factors in the development of atherosclerosis

Consistent with their central role in regulating SMC and EC homeostasis, the FoxO transcription factors have been shown to play a crucial role in the development of atherosclerosis. Notably, simultaneous deletion of three isoforms, FoxO1, FoxO3a and FoxO4 in the endothelial cells (vascular endothelial cell knockout mice; VECKO) has been shown to protect against atherosclerosis in the Ldl receptor knockout mice ($Ldlr^{-/-}$) fed with western type diet (WTD) (Tsuchiya et al., 2012). The VECKO $Ldlr^{-/-}$ mice fed with a WTD show increased eNOS expression and NO-dependent vasodilatation, when compared to WTD-fed $Ldlr^{-/-}$ mice. Also, these mice show decreased inflammation and reduced macrophage accumulation in atherosclerotic lesions. Importantly, the atheroprotective effects of triple FoxO deletion was accompanied by multiple beneficial effects including reduced expression of cell adhesion molecules and decreased levels of ROS, inflammation and apoptosis in ECs derived from VECKO mice, suggesting that FoxO factors regulate multiple atherogenic pathways in ECs (Tsuchiya et al., 2012). In contrast to the triple knockout of FoxO in endothelial cells, when FoxO4 null mice (*FoxO4*$^{-/-}$) are cross-bred with apolipoprotein *E*-deficient mice (*apoE*$^{-/-}$), a well-established mouse model of atherosclerosis, the resulting double knockout mice (FoxO4$^{-/-}$ apoE$^{-/-}$) display enhanced atherosclerosis upon a high-fat diet (HFD), when compared with *apoE* $^{-/-}$ mice (Zhu et al., 2011). Further, the FoxO4$^{-/-}$ apoE$^{-/-}$ mice show an increased presence of macrophages and T-cells in the atherosclerotic lesions and elevated serum level of IL-6 when compared to *apoE* $^{-/-}$ mice. Notably, transplantation of a FoxO4$^{-/-}$ bone marrow to C57B/6 mice was sufficient to increase the burden of atherosclerotic lesions following a HFD, indicating a key role for macrophages derived from FoxO4 deficient bone marrow (Zhu et al., 2011). In addition, it has been shown that inhibition of FoxO1 and FoxO3a in SMCs, either through direct genetic depletion or through sustained AKT activation by PTEN deletion, can promote vascular calcification (Deng et al., 2015). Mechanistically, inhibition of FoxO reduces the ubiquitination, which stabilizes the Runx2 transcription factor, an osteogenic regulator promoting osteogenic differentiation and vascular calcification. Consistent with this, the development of atherosclerosis in apoE$^{-/-}$ mice model was associated with increased Runx2 levels, which is correlated with increased AKT/FoxO phosphorylation and reduced Runx2 ubiquitination (Deng et al., 2015). Overall, the studies indicate distinct role of the FoxO transcription factors in different cell types that contribute to the development of atherosclerosis.

4. Role of FoxO in myocardial ischemia and ischemia-reperfusion injury

The myocardial oxygen demand is increased by aging-associated changes in the heart such as left ventricular hypertrophy. Combined with hypertrophy, the coronary atherosclerosis and myocardial microvascular defects increase with age, resulting in increased risk of myocardial ischemia (Paneni et al., 2017). Importantly, restoration of blood flow to the heart following myocardial ischemia, i.e., reperfusion, causes severe oxidative stress and cell death (Murphy & Steenbergen, 2008). In this context, the FoxO transcription factors play a crucial role in protecting the cardiomyocytes against oxidative stress-induced cell death. It has been shown that induction of oxidative stress promotes nuclear localization of FoxO1 and FoxO3a, resulting in activation of its target genes, which reduce ROS levels and promote cardiomyocyte cell survival (Sengupta et al., 2011). In line with this, mice deficient for FoxO1 and FoxO3a in the cardiomyocytes show increased susceptibility to myocardial infarction (MI) or ischemia-reperfusion (I/R) injury, which is accompanied by reduced expression of anti-apoptotic molecules, antioxidants and autophagy-related genes such as PINK1, CITED2, SOD2, catalase, LC3, Gabarapl1 and Gadd45a (Sengupta et al., 2011). Moreover, studies have shown that treatment with HDAC inhibitors such as Trichostatin and entinostat, a class I specific HDAC inhibitor, protects the heart against I/R injury through increased expression and enhanced nuclear FoxO3a activity. The increased FoxO3a activity promotes the expression of downstream antioxidant genes such as SOD2 and catalase and thus protects the heart against oxidative damage (Aune, Herr, Mani, & Menick, 2014; Guo et al., 2017). Similar to HDAC inhibitors, treatment with fibroblast growth factor-8 has been shown to protect against H_2O_2-induced oxidative damage in cardiomyocytes through elevated expression of FoxO1 and its downstream targets catalase and SOD2 (Singla, Wang, & Singla, 2017). Further, SIRT1 protects the heart against I/R by increasing nuclear localization and activity of FoxO1. Notably, the protective action of SIRT1 against oxidative stress in cardiomyocytes is reversed by downregulating FoxO1, suggesting SIRT1 functions via FoxO to exert its antioxidant effect (Hsu et al., 2010). In contrast to the protective role of FoxO1 and FoxO3a factors in the cardiomyocytes following myocardial ischemia, FoxO4 in the endothelial cells promote inflammatory reaction following MI (Zhu et al., 2015). Notably, endothelial-specific deletion of FoxO4, but not cardiomyocyte-specific deletion, reduces neutrophil infiltration and cytokine expression post MI along with significant improvement in cardiac function. Mechanistically, FoxO4 promotes the transcription of arginase1 which competes with eNOS for the common substrate, L-arginine and thus reduces the production of NO. The nitric oxide produced by NOS protects against enhanced inflammatory response, by inhibiting lymphocyte adhesion and migration across the endothelial barrier. Consistently, NOS inhibitor abrogates the cardioprotective effects of FoxO4 deficient mice upon MI, suggesting FoxO4 regulates inflammatory response and cardiac function during MI injury in NO-dependent manner (Zhu et al., 2015). Taken together, these data suggest that the effect of FoxO factors on ischemic processes in the heart is both isoform and cell type specific.

5. Role of FoxO in regulating metabolic homeostasis of heart

Aging-related changes in the heart can induce profound metabolic alterations in the cardiomyocytes. Importantly, failing heart heavily depends on the utilization of glucose for the production of ATP, unlike healthy hearts which predominantly utilize fatty-acids for energy production (Nagoshi, Yoshimura, Rosano, Lopaschuk, & Mochizuki, 2011). It has been shown that the transition from compensatory hypertrophy to heart failure is accompanied by an increase in AKT signaling and reduced FoxO1 levels (Wende et al., 2015). The sustained activation of AKT in the heart leads to suppression of mitochondrial fatty acid oxidation pathways at least in part through reduced FoxO-mediated transcription of nuclear-encoded mitochondrial genes, in concert with other transcriptional regulators such as PPARα and PGC-1α. Consistently, activation of AKT in cultured cardiomyocytes inhibits mitochondrial biogenesis and mitochondrial metabolic gene expression, which can be ameliorated by increased FoxO nuclear localization and target gene expression (Wende et al., 2015). Further, given the importance of glucose metabolism in failing hearts, defects in glucose oxidation can exacerbate the development of heart failure. Notably, FoxO1 plays an important role in regulating glucose oxidation in the cardiomyocytes through increasing the transcription of PDK4, which inhibits pyruvate dehydrogenase complex and prevents pyruvate from undergoing mitochondrial oxidation (Gopal et al., 2017). The synthetic glucocorticoid dexamethasone has been shown to increase PDK4 expression by enhancing the nuclear levels of FoxO1, through the action of multiple regulators, including phospho-p38 MAPK, Akt and SIRT1 (Puthanveetil et al., 2010). In addition, we recently identified that increased FoxO1 activity at the PDK4 promoter is responsible for defective glucose oxidation in the heart under SIRT6 deficient conditions (Khan et al., 2018). These evidences indicate a key role for FoxO in regulating substrate utilization as well as their oxidation in the heart and the potential to be regulated by different pathways.

Besides acting downstream of AKT, FoxO has been shown to regulate the phosphorylation of AKT itself, thus controlling insulin sensitivity and glucose uptake in the heart. It has been shown that FoxO1 and FoxO3a increase the phosphorylation and activation of AKT, by suppressing the activity of calcineurin and PP2A phosphatases, which target AKT for dephosphorylation (Ni et al., 2007). Consistent with this, treatment with the SIRT1 activator resveratrol, which promotes FoxO1 transcriptional activity, leads to increased phosphorylation of AKT in the heart. Under normal conditions, the FoxO mediated increase in AKT activity leads to phosphorylation of FoxO and feedback inhibition. However, sustained activation of FoxO1 or FoxO3a leads to increased basal level phosphorylation and activation of AKT and diminishes the insulin-stimulated increase in AKT activity and glucose uptake in cardiomyocytes. Remarkably, inactivation of FoxO3a enhances insulin-dependent phosphorylation of AKT, indicating a central role for FoxO in regulating insulin signaling in the heart (Ni et al., 2007).

Studies indicate that the FoxO transcription factors play an important role in the development of diabetic cardiomyopathy in the heart. In *db/db* as well as HFD-fed models of diabetes, the development of diabetic cardiomyopathy is accompanied by reduced phosphorylation, increased nuclear localization and activation of FoxO1 and FoxO3a (Battiprolu et al., 2012). Remarkably, mice with cardiomyocyte-specific deletion of both FoxO1 and FoxO3a do not develop cardiac dysfunction on being fed a HFD, despite the development of systemic insulin resistance in these mice. Further, deletion of FoxO1 alone suffices to rescue cardiac function in HFD-fed conditions, indicating a pivotal role for FoxO1 in the development of diabetic cardiomyopathy. Mechanistically, the cardiomyocyte-specific deletion of FoxO1 abrogates the increased lipid accumulation and a PDK4 induced shift from glycolytic to fatty-acid oxidation pathways observed under HFD-fed conditions. In addition, FoxO1 deletion maintains the responsiveness to insulin by preventing the inactivation of IRS-1 due to persistent FoxO activation under HFD-fed conditions (Battiprolu et al., 2012). Further, reminiscent of diabetic cardiomyopathy in mice, HFD-fed flies display accumulation of lipids in the heart, reduced cardiac contractility, conduction blocks and severe structural defects (Birse et al., 2010). Interestingly, attenuation of insulin signaling by expressing dFOXO in the myocardial cells protected the heart against the detrimental effects of HFD (Birse et al., 2010). These results suggest FoxO as a major target for treating the metabolic alterations induced in cardiomyocytes under diabetic conditions and thus prevent cardiac dysfunction.

6. Role of FoxO in regulating diet-induced alterations in heart

Calorie-restriction (CR) has been shown to extend organismal lifespan by protecting against age-related diseases, including cardiac dysfunction. However, the timing of the CR during the lifetime of the organism is critical, as CR initiated in 3 month old young mice leads to exacerbated cardiac hypertrophy and fibrosis along with an increased expression of inflammatory and aging-associated markers following CR (Sheng et al., 2017). On the other hand, CR in middle-aged and old mice prevented adverse cardiac remodeling as well as aging-associated changes in the heart. Importantly, the differential effects of CR on the cardiac phenotype was linked to the differential expression of FoxO factors and its downstream targets in different ages of mice. While the levels of FoxO1 and FoxO4 were significantly decreased in young mice following CR, the expression of FoxO1, FoxO3a, and FoxO4 were increased in middle-aged mice and FoxO3a was enhanced in old mice. In addition, the changes in FoxO expression was linked to differential changes in AMPK activity, as AMPK activity was upregulated in middle and old aged mice while it was downregulated in young mice following CR diet (Sheng et al., 2017). These findings highlight an important role for FoxO in mediating the effects of CR on cardiac function in an age-dependent manner.

HFD-induced obesity is a key contributor to the development of diabetes and cardiovascular diseases. A prolonged HFD feeding induces the development of cardiac hypertrophy and cardiac dysfunction characterized by reduced fractional shortening (Fang et al., 2008). The cardiomyocytes isolated from HFD-fed animals show contractile defects, intracellular Ca^{2+} defects and increased expression of hypertrophic and apoptotic markers (Fang et al., 2008; Relling et al., 2006). Notably, the HFD reduces the FoxO3a activity through increased phosphorylation of AKT and FoxO3a. Thus, the decreased FoxO3a activity leads to reduced expression of the downstream atrophic genes such as atrogin-1 and MuRF-1, leading to development of hypertrophy. Indeed, expression of a dominant-negative FoxO3a was sufficient to induce the expression of hypertrophic markers, suggesting that the AKT/FoxO signaling plays an important role in development of obesity-induced cardiac hypertrophy (Fang et al., 2008). Therefore, the FoxO factors play a key role in mediating the diet-induced changes in cardiac structure and function.

7. Role of FoxO in modulating autophagy and proteasomal degradation

Dysregulation of protein quality control due to defective autophagy or ubiquitin-proteasome system is emerging as a key contributor to the development of aging-related cardiac dysfunction (Ghosh, Vinod, Symons, & Boudina, 2020). FoxO has been shown to regulate both these processes in response to alterations in nutrient availability or stress stimuli. Cardiac specific overexpression of constitutively active FoxO3a has been shown to reduce cardiomyocyte size and decrease heart weight (Schips et al., 2011). Specifically, the constitutively active FoxO3a stimulates the transcription of autophagy related genes, leading to destruction of sarcomeres. Turning off the expression of constitutively active FoxO3a reverses the atrophic phenotype of heart (Schips et al., 2011). In addition, FoxO factors have been shown to play an important role in regulating cardiomyocyte cell size during starvation, by transcriptionally regulating the expression of autophagy related genes such as Gaparapl1 and Atg12 (Sengupta, Molkentin, & Yutzey, 2009). Indeed, overexpression of FoxO1 or FoxO3a is sufficient to induce a decrease in cardiomyocyte cell size with a concomitant increase in the expression of autophagy related genes. On the other hand, expression of a dominant negative FoxO1 attenuates glucose starvation induced decrease in cell size, suggesting an important role for FoxO in regulating cardiomyocyte cell size during starvation. Consistent with this, conditions of starvation or ischemic stress under in vivo conditions coincides with dephosphorylation and activation of FoxO factors which increases autophagy (Sengupta et al., 2009). Similarly, it has been shown that the activity of FoxO1 is increased upon glucose deprivation through an increase in the SIRT1 mediated deacetylation of FoxO1 (Hariharan et al., 2010). The increased FoxO1 activity enhances the expression of Rab7, induces lysosome biogenesis and promotes fusion of autophagic vesicles with endosomes and lysosomes (Hariharan et al., 2010). In agreement with this, the induction of autophagy upon starvation is abolished in cardiac specific FoxO1 knockout mice as well as in a mice expressing a FoxO1 mutant, which is incapable of interacting with SIRT1. Importantly, these mice showed exacerbated decline in cardiac function upon starvation, suggesting that FoxO1 is required for preserving the cardiac function during stressed conditions possibly through induction of autophagy (Hariharan et al., 2010). Besides SIRT1, another NAD^+-dependent enzyme PARP1 is also involved in the regulation of starvation induced autophagy in the cardiomyocytes. Notably, starvation induces PARP1 mediated PARylation of FoxO3a leading to the nuclear translocation of FoxO3a and activation of autophagy related genes such as Gaparapl1 and Atg12 (Wang, Xu, Zhang, Zhang, & Huang, 2018).

In addition to autophagy, FoxO also plays an important role in proteasome mediated degradation by activating the transcription of the muscle specific ubiquitin ligases such as atrogin-1 and MuRF-1. Notably, analysis of human heart failure patient samples display reduced AKT phosphorylation and elevated levels of FoxO3a and its downstream target atrogin-1, which might possibly lead to heart muscle loss and subsequent cardiac dysfunction (Galasso et al., 2010). Furthermore, abnormal calcium handling has been shown to increase FoxO-mediated expression of MuRF-1, which compromises myofibrillar integrity in zebrafish hearts (Shimizu et al., 2017). On the other hand, cardiac specific modest overexpression of dFOXO in the flies has been shown to reduce the age-related decline in cardiac function through improving the expression of genes associated with the ubiquitin-proteasome system. dFOXO overexpression in the heart significantly reduces the levels of ubiquitinated proteins in 7-week old flies, when compared to similarly aged controls. However, the beneficial effect of dFOXO overexpression was dose-dependent as excessive doses of dFOXO resulted in a maladaptive decrease in cardiac function (Blice-Baum et al., 2017). Collectively, these studies suggest a critical role for FoxO in the tight regulation of autophagy and proteasomal degradation in the heart in response to diverse stress conditions.

8. Role of FoxO in regulating age-dependent cardiac functions in flies

Cardiac performance has been shown to decline with aging in different organisms ranging from flies to higher order mammals, including humans (Chiao & Rabinovitch, 2015; Paternostro et al., 2001). In this context, flies with cardiac-specific overexpression of dFOXO or dPTEN which promotes FOXO activity through inhibition of AKT signaling, are shown to be protected against aging-associated decline in heart rate and stress-induced heart failure, suggesting a protective effect of dFOXO against age-dependent decline in cardiac function (Wessells et al., 2004). Moreover, dFOXO activates the transcription of d4EBP, which acts as a key downstream effector of dFOXO in conferring stress resistance in flies (Tettweiler, Miron, Jenkins, Sonenberg, & Lasko, 2005). In line with this, cardiac-specific RNAi mediated depletion of dFOXO or its downstream target d4EBP enhances stress induced heart failure rates in younger flies (Wessells et al., 2009). On the other hand, combined overexpression of dFOXO and d4EBP leads to slowed functional aging of the heart similar to flies overexpressing either gene alone. However, overexpressing dFOXO, along with d4EBP binding partner deIF4E which inhibits deIF4E function, overrides the beneficial effect of dFOXO and increases stress induced heart failure rates in younger flies (Wessells et al., 2009), suggesting that d4EBP acts as a critical effector for dFOXO mediated stress resistance that preserves cardiac function in flies.

Consistent with the protective effects of FoxO on cardiac aging, various cellular modulators of FoxO activity have been shown to regulate age-dependent changes in cardiac function. Interestingly, cardiac-specific over-expression of dSir2, the drosophila homolog of SIRT1, delays functional aging of the heart at least in part by mitigating oxidative stress through increasing dFOXO expression and enhancement of the superoxide dismutase activity (Wen, Zheng, Li, Lu, & Hou, 2019). In addition, cardiac-specific downregulation of the HDAC1 homolog Rpd3 confers stress resistance and improves cardiac function in flies, through upregulation of dFOXO as well as dSir2 (Kopp et al., 2015). Finally, increased ROS levels due to chronic activation of TOR signaling has been shown to induce dFoxO expression which in turn increases the expression of sestrins leading to feedback inhibition of TOR (Lee et al., 2010). These findings suggest that FoxO activation acts as an adaptive response to increased ROS levels and protects the heart from oxidative damage in flies.

9. Conclusion

In summary, FoxO transcription factors regulate a plethora of cellular processes such as stress-response, proliferation, migration, differentiation, apoptosis, autophagy and proteasomal degradation, which play a key role in the development of CVDs. Importantly, the effects of FoxO factors are cell, context and isoform specific and also dose-dependent. Therefore, careful modulation of FoxO activity is essential, while targeting FoxO for treating aging-related cardiovascular diseases.

References

Abid, M. R., Yano, K., Guo, S., Patel, V. I., Shrikhande, G., Spokes, K. C., et al. (2005). Forkhead transcription factors inhibit vascular smooth muscle cell proliferation and neointimal hyperplasia. *The Journal of Biological Chemistry*, *280*(33), 29864–29873. https://doi.org/10.1074/jbc.M502149200.

Alcendor, R. R., Gao, S., Zhai, P., Zablocki, D., Holle, E., Yu, X., et al. (2007). Sirt1 regulates aging and resistance to oxidative stress in the heart. *Circulation Research*, *100*(10), 1512–1521. https://doi.org/10.1161/01.RES.0000267723.65696.4a.

Aune, S. E., Herr, D. J., Mani, S. K., & Menick, D. R. (2014). Selective inhibition of class I but not class IIb histone deacetylases exerts cardiac protection from ischemia reperfusion. *Journal of Molecular and Cellular Cardiology*, *72*, 138–145. https://doi.org/10.1016/j.yjmcc.2014.03.005.

Basatemur, G. L., Jorgensen, H. F., Clarke, M. C. H., Bennett, M. R., & Mallat, Z. (2019). Vascular smooth muscle cells in atherosclerosis. *Nature Reviews. Cardiology*, *16*(12), 727–744. https://doi.org/10.1038/s41569-019-0227-9.

Battiprolu, P. K., Hojayev, B., Jiang, N., Wang, Z. V., Luo, X., Iglewski, M., et al. (2012). Metabolic stress-induced activation of FoxO1 triggers diabetic cardiomyopathy in mice. *The Journal of Clinical Investigation*, *122*(3), 1109–1118. https://doi.org/10.1172/JCI60329.

Birse, R. T., Choi, J., Reardon, K., Rodriguez, J., Graham, S., Diop, S., et al. (2010). High-fat-diet-induced obesity and heart dysfunction are regulated by the TOR pathway in Drosophila. *Cell Metabolism, 12*(5), 533–544. https://doi.org/10.1016/j.cmet.2010.09.014.

Blice-Baum, A. C., Zambon, A. C., Kaushik, G., Viswanathan, M. C., Engler, A. J., Bodmer, R., et al. (2017). Modest overexpression of FOXO maintains cardiac proteostasis and ameliorates age-associated functional decline. *Aging Cell, 16*(1), 93–103. https://doi.org/10.1111/acel.12543.

Burgering, B. M. (2008). A brief introduction to FOXOlogy. *Oncogene, 27*(16), 2258–2262. https://doi.org/10.1038/onc.2008.29.

Calnan, D. R., & Brunet, A. (2008). The FoxO code. *Oncogene, 27*(16), 2276–2288. https://doi.org/10.1038/onc.2008.21.

Chiao, Y. A., & Rabinovitch, P. S. (2015). The aging heart. *Cold Spring Harbor Perspectives in Medicine, 5*(9), a025148. https://doi.org/10.1101/cshperspect.a025148.

Chiribau, C. B., Cheng, L., Cucoranu, I. C., Yu, Y. S., Clempus, R. E., & Sorescu, D. (2008). FOXO3A regulates peroxiredoxin III expression in human cardiac fibroblasts. *The Journal of Biological Chemistry, 283*(13), 8211–8217. https://doi.org/10.1074/jbc.M710610200.

Dai, D. F., Chen, T., Johnson, S. C., Szeto, H., & Rabinovitch, P. S. (2012). Cardiac aging: From molecular mechanisms to significance in human health and disease. *Antioxidants & Redox Signaling, 16*(12), 1492–1526. https://doi.org/10.1089/ars.2011.4179.

Deng, L., Huang, L., Sun, Y., Heath, J. M., Wu, H., & Chen, Y. (2015). Inhibition of FOXO1/3 promotes vascular calcification. *Arteriosclerosis, Thrombosis, and Vascular Biology, 35*(1), 175–183. https://doi.org/10.1161/ATVBAHA.114.304786.

Diebold, I., Petry, A., Burger, M., Hess, J., & Gorlach, A. (2011). NOX4 mediates activation of FoxO3a and matrix metalloproteinase-2 expression by urotensin-II. *Molecular Biology of the Cell, 22*(22), 4424–4434. https://doi.org/10.1091/mbc.E10-12-0971.

Fang, C. X., Dong, F., Thomas, D. P., Ma, H., He, L., & Ren, J. (2008). Hypertrophic cardiomyopathy in high-fat diet-induced obesity: Role of suppression of forkhead transcription factor and atrophy gene transcription. *American Journal of Physiology. Heart and Circulatory Physiology, 295*(3), H1206–H1215. https://doi.org/10.1152/ajpheart.00319.2008.

Furuyama, T., Nakazawa, T., Nakano, I., & Mori, N. (2000). Identification of the differential distribution patterns of mRNAs and consensus binding sequences for mouse DAF-16 homologues. *The Biochemical Journal, 349*(Pt. 2), 629–634. https://doi.org/10.1042/0264-6021:3490629.

Galasso, G., De Rosa, R., Piscione, F., Iaccarino, G., Vosa, C., Sorriento, D., et al. (2010). Myocardial expression of FOXO3a-Atrogin-1 pathway in human heart failure. *European Journal of Heart Failure, 12*(12), 1290–1296. https://doi.org/10.1093/eurjhf/hfq102.

Galley, J. C., Durgin, B. G., Miller, M. P., Hahn, S. A., Yuan, S., Wood, K. C., et al. (2019). Antagonism of forkhead box subclass O transcription factors elicits loss of soluble guanylyl cyclase expression. *Molecular Pharmacology, 95*(6), 629–637. https://doi.org/10.1124/mol.118.115386.

Ghosh, R., Vinod, V., Symons, J. D., & Boudina, S. (2020). Protein and mitochondria quality control mechanisms and cardiac aging. *Cells, 9*(4), 933. https://doi.org/10.3390/cells9040933.

Gopal, K., Saleme, B., Al Batran, R., Aburasayn, H., Eshreif, A., Ho, K. L., et al. (2017). FoxO1 regulates myocardial glucose oxidation rates via transcriptional control of pyruvate dehydrogenase kinase 4 expression. *American Journal of Physiology. Heart and Circulatory Physiology, 313*(3), H479–H490. https://doi.org/10.1152/ajpheart.00191.2017.

Guo, Y., Li, Z., Shi, C., Li, J., Yao, M., & Chen, X. (2017). Trichostatin a attenuates oxidative stress-mediated myocardial injury through the FoxO3a signaling pathway. *International Journal of Molecular Medicine, 40*(4), 999–1008. https://doi.org/10.3892/ijmm.2017.3101.

Hariharan, N., Maejima, Y., Nakae, J., Paik, J., Depinho, R. A., & Sadoshima, J. (2010). Deacetylation of FoxO by Sirt1 plays an essential role in mediating starvation-induced autophagy in cardiac myocytes. *Circulation Research*, *107*(12), 1470–1482. https://doi.org/10.1161/CIRCRESAHA.110.227371.

Hsu, C. P., Zhai, P., Yamamoto, T., Maejima, Y., Matsushima, S., Hariharan, N., et al. (2010). Silent information regulator 1 protects the heart from ischemia/reperfusion. *Circulation*, *122*(21), 2170–2182. https://doi.org/10.1161/CIRCULATIONAHA.110.958033.

Huang, K., Yan, Z. Q., Zhao, D., Chen, S. G., Gao, L. Z., Zhang, P., et al. (2015). SIRT1 and FOXO mediate contractile differentiation of vascular smooth muscle cells under cyclic stretch. *Cellular Physiology and Biochemistry*, *37*(5), 1817–1829. https://doi.org/10.1159/000438544.

Kaestner, K. H., Knochel, W., & Martinez, D. E. (2000). Unified nomenclature for the winged helix/forkhead transcription factors. *Genes Dev*, *14*(2), 142–146. Retrieved from https://www.ncbi.nlm.nih.gov/pubmed/10702024.

Kawashima, S., & Yokoyama, M. (2004). Dysfunction of endothelial nitric oxide synthase and atherosclerosis. *Arteriosclerosis, Thrombosis, and Vascular Biology*, *24*(6), 998–1005. https://doi.org/10.1161/01.ATV.0000125114.88079.96.

Khan, D., Sarikhani, M., Dasgupta, S., Maniyadath, B., Pandit, A. S., Mishra, S., et al. (2018). SIRT6 deacetylase transcriptionally regulates glucose metabolism in heart. *Journal of Cellular Physiology*, *233*(7), 5478–5489. https://doi.org/10.1002/jcp.26434.

Kopp, Z. A., Hsieh, J. L., Li, A., Wang, W., Bhatt, D. T., Lee, A., et al. (2015). Heart-specific Rpd3 downregulation enhances cardiac function and longevity. *Aging (Albany NY)*, *7*(9), 648–663. https://doi.org/10.18632/aging.100806.

Lahteenvuo, J., & Rosenzweig, A. (2012). Effects of aging on angiogenesis. *Circulation Research*, *110*(9), 1252–1264. https://doi.org/10.1161/CIRCRESAHA.111.246116.

Lee, J. H., Budanov, A. V., Park, E. J., Birse, R., Kim, T. E., Perkins, G. A., et al. (2010). Sestrin as a feedback inhibitor of TOR that prevents age-related pathologies. *Science*, *327*(5970), 1223–1228. https://doi.org/10.1126/science.1182228.

Lee, H. Y., Chung, J. W., Youn, S. W., Kim, J. Y., Park, K. W., Koo, B. K., et al. (2007). Forkhead transcription factor FOXO3a is a negative regulator of angiogenic immediate early gene CYR61, leading to inhibition of vascular smooth muscle cell proliferation and neointimal hyperplasia. *Circulation Research*, *100*(3), 372–380. https://doi.org/10.1161/01.RES.0000257945.97958.77.

Lee, W. S., & Kim, J. (2018). Insulin-like growth factor-1 signaling in cardiac aging. *Biochimica et Biophysica Acta-Molecular Basis of Disease*, *1864*(5), 1931–1938. https://doi.org/10.1016/j.bbadis.2017.08.029.

Li, M., & Fukagawa, N. K. (2010). Age-related changes in redox signaling and VSMC function. *Antioxidants & Redox Signaling*, *12*(5), 641–655. https://doi.org/10.1089/ars.2009.2854.

Li, H. H., Kedar, V., Zhang, C., McDonough, H., Arya, R., Wang, D. Z., et al. (2004). Atrogin-1/muscle atrophy F-box inhibits calcineurin-dependent cardiac hypertrophy by participating in an SCF ubiquitin ligase complex. *The Journal of Clinical Investigation*, *114*(8), 1058–1071. https://doi.org/10.1172/JCI22220.

Li, H., Liang, J., Castrillon, D. H., DePinho, R. A., Olson, E. N., & Liu, Z. P. (2007). FoxO4 regulates tumor necrosis factor alpha-directed smooth muscle cell migration by activating matrix metalloproteinase 9 gene transcription. *Molecular and Cellular Biology*, *27*(7), 2676–2686. https://doi.org/10.1128/MCB.01748-06.

Li, H. H., Willis, M. S., Lockyer, P., Miller, N., McDonough, H., Glass, D. J., et al. (2007). Atrogin-1 inhibits Akt-dependent cardiac hypertrophy in mice via ubiquitin-dependent coactivation of Forkhead proteins. *The Journal of Clinical Investigation*, *117*(11), 3211–3223. https://doi.org/10.1172/JCI31757.

Liu, Z. P., Wang, Z., Yanagisawa, H., & Olson, E. N. (2005). Phenotypic modulation of smooth muscle cells through interaction of Foxo4 and myocardin. *Developmental Cell*, *9*(2), 261–270. https://doi.org/10.1016/j.devcel.2005.05.017.

Lu, L., Guo, J., Hua, Y., Huang, K., Magaye, R., Cornell, J., et al. (2017). Cardiac fibrosis in the ageing heart: Contributors and mechanisms. *Clinical and Experimental Pharmacology & Physiology*, *44*(Suppl. 1), 55–63. https://doi.org/10.1111/1440-1681.12753.

Martins, R., Lithgow, G. J., & Link, W. (2016). Long live FOXO: Unraveling the role of FOXO proteins in aging and longevity. *Aging Cell*, *15*(2), 196–207. https://doi.org/10.1111/acel.12427.

Mortuza, R., Chen, S., Feng, B., Sen, S., & Chakrabarti, S. (2013). High glucose induced alteration of SIRTs in endothelial cells causes rapid aging in a p300 and FOXO regulated pathway. *PLoS One*, *8*(1), e54514. https://doi.org/10.1371/journal.pone.0054514.

Murphy, E., & Steenbergen, C. (2008). Mechanisms underlying acute protection from cardiac ischemia-reperfusion injury. *Physiological Reviews*, *88*(2), 581–609. https://doi.org/10.1152/physrev.00024.2007.

Nagoshi, T., Yoshimura, M., Rosano, G. M., Lopaschuk, G. D., & Mochizuki, S. (2011). Optimization of cardiac metabolism in heart failure. *Current Pharmaceutical Design*, *17*(35), 3846–3853. https://doi.org/10.2174/138161211798357773.

Nakamura, M., & Sadoshima, J. (2018). Mechanisms of physiological and pathological cardiac hypertrophy. *Nature Reviews. Cardiology*, *15*(7), 387–407. https://doi.org/10.1038/s41569-018-0007-y.

Ni, Y. G., Berenji, K., Wang, N., Oh, M., Sachan, N., Dey, A., et al. (2006). Foxo transcription factors blunt cardiac hypertrophy by inhibiting calcineurin signaling. *Circulation*, *114*(11), 1159–1168. https://doi.org/10.1161/CIRCULATIONAHA.106.637124.

Ni, Y. G., Wang, N., Cao, D. J., Sachan, N., Morris, D. J., Gerard, R. D., et al. (2007). FoxO transcription factors activate Akt and attenuate insulin signaling in heart by inhibiting protein phosphatases. *Proceedings of the National Academy of Sciences of the United States of America*, *104*(51), 20517–20522. https://doi.org/10.1073/pnas.0610290104.

North, B. J., & Sinclair, D. A. (2012). The intersection between aging and cardiovascular disease. *Circulation Research*, *110*(8), 1097–1108. https://doi.org/10.1161/CIRCRESAHA.111.246876.

Olmos, Y., Valle, I., Borniquel, S., Tierrez, A., Soria, E., Lamas, S., et al. (2009). Mutual dependence of Foxo3a and PGC-1alpha in the induction of oxidative stress genes. *The Journal of Biological Chemistry*, *284*(21), 14476–14484. https://doi.org/10.1074/jbc.M807397200.

Orlandi, A., Bochaton-Piallat, M. L., Gabbiani, G., & Spagnoli, L. G. (2006). Aging, smooth muscle cells and vascular pathobiology: Implications for atherosclerosis. *Atherosclerosis*, *188*(2), 221–230. https://doi.org/10.1016/j.atherosclerosis.2006.01.018.

Paneni, F., Diaz Canestro, C., Libby, P., Luscher, T. F., & Camici, G. G. (2017). The aging cardiovascular system: Understanding it at the cellular and clinical levels. *Journal of the American College of Cardiology*, *69*(15), 1952–1967. https://doi.org/10.1016/j.jacc.2017.01.064.

Paternostro, G., Vignola, C., Bartsch, D. U., Omens, J. H., McCulloch, A. D., & Reed, J. C. (2001). Age-associated cardiac dysfunction in Drosophila melanogaster. *Circulation Research*, *88*(10), 1053–1058. https://doi.org/10.1161/hh1001.090857.

Potente, M., Fisslthaler, B., Busse, R., & Fleming, I. (2003). 11,12-Epoxyeicosatrienoic acid-induced inhibition of FOXO factors promotes endothelial proliferation by down-regulating p27Kip1. *The Journal of Biological Chemistry*, *278*(32), 29619–29625. https://doi.org/10.1074/jbc.M305385200.

Potente, M., Urbich, C., Sasaki, K., Hofmann, W. K., Heeschen, C., Aicher, A., et al. (2005). Involvement of Foxo transcription factors in angiogenesis and postnatal neovascularization. *The Journal of Clinical Investigation*, *115*(9), 2382–2392. https://doi.org/10.1172/JCI23126.

Pramod, S., & Shivakumar, K. (2014). Mechanisms in cardiac fibroblast growth: An obligate role for Skp2 and FOXO3a in ERK1/2 MAPK-dependent regulation of p27kip1. *American Journal of Physiology. Heart and Circulatory Physiology, 306*(6), H844–H855. https://doi.org/10.1152/ajpheart.00933.2013.

Puthanveetil, P., Wang, Y., Wang, F., Kim, M. S., Abrahani, A., & Rodrigues, B. (2010). The increase in cardiac pyruvate dehydrogenase kinase-4 after short-term dexamethasone is controlled by an Akt-p38-forkhead box other factor-1 signaling axis. *Endocrinology, 151*(5), 2306–2318. https://doi.org/10.1210/en.2009-1072.

Relling, D. P., Esberg, L. B., Fang, C. X., Johnson, W. T., Murphy, E. J., Carlson, E. C., et al. (2006). High-fat diet-induced juvenile obesity leads to cardiomyocyte dysfunction and upregulation of Foxo3a transcription factor independent of lipotoxicity and apoptosis. *Journal of Hypertension, 24*(3), 549–561. https://doi.org/10.1097/01.hjh.0000203846.34314.94.

Sang, T., Cao, Q., Wang, Y., Liu, F., & Chen, S. (2014). Overexpression or silencing of FOXO3a affects proliferation of endothelial progenitor cells and expression of cell cycle regulatory proteins. *PLoS One, 9*(8), e101703. https://doi.org/10.1371/journal.pone.0101703.

Savai, R., Al-Tamari, H. M., Sedding, D., Kojonazarov, B., Muecke, C., Teske, R., et al. (2014). Pro-proliferative and inflammatory signaling converge on FoxO1 transcription factor in pulmonary hypertension. *Nature Medicine, 20*(11), 1289–1300. https://doi.org/10.1038/nm.3695.

Schips, T. G., Wietelmann, A., Hohn, K., Schimanski, S., Walther, P., Braun, T., et al. (2011). FoxO3 induces reversible cardiac atrophy and autophagy in a transgenic mouse model. *Cardiovascular Research, 91*(4), 587–597. https://doi.org/10.1093/cvr/cvr144.

Sengupta, A., Molkentin, J. D., Paik, J. H., DePinho, R. A., & Yutzey, K. E. (2011). FoxO transcription factors promote cardiomyocyte survival upon induction of oxidative stress. *The Journal of Biological Chemistry, 286*(9), 7468–7478. https://doi.org/10.1074/jbc.M110.179242.

Sengupta, A., Molkentin, J. D., & Yutzey, K. E. (2009). FoxO transcription factors promote autophagy in cardiomyocytes. *The Journal of Biological Chemistry, 284*(41), 28319–28331. https://doi.org/10.1074/jbc.M109.024406.

Sheng, Y., Lv, S., Huang, M., Lv, Y., Yu, J., Liu, J., et al. (2017). Opposing effects on cardiac function by calorie restriction in different-aged mice. *Aging Cell, 16*(5), 1155–1167. https://doi.org/10.1111/acel.12652.

Shi, J., Wang, A., Sen, S., Wang, Y., Kim, H. J., Mitts, T. F., et al. (2012). Insulin induces production of new elastin in cultures of human aortic smooth muscle cells. *The American Journal of Pathology, 180*(2), 715–726. https://doi.org/10.1016/j.ajpath.2011.10.022.

Shimizu, H., Langenbacher, A. D., Huang, J., Wang, K., Otto, G., Geisler, R., et al. (2017). The Calcineurin-FoxO-MuRF1 signaling pathway regulates myofibril integrity in cardiomyocytes. *eLife, 6*, e27955. https://doi.org/10.7554/eLife.27955.

Singla, R. D., Wang, J., & Singla, D. K. (2017). Fibroblast growth factor-8 inhibits oxidative stress-induced apoptosis in H9c2 cells. *Molecular and Cellular Biochemistry, 425*(1–2), 77–84. https://doi.org/10.1007/s11010-016-2863-2.

Skurk, C., Izumiya, Y., Maatz, H., Razeghi, P., Shiojima, I., Sandri, M., et al. (2005). The FOXO3a transcription factor regulates cardiac myocyte size downstream of AKT signaling. *The Journal of Biological Chemistry, 280*(21), 20814–20823. https://doi.org/10.1074/jbc.M500528200.

Spurthi, K. M., Sarikhani, M., Mishra, S., Desingu, P. A., Yadav, S., Rao, S., et al. (2018). Toll-like receptor 2 deficiency hyperactivates the FoxO1 transcription factor and induces aging-associated cardiac dysfunction in mice. *The Journal of Biological Chemistry, 293*(34), 13073–13089. https://doi.org/10.1074/jbc.RA118.001880.

Steenman, M., & Lande, G. (2017). Cardiac aging and heart disease in humans. *Biophysical Reviews*, *9*(2), 131–137. https://doi.org/10.1007/s12551-017-0255-9.

Tan, W. Q., Wang, K., Lv, D. Y., & Li, P. F. (2008). Foxo3a inhibits cardiomyocyte hypertrophy through transactivating catalase. *The Journal of Biological Chemistry*, *283*(44), 29730–29739. https://doi.org/10.1074/jbc.M805514200.

Tettweiler, G., Miron, M., Jenkins, M., Sonenberg, N., & Lasko, P. F. (2005). Starvation and oxidative stress resistance in Drosophila are mediated through the eIF4E-binding protein, d4E-BP. *Genes & Development*, *19*(16), 1840–1843. https://doi.org/10.1101/gad.1311805.

Tsuchiya, K., Tanaka, J., Shuiqing, Y., Welch, C. L., DePinho, R. A., Tabas, I., et al. (2012). FoxOs integrate pleiotropic actions of insulin in vascular endothelium to protect mice from atherosclerosis. *Cell Metabolism*, *15*(3), 372–381. https://doi.org/10.1016/j.cmet.2012.01.018.

Tzivion, G., Dobson, M., & Ramakrishnan, G. (2011). FoxO transcription factors; regulation by AKT and 14-3-3 proteins. *Biochimica et Biophysica Acta*, *1813*(11), 1938–1945. https://doi.org/10.1016/j.bbamcr.2011.06.002.

Ungvari, Z., Kaley, G., de Cabo, R., Sonntag, W. E., & Csiszar, A. (2010). Mechanisms of vascular aging: New perspectives. *The Journals of Gerontology. Series A, Biological Sciences and Medical Sciences*, *65*(10), 1028–1041. https://doi.org/10.1093/gerona/glq113.

Vivar, R., Humeres, C., Anfossi, R., Bolivar, S., Catalan, M., Hill, J., et al. (2020). Role of FoxO3a as a negative regulator of the cardiac myofibroblast conversion induced by TGF-beta1. *Biochimica et Biophysica Acta, Molecular Cell Research*, *1867*(7), 118695. https://doi.org/10.1016/j.bbamcr.2020.118695.

Vivar, R., Humeres, C., Munoz, C., Boza, P., Bolivar, S., Tapia, F., et al. (2016). FoxO1 mediates TGF-beta1-dependent cardiac myofibroblast differentiation. *Biochimica et Biophysica Acta*, *1863*(1), 128–138. https://doi.org/10.1016/j.bbamcr.2015.10.019.

Wang, D., Uhrin, P., Mocan, A., Waltenberger, B., Breuss, J. M., Tewari, D., et al. (2018). Vascular smooth muscle cell proliferation as a therapeutic target. Part 1: Molecular targets and pathways. *Biotechnology Advances*, *36*(6), 1586–1607. https://doi.org/10.1016/j.biotechadv.2018.04.006.

Wang, C., Xu, W., Zhang, Y., Zhang, F., & Huang, K. (2018). PARP1 promote autophagy in cardiomyocytes via modulating FoxO3a transcription. *Cell Death & Disease*, *9*(11), 1047. https://doi.org/10.1038/s41419-018-1108-6.

Wang, Z., Yu, T., & Huang, P. (2016). Post-translational modifications of FOXO family proteins (review). *Molecular Medicine Reports*, *14*(6), 4931–4941. https://doi.org/10.3892/mmr.2016.5867.

Wang, Y., Zhou, Y., & Graves, D. T. (2014). FOXO transcription factors: Their clinical significance and regulation. *BioMed Research International*, *2014*, 925350. https://doi.org/10.1155/2014/925350.

Wen, D. T., Zheng, L., Li, J. X., Lu, K., & Hou, W. Q. (2019). The activation of cardiac dSir2-related pathways mediates physical exercise resistance to heart aging in old Drosophila. *Aging (Albany NY)*, *11*(17), 7274–7293. https://doi.org/10.18632/aging.102261.

Wende, A. R., O'Neill, B. T., Bugger, H., Riehle, C., Tuinei, J., Buchanan, J., et al. (2015). Enhanced cardiac Akt/protein kinase B signaling contributes to pathological cardiac hypertrophy in part by impairing mitochondrial function via transcriptional repression of mitochondrion-targeted nuclear genes. *Molecular and Cellular Biology*, *35*(5), 831–846. https://doi.org/10.1128/MCB.01109-14.

Wessells, R. J., Fitzgerald, E., Cypser, J. R., Tatar, M., & Bodmer, R. (2004). Insulin regulation of heart function in aging fruit flies. *Nature Genetics*, *36*(12), 1275–1281. https://doi.org/10.1038/ng1476.

Wessells, R., Fitzgerald, E., Piazza, N., Ocorr, K., Morley, S., Davies, C., et al. (2009). d4eBP acts downstream of both dTOR and dFoxo to modulate cardiac functional aging in Drosophila. *Aging Cell*, *8*(5), 542–552. https://doi.org/10.1111/j.1474-9726.2009.00504.x.

Wilhelm, K., Happel, K., Eelen, G., Schoors, S., Oellerich, M. F., Lim, R., et al. (2016). FOXO1 couples metabolic activity and growth state in the vascular endothelium. *Nature*, *529*(7585), 216–220. https://doi.org/10.1038/nature16498.

Yang, X., Gong, Y., Tang, Y., Li, H., He, Q., Gower, L., et al. (2013). Spry1 and Spry4 differentially regulate human aortic smooth muscle cell phenotype via Akt/FoxO/myocardin signaling. *PLoS One*, *8*(3), e58746. https://doi.org/10.1371/journal.pone.0058746.

Zhang, S., Zhao, Y., Xu, M., Yu, L., Zhao, Y., Chen, J., et al. (2013). FoxO3a modulates hypoxia stress induced oxidative stress and apoptosis in cardiac microvascular endothelial cells. *PLoS One*, *8*(11), e80342. https://doi.org/10.1371/journal.pone.0080342.

Zhu, M., Goetsch, S. C., Wang, Z., Luo, R., Hill, J. A., Schneider, J., et al. (2015). FoxO4 promotes early inflammatory response upon myocardial infarction via endothelial Arg1. *Circulation Research*, *117*(11), 967–977. https://doi.org/10.1161/CIRCRESAHA.115.306919.

Zhu, M., Zhang, Q. J., Wang, L., Li, H., & Liu, Z. P. (2011). FoxO4 inhibits atherosclerosis through its function in bone marrow derived cells. *Atherosclerosis*, *219*(2), 492–498. https://doi.org/10.1016/j.atherosclerosis.2011.09.038.

CHAPTER EIGHTEEN

Ghrelin and adipokines: An overview of their physiological role, antimicrobial activity and impact on cardiovascular conditions

Luís Perpétuo[a], Pierre Marie Voisin[e], Francisco Amado[c], Christophe Hirtz[d], and Rui Vitorino[a,b,c,*]

[a]iBiMED, Department of Medical Sciences, University of Aveiro, Aveiro, Portugal
[b]UnIC, Departamento de Cirurgia e Fisiologia, Faculdade de Medicina da Universidade do Porto, Porto, Portugal
[c]LAQV-REQUIMTE, Departamento de Química, Universidade de Aveiro, Aveiro, Portugal
[d]IRMB, Univ Montpellier, INSERM, CHU Montpellier, CNRS, Montpellier, France
[e]Pierre Fabre Oral Care, Castre, France
*Corresponding author: e-mail address: rvitorino@ua.pt

Contents

Vitamins and Hormones, Volume 115
ISSN 0083-6729
https://doi.org/10.1016/bs.vh.2020.12.019

Abstract

The human body has many different hormones that interact with each other and with other factors such as proteins, cell receptors and metabolites. There is still a limited understanding of some of the underlying biological mechanisms of some hormones. In the past decades, science and technology have made major advancements in regard to innovation and knowledge in fields such as medicine. However, some conditions are complex and have many variables that their full picture is still unclear, even though some of these conditions have an alarming rate of incidence and serious health consequences. Conditions such as type 2 diabetes, obesity, nonalcoholic liver disease (NAFLD), cancer in its different forms and even mental conditions, such as Alzheimer's disease, are some of the most common diseases in the 21st century. These conditions are relevant not only because of their high incidence on the general population, but also because of their severity. In this chapter, we present an overview of cardiovascular (CV) diseases. According to the World Health Organization (WHO), cardiovascular diseases, such as coronary artery disease (CAD), heart attack, cardiomyopathy and heart failure (among others), are the number one cause of death worldwide. In 2016, it was estimated that 17.9 million people died from CV diseases, representing more than 30% of all global deaths. Approximately 95% of people who died from CV diseases were so-called "premature deaths" because were referenced to individuals under the age of 70 years old.

In this chapter we described some of the hormones that may have an impact on CV diseases, including ghrelin, a peptide that is mostly produced in the stomach, known to induce hunger. Ghrelin is linked to an increase in body fat, i.e., adipose tissue in animals. For this reason, we also included the adipokines leptin, adiponectin and resistin.

The main objectives of this chapter are to present the state of the art knowledge concerning the mechanisms of each hormone relevant to CV diseases; to compile data and results that further elucidate the relevance of these peptides for several physiological events, conditions and diseases; and to discuss the metabolic impact of each hormone. We established connections between multiple peptides and the underlying condition/disease with tools such as STRING, referring to research using databases, such as UniProt, DisGeNET and Proteomics DB. Fig. 1 shows a network that summarizes the information presented in this chapter, which serves as a visual representation.

1. Introduction

Ghrelin participates in energy homeostasis, having an impact in energy input and output. Energy input is influenced by inducing appetite, while energy output is influenced by controlling ATP production, fat storage, glycogen storage, and/or short-term heat loss. This hormone is therefore linked with body weight gain which is related with increased adipose tissue mass. Ghrelin is also known to have an impact in conditions such as cardiovascular diseases due to its impact on inflammatory responses.

This behavior is justified by the fact that increased hunger leads to an increased food intake, increasing body weight. Body weight is proportional to adipose tissue mass. Adipose tissue is composed by cells named adipocytes that are known to produce cytokines named adipokines. Each adipocyte produces a small amount of cytokines that is insignificant to in vivo concentrations but it adds up to be a very significant amount due to the extension of adipocytes throughout the adipose tissue, leading to relevant effects in inflammatory processes. Adipokines are also thought to play a role in cardiovascular diseases and in antimicrobial processes (Fig. 1).

In this chapter we focus on the adipokines leptin, adiponectin and resistin. Leptin acts in the brain and modulates satiety, and is also a modulator of various immune and inflammatory responses (Lee et al., 2004). Adiponectin circulatory levels are associated with conditions such as of type 2 diabetes, insulin resistance, NAFLD, cardiovascular diseases and coronary artery disease (CAD), and is responsible for increasing fatty acid oxidation

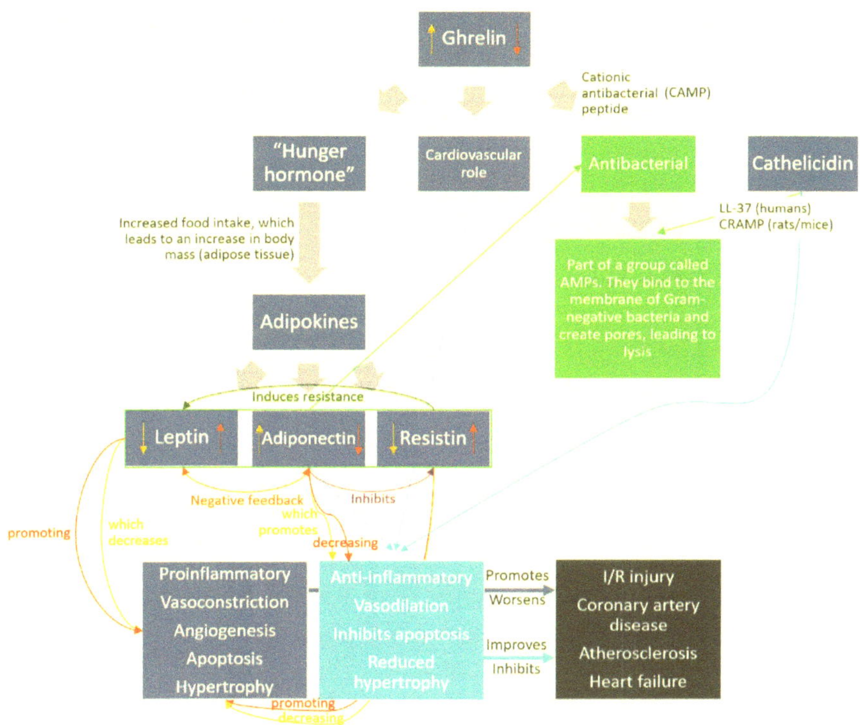

Fig. 1 Network that summarizes the interactions among the principal peptides mentioned in this chapter (ghrelin, leptin, adiponectin, resistin and cathelicidins) and their cardiovascular and antibacterial roles.

and reducing hepatic gluconeogenesis (Yamauchi et al., 2002). Resistin high levels are associated with insulin resistance, increased cardiovascular risk and endothelial dysfunction, among others. High levels of resistin inhibit cellular glucose uptake that are related to increase in triglyceride and cholesterol cellular mass in macrophages (Graveleau et al., 2005; Kusminski, McTernan, & Kumar, 2005). We also intend to clarify the existing information about the crosstalk between each of these hormones and, if available, with other relevant hormones/pathways. Furthermore, we want to focus on two aspects of these hormones (ghrelin and the mentioned adipokines): their role and potential as antibacterial agents, and their impact, both positive and negative, in cardiovascular conditions.

In this chapter, we compile scattered information presented in various articles and books regarding ghrelin and several adipokines, such as leptin, adiponectin and resistin. We describe the physical and chemical characteristics of each hormone, as well as its physiological and biological roles. Subsequently we focus on two aspects of these hormones: their role and potential as antibacterial agents and their impact on cardiovascular conditions. Our main objective is to bridge available information/data about the points mentioned previously resulting in a holistic view.

2. Ghrelin, the "hunger hormone"

2.1 Ghrelin

Ghrelin-obestatin preproprotein is a 117 amino acid peptide divided into 6 different sections: a signal peptide of 23 amino acids, a peptide sequence of 27/28 amino acids known as ghrelin-27/28, a propeptide, a 23 amino acid peptide known as obestatin, and a final propeptide. Both propeptides are removed in ghrelin's mature form, which originates from its self-cleavage behavior. This cleavage process also yields a second peptide, obestatin, a hormone that was first reported to be an inhibitor of appetite and food intake, but for which there is now controversy regarding its anorexic or orexic effects. Ghrelin only becomes active when a caprylic (octanoic) acid is linked post-translationally to the serine at the third position by the enzyme ghrelin O-acyltransferase (GOAT). The gene encoding ghrelin is located in the short (p) arm of chromosome 3 at position 25.3 (3p25.3).

Ghrelin is expressed mainly in the stomach and pancreas and is present in plasma and saliva at concentrations of approximately 100–200 pg/mL ($pg = 10^{-12} g$) (Aydin et al., 2005). Ghrelin is also secreted in lower amounts in the placenta (until the first trimester and after delivery), kidneys, pituitary, lungs and hypothalamus (Ghelardoni, Carnicelli, Frascarelli, Ronca-Testoni, & Zucchi, 2006; Gualillo et al., 2001; Korbonits et al., 2001; Mori et al., 2000). Ghrelin impacts the regulation of appetite, stimulating food intake, and metabolic pathways, increasing adiposity, triglyceride synthesis and gluconeogenesis (Barazzoni et al., 2007; Yamauchi et al., 2002). Ghrelin is also associated with increased circulating glucose levels and decreased insulin production in the pancreas, favoring the conservation of glucose for vital functions (such as brain metabolism) and stimulating fatty acid oxidation for energetic purposes (Broglio et al., 2001; Pinkney, 2014). Furthermore, some studies have shown that ghrelin stimulates osteoblast cell differentiation and bone mineral density (Fukushima et al., 2005). Studies have shown that there is specific stimulation of ghrelin secretion by the adrenergic hormones, norepinephrine and epinephrine, although it is a local phenomenon since experiments with intravenous infusions of these hormones did not alter ghrelin's plasma level (Zhao et al., 2010). Other studies have shown that ghrelin secretion changes in response to the stimulation of sympathetic nerves. Catecholamines, the classic product of sympathetic output, have been shown to correlate positively with ghrelin levels (Mundinger, Cummings, & Taborsky, 2006).

Ghrelin is known as the "hunger hormone" since it stimulates appetite and increases food intake when produced at elevated levels in the stomach. Ghrelin significantly inhibits adipocyte lipid accumulation and promotes lipolysis, which promote fat storage (Knerr, Herzog, Rauh, Rascher, & Horbach, 2006). Furthermore, studies have found that ghrelin levels in obese individuals are lower than in lean individuals and that ghrelin promotes obesity via its overproduction and decreased suppression after meals; in other words, this peptide's receptor has increased sensitivity (Makris et al., 2017).

On the other hand, a hormone called leptin is responsible for maintaining body weight. Leptin achieves this effect by sending signals to the brain that inhibit appetite and therefore decrease food intake. The lower the body weight, the less leptin is produced, increasing hunger and making it more difficult to maintain a low-calorie diet (Fig. 2).

3. Adipokines

Adipose tissue produces cytokines that, because of the location of their synthesis, are also known as adipokines or adipocytokines. Adipokines are bioactive molecules that are cell signaling proteins and play key roles in

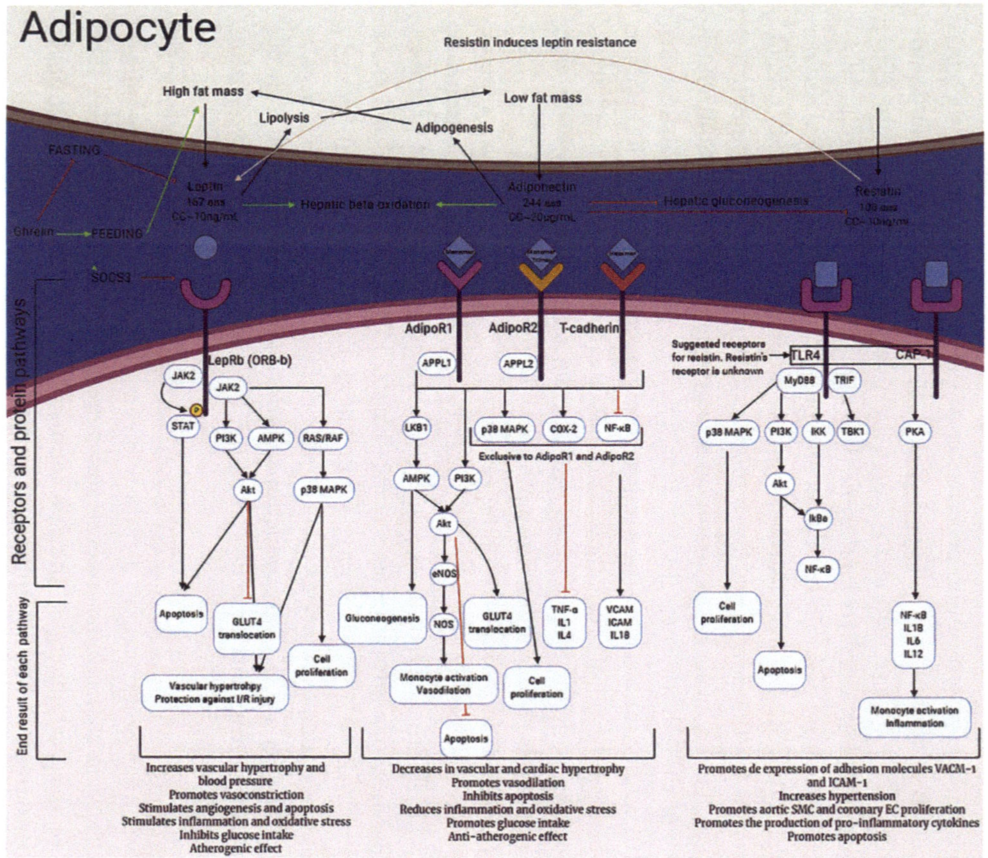

Fig. 2 Representation of the interactions among ghrelin, leptin, adiponectin and resistin; their pathways; and their impact on metabolism and cardiac and vascular conditions. CC means circulatory concentration, in this case, in plasma, and aa means amino acid (aas, plural). Note that the unidentified cell can be a myocyte cell, a vascular cell, etc. (see the above text for more references). All the information in this figure was previously described in detail in this chapter (Fig. 3).

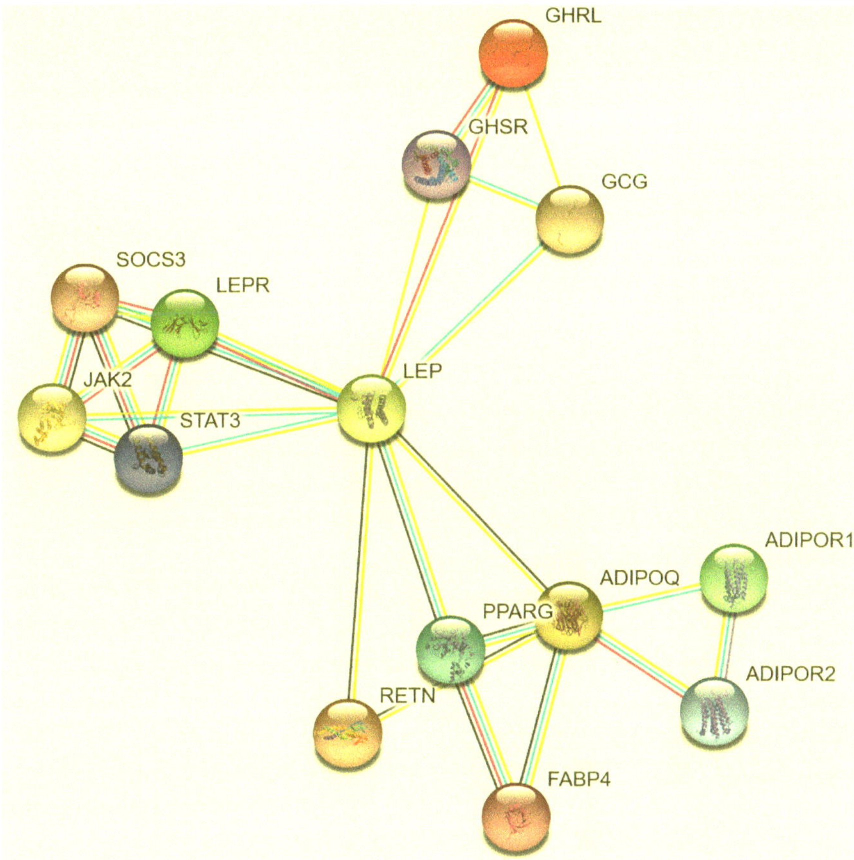

Fig. 3 Network of covered adipokines (leptin (LEP), adiponectin (ADIPOQ), resistin (RETN)) ghrelin (GHRL) and other factors, such as receptors (leptin receptor (LEPR), adiponectin receptors 1 and 2 (ADIPOR1 and 2) and peroxisome proliferator-activated receptor gamma (PPARG)) and signaling pathways: suppressor of cytokine signaling 3 (SOCS3), tyrosine-protein kinase (JAK2), signal transducer and activator of transcription 3 (STAT3), growth hormone secretagogue receptor type 1 (GHSR), glucagon (GCG) and fatty acid-binding protein (FABP4).

events such as tissue repair, homeostasis and immunity. The importance of adipokines is highlighted by the fact that adipose tissue is one of the largest organs in the body and even though the amount of adipokine produced by a single adipocyte is small, the overall amount produced by all adipose tissue is enough to affect body functions. Adipokines are also secreted by immune and endothelial cells and fibroblasts. When excessive amounts of adipokines are produced, they contribute to the pathogenesis of conditions such as cardiovascular complications. Adipokines can have autocrine, paracrine and endocrine effects: modification of insulin sensitivity, regulating glucose/lipid transportation; impact on cell differentiation and growth, immune cell attraction and by modifying insulin sensitivity since these processes can result from autocrine and paracrine mechanisms; impact of endothelial function, regulation of appetite, fat deposition, systemic insulin sensitivity and inflammation (respectively) (Blüher, 2012).

3.1 Leptin

Leptin is a 16-kDa non-glycosylated protein containing 167 amino acids; it is a member of the type I cytokine superfamily with a long chain four-helical bundle structure and is composed of a short signaling peptide sequence of 21 amino acids, followed by a sequence of 146 amino acids (Zhang & Gong, 2018). The OB gene, which is responsible for synthesizing leptin, is located on chromosome 7 (7q31.3) and is composed of three exons and two introns (Gong, Bi, Pratley, & Weintraub, 1996). Leptin release and mRNA expression are positively correlated with the size of adipocytes (Skurk, Alberti-Huber, Herder, & Hauner, 2007). Leptin acts through the leptin receptor, LEPR/OBR, and is generally produced by adipocytes in direct proportion to the mass of white adipose tissue (WAT). The secretion of leptin is greater from subcutaneous WAT than from visceral WAT (Fain, Bahouth, & Madan, 2004). Leptin is also produced in small amounts in other tissues, such as the stomach, epithelium, placenta and heart (Klok, Jakobsdottir, & Drent, 2007). Plasma leptin in humans exists at concentrations from 5 to 15 ng/mL (ng= 10^{-9} g) and circulates partially bound to plasma proteins. Protein-bound leptin exists in equilibrium with free leptin, and the latter is a bioactive hormone (Rajala et al., 2004; Sinha et al., 1996). Leptin mainly acts in the brain and modulates satiety and controls of reward and aversion. The classic primary physiologic role of leptin is to serve as a marker of long-term energy stores for the central nervous system (CNS). As the amount of adipose tissue decreases, the amount of leptin produced and that crosses the blood-brain barrier also decreases, increasing the sensation of hunger. Leptin plays other roles, such as modulating various immune and inflammatory responses, inducing activation of cytokines and modifying their production pattern toward the Th1 (T helper 1 cell) response and directly stimulating the expression and release of interleukins (IL-1a, IL-1b, and IL-6) (Lee et al., 2004; Shiver et al., 2002).

The physiological significance of leptin was emphasized by the successful administration of leptin to treat metabolic disorders of lipodystrophy in mice and humans (a disorder characterized by small amounts of adipose tissue accompanied by a severe resistance to insulin, which can lead to hyperinsulinemia, hyperglycemia, and an enlarged fatty liver). Fukushima et al. (2005) described an experiment in which two genetically engineered mouse models were conditioned to have lipodystrophy. The mice were observed to have very low plasma concentrations of leptin due to the loss of mature adipose tissue. Leptin supplementation overcame insulin resistance, diabetes, hyperlipidemia, and fatty liver in both models. These results support the theory that leptin exerts favorable effects on glucose and lipid metabolism. Human lipodystrophy patients have also been treated with leptin, resulting in an improvement in metabolic disorders and fatty liver disease (NAFLD). Leptin acts on target cells via plasma membrane receptors with at least six isoforms, from Ob-Ra through Ob-Rf (Ob-Rb is the predominant isoform), also known as LepRb. Leptin activity is mediated via Ob-R and is inhibited by cytokine signaling 3 receptors (SOCS3) (Frühbeck, 2006; Yamauchi et al., 2002).

It is known that ghrelin can inhibit leptin activity, but the specific mechanisms are still unclear. Some studies have shown that suppression of cytokine signaling 3 (SOCS3) plays a pivotal role in the modulation of leptin signaling by inhibiting leptin-activated Janus activated kinase 2 (JAK2), which means that SOCS3 negatively regulates hypothalamic leptin signaling. Ghrelin activates a protein pathway via cAMP (Epac), increasing SOCS3 expression, which affects leptin signal transduction. In other words, ghrelin can induce leptin resistance via SOCS3 (Heldsinger et al., 2014) (Fig. 2). Overexpression of SOCS3 is also responsible for insulin resistance, and its removal from skeletal muscle shows a protective effect against insulin resistance in obesity (Jorgensen et al., 2013). LepRb also activates other signaling pathways, such as the insulin receptor substrate-phosphatidylinositol 3-kinase (IRS-PI3K), SH2-containing protein tyrosine phosphatase 2-mitogen-activated protein kinase (SHP2-MAPK), and 5′ adenosine monophosphate-activated protein kinase—acetyl-CoA carboxylase (AMPK-ACC) pathways (Park & Ahima, 2015) (Fig. 2).

Leptin also causes conditions such as psoriasis (a skin disease that causes red, itchy scaly patches) and periodontitis (a serious gum infection that damages the soft tissue and the alveolar bone). The peptide plays a proinflammatory role, increasing Th-1 lymphocytes and cytokines, decreasing

Th-2 cytokines, and promoting the secretion of several proinflammatory factors, such as IL-6, IL-8 (CXCL8) and TNF-α (Furugen, Hayashida, Kitamura, & Saito, 2010; Kong et al., 2019) (Fig. 2). The proinflammatory characteristics of the peptide and the inflammatory characteristics of these two diseases share a link (Fig. 2 and Table 1). Leptin promotes increased vascular inflammation, stimulates apoptosis and increases vascular smooth muscle cell (VSMC) migration and proliferation. These characteristics are the link between the peptide and the two conditions mentioned.

3.2 Ghrelin and leptin

An overview of the dynamic between these two hormones is also of interest to highlight. Leptin is mostly secreted by adipose tissue, and ghrelin is mostly secreted by the stomach; both hormones can enter the brain through the bloodstream. Ghrelin and gastric leptin can reach the hypothalamus through the vagal nerve and nucleus tractus solitarus. Leptin and ghrelin both stimulate and suppress hypothalamic neurons, which contain various neuropeptides, resulting in anorexic or orexigenic effects, respectively. In other words, leptin can stimulate and inhibit anorexic effects, while ghrelin can stimulate and inhibit orexigenic effects. Both hormones can influence each other's levels; therefore, ghrelin can indirectly stimulate and inhibit anorexic effects, and leptin has the reciprocal effect for orexigenic effects (Klok et al., 2007) (Fig. 2). The extreme result of an unhealthy orexigenic effect is obesity. In obesity, there is decreased sensitivity to leptin, which leads to the inability to feel satiated even with high energy storage and high leptin levels. Obesity can be caused by leptin resistance, which is characterized by high leptin levels and decreased leptin sensitivity (Pan, Guo, & Su, 2014). Moreover, leptin can inhibit ghrelin by reducing ghrelin secretion by gastric cells and by suppressing the expression of ghrelin receptors in the neuropeptide Y system (NPY). NPY is a neuropeptide synthesized mainly by neurons in the sympathetic nervous system and not only acts as a strong vasoconstrictor but also promotes the growth of adipose tissue (Coiro et al., 2010). Obesity is characterized mainly as a condition in which there is an elevated amount of adipose tissue. Factors such as genetic predisposition, overeating and physical inactivity influence the expansion of adipose tissue. Obesity-associated diabetes, nonalcoholic fatty liver and metabolic syndrome are closely linked to adipose stress and dysfunction (Khan, Al-Goblan, & Al-Alfi, 2014; Paschos & Paletas, 2009). Moreover, adipose tissue has several functions, such as energy

Table 1 Summary of the cardiovascular effects of three different adipokines (adiponectin, leptin and resistin) and their respective mechanisms of action.

Adipocytokine	Action	Process	References
Adiponectin	Inhibition of apoptosis Suppress endothelial cells monocyte adhesion Promotion of angiogenesis	Stimulation of the interaction between Atk and AMPK on the endothelial cells (Atk is a second messenger in apoptosis and cell proliferation) that also reduces endothelial cells apoptosis by reducing TNF-α induced processes (involved in inflammation)	Kobayashi, (2004), Ouchi et al. (1999, 2004), and Hopkins, Ouchi, Shibata, and Walsh (2007)
	Anti-atherogenic	Regulation of the PI3K-Akt, endothelial NO synthase (eNOS) and AMPK signaling pathways	Shimada, Miyazaki, and Daida (2004)
	Reduction of vascular smooth muscle cells proliferation	Nitrous oxide is crucial of VSMC proliferation. Adiponectin regulates eNOS via PI3K and AMPK pathways, inhibiting the enzyme and reducing NO synthesis, therefore reducing VSMC proliferation	Arita et al. (2002) and Chen et al., (2006)
	Reduced myocardial damage and I/R injury	Activation of COX-2 (cyclooxygenase-2) and AMPK (5′ adenosine monophosphate-activated protein kinase) (anti-inflammatory enzymes) and suppression of TNF-α leading to a decrease in NO production	Takahashi et al., (2005), Ouchi, Shibata, & Walsh (2006), and Rojas et al. (2014)
	Reduction of hypertension	Similar mechanisms mentioned above like reduced atherosclerosis and monocyte adhesion	Ohashi et al., (2006) and Rojas et al. (2014)

Continued

Table 1 Summary of the cardiovascular effects of three different adipokines (adiponectin, leptin and resistin) and their respective mechanisms of action.—cont'd

Adipocytokine	Action	Process	References
Leptin	Stimulation of apoptosis	AMPK and MAPK pathways stimulate hypertrophic responses that lead to apoptosis	Wang et al. (1998)
	Proatherogenic	Induction of mitochondrial superoxide production and expression of MCP-1	Yamagishi et al. (2001)
	Increased VSMC migration and proliferation	Via signaling molecules involved are PI3K/Akt, p38 MAPK, mTOR, and RhoA/ROCK	Trovati et al. (2014)
	Cardiomyocytes hypertrophy and proliferation	Cardiomyocytes hypertrophy is stimulated via activation of p44/42 MAPK and p38 MAPK, while its proliferation is inhibited via inhibition of the PI3K-Atk and p44/42 MAPK signaling	Rajapurohitam, Gan, Kirshenbaum, and Karmazyn (2003) and Tajmir, Ceddia, Li, Coe, and Sweeney (2004)
	Increased vascular inflammation/EC dysfunction	Increase of inflammatory enzymes TNF-α and IL-6	Bouloumié et al. (1999)
	Reduced myocardial damage and I/R injury	Leptin mediates its actions via PI3-Akt and MAPK pathways that are involved in the reperfusion injury salvation kinase (RISK) pathway, protecting the myocardium against I/R injury	Yellon and Hausenloy (2007) and Schisler, Lang, and Willis (2016)

Resistin	Production of adhesion molecules	Induction of the expression of adhesion molecules vascular cell adhesion molecule 1 (VCAM-1) and intercellular adhesion molecule-1 (ICAM-1)	Kawanami et al. (2004)
	Increased levels of proatherogenic inflammatory markers	Increased production of interleukins like IL1, IL6, IL12 and TNF-α in an NF-κB-mediated process	Silswal et al. (2005)
	Protection against I/R injury	Via PI3K and PKC activation	Gao et al. (2007)
	Stimulation of apoptosis	Via PI3K/Akt pathway stimulation	Zuniga, Raghuraman, and Zhou (2018)
	Increased aortic SMC proliferation	Promotion of the proliferation of human aortic smooth muscle cells via PI3-Akt and p3B signaling (respectively). p38 is a class of MAPK, and is involved in proinflammatory responses	Calabro, Samudio, Willerson, and Yeh (2004) and Mu et al. (2006)
	Increased coronary EC proliferation		

storage, hormone production, thermal isolation (white adipose tissue) and thermogenesis (brown adipose tissue). Adipose tissue is also responsible for producing cytokines known as adipokines (Fig. 2).

3.3 Adiponectin

Another adipokine that plays a relevant biological role is adiponectin. Aside from adipose tissue, adiponectin can also be secreted by the placenta, cardiac myocytes, endothelial and skeletal cells. Adiponectin is a 30-kDa 244 amino acid polypeptide located on chromosome 3 (3q27.3). A mutation can occur at position 112, in which an arginine is replaced by a cysteine, leading to a condition known as adiponectin deficiency, which leads to impaired secretion of adiponectin from cells. There are four distinct regions of adiponectin: the first region is a short signal sequence that targets the hormone for secretion outside the cell, and the second region is a short variable region that varies between species. The first two regions account for a total of 41 amino acids. The third region is a 65-amino acid region with similarity to collagenous proteins, and the last region is a globular domain (136 amino acids). When the 3-dimensional structure of the globular region was determined, a striking similarity to tumor necrosis factor-α (TNF-α) was observed, despite their unrelated protein sequences (Shapiro & Scherer, 1998). Adiponectin is present in human plasma at concentrations ranging from 3 to 30 μg/mL ($\mu g = 10^{-6}$ g). These levels are reduced in obese individuals. Low adiponectin levels are associated with the prevalence of type 2 diabetes insulin resistance, NAFLD, and cardiovascular diseases, including coronary artery disease (CAD). Adiponectin is responsible for increasing fatty acid oxidation as a source of energy, reducing the synthesis of glucose in the liver and increasing insulin sensitivity via AMPK phosphorylation (Yamauchi et al., 2002). Studies have also shown that after percutaneous coronary intervention (PCI), low adiponectin levels are associated with exacerbation of cardiac damage and dysfunction and are an indicator of major adverse cardiac events at the 8–12-month follow-up (Shibata, Ouchi, Ohashi, & Murohara, 2017). The activity of adiponectin is mediated via the receptors AdipoR1 (mostly in skeletal muscle), AdipoR2 (in the liver) and T-cadherin (mostly expressed in the heart, spinal cord, neurons of the cerebral cortex and aortic walls), and all three receptors are expressed in cardiac tissue (Doyle et al., 1998; Hug et al., 2004; Yamauchi et al., 2003). AdipoR1 and AdipoR2 include an adaptative protein (APPL1) that is involved in several pathways. For AdipoR2, APPL1 starts a pathway that leads to the regulation of fatty acid oxidation. When this receptor is stimulated by a high adiponectin level, the fatty acid oxidation process is stimulated. For AdipoR1, APPL1 can regulate several processes, such as promoting GLUT-4 translocation (one of the several existing glucose transporters), which increases glucose uptake and can regulate AMPK-mediated processes, such as vasodilation, cytoprotection and fatty acid oxidation (common to AdipoR2 but through a different pathway). APPL1 from AdipoR1 can also present crosstalk with APPL1 from the insulin receptor, interfering with the PI3K/Akt pathway, resulting in metabolic effects, such as increased protein synthesis, glycogen synthesis, gluconeogenesis, lipogenesis, increased glucose uptake and utilization and reduced lipolysis (Achari & Jain, 2017) (Fig. 2).

Similar to leptin, adiponectin also plays a role in psoriasis and periodontitis, but with the opposite role. Adiponectin plays an anti-inflammatory role as it has an inverse correlation with both conditions. High levels of adiponectin suppress inflammation and responses from the immune system (Furugen et al., 2010; Kong et al., 2019) (Fig. 2).

3.4 Resistin

Resistin, also known as adipose tissue-specific secretory factor (ADSF), is a 12.5-kDa peptide with 108 amino acids that circulates in the plasma at concentrations of 2.5–21.5 ng/mL, and its gene is located on the 19th chromosome (19p13.2). This peptide is a cysteine-rich hormone that, similar to the previously mentioned peptides, is produced by adipose tissue and is composed of a short signal peptide (18 amino acids), followed by a sequence of 89 amino acids. Resistin was discovered in 2001 and was proposed to be a potential link between obesity and diabetes (Steppan & Lazar, 2002). In humans, resistin is produced in peripheral blood mononuclear cells (PBMCs), macrophages, and bone marrow cells (the primary source of circulating resistin). To a lesser extent, circulating resistin has also been observed in the pituitary gland, hypothalamus, epithelial cells from the gastrointestinal tract, adrenal glands, skeletal muscle, pancreas, spleen and placenta trophoblastic cells, bone marrow, trophoblastic cells in the placenta, pancreas, primary cell leukemia, synovial fluid, synovial tissue, and circulating blood (Codoñer-Franch & Alonso-Iglesias, 2015; Filková, Haluzík, Gay, & Šenolt, 2009; Jamaluddin, Weakley, Yao, & Chen, 2012; McTernan et al., 2002; Minn et al., 2003; Patel et al., 2003; Rajala, Obici, Scherer, & Rossetti, 2003; Satoh et al., 2004; Schäffler et al., 2003; Steppan & Lazar, 2002).

Resistin overexpression is associated with insulin resistance and dyslipidemia (Rajala et al., 2004; Sato et al., 2005; Silha et al., 2003). Resistin levels are increased in obesity, and resistin inhibits cellular glucose uptake, which is related to increases in triglycerides and cholesterol cellular mass in macrophages (Filková et al., 2009; Graveleau et al., 2005; Kusminski et al., 2005). In summary, high levels of resistin are associated with insulin resistance, an increased cardiovascular risk, endothelial dysfunction (via promotion of adhesion molecules, endothelial migration and proliferation), increased proatherogenic inflammatory markers, VSMC proliferation and poor prognosis in coronary artery disease (Calabro et al., 2004; Hausenloy & Yellon, 2006; Lubos et al., 2007; Pischon et al., 2005) (Fig. 2). Although the resistin receptor remains unknown, studies suggest that the proinflammatory effects of resistin may be mediated by its binding to an endotoxin receptor, Toll-like receptor 4 (TLR4) (Tarkowski, Bjersing, Shestakov, & Bokarewa, 2010). TLR4 can exert its effect via MyD88 (myeloid differentiation primary response gene 88) or TRIF (Toll-interleukin 1 receptor domain-containing adaptor inducing interferon-beta). In the MyD88 pathway, MAPK and IκB kinase (IKK) are activated, which ultimately promote the expression of proinflammatory cytokines, such as interleukins (IL1B, IL6 and IL12) and TNF-α. The TRIF pathway is different and ultimately promotes the expression of interferon-beta (IFN-β) and interferon-inducible protein 10 (IP-10). Interferon activates the innate immune response in infected cells and promotes cytokine production, presenting resistin with another antimicrobial factor that it must account for. It has also been suggested that resistin interacts with the receptor adenylyl cyclase-associated protein 1 (CAP-1), a known receptor that mediates inflammatory processes in human monocytes. CAP-1 activates adenylate cyclase, increasing intracellular levels of adenosine monophosphate, which increases protein kinase A (PKA) activity, inducing the expression of inflammatory cytokines (Hinshaw, 2012; Kopitar-Jerala, 2017) (Fig. 2).

Resistin plays a role in periodontitis, promoting inflammation and the synthesis of proinflammatory factors, such as TNF-α and IL-12. In this specific condition, studies have shown that resistin also participates in inflammatory processes caused by bacterial infections (Furugen et al., 2010).

3.5 Crosstalk between adipocytes and other proteins

Adipocytes are responsible for synthesizing leptin and adiponectin, and although the mechanisms are still not fully understood, the relationship between the two peptides has been studied. Adipocytes with a low-fat mass increase production of adiponectin, inhibiting lipolysis, stimulating the translocation of GLUT4 to the cell membrane and promoting the entry of glucose. This process increases adipocyte fat mass by accumulating lipids. High adiponectin levels also stimulate adipogenesis, i.e., the formation of adipocytes from stem cells. When adipocytes reach a high percentage of fat mass, there is an increase in the production of leptin, which stimulates lipolysis. Leptin also inhibits the translocation of GLUT4, which ensures that the adipocyte fat mass decreases over time. Fasting inhibits the synthesis of leptin, but these two elements can act independently (Stern, Rutkowski, & Scherer, 2016). In short, adipocytes produce adiponectin when they have a low-fat mass and produce leptin when they have a high fat mass. These peptides can impact liver metabolism. Both leptin and adiponectin decrease hepatic lipogenesis and increase β-oxidation though the activation of AMPK via a phosphorylation process. Phospho-AMPK phosphorylases acetyl-CoA carboxylase-1, limiting the production of malonyl CoA, enhancing fatty acid transport to mitochondria and β-oxidation.

Increases in fatty acid oxidation and decreases in both hepatic glucose output and lipid accumulation in the liver increase insulin sensitivity (Kim et al., 2007). Moreover, leptin is thought to be responsible for several cardiovascular diseases derived from obesity, while adiponectin is considered to be cardioprotective (as described previously).

There appears to be a direct reciprocal effect between resistin and adiponectin with respect to the inflammation of vascular ECs. High plasma levels of resistin correlate with proatherogenic inflammatory markers (Kunnari, Ukkola, Päivänsalo, & Kesäniemi, 2006), an increased cardiovascular risk, unstable angina, poor prognosis in coronary artery disease and metabolic syndrome (Lubos et al., 2007; Norata et al., 2007). Adiponectin has been shown to inhibit resistin's actions by inhibiting VCAM-1 (Kawanami et al., 2004).

A study using resistin transgenic mice with elevated levels of circulating resistin hypothesized that resistin creates centrally mediated effects known as leptin resistance; experiments that showed that acute resistin treatment inhibited leptin-mediated STAT3 phosphorylation (Zieba, Biernat, & Barć, 2020).

Both TNF-α and resistin enhance inflammatory processes in adipose tissues, and both activate NF-kB in their intracellular signaling pathways, but resistin does not induce the suppression of adipose-specific markers (CEBPA, FABP4 and SLC2A4). This finding suggests that resistin's intracellular signaling is distinct from that of TNF-α. Although there is not yet a known cellular receptor for resistin, it is known that it interacts with MAPK cascades that activate the p38 and ERKs, PI3K/Atk and PLC/IP3 pathways, which increase Ca^{2+} influx from the extracellular environment and from the endoplasmic reticulum (a proinflammatory process) (Filková et al., 2009). All these crosstalk pathways and their respective effects are highlighted in Fig. 2.

4. Antibacterial activity of ghrelin and adipokines

Ghrelin exists mostly as deacylated ghrelin (DAG) and acylated ghrelin (AG). Nonacylated ghrelin (or des-octanoyl or des-acyl) circulates in greater amounts than the acylated (or n-octanoyl) form. AG has a hydrophobic moiety with which it binds as an endogenous ligand of growth hormone secretagogue receptor (GHSR1α) in pituitary cells. Once bound to the receptor, secretion of the growth hormone is activated and induces orexigenic signaling in the brain. DAG, on the other hand, does not have the ability to bind and/or activate GHSR1α (Bellone et al., 2004; Kojima & Kangawa, 2005; Lim, Kola, & Korbonits, 2011). Reports state that AG has a half-life of approximately 240 min in humans, which reflects the action of the enzyme butyrylcholinesterase in inactivating AG in humans, while in rodents, this desoctanoylation process is approximately 8 times faster due to the action of carboxylesterases (De Vriese et al., 2004). AG also plays physiological roles, such as in glycemia, thymopoiesis and the prevention of neuronal apoptosis (Komatsuzawa et al., 2007). The level of AG in plasma is influenced by the ratio of fat, protein and carbohydrate intake and is associated with its physiological role in hunger due to its increase in pre- and postprandial periods (Ghelardoni et al., 2006). Both major forms of ghrelin can act as cationic antimicrobial peptides (CAMPs) and have been shown to have the same degree of bactericidal activity against Gram-negative bacteria, while their bactericidal effects against Gram-positive bacteria are minimal or absent. The cationic charge of CAMP promotes the binding of both major forms of ghrelin to negatively charged bacterial cell surfaces, where they perform their bactericidal activities, leading to negative results when tested on Gram-positive bacteria. It has been suggested that ghrelin, with its positive net charge and putative amphipathic helix structure, forces its hydrophobic residues into positions that potentiate the formation of transmembrane pores that can eventually lead to bacterial lysis (Kahlenberg & Kaplan, 2013).

Previously in this chapter, we approached the subject of peptides with antibacterial properties (antimicrobial peptides, or AMPs), such as ghrelin and the mentioned adipokines (leptin, adiponectin and resistin). AMPs are produced by all life forms and are part of innate immunity in higher organisms, while in bacteria, AMPs are produced to kill bacteria that compete for the same ecological system (Hancock & Chapple, 1999; Hassan, Kjos, Nes, Diep, & Lotfipour, 2012). AMPs can either be nonribosomally (mainly by bacteria) or ribosomally synthesized (a common process to all organisms) (Hancock & Chapple, 1999). Apart from their already mentioned antimicrobial activity against Gram-positive and Gram-negative bacteria, AMPs also can modulate immune responses (Hancock & Sahl, 2006). AMPs, when in contact with the cell membrane, form an amphipathic structure promoting an interaction between their charged domains and the hydrophilic groups of phospholipids of the cell membrane. At the same time, the hydrophobic domains of the peptide interact with the hydrophobic moiety of the bilipid layer. These two simultaneous events lead to the formation of a transmembrane pore (Mahlapuu, Håkansson, Ringstad, & Björn, 2016). It has been suggested that AMP-promoted membrane permeabilization leads to membrane rupture and lysis, a result of ion leakage, followed by transmembrane depolarization and dysfunction (Eckert, 2011). The formation of pores can also be relevant as a mechanism through which some AMPs gain access to the cytoplasm, where they target processes involving DNA/RNA and protein synthesis, protein folding and several enzymatic activities (Mahlapuu et al., 2016).

Cathelicidins are part of a large group of peptides (AMPs). These peptides are expressed by many immune cells, genital cells, and epithelial cells of the intestine, airway, skin, and urinary tract. Cathelicidins have been shown to be present in many autoimmune diseases, such as systemic lupus erythematous (the most common form of lupus), arthritis, psoriasis, atherosclerosis and type 1 diabetes. Cathelicidins have also been shown to play other roles, such as in immune cell activation, angiogenesis and chemoattraction, which are relevant for cardiovascular conditions. One form of cathelicidin that occurs in humans is called LL-37 (as already mentioned in this chapter), while in mice and rats, there is an analog called CRAMP (cathelicidin-related antimicrobial peptide). In animal models, CRAMP levels are reduced in individuals with cardiac I/R injury, and the peptide plays a protective role against I/R injury in vivo. CRAMP is linked to cardiomyocyte apoptosis prevention via activation of the Akt and ERK1/2 pathways. The activation of these pathways is thought to increase

FoxO3a phosphorylation and nuclear export, inhibiting apoptosis (FoxO3a is a nuclear Forkhead transcription factor that mediates the transcription of proapoptotic genes and, if phosphorylated and exported from the nucleus, inhibits apoptosis). In humans, LL-37 levels are reduced in patients with myocardial infarction (Bei et al., 2019). CRAMP is also found to affect Toll-like receptors (TLRs). TLRs are related to cardiac conditions: TLR4 mediates hypertrophy, TLR7 deficiency suppresses the cardiac remodeling process induced by myocardial infarction and TLR9 is essential in the pathology of the cardiac remodeling process of a post-myocardial infarction. CRAMP increases TLR9 expression without increasing expression of either TLR4 or TLR7 (Wang et al., 2020).

5. The role of ghrelin and its impact on cardiovascular systems and conditions

Ghrelin also has an impact on the cardiovascular system and therefore on cardiovascular conditions. Ghrelin and its receptors are widely distributed in cardiovascular tissues, and studies have confirmed that the effects of ghrelin in the cardiovascular system are mediated not only via its growth hormone-releasing effect but also by its direct effects on the heart. Ghrelin has been shown to have protective effects against conditions such as coronary atherosclerotic heart disease (CAD), hypertension, cardiomyopathy, pulmonary hypertension, peripheral vascular disease and heart failure. Ghrelin can promote peripheral blood vessel dilation, coronary artery constriction, endothelial function improvement and myocardial cell apoptosis inhibition (Zhang et al., 2010). The previously mentioned study also suggests that ghrelin may exert paracrine effects in the cardiovascular system.

Coronary artery disease (CAD), also known as arteriosclerosis, is a condition that affects the arteries that supply the heart and the heart muscle (myocardium) with blood: the coronary arteries. This condition is characterized by a build-up of plaque inside the artery walls, which causes the arteries to become narrower and not only slows blood flow but also increases blood pressure. Ghrelin's role in patients with CAD falls into three major categories: regulation of atherosclerosis and providing protective effects against myocardial ischemia/reperfusion injury. The regulatory mechanism of ghrelin on atherosclerosis is not yet clear. Studies have found that ghrelin has immunoregulatory and anti-inflammatory actions, such as inhibition of proinflammatory cytokine production and mononuclear cell binding, which may play a modulatory role in the mentioned condition.

Nonetheless, endothelial dysfunction is considered one of the earliest events of atherosclerotic development, and during this process, there is a decrease in plasma ghrelin levels (Tajtakova et al., 2005). This decrease suggests a correlation between ghrelin levels and atherosclerosis. Administration of exogenous ghrelin has been shown to improve endothelial dysfunction, and as mentioned previously, its antioxidative effects as an immunoregulatory and anti-inflammatory mediator may attenuate atherosclerosis (Tesauro et al., 2005). However, elevated concentrations of ghrelin may increase endothelial cell adhesion molecule expression, which can counteract the desired effect and increase the atherosclerosis risk in subjects with Prader-Willi syndrome, type 2 diabetes and anorexia nervosa (Skilton, Nakhla, Sieveking, Caterson, & Celermajer, 2005).

Peripheral vascular disease (PVD) is a condition that causes the blood vessels (arteries and veins) surrounding the heart and brain to narrow, block or spasm. This condition can also be caused by CAD, as mentioned previously. The major factors involved in the pathogenesis of PVD include vascular calcification and the action of different vasoconstrictors. This process is attenuated by ghrelin due to its vasodilator capacity and because it inhibits angiotensin-II (a vasoconstrictor) by increasing the intracellular concentration of cAMP (Rossi, Bertone, Petricca, & Santiemma, 2007). Ghrelin has been reported to be related to a nitric oxide (NO) mechanism (Okumura et al., 2002). In summary, NO causes vasodilation by initiating a cascade of biological events that relax smooth muscle cells lining blood vessels, therefore acting as a vasodilator. The mechanism by which ghrelin interacts with this cascade of events is not yet clear.

Hypertension is a long-term medical condition characterized by elevated blood pressure in the arteries. High blood pressure usually does not cause symptoms, but because of its long-term character, hypertension is a major risk factor for coronary artery disease, heart failure, atrial fibrillation, stroke, among others. Pulmonary hypertension is a type of high blood pressure that affects the arteries similar to general hypertension but is specific to the lung arteries and those on the right side of the heart. While some studies show that hypertension is associated with low levels of ghrelin, other studies performed in obese women show a positive association between ghrelin levels and hypertension, i.e., high levels of ghrelin were found in hypertensive subjects (Öner-Iyidoğan et al., 2007; Poykko et al., 2003). Ghrelin can be used to counteract cases of hypertension by decreasing blood pressure by acting as a vasodilator and suppressing oxidative stress (Kawczynska-Drozdz et al., 2006; Nagaya et al., 2001), especially in cases of endothelial dysfunction.

The exact blood pressure regulatory mechanism used by ghrelin is unknown but may involve central and peripheral pathways. In the brain, ghrelin has been demonstrated to suppress sympathetic activity and decrease blood pressure.

Cardiomyopathy is a group of conditions that affect the heart muscle and, consequently, its ability to pump blood and effectively distribute blood through all tissues. The types of cardiomyopathy include hypertrophic cardiomyopathy (usually inherited), dilated cardiomyopathy, restrictive cardiomyopathy, arrhythmogenic right ventricular dysplasia, and takotsubo cardiomyopathy. Ghrelin increases the size of cardiomyocytes, prolongs their survival, and protects them against apoptosis and myocardial injury induced by endoplasmic reticulum stress (ERS) (Popovic et al., 2004). In patients with heart failure, ghrelin can inhibit cardiomyocyte apoptosis and inhibit the angiotensin-II receptor. Reports indicate that patients with heart failure present elevated levels of ghrelin to increase appetite and weight, but other reports show the opposite results. Administration of ghrelin lowers lactate dehydrogenase and myoglobin levels produced by cardiomyocytes, improving coronary flow, heart rate, left ventricular systolic pressure, and left ventricular end-diastolic pressure. Administration of ghrelin may also improve left ventricular dysfunction (Chang et al., 2004). Ghrelin has a protective effect on myocardial I/R injury by reducing myocardial release of lactate dehydrogenase, which in turn, improves heart rate and left ventricular systolic and diastolic pressure (Chang et al., 2004). In summary, there is an association between ghrelin and various cardiovascular conditions and some common processes, such as oxidative stress and inflammation, although the exact mechanisms are not fully understood.

6. Role of adipokines in cardiovascular diseases

6.1 Leptin

As mentioned previously, leptin acts via LepRb receptors and is inhibited by cytokine signaling 3 receptor (SOCS3). Via this receptor, leptin modulates pathways such as the JAK/STAT, AMPK, PI3K-Akt and MAPK pathways (Frühbeck, 2006; Ren, 2004; Yamauchi et al., 2002; Zabeau et al., 2003). These pathways, as well as their receptors, are present in cardiovascular cells, demonstrating their relevance in cardiovascular diseases. The mentioned pathways are involved in the reperfusion injury salvation kinase (RISK) pathway and protect the myocardium against I/R injury via mechanical and chemical pathways (Hausenloy & Yellon, 2006).

Leptin has been shown to upregulate various mediators of inflammation, such as TNF-α, IL2, IL6, MCP-1, ROS, Th1-type cytokines from ECs and peripheral blood mononuclear cells, which play an important role in vascular inflammation and endothelial cell dysfunction (Bouloumié, Marumo, Lafontan, & Busse, 1999). Although some studies show an association between leptin and blood pressure, suggesting that the peptide induces hypertension, other studies suggest that leptin is a potent vasodilator in patients with coronary artery disease (Bełtowski, 2006; Momin et al., 2006). Leptin has beneficial actions, such as reducing cardiac lipotoxicity (Lee et al., 2004). Lipotoxicity is a metabolic syndrome that results from the accumulation of lipid intermediates in nonadipose tissue, leading to cellular dysfunction and apoptosis. In other words, leptin is related to decreased cardiomyocyte apoptosis. Leptin also causes hypertrophy and prevents the proliferation of cardiomyocytes via activation of pathways such as the PI3K-Akt and p38 pathways, and p44/42 MAPK is common in both pathways and plays a role in promoting cell growth and has a protective effect against I/R injury (Rajapurohitam et al., 2003; Tajmir et al., 2004).

In summary, leptin, through the ORB-b receptor, stimulates vascular and cardiac signaling (stimulating pathways such as the JAK/STAT, PI3K/Akt and MAPK pathways (38 and 44/42)). Vascular signaling contributes to increased vascular hypertrophy, blood pressure, vasoconstriction, oxidative stress, angiogenesis, inflammation and apoptosis. Cardiac signaling stimulates increases in SNS activity and left ventricular hypertrophy (Table 1 and Fig. 2).

6.2 Adiponectin

As mentioned previously, the actions of adiponectin are mediated via the receptors AdipoR1, AdipoR2 and T-cadherin, and all of these receptors are present in cardiac tissue. Some of the processes that follow the stimulation of these receptors include modulation of AMPK, PI3K, p38 and p44/42 MAPK and cyclooxygenase-2 (COX-2) (Shibata et al., 2004; Yamauchi et al., 2003). These processes are involved, for example, in inflammatory responses. Adiponectin is associated with some cardiovascular diseases because it is suggested that it promotes angiogenesis (promoting crosstalk between Akt and AMPK), suppresses monocyte adhesion to arteries and is involved in the origin of atherosclerosis through the actions of AMPK, PI3K-Akt and the enzyme NO synthase located in endothelial cells (eNOS) (Shimada et al., 2004). Atk is a protein kinase that can act as a second messenger in cell proliferation and apoptosis, and its stimulation is linked to reduced apoptosis. Taking a closer look at each topic mentioned previously, adiponectin regulates eNOS, an enzyme crucial for the production of nitrous oxide (NO), via the PI3K and AMPK pathways. NO is crucial for vasoconstriction and vasodilation, growth of smooth muscle cells, VSMC proliferation, among other processes (Arita et al., 2002; Chen et al., 2003). This regulation process can occur via adiponectin levels or via knockdown of its receptors, which decreases NO production. The activation of COX-2 and AMPK, as well as the suppression of TNF-α, leads to a decrease in NO production. Reducing NO levels has an antiatherogenic effect, reduces VMCS proliferation, reduces myocardial damage and myocardial I/R injury and reduces hypertension.

In summary, adiponectin, via its three receptors, promotes vascular and cardiac signaling by stimulating the AMPK and COX-2 pathways and eNOS, specifically in vascular signaling. These processes lead to decreases in vascular and cardiac hypertrophy, inflammation, apoptosis, and oxidative stress, as well as an increase in vasodilation (Table 1 and Fig. 2).

6.3 Resistin

Resistin receptors have yet to be identified, but it is known that resistin overexpression is associated with insulin resistance, dyslipidemia and obesity (Kusminski et al., 2005; Rajala et al., 2004; Sato et al., 2005). High plasma levels of resistin correlate with proatherogenic inflammatory markers, an increased cardiovascular risk, angina and coronary artery disease (Kunnari et al., 2006; Lubos et al., 2007; Norata et al., 2007). The proinflammatory role of resistin in atherosclerosis is supported by its role in human vascular ECs. Resistin induces endothelin-1 (ET-1) release, leading to increased expression of several proinflammatory cytokines, including IL1, IL6, IL12 and TNF-α, in an NF-κB-mediated fashion (Milan et al., 2002; Silswal et al., 2005). Resistin also stimulates the expression of the adhesion molecule vascular cell adhesion molecule 1 (VCAM-1) and intercellular adhesion molecule 1 (ICAM-1), which play a role in atherosclerosis. Resistin promotes the proliferation of human aortic smooth muscle cells via PI3-Akt and p44/42 signaling and increases the migration and proliferation of ECs derived from human coronary arteries via stimulation of p38 and p44/42 signaling (Calabro et al., 2004; Mu et al., 2006). Resistin induces angiogenesis via phosphatidylinositol 3-kinase (PI3K)/AKT phosphorylation and NF-κB and β-catenin activation. Resistin also plays a protective role against I/R injury via PI3K and PKC activation (Hinshaw, 2012).

In summary, resistin, via an unknown receptor (although it is suggested that it may act via the TLR4 and/or CAP-1 receptors), promotes the expression of adhesion molecules VACM-1 and ICAM-1; increases hypertension, aortic SMC and coronary EC proliferation; and promotes the production of proinflammatory cytokines. It is also suggested that resistin has a protective effect against I/R injury (Table 1 and Fig. 2).

7. Conclusion

In the coming years, it will be of interest to further study resistin. Specifically, we aim to identify the specific receptor of resistin and to further understand its role in and impact on conditions such as CAD and other cardiovascular diseases. While there is considerable information about adiponectin and leptin, as well as the link between them, there is still a lack of information regarding resistin, the last adipokine discussed in this chapter.

Some of the specific impacts of the three mentioned adipokines are presented in Table 1, but some of these impacts have only been tested in animal models (nonhumans) and/or in animal cells. It will be crucial to translate the results of animal experiments into humans to further validate their roles as antimicrobial agents and in cardiovascular diseases.

In this chapter, we described the physiological role of ghrelin, the "hunger hormone." As intended, we collected and presented information regarding its role in metabolism and as an antimicrobial agent and of its impact on cardiovascular conditions, such as CAD. We linked ghrelin with adipokines and explored three different adipokines: leptin, adiponectin and resistin. These peptides were described, as well as their role in and impact on biological events, with a focus on cardiovascular conditions and the underlying pathways and signaling cascades. While some information about the role and mechanisms of these peptides is already well established and has been confirmed, many factors, such as resistin's receptor and some signaling cascades, have yet to be described and studied. Nonetheless, this chapter serves as a compendium to elucidate what is already known about these peptides, linking several articles and studies together to present a comprehensive picture (Fig. 2) and to provide an idea of what is still missing and what the next steps should be.

Acknowledgments

Pablo Mohaupt and Pr. Dominique Deville de Périère (Plateforme de Protéomqiue Clinique, IRMB, CHU, France).

This work was supported by Portuguese Foundation for Science and Technology (FCT), European Union, QREN, FEDER, and COMPETE for funding the Unidade de Investigação Cardiovascular (UID/IC/00051/2019), iBiMED (UIDB/04501/2020, POCI-01-0145-FEDER-007628), LAQV/REQUIMTE (UIDB/50006/2020), and RNEM, Portuguese Mass Spectrometry Network, (LISBOA-01-0145-FEDER-402-022125) research units and the research project R.V. are supported by individual fellowship grants (IF/00286/2015).

Conflicts of interest

The authors declare no conflicts of interest.

References

Achari, A. E., & Jain, S. K. (2017). Adiponectin, a therapeutic target for obesity, diabetes, and endothelial dysfunction. *International Journal of Molecular Sciences, 18*(6), 1321. https://doi.org/10.3390/ijms18061321.

Arita, Y., Kihara, S., Ouchi, N., Maeda, K., Kuriyama, H., Okamoto, Y., et al. (2002). Adipocyte-derived plasma protein adiponectin acts as a platelet-derived growth factor-BB-binding protein and regulates growth factor-induced common postreceptor signal in vascular smooth muscle cell. *Circulation, 105*(24), 2893–2898. https://doi.org/10.1161/01.CIR.0000018622.84402.FF.

Aydin, S., Halifeoglu, I., Ozercan, I. H., Erman, F., Kilic, N., Aydin, S., et al. (2005). A comparison of leptin and ghrelin levels in plasma and saliva of young healthy subjects. *Peptides, 26*(4), 647–652. https://doi.org/10.1016/j.peptides.2004.11.008.

Barazzoni, R., Zanetti, M., Ferreira, C., Vinci, P., Pirulli, A., Mucci, M. P., et al. (2007). Relationships between desacylated and acylated ghrelin and insulin sensitivity in the metabolic syndrome. *Journal of Clinical Endocrinology and Metabolism, 92*(10), 3935–3940. https://doi.org/10.1210/jc.2006-2527.

Bei, Y., Pan, L. L., Zhou, Q., Zhao, C., Xie, Y., Wu, C., et al. (2019). Cathelicidin-related antimicrobial peptide protects against myocardial ischemia/reperfusion injury. *BMC Medicine, 17*(1), 1–20. https://doi.org/10.1186/s12916-019-1268-y.

Bellone, S., Rapa, A., Vivenza, D., Vercellotti, A., Petri, A., Radetti, G., et al. (2004). Circulating ghrelin levels in the newborn are positively associated with gestational age. *Clinical Endocrinology, 60*(5), 613–617. https://doi.org/10.1111/j.1365-2265.2004.02014.x.

Bełtowski, J. (2006). Role of leptin in blood pressure regulation and arterial hypertension. *Journal of Hypertension, 24*(5), 789–801. https://doi.org/10.1097/01.hjh.0000222743.06584.66.

Blüher, M. (2012). Clinical relevance of adipokines. *Diabetes and Metabolism Journal, 36*(5), 317–327. https://doi.org/10.4093/dmj.2012.36.5.317.

Bouloumié, A., Marumo, T., Lafontan, M., & Busse, R. (1999). Leptin induces oxidative stress in human endothelial cells. *The FASEB Journal, 13*(10), 1231–1238. https://doi.org/10.1096/fasebj.13.10.1231.

Broglio, F., Arvat, E., Benso, A., Gottero, C., Muccioli, G., Papotti, M., et al. (2001). Ghrelin, a natural GH secretagogue produced by the stomach, induces hyperglycemia and reduces insulin secretion in humans. *The Journal of Clinical Endocrinology and Metabolism, 86*(10), 5083–5086. https://doi.org/10.1210/jcem.86.10.8098.

Calabro, P., Samudio, I., Willerson, J. T., & Yeh, E. T. H. (2004). Resistin promotes smooth muscle cell proliferation through activation of extracellular signal-regulated kinase 1/2

and phosphatidylinositol 3-kinase pathways. *Circulation, 110*(21), 3335–3340. https://doi.org/10.1161/01.CIR.0000147825.97879.E7.
Chang, L., Ren, Y., Liu, X., Li, W. G., Yang, J., Geng, B., et al. (2004). Protective effects of ghrelin on ischemia/reperfusion injury in the isolated rat heart. *Journal of Cardiovascular Pharmacology, 43*(2), 165–170. https://doi.org/10.1097/00005344-200402000-00001.
Chen, M. M., Ashley, E. A., Deng, D. X. F., Tsalenko, A., Deng, A., Tabibiazar, R., et al. (2003). Novel role for the potent endogenous inotrope apelin in human cardiac dysfunction. *Circulation, 108*(12), 1432–1439. https://doi.org/10.1161/01.CIR.0000091235.94914.75.
Chen, J., Tan, B., Karteris, E., Zervou, S., Digby, J., Hillhouse, E. W., et al. (2006). Secretion of adiponectin by human placenta: Differential modulation of adiponectin and its receptors by cytokines. *Diabetologia, 49*(6), 1292–1302. https://doi.org/10.1007/s00125-006-0194-7.
Codoñer-Franch, P., & Alonso-Iglesias, E. (2015). Resistin: Insulin resistance to malignancy. *Clinica Chimica Acta, 438*, 46–54. https://doi.org/10.1016/j.cca.2014.07.043.
Coiro, V., Casti, A., Volta, E., Melani, A., Rubino, P., Saccani-Jotti, G., et al. (2010). Effect of physical training on reduction of circulating neuropeptide Y levels in elderly humans. *Journal of Endocrinological Investigation, 33*(2), 132–133. https://doi.org/10.3275/6475.
De Vriese, C., Gregoire, F., Lema-Kisoka, R., Waelbroeck, M., Robberecht, P., & Delporte, C. (2004). Ghrelin degradation by serum and tissue homogenates: Identification of the cleavage sites. *Endocrinology, 145*(11), 4997–5005. https://doi.org/10.1210/en.2004-0569.
Doyle, D. D., Goings, G. E., Upshaw-Earley, J., Page, E., Ranscht, B., & Palfrey, H. C. (1998). T-cadherin is a major glycophosphoinositol-anchored protein associated with noncaveolar detergent-insoluble domains of the cardiac sarcolemma. *Journal of Biological Chemistry, 273*(12), 6937–6943. https://doi.org/10.1074/jbc.273.12.6937.
Eckert, R. (2011). Road to clinical efficacy: Challenges and novel strategies for antimicrobial peptide development. *Future Microbiology, 6*(6), 635–651. https://doi.org/10.2217/fmb.11.27.
Fain, J. N., Bahouth, S. W., & Madan, A. K. (2004). TNFα release by the nonfat cells of human adipose tissue. *International Journal of Obesity, 28*(4), 616–622. https://doi.org/10.1038/sj.ijo.0802594.
Filková, M., Haluzík, M., Gay, S., & Šenolt, L. (2009). The role of resistin as a regulator of inflammation: Implications for various human pathologies. *Clinical Immunology, 133*(2), 157–170. https://doi.org/10.1016/j.clim.2009.07.013.
Frühbeck, G. (2006). Intracellular signalling pathways activated by leptin. *Biochemical Journal, 393*(1), 7–20. https://doi.org/10.1042/BJ20051578.
Fukushima, N., Hanada, R., Teranishi, H., Fukue, Y., Tachibana, T., Ishikawa, H., et al. (2005). Ghrelin directly regulates bone formation. *Journal of Bone and Mineral Research, 20*(5), 790–798. https://doi.org/10.1359/JBMR.041237.
Furugen, R., Hayashida, H., Kitamura, M., & Saito, T. (2010). Relationship between adipokines and periodontitis. *Japanese Dental Science Review, 46*(2), 159–164. https://doi.org/10.1016/j.jdsr.2010.01.001.
Gao, J., Chang Chua, C., Chen, Z., Wang, H., Xu, X., Hamdy, R. C., et al. (2007). Resistin, an adipocytokine, offers protection against acute myocardial infarction. *Journal of Molecular and Cellular Cardiology, 43*(5), 601–609. https://doi.org/10.1016/j.yjmcc.2007.08.009.
Ghelardoni, S., Carnicelli, V., Frascarelli, S., Ronca-Testoni, S., & Zucchi, R. (2006). Ghrelin tissue distribution: Comparison between gene and protein expression. *Journal of Endocrinological Investigation, 29*(2), 115–121. https://doi.org/10.1007/BF03344083.
Gong, D. W., Bi, S., Pratley, R. E., & Weintraub, B. D. (1996). Genomic structure and promoter analysis of the human obese gene. *Journal of Biological Chemistry, 271*(8), 3971–3974. https://doi.org/10.1074/jbc.271.8.3971.

Graveleau, C., Zaha, V. G., Mohajer, A., Banerjee, R. R., Dudley-Rucker, N., Steppan, C. M., et al. (2005). Mouse and human resistins impair glucose transport in primary mouse cardiomyocytes, and oligomerization is required for this biological action. *Journal of Biological Chemistry*, *280*(36), 31679–31685. https://doi.org/10.1074/jbc.M504008200.

Gualillo, O., Caminos, J., Blanco, M., Garcìa-Caballero, T., Kojima, M., Kangawa, K., et al. (2001). Ghrelin, a novel placental-derived hormone. *Endocrinology*, *142*(2), 788–794. https://doi.org/10.1210/endo.142.2.7987.

Hancock, R. E. W., & Chapple, D. S. (1999). Peptide antibiotics. *Antimicrobial Agents and Chemotherapy*, *43*(6), 1317–1323. https://doi.org/10.1128/aac.43.6.1317.

Hancock, R. E. W., & Sahl, H. G. (2006). Antimicrobial and host-defense peptides as new anti-infective therapeutic strategies. *Nature Biotechnology*, *24*(12), 1551–1557. https://doi.org/10.1038/nbt1267.

Hassan, M., Kjos, M., Nes, I. F., Diep, D. B., & Lotfipour, F. (2012). Natural antimicrobial peptides from bacteria: Characteristics and potential applications to fight against antibiotic resistance. *Journal of Applied Microbiology*, *113*(4), 723–736. https://doi.org/10.1111/j.1365-2672.2012.05338.x.

Hausenloy, D. J., & Yellon, D. M. (2006). Survival kinases in ischemic preconditioning and postconditioning. *Cardiovascular Research*, *70*(2), 240–253. https://doi.org/10.1016/j.cardiores.2006.01.017.

Heldsinger, A., Grabauskas, G., Wu, X., Zhou, S., Lu, Y., Song, I., et al. (2014). Ghrelin induces leptin resistance by activation of suppressor of cytokine signaling 3 expression in male rats: Implications in satiety regulation. *Endocrinology*, *155*(10), 3956–3969. https://doi.org/10.1210/en.2013-2095.

Hinshaw, J. E. (2012). Neuro-cognitive mechanism of imagining future and its role in suicide. In *Vol. 16. Annual review of cellular and developmental biology*, Issue January. https://doi.org/10.16128/j.cnki.1005- 3611.2012.02.007.

Hopkins, T. A., Ouchi, N., Shibata, R., & Walsh, K. (2007). Adiponectin actions in the cardiovascular system. *Cardiovascular Research*, *74*(1), 11–18. https://doi.org/10.1016/j.cardiores.2006.10.009.

Hug, C., Wang, J., Ahmad, N. S., Bogan, J. S., Tsao, T. S., & Lodish, H. F. (2004). T-cadherin is a receptor for hexameric and high-molecular-weight forms of Acrp30/adiponectin. *Proceedings of the National Academy of Sciences of the United States of America*, *101*(28), 10308–10313. https://doi.org/10.1073/pnas.0403382101.

Jamaluddin, M. S., Weakley, S. M., Yao, Q., & Chen, C. (2012). Resistin: Functional roles and therapeutic considerations for cardiovascular disease. *British Journal of Pharmacology*, *165*(3), 622–632. https://doi.org/10.1111/j.1476-5381.2011.01369.x.

Jorgensen, S. B., O'Neill, H. M., Sylow, L., Honeyman, J., Hewitt, K. A., Palanivel, R., et al. (2013). Deletion of skeletal muscle SOCS3 prevents insulin resistance in obesity. *Diabetes*, *62*(1), 56–64. https://doi.org/10.2337/db12-0443.

Kahlenberg, J. M., & Kaplan, M. J. (2013). Little peptide, big effects: The role of LL-37 in inflammation and autoimmune disease. *The Journal of Immunology*, *191*(10), 4895–4901. https://doi.org/10.4049/jimmunol.1302005.

Kawanami, D., Maemura, K., Takeda, N., Harada, T., Nojiri, T., Imai, Y., et al. (2004). Direct reciprocal effects of resistin and adiponectin on vascular endothelial cells: A new insight into adipocytokine-endothelial cell interactions. *Biochemical and Biophysical Research Communications*, *314*(2), 415–419. https://doi.org/10.1016/j.bbrc.2003.12.104.

Kawczynska-Drozdz, A., Olszanecki, R., Jawien, J., Brzozowski, T., Pawlik, W. W., Korbut, R., et al. (2006). Ghrelin inhibits vascular superoxide production in spontaneously hypertensive rats. *American Journal of Hypertension*, *19*(7), 764–767. https://doi.org/10.1016/j.amjhyper.2006.01.022.

Khan, M. Z., Al-Goblan, A. S., & Al-Alfi, M. A. (2014). Mechanism linking diabetes mellitus and obesity. *Diabetes, Metabolic Syndrome and Obesity: Targets and Therapy*, 7, 587–591. https://doi.org/10.2147/DMSO.S67400.

Kim, J., Wall, E., Van De Laplante, M., Azzara, A., Trujillo, M. E., Hofmann, S. M., et al. (2007). Kim & Scherer 2007, JCI. Expansión TA y sensibilización ins.pdf. *Journal of Clinical Investigation*, *117*(9), 2621–2637. https://doi.org/10.1172/JCI31021.in.

Klok, M. D., Jakobsdottir, S., & Drent, M. L. (2007). The role of leptin and ghrelin in the regulation of food intake and body weight in humans: A review. *Obesity Reviews*, *8*(1), 21–34. https://doi.org/10.1111/j.1467-789X.2006.00270.x.

Knerr, I., Herzog, D., Rauh, M., Rascher, W., & Horbach, T. (2006). Leptin and ghrelin expression in adipose tissues and serum levels in gastric banding patients. *European Journal of Clinical Investigation*, *36*(6), 389–394. https://doi.org/10.1111/j.1365-2362.2006.01642.x.

Kobayashi, H. (2004). Selective suppression of endothelial cell apoptosis by the high molecular weight form of adiponectin. *Circulation Research*, *94*(4), 27e–31. https://doi.org/10.1161/01.res.0000119921.86460.37.

Kojima, M., & Kangawa, K. (2005). Ghrelin: Structure and function. *Physiological Reviews*, *85*(2), 495–522. https://doi.org/10.1152/physrev.00012.2004.

Komatsuzawa, H., Ouhara, K., Kawai, T., Yamada, S., Fujiwara, T., Shiba, H., et al. (2007). Susceptibility of periodontopathogenic and cariogenic bacteria to defensins and potential therapeutic use of defensins in oral diseases. *Current Pharmaceutical Design*, *13*(30), 3084–3095. https://doi.org/10.2174/138161207782110426.

Kong, Y., Zhang, S., Wu, R., Su, X., Peng, D., Zhao, M., et al. (2019). New insights into different adipokines in linking the pathophysiology of obesity and psoriasis. *Lipids in Health and Disease*, *18*(1), 1–12. https://doi.org/10.1186/s12944-019-1115-3.

Kopitar-Jerala, N. (2017). The role of interferons in inflammation and inflammasome activation. *Frontiers in Immunology*, *8*(JUL). https://doi.org/10.3389/fimmu.2017.00873.

Korbonits, M., Bustin, S. A., Kojima, M., Jordan, S., Adams, E. F., Lowe, D. G., et al. (2001). The expression of the growth hormone secretagogue receptor ligand ghrelin in normal and abnormal human pituitary and other neuroendocrine tumors. *Journal of Clinical Endocrinology and Metabolism*, *86*(2), 881–887. https://doi.org/10.1210/jc.86.2.881.

Kunnari, A., Ukkola, O., Päivänsalo, M., & Kesäniemi, Y. A. (2006). High plasma resistin level is associated with enhanced highly sensitive C-reactive protein and leukocytes. *Journal of Clinical Endocrinology and Metabolism*, *91*(7), 2755–2760. https://doi.org/10.1210/jc.2005-2115.

Kusminski, C. M., McTernan, P. G., & Kumar, S. (2005). Role of resistin in obesity, insulin resistance and type II diabetes. *Clinical Science*, *109*(3), 243–256. https://doi.org/10.1042/CS20050078.

Lee, Y., Naseem, R. H., Duplomb, L., Park, B. H., Carry, D. J., Richardson, J. A., et al. (2004). Hyperleptinemia prevents lipotoxic cardiomyopathy in acyl CoA synthase transgenic mice. *Proceedings of the National Academy of Sciences of the United States of America*, *101*(37), 13624–13629. https://doi.org/10.1073/pnas.0405499101.

Lim, C. T., Kola, B., & Korbonits, M. (2011). The ghrelin/GOAT/GHS-R system and energy metabolism. *Reviews in Endocrine & Metabolic Disorders*, *12*(3), 173–186. https://doi.org/10.1007/s11154-011-9169-1.

Lubos, E., Messow, C. M., Schnabel, R., Rupprecht, H. J., Espinola-Klein, C., Bickel, C., et al. (2007). Resistin, acute coronary syndrome and prognosis results from the AtheroGene study. *Atherosclerosis*, *193*(1), 121–128. https://doi.org/10.1016/j.atherosclerosis.2006.05.039.

Mahlapuu, M., Håkansson, J., Ringstad, L., & Björn, C. (2016). Antimicrobial peptides: An emerging category of therapeutic agents. *Frontiers in Cellular and Infection Microbiology*, *6*(Dec), 1–12. https://doi.org/10.3389/fcimb.2016.00194.

Makris, M. C., Alexandrou, A., Papatsoutsos, E. G., Malietzis, G., Tsilimigras, D. I., Guerron, A. D., et al. (2017). Ghrelin and obesity: Identifying gaps and dispelling myths. A reappraisal. *In Vivo*, *31*(6), 1047–1050. https://doi.org/10.21873/invivo.11168.

McTernan, C. L., McTernan, P. G., Harte, A. L., Levick, P. L., Barnett, A. H., & Kumar, S. (2002). Resistin, central obesity, and type 2 diabetes. *Lancet*, *359*(9300), 46–47. https://doi.org/10.1016/S0140-6736(02)07281-1.

Milan, G., Granzotto, M., Scarda, A., Calcagno, A., Pagano, C., Federspil, G., et al. (2002). Resistin and adiponectin expression in visceral fat of obese rats: Effect of weight loss. *Obesity Research*, *10*(11), 1095–1103. https://doi.org/10.1038/oby.2002.149.

Minn, A. H., Patterson, N. B., Pack, S., Hoffmann, S. C., Gavrilova, O., Vinson, C., et al. (2003). Resistin is expressed in pancreatic islets. *Biochemical and Biophysical Research Communications*, *310*(2), 641–645. https://doi.org/10.1016/j.bbrc.2003.09.061.

Momin, A. U., Melikian, N., Shah, A. M., Grieve, D. J., Wheatcroft, S. B., John, L., et al. (2006). Leptin is an endothelial-independent vasodilator in humans with coronary artery disease: Evidence for tissue specificity of leptin resistance. *European Heart Journal*, *27*(19), 2294–2299. https://doi.org/10.1093/eurheartj/ehi831.

Mori, K., Yoshimoto, A., Takaya, K., Hosoda, K., Ariyasu, H., Yahata, K., et al. (2000). Kidney produces a novel acylated peptide, ghrelin. *FEBS Letters*, *486*(3), 213–216. https://doi.org/10.1016/S0014-5793(00)02308-5.

Mu, H., Ohashi, R., Yan, S., Chai, H., Yang, H., Lin, P., et al. (2006). Adipokine resistin promotes in vitro angiogenesis of human endothelial cells. *Cardiovascular Research*, *70*(1), 146–157. https://doi.org/10.1016/j.cardiores.2006.01.015.

Mundinger, T. O., Cummings, D. E., & Taborsky, G. J. (2006). Direct stimulation of ghrelin secretion by sympathetic nerves. *Endocrinology*, *147*(6), 2893–2901. https://doi.org/10.1210/en.2005-1182.

Nagaya, N., Kojima, M., Uematsu, M., Yamagishi, M., Hosoda, H., Oya, H., et al. (2001). Hemodynamic and hormonal effects of human ghrelin in healthy volunteers. *American Journal of Physiology. Regulatory, Integrative and Comparative Physiology*, *280*(549–5), 1483–1487. https://doi.org/10.1152/ajpregu.2001.280.5.r1483.

Norata, G. D., Ongari, M., Garlaschelli, K., Raselli, S., Grigore, L., & Catapano, A. L. (2007). Plasma resistin levels correlate with determinants of the metabolic syndrome. *European Journal of Endocrinology*, *156*(2), 279–284. https://doi.org/10.1530/eje.1.02338.

Ohashi, K., Kihara, S., Ouchi, N., Kumada, M., Fujita, K., Hiuge, A., et al. (2006). Adiponectin replenishment ameliorates obesity-related hypertension. *Hypertension*, *47*(6), 1108–1116. https://doi.org/10.1161/01.hyp.0000222368.43759.a1.

Okumura, H., Nagaya, N., Enomoto, M., Nakagawa, E., Oya, H., & Kangawa, K. (2002). Vasodilatory effect of ghrelin, an endogenous peptide from the stomach. *Journal of Cardiovascular Pharmacology*, *39*(6), 779–783. https://doi.org/10.1097/00005344-200206000-00001.

Öner-Iyidoğan, Y., Koçak, H., Gürdöl, F., Öner, P., Işsever, H., & Esin, D. (2007). Circulating ghrelin levels in obese women: A possible association with hypertension. *Scandinavian Journal of Clinical and Laboratory Investigation*, *67*(5), 568–576. https://doi.org/10.1080/00365510701210186.

Ouchi, N., Kihara, S., Arita, Y., Maeda, K., Kuriyama, H., Okamoto, Y., et al. (1999). Novel modulator for endothelial adhesion molecules: Adipocyte-derived plasma protein adiponectin. *Circulation*, *100*(25), 2473–2476. https://doi.org/10.1161/01.cir.100.25.2473.

Ouchi, N., Kobayashi, H., Kihara, S., Kumada, M., Sato, K., Inoue, T., et al. (2004). Adiponectin stimulates angiogenesis by promoting cross-talk between AMP-activated protein kinase and Akt signaling in endothelial cells. *Journal of Biological Chemistry*, *279*(2), 1304–1309. https://doi.org/10.1074/jbc.m310389200.

Ouchi, N., Shibata, R., & Walsh, K. (2006). Cardioprotection by adiponectin. *Trends in Cardiovascular Medicine*, *16*(5), 141–146. https://doi.org/10.1016/j.tcm.2006.03.001.

Pan, H., Guo, J., & Su, Z. (2014). Advances in understanding the interrelations between leptin resistance and obesity. *Physiology and Behavior, 130*, 157–169. https://doi.org/10.1016/j.physbeh.2014.04.003.

Park, H. K., & Ahima, R. S. (2015). Physiology of leptin: Energy homeostasis, neuroendocrine function and metabolism. *Metabolism, Clinical and Experimental, 64*(1), 24–34. https://doi.org/10.1016/j.metabol.2014.08.004.

Paschos, P., & Paletas, K. (2009). Non alcoholic fatty liver disease and metabolic syndrome. *Hippokratia, 13*(1), 9–19.

Patel, L., Buckels, A. C., Kinghorn, I. J., Murdock, P. R., Holbrook, J. D., Plumpton, C., et al. (2003). Resistin is expressed in human macrophages and directly regulated by PPARγ activators. *Biochemical and Biophysical Research Communications, 300*(2), 472–476. https://doi.org/10.1016/S0006-291X(02)02841-3.

Pinkney, J. (2014). The role of ghrelin in metabolic regulation. *Current Opinion in Clinical Nutrition and Metabolic Care, 17*(6), 497–502. https://doi.org/10.1097/MCO.0000000000000101.

Pischon, T., Bamberger, C. M., Kratzsch, J., Zyriax, B. C., Algenstaedt, P., Boeing, H., et al. (2005). Association of plasma resistin levels with coronary heart disease in women. *Obesity Research, 13*(10), 1764–1771. https://doi.org/10.1038/oby.2005.215.

Popovic, V., Svetel, M., Djurovic, M., Petrovic, S., Doknic, M., Pekic, S., et al. (2004). Circulating and cerebrospinal fluid ghrelin and leptin: Potential role in altered body weight in Huntington's disease. *European Journal of Endocrinology, 151*(4), 451–455. https://doi.org/10.1530/eje.0.1510451.

Poykko, S., Kellokoski, E., Horkko, S., Kauma, H., Kesaniemi, Y., & Ukkola, O. (2003). Low plasma ghrelin is associated with insulin resistance. *Diebetes, 52*(10), 2546–2553.

Rajala, M. W., Obici, S., Scherer, P. E., & Rossetti, L. (2003). Adipose-derived resistin and molecule—β selectively impair. *Journal of Clinical Investigation, 111*(2), 225–230. https://doi.org/10.1172/JCI200316521.Introduction.

Rajala, M. W., Qi, Y., Patel, H. R., Takahashi, N., Banerjee, R., Pajvani, U. B., et al. (2004). Regulation of resistin expression and circulating levels in obesity, diabetes, and fasting. *Diabetes, 53*(7), 1671–1679. https://doi.org/10.2337/diabetes.53.7.1671.

Rajapurohitam, V., Gan, X. T., Kirshenbaum, L. A., & Karmazyn, M. (2003). The obesity-associated peptide leptin induces hypertrophy in neonatal rat ventricular myocytes. *Circulation Research, 93*(4), 277–279. https://doi.org/10.1161/01.RES.0000089255.37804.72.

Ren, J. (2004). Leptin and hyperleptinemia—From friend to foe for cardiovascular function. *Journal of Endocrinology, 181*(1), 1–10. https://doi.org/10.1677/joe.0.1810001.

Rojas, E., Rodríguez-Molina, D., Bolli, P., Israili, Z. H., Faría, J., Fidilio, E., et al. (2014). The role of adiponectin in endothelial dysfunction and hypertension. *Current Hypertension Reports, 16*(8), 463. https://doi.org/10.1007/s11906-014-0463-7.

Rossi, F., Bertone, C., Petricca, S., & Santiemma, V. (2007). Ghrelin inhibits angiotensin II-induced migration of human aortic endothelial cells. *Atherosclerosis, 192*(2), 291–297. https://doi.org/10.1016/j.atherosclerosis.2006.07.021.

Sato, N., Kobayashi, K., Inoguchi, T., Sonoda, N., Imamura, M., Sekiguchi, N., et al. (2005). Adenovirus-mediated high expression of resistin causes dyslipidemia in mice. *Endocrinology, 146*(1), 273–279. https://doi.org/10.1210/en.2004-0985.

Satoh, H., Nguyen, M. T. A., Miles, P. D. G., Imamura, T., Usui, I., & Olefsky, J. M. (2004). Adenovirus-mediated chronic "hyper-resistinemia" leads to in vivo insulin resistance in normal rats. *Journal of Clinical Investigation, 114*(2), 224–231. https://doi.org/10.1172/JCI20785.

Schäffler, A., Ehling, A., Neumann, E., Herfarth, H., Tarner, I., Schölmerich, J., et al. (2003). Adipocytokines in synovial fluid. *JAMA, 290*(13), 1709–1710. https://doi.org/10.1001/jama.290.13.1709-c.

Schisler, J., Lang, C. H., & Willis, M. (2016). Endocrinology of the heart in health and disease: Integrated, cellular, and molecular endocrinology of the heart. In *Endocrinology of the heart in health and disease: Integrated, cellular, and molecular endocrinology of the heart.*

Shapiro, L., & Scherer, P. E. (1998). The crystal structure of a complement-1q family protein suggests an evolutionary link to tumor necrosis factor. *Current Biology, 8*(6), 335–340. https://doi.org/10.1016/s0960-9822(98)70133-2.

Shibata, R., Ouchi, N., Ito, M., Kihara, S., Shiojima, I., Pimentel, D. R., et al. (2004). Adiponectin-mediated modulation of hypertrophic signals in the heart. *Nature Medicine, 10*(12), 1384–1389. https://doi.org/10.1038/nm1137.

Shibata, R., Ouchi, N., Ohashi, K., & Murohara, T. (2017). The role of adipokines in cardiovascular disease. *Journal of Cardiology, 70*(4), 329–334. https://doi.org/10.1016/j.jjcc.2017.02.006.

Shimada, K., Miyazaki, T., & Daida, H. (2004). Adiponectin and atherosclerotic disease. *Clinica Chimica Acta, 344*(1–2), 1–12. https://doi.org/10.1016/j.cccn.2004.02.020.

Shiver, J., Reimann, K., Lord, C., Miura, A., Khunkhun, R., Wagner, W., et al. (2002). Leptin stimulates fatty-acid oxidation by activating AMP-activated protein kinase. *Nature, 415*(January), 339–343.

Silha, J. V., Krsek, M., Skrha, J. V., Sucharda, P., Nyomba, B. L. G., & Murphy, L. J. (2003). Plasma resistin, adiponectin and leptin levels in lean and obese subjects: Correlations with insulin resistence. *European Journal of Endocrinology, 149*(4), 331–335. https://doi.org/10.1530/eje.0.1490331.

Silswal, N., Singh, A. K., Aruna, B., Mukhopadhyay, S., Ghosh, S., & Ehtesham, N. Z. (2005). Human resistin stimulates the pro-inflammatory cytokines TNF-α and IL-12 in macrophages by NF-κB-dependent pathway. *Biochemical and Biophysical Research Communications, 334*(4), 1092–1101. https://doi.org/10.1016/j.bbrc.2005.06.202.

Sinha, M. K., Opentanova, I., Ohannesian, J. P., Kolaczynski, J. W., Heiman, M. L., Hale, J., et al. (1996). Evidence of free and bound leptin in human circulation: Studies in lean and obese subjects and during short-term fasting. *Journal of Clinical Investigation, 98*(6), 1277–1282. https://doi.org/10.1172/JCI118913.

Skilton, M. R., Nakhla, S., Sieveking, D. P., Caterson, I. D., & Celermajer, D. S. (2005). Pathophysiological levels of the obesity related peptides resistin and ghrelin increase adhesion molecule expression on human vascular endothelial cells. *Clinical and Experimental Pharmacology and Physiology, 32*(10), 839–844. https://doi.org/10.1111/j.1440-1681.2005.04274.x.

Skurk, T., Alberti-Huber, C., Herder, C., & Hauner, H. (2007). Relationship between adipocyte size and adipokine expression and secretion. *Journal of Clinical Endocrinology and Metabolism, 92*(3), 1023–1033. https://doi.org/10.1210/jc.2006-1055.

Steppan, C. M., & Lazar, M. A. (2002). Resistin and obesity-associated insulin resistance. *Trends in Endocrinology and Metabolism, 13*(1), 18–23. https://doi.org/10.1016/S1043-2760(01)00522-7.

Stern, J. H., Rutkowski, J. M., & Scherer, P. E. (2016). Adiponectin, leptin, and fatty acids in the maintenance of metabolic homeostasis through adipose tissue crosstalk. *Cell Metabolism, 23*(5), 770–784. https://doi.org/10.1016/j.cmet.2016.04.011.

Tajmir, P., Ceddia, R. B., Li, R. K., Coe, I. R., & Sweeney, G. (2004). Leptin increases cardiomyocyte hyperplasia via extracellular signal-regulated kinase- and phosphatidylinositol 3-kinase-dependent signaling pathways. *Endocrinology, 145*(4), 1550–1555. https://doi.org/10.1210/en.2003-1128.

Tajtakova, M., Petrovicova, J., Spurny, P., Pytliak, M., Petrasova, D., & Zemberova, E. (2005). Selected hormones levels in individuals with endothelial dysfunction and insulin resistance. *Bratislavské Lekárske Listy, 106*(1), 37–40.

Takahashi, T., Saegusa, S., Sumino, H., Nakahashi, T., Iwai, K., Morimoto, S., et al. (2005). Adiponectin, T-cadherin and tumour necrosis factor-α in damaged cardiomyocytes from autopsy specimens. *Journal of International Medical Research, 33*(2), 236–244. https://doi.org/10.1177/147323000503300212.

Tarkowski, A., Bjersing, J., Shestakov, A., & Bokarewa, M. I. (2010). Resistin competes with lipopolysaccharide for binding to toll-like receptor 4. *Journal of Cellular and Molecular Medicine, 14*(6 B), 1419–1431. https://doi.org/10.1111/j.1582-4934.2009.00899.x.

Tesauro, M., Schinzari, F., Iantorno, M., Rizza, S., Melina, D., Lauro, D., et al. (2005). Ghrelin improves endothelial function in patients with metabolic syndrome. *Circulation, 112*(19), 2986–2992. https://doi.org/10.1161/CIRCULATIONAHA.105.553883.

Trovati, M., Doronzo, G., Barale, C., Vaccheris, C., Russo, I., & Cavalot, F. (2014). Leptin and vascular smooth muscle cells. *Current Pharmaceutical Design, 20*(4), 625–634. https://doi.org/10.2174/13816128113199990022.

Wang, X., Chen, L., Zhao, X., Xiao, L., Yi, S., Kong, Y., et al. (2020). A cathelicidin-related antimicrobial peptide suppresses cardiac hypertrophy induced by pressure overload by regulating IGFR1/PI3K/AKT and TLR9/AMPKα. *Cell Death & Disease, 11*(2), 6. https://doi.org/10.1038/s41419-020-2296-4.

Wang, M.-Y., Koyama, K., Shimabukuro, M., Mangelsdorf, D., Newgard, C. B., & Unger, R. H. (1998). Overexpression of leptin receptors in pancreatic islets of Zucker diabetic fatty rats restores GLUT-2, glucokinase, and glucose-stimulated insulin secretion. *Proceedings of the National Academy of Sciences, 95*(20), 11921–11926. https://doi.org/10.1073/pnas.95.20.11921.

Yamagishi, S., Edelstein, D., Du, X., Kaneda, Y., Guzmán, M., & Brownlee, M. (2001). Leptin induces mitochondrial superoxide production and monocyte chemoattractant protein-1 expression in aortic endothelial cells by increasing fatty acid oxidation via protein kinase A. *Journal of Biological Chemistry, 276*(27), 25096–25100. https://doi.org/10.1074/jbc.m007383200.

Yamauchi, T., Kamon, J., Minokoshi, Y., Ito, Y., Waki, H., Uchida, S., et al. (2002). Adiponectin stimulates glucose utilization and fatty-acid oxidation by activating AMP-activated protein kinase. *Nature Medicine, 8*(11), 1288–1295. https://doi.org/10.1038/nm788.

Yamauchi, T., Kamon, J., Waki, H., Imai, Y., Shimozawa, N., Hioki, K., et al. (2003). Globular adiponectin protected ob/ob mice from diabetes and ApoE-deficient mice from atherosclerosis. *Journal of Biological Chemistry, 278*(4), 2461–2468. https://doi.org/10.1074/jbc.M209033200.

Yellon, D. M., & Hausenloy, D. J. (2007). Myocardial reperfusion injury. *New England Journal of Medicine, 357*(11), 1121–1135. https://doi.org/10.1056/nejmra071667.

Zabeau, L., Lavens, D., Peelman, F., Eyckerman, S., Vandekerckhove, J., & Tavernier, J. (2003). The ins and outs of leptin receptor activation. *FEBS Letters, 546*(1), 45–50. https://doi.org/10.1016/S0014-5793(03)00440-X.

Zhang, J., & Gong, M. (2018). Review of the role of leptin in the regulation of male reproductive function. *Andrologia, 50*(4), 1–7. https://doi.org/10.1111/and.12965.

Zhang, G., Yin, X., Qi, Y., Pendyala, L., Chen, J., H, D., et al. (2010). Ghrelin and cardiovascular diseases. *Current Cardiology Reviews, 6*(1), 396–398. https://doi.org/10.2174/157340310790231662.

Zhao, T. J., Sakata, I., Li, R. L., Liang, G., Richardson, J. A., Brown, M. S., et al. (2010). Ghrelin secretion stimulated by β1-adrenergic receptors in cultured ghrelinoma cells and in fasted mice. *Proceedings of the National Academy of Sciences of the United States of America, 107*(36), 15868–15873. https://doi.org/10.1073/pnas.1011116107.

Zieba, D. A., Biernat, W., & Barć, J. (2020). Roles of leptin and resistin in metabolism, reproduction, and leptin resistance. *Domestic Animal Endocrinology, 73*, 106472. https://doi.org/10.1016/j.domaniend.2020.106472.

Zuniga, M. C., Raghuraman, G., & Zhou, W. (2018). Physiologic levels of resistin induce a shift from proliferation to apoptosis in macrophage and VSMC co-culture. *Surgery (United States), 163*(4), 906–911. https://doi.org/10.1016/j.surg.2017.10.051.

CHAPTER NINETEEN

Sex hormones and cognition in aging

Caroline Gurvich*, Jessica Le, Natalie Thomas, Elizabeth H.X. Thomas, and Jayashri Kulkarni
Monash Alfred Psychiatry Research Centre, Monash University Central Clinical School and The Alfred Hospital, Melbourne, VIC, Australia
*Corresponding author: e-mail address: caroline.gurvich@monash.edu

Contents

Abstract

Hormones of the hypothalamic-pituitary-gonadal axis that regulate reproductive function are also potent neurosteriods that have multiple effects on the development, maintenance and function of the brain. There is a growing body of evidence linking sex hormones to cognitive functioning across the lifespan. Both subjective and objective cognitive changes can occur with aging. For women, cognitive complains are commonly associated with the menopause transition—a time of significant hormone flux. Sex differences in neurodegenerative conditions associated with cognitive dysfunction, such as Alzheimer's disease and Parkinson's disease, suggest a potential link between sex hormones and cognitive decline. Evidence for the effects of hormone therapy on cognition is growing, but remains inconclusive. This chapter provides an overview of sex hormones and cognition in association with healthy aging, including a focus on the menopause

Vitamins and Hormones, Volume 115
ISSN 0083-6729
https://doi.org/10.1016/bs.vh.2020.12.020

transition, as well as reviewing findings linking sex hormones to cognitive decline associated with Alzheimer's disease and Parkinson's disease. An overview of hormone therapy and cognition is also provided.

1. Introduction

Hormones of the hypothalamic-pituitary-gonadal (HPG) axis that regulate reproductive function also have multiple effects on the development and function of the brain (Galea, Frick, Hampson, Sohrabji, & Choleris, 2017). Animal studies and theoretical models have strongly implicated a role for sex steroid hormones in cognitive functioning (Du & Hill, 2019; Gurvich et al., 2018). The aging process in men and women is associated with significant changes in the physiological levels of sex hormones. Both subjective and objective cognitive changes also occur with aging, and there is a body of evidence associating sex hormones with age-related cognitive changes as well as cognitive decline. This chapter outlines associations between sex hormones and different aspects of cognitive aging including healthy aging in men and women, the menopausal transition in women, and the potential role of sex hormones in neurodegenerative disorders involving cognitive decline. An overview of hormone therapy in treating cognitive decline and reducing the risk of dementia is also provided.

2. Sex hormones and their associations with cognition

Sex hormone production is regulated by the hypothalamus-pituitary-gonadal (HPG) axis (Meethal & Atwood, 2005). The hypothalamus in the brain secretes gonadotropin-releasing hormone (GnRH), which stimulates the anterior pituitary to produce luteinizing hormone (LH) and follicle stimulating hormone (FSH). LH and FSH stimulate gonads to produce sex hormones, including estrogens, progesterone and androgens such as testosterone. These hormones can negatively feedback to the hypothalamus and pituitary to decrease GnRH, LH and FSH secretion. In addition to their role regulating reproductive functions, a substantial body of literature supports the influence of sex hormones on cognitive processes in the brain. The clearest evidence for effects of sex hormones in the CNS are the presence of receptors for estrogens, progesterone and androgens in cognitively relevant areas of the brain such as the hippocampus and the prefrontal cortex (Brinton et al., 2008; Hara, Waters, McEwen, & Morrison, 2015; Olivier, 2006).

Sex steroid hormones are lipophilic, which means that peripheral levels can pass through the blood brain barrier. Much of the literature has focused on the neuroprotective effects of estrogen in animal studies and its role in synapse formation and transmission (Barth, Villringer, & Sacher, 2015; Hara et al., 2015). Estradiol acts in the central nervous system via fast acting, non-classical membrane associated receptors in addition to classical genomic actions to modulate the synthesis, release, and metabolism of different cognitively relevant neurotransmitter systems (e.g., glutamatergic, GABAergic, serotonergic, and dopaminergic synapses) (Barth et al., 2015) and permanently alter the structure of the nervous system through a variety of mechanisms, such as myelination, neural pruning, apoptosis, and dendritic spine remodeling (Del Río et al., 2018).

Progesterone is less studied. Progesterone receptors have been identified throughout the brain, outside the hypothalamus in cortical and subcortical regions (Woolley & McEwen, 1993). Progesterone and its metabolite, allopregnanolone have neuroprotective effects, promote nerve regeneration and myelination (see review; Schumacher et al., 2014). Testosterone can be converted into estradiol by aromatase or metabolized to dihydrotestosterone (DHT) and bind to androgen receptors (Janowsky, 2006). Therefore, testosterone can affect cognition via androgen or estrogen receptors. Testosterone has been demonstrated to have neuroprotective effects against serum deprivation-induced apoptosis, oxidative stress, and soluble amyloid β (Aβ) toxicity (Pike, Carroll, Rosario, & Barron, 2009).

Hence, there are numerous neuroprotective and neuromodulatory actions of estrogens and androgens that have direct relevance to cognition.

3. Endogenous hormones and cognition in women of reproductive age

3.1 Menstrual cycle studies

The menstrual cycle provides a convenient model of hormonal flux that has garnered interest in understanding endogenous hormonal influences on cognitive functioning in reproductive aged women. The hormonal fluctuations of the menstrual cycle are regulated by the hypothalamus–pituitary–gonadal (HPG) axis a 28-day cycle. Both hormone levels then gradually fall in the late luteal phase before menses. Most research into neurocognitive performance across the menstrual cycle has been conducted with a hypothesis that stems from research into cognitive sex differences. Historically, it is claimed that men perform better in visuospatial tasks, while women demonstrate an advantage in verbal skills (Sherman, 1967). This theory was applied to the menstrual cycle by assuming that performance in these "sex dimorphic" tasks differed, at least in part, due to levels of sex hormones. The hypothesis was that women would perform better in the follicular phase (when sex hormone levels were low) on visuospatial tasks that historically have been described as male favoring tasks. In contrast, performance on verbal tasks that are suggested to favor females would be improved in the luteal phase when levels of sex hormones were higher (Sundstrom Poromaa & Gingnell, 2014). However, most studies do not appear to demonstrate significant effects of menstrual cycle phase on cognitive performance (Le, Thomas, & Gurvich, 2020; Sundstrom Poromaa & Gingnell, 2014).

3.2 Polycystic ovarian syndrome (PCOS)

PCOS is the most common endocrine disorder in women, affecting 6–10% of reproductive-aged women worldwide (Mumusoglu & Yildiz, 2020). Clinical symptoms vary and can include irregular menses, infertility, obesity, and hirsutism. This complex heterogeneous syndrome is diagnosed by having two of the following three presentations: oligo- or anovulation, hyperandrogenism and polycystic ovaries (The Rotterdam ESHRE/ASRM-Sponsored PCOS Consensus Workshop Group, 2004). Hyperandrogenemia induces increases in testosterone, androstenedione, dehydroepiandrosterone (DHEA), DHEA-S, 17-hydroxyprogersetone and estrone and there is are abnormal LH/FSH ratios (Dumitrescu, Mehedintu, Briceag, Purcarea, & Hudita, 2015). Hence, there is a growing interest in the effect of the excess sex hormones on cognition in women with PCOS.

Studies investigating cognition in women with PCOS have had mixed findings. Higher free testosterone has been correlated with better verbal psychomotor speed but not processing speed, while higher androstenedione level has been associated with worse executive functions (Franik et al., 2019), suggesting that the influence of different androgens is complex and varies for different cognitive functions. In contrast, Rees, Udiawar, Berlot, Jones, and O'Sullivan (2016) observed no significant relationship between testosterone levels and cognitive performance. Despite this, the study did observe that women with PCOS performed better across a range of tasks measuring executive functioning, episodic memory and IQ (Rees et al., 2016). Women with PCOS have also been shown to have better visuospatial cognition than controls, associated with increased testosterone (Barry, Parekh, & Hardiman, 2013), and poorer performance in manual dexterity (Schattmann & Sherwin, 2007b).

Antiandrogenic treatment has been shown to improve working memory performance (Soleman et al., 2016) and manual dexterity (Schattmann & Sherwin, 2007a), though the latter is in conflict with findings that increased estradiol enhances manual dexterity in women (Hampson, 1990; Hampson & Kimura, 1988; Phillips & Sherwin, 1992). There have also been multiple studies that have observed no significant differences on cognitive performance between women with PCOS and healthy women (Barnard, Balen, Ferriday, Tiplady, & Dye, 2007; Castellano et al., 2015; Jarrett et al., 2019; Velusami & Sivasubramanian, 2019); these inconsistent findings in literature highlight the need for further exploration in this area.

4. Sex hormones and cognition during healthy aging

In the aging process, both men and women experience changes in the level and function of many hormones (Conrad & Bimonte-Nelson, 2010). Following menopause in women, levels of estrogen and progesterone decline substantially (Burger, Hale, Robertson, & Dennerstein, 2007). In estrogens, the decline is most marked in 17β-estradiol (E2) relative to estrone (E1) (Rannevik et al., 1995). Comparatively, testosterone levels in females have a more steady decline that starts in the reproductive years, which is unrelated to menopausal changes (Davison, Bell, Donath, Montalto, & Davis, 2005). In contrast to the abrupt changes in women, men overall have a slower, gradual decline in sex hormones. Testosterone decreases about 2% a year from the third decade onwards (Feldman et al., 2002), as well as a decrease in dehydroepiandrosterone (DHEA) and increases in LH, FSH, and sex hormone-binding globulin (Araujo & Wittert, 2011).

As people age, they also typically experience a decline in certain aspects of cognitive domains as part of normal aging including gradual declines in processing speed, attention, memory, language, visuospatial and executive function abilities (Harada, Natelson Love, & Triebel, 2013). Despite considerable heterogeneity in patterns of cognitive change both across and within individuals, broad differences in cognitive trajectories have been reported between men and women– women may have a specific decline related to menopausal changes (see Section 3) and men potentially have a steeper decline in visuospatial abilities and psychomotor speed than women (McCarrey, An, Kitner-Triolo, Ferrucci, & Resnick, 2016), even though as a group, men traditionally demonstrate an advantage in these domains compared to women (Levine, Foley, Lourenco, Ehrlich, & Ratliff, 2016; Li & Singh, 2014). As both the endocrine environments and cognitive abilities appear to be temporally related during the aging process, it is possible that decreases in hormone levels are directly impacting cognitive abilities. A review of 22 cross-sectional studies of older men found that 15 studies reported positive or curvilinear associations of serum testosterone with varying aspects of cognitive function (Hogervorst, 2013), especially in those who were healthy with no comorbidities. Contrastingly, in older women, negative or no associations have been found between testosterone levels and verbal memory performance (Hogervorst, 2013; Ryan et al., 2012; Thilers, Macdonald, & Herlitz, 2006). Estrogen is not positively related to cognitive performance in men (Hogervorst, 2013), but as discussed in Section 3, it has been heavily researched regarding the changes occurring in the female menopausal transition.

5. Sex hormones and cognitive changes in menopause

Menopause is a major reproductive event in the female lifetime that significantly changes the hormone environment. As the natural or surgical cessation of ovarian hormone production (and thereby reproductive ability), it is often accompanied by physical, behavioral and cognitive changes.

Perimenopause is defined as the time immediately preceding the menopause, beginning with endocrine, biological, and clinical changes, and ending a year after the final menstrual period (Brambilla, McKinlay, & Johannes, 1994). Perimenopause has a median age of onset of 47.5 years (Gold, 2011) and is diagnosed **clinically** on the basis of varying menstrual cycle lengths, and can have accompanying "classical" menopausal symptoms such as hot flushes, sleep disturbance and vaginal dryness (Harlow et al., 2012; Soules et al., 2001). The menopausal transition is associated with labile fluctuations and the eventual decline of ovarian estrogens, progesterone, as well as rising levels of FSH (Burger et al., 2007).

Women commonly describe subjective cognitive complaints of memory loss and "clouding" and "blocking" of thinking processes during the menopause transition (Schaafsma, Homewood, & Taylor, 2010). An association has been demonstrated between subjective cognitive complaints and scores on objective tests of attention, processing speed and memory (Schaafsma et al., 2010).

Four cross sectional studies that compared neuropsychological test performance in pre-, peri- and postmenopausal women were included in a meta-analysis that concluded peri- and postmenopausal stages were associated with decreases in delayed verbal memory compared to pre-menopause.

The post-menopausal stage was associated with decreases in phonemic verbal fluency compared to the perimenopause (Weber, Maki, & McDermott, 2014). Declines in attentional processes (Schaafsma et al., 2010) and executive function (Weber, Mapstone, Staskiewicz, & Maki, 2012) have also been implicated.

Longitudinal studies examining cognitive function across the menopause transition are limited. The Study of Women's Health Across the Nation (SWAN) longitudinal study followed 2, 362 women aged between 42 and 52 for 4 years and measured menopause stage (Greendale et al., 2009) and the Kinmen Women-Health Investigation (KIWI) longitudinal cohort study followed 495 Taiwanese women between the ages of 40–54 years for 18 months, of which 23% (114 women) transitioned to perimenopause (Fuh, Wang, Lee, Lu, & Juang, 2006). Neither study revealed significant cognitive *declines* associated with perimenopause; however both studies found that perimenopause was associated with a decrement in cognitive performance that was characterized by women not being able to learn as well as they had during premenopause, i.e., a lack of improvement over time rather than an overt decline. The KIWI study observed this cognitive decrement in verbal fluency (Fuh et al., 2006) and the SWAN study in the area of processing speed and verbal episodic memory (Greendale et al., 2009). The 14-year Penn Ovarian Aging Study (POAS) examined verbal memory and psychomotor processing speed annually in 403 women, who were all premenopausal at assessment 1 and mostly post-menopausal at the final assessment. The natural transition to menopause was associated with a clinically modest but statistically significant decline on immediate and delayed verbal recall (Epperson, Sammel, & Freeman, 2013).

The effect of menopausal status and serum levels of estradiol and FSH on verbal abilities has shown to be independent from age (Berent-Spillson et al., 2012). Furthermore, a later age of menopause or longer reproductive period has been associated with improved verbal memory performance (Kuh, Cooper, Moore, Richards, & Hardy, 2018), suggesting longer lifetime exposure to endogenous estrogen may benefit cognition. While it has been suggested that menopausal symptoms such as concurrent depressive (Greendale et al., 2010), sleep or vasomotor symptoms (Weber et al., 2012) may affect cognition, a more recent study by Weber, Rubin, and Maki (2013) has shown that cognitive decline may be an independent process to other menopausal symptoms.

Surgical menopause, the removal of ovaries via bilateral oophorectomy, has been associated with cognitive impairment and an increased risk of developing Alzheimer's disease later in life (Bove et al., 2014; Phung et al., 2010; Rocca et al., 2007). The increased risk particularly afflicts younger women (who are premenopausal at the time of surgical menopause), hypothesized to be due to the abrupt hormone disruption of the surgery, particularly estrogen deprivation (Rocca, Grossardt, & Shuster, 2014). This causal link is supported by findings which show that women who commenced and continued hormone therapy after oophorectomy did not have an increased risk of Alzheimer's disease (Bove et al., 2014; Rocca et al., 2007). Furthermore, a case-control study has demonstrated that women with oophorectomy prior to the age of natural menopause (before the age of 50) had structural abnormalities in the medial temporal lobe later in life (between ages 62 and 68) that were consistent with early biomarker changes associated with Alzheimer's disease pathology (Zeydan et al., 2019).

6. Sex hormones and their role in neurodegenerative disorders

The decline in sex hormones that occurs with aging has also been implicated in the risk of neurodegenerative disorders involving cognitive dysfunction, such as Alzheimer's disease (AD) and Parkinson's disease (PD).

6.1 Alzheimer's disease

The decline in sex hormones post-menopause has been implicated as a possible reason why the incidence of AD is higher in women than men. In 2010 US census data, up to two-thirds of AD incidence were women (Hebert, Weuve, Scherr, & Evans, 2013). Furthermore, when diagnosed, women have shown greater cognitive deficits than men across memory, verbal and visuospatial domains, which was not explained by dementia severity, age or education level (Irvine, Laws, Gale, & Kondel, 2012). The higher incidence may be because women form a larger part of the aged population due to their longer lifespans. However, it has also been hypothesized that the abrupt drop in estrogen levels from menopause and thus loss of neuroprotective properties makes women more vulnerable to developing AD (Pike, 2017). Furthermore, their longer lifespan does mean they spend a longer time (about a third of their lifetime) with decreased levels of estradiol and prosterone (as compared to pre-menopause levels) (Li & Singh, 2014). Evidence from post-mortem studies also show that it may be low estrogen in the brain itself which is key to AD. One study found that post-mortem brain levels of estrogen in women with AD were lower than that of women without AD even when serum levels were comparable (Yue et al., 2005), and another replicated this finding but found this difference only in women aged 80 and over (Rosario, Chang, Head, Stanczyk, & Pike, 2011).

Estrogen has been linked with the pathophysiology of AD in animal studies. Estrogen depletion by oophorectomy increased levels of soluble β-amyloid protein, the central protein involved in AD pathogenesis, which was then reversible with estradiol treatment (Petanceska, Nagy, Frail, & Gandy, 2000). Both estrogen and progesterone have also been implicated in tau phosphorylation in the brain, which is the key protein involved in neurofibrillary tangles formation characteristic in AD pathology (Muñoz-Mayorga, Guerra-Araiza, Torner, & Morales, 2018). In men, even though the decline in sex hormones is slower and more gradual over their lifetime, decreased testosterone levels are similarly linked with increased risk of Alzheimer's disease. Most studies show that serum levels of testosterone are lower in men with AD compared to men without AD (Pike, 2017), and this is also the case with brain levels (Rosario et al., 2011). Sex hormones may also interact with genetic disposition to AD. For example, the Apolipoprotein E (APOE) epsilon4 allele is a genetic risk factor developing sporadic or late-onset AD (Neu et al., 2017). In one study of 45 men, there was a significant interaction between low testosterone levels and having the allele on general cognitive performance (Burkhardt et al., 2006). In women, those without the allele responded better to hormone treatment than those with the allele (Yaffe, Haan, Byers, Tangen, & Kuller, 2000).

Aside from the main body of research for estrogen and testosterone loss in AD, emerging evidence has also suggested that luteinizing hormone (LH) is involved in cognitive decline and AD pathogenesis. Like sex steroids, there are LH receptors in areas in cognitively relevant areas of the brain such as the hippocampus (Blair, Bhatta, McGee, & Casadesus, 2015). LH increases up to threefold in women and twofold in men in the aging process (Blair et al., 2015), and women with AD tend to have higher LH serum levels than both healthy women and women with frontotemporal dementia (Short, Bowen, O'Brien, & Graff-Radford, 2001). Increases in LH have been correlated with cognitive deficits in men (Hyde et al., 2010) and women (Rodrigues et al., 2008), as well as plasma β-amyloid levels (Verdile et al., 2008). These observational findings have also been mirrored in animal studies where cerebral amyloid levels increased with chronic LH administration (Wahjoepramono et al., 2011). Therefore, it is important to consider the impact of both sex steroid hormones and gonadotropins in the development of AD.

6.2 Parkinson's disease

Parkinson's disease (PD) is a neurodegenerative movement disorder related to loss of cells in the substantia nigra in the brain, resulting in decreased dopamine in the brain. The cognitive symptoms of Parkinson's disease in patients without dementia reflect the fronto-striatal pattern of degeneration and involve disruptions to executive functions, attention, memory, and visuospatial cognition. Gender differences in cognitive profiles are not well studied. Available evidence suggests the occurrence of cognitive impairment in PD might be greater in males, and males may show an advantage on visuospatial tasks with females showing an advantage on verbal tasks (Heller, Dogan, Schulz, & Reetz, 2014; Miller & Cronin-Golomb, 2010). Women with PD may also show a slower decline of cognitive functioning compared to men with PD (Locascio, Corkin, & Growdon, 2003). It is not clear whether these patterns of findings are disease specific or a reflection of a general gender effect that is also observed in the healthy population. It also raises the possibility of estrogen playing a neuroprotective role, independently or via estrogen's interactions with dopaminergic neurotransmitter pathways.

Similar to the risks related to AD, longer estrogen exposure during lifetime is possibly linked to decreased PD risk and milder features (Picillo et al., 2017). Following menopause, reduced estrogen has also been suggested to increase the PD risk (Picillo et al., 2017). However, contrary to AD, men are more likely than women to have PD, have an increased prevalence and risk of PD-associated cognitive impairment as well as more rapid cognitive decline once afflicted (Meoni, Macerollo, & Moro, 2020). These findings suggest a role for estrogen's neuroprotective qualities in women on dopaminergic pathways. However, other factors have been implicated for the sex differences in PD—including sex-specific biomarkers, genetics and different environmental risk factors (Meoni et al., 2020).

Some studies have examined estrogen therapy in PD in post-menopausal women with some motor benefits (e.g., Saunders-Pullman et al., 1999). A recent meta-analysis on estrogen therapy and the risk of PD showed that estrogen therapy significantly reduced the risk of PD; however cognition was not specifically examined (Song et al., 2020). In an observational study of 10,145 elderly women with PD, Fernandez and Lapane (2000) found women who were estrogen users were less likely to experience cognitive decline; however, they were not able to assess the duration of estrogen use and the concomitant administration of progesterone in each patient. Further research is required to explore estrogen's potential beneficial effects on PD related cognitive decline.

7. Hormone therapy and cognition

7.1 Estrogen therapy

While biological studies show a clear neuroprotective role for estrogen, research in exogenous hormone therapy is controversial. The Women's Health Initiative Memory Study (WHIMS), an ancillary study of the Women's Health Initiative, was the first major randomized double-blinded clinical trial to investigate the therapeutic effects of conjugate equine estrogens (CEE) alone and plus medroxyprogesterone acetate (MPA) on the risk of all-cause dementia and mild cognitive impairment in postmenopausal women (Shumaker et al., 2003). The results were unexpected—in the study of 4532 postmenopausal women, the CEE plus progestin group had an increased risk (HR = 2.05) of all-cause probable dementia and no difference in effect on mild cognitive impairment. Alzheimer's disease was the most common classification in both groups. An extension of the study also showed a decrement in global cognitive function and several specific domains including verbal memory (Espeland et al., 2010). In addition to the medical side effects found with hormone therapy in this trial (heart disease, stroke, pulmonary embolism, breast cancer), the risks of hormone therapy seemed to be too high. However, researchers suggested that timing of initiation of treatment was key to the controversial results (Maki, 2013). As women in the study were all aged 65 and over, the short-term follow-up of an average of 4 years suggested that the results were reflective of a hastening of existing neuropathology instead of primary presentation.

A "critical window" hypothesis was developed, which proposed that hormonal therapy given earlier post-menopause could be beneficial (Maki, 2013). This was supported by small clinical trials which demonstrated either beneficial or neutral effect in younger age groups (Maki, 2013). Observational studies, such as the MIRAGE study (Henderson et al., 2005), showed that women who started taking hormone therapy between ages 50–63 compared to older age brackets were the only age group to have reduced risk of AD. Earlier instead of later initiation would allegedly reduce risk according to the "Healthy Cell Bias Theory," in which estrogens are neuroprotective for healthy neurones, but once neurons start to undergo pathological changes, they hasten pathological processes and can lead to cell death (Brinton, 2008). Therefore, it is potentially the health of the women and thus the health of their neurones which is critical to the efficacy of hormonal treatment (Hogervorst, 2013). Recent major clinical trials have failed to support the critical window hypothesis. The Kronos Early Estrogen

Prevention Study (KEEPS) explored early intervention with a large sample of 693 recently postmenopausal women within 3 years of their last menstrual period randomized into three groups: for oral CEE and micronized progesterone (m-P), a transdermal estradiol patch and m-P, and placebo (Gleason et al., 2015). Using the modified mini-mental state examination (3MS) which tests global cognitive performance, there was no difference between treatment groups from 4 years of follow-up. Similarly, another extension of the original WHIMS study (WHIMS-Young/WHIMSY) found that participants ages 50–55 who had been originally assigned to hormonal therapy or placebo showed no differences (Espeland et al., 2017). The Early vs Late Intervention Trial with Estradiol (ELITE) compared 643 women into groups by length from menopause (<6 years and > 10 years) with estradiol or placebo. When compared with placebo, both early or late estradiol therapy had no effect on verbal memory, executive function or global cognitive performance (Henderson et al., 2016). These trials fail to support the timing hypothesis, which may not be crucial as previously thought. They however do not assess the effect of hormone therapy on risk of dementia, and thus WHIMS remains the only major clinical trial to do so.

Timing is not the only factor that may influence estrogen efficacy. In the prospective Nurse's Health Study of 16,514 participants, women with the APOE allele and were current hormone therapy users had the fastest rate of cognitive decline (Kang & Grodstein, 2012). This also suggests that genetic risk factors can hinder the positive effects of estrogen. Furthermore, no benefits have been demonstrated in the few clinical trials investigating estrogen as a treatment once dementia is diagnosed in a Cochrane systematic review (Hogervorst, Yaffe, Richards, & Huppert, 2009). Overall, the research in hormone therapy highlights the complexities of understanding how estrogen affects cognitive performance.

Due to the possible medical risks of typical systematic estrogen therapy, using selective estrogen receptor modulators (SERMs) have been considered for both menopausal symptoms and cognition. SERMs such as Raloxifene may be useful as they have more tissue specificity, being agonists for estrogen receptors in bone and brain tissue, but antagonists in reproductive tissue (Yang, Yu, & Zhang, 2013). In their systematic review of randomized controlled trials, Yang et al. found four studies which had mixed results (Yang et al., 2013). However, the two studies with larger samples and robust study designs suggested that raloxifene could benefit verbal memory (Jacobsen, Samson, Emmelot-Vonk, & Verhaar, 2010), or that a large dose of 120 mg/day could reduce the risk of mild cognitive impairment (Yaffe et al., 2005).

7.2 Progesterone therapy

Postmenopausal hormone therapy use by women with intact uteri consists of a combined estrogen and progesterone regimen. Evidence is emerging to suggest that the cognitive effects of hormone formulations are heavily influenced by the progestin formulation (Berent-Spillson et al., 2015). The CEE/MPA arm of the WHIMS study showed increased dementia risk, while the CEE arm had no effect (Shumaker et al., 2003). Other studies have also demonstrated more positive results if CEE or estradiol is used alone compared to estradiol or CEE combined with MPA (Maki, 2012). While it must be considered that many women allocated to unopposed estrogen are those that have had a hysterectomy and therefore do not require progesten to protect the uterus from the risk of endometrial cancer. If these women also had an oophorectomy at the time of hysterectomy this may be associated with independent cognitive effects (see Section 3). Nevertheless, it has been suggested that synthetic progestins may reduce some of the neuroprotective effects of estrogen and several studies have demonstrated that synthetic progestins, particularly MPA, are associated with reductions in verbal abilities in postmenopausal women (Coker et al., 2010; Maki & Henderson, 2012; Resnick et al., 2006). Thus, the progesterone component of hormone therapy may have a crucial cognitive effect.

Only a few small studies have examined the short-term effects of progesterone treatment on cognition in postmenopausal women. One small study in involved 24 younger (50 to 55 years) postmenopausal women who all received 12 weeks of estrogen therapy (CEE) as well as being assigned to either MPA or micronized progesterone, or CEE plus placebo. Results showed that women in the CEE + micronized progesterone group had a small decline in delayed verbal memory scores as well as a significant increase (benefit) in working memory scores (Sherwin & Grigorova, 2011). Another study of recently postmenopausal women assessed the effects of 90 days of placebo, estrogen or progesterone treatment on cognitive functioning as well as functional MRI (Berent-Spillson et al., 2015). Progesterone, but not estrogen, was associated with improved neuropsychological measures of verbal working memory compared to placebo. These findings are considered preliminary and replication in larger samples are required to consider their implications. While there are some suggestions that MPA might have a deleterious effect of cognition in older postmenopausal women, it remains unclear whether progesterone has clinically important effects on cognition (Henderson, 2018).

7.3 Testosterone therapy

Some studies have shown that testosterone treatment in postmenopausal women to raise physiological levels to premenopausal levels has some benefits on verbal learning and memory (Davis & Wahlin-Jacobsen, 2015). The addition of transdermal testosterone to oral estrogen in an open label trial improved immediate and delayed verbal memory in 61 healthy postmenopausal women (Shah et al., 2006). Similarly, an open label pilot study in nine healthy postmenopausal women on estrogen therapy demonstrated improvement in verbal learning and memory, over 26 weeks of receiving transdermal testosterone spray (Davison et al., 2011). However, the findings from a randomized-controlled trial with 200 healthy postmenopausal women randomly assigned to receive estrogen, testosterone or placebo treatment for 1 month suggest that too much testosterone compared to estrogen may decrease verbal ability, and a specific balance between estrogen and testosterone is needed to achieve optimal effects in cognitive performance (Kocoska-Maras et al., 2013).

Studies have shown improvement in memory in cognitively normal older men with low testosterone after testosterone supplementation through gel (Cherrier, Craft, & Matsumoto, 2003; Resnick et al., 2017) and intramuscularly (Vaughan, Goldstein, & Tenover, 2007) and improvement in attention and executive functioning after transdermal delivery (Kenny, Bellantonio, Gruman, Acosta, & Prestwood, 2002). However, several studies looking at performance in cognitively normal older men with low-normal testosterone have also found no effect on cognitive performance (Emmelot-Vonk et al., 2008; Haren, Wittert, Chapman, Coates, & Morley, 2005; Sih et al., 1997). Testosterone supplementation has also been shown to improve cognition in older men with Alzheimer's Disease and normal testosterone levels (Cherrier et al., 2005; Fukai, Akishita, Yamada, Toba, & Ouchi, 2010; Lu et al., 2006), though this has not been observed in men with Alzheimer's Disease and low testosterone levels (Kenny, Fabregas, Song, Biskup, & Bellantonio, 2004; Tan & Pu, 2003). The inconsistent findings in older men appear to be dependent on endogenous T level, suggesting an optimal testosterone level at which older men improve cognitive performance (Hua, Hildreth, & Pelak, 2016).

8. Conclusion

Sex hormones appear to influence the cognitive changes that occur with aging, but more research is warranted to clearly understand their role.

In the overall cognitive decline in the normal aging process, it is unclear how significant the role of sex hormones is and if the process occurs only in parallel. However, sex differences in cognitive decline may be affected by the sex differences in hormone changes. In particular, verbal memory and fluency may be specifically affected by the female menopausal transition. This unique menopausal decline in estrogen as well as the female lifetime exposure to estrogen also appears to mediate AD risk. In aging men, decreased testosterone is also related to decreased cognitive performance and AD. For PD, the sex hormone link is less explored, but the neuroprotective properties of estrogen may possible affect the dopaminergic physiology of the disease. However, the controversial research in hormone therapy demonstrates that estrogen or testosterone depletion cannot simply be the mechanisms underlying cognitive decline. This may be because the effect of sex hormones may be dependent on other factors such as physical and neurological health. The timing of hormone therapy was initially considered a crucial factor, however, most recent evidence does not appear to support the idea that timing of hormone therapy impacts cognitive outcomes. Aside from clinical trials for hormone therapy, the observational studies investigating sex hormones and cognition in aging are unable to fully consider the multifactorial nature of cognitive changes—genetics, environment, education and health likely all play a role. Further research is required to explore these complex biological mechanisms, and the effect of hormone therapy on dementia risk demands further attention.

References

Araujo, A. B., & Wittert, G. A. (2011). Endocrinology of the aging male. *Best Practice & Research. Clinical Endocrinology & Metabolism*, *25*(2), 303–319. https://doi.org/10.1016/j.beem.2010.11.004.

Barnard, L., Balen, A. H., Ferriday, D., Tiplady, B., & Dye, L. (2007). Cognitive functioning in polycystic ovary syndrome. *Psychoneuroendocrinology*, *32*(8–10), 906–914.

Barry, J. A., Parekh, H. S. K., & Hardiman, P. J. (2013). Visual-spatial cognition in women with polycystic ovarian syndrome: The role of androgens. *Human Reproduction*, *28*(10), 2832–2837. https://doi.org/10.1093/humrep/det335.

Barth, C., Villringer, A., & Sacher, J. (2015). Sex hormones affect neurotransmitters and shape the adult female brain during hormonal transition periods. *Frontiers in Neuroscience*, *9*, 37. https://doi.org/10.3389/fnins.2015.00037.

Berent-Spillson, A., Briceno, E., Pinsky, A., Simmen, A., Persad, C. C., Zubieta, J. K., et al. (2015). Distinct cognitive effects of estrogen and progesterone in menopausal women. *Psychoneuroendocrinology*, *59*, 25–36. https://doi.org/10.1016/j.psyneuen.2015.04.020.

Berent-Spillson, A., Persad, C. C., Love, T., Sowers, M., Randolph, J. F., Zubieta, J.-K., et al. (2012). Hormonal environment affects cognition independent of age during the menopause transition. *The Journal of Clinical Endocrinology and Metabolism*, *97*(9), E1686–E1694. https://doi.org/10.1210/jc.2012-1365.

Blair, J. A., Bhatta, S., McGee, H., & Casadesus, G. (2015). Luteinizing hormone: Evidence for direct action in the CNS. *Hormones and Behavior, 76*, 57–62. https://doi.org/10.1016/j.yhbeh.2015.06.020.

Bove, R., Secor, E., Chibnik, L. B., Barnes, L. L., Schneider, J. A., Bennett, D. A., et al. (2014). Age at surgical menopause influences cognitive decline and Alzheimer pathology in older women. *Neurology, 82*(3), 222–229. https://doi.org/10.1212/WNL.0000000000000033.

Brambilla, D. J., McKinlay, S. M., & Johannes, C. B. (1994). Defining the perimenopause for application in epidemiologic investigations. *American Journal of Epidemiology, 140*(12), 1091–1095. https://doi.org/10.1093/oxfordjournals.aje.a117209.

Brinton, R. D. (2008). The healthy cell bias of estrogen action: Mitochondrial bioenergetics and neurological implications. *Trends in Neurosciences, 31*(10), 529–537. https://doi.org/10.1016/j.tins.2008.07.003.

Brinton, R. D., Thompson, R. F., Foy, M. R., Baudry, M., Wang, J., Finch, C. E., et al. (2008). Progesterone receptors: Form and function in brain. *Frontiers in Neuroendocrinology, 29*(2), 313–339. https://doi.org/10.1016/j.yfrne.2008.02.001.

Burger, H. G., Hale, G. E., Robertson, D. M., & Dennerstein, L. (2007). A review of hormonal changes during the menopausal transition: Focus on findings from the Melbourne Women's midlife health project. *Human Reproduction Update, 13*(6), 559–565. https://doi.org/10.1093/humupd/dmm020.

Burkhardt, M. S., Foster, J. K., Clarnette, R. M., Chubb, S. A., Bruce, D. G., Drummond, P. D., et al. (2006). Interaction between testosterone and apolipoprotein E epsilon4 status on cognition in healthy older men. *The Journal of Clinical Endocrinology and Metabolism, 91*(3), 1168–1172. https://doi.org/10.1210/jc.2005-1072.

Castellano, C. A., Baillargeon, J. P., Nugent, S., Tremblay, S., Fortier, M., Imbeault, H., et al. (2015). Regional brain glucose hypometabolism in young women with polycystic ovary syndrome: Possible link to mild insulin resistance. *PLoS One, 10*(12), e0144116, (Electronic Resource). https://doi.org/10.1371/journal.pone.0144116.

Cherrier, M. M., Craft, S., & Matsumoto, A. H. (2003). Cognitive changes associated with supplementation of testosterone or dihydrotestosterone in mildly hypogonadal men: A preliminary report. *Journal of Andrology, 24*(4), 568–576.

Cherrier, M., Matsumoto, A., Amory, J., Asthana, S., Bremner, W., Peskind, E., et al. (2005). Testosterone improves spatial memory in men with Alzheimer disease and mild cognitive impairment. *Neurology, 64*(12), 2063–2068.

Coker, L. H., Espeland, M. A., Rapp, S. R., Legault, C., Resnick, S. M., Hogan, P., et al. (2010). Postmenopausal hormone therapy and cognitive outcomes: The Women's Health Initiative memory study (WHIMS). *The Journal of Steroid Biochemistry and Molecular Biology, 118*(4–5), 304–310. https://doi.org/10.1016/j.jsbmb.2009.11.007.

Conrad, C. D., & Bimonte-Nelson, H. A. (2010). Impact of the hypothalamic-pituitary-adrenal/gonadal axes on trajectory of age-related cognitive decline. *Progress in Brain Research, 182*, 31–76. https://doi.org/10.1016/S0079-6123(10)82002-3.

Davis, S. R., & Wahlin-Jacobsen, S. (2015). Testosterone in women—The clinical significance. *The Lancet Diabetes & Endocrinology, 3*(12), 980–992. https://doi.org/10.1016/S2213-8587(15)00284-3.

Davison, S. L., Bell, R., Donath, S., Montalto, J. G., & Davis, S. R. (2005). Androgen levels in adult females: Changes with age, menopause, and oophorectomy. *The Journal of Clinical Endocrinology and Metabolism, 90*(7), 3847–3853. https://doi.org/10.1210/jc.2005-0212.

Davison, S. L., Bell, R. J., Gavrilescu, M., Searle, K., Maruff, P., Gogos, A., et al. (2011). Testosterone improves verbal learning and memory in postmenopausal women: Results from a pilot study. *Maturitas, 70*(3), 307–311.

Del Río, J. P., Alliende, M. I., Molina, N., Serrano, F. G., Molina, S., & Vigil, P. (2018). Steroid hormones and their action in Women's brains: The importance of hormonal balance. *Frontiers in Public Health, 6*, 141. https://doi.org/10.3389/fpubh.2018.00141.

Du, X., & Hill, R. A. (2019). Hypothalamic-pituitary-gonadal axis dysfunction: An innate pathophysiology of schizophrenia? *General and Comparative Endocrinology, 275*, 38–43. https://doi.org/10.1016/j.ygcen.2019.02.009.

Dumitrescu, R., Mehedintu, C., Briceag, I., Purcarea, V. L., & Hudita, D. (2015). The polycystic ovary syndrome: an update on metabolic and hormonal mechanisms. *Journal of Medicine and Life, 8*(2), 142–145. Retrieved from https://www.ncbi.nlm.nih.gov/pubmed/25866568.

Emmelot-Vonk, M. H., Verhaar, H. J., Pour, H. R. N., Aleman, A., Lock, T. M., Bosch, J. R., et al. (2008). Effect of testosterone supplementation on functional mobility, cognition, and other parameters in older men: A randomized controlled trial. *JAMA, 299*(1), 39–52.

Epperson, C. N., Sammel, M. D., & Freeman, E. W. (2013). Menopause effects on verbal memory: Findings from a longitudinal community cohort. *The Journal of Clinical Endocrinology and Metabolism, 98*(9), 3829–3838. https://doi.org/10.1210/jc.2013-1808.

Espeland, M. A., Brunner, R. L., Hogan, P. E., Rapp, S. R., Coker, L. H., Legault, C., et al. (2010). Long-term effects of conjugated equine estrogen therapies on domain-specific cognitive function: Results from the Women's Health Initiative study of cognitive aging extension. *Journal of the American Geriatrics Society, 58*(7), 1263–1271. https://doi.org/10.1111/j.1532-5415.2010.02953.x.

Espeland, M. A., Rapp, S. R., Manson, J. E., Goveas, J. S., Shumaker, S. A., Hayden, K. M., et al. (2017). Long-term effects on cognitive trajectories of postmenopausal hormone therapy in two age Groups. *The Journals of Gerontology. Series A, Biological Sciences and Medical Sciences, 72*(6), 838–845. https://doi.org/10.1093/gerona/glw156.

Feldman, H. A., Longcope, C., Derby, C. A., Johannes, C. B., Araujo, A. B., Coviello, A. D., et al. (2002). Age trends in the level of serum testosterone and other hormones in middle-aged men: Longitudinal results from the Massachusetts male aging study. *The Journal of Clinical Endocrinology and Metabolism, 87*(2), 589–598. https://doi.org/10.1210/jcem.87.2.8201.

Fernandez, H. H., & Lapane, K. L. (2000). Estrogen use among nursing home residents with a diagnosis of Parkinson's disease. *Movement Disorders, 15*(6), 1119–1124. doi:10.1002/1531-8257(200011)15:6<1119::aid-mds1009>3.0.co;2-0.

Franik, G., Krysta, K., Witkowska, A., Dudek, A., Krzystanek, M., & Madej, P. (2019). The impact of sex hormones and metabolic markers on depressive symptoms and cognitive functioning in PCOS patients. *Gynecological Endocrinology, 35*(11), 965–969. https://doi.org/10.1080/09513590.2019.1613359.

Fuh, J. L., Wang, S. J., Lee, S. J., Lu, S. R., & Juang, K. D. (2006). A longitudinal study of cognition change during early menopausal transition in a rural community. *Maturitas, 53*(4), 447–453. https://doi.org/10.1016/j.maturitas.2005.07.009.

Fukai, S., Akishita, M., Yamada, S., Toba, K., & Ouchi, Y. (2010). Effects of testosterone in older men with mild-to-moderate cognitive impairment. *Journal of the American Geriatrics Society, 58*(7), 1419–1421.

Galea, L. A. M., Frick, K. M., Hampson, E., Sohrabji, F., & Choleris, E. (2017). Why estrogens matter for behavior and brain health. *Neuroscience and Biobehavioral Reviews, 76*(Pt B), 363–379. https://doi.org/10.1016/j.neubiorev.2016.03.024.

Gleason, C. E., Dowling, N. M., Wharton, W., Manson, J. E., Miller, V. M., Atwood, C. S., et al. (2015). Effects of hormone therapy on cognition and mood in recently postmenopausal women: Findings from the randomized, controlled KEEPS-cognitive and affective study. *PLoS Medicine, 12*(6). https://doi.org/10.1371/journal.pmed.1001833.

Gold, E. B. (2011). The timing of the age at which natural menopause occurs. *Obstetrics and Gynecology Clinics of North America, 38*(3), 425–440. https://doi.org/10.1016/j.ogc.2011.05.002.

Greendale, G. A., Huang, M. H., Wight, R. G., Seeman, T., Luetters, C., Avis, N. E., et al. (2009). Effects of the menopause transition and hormone use on cognitive performance in midlife women. *Neurology, 72*(21), 1850–1857. https://doi.org/10.1212/WNL.0b013e3181a71193.

Greendale, G. A., Wight, R. G., Huang, M.-H., Avis, N., Gold, E. B., Joffe, H., et al. (2010). Menopause-associated symptoms and cognitive performance: Results from the study of women's health across the nation. *American Journal of Epidemiology, 171*(11), 1214–1224. https://doi.org/10.1093/aje/kwq067.

Gurvich, C., Gavrilidis, E., Worsley, R., Hadaib, A., Thomas, N., & Kulkarni, J. (2018). Menstrual cycle irregularity and menopause status influence cognition in women with schizophrenia. *Psychoneuroendocrinology, 96*, 173–178. https://doi.org/10.1016/j.psyneuen.2018.06.022.

Hampson, E. (1990). Estrogen-related variations in human spatial and articulatory-motor skills. *Psychoneuroendocrinology, 15*(2), 97–111.

Hampson, E., & Kimura, D. (1988). Reciprocal effects of hormonal fluctuations on human motor and perceptual-spatial skills. *Behavioral Neuroscience, 102*(3), 456.

Hara, Y., Waters, E. M., McEwen, B. S., & Morrison, J. H. (2015). Estrogen effects on cognitive and synaptic health over the Lifecourse. *Physiological Reviews, 95*(3), 785–807. https://doi.org/10.1152/physrev.00036.2014.

Harada, C. N., Natelson Love, M. C., & Triebel, K. L. (2013). Normal cognitive aging. *Clinics in Geriatric Medicine, 29*(4), 737–752. https://doi.org/10.1016/j.cger.2013.07.002.

Haren, M. T., Wittert, G. A., Chapman, I. M., Coates, P., & Morley, J. E. (2005). Effect of oral testosterone undecanoate on visuospatial cognition, mood and quality of life in elderly men with low-normal gonadal status. *Maturitas, 50*(2), 124–133.

Harlow, S. D., Gass, M., Hall, J. E., Lobo, R., Maki, P., Rebar, R. W., et al. (2012). Executive summary of the stages of reproductive aging workshop + 10: Addressing the unfinished agenda of staging reproductive aging. *The Journal of Clinical Endocrinology and Metabolism, 97*(4), 1159–1168. https://doi.org/10.1210/jc.2011-3362.

Hebert, L. E., Weuve, J., Scherr, P. A., & Evans, D. A. (2013). Alzheimer disease in the United States (2010-2050) estimated using the 2010 census. *Neurology, 80*(19), 1778–1783. https://doi.org/10.1212/WNL.0b013e31828726f5.

Heller, J., Dogan, I., Schulz, J. B., & Reetz, K. (2014). Evidence for gender differences in cognition, emotion and quality of life in Parkinson's disease? *Aging and Disease, 5*(1), 63–75. https://doi.org/10.14366/AD.2014.050063.

Henderson, V. W. (2018). Progesterone and human cognition. *Climacteric, 21*(4), 333–340. https://doi.org/10.1080/13697137.2018.1476484.

Henderson, V. W., Benke, K. S., Green, R. C., Cupples, L. A., Farrer, L. A., & MIRAGE Study Group. (2005). Postmenopausal hormone therapy and Alzheimer's disease risk: Interaction with age. *Journal of Neurology, Neurosurgery, and Psychiatry, 76*(1), 103–105. https://doi.org/10.1136/jnnp.2003.024927.

Henderson, V. W., St John, J. A., Hodis, H. N., McCleary, C. A., Stanczyk, F. Z., Shoupe, D., et al. (2016). Cognitive effects of estradiol after menopause: A randomized trial of the timing hypothesis. *Neurology, 87*(7), 699–708. https://doi.org/10.1212/WNL.0000000000002980.

Hogervorst, E. (2013). Effects of gonadal hormones on cognitive behaviour in elderly men and women. *Journal of Neuroendocrinology, 25*(11), 1182–1195. https://doi.org/10.1111/jne.12080.

Hogervorst, E., Yaffe, K., Richards, M., & Huppert, F. A. (2009). Hormone replacement therapy to maintain cognitive function in women with dementia. *Cochrane Database of*

Systematic Reviews, *2009*(1), CD003799. https://doi.org/10.1002/14651858.CD003799.pub2. PMID: 19160224. PMCID: PMC7156885.

Hua, J. T., Hildreth, K. L., & Pelak, V. S. (2016). Effects of testosterone therapy on cognitive function in aging: A systematic review. *Cognitive and Behavioral Neurology: Official Journal of the Society for Behavioral and Cognitive Neurology*, *29*(3), 122.

Hyde, Z., Flicker, L., Almeida, O. P., McCaul, K. A., Jamrozik, K., Hankey, G. J., et al. (2010). Higher luteinizing hormone is associated with poor memory recall: The health in men study. *Journal of Alzheimer's Disease*, *19*(3), 943–951. https://doi.org/10.3233/jad-2010-1342.

Irvine, K., Laws, K. R., Gale, T. M., & Kondel, T. K. (2012). Greater cognitive deterioration in women than men with Alzheimer's disease: A meta analysis. *Journal of Clinical and Experimental Neuropsychology*, *34*(9), 989–998. https://doi.org/10.1080/13803395.2012.712676.

Jacobsen, D. E., Samson, M. M., Emmelot-Vonk, M. H., & Verhaar, H. J. J. (2010). Raloxifene improves verbal memory in late postmenopausal women: A randomized, double-blind, placebo-controlled trial. *Menopause*, *17*(2), 309–314. https://doi.org/10.1097/gme.0b013e3181bd54df.

Janowsky, J. S. (2006). Thinking with your gonads: Testosterone and cognition. *Trends in Cognitive Sciences*, *10*(2), 77–82. https://doi.org/10.1016/j.tics.2005.12.010.

Jarrett, B. Y., Vantman, N., Mergler, R. J., Brooks, E. D., Pierson, R. A., Chizen, D. R., et al. (2019). Dysglycemia, not altered sex steroid hormones, affects cognitive function in polycystic ovary syndrome. *Journal of the Endocrine Society*, *3*(10), 1858–1868.

Kang, J. H., & Grodstein, F. (2012). Postmenopausal hormone therapy, timing of initiation, APOE and cognitive decline. *Neurobiology of Aging*, *33*(7), 1129–1137. https://doi.org/10.1016/j.neurobiolaging.2010.10.007.

Kenny, A. M., Bellantonio, S., Gruman, C. A., Acosta, R. D., & Prestwood, K. M. (2002). Effects of transdermal testosterone on cognitive function and health perception in older men with low bioavailable testosterone levels. *The Journals of Gerontology Series A: Biological Sciences and Medical Sciences*, *57*(5), M321–M325.

Kenny, A. M., Fabregas, G., Song, C., Biskup, B., & Bellantonio, S. (2004). Effects of testosterone on behavior, depression, and cognitive function in older men with mild cognitive loss. *The Journals of Gerontology Series A: Biological Sciences and Medical Sciences*, *59*(1), M75–M78.

Kocoska-Maras, L., Rådestad, A. F., Carlström, K., Bäckström, T., Von Schoultz, B., & Hirschberg, A. L. (2013). Cognitive function in association with sex hormones in postmenopausal women. *Gynecological Endocrinology*, *29*(1), 59–62.

Kuh, D., Cooper, R., Moore, A., Richards, M., & Hardy, R. (2018). Age at menopause and lifetime cognition: Findings from a British birth cohort study. *Neurology*, *90*(19), e1673–e1681. https://doi.org/10.1212/WNL.0000000000005486.

Le, J., Thomas, N., & Gurvich, C. (2020). Cognition, the menstrual cycle, and premenstrual disorders: A review. *Brain Sciences*, *10*(4). https://doi.org/10.3390/brainsci10040198.

Levine, S. C., Foley, A., Lourenco, S., Ehrlich, S., & Ratliff, K. (2016). Sex differences in spatial cognition: Advancing the conversation. *Wiley Interdisciplinary Reviews: Cognitive Science*, 7(2), 127–155. https://doi.org/10.1002/wcs.1380.

Li, R., & Singh, M. (2014). Sex differences in cognitive impairment and Alzheimer's disease. *Frontiers in Neuroendocrinology*, *35*(3), 385–403. https://doi.org/10.1016/j.yfrne.2014.01.002.

Locascio, J. J., Corkin, S., & Growdon, J. H. (2003). Relation between clinical characteristics of Parkinson's disease and cognitive decline. *Journal of Clinical and Experimental Neuropsychology*, *25*(1), 94–109. https://doi.org/10.1076/jcen.25.1.94.13624.

Lu, P. H., Masterman, D. A., Mulnard, R., Cotman, C., Miller, B., Yaffe, K., et al. (2006). Effects of testosterone on cognition and mood in male patients with mild Alzheimer disease and healthy elderly men. *Archives of Neurology*, *63*(2), 177–185.

Maki, P. M. (2012). Minireview: Effects of different HT formulations on cognition. *Endocrinology, 153*(8), 3564–3570. https://doi.org/10.1210/en.2012-1175.

Maki, P. M. (2013). Critical window hypothesis of hormone therapy and cognition: A scientific update on clinical studies. *Menopause (New York, N.Y.), 20*(6), 695–709. https://doi.org/10.1097/GME.0b013e3182960cf8.

Maki, P. M., & Henderson, V. W. (2012). Hormone therapy, dementia, and cognition: The Women's Health Initiative 10 years on. *Climacteric, 15*(3), 256–262. https://doi.org/10.3109/13697137.2012.660613.

McCarrey, A. C., An, Y., Kitner-Triolo, M. H., Ferrucci, L., & Resnick, S. M. (2016). Sex differences in cognitive trajectories in clinically normal older adults. *Psychology and Aging, 31*(2), 166–175. https://doi.org/10.1037/pag0000070.

Meethal, S. V., & Atwood, C. S. (2005). The role of hypothalamic-pituitary-gonadal hormones in the normal structure and functioning of the brain. *Cellular and Molecular Life Sciences, 62*(3), 257–270. https://doi.org/10.1007/s00018-004-4381-3.

Meoni, S., Macerollo, A., & Moro, E. (2020). Sex differences in movement disorders. *Nature Reviews Neurology, 16*(2), 84–96. https://doi.org/10.1038/s41582-019-0294-x.

Miller, I. N., & Cronin-Golomb, A. (2010). Gender differences in Parkinson's disease: Clinical characteristics and cognition. *Movement Disorders, 25*(16), 2695–2703. https://doi.org/10.1002/mds.23388.

Mumusoglu, S., & Yildiz, B. O. (2020). PCOS phenotypes and prevalence: Differential impact of diagnostic criteria and clinical versus unselected population. *Current Opinion in Endocrine and Metabolic Research, 12*, 66–71.

Muñoz-Mayorga, D., Guerra-Araiza, C., Torner, L., & Morales, T. (2018). Tau phosphorylation in female neurodegeneration: Role of Estrogens, progesterone, and prolactin. *Frontiers in Endocrinology, 9*, 133. https://doi.org/10.3389/fendo.2018.00133.

Neu, S. C., Pa, J., Kukull, W., Beekly, D., Kuzma, A., Gangadharan, P., et al. (2017). Apolipoprotein E genotype and sex risk factors for Alzheimer disease: A meta-analysis. *JAMA Neurology, 74*(10), 1178–1189. https://doi.org/10.1001/jamaneurol.2017.2188.

Olivier, B. (2006). Testosterone and cognitive function: Current clinical evidence of a relationship. *European Journal of Endocrinology, 155*(6), 773–781. https://doi.org/10.1530/eje.1.02306.

Petanceska, S. S., Nagy, V., Frail, D., & Gandy, S. (2000). Ovariectomy and 17beta-estradiol modulate the levels of Alzheimer's amyloid beta peptides in brain. *Experimental Gerontology, 35*(9–10), 1317–1325. https://doi.org/10.1016/s0531-5565(00)00157-1.

Phillips, S. M., & Sherwin, B. B. (1992). Effects of estrogen on memory function in surgically menopausal women. *Psychoneuroendocrinology, 17*(5), 485–495.

Phung, T. K., Waltoft, B. L., Laursen, T. M., Settnes, A., Kessing, L. V., Mortensen, P. B., et al. (2010). Hysterectomy, oophorectomy and risk of dementia: A nationwide historical cohort study. *Dementia and Geriatric Cognitive Disorders, 30*(1), 43–50. https://doi.org/10.1159/000314681.

Picillo, M., Nicoletti, A., Fetoni, V., Garavaglia, B., Barone, P., & Pellecchia, M. T. (2017). The relevance of gender in Parkinson's disease: A review. *Journal of Neurology, 264*(8), 1583–1607. https://doi.org/10.1007/s00415-016-8384-9.

Pike, C. J. (2017). Sex and the development of Alzheimer's disease. *Journal of Neuroscience Research, 95*(1–2), 671–680. https://doi.org/10.1002/jnr.23827.

Pike, C. J., Carroll, J. C., Rosario, E. R., & Barron, A. M. (2009). Protective actions of sex steroid hormones in Alzheimer's disease. *Frontiers in Neuroendocrinology, 30*(2), 239–258. https://doi.org/10.1016/j.yfrne.2009.04.015.

Rannevik, G., Jeppsson, S., Johnell, O., Bjerre, B., Laurell-Borulf, Y., & Svanberg, L. (1995). A longitudinal study of the perimenopausal transition: Altered profiles of steroid and pituitary hormones, SHBG and bone mineral density. *Maturitas, 21*(2), 103–113. https://doi.org/10.1016/0378-5122(94)00869-9.

Rees, D. A., Udiawar, M., Berlot, R., Jones, D. K., & O'Sullivan, M. J. (2016). White matter microstructure and cognitive function in young women with polycystic ovary syndrome. *The Journal of Clinical Endocrinology and Metabolism, 101*(1), 314–323. https://doi.org/10.1210/jc.2015-2318.

Resnick, S. M., Maki, P. M., Rapp, S. R., Espeland, M. A., Brunner, R., Coker, L. H., et al. (2006). Effects of combination estrogen plus progestin hormone treatment on cognition and affect. *The Journal of Clinical Endocrinology and Metabolism, 91*(5), 1802–1810. https://doi.org/10.1210/jc.2005-2097.

Resnick, S. M., Matsumoto, A. M., Stephens-Shields, A. J., Ellenberg, S. S., Gill, T. M., Shumaker, S. A., et al. (2017). Testosterone treatment and cognitive function in older men with low testosterone and age-associated memory impairment. *JAMA, 317*(7), 717–727.

Rocca, W. A., Bower, J. H., Maraganore, D. M., Ahlskog, J. E., Grossardt, B. R., de Andrade, M., et al. (2007). Increased risk of cognitive impairment or dementia in women who underwent oophorectomy before menopause. *Neurology, 69*(11), 1074–1083. https://doi.org/10.1212/01.wnl.0000276984.19542.e6.

Rocca, W. A., Grossardt, B. R., & Shuster, L. T. (2014). Oophorectomy, estrogen, and dementia: A 2014 update. *Molecular and Cellular Endocrinology, 389*(1–2), 7–12. https://doi.org/10.1016/j.mce.2014.01.020.

Rodrigues, M. A., Verdile, G., Foster, J. K., Hogervorst, E., Joesbury, K., Dhaliwal, S., et al. (2008). Gonadotropins and cognition in older women. *Journal of Alzheimer's Disease, 13*(3), 267–274. https://doi.org/10.3233/jad-2008-13304.

Rosario, E. R., Chang, L., Head, E. H., Stanczyk, F. Z., & Pike, C. J. (2011). Brain levels of sex steroid hormones in men and women during normal aging and in Alzheimer's disease. *Neurobiology of Aging, 32*(4), 604–613. https://doi.org/10.1016/j.neurobiolaging.2009.04.008.

Ryan, J., Stanczyk, F. Z., Dennerstein, L., Mack, W. J., Clark, M. S., Szoeke, C., et al. (2012). Hormone levels and cognitive function in postmenopausal midlife women. *Neurobiology of Aging, 33*(7), 1138–1147. https://doi.org/10.1016/j.neurobiolaging.2012.04.011.

Saunders-Pullman, R., Gordon-Elliott, J., Parides, M., Fahn, S., Saunders, H. R., & Bressman, S. (1999). The effect of estrogen replacement on early Parkinson's disease. *Neurology, 52*(7), 1417–1421. https://doi.org/10.1212/wnl.52.7.1417.

Schaafsma, M., Homewood, J., & Taylor, A. (2010). Subjective cognitive complaints at menopause associated with declines in performance of verbal memory and attentional processes. *Climacteric, 13*(1), 84–98. https://doi.org/10.3109/13697130903009187.

Schattmann, L., & Sherwin, B. B. (2007a). Effects of the pharmacologic manipulation of testosterone on cognitive functioning in women with polycystic ovary syndrome: A randomized, placebo-controlled treatment study. *Hormones and Behavior, 51*(5), 579–586. https://doi.org/10.1016/j.yhbeh.2007.02.002.

Schattmann, L., & Sherwin, B. B. (2007b). Testosterone levels and cognitive functioning in women with polycystic ovary syndrome and in healthy young women. *Hormones and Behavior, 51*(5), 587–596.

Schumacher, M., Mattern, C., Ghoumari, A., Oudinet, J. P., Liere, P., Labombarda, F., et al. (2014). Revisiting the roles of progesterone and allopregnanolone in the nervous system: Resurgence of the progesterone receptors. *Progress in Neurobiology, 113*, 6–39. https://doi.org/10.1016/j.pneurobio.2013.09.004.

Shah, S., Bell, R. J., Savage, G., Goldstat, R., Papalia, M.-A., Kulkarni, J., et al. (2006). Testosterone aromatization and cognition in women: A randomized, placebo-controlled trial. *Menopause, 13*(4), 600–608.

Sherman, J. A. (1967). Problem of sex differences in space perception and aspects of intellectual functioning. *Psychological Review, 74*(4), 290–299. https://doi.org/10.1037/h0024723.

Sherwin, B. B., & Grigorova, M. (2011). Differential effects of estrogen and micronized progesterone or medroxyprogesterone acetate on cognition in postmenopausal women. *Fertility and Sterility*, *96*(2), 399–403. https://doi.org/10.1016/j.fertnstert.2011.05.079.

Short, R. A., Bowen, R. L., O'Brien, P. C., & Graff-Radford, N. R. (2001). Elevated gonadotropin levels in patients with Alzheimer disease. *Mayo Clinic Proceedings*, *76*(9), 906–909. https://doi.org/10.4065/76.9.906.

Shumaker, S. A., Legault, C., Rapp, S. R., Thal, L., Wallace, R. B., Ockene, J. K., et al. (2003). Estrogen plus progestin and the incidence of dementia and mild cognitive impairment in Postmenopausal WomenThe Women's Health Initiative Memory Study: A randomized controlled trial. *JAMA*, *289*(20), 2651–2662. https://doi.org/10.1001/jama.289.20.2651.

Sih, R., Morley, J. E., Kaiser, F. E., Perry, H. M., III, Patrick, P., & Ross, C. (1997). Testosterone replacement in older hypogonadal men: A 12-month randomized controlled trial. *The Journal of Clinical Endocrinology & Metabolism*, *82*(6), 1661–1667.

Soleman, R. S., Kreukels, B. P. C., Veltman, D. J., Cohen-Kettenis, P. T., Hompes, P. G. A., Drent, M. L., et al. (2016). Does polycystic ovary syndrome affect cognition? A functional magnetic resonance imaging study exploring working memory. *Fertility and Sterility*, *105*(5), 1314–1321e1. https://doi.org/10.1016/j.fertnstert.2016.01.034.

Song, Y. J., Li, S. R., Li, X. W., Chen, X., Wei, Z. X., Liu, Q. S., et al. (2020). The effect of Estrogen replacement therapy on Alzheimer's disease and Parkinson's disease in postmenopausal women: A meta-analysis. *Frontiers in Neuroscience*, *14*, 157. https://doi.org/10.3389/fnins.2020.00157.

Soules, M. R., Sherman, S., Parrott, E., Rebar, R., Santoro, N., Utian, W., et al. (2001). Executive summary: Stages of reproductive aging workshop (STRAW). *Climacteric*, *4*(4), 267–272. Retrieved from https://www.ncbi.nlm.nih.gov/pubmed/11770182.

Sundstrom Poromaa, I., & Gingnell, M. (2014). Menstrual cycle influence on cognitive function and emotion processing-from a reproductive perspective. *Frontiers in Neuroscience*, *8*, 380. https://doi.org/10.3389/fnins.2014.00380.

Tan, R., & Pu, S. (2003). A pilot study on the effects of testosterone in hypogonadal aging male patients with Alzheimer's disease. *The Aging Male*, *6*(1), 13–17.

The Rotterdam ESHRE/ASRM-Sponsored PCOS Consensus Workshop Group. (2004). Revised 2003 consensus on diagnostic criteria and long-term health risks related to polycystic ovary syndrome. *Fertility and Sterility*, *81*(1), 19–25.

Thilers, P. P., Macdonald, S. W., & Herlitz, A. (2006). The association between endogenous free testosterone and cognitive performance: A population-based study in 35 to 90 year-old men and women. *Psychoneuroendocrinology*, *31*(5), 565–576. https://doi.org/10.1016/j.psyneuen.2005.12.005.

Vaughan, C., Goldstein, F. C., & Tenover, J. L. (2007). Exogenous testosterone alone or with finasteride does not improve measurements of cognition in healthy older men with low serum testosterone. *Journal of Andrology*, *28*(6), 875–882.

Velusami, D., & Sivasubramanian, S. (2019). Sympathovagal imbalance and neurophysiologic cognitive assessment using evoked potentials in polycystic ovary syndrome in young adolescents—A cross-sectional study. *Journal of Basic and Clinical Physiology and Pharmacology*, *30*(2), 233–237. https://doi.org/10.1515/jbcpp-2018-0081.

Verdile, G., Yeap, B. B., Clarnette, R. M., Dhaliwal, S., Burkhardt, M. S., Chubb, S. A., et al. (2008). Luteinizing hormone levels are positively correlated with plasma amyloid-beta protein levels in elderly men. *Journal of Alzheimer's Disease*, *14*(2), 201–208. https://doi.org/10.3233/jad-2008-14208.

Wahjoepramono, E. J., Wijaya, L. K., Taddei, K., Bates, K. A., Howard, M., Martins, G., et al. (2011). Direct exposure of Guinea pig CNS to human luteinizing hormone increases cerebrospinal fluid and cerebral beta amyloid levels. *Neuroendocrinology*, *94*(4), 313–322. https://doi.org/10.1159/000330812.

Weber, M. T., Maki, P. M., & McDermott, M. P. (2014). Cognition and mood in perimenopause: A systematic review and meta-analysis. *The Journal of Steroid Biochemistry and Molecular Biology*, *142*, 90–98. https://doi.org/10.1016/j.jsbmb.2013.06.001.

Weber, M. T., Mapstone, M., Staskiewicz, J., & Maki, P. M. (2012). Reconciling subjective memory complaints with objective memory performance in the menopausal transition. *Menopause (New York, N.Y.)*, *19*(7), 735–741. https://doi.org/10.1097/gme.0b013e318241fd22.

Weber, M. T., Rubin, L. H., & Maki, P. M. (2013). Cognition in perimenopause: The effect of transition stage. *Menopause (New York, N.Y.)*, *20*(5), 511–517. https://doi.org/10.1097/gme.0b013e31827655e5.

Woolley, C. S., & McEwen, B. S. (1993). Roles of estradiol and progesterone in regulation of hippocampal dendritic spine density during the estrous cycle in the rat. *Journal of Comparative Neurology*, *336*(2), 293–306. https://doi.org/10.1002/cne.903360210.

Yaffe, K., Haan, M., Byers, A., Tangen, C., & Kuller, L. (2000). Estrogen use, APOE, and cognitive decline: Evidence of gene-environment interaction. *Neurology*, *54*(10), 1949–1954. https://doi.org/10.1212/wnl.54.10.1949.

Yaffe, K., Krueger, K., Cummings, S. R., Blackwell, T., Henderson, V. W., Sarkar, S., et al. (2005). Effect of Raloxifene on prevention of dementia and cognitive impairment in older women: The multiple outcomes of Raloxifene evaluation (MORE) randomized trial. *American Journal of Psychiatry*, *162*(4), 683–690. https://doi.org/10.1176/appi.ajp.162.4.683.

Yang, Z. D., Yu, J., & Zhang, Q. (2013). Effects of raloxifene on cognition, mental health, sleep and sexual function in menopausal women: A systematic review of randomized controlled trials. *Maturitas*, *75*(4), 341–348. https://doi.org/10.1016/j.maturitas.2013.05.010.

Yue, X., Lu, M., Lancaster, T., Cao, P., Honda, S.-I., Staufenbiel, M., et al. (2005). Brain estrogen deficiency accelerates Abeta plaque formation in an Alzheimer's disease animal model. *Proceedings of the National Academy of Sciences of the United States of America*, *102*(52), 19198–19203. https://doi.org/10.1073/pnas.0505203102.

Zeydan, B., Tosakulwong, N., Schwarz, C. G., Senjem, M. L., Gunter, J. L., Reid, R. I., et al. (2019). Association of Bilateral Salpingo-Oophorectomy before Menopause Onset with Medial Temporal Lobe Neurodegeneration. *JAMA Neurology*, *76*(1), 95–100. https://doi.org/10.1001/jamaneurol.2018.3057.

CHAPTER TWENTY

Role of hormones in sarcopenia

T. Priego[a], A.I. Martín[a], D. González-Hedström[b,c], M. Granado[b,d], and A. López-Calderón[a,*]

[a]Departamento de Fisiología, Facultad de Medicina, Universidad Complutense de Madrid, Madrid, Spain
[b]Departamento de Fisiología, Facultad de Medicina, Universidad Autónoma de Madrid, Madrid, Spain
[c]Pharmactive Biotech Products S.L. Parque Científico de Madrid. Avenida del Doctor Severo Ochoa, 37 Local 4J, 28108 Alcobendas, Madrid, Spain
[d]CIBER Fisiopatología de la Obesidad y Nutrición. Instituto de Salud Carlos III, Madrid, Spain
[*]Corresponding author: e-mail address: ALC@med.ucm.es

Contents

Abstract

Aging involves numerous changes in body composition that include a decrease in skeletal muscle mass. The gradual reduction in muscle mass is associated with a simultaneous decrease in muscle strength, which leads to reduced mobility, fragility and loss of independence. This process called sarcopenia is secondary to several factors such as sedentary lifestyle, inadequate nutrition, chronic inflammatory state and neurological alterations. However, the endocrine changes associated with aging seem to be of special importance in the development of sarcopenia. On one hand, advancing age is associated with a decreased secretion of the main hormones that stimulate skeletal muscle mass and function (growth hormone, insulin-like growth factor 1 (IGF-I), testosterone

Vitamins and Hormones, Volume 115
ISSN 0083-6729
https://doi.org/10.1016/bs.vh.2020.12.021

and estradiol). On the other hand, the alteration of the IGF-I signaling along with decreased insulin sensitivity also have an important impact on myogenesis. Other hormones that decline with aging such as the adrenal-derived dehydroepiandrosterone, thyroid hormones and vitamin D seem to also be involved in sarcopenia. Adipokines released by adipose tissue show important changes during aging and can affect muscle physiology and metabolism. In addition, catabolic hormones such as cortisol and angiotensin II can accelerate aged-induced muscle atrophy, as they are involved in muscle wasting and their levels increase with age. The role played by all of these hormones and the possible use of some of them as therapeutic tools for treating sarcopenia will be discussed.

1. Introduction

Sarcopenia is a term derived from the Greek words sarx (flesh) and penia (deficiency). It is defined as "a syndrome characterized by progressive and generalized loss of skeletal muscle mass and strength with a risk of physical disability" (Volpato et al., 2014). More than 50 million people suffer from sarcopenia, and it is estimated that >200 million older adults will be affected in the next 40 years (Cruz-Jentoft et al., 2010). The rate of annual skeletal muscle loss appears to be approximately 0.1–0.5% starting from age 30, with a dramatic acceleration post age 65 (Hughes, Frontera, Roubenoff, Evans, & Singh, 2002). The reduction in muscle mass associated with weakness and loss of independence represents an important public health problem, linked to economic and social implications, including hospitalization and death (Morley, Anker, & von Haehling, 2014). Several studies have yielded higher reduction of muscle mass in men compared to women, but other studies have reported high rates of sarcopenia among women or even similar prevalence in both genders (Shafiee et al., 2017). Differences between studies are a consequence of the different methods used to assess muscle mass, but it seems that women suffer more muscle atrophy because they initially have smaller amounts of muscle and live longer.

The decrease in muscle mass with aging involves a decline in muscle strength and muscle efficiency. The relationship between muscle strength and muscle mass does not seem to be linear, since strength reduction was more rapid than muscle mass loss (Goodpaster et al., 2006). In addition, all muscles are not uniformly affected by aging, since the rate of loss of the lower limb muscles is higher than that of upper limbs (Mitchell et al., 2012). A decrease in the number of muscle fibers and a shortening of them has been observed with age (Tieland, Trouwborst, & Clark, 2018). Atrophy of muscle fibers mostly affects type IIa (fast, glycolytic fibers) muscle fibers (Nilwik et al., 2013), and it is accompanied by an increase in extra and intra-myocellular lipid deposition (Mitchell et al., 2012). A decrease in muscle mass occurs if the rate of muscle protein synthesis is lower than the rate of muscle proteolysis. In older persons, whole body protein synthesis was 19% lower when compared to that of younger persons, but skeletal muscle protein synthesis was 55% lower (Balagopal, Rooyackers, Adey, Ades, & Nair, 1997). Furthermore, defects in protein trafficking, degradation, and removal such as, alterations in autophagy, ubiquitination, lysosomal degradation, and decreased proteasome activity have been reported in advancing age (Carnio et al., 2014; Sakuma et al., 2016; Wohlgemuth, Seo, Marzetti, Lees, & Leeuwenburgh, 2010). The decrease in proteolytic activity and the inefficient removal of nonfunctional proteins contribute to rapid accumulation of oxidized proteins, cellular dysfunction and senescence. In addition, sarcopenia is associated with mitochondrial dysfunction, which includes loss of muscle mitochondria, mitochondrial DNA mutations and an increase in mitochondrial reactive oxygen species (ROS) emission (Coen, Musci, Hinkley, & Miller, 2018). These mitochondrial abnormalities not only affect muscle fibers but also motor neurons and neuromuscular junction (Rygiel, Picard, & Turnbull, 2016). Several data suggest that oxidative damage could be involved in neuromuscular junction degeneration (McDonagh et al., 2016). Aging is not only accompanied by a decline in muscle protein turnover, but is also associated with a decrease in the number and the proliferative and regenerative capacity of satellite cells after different stimuli (Verdijk et al., 2007). In this sense, exposure of satellite cells from old mice to young serum by parabiotic pairings (a shared circulatory system) restored the proliferation and regenerative capacity of aged satellite cells (Conboy et al., 2005). These results suggest that the age-related decline of skeletal muscle progenitor cell activity can be modulated by systemic factors such as hormones, which change with age.

1.1 Nonendocrine factors promoting sarcopenia

Sarcopenia is multifactorial in its development, and several promoting factors have been identified (Fig. 1) (Tournadre, Vial, Capel, Soubrier, & Boirie, 2019).

Sedentary lifestyle. The decline in muscle fiber and strength is more pronounced in patients with physical inactivity. For this reason, several

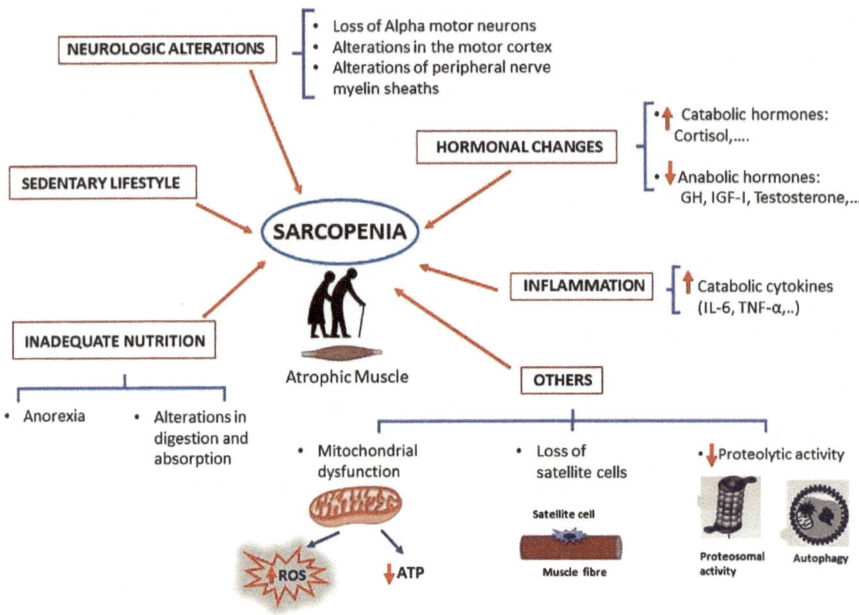

Fig. 1 Major causes of sarcopenia. GH, growth hormone; IGF-I, insulin-like growth factor; IL-6, interleukin 6; ROS, reactive oxygen species; TNF-α, tumor necrosis factor.

physical training programs have been proposed to mitigate sarcopenia (Montero-Fernandez & Serra-Rexach, 2013). There are various mechanisms that mediate the effects of exercise: proliferation of satellite cells, increase in energy production by mitochondria and capillary density, down-regulation of chronic inflammation and decreased apoptosis of skeletal myocytes (Marzetti et al., 2017).

Inadequate nutrition. The average food intake falls by about 25% between 40 and 70 years of age. This phenomenon called "anorexia of ageing" comes with several disabilities like poor dentition, loss of taste and olfaction, alterations in digestion and absorption. In addition, impairment in the regulation of food intake can also contribute to malnutrition (Nieuwenhuizen, Weenen, Rigby, & Hetherington, 2010). Many researchers propose nutritional intervention to prevent sarcopenia. Bo et al. (2019) reported that combined supplementation of whey protein and vitamin D and E can significantly improve muscle mass and strength in older adults with sarcopenia.

Chronic inflammation. The aging process is accompanied by an increased release of pro-inflammatory cytokines (Wahlin-Larsson, Carnac, & Kadi, 2014). The rise in cytokines, such as interleukin (IL)-6 and tumor necrosis factor (TNF)-α, IL-1 as well as C-reactive protein, has been associated with lower muscle function (Visser et al., 2002). Cytokines may induce sarcopenia by activating the ubiquitin-proteasome pathway (Mukund & Subramaniam, 2020), decreasing IGF-1 production, and by inducing insulin resistance.

Loss of alpha motor neurons and neurological alterations. There is a 50% loss of alpha motor-neurons after the seventh decade (Drey et al., 2014). The remaining alpha-motor-neurons reinnervate the denerved fibers, leading to increased motor unit size and decreased muscle strength. This reinnervation contributes to the final differentiation of type II fibers to type I fibers. This way, type II fibers are recruited into slow motor units (Lang et al., 2010). All of these changes decrease muscle power, strength and function. In addition, alterations in the motor cortex, spinal cord (decrease in the number of alpha motor neurons) and peripheral neurons (decline in peripheral nerve fibers and alterations of their myelin sheaths) may significantly contribute to the effects of age on muscle.

2. The GH-IGF-I axis

The term somatopause refers to the decrease in serum concentrations of growth hormone (GH) and insulin-like growth factor-I (IGF-I) with age. In humans, the decrease in circulating GH starts in an exponential mode after reaching final height (Rudman et al., 1981). However, the decrease in circulating IGF-I is not as great as the decrease in circulating GH (Maggio, Ble, Ceda, & Metter, 2006). The progressive decline in plasma GH and IGF-I with age is mainly due to a decrease in pituitary GH secretion and in GH pulses rather than to a decrease in GH pulse frequency and GH half-life. However, aging does not decrease hypoglycemia-induced GH release or the increase in hepatic IGF-I synthesis after exogenous GH administration (Veldhuis, 2008). Together with age, decreased secretion of gonadal steroids and increased abdominal visceral fat are factors that are very closely associated with the decline in pituitary GH secretion (Veldhuis et al., 1995).

The relationship between somatopause and sarcopenia has been studied in several GH disorders that lead to GH hyposecretion. One of these is congenital isolated GH deficiency, secondary to a mutation in the GH-releasing hormone receptor (GHRH), present in a large cohort of individuals in northeastern Brazil (Aguiar-Oliveira & Bartke, 2019). These individuals are dwarves, due to very low GH and IGF-I levels; they show increased adiposity and their skeletal muscle mass is reduced. However, muscle function in these patients is adequate with increased muscle strength parameters, adjusted for body weight and composition, as well as the capacity for vigorous physical activity with improved resistance to fatigue (Aguiar-Oliveira & Bartke, 2019). A possible explanation of this apparent contradictory data can be the fact that these patients have greater insulin sensitivity than individuals with normal GH-IGF-I secretion (Aguiar-Oliveira & Bartke, 2019). Similarly, in Laron patients, which have GH resistance and low IGF-I levels, an increase in insulin sensitivity has also been observed (Guevara-Aguirre et al., 2015).

In healthy elderly men, although GH administration modestly increases muscle mass, it does not improve physical strength or endurance (Sattler, 2013). Therefore, in contrast to the well-known role of IGF-I in increasing muscle stretch, a direct effect of GH on muscle stretch and function is not clear. The lack of effect of GH treatment on muscle function in elderly can be due to the induction of GH-related insulin resistance. One of the mechanisms by which GH administration can reduce insulin sensitivity is by inducting suppression of the cytokine signaling (SOCS) protein family, which down-regulates insulin signaling (Sattler, 2013). In addition to early insulin resistance, it should be noted that GH administration frequently causes other adverse events including edema, arthralgias, carpal tunnel syndrome and an increased risk of cancer, which lead to safety concerns (Sattler, 2013).

In skeletal muscle, the IGF-I/Akt/mTOR pathway plays an important role, increasing myofibrillar protein biosynthesis (Martín, Priego, & López-Calderón, 2018). Another signaling pathway activated by IGF-I is the Ras/Raf/ERK pathway, which is able to increase cell proliferation, cell differentiation and cell survival (Matheny, Merritt, Zannikos, Farrar, & Adamo, 2009).

There are contradictory reports regarding the role of circulating IGF-I in sarcopenia and frailty. A cross-sectional study in healthy elderly women showed that plasma IGF-I levels did not explain the age-associated decline in muscle function (Boonen et al., 1998). In contrast, it has been reported that aging-induced decrease in myosin heavy-chain synthesis correlates with muscle mass and strength, as well as with circulating levels of IGF-I, both in men and women (Balagopal et al., 1997). Furthermore, in old healthy women that are functionally active, lower circulating levels of IGF-I have been found to be related to lower values of quadricep maximal muscle power and optimal shortening velocity and activity (Kostka et al., 2000).

2.1 Local IGF-I

Skeletal muscle also produces and secretes IGF-I containing autocrine and paracrine actions and is an important regulator of muscle function. In liver-specific IGF-I knockout mice, which have a 75% reduction of circulating IGF-I, normal muscle development has been reported (Yakar et al., 1999). Furthermore, several data indicate that IGF-I overexpression in skeletal muscle has the capacity to maintain muscle mass and function during aging (Musaro et al., 2001) and in animal models of neuromuscular diseases (Bosch-Marce et al., 2011). In addition, in old mice, local overexpression of IGF-I in muscle is able to induce hypertrophy and regeneration of skeletal muscle (Musaro et al., 2001). All these data suggest that IGF-I synthetized in skeletal muscles is an important factor in the regulation of muscle mass and function.

The different roles of both circulating and muscle IGF-I in preserving skeletal muscle mass and function during the aging process can be a positive response, as has been pointed out by Ascenzi et al. (2019). Whereas circulating IGF-I can induce hyperplasia and spontaneous tumor formation, locally synthetized IGF-I control growth, survival, and differentiation after injury or physical exercise only locally. One of the mechanisms by which IGF-I counteracts sarcopenia is by activating two pathways normally affected by aging, autophagy and PGC-1-mediated signaling, as well as decreasing the basal inflammatory state and preserving the integrity of neuromuscular junctions (Ascenzi et al., 2019).

When local IGF-I is released it stimulates satellite cell proliferation, promotes amino acid uptake by muscle, decreases proteolysis and facilitates muscle recovery after damage or injury. The decrease in resident satellite cells is attenuated in mice overexpressing IGF-I in skeletal muscle (Musaro et al., 2001). However, as mentioned above, lower response of satellite cells in aged muscle is secondary to the systemic environment, rather than to the intrinsic activity of muscle stem cells (Conboy et al., 2005). Therefore, changes in circulating IGF-I concentrations may also contribute to aging-induced decrease in the number of satellite cells.

IGF-I synthesis in skeletal muscle can be increased by anabolic hormones such as GH and testosterone, and decreased by cortisol, angiotensin II and cytokines (López-Menduiña, Martín, Castillero, Villanúa, & López-Calderón, 2012; Martin et al., 2018). However, local production of IGF-I may only be necessary, as in the case of physical exercise, where the induced-muscle hypertrophy is secondary to skeletal muscle load and mechanical-stretch mechanisms, indicating that muscle hypertrophy is dependent on local growth factor production, rather than secondary to systemic GH or IGF-I (Hameed, Harridge, & Goldspink, 2002).

2.2 Effects of physical exercise on muscle IGF-I

Inactivity is one of the major causes of sarcopenia. In humans aged 60–70 years, 10 days of bed rest can induce 1 kg loss. On the contrary, physical exercise is one of the stimuli that increases skeletal muscle mass. Therefore, sarcopenia can be due to the fact that aging decreases skeletal muscle plasticity in response to exercise.

Resistance exercise increases local IGF-I expression in skeletal muscle together with an increase in skeletal muscle strength and growth, a process that is independent of circulating IGF-I levels (Matheny et al., 2009). Furthermore, the increase in circulating IGF-I levels does not modify skeletal muscle protein synthesis or hypertrophy during resistance exercise training, whereas the upregulation of the skeletal muscle IGF-I expression induces muscle hypertrophy (Harridge, 2007). All of these data indicate that local, but not systemic, IGF-I may be involved in compensatory growth of muscle after resistance training and increased mechanical load. However, in addition to IGF-I, activation of the mTOR signaling pathway in muscle after acute mechanical loading can also be IGF-I-independent (O'Neil, Duffy, Frey, & Hornberger, 2009).

Aging decreases growth factors, but skeletal muscle is able to respond to strength training by increasing local IGF-I expression (Hameed et al., 2002). However, this response seems to be reduced by aging. In this sense, older individuals have a reduced capacity to induce muscle hypertrophy with resistance exercise (Rivas et al., 2014). Small noncoding RNAs, called microRNAs (miRNAs), regulate muscle phenotype via alterations in gene expression, and they are potential regulators of exercise-induced adaptations in skeletal muscle. One of these is miR-126, which has a reduced response to exercise-induced downregulation in old men (Rivas et al., 2014).

3. Insulin resistance and glucose regulation

Type 2 diabetes mellitus is one of the most prevalent chronic diseases in older adults, and aging-induced muscle atrophy is associated with decreased insulin sensitivity. Skeletal muscle not only assures locomotion, but it also plays a key role in homeostasis by regulating metabolism and insulin sensitivity. In addition, skeletal muscle released muscle-derived cytokines, called myokines, which exert endocrine effects on visceral fat and have anti-inflammatory effects (Iizuka, Machida, & Hirafuji, 2014). Therefore, a decrease in muscle mass can result in myokine dysregulation as well as in metabolic and endocrine alterations. Sarcopenia is associated with the onset of metabolic diseases, such as anabolic/insulin resistance, type 2 diabetes, dyslipidemia and obesity (Moon, 2014). Several factors contribute to those changes: the effects of aging on insulin secretion, a decrease in both physical activity and muscle mass, as well as, an increase in adiposity. The important role of muscle mass on insulin-resistance is shown by the fact that myostatin inhibition in old rats, not only increases muscle mass, but it also improves insulin sensitivity (Shou, Chen, & Xiao, 2020). In aged rats, it has been shown that an increase in muscle mass raises skeletal muscle glucose uptake and improves insulin sensitivity (Shou et al., 2020). All of these data indicate a relationship between diabetes and sarcopenia, and this relationship seems to be bidirectional.

Although aging does not impair proinsulin synthesis, it alters beta-cell response to hyperglycemia and to other secretagogues in rats (Li et al., 2019). Similarly, insulin sensitivity to oral glucose and to incretins is decreased in old people, whereas insulin clearance is decreased (Chang & Halter, 2003). It has been reported that, compared with younger male subjects (30 years old), older ones (65–70 years) have reduced glucose metabolism and decreased expression of skeletal muscle glucose transporter 4 (GLUT4) in type II fibers (Gaster, Poulsen, Handberg, Schroder, & Beck-Nielsen, 2000). In healthy, lean, elderly individuals, without type 2 diabetes, an impairment of insulin-stimulated protein kinase B activity in skeletal muscle has been reported (Petersen et al., 2015). Authors indicate that the decrease in muscle mitochondrial function is an important predisposing factor for ectopic lipid accumulation in muscle, which would result in defects in insulin signaling and ultimately lead to insulin resistance. In turn, muscle insulin resistance induces a decreased trafficking of GLUT4 from intracellular GLUT4 storage vesicles to the plasma membrane, causing reduced glucose utilization by muscles (Consitt, Dudley, & Saxena, 2019). All of the aforementioned increase the risk of insulin resistance or be secondary to insulin resistance.

In elderly diabetic type II patients the increase in incretin GLP-1 levels, by dipeptidyl peptidase 4 inhibitor administration, ameliorates glycemic control, decreases inflammation and increases skeletal muscle mass, strength and gait speed (Rizzo et al., 2016). In addition, a GLP-1 agonist is able to directly decrease muscle proteolysis in muscle cells in culture (Hong, Lee,

Jeong, Choi, & Jun, 2019). These data suggest that the beneficial effects of this gastrointestinal hormone on muscle mass are exerted through both increasing insulin secretion and a direct action at the skeletal muscle level.

Aging not only decreases muscle mass, but it also increases fat mass. The increase in intracellular muscle fat contributes to insulin resistance, and in turn, impacts muscle protein turnover and contractile dysfunction, due to structural and functional alterations (Gregorich et al., 2016). The increase in visceral fat mass induces the release of proinflammatory cytokines, which leads to a state of low-grade inflammation. These cytokines are one of the factors responsible for the decrease in muscle mass and function. In addition, obesity decreases adiponectin secretion and, as will be discussed below, the decrease in adiponectin levels decreases insulin sensitivity. Obesity will become an important challenge for the health system in the near future.

4. Adrenal hormones

In humans, aging affects steroid biosynthesis in a different way. Secretion of cortisol by zona fasciculata of the adrenal cortex is increased, as a result of diminished negative feedback due to impaired sensitivity of the hypothalamic-pituitary-adrenal axis. However, secretion of the main androgen dehydroepiandrosterone (DHEA) by the zona reticularis of the adrenal cortex is decreased (Yiallouris et al., 2019).

4.1 Cortisol

The secretion of glucocorticoids (cortisol in humans) is activated by stressful stimuli and displays a marked circadian rhythm in humans with higher levels in the morning (7:00 am) and a nadir in the late evening. Cortisol secretion is controlled by the hypothalamic-pituitary-adrenal axis and locally by the action of 11b-hydroxysteroid dehydrogenase 1 (11β-HSD1), an enzyme which converts cortisone into active cortisol. As a catabolic hormone, higher cortisol levels are linked with muscle mass reduction and has been proposed as a mechanism for sarcopenia (Marcell, 2003). Glucocorticoids affect muscle mass through a combination of mechanisms (Schakman, Kalista, Barbe, Loumaye, & Thissen, 2013) such as inducing muscle proteolysis (by the ubiquitin–proteasome and lysosomal systems), decreasing protein synthesis, inhibiting the insulin/IGF-I signaling pathway, and altering the local production of growth factors that control muscle mass (IGF-I and myostatin). Several functional changes in the hypothalamic-pituitary-adrenal axis have been described in the elderly:

increased cortisol concentrations, diminished axis sensitivity and alterations in cortisol secretion pattern (Yiallouris et al., 2019). Additionally, an increased cortisol response to stress in aging has been reported (Otte et al., 2005). Moreover, 11β-HSD shows increased activity during aging, increasing cortisol availability (Kilgour et al., 2013).

Several studies have investigated the relationship between cortisol age-related alterations and muscle mass or strength. Higher cortisol was associated with lower appendicular lean mass in older people (Westbury et al., 2018). Likewise, an association between higher salivary, but not serum cortisol, with lower skeletal mass and function has been reported (Peeters, van Schoor, van Rossum, Visser, & Lips, 2008). Baylis et al. (2013) proposed that high serum cortisol/DHEA ratio is significantly associated with an increased likelihood of frailty in a cohort of community-dwelling older people. In agreement with these results, Yanagita et al. (2019) reported this relationship in elderly patients with type 2 diabetes mellitus. In contrast, Kilgour et al. (2013) found no significant relationships between plasma cortisol, urinary glucocorticoid metabolites or cortisol receptor expression and muscle mass or strength. In spite of these results, they found a significant association between increased muscle 11β-HSD1 expression and lower quadricep strength. Increased 11β-HSD1 expression is likely to increase myocellular cortisol levels and it may contribute to age-related loss of muscle mass. In this sense, skeletal muscle 11β-HSD1 expression is upregulated in aging women, and this situation is associated with reduced grip strength (Hassan-Smith et al., 2015). The age-related decline in IGF-I and/or estrogen levels could be responsible for the increased expression of this enzyme (Frajese, Taylor, Jenkins, Besser, & Monson, 2004; McInnes et al., 2012), although the increased cytokine release might also be involved (Tomlinson et al., 2004). In summary, most of the available evidence supports that cortisol alterations during aging are involved in sarcopenia and frailty.

4.2 Dehydroepiandrosterone (DHEA) and its sulfate (DHEAs)

Over time DHEA levels gradually decrease in both men and women, and it is possible that this decrease plays a role in age-related muscle loss. Serum DHEA concentrations drop progressively from the age of 30 years. Thus, at 70–80 years, DHEA levels reach approximately 20% of peak values in men, and 30% of peak values in women versus people younger than 40 years (van den Beld et al., 2018).

Since no specific DHEA receptor has been identified, its effects in muscle could be mediated by its conversion into more active androgens or into estrogens. DHEA increases IGF-I levels and its bioavailability in muscles by decreasing the levels of IGF-I binding protein-1 (Vitale, Cesari, & Mari, 2016). In addition, DHEA supplementation increases plasma testosterone levels in women and IGF-I in men. Nevertheless, in this study DHEA treatment did not produce clear changes in muscle mass, strength or function (Samaras, Samaras, Frangos, Forster, & Philippe, 2013). However, a positive effect of DHEA administration on muscle function and composition has been reported in older subjects (Maggio, Lauretani, & Ceda, 2013), while a systematic review failed to confirm such effects (Baker, Karan, & Kenny, 2011). Discrepancies between results could be related to the variety of study designs, treatment durations and doses, and sample population size. Accordingly, to date, the relationship between DHEA replacement and sarcopenia remains inconclusive. Additional large clinical trials are necessary to identify the clinical role of DHEA replacement in this population.

5. Gonadal steroids

5.1 Testosterone

Testosterone (T) is the most important androgen in plasma. T has important metabolic effects on protein, carbohydrate, and fat metabolism, and therefore is critical to maintain muscle mass and function, bone mass and body composition. A decrease in T levels with aging is referred to as andropause and is characterized by a decreased libido and erectile difficulties, increased fatigability, and decreased muscle strength. T decline does not appear to be exclusively attributable to aging, since risk factors operate independently and interact with age, including health and lifestyle factors such as obesity, stress and smoking (Travison, Araujo, O'Donnell, Kupelian, & McKinlay, 2007).

Reduced T levels could contribute to muscle wasting and to an age-related decrease in muscle strength. Several mechanisms are involved in the positive effect of T on skeletal muscle. On one hand, T increases protein synthesis by increasing amino acid utilization and subsequently enhances muscle fiber size (Ferrando et al., 2002). On the other hand, T treatment increases both type I and type II muscle fibers and the number of myonuclear per fiber (Sinha-Hikim et al., 2002). It has also been reported that T stimulates satellite cell activation and proliferation, as well as the regeneration of both young and aged mouse muscle (Serra et al., 2013). In addition, T administration improves muscle mitochondrial biogenesis and function (Pronsato, Milanesi, & Vasconsuelo, 2020). T also increases intramuscular IGF-I concentrations (Ferrando et al., 2002) and downregulates myostatin gene expression (Dubois et al., 2014), as well as, counteracts the proteolytic action of glucocorticoids in muscle (Martin et al., 2018).

Numerous studies have shown the efficacy of treatment with T in the elderly, but the results varied based on study characteristics such as subjects, dosages and treatment type (Shin, Jeon, & Kim, 2018). Analysis of the literature has concluded that T supplementation attenuates aging-induced decreases in muscle mass (Corona et al., 2016; Ferrando et al., 2002; Srinivas-Shankar et al., 2010). A systematic review and meta-analysis of the research on testosterone replacement therapy (TRT), of 31 relevant double-blind placebo-controlled randomized clinical trials, concluded that intramuscular TRT produces greater muscular benefit than transdermal administration (Skinner et al., 2018). Intramuscular TRT improves total body strength more than transdermal TRT, with the effect being more pronounced for lower-extremities than for upper-extremities. These better results may be a consequence of the fact that intramuscular TRT produces higher circulating T levels. Nevertheless, conflicting results are also present among meta-analyses (Hildreth et al., 2013; Storer et al., 2017) which reported very minimal changes in fat-free mass and in strength improvement with TRT. Discrepancies between studies result from the different methods used to assess muscle mass and to quantify muscle strength. In this sense, Nam, Lee, Yun, and Cho (2018) conducted a review to evaluate the effects of TRT on muscle strength and physical function. The analysis concluded that TRT is not associated with increased muscle strength (evaluated by hand-grip strength and leg muscle strength), whereas it does increase physical function (assessed by the 6-min walk test, Physical Activity Scale for the Elderly score, and other physical performance tests). This study highlights the difficulty to compare the results as a consequence of the heterogeneity of protocol to assess both muscle strength and physical function. In summary, although considerable evidence suggests that T treatment improves sarcopenia by increasing muscle mass, its beneficial effects on muscle function are not conclusive.

5.2 Estrogens

Circulating levels of estrogens (17-β estradiol and estrone) change throughout life. In their mid-40s most women experience their menopausal transition, where the level of female hormones declines and menstrual cycle becomes irregular. Taking into account that the life expectancy of women in European countries is 82 years, women spend more than one-third of their life in the postmenopausal state. During menopause a decline in muscle mass and strength together with a reduction in the amount of contractile protein per cell unit, mainly in type II fibers is frequently observed (Martel et al., 2006). Therefore, it has been postulated that the reduction of estrogen production may partially explain this accelerated muscle loss. In fact, estrogens have an important role in skeletal muscle physiology. It has been described that both alpha and beta estrogen receptors (ERα and ERβ) are expressed in skeletal muscle cells of both genders (Wiik, Ekman, Johansson, Jansson, & Esbjornsson, 2009).

In animal models, estrogens improve muscle contractile properties and exercise endurance (Moran, Nelson, Landisch, Warren, & Lowe, 2007). Specifically in females, estrogens induce changes in myosin relaxation states and this is affected by the age-dependent decrease of estradiol (Colson, Petersen, Collins, Lowe, & Thomas, 2015). In addition, estrogen seems to prevent the decrease in the satellite cell pool associated with aging. Estradiol binding to ERα regulates satellite cell maintenance and function, thereby promoting muscle regeneration (Collins, Laakkonen, & Lowe, 2019). ERα has been observed in the mitochondria and perinuclear compartments (Milanesi, Russo de Boland, & Boland, 2008), and estrogens maintain mitochondrial health (Ribas et al., 2016) and metabolic homeostasis (Ikeda, Horie-Inoue, & Inoue, 2019). Estrogen also prevents peroxidation of polyunsaturated fatty acids from membrane phospholipids, preserving an effective electron transport (Irwin et al., 2008). In addition estradiol restores redox balance and insulin sensitivity in skeletal muscle (Torres et al., 2018). These estradiol effects contribute to the higher risk for type 2 diabetes that develops in menopause.

Throughout binding to ERβ, estrogens can regulate the expression of myogenic regulatory factors (Dieli-Conwright, Spektor, Rice, Sattler, & Schroeder, 2009) and activate the IGF-I signaling pathway in muscle (Ahtiainen et al., 2012). In this sense, estrogens upregulate MyoD expression (Pedraza-Alva, Zingg, Donda, & Perez-Martinez, 2009), whereas they suppress myostatin expression, a negative regulator of myogenesis (Dieli-Conwright et al., 2009). Other authors have reported that estrogen protects against increased cellular apoptosis associated with aging (Kerksick, Taylor, Harvey, & Willoughby, 2008). However, the apoptotic signaling pathways involved in the response to estrogen are not fully determined.

Although various studies have found a link between the loss of estrogen at menopause and sarcopenia (Cooper, Mishra, Clennell, Guralnik, & Kuh, 2008; van Geel, Geusens, Winkens, Sels, & Dinant, 2009), controversial results in relation to estrogen administration for maintaining skeletal muscle function have been reported (Brown, 2008; Taaffe et al., 2005). In postmenopausal women, no effect of estrogen treatment in muscle mass or function has been observed (Javed, Mayhew, Shea, & Raina, 2019). However, other studies support the hypothesis that hormone replacement therapy (HRT) has a positive and measurable impact on muscle strength (Taaffe et al., 2005; Tiidus, 2011). A meta-analysis comparing data from 23 studies concluded that HRT is beneficial on skeletal muscle strength in postmenopausal women (Greising, Baltgalvis, Lowe, & Warren, 2009). Furthermore, in postmenopausal monozygotic twins, twins receiving HRT had better mobility, greater muscle body mass, less fat mass and increased involuntary force than their sisters (Finni et al., 2011). In summary, there is enough evidence pointing out that HRT is beneficial for muscle strength without affecting muscle size. Despite the benefits of prolonged postmenopausal estrogen replacement therapy on muscle strength, undesirable effects like increased risks of cardiovascular disease, breast cancer, and pulmonary embolism have been reported (Ettinger, Quesenberry, Schroeder, & Friedman, 2018). Although this is clearly an area that requires further investigation, estrogen therapy is not recommended for preventing or treating sarcopenia in postmenopausal women.

6. Adipokines

It is well established that adipose tissue plays an active role in energy homeostasis as well as in the control of neuroendocrine, autonomic and immune functions (Ahima & Flier, 2000). To coordinate all these actions, adipose tissue produces a wide variety of hormones called adipokines, such as adiponectin, leptin, resistin, visfatin, among others. Through these hormones there is an important crosstalk between organs, including skeletal muscle (Dyck, 2009). With aging, these pathways are significantly altered, and this alteration is related with the increase in metabolic disorders such as type 2 diabetes mellitus (T2DM) and cardiovascular disease. There is also increasing evidence that adipokine dysregulation is associated with wasting syndromes such as cachexia and sarcopenia (Unger, 2005).

The increase in adipose tissue mass with age is accompanied by an increase in the main adipokines; leptin and adiponectin, and with a higher ratio between resistin/IGF-I (Bucci et al., 2013). These parameters have been inversely associated with muscle strength in old subjects (Bucci et al., 2013), reinforcing the important role of these adipokines in sarcopenia. Remarkably, in obese subjects the prevalence of sarcopenia is greater than in non-obese ones (Srikanthan, Hevener, & Karlamangla, 2010). Individuals with sarcopenic obesity show increased levels of adipocyte-derived pro-inflammatory markers (Batsis & Villareal, 2018), which have negative effects on muscle regeneration (Akhmedov & Berdeaux, 2013) and promote atrophy (Adams et al., 2008). Recently, O'Leary et al. (2018) reported direct negative effects of the obese human adipose tissue secretome (milieu secreted by the subcutaneous adipose tissue) on myocyte myogenesis from human skeletal muscle, particularly in those myocytes from older subjects. The fact that the secretome from lean adipose tissue has no detrimental effect on myogenesis (Pellegrinelli et al., 2015) indicates that anti-myogenic adipokines are pro-inflammatory, which are elevated in obese individuals.

Leptin has an important role controlling adipose tissue mass, energy expenditure and food intake (Friedman, 2011). Leptin is mainly produced by adipose tissue, but it is also produced by other tissues such as skeletal muscle and bone cells (Hamrick, 2017). Specifically, in skeletal muscle leptin has a range of physiological actions, such as increasing insulin sensitivity, glucose utilization and lipolysis (Dyck, 2009). The lack of leptin in the ob/ob mice model causes alterations in muscle physiology that resemble those of aged rodents (Warmington, Tolan, & McBennett, 2000). Thus, leptin may also have important effects in myogenesis. In humans, lower levels of leptin have been observed in sarcopenic individuals, while centenarians showed increased levels of this hormone (Pareja-Galeano et al., 2017). An association between low leptin levels and high TNF-α and higher mortality risk has also been described (Arai, Kamide, & Hirose, 2019).

Remarkably, leptin treatment in aged mice improves skeletal muscle mass and muscle fiber size (Hamrick et al., 2010), showing the important role of this adipokine on sarcopenia. The mechanism underlying this effect seems to be by increasing muscle mass and decreasing the expression of atrophy-related factors such as myostatin, muscle RING-finger protein-1 (MuRF1), and muscle atrophy F-box (MAFbx) (Sainz et al., 2009). Leptin may also act via the IGF-I system as both circulating and muscle-derived IGF-I are increased by leptin treatment in aged mice (Hamrick et al., 2015). Leptin has also proved to indirectly mediate muscle regeneration by suppressing miR-489 (Hamrick et al., 2010), which maintains muscle satellite cells in a quiescent state (Cheung et al., 2012). Thus, treatment with leptin will not only be beneficial for increasing muscle mass, but it will also enhance muscle regeneration and repair.

Loss of positive leptin actions on muscle in obese individuals despite increased leptin levels shown by these subjects could be due to leptin resistance (Myers, Cowley, & Munzberg, 2008). In this sense, the decrease in leptin sensitivity with obesity at central level (Liu, Bian, Liu, & Endoh, 2007) and also in skeletal muscle has been reported (Fuentes et al., 2010). In fact, as previously mentioned, sarcopenia is more severe in obese individuals than in lean ones, and the role of leptin resistance in this circumstance has been hypothesized (Kohara et al., 2011). Leptin receptor is expressed in large quantities in musculoskeletal tissue, and the lack of leptin receptors specifically in skeletal muscle causes muscle atrophy (Arounleut et al., 2013). Thus, leptin resistance may worsen sarcopenia in obese, aged patients.

In addition to leptin, the other important hormone produced by adipose tissue is adiponectin. This adipokine has a key role in energy homeostasis, insulin sensitivity, and has a potent anti-inflammatory effect. Adiponectin is also associated with positive actions on muscle metabolism (Dyck, 2009). The insulin-sensitizing effect of adiponectin on muscle results in an increase in the use of glucose and fatty acid resources by skeletal muscle fibers and adiponectin also increases myogenesis (Liu & Sweeney, 2014). In addition, this hormone has a protective effect on muscle protein degradation by upregulating the IRS-1/Akt signaling pathway (Zhou, Du, Hu, Walsh, & Wang, 2007). However, in the elderly, the levels of serum adiponectin increase and are correlated with low lean mass (Santoro et al., 2019). Furthermore, adiponectin has been described as a biomarker of adverse body composition during aging (Baker et al., 2019). In a recent prospective study, Biercewicz et al. (2020) found in a population of 88 elderly women, that adiponectin levels were higher and leptin levels lower in those women with a risk of malnutrition and sarcopenia in comparison with well-nourished women. However, the parameter that was found as a more stable indicator of nutritional status was the leptin to adiponectin ratio. This ratio decreases as undernourishment, and therefore sarcopenia, becomes more pronounced. Other authors have also described the association between leptin to adiponectin ratio and parameters of muscle physiology such as muscle strength. This association was positive. Thus, high adiponectin levels may be negative for sarcopenia (Bucci et al., 2013). These results do not agree with the positive effects of adiponectin on muscle. This paradox may be explained because this hormone increases with weight loss. Thus, the relationship between sarcopenia and higher levels of adiponectin may be the consequence and not the cause; the loss of muscle with age may be related with the higher secretion of adiponectin (Baker et al., 2019). It has also been proposed that the age-induced increased levels of adiponectin may be the consequence of an adiponectin resistance (Kizer et al., 2010). Curiously, other authors have reported low levels of adiponectin in sarcopenic individuals (Can et al., 2017). Li et al. (2019) described an inverse correlation between adiponectin and sarcopenia risk. The same authors showed a significant increase in adiponectin in sarcopenic subjects subjected to a positive lifestyle intervention (whey supplementation with resistance exercise training). Similarly, high levels of adiponectin have been observed in centenarians and has been related with a compensatory response to maintain metabolic homeostasis, reduce inflammation and oxidative stress, which will eventually counteract catabolic states such as sarcopenia (Arai et al., 2019). In this sense, adiponectin seems to be protective against sarcopenia. In fact, it has been described in an animal model of postmenopausal sarcopenia, with aged ovarectomized rats, that treatment with adiponectin resulted beneficial in muscle mass. This treatment reversed alterations in atrogenes and anti-atrophy factors that were increased in this sarcopenic model (China et al., 2017).

Resistin is another hormone produced by adipose tissue, with pro-inflammatory properties (Butler et al., 2009) and effects impairing the activating pathways of insulin. In muscle, resistin inhibits myogenic differentiation of myoblasts (Sheng et al., 2013). In older individuals, the resistin/IGF-I ratio increases and is associated with insulin resistance and lower quadricep strength in old men but not in old women (Bucci et al., 2013). Likewise, an inverse association between resistin concentrations and abdominal skeletal muscle density has been reported (Van Hollebeke, Cushman, Schlueter, & Allison, 2018). In myotube culture assays from old subjects, incubation with resistin at a concentration similar to that of older individuals had adverse effects on myogenesis (O'Leary et al., 2018). The fact that myocytes from young individuals were less affected indicates an age-dependent effect of resistin. The way that resistin exerts its detrimental effects on muscle is via NF-κB (O'Leary et al., 2018), a pathway that is increased in aged individuals (Tilstra, Clauson, Niedernhofer, & Robbins, 2011). This may be this the reason for the higher sensibility of old muscle to the unfavorable effects of resistin.

7. Angiotensin II

Angiotensin II is a hormone that regulates a variety of physiological functions related with body fluid balance, blood pressure and sodium homeostasis. The angiotensin-converting enzyme 1 (ACE1) converts angiotensin I (inactive) into the active form, angiotensin II, which is an important component in blood pressure control. Pharmacological treatment with ACE1 inhibitors, widely used in aged people to control high blood pressure, has revealed the possible role of angiotensin II in muscle maintenance and sarcopenia (Carter & Groban, 2008). In this sense, it has been described that the treatment with ACE1 inhibitors is associated with greater muscle mass in the lower extremities of older persons in comparison with other treatments for hypertension (Di Bari et al., 2004). These drugs have also been shown to improve mitochondrial function, increase IGF-I levels and enhance glucose uptake by the skeletal muscle (Dietze & Henriksen, 2008).

On the other hand, high levels of angiotensin II are negatively associated with skeletal muscle strength (Delafontaine & Yoshida, 2016). Angiotensin II acts on IGF-I/insulin signaling pathways, by decreasing Akt phosphorylation and activating muscle proteolysis by the ubiquitin-proteasome system (Song et al., 2005) and caspase-3 apoptotic pathways in muscle (Delafontaine & Yoshida, 2016). Angiotensin II also directly acts on the IGF-I system, since infusion of this hormone in rats decreases IGF-I, IGFBP-3 and BP-5 levels in skeletal muscle (Brink, Wellen, & Delafontaine, 1996). Thus, it seems that one of the ways by which angiotensin II induces muscle atrophy is by disrupting the IGF-I system. In addition, activation of the IGF-I/insulin pathway can prevent these adverse effects of angiotensin II on skeletal muscle (Delafontaine & Yoshida, 2016).

Treatment with angiotensin II receptor blockers seems also to be protective against muscle loss and strength decrease (Lin et al., 2019). In animal models, blockage of angiotensin II receptor restores muscle remodeling and improves sarcopenia (Burks et al., 2011). Other evidence of the importance of the angiotensin system on muscle physiology is the recent result on the involvement of ACE2 enzyme. This enzyme transforms active angiotensin II into an inactive compound called angiotensin 1–7. When this enzyme is deleted in animal models, muscle weakness and early sarcopenia develops. Thus, this enzyme may be beneficial in muscle function in two ways; by lowering angiotensin II levels and by the direct effect of angiotensin 1–7 on muscle. In this sense, some studies have shown that angiotensin 1–7 has protective effects on skeletal muscle. It has also been reported that angiotensin 1–7 infusion improves muscle strength in mice lacking ACE2. However, it does not completely restore the sarcopenic effects (Takeshita et al., 2018). All these studies indicate that angiotensin II is a new important regulator of skeletal muscle, as reviewed by Cabello-Verrugio, Morales, Rivera, Cabrera, and Simon (2015). Thus, pharmacological treatments for hypertension may be also beneficial against muscle loss, enhancing cost-effectiveness of these therapies for older patients with sarcopenia.

8. Vitamin D

Vitamin D is other important hormone related with muscle physiology (Pojednic & Ceglia, 2014). This hormone is a fat-soluble molecule with pleiotropic actions in the organism. Specifically in skeletal muscle, it increases cell proliferation, differentiation and regeneration and prevents muscle wasting (Bischoff-Ferrari, 2012). With aging, the prevalence of vitamin D deficiency increases (Lips, 2001), and is predictive of low muscle mass and strength in elderly adults (Visser, Deeg, Lips, & Longitudinal Aging Study, 2003). Vitamin D levels have been negatively correlated with sarcopenia in several cross-sectional and prospective studies (Scott et al., 2010; Seo et al., 2012; Tieland, Brouwer-Brolsma, Nienaber-Rousseau, van Loon, & De Groot, 2013). Hirani et al. (2017) in a longitudinal study described a significant increased risk in the rate of sarcopenia in older subjects with low levels of vitamin D. However, other authors did not find the same association (Gielen et al., 2015).

Type II fiber reduction is a characteristic of sarcopenia. Vitamin D increases the number and diameters of fiber II cells (Agergaard et al., 2015; Knutsen, Brekke, Gjelstad, & Lagerlov, 2010). In this sense, there is an association between vitamin D deficiency and type II fiber atrophy, together with other muscle abnormalities such as fat cell infiltration, muscle glycogen granules, fibrosis and interfibrillary gaps (Girgis, 2020). Other characteristics of sarcopenia are low muscle strength, frailty and falls and all of these characteristics are associated with low levels of 25-hydroxyvitamin D (Bischoff-Ferrari, 2012; Visser et al., 2003). However, it seems that this association depends on the sex of the individual, as it has been described that vitamin D insufficiency and low lean mass have a strong association in men but not in women (Spira et al., 2019).

The vitamin D receptor is also affected by aging; as its expression as has been reported to decrease with age (Bischoff-Ferrari et al., 2004). In fact, if the vitamin D receptor is deleted in mice, animals show sarcopenia and impaired muscle function (Girgis et al., 2019). Therefore, the effects of vitamin D deficiency in the development of sarcopenia can be intensified by the age-dependent loss of its receptor. Vitamin D treatment could be a very useful tool for preventing aging-induced muscle atrophy. There is an increasing number of studies on the beneficial role of vitamin D in preventing sarcopenia (reviewed in Remelli, Vitali, Zurlo, & Volpato, 2019). Conversely, the meta-analysis carried out by Beaudart et al. (2014) pointed out that the effect of supplementation with this hormone has effects on muscle strength but little effect on muscle mass. Other meta-analyses (Antoniak & Greig, 2017) and animal studies (Yang et al., 2020) have reported that the efficiency of the vitamin D treatment on muscle depends on its combination with physical activity. Therefore, more studies must be performed in order to better know the role of vitamin D on sarcopenia and the possible benefits of its supplementation.

9. Thyroid hormones

Thyroid hormones (TH) have important actions in muscle physiology in different aspects such as myogenesis and mitochondrial function (Bloise, Cordeiro, & Ortiga-Carvalho, 2018). During aging TH decrease, and this decrease seems to be related with the reduction of myogenesis and the number of type II fibers (Bloise, Oliveira, Cordeiro, & Ortiga-Carvalho, 2018). Elderly patients with subclinical hypothyroidism showed smaller muscle cross-sectional area compared to euthyroid age-matched controls (Moon et al., 2010). Likewise, a positive association between free T3 (FT3) plasma levels and skeletal muscle mass in a population of euthyroid subjects has also been reported (Sheng et al., 2019), and low FT3 was associated with higher frailty score (Bertoli et al., 2017). In addition, in a recent large-scale longitudinal study, authors reported that old euthyroid subjects with attenuated loss of muscle strength (compared with others) were those with higher FT3 levels and a higher FT3/FT4 ratio, after adjusting for multiple confounding factors (Gu et al., 2019). All of these data indicate that TH have a protective effect against the progression of sarcopenia. Furthermore, if these patients are treated with TH, their muscle strength improves and their muscle cross-sectional area increases.

Recently, Szlejf et al. (2020) have found new evidence between the thyroid axis and sarcopenia. They found a U-shaped association between sarcopenia and thyroid stimulating hormone (TSH), the main regulator of TH synthesis and secretion. In this sense, either high or low levels of TSH were associated with low body mass in older adults without overt thyroid dysfunction. The fact that TSH associations with sarcopenia were more potent that the ones found with TH levels may be due to the robustness of the TSH assays. Similarly, other authors have described an association between TSH and muscle strength in cross-sectional studies (Kim et al., 2018; Spira et al., 2019). However, other authors did not find this association when a cohort of old euthyroid adults was studied (Gu et al., 2019).

All in all, there is evidence indicating that different levels of the thyroid system may be altered in the aging process and be responsible, at least in part, for the loss of muscle and the alteration of its physiology leading to sarcopenia.

10. Concluding remarks

Sarcopenia represents an important public health problem, affecting the health and quality of life of older people. Sarcopenia is characterized by a loss of muscle mass, which ultimately leads to a loss of strength and muscle efficiency. Origin and development of sarcopenia is multifactorial. Available epidemiological evidence points to the endocrine environment as an element of key importance (Fig. 2). The GH-IGF-I axis, although it has an important anabolic role throughout life and decreases during aging, it does not seem to have an important role in sarcopenia development. However, the IGF-I produced locally in the muscle, which plays important autocrine and paracrine roles, has been reported to be crucial for preventing age-associated muscle atrophy. Local IGF-I is dependent on anabolic hormones such as testosterone, estrogens and DHEA, and all of them are negatively affected by age. The decrease in these anabolic hormones together with the increase in catabolic hormones such as cortisol has an important impact on decreasing muscle myogenesis and increasing muscle wasting. Metabolic pathways are also affected; muscle use of glucose and fatty acids decreases as a result of the age-induced insulin resistance together with the decrease in leptin sensitivity and the increase in resistin. These conditions are worse in those individuals with obesity and sedentarism and have an important impact on the type of fibers (with the loss of type II fibers) and the accumulation of fatty acids in muscle fiber, aggravating the diabetic state. The alteration of the thyroid axis and vitamin D deficiency, very prevalent

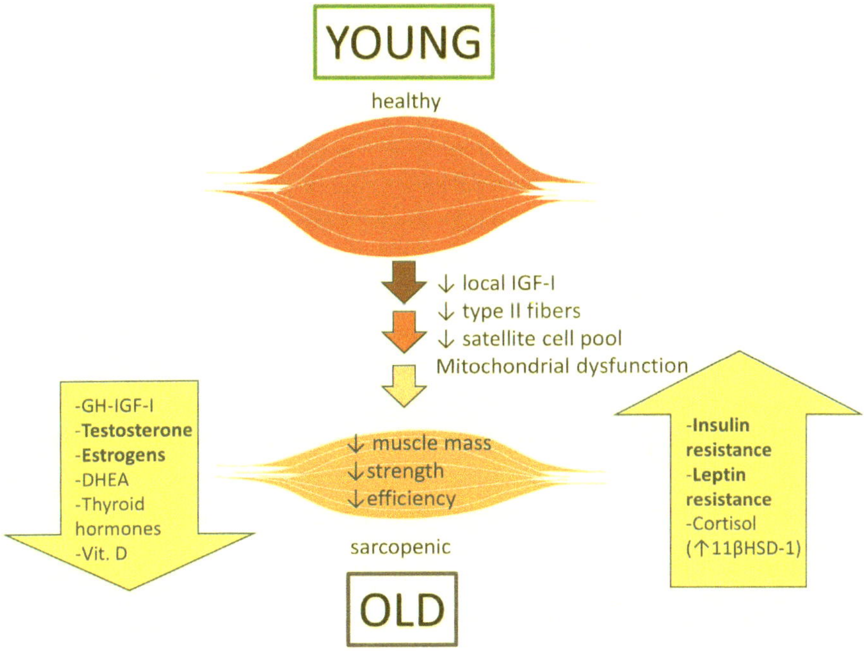

Fig. 2 Main age-related hormonal alterations involved in the development of sarcopenia. Key factors are highlighted in bold font. DHEA, dehydroepiandrosterone; GH, growth hormone; 11β-HSD1, 11b-hydroxysteroid dehydrogenase 1; IGF-I, insulin-like growth factor; Vit. D, vitamin D.

in aging, can also affect these metabolic pathways. The beneficial effects of insulin sensitizers and supplementation with vitamin D have been reported in preventing the loss of muscle mass and strength without the adverse effects of other anabolic hormones. All of these treatments have been proven to be more effective if they are combined with physical exercise. In addition, important new regulators such as angiotensin II have been revealed, opening up new therapeutic windows. Although many studies have been performed, it is still necessary to increase evidence with more studies in order to establish the efficacy and long-term safety of potential therapeutic interventions to ameliorate muscle function.

References

Adams, V., Mangner, N., Gasch, A., Krohne, C., Gielen, S., Hirner, S., et al. (2008). Induction of MuRF1 is essential for TNF-alpha-induced loss of muscle function in mice. *Journal of Molecular Biology*, *384*(1), 48–59. https://doi.org/10.1016/j.jmb.2008.08.087.

Agergaard, J., Trostrup, J., Uth, J., Iversen, J. V., Boesen, A., Andersen, J. L., et al. (2015). Does vitamin-D intake during resistance training improve the skeletal muscle hypertrophic and strength response in young and elderly men?—A randomized controlled trial. *Nutrition & Metabolism (London)*, *12*, 32. https://doi.org/10.1186/s12986-015-0029-y.

Aguiar-Oliveira, M. H., & Bartke, A. (2019). Growth hormone deficiency: Health and longevity. *Endocrine Reviews*, *40*(2), 575–601. https://doi.org/10.1210/er.2018-00216.

Ahima, R. S., & Flier, J. S. (2000). Adipose tissue as an endocrine organ. *Trends in Endocrinology and Metabolism*, *11*(8), 327–332. https://doi.org/10.1016/s1043-2760(00)00301-5.

Ahtiainen, M., Pollanen, E., Ronkainen, P. H., Alen, M., Puolakka, J., Kaprio, J., et al. (2012). Age and estrogen-based hormone therapy affect systemic and local IL-6 and IGF-1 pathways in women. *Age (Dordrecht, Netherlands)*, *34*(5), 1249–1260. https://doi.org/10.1007/s11357-011-9298-1.

Akhmedov, D., & Berdeaux, R. (2013). The effects of obesity on skeletal muscle regeneration. *Frontiers in Physiology*, *4*, 371. https://doi.org/10.3389/fphys.2013.00371.

Antoniak, A. E., & Greig, C. A. (2017). The effect of combined resistance exercise training and vitamin D3 supplementation on musculoskeletal health and function in older adults: A systematic review and meta-analysis. *BMJ Open*, 7(7), e014619. https://doi.org/10.1136/bmjopen-2016-014619.

Arai, Y., Kamide, K., & Hirose, N. (2019). Adipokines and aging: Findings from centenarians and the very old. *Frontiers in Endocrinology*, *10*, 142. https://doi.org/10.3389/fendo.2019.00142.

Arounleut, P., Bowser, M., Upadhyay, S., Shi, X. M., Fulzele, S., Johnson, M. H., et al. (2013). Absence of functional leptin receptor isoforms in the POUND (Lepr(db/lb)) mouse is associated with muscle atrophy and altered myoblast proliferation and differentiation. *PLoS One*, *8*(8), e72330. https://doi.org/10.1371/journal.pone.0072330.

Ascenzi, F., Barberi, L., Dobrowolny, G., Villa Nova Bacurau, A., Nicoletti, C., Rizzuto, E., et al. (2019). Effects of IGF-1 isoforms on muscle growth and sarcopenia. *Aging Cell*, *18*(3), e12954. https://doi.org/10.1111/acel.12954.

Baker, W. L., Karan, S., & Kenny, A. M. (2011). Effect of dehydroepiandrosterone on muscle strength and physical function in older adults: A systematic review. *Journal of the American Geriatrics Society*, *59*(6), 997–1002. https://doi.org/10.1111/j.1532-5415.2011.03410.x.

Baker, J. F., Newman, A. B., Kanaya, A., Leonard, M. B., Zemel, B., Miljkovic, I., et al. (2019). The adiponectin paradox in the elderly: Associations with body composition, physical functioning, and mortality. *The Journals of Gerontology. Series A, Biological Sciences and Medical Sciences*, *74*(2), 247–253. https://doi.org/10.1093/gerona/gly017.

Balagopal, P., Rooyackers, O. E., Adey, D. B., Ades, P. A., & Nair, K. S. (1997). Effects of aging on in vivo synthesis of skeletal muscle myosin heavy-chain and sarcoplasmic protein in humans. *The American Journal of Physiology*, *273*(4), E790–E800. https://doi.org/10.1152/ajpendo.1997.273.4.E790.

Batsis, J. A., & Villareal, D. T. (2018). Sarcopenic obesity in older adults: Aetiology, epidemiology and treatment strategies. *Nature Reviews. Endocrinology*, *14*(9), 513–537. https://doi.org/10.1038/s41574-018-0062-9.

Baylis, D., Bartlett, D. B., Syddall, H. E., Ntani, G., Gale, C. R., Cooper, C., et al. (2013). Immune-endocrine biomarkers as predictors of frailty and mortality: A 10-year longitudinal study in community-dwelling older people. *Age (Dordrecht, Netherlands)*, *35*(3), 963–971. https://doi.org/10.1007/s11357-012-9396-8.

Beaudart, C., Buckinx, F., Rabenda, V., Gillain, S., Cavalier, E., Slomian, J., et al. (2014). The effects of vitamin D on skeletal muscle strength, muscle mass, and muscle power: A systematic review and meta-analysis of randomized controlled trials. *The Journal of Clinical Endocrinology and Metabolism*, *99*(11), 4336–4345. https://doi.org/10.1210/jc.2014-1742.

Bertoli, A., Valentini, A., Cianfarani, M. A., Gasbarra, E., Tarantino, U., & Federici, M. (2017). Low FT3: A possible marker of frailty in the elderly. *Clinical Interventions in Aging, 12*, 335–341. https://doi.org/10.2147/CIA.S125934.

Biercewicz, M., Slusarz, R., Kedziora-Kornatowska, K., Filipska, K., Bielawski, K., & Ruszkowska-Ciastek, B. (2020). Assessment of leptin-to-adiponectin ratio in prediction of insulin resistance and nutrition status in a geriatric female population. *Journal of Physiology and Pharmacology, 71*(1), 35–46. https://doi.org/10.26402/jpp.2020.1.02.

Bischoff-Ferrari, H. A. (2012). Relevance of vitamin D in muscle health. *Reviews in Endocrine & Metabolic Disorders, 13*(1), 71–77. https://doi.org/10.1007/s11154-011-9200-6.

Bischoff-Ferrari, H. A., Borchers, M., Gudat, F., Durmuller, U., Stahelin, H. B., & Dick, W. (2004). Vitamin D receptor expression in human muscle tissue decreases with age. *Journal of Bone and Mineral Research, 19*(2), 265–269. https://doi.org/10.1359/jbmr.2004.19.2.265.

Bloise, F. F., Cordeiro, A., & Ortiga-Carvalho, T. M. (2018). Role of thyroid hormone in skeletal muscle physiology. *The Journal of Endocrinology, 236*(1), R57–R68. https://doi.org/10.1530/JOE-16-0611.

Bloise, F. F., Oliveira, T. S., Cordeiro, A., & Ortiga-Carvalho, T. M. (2018). Thyroid hormones play role in sarcopenia and myopathies. *Frontiers in Physiology, 9*, 560. https://doi.org/10.3389/fphys.2018.00560.

Bo, Y., Liu, C., Ji, Z., Yang, R., An, Q., Zhang, X., et al. (2019). A high whey protein, vitamin D and E supplement preserves muscle mass, strength, and quality of life in sarcopenic older adults: A double-blind randomized controlled trial. *Clinical Nutrition, 38*(1), 159–164. https://doi.org/10.1016/j.clnu.2017.12.020.

Boonen, S., Lysens, R., Verbeke, G., Joosten, E., Dejaeger, E., Pelemans, W., et al. (1998). Relationship between age-associated endocrine deficiencies and muscle function in elderly women: A cross-sectional study. *Age and Ageing, 27*(4), 449–454. https://doi.org/10.1093/ageing/27.4.449.

Bosch-Marce, M., Wee, C. D., Martinez, T. L., Lipkes, C. E., Choe, D. W., Kong, L., et al. (2011). Increased IGF-1 in muscle modulates the phenotype of severe SMA mice. *Human Molecular Genetics, 20*(9), 1844–1853. https://doi.org/10.1093/hmg/ddr067.

Brink, M., Wellen, J., & Delafontaine, P. (1996). Angiotensin II causes weight loss and decreases circulating insulin-like growth factor I in rats through a pressor-independent mechanism. *The Journal of Clinical Investigation, 97*(11), 2509–2516. https://doi.org/10.1172/JCI118698.

Brown, M. (2008). Skeletal muscle and bone: Effect of sex steroids and aging. *Advances in Physiology Education, 32*(2), 120–126. https://doi.org/10.1152/advan.90111.2008.

Bucci, L., Yani, S. L., Fabbri, C., Bijlsma, A. Y., Maier, A. B., Meskers, C. G., et al. (2013). Circulating levels of adipokines and IGF-1 are associated with skeletal muscle strength of young and old healthy subjects. *Biogerontology, 14*(3), 261–272. https://doi.org/10.1007/s10522-013-9428-5.

Burks, T. N., Andres-Mateos, E., Marx, R., Mejias, R., Van Erp, C., Simmers, J. L., et al. (2011). Losartan restores skeletal muscle remodeling and protects against disuse atrophy in sarcopenia. *Science Translational Medicine, 3*(82), 82ra37. https://doi.org/10.1126/scitranslmed.3002227.

Butler, J., Kalogeropoulos, A., Georgiopoulou, V., de Rekeneire, N., Rodondi, N., Smith, A. L., et al. (2009). Serum resistin concentrations and risk of new onset heart failure in older persons: The health, aging, and body composition (Health ABC) study. *Arteriosclerosis, Thrombosis, and Vascular Biology, 29*(7), 1144–1149. https://doi.org/10.1161/ATVBAHA.109.186783.

Cabello-Verrugio, C., Morales, M. G., Rivera, J. C., Cabrera, D., & Simon, F. (2015). Renin-angiotensin system: An old player with novel functions in skeletal muscle. *Medicinal Research Reviews, 35*(3), 437–463. https://doi.org/10.1002/med.21343.

Can, B., Kara, O., Kizilarslanoglu, M. C., Arik, G., Aycicek, G. S., Sumer, F., et al. (2017). Serum markers of inflammation and oxidative stress in sarcopenia. *Aging Clinical and Experimental Research*, *29*(4), 745–752. https://doi.org/10.1007/s40520-016-0626-2.

Carnio, S., LoVerso, F., Baraibar, M. A., Longa, E., Khan, M. M., Maffei, M., et al. (2014). Autophagy impairment in muscle induces neuromuscular junction degeneration and precocious aging. *Cell Reports*, *8*(5), 1509–1521. https://doi.org/10.1016/j.celrep.2014.07.061.

Carter, C. S., & Groban, L. (2008). Role of the renin-angiotensin system in age-related sarcopenia and diastolic dysfunction. *Aging Health*, *4*(1), 37–46. https://doi.org/10.2217/1745509X.4.1.37.

Chang, A. M., & Halter, J. B. (2003). Aging and insulin secretion. *American Journal of Physiology. Endocrinology and Metabolism*, *284*(1), E7–12. https://doi.org/10.1152/ajpendo.00366.2002.

Cheung, T. H., Quach, N. L., Charville, G. W., Liu, L., Park, L., Edalati, A., et al. (2012). Maintenance of muscle stem-cell quiescence by microRNA-489. *Nature*, *482*(7386), 524–528. https://doi.org/10.1038/nature10834.

China, S. P., Pal, S., Chattopadhyay, S., Porwal, K., Kushwaha, S., Bhattacharyya, S., et al. (2017). Globular adiponectin reverses osteo-sarcopenia and altered body composition in ovariectomized rats. *Bone*, *105*, 75–86. https://doi.org/10.1016/j.bone.2017.08.005.

Coen, P. M., Musci, R. V., Hinkley, J. M., & Miller, B. F. (2018). Mitochondria as a target for mitigating sarcopenia. *Frontiers in Physiology*, *9*, 1883. https://doi.org/10.3389/fphys.2018.01883.

Collins, B. C., Laakkonen, E. K., & Lowe, D. A. (2019). Aging of the musculoskeletal system: How the loss of estrogen impacts muscle strength. *Bone*, *123*, 137–144. https://doi.org/10.1016/j.bone.2019.03.033.

Colson, B. A., Petersen, K. J., Collins, B. C., Lowe, D. A., & Thomas, D. D. (2015). The myosin super-relaxed state is disrupted by estradiol deficiency. *Biochemical and Biophysical Research Communications*, *456*(1), 151–155. https://doi.org/10.1016/j.bbrc.2014.11.050.

Conboy, I. M., Conboy, M. J., Wagers, A. J., Girma, E. R., Weissman, I. L., & Rando, T. A. (2005). Rejuvenation of aged progenitor cells by exposure to a young systemic environment. *Nature*, *433*(7027), 760–764. https://doi.org/10.1038/nature03260.

Consitt, L. A., Dudley, C., & Saxena, G. (2019). Impact of endurance and resistance training on skeletal muscle glucose metabolism in older adults. *Nutrients*, *11*(11), 2636. https://doi.org/10.3390/nu11112636.

Cooper, R., Mishra, G., Clennell, S., Guralnik, J., & Kuh, D. (2008). Menopausal status and physical performance in midlife: Findings from a British birth cohort study. *Menopause*, *15*(6), 1079–1085. https://doi.org/10.1097/gme.0b013e31816f63a3.

Corona, G., Giagulli, V. A., Maseroli, E., Vignozzi, L., Aversa, A., Zitzmann, M., et al. (2016). Testosterone supplementation and body composition: Results from a meta-analysis of observational studies. *Journal of Endocrinological Investigation*, *39*(9), 967–981. https://doi.org/10.1007/s40618-016-0480-2.

Cruz-Jentoft, A. J., Baeyens, J. P., Bauer, J. M., Boirie, Y., Cederholm, T., Landi, F., et al. (2010). Sarcopenia: European consensus on definition and diagnosis: Report of the European Working Group on Sarcopenia in Older People. *Age and Ageing*, *39*(4), 412–423. https://doi.org/10.1093/ageing/afq034.

Delafontaine, P., & Yoshida, T. (2016). The renin-angiotensin system and the biology of skeletal muscle: Mechanisms of muscle wasting in chronic disease states. *Transactions of the American Clinical and Climatological Association*, *127*, 245–258. Retrieved from https://www.ncbi.nlm.nih.gov/pubmed/28066057.

Di Bari, M., van de Poll-Franse, L. V., Onder, G., Kritchevsky, S. B., Newman, A., Harris, T. B., et al. (2004). Antihypertensive medications and differences in muscle mass in older persons: The Health, Aging and Body Composition Study. *Journal of the American Geriatrics Society*, *52*(6), 961–966. https://doi.org/10.1111/j.1532-5415.2004.52265.x.

Dieli-Conwright, C. M., Spektor, T. M., Rice, J. C., Sattler, F. R., & Schroeder, E. T. (2009). Influence of hormone replacement therapy on eccentric exercise induced myogenic gene expression in postmenopausal women. *Journal of Applied Physiology (Bethesda, MD: 1985)*, *107*(5), 1381–1388. https://doi.org/10.1152/japplphysiol.00590.2009.

Dietze, G. J., & Henriksen, E. J. (2008). Angiotensin-converting enzyme in skeletal muscle: Sentinel of blood pressure control and glucose homeostasis. *Journal of the Renin-Angiotensin-Aldosterone System*, *9*(2), 75–88. https://doi.org/10.3317/jraas.2008.011.

Drey, M., Krieger, B., Sieber, C. C., Bauer, J. M., Hettwer, S., Bertsch, T., et al. (2014). Motoneuron loss is associated with sarcopenia. *Journal of the American Medical Directors Association*, *15*(6), 435–439. https://doi.org/10.1016/j.jamda.2014.02.002.

Dubois, V., Laurent, M. R., Sinnesael, M., Cielen, N., Helsen, C., Clinckemalie, L., et al. (2014). A satellite cell-specific knockout of the androgen receptor reveals myostatin as a direct androgen target in skeletal muscle. *The FASEB Journal*, *28*(7), 2979–2994. https://doi.org/10.1096/fj.14-249748.

Dyck, D. J. (2009). Adipokines as regulators of muscle metabolism and insulin sensitivity. *Applied Physiology, Nutrition, and Metabolism*, *34*(3), 396–402. https://doi.org/10.1139/H09-037.

Ettinger, B., Quesenberry, C., Schroeder, D. A., & Friedman, G. (2018). Long-term postmenopausal estrogen therapy may be associated with increased risk of breast cancer: A cohort study. *Menopause*, *25*(11), 1191–1194. https://doi.org/10.1097/GME.0000000000001216.

Ferrando, A. A., Sheffield-Moore, M., Yeckel, C. W., Gilkison, C., Jiang, J., Achacosa, A., et al. (2002). Testosterone administration to older men improves muscle function: Molecular and physiological mechanisms. *American Journal of Physiology. Endocrinology and Metabolism*, *282*(3), E601–E607. https://doi.org/10.1152/ajpendo.00362.2001.

Finni, T., Noorkoiv, M., Pollanen, E., Ronkainen, P. H., Alen, M., Kaprio, J., et al. (2011). Muscle function in monozygotic female twin pairs discordant for hormone replacement therapy. *Muscle & Nerve*, *44*(5), 769–775. https://doi.org/10.1002/mus.22162.

Frajese, G. V., Taylor, N. F., Jenkins, P. J., Besser, G. M., & Monson, J. P. (2004). Modulation of cortisol metabolism during treatment of acromegaly is independent of body composition and insulin sensitivity. *Hormone Research*, *61*(5), 246–251. https://doi.org/10.1159/000077135.

Friedman, J. M. (2011). Leptin and the regulation of body weigh. *The Keio Journal of Medicine*, *60*(1), 1–9. https://doi.org/10.2302/kjm.60.1.

Fuentes, T., Ara, I., Guadalupe-Grau, A., Larsen, S., Stallknecht, B., Olmedillas, H., et al. (2010). Leptin receptor 170 kDa (OB-R170) protein expression is reduced in obese human skeletal muscle: A potential mechanism of leptin resistance. *Experimental Physiology*, *95*(1), 160–171. https://doi.org/10.1113/expphysiol.2009.049270.

Gaster, M., Poulsen, P., Handberg, A., Schroder, H. D., & Beck-Nielsen, H. (2000). Direct evidence of fiber type-dependent GLUT-4 expression in human skeletal muscle. *American Journal of Physiology. Endocrinology and Metabolism*, *278*(5), E910–E916. https://doi.org/10.1152/ajpendo.2000.278.5.E910.

Gielen, E., O'Neill, T. W., Pye, S. R., Adams, J. E., Wu, F. C., Laurent, M. R., et al. (2015). Endocrine determinants of incident sarcopenia in middle-aged and elderly European men. *Journal of Cachexia, Sarcopenia and Muscle*, *6*(3), 242–252. https://doi.org/10.1002/jcsm.12030.

Girgis, C. M. (2020). Vitamin D and skeletal muscle: Emerging roles in development, anabolism and repair. *Calcified Tissue International*, *106*(1), 47–57. https://doi.org/10.1007/s00223-019-00583-4.

Girgis, C. M., Cha, K. M., So, B., Tsang, M., Chen, J., Houweling, P. J., et al. (2019). Mice with myocyte deletion of vitamin D receptor have sarcopenia and impaired muscle function. *Journal of Cachexia, Sarcopenia and Muscle, 10*(6), 1228–1240. https://doi.org/10.1002/jcsm.12460.

Goodpaster, B. H., Park, S. W., Harris, T. B., Kritchevsky, S. B., Nevitt, M., Schwartz, A. V., et al. (2006). The loss of skeletal muscle strength, mass, and quality in older adults: The health, aging and body composition study. *The Journals of Gerontology. Series A, Biological Sciences and Medical Sciences, 61*(10), 1059–1064. https://doi.org/10.1093/gerona/61.10.1059.

Gregorich, Z. R., Peng, Y., Cai, W., Jin, Y., Wei, L., Chen, A. J., et al. (2016). Top-down targeted proteomics reveals decrease in myosin regulatory light-chain phosphorylation that contributes to sarcopenic muscle dysfunction. *Journal of Proteome Research, 15*(8), 2706–2716. https://doi.org/10.1021/acs.jproteome.6b00244.

Greising, S. M., Baltgalvis, K. A., Lowe, D. A., & Warren, G. L. (2009). Hormone therapy and skeletal muscle strength: A meta-analysis. *The Journals of Gerontology. Series A, Biological Sciences and Medical Sciences, 64*(10), 1071–1081. https://doi.org/10.1093/gerona/glp082.

Gu, Y., Meng, G., Wu, H., Zhang, Q., Liu, L., Bao, X., et al. (2019). Thyroid function as a predictor of handgrip strength among middle-aged and older euthyroid adults: The TCLSIH Cohort Study. *Journal of the American Medical Directors Association, 20*(10), 1236–1241. https://doi.org/10.1016/j.jamda.2019.06.014.

Guevara-Aguirre, J., Rosenbloom, A. L., Balasubramanian, P., Teran, E., Guevara-Aguirre, M., Guevara, C., et al. (2015). GH receptor deficiency in Ecuadorian adults is associated with obesity and enhanced insulin sensitivity. *The Journal of Clinical Endocrinology and Metabolism, 100*(7), 2589–2596. https://doi.org/10.1210/jc.2015-1678.

Hameed, M., Harridge, S. D., & Goldspink, G. (2002). Sarcopenia and hypertrophy: A role for insulin-like growth factor-1 in aged muscle? *Exercise and Sport Sciences Reviews, 30*(1), 15–19. https://doi.org/10.1097/00003677-200201000-00004.

Hamrick, M. W. (2017). Role of the cytokine-like hormone leptin in muscle-bone crosstalk with aging. *Journal of Bone Metabolism, 24*(1), 1–8. https://doi.org/10.11005/jbm.2017.24.1.1.

Hamrick, M. W., Dukes, A., Arounleut, P., Davis, C., Periyasamy-Thandavan, S., Mork, S., et al. (2015). The adipokine leptin mediates muscle- and liver-derived IGF-1 in aged mice. *Experimental Gerontology, 70*, 92–96. https://doi.org/10.1016/j.exger.2015.07.014.

Hamrick, M. W., Herberg, S., Arounleut, P., He, H. Z., Shiver, A., Qi, R. Q., et al. (2010). The adipokine leptin increases skeletal muscle mass and significantly alters skeletal muscle miRNA expression profile in aged mice. *Biochemical and Biophysical Research Communications, 400*(3), 379–383. https://doi.org/10.1016/j.bbrc.2010.08.079.

Harridge, S. D. (2007). Plasticity of human skeletal muscle: Gene expression to in vivo function. *Experimental Physiology, 92*(5), 783–797. https://doi.org/10.1113/expphysiol.2006.036525.

Hassan-Smith, Z. K., Morgan, S. A., Sherlock, M., Hughes, B., Taylor, A. E., Lavery, G. G., et al. (2015). Gender-specific differences in skeletal muscle 11beta-HSD1 expression across healthy aging. *The Journal of Clinical Endocrinology and Metabolism, 100*(7), 2673–2681. https://doi.org/10.1210/jc.2015-1516.

Hildreth, K. L., Barry, D. W., Moreau, K. L., Vande Griend, J., Meacham, R. B., Nakamura, T., et al. (2013). Effects of testosterone and progressive resistance exercise in healthy, highly functioning older men with low-normal testosterone levels. *The Journal of Clinical Endocrinology and Metabolism, 98*(5), 1891–1900. https://doi.org/10.1210/jc.2013-2227.

Hirani, V., Cumming, R. G., Naganathan, V., Blyth, F., Le Couteur, D. G., Hsu, B., et al. (2017). Longitudinal associations between vitamin D metabolites and sarcopenia in older

Australian men: The Concord Health and Aging in Men Project. *The Journals of Gerontology. Series A, Biological Sciences and Medical Sciences*, *73*(1), 131–138. https://doi.org/10.1093/gerona/glx086.

Hong, Y., Lee, J. H., Jeong, K. W., Choi, C. S., & Jun, H. S. (2019). Amelioration of muscle wasting by glucagon-like peptide-1 receptor agonist in muscle atrophy. *Journal of Cachexia, Sarcopenia and Muscle*, *10*(4), 903–918. https://doi.org/10.1002/jcsm.12434.

Hughes, V. A., Frontera, W. R., Roubenoff, R., Evans, W. J., & Singh, M. A. (2002). Longitudinal changes in body composition in older men and women: Role of body weight change and physical activity. *The American Journal of Clinical Nutrition*, *76*(2), 473–481. https://doi.org/10.1093/ajcn/76.2.473.

Iizuka, K., Machida, T., & Hirafuji, M. (2014). Skeletal muscle is an endocrine organ. *Journal of Pharmacological Sciences*, *125*(2), 125–131. https://doi.org/10.1254/jphs.14r02cp.

Ikeda, K., Horie-Inoue, K., & Inoue, S. (2019). Functions of estrogen and estrogen receptor signaling on skeletal muscle. *The Journal of Steroid Biochemistry and Molecular Biology*, *191*, 105375. https://doi.org/10.1016/j.jsbmb.2019.105375.

Irwin, R. W., Yao, J., Hamilton, R. T., Cadenas, E., Brinton, R. D., & Nilsen, J. (2008). Progesterone and estrogen regulate oxidative metabolism in brain mitochondria. *Endocrinology*, *149*(6), 3167–3175. https://doi.org/10.1210/en.2007-1227.

Javed, A. A., Mayhew, A. J., Shea, A. K., & Raina, P. (2019). Association between hormone therapy and muscle mass in postmenopausal women: A systematic review and meta-analysis. *JAMA Network Open*, *2*(8), e1910154. https://doi.org/10.1001/jamanetworkopen.2019.10154.

Kerksick, C., Taylor, L.t., Harvey, A., & Willoughby, D. (2008). Gender-related differences in muscle injury, oxidative stress, and apoptosis. *Medicine and Science in Sports and Exercise*, *40*(10), 1772–1780. https://doi.org/10.1249/MSS.0b013e31817d1cce.

Kilgour, A. H., Gallagher, I. J., MacLullich, A. M., Andrew, R., Gray, C. D., Hyde, P., et al. (2013). Increased skeletal muscle 11betaHSD1 mRNA is associated with lower muscle strength in ageing. *PLoS One*, *8*(12), e84057. https://doi.org/10.1371/journal.pone.0084057.

Kim, B. J., Lee, S. H., Isales, C. M., Hamrick, M. W., Kwak, M. K., & Koh, J. M. (2018). Association of serum TSH with handgrip strength in community-dwelling euthyroid elderly. *The Journal of Clinical Endocrinology and Metabolism*, *103*(11), 3986–3992. https://doi.org/10.1210/jc.2018-01095.

Kizer, J. R., Arnold, A. M., Strotmeyer, E. S., Ives, D. G., Cushman, M., Ding, J., et al. (2010). Change in circulating adiponectin in advanced old age: Determinants and impact on physical function and mortality. The cardiovascular Health Study All Stars Study. *The Journals of Gerontology. Series A, Biological Sciences and Medical Sciences*, *65*(11), 1208–1214. https://doi.org/10.1093/gerona/glq122.

Knutsen, K. V., Brekke, M., Gjelstad, S., & Lagerlov, P. (2010). Vitamin D status in patients with musculoskeletal pain, fatigue and headache: A cross-sectional descriptive study in a multi-ethnic general practice in Norway. *Scandinavian Journal of Primary Health Care*, *28*(3), 166–171. https://doi.org/10.3109/02813432.2010.505407.

Kohara, K., Ochi, M., Tabara, Y., Nagai, T., Igase, M., & Miki, T. (2011). Leptin in sarcopenic visceral obesity: Possible link between adipocytes and myocytes. *PLoS One*, *6*(9), e24633. https://doi.org/10.1371/journal.pone.0024633.

Kostka, T., Arsac, L. M., Patricot, M. C., Berthouze, S. E., Lacour, J. R., & Bonnefoy, M. (2000). Leg extensor power and dehydroepiandrosterone sulfate, insulin-like growth factor-I and testosterone in healthy active elderly people. *European Journal of Applied Physiology*, *82*(1–2), 83–90. https://doi.org/10.1007/s004210050655.

Lang, T., Streeper, T., Cawthon, P., Baldwin, K., Taaffe, D. R., & Harris, T. B. (2010). Sarcopenia: Etiology, clinical consequences, intervention, and assessment. *Osteoporosis International*, *21*(4), 543–559. https://doi.org/10.1007/s00198-009-1059-y.

Li, N., Liu, F., Yang, P., Xiong, F., Yu, Q., Li, J., et al. (2019). Aging and stress induced beta cell senescence and its implication in diabetes development. *Aging (Albany NY)*, *11*(21), 9947–9959. https://doi.org/10.18632/aging.102432.

Li, C. W., Yu, K., Shyh-Chang, N., Li, G. X., Jiang, L. J., Yu, S. L., et al. (2019). Circulating factors associated with sarcopenia during ageing and after intensive lifestyle intervention. *Journal of Cachexia, Sarcopenia and Muscle*, *10*(3), 586–600. https://doi.org/10.1002/jcsm.12417.

Lin, Y. L., Chen, S. Y., Lai, Y. H., Wang, C. H., Kuo, C. H., Liou, H. H., et al. (2019). Angiotensin II receptor blockade is associated with preserved muscle strength in chronic hemodialysis patients. *BMC Nephrology*, *20*(1), 54. https://doi.org/10.1186/s12882-019-1223-3.

Lips, P. (2001). Vitamin D deficiency and secondary hyperparathyroidism in the elderly: Consequences for bone loss and fractures and therapeutic implications. *Endocrine Reviews*, *22*(4), 477–501. https://doi.org/10.1210/edrv.22.4.0437.

Liu, Z. J., Bian, J., Liu, J., & Endoh, A. (2007). Obesity reduced the gene expressions of leptin receptors in hypothalamus and liver. *Hormone and Metabolic Research*, *39*(7), 489–494. https://doi.org/10.1055/s-2007-981680.

Liu, Y., & Sweeney, G. (2014). Adiponectin action in skeletal muscle. *Best Practice & Research. Clinical Endocrinology & Metabolism*, *28*(1), 33–41. https://doi.org/10.1016/j.beem.2013.08.003.

López-Menduina, M., Martin, A. I., Castillero, E., Villanúa, M. A., & López-Calderón, A. (2012). Short-term growth hormone or IGF-I administration improves the IGF-IGFBP system in arthritic rats. *Growth Hormone & IGF Research*, *22*(1), 22–29. https://doi.org/10.1016/j.ghir.2011.12.003.

Maggio, M., Ble, A., Ceda, G. P., & Metter, E. J. (2006). Decline in insulin-like growth factor-I levels across adult life span in two large population studies. *The Journals of Gerontology. Series A, Biological Sciences and Medical Sciences*, *61*(2), 182–183. https://doi.org/10.1093/gerona/61.2.182.

Maggio, M., Lauretani, F., & Ceda, G. P. (2013). Sex hormones and sarcopenia in older persons. *Current Opinion in Clinical Nutrition and Metabolic Care*, *16*(1), 3–13. https://doi.org/10.1097/MCO.0b013e32835b6044.

Marcell, T. J. (2003). Sarcopenia: Causes, consequences, and preventions. *The Journals of Gerontology. Series A, Biological Sciences and Medical Sciences*, *58*(10), M911–M916. https://doi.org/10.1093/gerona/58.10.m911.

Martel, G. F., Roth, S. M., Ivey, F. M., Lemmer, J. T., Tracy, B. L., Hurlbut, D. E., et al. (2006). Age and sex affect human muscle fibre adaptations to heavy-resistance strength training. *Experimental Physiology*, *91*(2), 457–464. https://doi.org/10.1113/expphysiol.2005.032771.

Martín, A. I., Priego, T., & López-Calderón, A. (2018). Hormones and muscle atrophy. *Advances in Experimental Medicine and Biology*, *1088*, 207–233. https://doi.org/10.1007/978-981-13-1435-3_9.

Marzetti, E., Calvani, R., Tosato, M., Cesari, M., Di Bari, M., Cherubini, A., et al. (2017). Physical activity and exercise as countermeasures to physical frailty and sarcopenia. *Aging Clinical and Experimental Research*, *29*(1), 35–42. https://doi.org/10.1007/s40520-016-0705-4.

Matheny, R. W., Merritt, E., Zannikos, S. V., Farrar, R. P., & Adamo, M. L. (2009). Serum IGF-I-deficiency does not prevent compensatory skeletal muscle hypertrophy in resistance exercise. *Experimental Biology and Medicine (Maywood, N.J.)*, *234*(2), 164–170. https://doi.org/10.3181/0808-RM-251.

McDonagh, B., Scullion, S. M., Vasilaki, A., Pollock, N., McArdle, A., & Jackson, M. J. (2016). Ageing-induced changes in the redox status of peripheral motor nerves imply an effect on redox signalling rather than oxidative damage. *Free Radical Biology & Medicine*, *94*, 27–35. https://doi.org/10.1016/j.freeradbiomed.2016.02.008.

McInnes, K. J., Andersson, T. C., Simonyte, K., Soderstrom, I., Mattsson, C., Seckl, J. R., et al. (2012). Association of 11beta-hydroxysteroid dehydrogenase type I expression and activity with estrogen receptor beta in adipose tissue from postmenopausal women. *Menopause*, *19*(12), 1347–1352. https://doi.org/10.1097/gme.0b013e318258aad7.

Milanesi, L., Russo de Boland, A., & Boland, R. (2008). Expression and localization of estrogen receptor alpha in the C2C12 murine skeletal muscle cell line. *Journal of Cellular Biochemistry*, *104*(4), 1254–1273. https://doi.org/10.1002/jcb.21706.

Mitchell, W. K., Williams, J., Atherton, P., Larvin, M., Lund, J., & Narici, M. (2012). Sarcopenia, dynapenia, and the impact of advancing age on human skeletal muscle size and strength; a quantitative review. *Frontiers in Physiology*, *3*, 260. https://doi.org/10.3389/fphys.2012.00260.

Montero-Fernandez, N., & Serra-Rexach, J. A. (2013). Role of exercise on sarcopenia in the elderly. *European Journal of Physical and Rehabilitation Medicine*, *49*(1), 131–143. Retrieved from https://www.ncbi.nlm.nih.gov/pubmed/23575207.

Moon, S. S. (2014). Low skeletal muscle mass is associated with insulin resistance, diabetes, and metabolic syndrome in the Korean population: The Korea National Health and Nutrition Examination Survey (KNHANES) 2009-2010. *Endocrine Journal*, *61*(1), 61–70. https://doi.org/10.1507/endocrj.ej13-0244.

Moon, M. K., Lee, Y. J., Choi, S. H., Lim, S., Yang, E. J., Lim, J. Y., et al. (2010). Subclinical hypothyroidism has little influences on muscle mass or strength in elderly people. *Journal of Korean Medical Science*, *25*(8), 1176–1181. https://doi.org/10.3346/jkms.2010.25.8.1176.

Moran, A. L., Nelson, S. A., Landisch, R. M., Warren, G. L., & Lowe, D. A. (2007). Estradiol replacement reverses ovariectomy-induced muscle contractile and myosin dysfunction in mature female mice. *Journal of Applied Physiology (Bethesda, MD: 1985)*, *102*(4), 1387–1393. https://doi.org/10.1152/japplphysiol.01305.2006.

Morley, J. E., Anker, S. D., & von Haehling, S. (2014). Prevalence, incidence, and clinical impact of sarcopenia: Facts, numbers, and epidemiology-update 2014. *Journal of Cachexia, Sarcopenia and Muscle*, *5*(4), 253–259. https://doi.org/10.1007/s13539-014-0161-y.

Mukund, K., & Subramaniam, S. (2020). Skeletal muscle: A review of molecular structure and function, in health and disease. *Wiley Interdisciplinary Reviews. Systems Biology and Medicine*, *12*(1), e1462. https://doi.org/10.1002/wsbm.1462.

Musaro, A., McCullagh, K., Paul, A., Houghton, L., Dobrowolny, G., Molinaro, M., et al. (2001). Localized Igf-1 transgene expression sustains hypertrophy and regeneration in senescent skeletal muscle. *Nature Genetics*, *27*(2), 195–200. https://doi.org/10.1038/84839.

Myers, M. G., Cowley, M. A., & Munzberg, H. (2008). Mechanisms of leptin action and leptin resistance. *Annual Review of Physiology*, *70*, 537–556. https://doi.org/10.1146/annurev.physiol.70.113006.100707.

Nam, Y. S., Lee, G., Yun, J. M., & Cho, B. (2018). Testosterone replacement, muscle strength, and physical function. *The World Journal of Men's Health*, *36*(2), 110–122. https://doi.org/10.5534/wjmh.182001.

Nieuwenhuizen, W. F., Weenen, H., Rigby, P., & Hetherington, M. M. (2010). Older adults and patients in need of nutritional support: Review of current treatment options and factors influencing nutritional intake. *Clinical Nutrition*, *29*(2), 160–169. https://doi.org/10.1016/j.clnu.2009.09.003.

Nilwik, R., Snijders, T., Leenders, M., Groen, B. B., van Kranenburg, J., Verdijk, L. B., et al. (2013). The decline in skeletal muscle mass with aging is mainly attributed to a reduction in type II muscle fiber size. *Experimental Gerontology*, *48*(5), 492–498. https://doi.org/10.1016/j.exger.2013.02.012.

O'Leary, M. F., Wallace, G. R., Davis, E. T., Murphy, D. P., Nicholson, T., Bennett, A. J., et al. (2018). Obese subcutaneous adipose tissue impairs human myogenesis, particularly in old skeletal muscle, via resistin-mediated activation of NFkappaB. *Scientific Reports, 8*(1), 15360. https://doi.org/10.1038/s41598-018-33840-x.

O'Neil, T. K., Duffy, L. R., Frey, J. W., & Hornberger, T. A. (2009). The role of phosphoinositide 3-kinase and phosphatidic acid in the regulation of mammalian target of rapamycin following eccentric contractions. *The Journal of Physiology, 587*(Pt. 14), 3691–3701. https://doi.org/10.1113/jphysiol.2009.173609.

Otte, C., Hart, S., Neylan, T. C., Marmar, C. R., Yaffe, K., & Mohr, D. C. (2005). A meta-analysis of cortisol response to challenge in human aging: Importance of gender. *Psychoneuroendocrinology, 30*(1), 80–91. https://doi.org/10.1016/j.psyneuen.2004.06.002.

Pareja-Galeano, H., Santos-Lozano, A., Sanchis-Gomar, F., Fiuza-Luces, C., Garatachea, N., Galvez, B. G., et al. (2017). Circulating leptin and adiponectin concentrations in healthy exceptional longevity. *Mechanisms of Ageing and Development, 162*, 129–132. https://doi.org/10.1016/j.mad.2016.02.014.

Pedraza-Alva, G., Zingg, J. M., Donda, A., & Perez-Martinez, L. (2009). Estrogen receptor regulates MyoD gene expression by preventing AP-1-mediated repression. *Biochemical and Biophysical Research Communications, 389*(2), 360–365. https://doi.org/10.1016/j.bbrc.2009.08.153.

Peeters, G. M., van Schoor, N. M., van Rossum, E. F., Visser, M., & Lips, P. (2008). The relationship between cortisol, muscle mass and muscle strength in older persons and the role of genetic variations in the glucocorticoid receptor. *Clinical Endocrinology, 69*(4), 673–682. https://doi.org/10.1111/j.1365-2265.2008.03212.x.

Pellegrinelli, V., Rouault, C., Rodriguez-Cuenca, S., Albert, V., Edom-Vovard, F., Vidal-Puig, A., et al. (2015). Human adipocytes induce inflammation and atrophy in muscle cells during obesity. *Diabetes, 64*(9), 3121–3134. https://doi.org/10.2337/db14-0796.

Petersen, K. F., Morino, K., Alves, T. C., Kibbey, R. G., Dufour, S., Sono, S., et al. (2015). Effect of aging on muscle mitochondrial substrate utilization in humans. *Proceedings of the National Academy of Sciences of the United States of America, 112*(36), 11330–11334. https://doi.org/10.1073/pnas.1514844112.

Pojednic, R. M., & Ceglia, L. (2014). The emerging biomolecular role of vitamin D in skeletal muscle. *Exercise and Sport Sciences Reviews, 42*(2), 76–81. https://doi.org/10.1249/JES.0000000000000013.

Pronsato, L., Milanesi, L., & Vasconsuelo, A. (2020). Testosterone induces up-regulation of mitochondrial gene expression in murine C2C12 skeletal muscle cells accompanied by an increase of nuclear respiratory factor-1 and its downstream effectors. *Molecular and Cellular Endocrinology, 500*, 110631. https://doi.org/10.1016/j.mce.2019.110631.

Remelli, F., Vitali, A., Zurlo, A., & Volpato, S. (2019). Vitamin D deficiency and sarcopenia in older persons. *Nutrients, 11*(12), 2861. https://doi.org/10.3390/nu11122861.

Ribas, V., Drew, B. G., Zhou, Z., Phun, J., Kalajian, N. Y., Soleymani, T., et al. (2016). Skeletal muscle action of estrogen receptor alpha is critical for the maintenance of mitochondrial function and metabolic homeostasis in females. *Science Translational Medicine, 8*(334), 334ra354. https://doi.org/10.1126/scitranslmed.aad3815.

Rivas, D. A., Lessard, S. J., Rice, N. P., Lustgarten, M. S., So, K., Goodyear, L. J., et al. (2014). Diminished skeletal muscle microRNA expression with aging is associated with attenuated muscle plasticity and inhibition of IGF-1 signaling. *The FASEB Journal, 28*(9), 4133–4147. https://doi.org/10.1096/fj.14-254490.

Rizzo, M. R., Barbieri, M., Fava, I., Desiderio, M., Coppola, C., Marfella, R., et al. (2016). Sarcopenia in elderly diabetic patients: Role of dipeptidyl peptidase 4 inhibitors. *Journal of the American Medical Directors Association, 17*(10), 896–901. https://doi.org/10.1016/j.jamda.2016.04.016.

Rudman, D., Kutner, M. H., Rogers, C. M., Lubin, M. F., Fleming, G. A., & Bain, R. P. (1981). Impaired growth hormone secretion in the adult population: Relation to age and adiposity. *The Journal of Clinical Investigation, 67*(5), 1361–1369. https://doi.org/10.1172/jci110164.

Rygiel, K. A., Picard, M., & Turnbull, D. M. (2016). The ageing neuromuscular system and sarcopenia: A mitochondrial perspective. *The Journal of Physiology, 594*(16), 4499–4512. https://doi.org/10.1113/JP271212.

Sainz, N., Rodriguez, A., Catalan, V., Becerril, S., Ramirez, B., Gomez-Ambrosi, J., et al. (2009). Leptin administration favors muscle mass accretion by decreasing FoxO3a and increasing PGC-1alpha in ob/ob mice. *PLoS One, 4*(9), e6808. https://doi.org/10.1371/journal.pone.0006808.

Sakuma, K., Kinoshita, M., Ito, Y., Aizawa, M., Aoi, W., & Yamaguchi, A. (2016). p62/SQSTM1 but not LC3 is accumulated in sarcopenic muscle of mice. *Journal of Cachexia, Sarcopenia and Muscle*, 7(2), 204–212. https://doi.org/10.1002/jcsm.12045.

Samaras, N., Samaras, D., Frangos, E., Forster, A., & Philippe, J. (2013). A review of age-related dehydroepiandrosterone decline and its association with well-known geriatric syndromes: Is treatment beneficial? *Rejuvenation Research, 16*(4), 285–294. https://doi.org/10.1089/rej.2013.1425.

Santoro, A., Guidarelli, G., Ostan, R., Giampieri, E., Fabbri, C., Bertarelli, C., et al. (2019). Gender-specific association of body composition with inflammatory and adipose-related markers in healthy elderly Europeans from the NU-AGE study. *European Radiology, 29*(9), 4968–4979. https://doi.org/10.1007/s00330-018-5973-2.

Sattler, F. R. (2013). Growth hormone in the aging male. *Best Practice & Research. Clinical Endocrinology & Metabolism, 27*(4), 541–555. https://doi.org/10.1016/j.beem.2013.05.003.

Schakman, O., Kalista, S., Barbe, C., Loumaye, A., & Thissen, J. P. (2013). Glucocorticoid-induced skeletal muscle atrophy. *The International Journal of Biochemistry & Cell Biology, 45*(10), 2163–2172. https://doi.org/10.1016/j.biocel.2013.05.036.

Scott, D., Blizzard, L., Fell, J., Ding, C., Winzenberg, T., & Jones, G. (2010). A prospective study of the associations between 25-hydroxy-vitamin D, sarcopenia progression and physical activity in older adults. *Clinical Endocrinology, 73*(5), 581–587. https://doi.org/10.1111/j.1365-2265.2010.03858.x.

Seo, J. A., Cho, H., Eun, C. R., Yoo, H. J., Kim, S. G., Choi, K. M., et al. (2012). Association between visceral obesity and sarcopenia and vitamin D deficiency in older Koreans: The Ansan Geriatric Study. *Journal of the American Geriatrics Society, 60*(4), 700–706. https://doi.org/10.1111/j.1532-5415.2012.03887.x.

Serra, C., Tangherlini, F., Rudy, S., Lee, D., Toraldo, G., Sandor, N. L., et al. (2013). Testosterone improves the regeneration of old and young mouse skeletal muscle. *The Journals of Gerontology. Series A, Biological Sciences and Medical Sciences, 68*(1), 17–26. https://doi.org/10.1093/gerona/gls083.

Shafiee, G., Keshtkar, A., Soltani, A., Ahadi, Z., Larijani, B., & Heshmat, R. (2017). Prevalence of sarcopenia in the world: A systematic review and meta-analysis of general population studies. *Journal of Diabetes and Metabolic Disorders, 16*, 21. https://doi.org/10.1186/s40200-017-0302-x.

Sheng, C. H., Du, Z. W., Song, Y., Wu, X. D., Zhang, Y. C., Wu, M., et al. (2013). Human resistin inhibits myogenic differentiation and induces insulin resistance in myocytes. *BioMed Research International, 2013*, 804632. https://doi.org/10.1155/2013/804632.

Sheng, Y., Ma, D., Zhou, Q., Wang, L., Sun, M., Wang, S., et al. (2019). Association of thyroid function with sarcopenia in elderly Chinese euthyroid subjects. *Aging Clinical and Experimental Research, 31*(8), 1113–1120. https://doi.org/10.1007/s40520-018-1057-z.

Shin, M. J., Jeon, Y. K., & Kim, I. J. (2018). Testosterone and sarcopenia. *The World Journal of Men's Health, 36*(3), 192–198. https://doi.org/10.5534/wjmh.180001.

Shou, J., Chen, P. J., & Xiao, W. H. (2020). Mechanism of increased risk of insulin resistance in aging skeletal muscle. *Diabetology and Metabolic Syndrome, 12*, 14. https://doi.org/10.1186/s13098-020-0523-x.

Sinha-Hikim, I., Artaza, J., Woodhouse, L., Gonzalez-Cadavid, N., Singh, A. B., Lee, M. I., et al. (2002). Testosterone-induced increase in muscle size in healthy young men is associated with muscle fiber hypertrophy. *American Journal of Physiology. Endocrinology and Metabolism, 283*(1), E154–E164. https://doi.org/10.1152/ajpendo.00502.2001.

Skinner, J. W., Otzel, D. M., Bowser, A., Nargi, D., Agarwal, S., Peterson, M. D., et al. (2018). Muscular responses to testosterone replacement vary by administration route: A systematic review and meta-analysis. *Journal of Cachexia, Sarcopenia and Muscle, 9*(3), 465–481. https://doi.org/10.1002/jcsm.12291.

Song, Y. H., Li, Y., Du, J., Mitch, W. E., Rosenthal, N., & Delafontaine, P. (2005). Muscle-specific expression of IGF-1 blocks angiotensin II-induced skeletal muscle wasting. *The Journal of Clinical Investigation, 115*(2), 451–458. https://doi.org/10.1172/JCI22324.

Spira, D., Buchmann, N., Konig, M., Rosada, A., Steinhagen-Thiessen, E., Demuth, I., et al. (2019). Sex-specific differences in the association of vitamin D with low lean mass and frailty: Results from the Berlin Aging Study II. *Nutrition, 62*, 1–6. https://doi.org/10.1016/j.nut.2018.11.020.

Srikanthan, P., Hevener, A. L., & Karlamangla, A. S. (2010). Sarcopenia exacerbates obesity-associated insulin resistance and dysglycemia: Findings from the National Health and Nutrition Examination Survey III. *PLoS One, 5*(5), e10805. https://doi.org/10.1371/journal.pone.0010805.

Srinivas-Shankar, U., Roberts, S. A., Connolly, M. J., O'Connell, M. D., Adams, J. E., Oldham, J. A., et al. (2010). Effects of testosterone on muscle strength, physical function, body composition, and quality of life in intermediate-frail and frail elderly men: A randomized, double-blind, placebo-controlled study. *The Journal of Clinical Endocrinology and Metabolism, 95*(2), 639–650. https://doi.org/10.1210/jc.2009-1251.

Storer, T. W., Basaria, S., Traustadottir, T., Harman, S. M., Pencina, K., Li, Z., et al. (2017). Effects of testosterone supplementation for 3 years on muscle performance and physical function in older men. *The Journal of Clinical Endocrinology and Metabolism, 102*(2), 583–593. https://doi.org/10.1210/jc.2016-2771.

Szlejf, C., Suemoto, C. K., Janovsky, C., Barreto, S. M., Diniz, M., Lotufo, P. A., et al. (2020). Thyroid function and sarcopenia: Results from the ELSA-Brasil Study. *Journal of the American Geriatrics Society, 68*, 1545–1553. https://doi.org/10.1111/jgs.16416.

Taaffe, D. R., Newman, A. B., Haggerty, C. L., Colbert, L. H., de Rekeneire, N., Visser, M., et al. (2005). Estrogen replacement, muscle composition, and physical function: The Health ABC Study. *Medicine and Science in Sports and Exercise, 37*(10), 1741–1747. https://doi.org/10.1249/01.mss.0000181678.28092.31.

Takeshita, H., Yamamoto, K., Nozato, S., Takeda, M., Fukada, S. I., Inagaki, T., et al. (2018). Angiotensin-converting enzyme 2 deficiency accelerates and angiotensin 1-7 restores age-related muscle weakness in mice. *Journal of Cachexia, Sarcopenia and Muscle, 9*(5), 975–986. https://doi.org/10.1002/jcsm.12334.

Tieland, M., Brouwer-Brolsma, E. M., Nienaber-Rousseau, C., van Loon, L. J., & De Groot, L. C. (2013). Low vitamin D status is associated with reduced muscle mass and impaired physical performance in frail elderly people. *European Journal of Clinical Nutrition, 67*(10), 1050–1055. https://doi.org/10.1038/ejcn.2013.144.

Tieland, M., Trouwborst, I., & Clark, B. C. (2018). Skeletal muscle performance and ageing. *Journal of Cachexia, Sarcopenia and Muscle, 9*(1), 3–19. https://doi.org/10.1002/jcsm.12238.

Tiidus, P. M. (2011). Benefits of estrogen replacement for skeletal muscle mass and function in post-menopausal females: Evidence from human and animal studies. *The Eurasian Journal of Medicine, 43*(2), 109–114. https://doi.org/10.5152/eajm.2011.24.

Tilstra, J. S., Clauson, C. L., Niedernhofer, L. J., & Robbins, P. D. (2011). NF-kappaB in aging and disease. *Aging and Disease, 2*(6), 449–465. Retrieved from https://www.ncbi.nlm.nih.gov/pubmed/22396894.

Tomlinson, J. W., Walker, E. A., Bujalska, I. J., Draper, N., Lavery, G. G., Cooper, M. S., et al. (2004). 11beta-hydroxysteroid dehydrogenase type 1: A tissue-specific regulator of glucocorticoid response. *Endocrine Reviews*, *25*(5), 831–866. https://doi.org/10.1210/er.2003-0031.

Torres, M. J., Kew, K. A., Ryan, T. E., Pennington, E. R., Lin, C. T., Buddo, K. A., et al. (2018). 17beta-estradiol directly lowers mitochondrial membrane microviscosity and improves bioenergetic function in skeletal muscle. *Cell Metabolism*, *27*(1), 167–179.e167. https://doi.org/10.1016/j.cmet.2017.10.003.

Tournadre, A., Vial, G., Capel, F., Soubrier, M., & Boirie, Y. (2019). Sarcopenia. *Joint, Bone, Spine*, *86*(3), 309–314. https://doi.org/10.1016/j.jbspin.2018.08.001.

Travison, T. G., Araujo, A. B., O'Donnell, A. B., Kupelian, V., & McKinlay, J. B. (2007). A population-level decline in serum testosterone levels in American men. *The Journal of Clinical Endocrinology and Metabolism*, *92*(1), 196–202. https://doi.org/10.1210/jc.2006-1375.

Unger, R. H. (2005). Longevity, lipotoxicity and leptin: The adipocyte defense against feasting and famine. *Biochimie*, *87*(1), 57–64. https://doi.org/10.1016/j.biochi.2004.11.014.

van den Beld, A. W., Kaufman, J. M., Zillikens, M. C., Lamberts, S. W. J., Egan, J. M., & van der Lely, A. J. (2018). The physiology of endocrine systems with ageing. *The Lancet Diabetes and Endocrinology*, *6*(8), 647–658. https://doi.org/10.1016/S2213-8587(18)30026-3.

van Geel, T. A., Geusens, P. P., Winkens, B., Sels, J. P., & Dinant, G. J. (2009). Measures of bioavailable serum testosterone and estradiol and their relationships with muscle mass, muscle strength and bone mineral density in postmenopausal women: A cross-sectional study. *European Journal of Endocrinology*, *160*(4), 681–687. https://doi.org/10.1530/EJE-08-0702.

Van Hollebeke, R. B., Cushman, M., Schlueter, E. F., & Allison, M. A. (2018). Abdominal muscle density is inversely related to adiposity inflammatory mediators. *Medicine and Science in Sports and Exercise*, *50*(7), 1495–1501. https://doi.org/10.1249/MSS.0000000000001570.

Veldhuis, J. D. (2008). Aging and hormones of the hypothalamo-pituitary axis: Gonadotropic axis in men and somatotropic axes in men and women. *Ageing Research Reviews*, 7(3), 189–208. https://doi.org/10.1016/j.arr.2007.12.005.

Veldhuis, J. D., Liem, A. Y., South, S., Weltman, A., Weltman, J., Clemmons, D. A., et al. (1995). Differential impact of age, sex steroid hormones, and obesity on basal versus pulsatile growth hormone secretion in men as assessed in an ultrasensitive chemiluminescence assay. *The Journal of Clinical Endocrinology and Metabolism*, *80*(11), 3209–3222. https://doi.org/10.1210/jcem.80.11.7593428.

Verdijk, L. B., Koopman, R., Schaart, G., Meijer, K., Savelberg, H. H., & van Loon, L. J. (2007). Satellite cell content is specifically reduced in type II skeletal muscle fibers in the elderly. *American Journal of Physiology. Endocrinology and Metabolism*, *292*(1), E151–E157. https://doi.org/10.1152/ajpendo.00278.2006.

Visser, M., Deeg, D. J., Lips, P., & Longitudinal Aging Study, A. (2003). Low vitamin D and high parathyroid hormone levels as determinants of loss of muscle strength and muscle mass (sarcopenia): The Longitudinal Aging Study Amsterdam. *The Journal of Clinical Endocrinology and Metabolism*, *88*(12), 5766–5772. https://doi.org/10.1210/jc.2003-030604.

Visser, M., Pahor, M., Taaffe, D. R., Goodpaster, B. H., Simonsick, E. M., Newman, A. B., et al. (2002). Relationship of interleukin-6 and tumor necrosis factor-alpha with muscle mass and muscle strength in elderly men and women: The Health ABC Study. *The Journals of Gerontology. Series A, Biological Sciences and Medical Sciences*, *57*(5), M326–M332. https://doi.org/10.1093/gerona/57.5.m326.

Vitale, G., Cesari, M., & Mari, D. (2016). Aging of the endocrine system and its potential impact on sarcopenia. *European Journal of Internal Medicine, 35*, 10–15. https://doi.org/10.1016/j.ejim.2016.07.017.

Volpato, S., Bianchi, L., Cherubini, A., Landi, F., Maggio, M., Savino, E., et al. (2014). Prevalence and clinical correlates of sarcopenia in community-dwelling older people: Application of the EWGSOP definition and diagnostic algorithm. *The Journals of Gerontology. Series A, Biological Sciences and Medical Sciences, 69*(4), 438–446. https://doi.org/10.1093/gerona/glt149.

Wahlin-Larsson, B., Carnac, G., & Kadi, F. (2014). The influence of systemic inflammation on skeletal muscle in physically active elderly women. *Age (Dordrecht, Netherlands), 36*(5), 9718. https://doi.org/10.1007/s11357-014-9718-0.

Warmington, S. A., Tolan, R., & McBennett, S. (2000). Functional and histological characteristics of skeletal muscle and the effects of leptin in the genetically obese (ob/ob) mouse. *International Journal of Obesity and Related Metabolic Disorders, 24*(8), 1040–1050. https://doi.org/10.1038/sj.ijo.0801357.

Westbury, L. D., Fuggle, N. R., Syddall, H. E., Duggal, N. A., Shaw, S. C., Maslin, K., et al. (2018). Relationships between markers of inflammation and muscle mass, strength and function: Findings from the Hertfordshire Cohort Study. *Calcified Tissue International, 102*(3), 287–295. https://doi.org/10.1007/s00223-017-0354-4.

Wiik, A., Ekman, M., Johansson, O., Jansson, E., & Esbjornsson, M. (2009). Expression of both oestrogen receptor alpha and beta in human skeletal muscle tissue. *Histochemistry and Cell Biology, 131*(2), 181–189. https://doi.org/10.1007/s00418-008-0512-x.

Wohlgemuth, S. E., Seo, A. Y., Marzetti, E., Lees, H. A., & Leeuwenburgh, C. (2010). Skeletal muscle autophagy and apoptosis during aging: Effects of calorie restriction and life-long exercise. *Experimental Gerontology, 45*(2), 138–148. https://doi.org/10.1016/j.exger.2009.11.002.

Yakar, S., Liu, J. L., Stannard, B., Butler, A., Accili, D., Sauer, B., et al. (1999). Normal growth and development in the absence of hepatic insulin-like growth factor I. *Proceedings of the National Academy of Sciences of the United States of America, 96*(13), 7324–7329. https://doi.org/10.1073/pnas.96.13.7324.

Yanagita, I., Fujihara, Y., Kitajima, Y., Tajima, M., Honda, M., Kawajiri, T., et al. (2019). A high serum cortisol/DHEA-S ratio is a risk factor for sarcopenia in elderly diabetic patients. *Journal of the Endocrine Society, 3*(4), 801–813. https://doi.org/10.1210/js.2018-00271.

Yang, A., Lv, Q., Chen, F., Wang, Y., Liu, Y., Shi, W., et al. (2020). The effect of vitamin D on sarcopenia depends on the level of physical activity in older adults. *Journal of Cachexia, Sarcopenia and Muscle, 11*, 678–689. https://doi.org/10.1002/jcsm.12545.

Yiallouris, A., Tsioutis, C., Agapidaki, E., Zafeiri, M., Agouridis, A. P., Ntourakis, D., et al. (2019). Adrenal aging and its implications on stress responsiveness in humans. *Frontiers in Endocrinology, 10*, 54. https://doi.org/10.3389/fendo.2019.00054.

Zhou, Q., Du, J., Hu, Z., Walsh, K., & Wang, X. H. (2007). Evidence for adipose-muscle cross talk: Opposing regulation of muscle proteolysis by adiponectin and fatty acids. *Endocrinology, 148*(12), 5696–5705. https://doi.org/10.1210/en.2007-0183.

CHAPTER TWENTY-ONE

Sources of lumbar back pain during aging and potential therapeutic targets

T. Peter Li, Amit Jain, and Xu Cao*
From the Department of Orthopedic Surgery, The Johns Hopkins University School of Medicine, Baltimore, MD, United States
*Corresponding author: e-mail address: xcao11@jhmi.edu

Contents

Abstract

Lumbar back pain during aging is a major clinical problem, the origins and underlying mechanisms of which are challenging to study. Degenerative changes occur in various parts of the functional spinal unit, such the vertebral endplate and intervertebral disc. The homeostasis of these structural components is regulated by signaling molecules, such as transforming growth factor-β and parathyroid hormone. Previous efforts to understand sources of lumbar back pain focused on sensory innervation in the degenerative intervertebral disc, but intervertebral disc degeneration is frequently asymptomatic. An *in vivo* mouse model of lumbar spine aging and degeneration, combined with genetic technology, has identified endplate innervation as a major source of lumbar back pain and a potential therapeutic target. In this review, we consider how each structural component of the functional spinal unit contributes to lumbar back pain, how the homeostasis of each component is regulated, and how these findings can be used to develop potential therapies.

Vitamins and Hormones, Volume 115
ISSN 0083-6729
https://doi.org/10.1016/bs.vh.2020.12.022

1. Introduction

Lumbar back pain affects 540 million people worldwide and is the most common cause of job-related disability and the main reason for lost productivity at work (Buchbinder et al., 2018; Hartvigsen et al., 2018). Estimated lifetime prevalence of lumbar back pain in adults is 84% (Cassidy, Carroll, & Cote, 1998; NINDS, 2020); this figure is rising in the setting of a growing aging population. Acute, nonspecific back pain has a 1-year recurrence rate of 20–44% and a lifetime recurrence rate of 72% (Andersson, 1999). The cost of lumbar back pain exceeds $100 billion per year in the United States; two-thirds of this cost represents lost wages and reduced productivity, and one-third represents direct costs to the health care system (Katz, 2006). The cost of lumbar back pain is steadily increasing (Martin et al., 2008); yet, there is no disease-modifying medication to cure or slow the progression of spine-related back pain (Qaseem, Wilt, McLean, & Forciea, 2017).

An important mechanical function of the lumbar spine is to support the upper body by transmitting compressive and shearing forces to the lower body during the performance of everyday activities. Mechanical stability of the spinal system must be maintained to enable painless transmission of these forces. Namely, the vertebrae must maintain their anatomic relationship to limit relative displacement during physiologic postures and loading (Guillot et al., 1990). Spinal instability is considered one of the major causes of lumbar back pain.

Stability of the lumbar spine is maintained by the coordination of muscles and multiple sets of functional spinal units (FSUs), which are defined as the smallest physiologic motion unit of the spine consisting of two adjacent vertebrae, vertebral endplates, and an intervertebral disc (Panjabi, 1992a, 1992b) (Fig. 1A). Increased mechanical instability in the FSU during aging accelerates the degeneration of the FSU (Bian et al., 2016, 2017; Cao et al., 2018) (Fig. 1B), resulting in back pain. In this chapter, we consider how each component of the FSU contributes to lumbar back pain, how the homeostasis of each component is regulated, and how these findings can be used to develop potential therapies.

2. Paravertebral muscles and ligaments stabilize the FSU

Paraspinal muscles, although by definition not part of the FSU, are critical to stabilizing the spinal column and play a major role in the

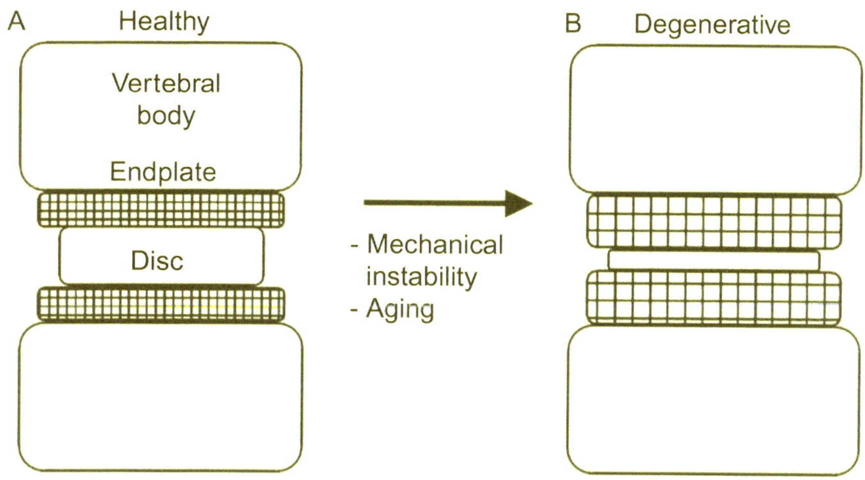

Fig. 1 Mechanical instability associated with aging accelerates the degeneration of the functional spinal unit. (A) Normal functional spinal unit with healthy vertebral endplates and intervertebral disc. (B) Degenerative changes of the functional spinal unit involve endplate sclerosis and hypertrophy with an increase in porosity, as well as loss of height and volume of the intervertebral disc.

pathophysiology of lumbar back pain. Paraspinal muscles support the spine akin to guy-wires supporting a bending mast. Clinical studies support the hypothesis that musculoligamentous disruption around the FSU leads to instability and accelerates degeneration. For example, spine stabilization exercises benefit patients with lumbar back pain (Hebert, Koppenhaver, Magel, & Fritz, 2010; Hicks, Fritz, Delitto, & McGill, 2005; Rasmussen-Barr, Nilsson-Wikmar, & Arvidsson, 2003). Paraspinal muscular atrophy is associated with higher incidence and severity of lumbar back pain (Barker, Shamley, & Jackson, 2004; Beneck & Kulig, 2012; Kang, Shin, Kim, Lee, & Lee, 2007). Further supporting the hypothesis, detachment of muscles and ligaments from the FSU in mice accelerates the development of spondylosis (Miyamoto, Yonenobu, & Ono, 1991). This spinal instability mouse model mimics degenerative spinal disease seen in human aging, causing sclerosis of endplates and narrowing of intervertebral discs, resulting in pain behavior (Bian et al., 2016, 2017). This spinal instability mouse model of spinal degeneration and pain has been a reliable model for investigating mechanisms of back pain.

3. Pathophysiology of lumbar back pain

Sources of back pain are multifaceted and challenging to diagnose, partly because of the complex anatomy and function of the spine. Spinal pain involves the peripheral nervous system, central nervous system, and their interaction. Sensory innervation of the spine has been reviewed extensively elsewhere (Allegri et al., 2016; Izzo, Popolizio, D'Aprile, & Muto, 2015). In brief, intervertebral discs, vertebral bodies, and facet joints are innervated by somatic and autonomic sensory systems. Referred pain is triggered by stimulation of nociceptive terminals, which converge onto higher-order neurons in the dorsal horn of the spinal cord and ultimately in the thalamus (Bogduk, 2009). Lumbar back pain can be caused by the stimulation of nociceptive nerve endings, which can result in a hypersensitive state or imbalance of inhibitory and excitatory neurotransmission in central pain centers (Moskowitz, 2008). Alternatively, lumbar back pain can be caused by changes in nerve density in various structures of the FSU.

In a healthy FSU, sensory nerve endings are limited to the outermost external lamellae of the annulus fibrosus, responsible for nociception and proprioception (Freemont et al., 1997). High hydrostatic pressures in the inner portion of a healthy disc and the dense trabeculation of endplates preclude the penetration of high-density nerve endings. During aging, endplates become porous (Bian et al., 2016, 2017; Papadakis, Sapkas, Papadopoulos, & Katonis, 2011; Rodriguez et al., 2012), the annulus of intervertebral discs becomes more disrupted (Setton & Chen, 2006), intervertebral disc spaces become more narrowed (Rodriguez et al., 2012; Taher et al., 2012), and facet joints become arthritic (Kim et al., 2015). Nociceptive nerve density increases in porous endplates and degenerative intervertebral discs (Peng, 2013; Setton & Chen, 2006). Sensory neural filaments also increase in degenerative facet joints (Kim et al., 2015). Interestingly, nerve density is greater in porous endplates than in healthy endplates or degenerative intervertebral discs (Fields, Liebenberg, & Lotz, 2014). Attenuating sensory innervation in endplates by inhibiting osteoclast activity reduces pain sensitivity (Ni et al., 2019), indicating that prevention or reversal of degenerative changes in endplates could mitigate lumbar back pain.

Pathways to spinal hypersensitivity involve the concerted actions of neurotransmitters, neuromodulators, and inflammatory mediators, such as prostaglandins (Coderre, 1999; Crain & Shen, 2000; DeLeo & Yezierski, 2001; Moskowitz, 2008; Rutkowski, Winkelstein, Hickey, Pahl, & DeLeo, 2002). Prostaglandin E_2 (PGE_2) is involved in the pathophysiology of inflammation and the generation of spinal hypersensitivity (Schaible, Ebersberger, & Von Banchet, 2002). PGE_2 is elevated in porous endplates to stimulate nociceptive nerves by increasing sodium influx in dorsal root ganglion neurons (Ni et al., 2019). The E-type prostanoid receptor 4 (EP4), a primary mediator of PGE_2-evoked inflammatory pain hypersensitivity (Lin et al., 2006), is elevated in human degenerative facet joint tissue (Kim et al., 2015). However, it remains unclear how PGE_2 directly contributes to pain associated with degenerative facet joints. Nonetheless, prostaglandins are important mediators of lumbar back pain in diverse structural regions of the FSU.

4. Facet joint degeneration as a source of pain

Facet joint osteoarthritic degeneration is a source of lumbar back pain in 15–40% of patients with chronic lumbar back pain (Manchikanti et al., 2001; Schwarzer, Wang, Bogduk, McNaught, & Laurent, 1995). Facet joints are synovial joints between two adjacent vertebrae; they consist of hyaline cartilage, synovium, and joint capsule. The medial branch of the dorsal ramus provides rich sensory innervation in the capsule, synovial membranes, and subchondral bone of any given facet joint. Lumbar back pain is associated with facet hypertrophy, inflammation, and degradation of facet joints (Jackson, 1992; van Kleef et al., 2010; Varlotta et al., 2011a, 2011b); however, the causal relationship between the degenerative facet joint and lumbar back pain remains unclear. Studying isolated facet joint degeneration *in vivo* is challenging because joint degeneration is frequently associated with degeneration in other parts of the FSU, such as the intervertebral disc and endplates. There is a need to create an animal model with isolated symptomatic facet joint degeneration to understand the pain mechanism associated with facet joint degeneration alone without confounding from other structures of the FSU.

5. Intervertebral disc disruption as a source of pain

Intervertebral disc degeneration occurs in 58% of asymptomatic patients, suggesting that disc degeneration alone is not painful (Borenstein et al., 2001; Hurri & Karppinen, 2004). In fact, patients with degenerative disc disease have marked enrichment of nociceptive nerve fibers in the endplate region as opposed to the disc (Brown et al., 1997). However, disruptions such as tears or fissures in the disc can result in the development of inflammatory granulation tissue and centralization of nociceptive nerve fibers (Freemont et al., 1997; Peng et al., 2005), which may contribute to discogenic pain (Peng, 2013). Interestingly, internal disc disruption is often preceded and surrounded by endplate abnormality (Peng, 2013), suggesting that disc abnormality could be a structural change secondary to endplate disease as opposed to a primary contributor to lumbar back pain.

6. Vertebral endplate degeneration as a source of pain

Vertebral endplates are positioned at the cranial and caudal ends of each intervertebral disc. Endplates transmit mechanical loads produced by bodyweight and muscle activity between vertebrae (Broberg, 1983; Paesold, Nerlich, & Boos, 2007). With aging, endplates become sclerotic (Moore, 2000). The severity of vertebral endplate degeneration is detectable on magnetic resonance imaging as Modic changes, ranging from inflammatory edema to extensive sclerotic changes in the subchondral bone (Dudli, Fields, Samartzis, Karppinen, & Lotz, 2016). The magnitude of Modic change is directly associated with the severity of lumbar back pain (Jarvinen et al., 2015). Sclerotic endplates are highly porous (Bian et al., 2016, 2017; Papadakis et al., 2011; Rodriguez et al., 2012). Nerve density in porous endplates is greater than in healthy endplates and in degenerative nucleus pulposus (Fields et al., 2014), suggesting that the main source of lumbar back pain is the endplate in degenerative disc disease with associated endplate degeneration.

In a mouse model of lumbar spine instability and aging (Ariga et al., 2001; Bian et al., 2016; Miyamoto et al., 1991), osteoclasts at the endplates facilitated lumbar back pain in multiple ways (Ni et al., 2019). Osteoclasts resorb endplate bone to increase porous space, and they secrete Netrin-1, an axon guidance factor (Moore, Zhang, Lynch, & Sheetz, 2012; Serafini et al., 1996), to attract nociceptive sensory nerve fibers to innervate porous space, contributing to spinal hypersensitivity (Ni et al., 2019) (Fig. 2). These results suggest that the inhibition osteoclastic function to reduce endplate porosity could be a therapeutic target for lumbar back pain. Consistent with this idea, zoledronic acid and denosumab, which inhibit osteoclast activity, have analgesic effects in patients with lumbar back pain associated with Modic changes (Cai et al., 2018; Koivisto et al., 2014). In fact, inhibiting osteoclast formation in a mouse model of lumbar spine instability and aging significantly reduced spinal hypersensitivity (Ni et al., 2019). Importantly, endplate porosity is believed to result from an imbalance of exaggerated osteoblastic response and osteoclastic resorption (Wolf & Denko, 1958). Therefore, altering endplate porosity by modulating the balance of osteoblastic stimulation and osteoclastic inhibition, may be a therapeutic target for lumbar back pain.

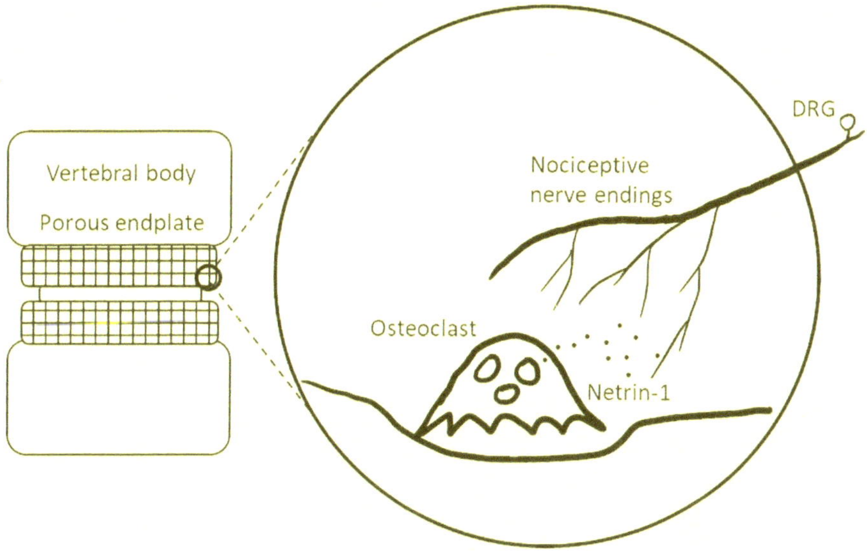

Fig. 2 Osteoclasts contribute to spinal pain by increasing porous space in endplates and attracting nociceptive sensory nerve fibers. In a degenerative functional spinal unit, osteoclasts increase the porosity of hypertrophic endplates and the density of nociceptive nerve endings through an axon-guiding signaling molecule, Netrin-1, which induces increased nociceptive innervation in the more porous endplate. DRG: dorsal root ganglion cell.

7. Roles of transforming growth factor–β (TGFβ) and parathyroid hormone (PTH) in maintaining the homeostasis of the FSU

TGFβ is a cytokine with multiple functions, including modulating chondrocyte anabolic activity in articular cartilage, maintaining bone homeostasis during bone remodeling, and contributing to tissue repair (Wipff, Rifkin, Meister, & Hinz, 2007; Zhen et al., 2013). Intervertebral disc homeostasis requires a fine balance in TGFβ activation. Overactivation of TGFβ from pathologic mechanical loading of the spine results in disc degeneration, yet complete inhibition of TGFβ signaling also results in disc degeneration (Bian et al., 2017). During aging and aberrant mechanical loading, endplates become sclerotic and hypertrophic, resulting in less efficient diffusion of nutrients to support intervertebral disc homeostasis (Bian et al., 2016; Moore, 2000). Excessive activation of TGFβ in aberrant mechanical loading causes endplate sclerosis (Bian et al., 2016). These findings suggest that reduction of supraphysiologic TGFβ signaling is a potential therapeutic target in degenerative endplate and disc diseases.

PTH, secreted by the parathyroid gland, stimulates bone remodeling and coordinates signaling of multiple cytokines and growth factors, including TGFβ, Wnt, bone morphogenetic protein, and insulin-like growth factor (Crane & Cao, 2014; Qiu et al., 2018, 2010; Wan et al., 2008; Yu et al., 2012). PTH modulates downstream gene expression through type-1 PTH receptor (PTH1R), a G-protein-coupled receptor (Kawane, Mimura, Yanagawa, Fujii-Kuriyama, & Horiuchi, 2003). PTH1R is expressed in the chondrocytes of endplates, osteoblasts and osteocytes of bony endplates, and nucleus pulposus cells in intervertebral discs (Cao et al., 2018), which suggests that PTH may coordinate the homeostasis of multiple regions of the FSU. In fact, intermittent injection of PTH in aged mice attenuates intervertebral disc degeneration by direct signaling through nucleus pulposus cells to modulate TGFβ activity (Cao et al., 2018). Furthermore, intermittent injection of PTH restores native endplate structure (Jia et al., 2016), a process that is believed to involve the Wnt/β-catenin pathway. In addition to the finding that endplate porosity is directly associated with nociceptive innervation density (Ni et al., 2019), these studies suggest that restoring endplate structure by modulating PTH signaling is a potential therapeutic target for degenerative lumbar back pain.

8. Conclusion

With aging and spinal instability, degenerative changes of the FSU are evident in multiple structural regions, such as the intervertebral disc and the neighboring vertebral endplates. Spinal pain may come from several structural sources, but emerging evidence suggests a more focal source in the endplate. In an *in vivo* mouse spine degeneration model, most spinal pain comes from vertebral endplates secondary to an osteoclast-mediated increase in endplate porosity and subsequent nociceptive innervation (Ni et al., 2019), which suggests that restoring the endplate structure could alleviate most spinal pain. Given that the modulation of endplate structure involves an intricate interplay of osteoblasts, osteoclasts, and several key molecules, such as PTH and TGFβ, several therapeutic targets for back pain remain to be explored.

Acknowledgment

The authors thank Jenni Weems and Rachel Box in the editorial office at the Department of Orthopedic Surgery, The Johns Hopkins University, for editing the manuscript.

References

Allegri, M., Montella, S., Salici, F., Valente, A., Marchesini, M., Compagnone, C., et al. (2016). Mechanisms of low back pain: A guide for diagnosis and therapy. *F1000Res*, *5*, 1530. https://doi.org/10.12688/f1000research.8105.2.

Andersson, G. B. (1999). Epidemiological features of chronic low-back pain. *Lancet*, *354*(9178), 581–585. https://doi.org/10.1016/s0140-6736(99)01312-4.

Ariga, K., Miyamoto, S., Nakase, T., Okuda, S., Meng, W., Yonenobu, K., et al. (2001). The relationship between apoptosis of endplate chondrocytes and aging and degeneration of the intervertebral disc. *Spine (Phila Pa 1976)*, *26*(22), 2414–2420. https://doi.org/10.1097/00007632-200111150-00004.

Barker, K. L., Shamley, D. R., & Jackson, D. (2004). Changes in the cross-sectional area of multifidus and psoas in patients with unilateral back pain: The relationship to pain and disability. *Spine (Phila Pa 1976)*, *29*(22), E515–E519. https://doi.org/10.1097/01.brs.0000144405.11661.eb.

Beneck, G. J., & Kulig, K. (2012). Multifidus atrophy is localized and bilateral in active persons with chronic unilateral low back pain. *Archives of Physical Medicine and Rehabilitation*, *93*(2), 300–306. https://doi.org/10.1016/j.apmr.2011.09.017.

Bian, Q., Jain, A., Xu, X., Kebaish, K., Crane, J. L., Zhang, Z., et al. (2016). Excessive activation of TGFbeta by spinal instability causes vertebral endplate sclerosis. *Scientific Reports*, *6*, 27093. https://doi.org/10.1038/srep27093.

Bian, Q., Ma, L., Jain, A., Crane, J. L., Kebaish, K., Wan, M., et al. (2017). Mechanosignaling activation of TGFbeta maintains intervertebral disc homeostasis. *Bone Research*, *5*, 17008. https://doi.org/10.1038/boneres.2017.8.

Bogduk, N. (2009). On the definitions and physiology of back pain, referred pain, and radicular pain. *Pain*, *147*(1–3), 17–19. https://doi.org/10.1016/j.pain.2009.08.020.

Borenstein, D. G., O'Mara, J. W., Jr., Boden, S. D., Lauerman, W. C., Jacobson, A., Platenberg, C., et al. (2001). The value of magnetic resonance imaging of the lumbar spine to predict low-back pain in asymptomatic subjects : A seven-year follow-up study. *The Journal of Bone and Joint Surgery. American Volume*, *83*(9), 1306–1311. https://doi.org/10.2106/00004623-200109000-00002.

Broberg, K. B. (1983). On the mechanical behaviour of intervertebral discs. *Spine (Phila Pa 1976)*, *8*(2), 151–165. https://doi.org/10.1097/00007632-198303000-00006.

Brown, M. F., Hukkanen, M. V., McCarthy, I. D., Redfern, D. R., Batten, J. J., Crock, H. V., et al. (1997). Sensory and sympathetic innervation of the vertebral endplate in patients with degenerative disc disease. *Journal of Bone and Joint Surgery. British Volume (London)*, *79*(1), 147–153. https://doi.org/10.1302/0301-620x.79b1.6814.

Buchbinder, R., van Tulder, M., Oberg, B., Costa, L. M., Woolf, A., Schoene, M., et al. (2018). Low back pain: A call for action. *Lancet*, *391*(10137), 2384–2388. https://doi.org/10.1016/s0140-6736(18)30488-4.

Cai, G., Laslett, L. L., Aitken, D., Halliday, A., Pan, F., Otahal, P., et al. (2018). Effect of Zoledronic acid and Denosumab in patients with low Back pain and Modic change: A proof-of-principle trial. *Journal of Bone and Mineral Research*, *33*(5), 773–782. https://doi.org/10.1002/jbmr.3376.

Cassidy, J. D., Carroll, L. J., & Cote, P. (1998). The Saskatchewan health and back pain survey. The prevalence of low back pain and related disability in Saskatchewan adults. *Spine (Phila Pa 1976)*, *23*(17), 1860–1866, discussion 1867. https://doi.org/10.1097/00007632-199809010-00012.

Coderre, C. (1999). In S. Cowan (Ed.), *Excitatory amino acid antagonists: Potential analgesics for persistent pain*. Wiley Liss.

Crain, S. M., & Shen, K. F. (2000). Antagonists of excitatory opioid receptor functions enhance morphine's analgesic potency and attenuate opioid tolerance/dependence liability. *Pain*, *84*(2–3), 121–131. https://doi.org/10.1016/s0304-3959(99)00223-7.

Crane, J. L., & Cao, X. (2014). Bone marrow mesenchymal stem cells and TGF-beta signaling in bone remodeling. *The Journal of Clinical Investigation*, *124*(2), 466–472. https://doi.org/10.1172/jci70050.

DeLeo, J. A., & Yezierski, R. P. (2001). The role of neuroinflammation and neuroimmune activation in persistent pain. *Pain*, *90*(1–2), 1–6. https://doi.org/10.1016/s0304-3959(00)00490-5.

Dudli, S., Fields, A. J., Samartzis, D., Karppinen, J., & Lotz, J. C. (2016). Pathobiology of Modic changes. *European Spine Journal*, *25*(11), 3723–3734. https://doi.org/10.1007/s00586-016-4459-7.

Fields, A. J., Liebenberg, E. C., & Lotz, J. C. (2014). Innervation of pathologies in the lumbar vertebral end plate and intervertebral disc. *The Spine Journal*, *14*(3), 513–521. https://doi.org/10.1016/j.spinee.2013.06.075.

Freemont, A. J., Peacock, T. E., Goupille, P., Hoyland, J. A., O'Brien, J., & Jayson, M. I. (1997). Nerve ingrowth into diseased intervertebral disc in chronic back pain. *Lancet*, *350*(9072), 178–181. https://doi.org/10.1016/s0140-6736(97)02135-1.

Guillot, M., Fournier, J., Scheye, T., Escande, G., Chazal, J., Tanguy, A., et al. (1990). Mechanics of the characteristic geometry of the human spine undergoing vertical pressure. *Bulletin de l'Association des Anatomistes*, *74*(225), 7–8, (Justification mecanique approchee de la geometrie particulaiere du rachis humain soumis a contrainte verticale.).

Hartvigsen, J., Hancock, M. J., Kongsted, A., Louw, Q., Ferreira, M. L., Genevay, S., et al. (2018). What low back pain is and why we need to pay attention. *Lancet*, *391*(10137), 2356–2367. https://doi.org/10.1016/s0140-6736(18)30480-x.

Hebert, J. J., Koppenhaver, S. L., Magel, J. S., & Fritz, J. M. (2010). The relationship of transversus abdominis and lumbar multifidus activation and prognostic factors for clinical success with a stabilization exercise program: A cross-sectional study. *Archives of Physical Medicine and Rehabilitation*, *91*(1), 78–85. https://doi.org/10.1016/j.apmr.2009.08.146.

Hicks, G. E., Fritz, J. M., Delitto, A., & McGill, S. M. (2005). Preliminary development of a clinical prediction rule for determining which patients with low back pain will respond to a stabilization exercise program. *Archives of Physical Medicine and Rehabilitation*, *86*(9), 1753–1762. https://doi.org/10.1016/j.apmr.2005.03.033.

Hurri, H., & Karppinen, J. (2004). Discogenic pain. *Pain*, *112*(3), 225–228. https://doi.org/10.1016/j.pain.2004.08.016.

Izzo, R., Popolizio, T., D'Aprile, P., & Muto, M. (2015). Spinal pain. *European Journal of Radiology*, *84*(5), 746–756. https://doi.org/10.1016/j.ejrad.2015.01.018.

Jackson, R. P. (1992). The facet syndrome. Myth or reality? *Clinical Orthopaedics and Related Research*, *279*, 110–121.

Jarvinen, J., Karppinen, J., Niinimaki, J., Haapea, M., Gronblad, M., Luoma, K., et al. (2015). Association between changes in lumbar Modic changes and low back symptoms over a two-year period. *BMC Musculoskeletal Disorders*, *16*, 98. https://doi.org/10.1186/s12891-015-0540-3.

Jia, H., Ma, J., Lv, J., Ma, X., Xu, W., Yang, Y., et al. (2016). Oestrogen and parathyroid hormone alleviate lumbar intervertebral disc degeneration in ovariectomized rats and enhance Wnt/beta-catenin pathway activity. *Scientific Reports*, *6*, 27521. https://doi.org/10.1038/srep27521.

Kang, C. H., Shin, M. J., Kim, S. M., Lee, S. H., & Lee, C. S. (2007). MRI of paraspinal muscles in lumbar degenerative kyphosis patients and control patients with chronic low back pain. *Clinical Radiology*, *62*(5), 479–486. https://doi.org/10.1016/j.crad.2006.12.002.

Katz, J. N. (2006). Lumbar disc disorders and low-back pain: Socioeconomic factors and consequences. *The Journal of Bone and Joint Surgery. American Volume*, *88*(Suppl 2), 21–24. https://doi.org/10.2106/jbjs.e.01273.

Kawane, T., Mimura, J., Yanagawa, T., Fujii-Kuriyama, Y., & Horiuchi, N. (2003). Parathyroid hormone (PTH) down-regulates PTH/PTH-related protein receptor gene expression in UMR-106 osteoblast-like cells via a 3′,5′-cyclic adenosine monophosphate-dependent, protein kinase A-independent pathway. *The Journal of Endocrinology*, *178*(2), 247–256. https://doi.org/10.1677/joe.0.1780247.

Kim, J. S., Ali, M. H., Wydra, F., Li, X., Hamilton, J. L., An, H. S., et al. (2015). Characterization of degenerative human facet joints and facet joint capsular tissues. *Osteoarthritis and Cartilage*, *23*(12), 2242–2251. https://doi.org/10.1016/j.joca.2015.06.009.

Koivisto, K., Kyllonen, E., Haapea, M., Niinimaki, J., Sundqvist, K., Pehkonen, T., et al. (2014). Efficacy of zoledronic acid for chronic low back pain associated with Modic changes in magnetic resonance imaging. *BMC Musculoskeletal Disorders*, *15*, 64. https://doi.org/10.1186/1471-2474-15-64.

Lin, C. R., Amaya, F., Barrett, L., Wang, H., Takada, J., Samad, T. A., et al. (2006). Prostaglandin E2 receptor EP4 contributes to inflammatory pain hypersensitivity. *The Journal of Pharmacology and Experimental Therapeutics*, *319*(3), 1096–1103. https://doi.org/10.1124/jpet.106.105569.

Manchikanti, L., Singh, V., Pampati, V., Damron, K. S., Barnhill, R. C., Beyer, C., et al. (2001). Evaluation of the relative contributions of various structures in chronic low back pain. *Pain Physician*, *4*(4), 308–316.

Martin, B. I., Deyo, R. A., Mirza, S. K., Turner, J. A., Comstock, B. A., Hollingworth, W., et al. (2008). Expenditures and health status among adults with back and neck problems. *JAMA*, *299*(6), 656–664. https://doi.org/10.1001/jama.299.6.656.

Miyamoto, S., Yonenobu, K., & Ono, K. (1991). Experimental cervical spondylosis in the mouse. *Spine (Phila Pa 1976)*, *16*(10 Suppl), S495–S500. https://doi.org/10.1097/00007632-199110001-00008.

Moore, R. J. (2000). The vertebral end-plate: What do we know? *European Spine Journal*, *9*(2), 92–96. https://doi.org/10.1007/s005860050217.

Moore, S. W., Zhang, X., Lynch, C. D., & Sheetz, M. P. (2012). Netrin-1 attracts axons through FAK-dependent mechanotransduction. *The Journal of Neuroscience*, *32*(34), 11574–11585. https://doi.org/10.1523/JNEUROSCI.0999-12.2012.

Moskowitz, M. (2008). In C. Slipman, R. Derby, & F. Simeone (Eds.), *Central influences on pain*. Philadelphia: Saunders Elsevier.

Ni, S., Ling, Z., Wang, X., Cao, Y., Wu, T., Deng, R., et al. (2019). Sensory innervation in porous endplates by Netrin-1 from osteoclasts mediates PGE2-induced spinal hypersensitivity in mice. *Nature Communications*, *10*(1), 5643. https://doi.org/10.1038/s41467-019-13476-9.

NINDS. (2020). *Low back pain fact sheet*. Retrieved 2/10 from https://www.ninds.nih.gov/Disorders/Patient-Caregiver-Education/Fact-Sheets/Low-Back-Pain-Fact-Sheet.

Paesold, G., Nerlich, A. G., & Boos, N. (2007). Biological treatment strategies for disc degeneration: Potentials and shortcomings. *European Spine Journal*, *16*(4), 447–468. https://doi.org/10.1007/s00586-006-0220-y.

Panjabi, M. M. (1992a). The stabilizing system of the spine. Part I. function, dysfunction, adaptation, and enhancement. *Journal of Spinal Disorders*, *5*(4), 383–389. discussion 397 https://doi.org/10.1097/00002517-199212000-00001.

Panjabi, M. M. (1992b). The stabilizing system of the spine. Part II. Neutral zone and instability hypothesis. *Journal of Spinal Disorders*, *5*(4), 390–396, discussion 397. https://doi.org/10.1097/00002517-199212000-00002.

Papadakis, M., Sapkas, G., Papadopoulos, E. C., & Katonis, P. (2011). Pathophysiology and biomechanics of the aging spine. *The Open Orthopaedics Journal*, *5*, 335–342. https://doi.org/10.2174/1874325001105010335.

Peng, B. G. (2013). Pathophysiology, diagnosis, and treatment of discogenic low back pain. *World Journal of Orthopedics, 4*(2), 42–52. https://doi.org/10.5312/wjo.v4.i2.42.

Peng, B., Wu, W., Hou, S., Li, P., Zhang, C., & Yang, Y. (2005). The pathogenesis of discogenic low back pain. *Journal of Bone and Joint Surgery. British Volume (London), 87*(1), 62–67.

Qaseem, A., Wilt, T. J., McLean, R. M., & Forciea, M. A. (2017). Noninvasive treatments for acute, subacute, and chronic low back pain: A clinical practice guideline from the American College of Physicians. *Annals of Internal Medicine, 166*(7), 514–530. https://doi.org/10.7326/m16-2367.

Qiu, T., Crane, J. L., Xie, L., Xian, L., Xie, H., & Cao, X. (2018). IGF-I induced phosphorylation of PTH receptor enhances osteoblast to osteocyte transition. *Bone Research, 6*, 5. https://doi.org/10.1038/s41413-017-0002-7.

Qiu, T., Wu, X., Zhang, F., Clemens, T. L., Wan, M., & Cao, X. (2010). TGF-beta type II receptor phosphorylates PTH receptor to integrate bone remodelling signalling. *Nature Cell Biology, 12*(3), 224–234. https://doi.org/10.1038/ncb2022.

Rasmussen-Barr, E., Nilsson-Wikmar, L., & Arvidsson, I. (2003). Stabilizing training compared with manual treatment in sub-acute and chronic low-back pain. *Manual Therapy, 8*(4), 233–241. https://doi.org/10.1016/s1356-689x(03)00053-5.

Rodriguez, A. G., Rodriguez-Soto, A. E., Burghardt, A. J., Berven, S., Majumdar, S., & Lotz, J. C. (2012). Morphology of the human vertebral endplate. *Journal of Orthopaedic Research, 30*(2), 280–287. https://doi.org/10.1002/jor.21513.

Rutkowski, M. D., Winkelstein, B. A., Hickey, W. F., Pahl, J. L., & DeLeo, J. A. (2002). Lumbar nerve root injury induces central nervous system neuroimmune activation and neuroinflammation in the rat: Relationship to painful radiculopathy. *Spine (Phila Pa 1976), 27*(15), 1604–1613. https://doi.org/10.1097/00007632-200208010-00003.

Schaible, H. G., Ebersberger, A., & Von Banchet, G. S. (2002). Mechanisms of pain in arthritis. *Annals of the New York Academy of Sciences, 966*, 343–354. https://doi.org/10.1111/j.1749-6632.2002.tb04234.x.

Schwarzer, A. C., Wang, S. C., Bogduk, N., McNaught, P. J., & Laurent, R. (1995). Prevalence and clinical features of lumbar zygapophysial joint pain: A study in an Australian population with chronic low back pain. *Annals of the Rheumatic Diseases, 54*(2), 100–106. https://doi.org/10.1136/ard.54.2.100.

Serafini, T., Colamarino, S. A., Leonardo, E. D., Wang, H., Beddington, R., Skarnes, W. C., et al. (1996). Netrin-1 is required for commissural axon guidance in the developing vertebrate nervous system. *Cell, 87*(6), 1001–1014. https://doi.org/10.1016/s0092-8674(00)81795-x.

Setton, L. A., & Chen, J. (2006). Mechanobiology of the intervertebral disc and relevance to disc degeneration. *The Journal of Bone and Joint Surgery. American Volume, 88*(Suppl 2), 52–57. https://doi.org/10.2106/jbjs.f.00001.

Taher, F., Essig, D., Lebl, D. R., Hughes, A. P., Sama, A. A., Cammisa, F. P., et al. (2012). Lumbar degenerative disc disease: Current and future concepts of diagnosis and management. *Advances in Orthopedics, 2012*, 970752. https://doi.org/10.1155/2012/970752.

van Kleef, M., Vanelderen, P., Cohen, S. P., Lataster, A., Van Zundert, J., & Mekhail, N. (2010). 12. Pain originating from the lumbar facet joints. *Pain Practice, 10*(5), 459–469. https://doi.org/10.1111/j.1533-2500.2010.00393.x.

Varlotta, G. P., Lefkowitz, T. R., Schweitzer, M., Errico, T. J., Spivak, J., Bendo, J. A., et al. (2011a). The lumbar facet joint: A review of current knowledge: Part 1: Anatomy, biomechanics, and grading. *Skeletal Radiology, 40*(1), 13–23. https://doi.org/10.1007/s00256-010-0983-4.

Varlotta, G. P., Lefkowitz, T. R., Schweitzer, M., Errico, T. J., Spivak, J., Bendo, J. A., et al. (2011b). The lumbar facet joint: A review of current knowledge: Part II: Diagnosis and management. *Skeletal Radiology, 40*(2), 149–157. https://doi.org/10.1007/s00256-010-0984-3.

Wan, M., Yang, C., Li, J., Wu, X., Yuan, H., Ma, H., et al. (2008). Parathyroid hormone signaling through low-density lipoprotein-related protein 6. *Genes & Development, 22*(21), 2968–2979. https://doi.org/10.1101/gad.1702708.

Wipff, P. J., Rifkin, D. B., Meister, J. J., & Hinz, B. (2007). Myofibroblast contraction activates latent TGF-beta1 from the extracellular matrix. *The Journal of Cell Biology, 179*(6), 1311–1323. https://doi.org/10.1083/jcb.200704042.

Wolf, H. L., & Denko, J. V. (1958). Osteosclerosis in chronic renal disease. *The American Journal of the Medical Sciences, 235*(1), 33–42. https://doi.org/10.1097/00000441-195801000-00004.

Yu, B., Zhao, X., Yang, C., Crane, J., Xian, L., Lu, W., et al. (2012). Parathyroid hormone induces differentiation of mesenchymal stromal/stem cells by enhancing bone morphogenetic protein signaling. *Journal of Bone and Mineral Research, 27*(9), 2001–2014. https://doi.org/10.1002/jbmr.1663.

Zhen, G., Wen, C., Jia, X., Li, Y., Crane, J. L., Mears, S. C., et al. (2013). Inhibition of TGF-β signaling in mesenchymal stem cells of subchondral bone attenuates osteoarthritis. *Nature Medicine, 19*(6), 704–712. https://doi.org/10.1038/nm.3143.

Zheng, L., Cao, Y., Ni, S., Qi, H., Ling, Z., Xu, X., et al. (2018). Ciliary parathyroid hormone signaling activates transforming growth factor-beta to maintain intervertebral disc homeostasis during aging. *Bone Research, 6*, 21. https://doi.org/10.1038/s41413-018-0022-y.

CHAPTER TWENTY-TWO

Leydig cell aging: Molecular mechanisms and treatments

V. Papadopoulos[a,*] and B.R. Zirkin[b]

[a]Department of Pharmacology and Pharmaceutical Sciences, School of Pharmacy, University of Southern California, Los Angeles, CA, United States
[b]Department of Biochemistry and Molecular Biology, Johns Hopkins Bloomberg School of Public Health, Baltimore, MD, United States
*Corresponding author: e-mail address: vpapadop@usc.edu

Contents

Abstract

Late-onset hypogonadism, resulting from deficiency in serum testosterone (T), affects the health and quality of life of millions of aging men. T is synthesized by Leydig cells (LCs) in response to luteinizing hormone (LH). LH binds LC plasma membrane receptors, inducing the formation of a supramolecular complex of cytosolic and mitochondrial proteins, the Steroidogenic InteracTomE (SITE). SITE proteins are involved in targeting cholesterol to CYP11A1 in the mitochondria, the first enzyme of the steroidogenic cascade. Cholesterol translocation is the rate-determining step in T formation. With aging, LC defects occur that include changes in SITE, an increasingly oxidative intracellular environment, and reduced androgen formation and serum T levels. T replacement therapy (TRT) will restore T levels, but reported side effects make it desirable to develop additional strategies for increasing T. One approach is to target LC protein-protein interactions and thus increase T production by the hypofunctional Leydig cells themselves.

Vitamins and Hormones, Volume 115
ISSN 0083-6729
https://doi.org/10.1016/bs.vh.2020.12.023

Abbreviations

3β-HSD	3β-hydroxysteroid dehydrogenase
17β-HSD3	17β-hydroxysteroid dehydrogenase
ACBD1/DBI	acyl-CoA-binding protein 1 or diazepam binding inhibitor
ACBD3	acyl-CoA binding domain-containing 3
ATAD3	AAA+ ATPase domain 3
cAMP	cyclic adenosine monophosphate
CYP11A1	cholesterol side-chain cleavage enzyme
CYP17A1	17α-hydroxylase, 17,20-lyase
ER	endoplasmic reticulum
GnRH	gonadotropin-releasing hormone
hCG	human chorionic gonadotropin
IMM	inner mitochondrial membrane
LC	Leydig cell
LH	luteinizing hormone
LOH	late-onset hypogonadism
MAM	mitochondria-associated membranes
Nrf2	nuclear factor erythroid-2-related factor
OMM	outer mitochondrial membrane
PKARIα	protein kinase A regulatory subunit I alpha
PPI	protein-protein interaction
ROS	reactive oxygen species
SNARE	soluble *N*-ethylmaleimide-sensitive factor attachment receptor
SITE	steroidogenic InteracTomE
STAR	steroidogenic acute regulatory protein
T	testosterone
TRT	testosterone replacement therapy
TSPO	translocator protein
VDAC	voltage-dependent anion channel

1. Introduction: Testosterone production and the aging testis

Testosterone (T), made by the Leydig cells (LCs) of the testis in response to luteinizing hormone (LH) from the anterior pituitary gland, drives the establishment and function of the male reproductive system from gestation to adulthood (Payne & Hardy, 2007; Zirkin & Papadopoulos, 2018). T also exerts organizational and morphogenic effects on non-reproductive organs, including the brain (Payne & Hardy, 2007; Zirkin & Papadopoulos, 2018). During gestation, fetal LCs arise from mesenchymal precursors and produce high levels of T before acquiring luteinizing hormone (LH) receptors (Habert, Lejeune, & Saez, 2001; Zirkin & Papadopoulos, 2018). T production declines with the loss of fetal LCs and then gradually increases to high levels with the development of adult LCs from stem cells (Habert et al., 2001; Zirkin & Papadopoulos, 2018). With aging, T levels decrease at a rate of 0.4–2% per year starting at age 30 (Harman, Metter, Tobin, Pearson, & Blackman, 2001; Kaufman & Vermeulen, 2005; Wu et al., 2008), and this can result in pathologically low levels of serum T, termed hypogonadism, at advanced ages. Primary hypogonadism is defined as reduced LC androgen production when LH levels have not changed or have increased. In secondary hypogonadism, LH levels are reduced and become inadequate to maintain T levels (Tajar et al., 2010). In some men, there is a mix of central (hypothalamic and/or pituitary) and gonadal deficiencies (Khera et al., 2016; McBride, Carson, & Coward, 2016; Zirkin & Tenover, 2012).

Progressive decline of T with aging results in 20–50% of men over age 60 with significantly reduced T levels (Araujo et al., 2007; Harman et al., 2001; Mohr, Guay, O'Donnell, & McKinlay, 2005; Travison, Araujo, O'Donnell, Kupelian, & McKinlay, 2007). Age-related decline in T along with associated symptoms is now referred to as late-onset hypogonadism (LOH) (Nieschlag, 2019; Pye et al., 2014). LOH is characterized by symptoms that include loss of libido, erectile dysfunction, loss of muscle mass, increased body fat, anemia, osteoporosis, depressed mood, decreased vitality, sweating, and hot flushes, as well as greater likelihood of both metabolic syndrome and cardiovascular disease (Nieschlag, 2019; Pye et al., 2014). Leydig cells in aged males (> 65 years of age) were found to be less responsive to LH stimulation as compared to younger males (<50 years of age) (Rubens, Dhont, & Vermeulen, 1974). This is also true of old Brown Norway rats, the most well studied animal model of aging, though the number of LCs per testis is the same in both young and old animals (Chen, Hardy, Huhtaniemi, & Zirkin, 1994; Midzak, Chen, Papadopoulos, & Zirkin, 2009). Gonadotropin-releasing hormone (GnRH) is released from the hypothalamus in a pulsatile manner, the periodicity and amplitude of which determine the pattern of secretion of the gonadotropins, LH and FSH, from the anterior pituitary (Araujo & Wittert, 2011). As some men age, there is a decrease in GnRH secretion from the hypothalamus and alterations in the pulse amplitude and frequencies of gonadotropins, either due to increased sensitivity to negative feedback or innate production defects (Urban, Veldhuis, Blizzard, & Dufau, 1988; Veldhuis, Urban, Lizarralde, Johnson, & Iranmanesh, 1992). Decreased GnRH gene expression and reduced release of LH from the pituitary have been found in older rats of some strains (e.g., Sprague-Dawley) (Kaler & Critchlow, 1984), but in other strains (e.g., Brown Norway rats) serum LH levels do not decrease with aging (Midzak et al., 2009). These observations, together with clinical studies, e.g., European Male Aging Study, suggest that LOH may occur in response to primary testicular failure, secondary disruption of the hypothalamic-pituitary-gonadal axis, or a combination of both (Golan, Scovell, & Ramasamy, 2015; Huhtaniemi, 2014).

T replacement therapy (TRT) is currently the only approved therapy to treat LOH. TRT involves the use of synthetic T analogs to improve patient symptoms (Bhasin et al., 2010; Carruthers, 2009; Corona, Sforza, & Maggi, 2017; Goodman et al., 2015; Guay et al., 2003). Side effects of TRT include polycythemia (Bachman et al., 2014; Ip et al., 2010), gynecomastia (Borst & Mulligan, 2007), and infertility (Samplaski et al., 2014). Additionally, cardiovascular concerns were raised in retrospective (Finkle et al., 2014; Vigen et al., 2013) and prospective (Basaria et al., 2010) studies. It should be noted, however, that some studies showed no effects of TRT or even cardiovascular improvements (Baillargeon et al., 2014; Shores, Smith, Forsberg, Anawalt, & Matsumoto, 2012). The use of hCG (Coviello et al., 2005; Depenbusch, von Eckardstein, Simoni, & Nieschlag, 2002; Hsieh, Pastuszak, Hwang, & Lipshultz, 2013; Turek, Williams, Gilbaugh 3rd, & Lipshultz, 1995), modulators of the estrogen receptor (Katz, Nabulsi, Tal, & Mulhall, 2012; Lim & Fang, 1976; Moskovic, Katz, Akhavan, Park, & Mulhall, 2012; Tenover & Bremner, 1991), and aromatase inhibitors (Leder, Rohrer, Rubin, Gallo, & Longcope, 2004; Veldhuis & Iranmanesh, 2005) have been proposed as alternatives to TRT. Methods to increase serum T in addition to TRT are clearly desirable (Lo, Rodriguez, Pastuszak, & Khera, 2017; McCullough, 2015).

Detailed knowledge of how T formation is regulated is essential for understanding the molecular basis of T decline in LOH and how this might be reversed by pharmacological means to restore T production by old LCs. Pituitary LH is the signal triggering adult LCs to produce T (Thompson & Kaiser, 2014). Control of T output is at two levels: substrate (cholesterol) availability and trafficking and enzymatic activities. Previous studies in Brown Norway rats indicated that both cholesterol transfer and enzyme expression levels are reduced in LCs from old rats (Beattie, Adekola, Papadopoulos, Chen, & Zirkin, 2015; Culty et al., 2002). However, enzyme availability is not rate-limiting since there is excess of bioactive steroidogenic enzymes present to metabolize large amounts of substrates. What seems to be limiting in the aging LC is the amount of cholesterol available to mitochondrial CYP11A1 for steroidogenesis (Culty et al., 2002).

2. Cell biology of Leydig cell steroidogenesis

In response to LH, cAMP stimulates the transport of free cholesterol from the plasma membrane and intracellular stores to mitochondria, and its subsequent transfer from the outer mitochondrial membrane (OMM) to the inner mitochondrial membrane (IMM), leading to steroidogenesis (Jefcoate, 2002; Miller & Bose, 2011; Rone, Fan, & Papadopoulos, 2009) (Fig. 1). The initial rapid response to LH and cAMP is mediated by hormone-sensitive protein synthesis, protein-protein interactions (PPIs), and organelle communications (Issop, Rone, & Papadopoulos, 2013; Miller & Bose, 2011). Chronic stimulation by LH is required for optimal expression of steroidogenic enzymes and thus sustainable steroid formation (Payne & Hales, 2004). Cholesterol transport may occur through either a vesicular or non-vesicular transport pathway. There is evidence that vesicular transport occurs through increased association of mitochondria with the endoplasmic reticulum in LCs in response to hormone, mediated via AAA+ ATPase domain 3 (ATAD3) protein (Issop et al., 2015). It is also possible that soluble N- ethylmaleimide-sensitive factor attachment receptor (SNARE) complexes facilitate the delivery of cholesterol by promoting functional PPIs between lipid droplets and mitochondria (Lin et al., 2016). Indeed, strong interorganelle interactions facilitate free cholesterol exchange in response to LH in LCs (Issop et al., 2013). The non-vesicular transport pathway may be mediated by proteins such as the steroidogenic acute regulatory protein (STAR) (Prinz, 2007).

Great understanding of steroidogenesis has come from the identification of the Steroidogenic InteracTomE (SITE), a protein scaffold (Fig. 2) created by PPIs of cytosolic and mitochondrial proteins that mediate free cholesterol targeting to OMM (Rone et al., 2009, 2012). It is important to clarify that initially a hormone-induced complex of proteins derived from the cytosol and OMM, referred to as the transduceosome, was shown to facilitate cholesterol movement to mitochondria, (Issop et al., 2013; Papadopoulos, Liu, & Culty, 2007). The formation of the transduceosome confers cAMP signaling specificity, transduction, and amplification, leading to increased cholesterol transport into mitochondria. Cholesterol reaches CYP11A1 through outer/inner mitochondrial membrane contact sites. SITE encompasses the entire protein complex composed of cytosolic, outer and inner mitochondrial membrane proteins.

This protein complex is composed of the cytosolic proteins acyl-CoA binding domain-containing 1 (ACBDI, also known as diazepam binding inhibitor, DBI) and 3 (ACBD3), the latter an A-kinase anchoring protein;

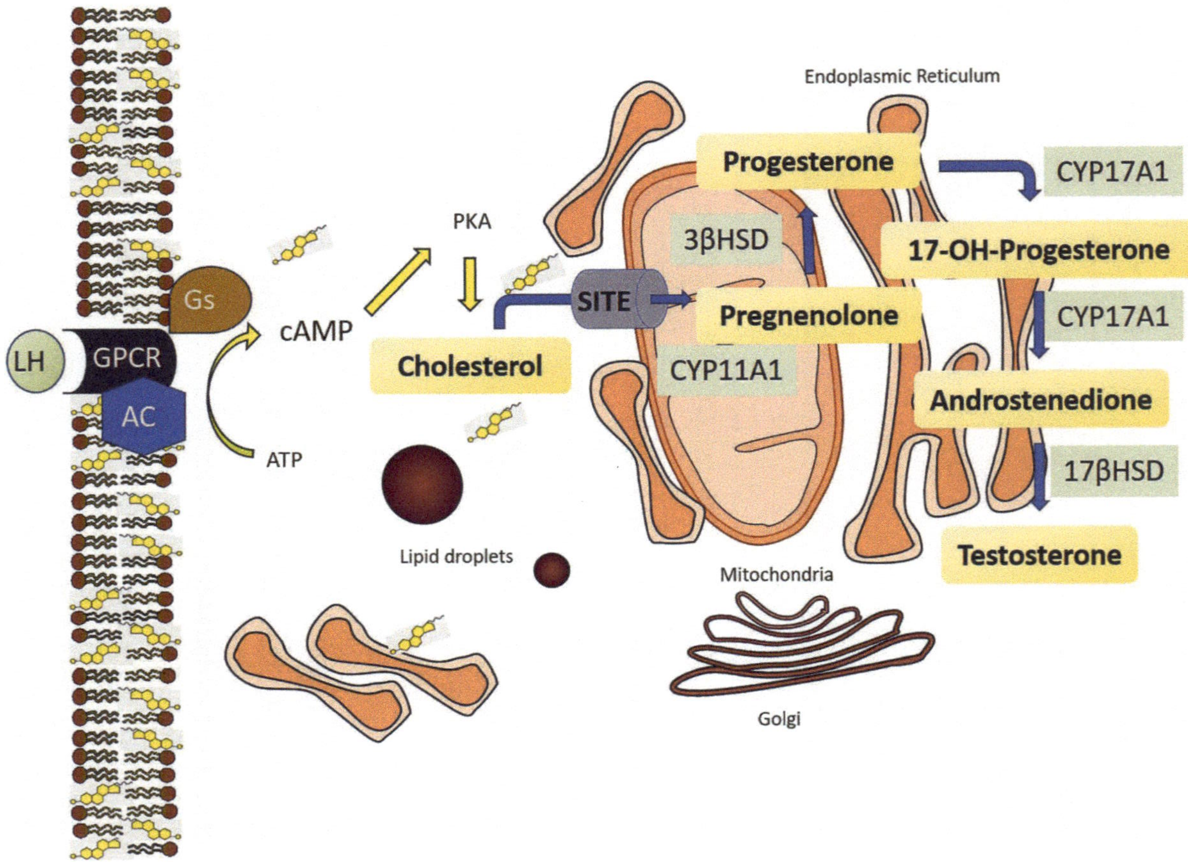

Fig. 1 See figure legend on opposite page.

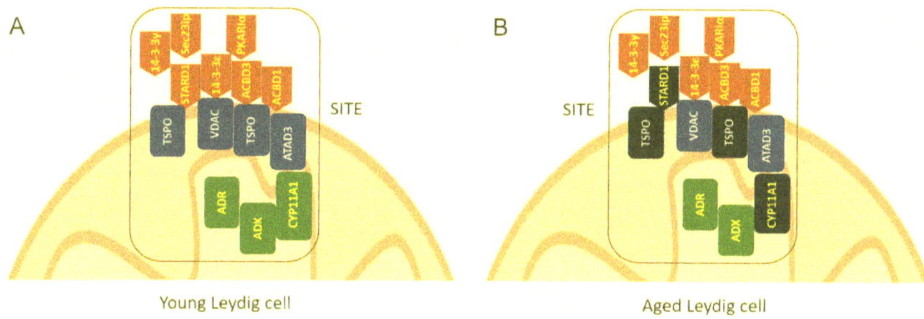

Fig. 2 SITE formation driving cholesterol import into mitochondrla. Cholesterol import into mitochondria is the result of series of protein-protein interactions. The involved proteins include: (A) cytosolic proteins ACBD1, ACBD3, Sec23ip, PKARIα, STARD1, and 14-3-3 adaptor proteins, 14-3-3γ and 14-3-3ε (red); (B) OMM proteins TSPO, VDAC and ATAD3 (blue); and (c) IMM proteins CYP11A1, ADX-adrenodoxin and ADR-adrenodoxin reductase (green). ACBD1/DBI is a TSPO endogenous ligand. The TSPO and VDAC complex located at the contact sites between outer and inner mitochondrial membranes recruits ACBD3 protein, a PKA anchoring protein, to mitochondria. The hormone-induced STARD1 protein is targeted to OMM where it interacts with VDAC and is locally phosphorylated by PKA for maximal activity. 14-3-3 adaptor proteins, binding to either STAR (14-3-3γ) or VDAC1 (14-3-3ε), provide negative control of steroid formation, thus allowing for sustainable steroid formation. Sec23pi binds to and controls 14-3-3γ activity. The formation of SITE allows for the translocation of cholesterol directly across mitochondrial membranes and intermembrane space to CYP11A1 for metabolism. In aged Leydig cells the levels of TSPO, STARD1 and CYP11A1 are reduced.

protein kinase A regulatory subunit I alpha (PKARIα); 14-3-3 adaptor proteins; STAR; and the outer/inner mitochondrial membrane contact site proteins translocator protein (TSPO), voltage-dependent anion channel (VDAC), ATAD3, and the IMM CYP11A1, the enzyme that metabolizes cholesterol to pregnenolone. The assembly of SITE allows for the separation

---

Fig. 1 Cell biology of Leydig cell T formation. LH binds its receptor on the Leydig cell membrane. LH receptor/G protein coupling results in increased cAMP production. cAMP stimulates the mobilization and transport of cholesterol to and into the mitochondria in part by activating cAMP-dependent protein kinase (PKA). Free cholesterol for steroidogenesis originates from various intracellular locations, including plasma membrane, endoplasmic reticulum and lipid droplets. LH binding to LC plasma membrane receptors induces the formation of a supramolecular complex of cytosolic and mitochondrial proteins, referred to as the Steroidogenic InteracTomE (SITE). Cholesterol movement into mitochondria is accomplished by a series of protein-protein interactions at SITE ending at active site of CYP11A1 in the matrix side of IMM. Cholesterol is converted to pregnenolone by CYP11A1, pregnenolone is converted into progesterone by enzymes present in mitochondria and endoplasmic reticulum (3βHSD), and progesterone is metabolized to T by enzymes exclusively present in the endoplasmic reticulum (CYP17A1 and 17βHSD).

of the steroidogenic cholesterol pool from structural membrane cholesterol and prevents unwanted crosstalk of cholesterol with other pathways, while optimizing substrate concentration and targeting to CYP11A1 (Rone et al., 2012). The rapid induction of STAR formation by LH and its subsequent targeting to SITE and insertion into OMM lead to accelerated cholesterol delivery to CYP11A1 (Midzak, Rammouz, & Papadopoulos, 2012; Rone et al., 2012). The resulting pregnenolone is further metabolized to progesterone by 3β-hydroxysteroid dehydrogenase (3β-HSD), present in mitochondria but predominantly found in the endoplasmic reticulum (ER), and then by CYP17A1 and 17β-HSD3, found only in the ER (Payne & Hales, 2004). Thus, cholesterol transfer is likely achieved through a hormone-dependent organelle communication network mediated by inter-organelle trafficking and PPIs, resulting in the efficient and timely delivery of cholesterol into mitochondria for steroid synthesis. Despite the progress made in understanding this process in steroidogenic cells, we do not know whether this mechanism is functional in old LCs where reduced levels of expression of SITE proteins and defects in specific PPIs are likely to be the basis for reduced cholesterol delivery and T production.

As with the steroidogenic enzymes, most of the proteins forming SITE are present in other steroidogenic tissues, such as the adrenal. Inborn errors in steroid biosynthesis are generally linked to steroidogenic enzyme or *STAR* mutations that affect both adrenal and gonadal function. However, in the case of LOH, T levels are reduced (Harman, et al., 2001; Khera et al., 2016; Tajar et al., 2010; Zirkin & Tenover, 2012) but there are no changes in circulating cortisol levels (Kobori et al., 2009; Lupien et al., 1999; Parker et al., 2000; Yen & Laughlin, 1998; Yiallouris et al., 2019), suggesting that there are testis-specific defects in T formation. Moreover, there are no genetic determinants of LOH identified as yet. It is our hypothesis that aging is associated with LC-specific changes in interorganelle trafficking and PPIs at the SITE that lead to reduced cholesterol transfer to CYP11A1 and T production in late-onset hypogonadism.

3. Steroidogenic cholesterol: Sources, trafficking and targeting to CYP11A1

De-esterification of cholesterol esters stored in lipid droplets may provide the bulk of cholesterol that sustains T formation (Rone et al., 2009). The physical association of the mitochondria and endoplasmic reticulum

(ER) seems likely to play a role in regulating the rate of steroid formation (Issop et al., 2015). Confocal microscopy, using antisera against organelle-specific antigens, and electron microscopy studies, demonstrated that there is an increase in the number of mitochondria–ER contact sites in response to hormone treatment in MA-10 cells (Issop et al., 2015). Electron tomography and 3D reconstruction allowed for the visualization of mitochondria-associated membranes (MAMs) (Issop et al., 2015). MAMs were isolated and found to contain the 67-kDa long isoform of ATAD3, a protein anchored in IMM and enriched in OMM/IMM contact sites. Deletion of the ATAD3 N-terminus resulted in the reduction of hormone-stimulated progesterone biosynthesis, suggesting a role of ATAD3 in mitochondria–ER contact site formation (Issop et al., 2015). These results demonstrated that the hormone-induced, ATAD3-mediated MAM formation participates in the optimal transfer of cholesterol from the ER into mitochondria for steroidogenesis.

While studying the mechanism of TSPO integration into OMM, we observed the formation of a large 800-kDa protein complex, which included TSPO and VDAC (Rone et al., 2012). Mass spectrometry analysis indicated that the 800-kDa mitochondrial complex contained OMM and IMM proteins, including CYP11A1. Further studies indicated that TSPO and CYP11A1 levels in this complex increased in MA-10 cells in response to cAMP. This finding suggests that this complex represents an OMM/IMM contact site where the conversion of cholesterol to pregnenolone occurs. To identify the cholesterol entry site, mitochondria were isolated from MA-10 cells, incubated with the photoactivatable ^{3}H-photocholesterol, and exposed to UV. This approach showed that the complex offers a microenvironment needed for CYP11A1 activity. These results identified SITE, a bioactive protein complex spanning the OMM/IMM, to be responsible for hormone-induced import, segregation, targeting, and metabolism of cholesterol (Fig. 2).

Cholesterol movement in MA-10 and primary rat LCs was tracked, using the fourth domain of perfringolysin O (D4) that binds cholesterol (Venugopal et al., 2016). Staining was primarily localized to the inner leaflet of the plasma membrane and was partially released upon treatment with a cAMP analogue. These results suggest that much of the steroidogenic pool of cholesterol, mobilized by cAMP for acute steroidogenesis, originates from the plasma membrane, and provide evidence that the plasma membrane is an important source of cholesterol for acute steroidogenesis in these cells as well as in MA-10 cells. These findings together with the fact that mitochondria are cholesterol-poor organelles whereas the plasma membrane contains most of the free cholesterol in the cell, suggest that plasma membrane cholesterol is readily available for steroidogenesis, whereas lipid droplet-associated cholesterol is used at later times. These steroidogenic pool of cholesterol will be targeted to mitochondria through MAMs and the 800-kDa SITE complex to CYP11A1 in IMM.

4. SITE proteins in cholesterol import machinery and steroidogenesis

4.1. *ACBD1/DBI* was identified in search of proteins mediating cholesterol transfer into mitochondria (Besman et al., 1989). DBI is important for hormone-dependent steroidogenesis (Besman et al., 1989; Boujrad, Hudson, & Papadopoulos, 1993; Schultz, Pelto-Huikko, & Alho, 1992). It displaces the benzodiazepine diazepam from brain $GABA_A$ receptors (Costa & Guidotti, 1991) and from the peripheral-type benzodiazepine receptor, now known as TSPO (Papadopoulos, Berkovich, Krueger, Costa, & Guidotti, 1991).

4.2. *TSPO* (18-kDa) is a mitochondrial high affinity cholesterol binding protein localized in the OMM (Jaipuria et al., 2017; Papadopoulos et al., 2006), A number of studies have provided evidence that TSPO is involved in inducing the translocation of cholesterol from OMM to IMM (Krueger & Papadopoulos, 1990; Yanagibashi et al., 1989) and that binding of TSPO ligands stimulates adrenal, brain and testicular steroid formation in vivo (Lacapere & Papadopoulos, 2003; Papadopoulos et al., 2015, 2006; Rupprecht et al., 2010). Blocking the cholesterol recognition amino acid consensus domain of TSPO was shown to block hormone-induced steroid formation in cells both in vitro and in vivo (Chung et al., 2013; Midzak, Akula, Lecanu, & Papadopoulos, 2011; Midzak, Akula, Rone, & Papadopoulos, 2015; Midzak, Rammouz, & Papadopoulos, 2012). CRISPR/Cas9-induced *Tspo* knockout in MA-10 cells resulted in reduced mitochondrial membrane potential, and the cells made little to no steroids (Fan, Wang, Zirkin, & Papadopoulos, 2018). Additionally, TSPO-specific drug ligands have been shown to stimulate steroid formation in steroidogenic cells of hypogonadal aged and young rats and in isolated mitochondria (Chung et al., 2013; Lacapere & Papadopoulos, 2003; Papadopoulos et al., 2006; Papadopoulos, Mukhin, Costa, & Krueger, 1990). For example, despite the reduced amount of TSPO in LCs of aged Brown Norway rats (Culty et al., 2002), pharmacological activation of TSPO with drug ligands elevated T production by the LCs back to the higher levels produced by young cells (Chung et al., 2013). In agreement with these findings, an age-dependent androgen deficiency was observed in *Tspo* knockout mice (Barron, Ji, Kito, Suhara, & Higuchi, 2018). TSPO drug ligands also increased T formation in LH-suppressed Sprague-Dawley rats, a model of secondary hypogonadism that was generated by treating the animals with the GnRH antagonist cetrorelix (Aghazadeh, Martinez-Arguelles, Fan, Culty, & Papadopoulos, 2014). Taken together, these studies strongly support the conclusion that TSPO plays an important role in cholesterol import into mitochondria and thus in steroidogenesis (Papadopoulos, Fan, & Zirkin, 2018).

4.3. *VDAC* has been shown to associate with TSPO (Liu, Rone, & Papadopoulos, 2006; McEnery, Snowman, Trifiletti, & Snyder, 1992). VDAC is an OMM protein that functions as a gatekeeper for the entry and exit of mitochondrial metabolites, thereby controlling crosstalk between mitochondria and the rest of the cell. Knocking down VDAC1, the major VDAC in MA-10 cells, resulted in decrease in hCG-induced progesterone production (Rone et al., 2012). Erastin, a drug targeting VDAC permeability, also resulted in reduced steroidogenesis, confirming that VDAC is important for contact site formation and steroid biosynthesis (Rone et al., 2012).

4.4. *STAR* is a LH-responsive, 37-kDa protein also involved in mitochondrial cholesterol transport (Bose, Lingappa, & Miller, 2002a; Clark, Wells, King, & Stocco, 1994; Miller, 2007), though its specific mechanisms of action remains uncertain. In the testis, STAR de novo synthesis parallels LH-induced steroid formation (Clark et al., 1995). There is now solid evidence that STAR functions at the OMM by activating a mitochondrial receptor or transport mechanism (Arakane et al., 1996; Baker, Epand, Epand, & Miller, 2007; Bose, Lingappa, & Miller, 2002b). The 37-kDa STAR was not processed to the mature 30-kDa protein in TSPO-depleted cells or in the presence of a peptide antagonist of TSPO (Gazouli, Han, & Papadopoulos, 2002; Hauet et al., 2005), suggesting that TSPO may play a direct role in the activity of STAR. These data were further corroborated by results showing that *Tspo* deletion from MA-10 cells using the CRISPR/Cas9 system affects hormone-induced steroid formation and STAR processing (Fan et al., 2018). STAR was also shown to directly interact with both VDAC1 and VDAC2 (Bose, Whittal, Miller, & Bose, 2008; Prasad et al., 2015). These findings suggest that the absence of STAR disrupts TSPO-mediated steroid formation. Collectively, these results show that while STAR is dependent on TSPO and VDAC for proper function, TSPO is also dependent on STAR for steroidogenesis.

4.5. *14-3-3 proteins.* The 14-3-3 family of proteins has been shown to mediate crosstalk among multiple protein networks (Aghazadeh & Papadopoulos, 2015). Proteomic analysis of proteins associated with mitochondria in control and hormone-treated MA-10 cells identified 14-3-3γ in the cytoplasm and 14-3-3ε at OMM (Aghazadeh et al., 2012; Aghazadeh, et al., 2014). 14-3-3ε and γ were found to interact with STAR, TSPO and VDAC1 and to control the rate of cholesterol import. Sec23ip, a protein involved in vesicle trafficking, interacts with 14-3-3γ during steroidogenesis (Aghazadeh et al., 2020) and is involved in cholesterol trafficking to supply cholesterol for steroidogenesis through interactions with 14-3-3γ (Aghazadeh et al., 2020). 14-3-3ε was also found to interact with VDAC1 forming a scaffold that limits the availability of cholesterol for steroidogenesis (Aghazadeh et al., 2014). This interaction results in multiple downstream effects, including localization of 14-3-3ε to the mitochondria, negative regulation of steroidogenesis by buffering cholesterol import through intercalation between TSPO and VDAC1, and regulation of cholesterol and ligand binding to TSPO.

4.6. *Other proteins of SITE.* ACBD3 (PAP7) has been identified as a protein that binds both TSPO and the regulatory subunit RIα of PKA (Liu et al., 2003a; Liu, Li, & Papadopoulos, 2003). Transfection of the TSPO- and PKARIα-binding domain of ACBD3 significantly inhibited steroidogenesis. These results confirmed ACBD3's role in steroidogenesis via interaction with TSPO and PKARIα. ACBD3 functions as an A-kinase anchoring protein, a family of proteins that recruit the PKA holoenzyme into proximity of its substrate (Torres-Quesada, Mayrhofer, & Stefan, 2017). Thus, the role of ACBD3 might be to anchor PKARIα close to mitochondrial TSPO and STAR, favoring their co-localization and phosphorylation by PKA. When TSPO, ACBD3, and PKARIα PPIs were disrupted by mutant proteins or competing peptides, steroidogenesis was inhibited, confirming the involvement of this complex in cholesterol transport into mitochondria (Liu et al., 2006). These observations show that the regulation of steroidogenesis is not under the control of any single protein, but rather results from inter-organelle interactions with the mitochondria mediated by PPIs at the OMM. Changes in these interactions may mediate the age-dependent reductions of T.

5. Oxidant/antioxidant imbalance and reductions in testosterone production

Numerous studies have reported the association of age-dependent oxidant/antioxidant imbalance with cellular DNA, protein and lipid damage (Jang, Blum, Liu, & Finkel, 2018). In steroidogenic cells, ROS are produced by the mitochondrial electron transport system but also from the P450 reactions in the mitochondria and smooth endoplasmic reticulum (Hornsby, 1989; Peltola, Huhtaniemi, Metsa-Ketela, & Ahotupa, 1996). Aging in Leydig cells has been shown to be accompanied by an increasingly oxidative intracellular environment (Cao, Leers-Sucheta, & Azhar, 2004; Chen et al., 2001; Luo et al., 2006; Zirkin & Papadopoulos, 2018). An altered oxidant/antioxidant environment might result from increased oxidant production, decreases in protective antioxidants, or both. Mitochondrial superoxide production was shown to increase in Leydig cells of aged as compared to young rats (Chen et al., 2001), and ROS was stimulated in both young and aged cells in response to LH (Beattie et al., 2013). Long-term suppression of LH and thus of steroidogenesis prevented or delayed age-dependent reduction in T production (Chen & Zirkin, 1999). Additionally, the antioxidants superoxide dismutase-1 and -2, glutathione peroxidase, and glutathione (GSH) were found to decrease with aging (Cao et al., 2004; Chen et al., 2001; Luo et al., 2006). These data, taken together, support a causative role for an altered oxidant/antioxidant environment in the reduced ability of aging Leydig cells to produce T (Fig. 3).

The increasingly oxidative environment and decreases in antioxidants in old Leydig cells raise the question of how levels of antioxidants, which protect cells from oxidative stress, are regulated. NRF2 is a transcription factor that regulates phase 2 antioxidant genes in numerous cells. Stimulation of Nrf2 activates an array of Nrf2-responsive genes and of their protein products, thus eliciting protection against a broad range of endogenous and exogenous factors that otherwise might disrupt cellular homeostasis through oxidative damage (Cho, Reddy, Debiase, Yamamoto, & Kleeberger, 2005). Upon increased intracellular oxidative stress, NRF2 translocates to the nucleus, there interacting with the androgen response

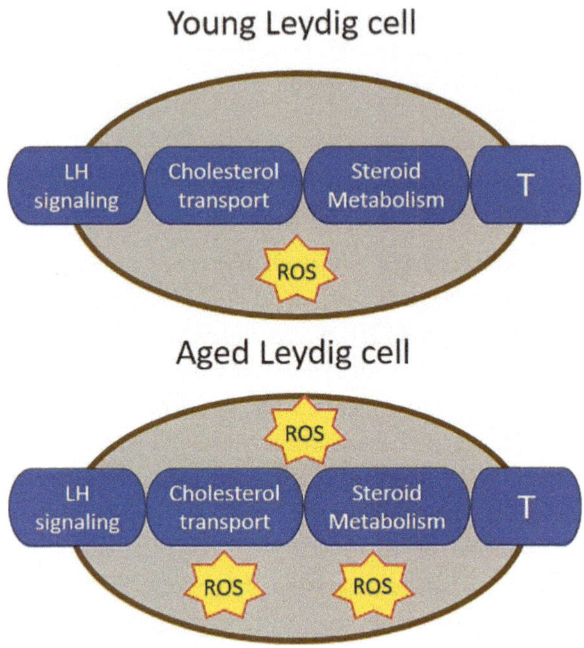

Fig. 3 Role of ROS in age-dependent Leydig cell steroidogenesis. Aging impacts steroidogenesis by affecting the intracellular redox balance in part through increased ROS production. This has significant effects on LH signaling, cholesterol transport into the mitochondria, steroid metabolism, and thus T formation.

element in the promoter region of antioxidant genes to promote the transcription of target antioxidant genes (Cho, Reddy, & Kleeberger, 2006; Giudice & Montella, 2006). Knocking out *Nrf2* was found to have no significant effect on serum T level or on T production by Leydig cells in young (3-month-old) mice, but by middle age (8 months), significant reductions were seen in serum T concentration and in Leydig cell T formation (Chen et al., 2015). By 24 months, both the wild-type and knockout mice had reduced serum T and Leydig cell T production, but with reductions more extensive in the knockouts. These observations support the hypothesis that suppression of the Leydig cell antioxidant defense capacity, in this case via *Nrf2* knockout, had subsequent detrimental effects on T formation. As yet, however, there have been few cause-effect studies directed to understanding the mechanism by which the intracellular environment of aging Leydig cells changes with aging, and how the changes, over time, elicit reduced T formation.

6. Existing treatments to increase serum T levels

In men with secondary hypogonadism, serum T can be elevated by TRT, administering LH or hCG to stimulate Leydig cell T production, or indirectly with clomiphene or aromatase inhibitor (Mulhall et al., 2018). However, hypogonadism in most aging patients is not the result of central deficiencies, but rather (or in addition) is the result of primary hypogonadism. In such men, attempts to increase serum T levels by LH administration typically are not effective. Rather, these men are treated by TRT. TRT involves the use of synthetic T analogs to increase serum T levels into the eugonadal range and thus improve patient symptoms (Bhasin et al., 2010; Carruthers, 2009; Corona et al., 2017; Goodman et al., 2015; Guay et al., 2003). Prescription T sales have dramatically increased in the U.S. since 2000 (Kolettis, Purcell, Parker, Poston, & Nangia, 2015). T is delivered via injections, scrotal and non-scrotal transdermal patches, and oral, buccal and gel preparations (Abadilla & Dobs, 2012; Gooren & Bunck, 2003; Wang et al., 2011).

As noted above, recent studies suggest that there may be increased risk of cardiovascular disease in older men after TRT (Basaria et al., 2010; Finkle et al., 2014; Vigen et al., 2013; Xu, Freeman, Cowling, & Schooling, 2013). These studies led to the FDA cautioning that aging men who take exogenous T face increased risk of stroke and heart attack (https://www.fda.gov/drugs/drug-safety-and-availability/fda-drug-safety-communication-fda-cautions-about-using-testosterone-products-low-testosterone-due) and medical societies adjusting the guidelines for use of TRT (Bhasin et al., 2018; Goodman et al., 2015). There also are reports suggesting that exogenous T treatment might increase the risk of prostate cancer (Bosland, 2014). Thus, the availability of additional methods to TRT by which to increase serum T clearly is desirable.

7. Targeting critical protein-protein interactions to reverse LOH

Knowledge of the steps and mechanisms in testosterone formation has made it possible to consider the use of pharmacological means to increase serum (and intratesticular) T by stimulating the Leydig cells themselves.

In particular, manipulation of SITE components might lead to first-in-class biologics and new chemical entities able to restore the ability of the Leydig cells to form T in aged testes. An approach that targets the Leydig cells themselves could be beneficial to many among the millions of men who currently use TRT to elevate serum T (Goodman et al., 2015).

We initially used TSPO as the target by using TSPO drug ligands to induce T formation. Although this approach did increase T production by aged Leydig cells in vitro and therefore resulted in elevated serum T levels both in hypogonadal aged rats (Chung et al., 2013) and in young rats rendered hypogonadal in response to the GnRH antagonist cetrorelix (Aghazadeh et al., 2014), adrenal and brain steroids also were affected. TSPO drug ligands can increase steroid production in various states of reduced endogenous steroidogenesis and for hormone-independent steroid formation, including neuropsychiatric disorders and hypogonadism (Papadopoulos et al., 2018). These results notwithstanding, the very high levels of TSPO in Leydig cells hold promise that an appropriate dose of administered TSPO drug ligand might elevate T production by Leydig cells specifically, with minor if any effects on the adrenal and/or brain in normal in vivo settings.

We have also begun to evaluate other possible targets in SITE. As noted earlier, 14-3-3ε protein adaptor binding to VDAC1 is a negative regulator of cholesterol import into mitochondria and testicular steroidogenesis. Cell penetrating peptides that block 14-3-3ε-VDAC1 interactions have been shown to induce T formation in vitro in Leydig cell and rat testis (Aghazadeh et al., 2014). In addition, the identified lead peptide TVS167 rescued intratesticular and serum T formation in adult male rats treated with GnRH antagonist, which dampened LH and T production (Aghazadeh et al., 2014). The VDAC1 14-3-3ε binding motif that contains S167 has a 100% across-species homology, including human. Thus, it is highly probable that bioactive peptides can be developed that will also induce T formation in humans. In fact, surface mapping of 14-3-3ε indicated that TVS167 binds to the open structure of 14-3-3ε and blocks docking of 14-3-3ε to VDAC1 in all species (Aghazadeh et al., 2014). This suggests the exciting possibility of recovering the ability of old LCs to form T without flooding the body with a long-acting synthetic T derivative and without administering LH-suppressive T.

8. Conclusions and future directions

LOH resulting from age-dependent deficiency in serum T affects the quality of life and well-being of millions of aging men worldwide via its association with numerous chronic and other diseases. Understanding the molecular determinants of LOH would allow us to identify new therapeutic targets that could be used to restore the ability of the testis to form T and to design preventive approaches to maintain T formation. The discovery of specialized protein networks forming the SITE, a hormonally regulated multiprotein complex that drives the transfer of cholesterol from storage sites across membranes and aqueous spaces into the IMM, is a new way of looking at complex processes. Understanding the differences in the organization and regulation of the SITE in old versus young LCs should explain differences in the ability of these cells to form T. Moreover, it is now clear that, despite their perceived similarities, the aging adrenal and testis present critical differences in the regulation of their respective steroidogenic pathways that may be key for developing tissue-specific therapies. These tissue-specific differences also indicate that the age-dependent decline in T by LCs is not simply a general phenomenon associated with aging that is common to all steroidogenic and other tissues. Identifying LC-specific and age-dependent PPIs are expected to provide targets for drug development to restore T formation without affecting adrenal corticosteroid production. Moreover, by understanding cause-effect relationships between oxidative stress and steroid formation, we will better understand the cause(s) of reduced T formation with aging, and this might lead to interventions designed to extend years of healthy life.

Exploiting these differences is a new concept that should lead to testis-specific treatments for LOH.

In addition to providing potential benefit to aging men, the design of new therapies that increase intratesticular bioactive androgen levels without affecting the hypothalamic-pituitary axis could be of importance for subfertile and infertile young men, including the many men diagnosed with idiopathic infertility who present with reduced circulating T, men with orchitis, and men following trauma [injury to genitalia, spinal cord injury], torsion, surgery, chemotherapy, irradiation, and in response to some medications (acquired hypogonadism).

Acknowledgments

This work was supported by grants from the National Institutes of Health [R01 AG021092] and the John Stauffer Dean's Chair in Pharmaceutical Sciences [University of Southern California].

References

Abadilla, K. A., & Dobs, A. S. (2012). Topical testosterone supplementation for the treatment of male hypogonadism. *Drugs*, *72*, 1591–1603.

Aghazadeh, Y., Martinez-Arguelles, D. B., Fan, J., Culty, M., & Papadopoulos, V. (2014). Induction of androgen formation in the male by a TAT-VDAC1 fusion peptide blocking 14-3-3varepsilon protein adaptor and mitochondrial VDAC1 interactions. *Molecular Therapy: The journal of the American Society of Gene Therapy*, *22*, 1779–1791.

Aghazadeh, Y., & Papadopoulos, V. (2015). The role of the 14-3-3 protein family in health, disease, and drug development. *Drug Discovery Today*, *21*, 278.

Aghazadeh, Y., Rone, M. B., Blonder, J., Ye, X., Veenstra, T. D., Hales, D. B., et al. (2012). Hormone-induced 14-3-3gamma adaptor protein regulates steroidogenic acute regulatory protein activity and steroid biosynthesis in MA-10 Leydig cells. *The Journal of Biological Chemistry*, *287*, 15380–15394.

Aghazadeh, Y., Venugopal, S., Martinez-Arguelles, D. B., Boisvert, A., Blonder, J., & Papadopoulos, V. (2020). Identification of Sec23ip, Part of 14-3-3gamma protein network, as a regulator of acute steroidogenesis in MA-10 leydig cells. *Endocrinology*, *161*, bqz036.

Arakane, F., Sugawara, T., Nishino, H., Liu, Z., Holt, J. A., Pain, D., et al. (1996). Steroidogenic acute regulatory protein (StAR) retains activity in the absence of its mitochondrial import sequence: Implications for the mechanism of StAR action. *Proceedings of the National Academy of Sciences of the United States of America*, *93*, 13731–13736.

Araujo, A. B., Esche, G. R., Kupelian, V., O'Donnell, A. B., Travison, T. G., Williams, R. E., et al. (2007). Prevalence of symptomatic androgen deficiency in men. *The Journal of Clinical Endocrinology and Metabolism*, *92*, 4241–4247.

Araujo, A. B., & Wittert, G. A. (2011). Endocrinology of the aging male. *Best Practice & Research. Clinical Endocrinology & Metabolism*, *25*, 303–319.

Bachman, E., Travison, T. G., Basaria, S., Davda, M. N., Guo, W., Li, M., et al. (2014). Testosterone induces erythrocytosis via increased erythropoietin and suppressed hepcidin: Evidence for a new erythropoietin/hemoglobin set point. *The Journals of Gerontology. Series A, Biological Sciences and Medical Sciences*, *69*, 725–735.

Baillargeon, J., Urban, R. J., Kuo, Y. F., Ottenbacher, K. J., Raji, M. A., Du, F., et al. (2014). Risk of myocardial infarction in older men receiving testosterone therapy. *The Annals of Pharmacotherapy*, *48*, 1138–1144.

Baker, B. Y., Epand, R. F., Epand, R. M., & Miller, W. L. (2007). Cholesterol binding does not predict activity of the steroidogenic acute regulatory protein, StAR. *The Journal of Biological Chemistry*, *282*, 10223–10232.

Barron, A. M., Ji, B., Kito, S., Suhara, T., & Higuchi, M. (2018). Steroidogenic abnormalities in translocator protein knockout mice and significance in the aging male. *The Biochemical Journal*, *475*, 75–85.

Basaria, S., Coviello, A. D., Travison, T. G., Storer, T. W., Farwell, W. R., Jette, A. M., et al. (2010). Adverse events associated with testosterone administration. *The New England Journal of Medicine*, *363*, 109–122.

Beattie, M. C., Adekola, L., Papadopoulos, V., Chen, H., & Zirkin, B. R. (2015). Leydig cell aging and hypogonadism. *Experimental Gerontology*, *68*, 87–91.

Beattie, M. C., Chen, H., Fan, J., Papadopoulos, V., Miller, P., & Zirkin, B. R. (2013). Aging and luteinizing hormone effects on reactive oxygen species production and DNA damage in rat Leydig cells. *Biology of Reproduction, 88*, 100.

Besman, M. J., Yanagibashi, K., Lee, T. D., Kawamura, M., Hall, P. F., & Shively, J. E. (1989). Identification of des-(Gly-Ile)-endozepine as an effector of corticotropin-dependent adrenal steroidogenesis: Stimulation of cholesterol delivery is mediated by the peripheral benzodiazepine receptor. *Proceedings of the National Academy of Sciences of the United States of America, 86*, 4897–4901.

Bhasin, S., Brito, J. P., Cunningham, G. R., Hayes, F. J., Hodis, H. N., Matsumoto, A. M., et al. (2018). Testosterone therapy in men with hypogonadism: An endocrine society clinical practice guideline. *The Journal of Clinical Endocrinology and Metabolism, 103*, 1715–1744.

Bhasin, S., Cunningham, G. R., Hayes, F. J., Matsumoto, A. M., Snyder, P. J., Swerdloff, R. S., et al. (2010). Testosterone therapy in men with androgen deficiency syndromes: an Endocrine Society clinical practice guideline. *The Journal of Clinical Endocrinology and Metabolism, 95*, 2536–2559.

Borst, S. E., & Mulligan, T. (2007). Testosterone replacement therapy for older men. *Clinical Interventions in Aging, 2*, 561–566.

Bose, H. S., Lingappa, V. R., & Miller, W. L. (2002a). Rapid regulation of steroidogenesis by mitochondrial protein import. *Nature, 417*, 87–91.

Bose, H. S., Lingappa, V. R., & Miller, W. L. (2002b). The steroidogenic acute regulatory protein, StAR, works only at the outer mitochondrial membrane. *Endocrine Research, 28*, 295–308.

Bose, M., Whittal, R. M., Miller, W. L., & Bose, H. S. (2008). Steroidogenic activity of StAR requires contact with mitochondrial VDAC1 and phosphate carrier protein. *The Journal of Biological Chemistry, 283*, 8837–8845.

Bosland, M. C. (2014). Testosterone treatment is a potent tumor promoter for the rat prostate. *Endocrinology, 155*, 4629–4633.

Boujrad, N., Hudson, J. R., Jr., & Papadopoulos, V. (1993). Inhibition of hormone-stimulated steroidogenesis in cultured Leydig tumor cells by a cholesterol-linked phosphorothioate oligodeoxynucleotide antisense to diazepam-binding inhibitor. *Proceedings of the National Academy of Sciences of the United States of America, 90*, 5728–5731.

Cao, L., Leers-Sucheta, S., & Azhar, S. (2004). Aging alters the functional expression of enzymatic and non-enzymatic anti-oxidant defense systems in testicular rat Leydig cells. *The Journal of Steroid Biochemistry and Molecular Biology, 88*, 61–67.

Carruthers, M. (2009). Time for international action on treating testosterone deficiency syndrome. *The Aging Male, 12*, 21–28.

Chen, H., Cangello, D., Benson, S., Folmer, J., Zhu, H., Trush, M. A., et al. (2001). Age-related increase in mitochondrial superoxide generation in the testosterone-producing cells of Brown Norway rat testes: Relationship to reduced steroidogenic function? *Experimental Gerontology, 36*, 1361–1373.

Chen, H., Hardy, M. P., Huhtaniemi, I., & Zirkin, B. R. (1994). Age-related decreased Leydig cell testosterone production in the brown Norway rat. *Journal of Andrology, 15*, 551–557.

Chen, H., Jin, S., Guo, J., Kombairaju, P., Biswal, S., & Zirkin, B. R. (2015). Knockout of the transcription factor Nrf2: Effects on testosterone production by aging mouse Leydig cells. *Molecular and Cellular Endocrinology, 409*, 113–120.

Chen, H., & Zirkin, B. R. (1999). Long-term suppression of Leydig cell steroidogenesis prevents Leydig cell aging. *Proceedings of the National Academy of Sciences of the United States of America, 96*, 14877–14881.

Cho, H. Y., Reddy, S. P., Debiase, A., Yamamoto, M., & Kleeberger, S. R. (2005). Gene expression profiling of NRF2-mediated protection against oxidative injury. *Free Radical Biology & Medicine, 38*, 325–343.
Cho, H. Y., Reddy, S. P., & Kleeberger, S. R. (2006). Nrf2 defends the lung from oxidative stress. *Antioxidants & Redox Signaling, 8*, 76–87.
Chung, J. Y., Chen, H., Midzak, A., Burnett, A. L., Papadopoulos, V., & Zirkin, B. R. (2013). Drug ligand-induced activation of translocator protein (TSPO) stimulates steroid production by aged brown Norway rat Leydig cells. *Endocrinology, 154*, 2156–2165.
Clark, B. J., Soo, S. C., Caron, K. M., Ikeda, Y., Parker, K. L., & Stocco, D. M. (1995). Hormonal and developmental regulation of the steroidogenic acute regulatory protein. *Molecular Endocrinology, 9*, 1346–1355.
Clark, B. J., Wells, J., King, S. R., & Stocco, D. M. (1994). The purification, cloning, and expression of a novel luteinizing hormone-induced mitochondrial protein in MA-10 mouse Leydig tumor cells. Characterization of the steroidogenic acute regulatory protein (StAR). *The Journal of Biological Chemistry, 269*, 28314–28322.
Corona, G., Sforza, A., & Maggi, M. (2017). Testosterone replacement therapy: Long-term safety and efficacy. *The World Journal of Mens Health, 35*, 65–76.
Costa, E., & Guidotti, A. (1991). Diazepam binding inhibitor (DBI): A peptide with multiple biological actions. *Life Sciences, 49*, 325–344.
Coviello, A. D., Matsumoto, A. M., Bremner, W. J., Herbst, K. L., Amory, J. K., Anawalt, B. D., et al. (2005). Low-dose human chorionic gonadotropin maintains intratesticular testosterone in normal men with testosterone-induced gonadotropin suppression. *The Journal of Clinical Endocrinology and Metabolism, 90*, 2595–2602.
Culty, M., Luo, L., Yao, Z. X., Chen, H., Papadopoulos, V., & Zirkin, B. R. (2002). Cholesterol transport, peripheral benzodiazepine receptor, and steroidogenesis in aging Leydig cells. *Journal of Andrology, 23*, 439–447.
Depenbusch, M., von Eckardstein, S., Simoni, M., & Nieschlag, E. (2002). Maintenance of spermatogenesis in hypogonadotropic hypogonadal men with human chorionic gonadotropin alone. *European Journal of Endocrinology, 147*, 617–624.
Fan, J., Wang, K., Zirkin, B., & Papadopoulos, V. (2018). CRISPR/Cas9 mediated Tspo gene mutations lead to reduced mitochondrial membrane potential and steroid formation in MA-10 mouse tumor Leydig cells. *Endocrinology, 159*, 1130–1146.
Finkle, W. D., Greenland, S., Ridgeway, G. K., Adams, J. L., Frasco, M. A., Cook, M. B., et al. (2014). Increased risk of non-fatal myocardial infarction following testosterone therapy prescription in men. *PLoS One, 9*, e85805.
Gazouli, M., Han, Z., & Papadopoulos, V. (2002). Identification of a peptide antagonist to the peripheral-type benzodiazepine receptor (PBR) that inhibits hormone-stimulated Leydig cell steroid formation. *Journal of Pharmacology and Experimental Therapeutics, 303*, 627–632.
Giudice, A., & Montella, M. (2006). Activation of the Nrf2-ARE signaling pathway: A promising strategy in cancer prevention. *BioEssays, 28*, 169–181.
Golan, R., Scovell, J. M., & Ramasamy, R. (2015). Age-related testosterone decline is due to waning of both testicular and hypothalamic-pituitary function. *The Aging Male, 18*, 201–204.
Goodman, N., Guay, A., Dandona, P., Dhindsa, S., Faiman, C., Cunningham, G. R., et al. (2015). American Association of Clinical Endocrinologists and American College of Endocrinology Position Statement on the Association of Testosterone and Cardiovascular Risk. *Endocrine Practice, 21*, 1066–1073.
Gooren, L. J., & Bunck, M. C. (2003). Transdermal testosterone delivery: Testosterone patch and gel. *World Journal of Urology, 21*, 316–319.

Guay, A. T., Spark, R. F., Bansal, S., Cunningham, G. R., Goodman, N. F., Nankin, H. R., et al. (2003). American Association of Clinical Endocrinologists medical guidelines for clinical practice for the evaluation and treatment of male sexual dysfunction: a couple's problem—2003 update. *Endocrine Practice*, *9*, 77–95.

Habert, R., Lejeune, H., & Saez, J. M. (2001). Origin, differentiation and regulation of fetal and adult Leydig cells. *Molecular and Cellular Endocrinology*, *179*, 47–74.

Harman, S. M., Metter, E. J., Tobin, J. D., Pearson, J., & Blackman, M. R. (2001). Longitudinal effects of aging on serum total and free testosterone levels in healthy men. Baltimore Longitudinal Study of Aging. *The Journal of Clinical Endocrinology and Metabolism*, *86*, 724–731.

Hauet, T., Yao, Z., Bose, S., Wall, C. T., Han, Z., Hales, B. D., et al. (2005). Peripheral-type benzodiazepine receptor-mediated action of steroidogenic acute regulatory protein on cholesterol entry into Leydig cell mitochondria. *Molecular Endocrinology*, *19*, 540–554.

Hornsby, P. J. (1989). Steroid and xenobiotic effects on the adrenal cortex: Mediation by oxidative and other mechanisms. *Free Radical Biology & Medicine*, *6*, 103–115.

Hsieh, T. C., Pastuszak, A. W., Hwang, K., & Lipshultz, L. I. (2013). Concomitant intramuscular human chorionic gonadotropin preserves spermatogenesis in men undergoing testosterone replacement therapy. *The Journal of Urology*, *189*, 647–650.

Huhtaniemi, I. T. (2014). Andropause—Lessons from the European Male Ageing Study. *Annales d'Endocrinologie*, *75*, 128–131.

Ip, F. F., di Pierro, I., Brown, R., Cunningham, I., Handelsman, D. J., & Liu, P. Y. (2010). Trough serum testosterone predicts the development of polycythemia in hypogonadal men treated for up to 21 years with subcutaneous testosterone pellets. *European Journal of Endocrinology*, *162*, 385–390.

Issop, L., Fan, J., Lee, S., Rone, M. B., Basu, K., Mui, J., et al. (2015). Mitochondria-associated membrane formation in hormone-stimulated Leydig cell steroidogenesis: Role of ATAD3. *Endocrinology*, *156*, 334–345.

Issop, L., Rone, M. B., & Papadopoulos, V. (2013). Organelle plasticity and interactions in cholesterol transport and steroid biosynthesis. *Molecular and Cellular Endocrinology*, *371*, 34–46.

Jaipuria, G., Leonov, A., Giller, K., Vasa, S. K., Jaremko, L., Jaremko, M., et al. (2017). Cholesterol-mediated allosteric regulation of the mitochondrial translocator protein structure. *Nature Communications*, *8*, 14893.

Jang, J. Y., Blum, A., Liu, J., & Finkel, T. (2018). The role of mitochondria in aging. *The Journal of Clinical Investigation*, *128*, 3662–3670.

Jefcoate, C. (2002). High-flux mitochondrial cholesterol trafficking, a specialized function of the adrenal cortex. *The Journal of Clinical Investigation*, *110*, 881–890.

Kaler, L. W., & Critchlow, V. (1984). Anterior pituitary luteinizing hormone secretion during continuous perifusion in aging male rats. *Mechanisms of Ageing and Development*, *25*, 103–115.

Katz, D. J., Nabulsi, O., Tal, R., & Mulhall, J. P. (2012). Outcomes of clomiphene citrate treatment in young hypogonadal men. *BJU International*, *110*, 573–578.

Kaufman, J. M., & Vermeulen, A. (2005). The decline of androgen levels in elderly men and its clinical and therapeutic implications. *Endocrine Reviews*, *26*, 833–876.

Khera, M., Broderick, G. A., Carson, C. C., 3rd, Dobs, A. S., Faraday, M. M., Goldstein, I., et al. (2016). Adult-onset hypogonadism. *Mayo Clinic Proceedings*, *91*, 908–926.

Kobori, Y., Koh, E., Sugimoto, K., Izumi, K., Narimoto, K., Maeda, Y., et al. (2009). The relationship of serum and salivary cortisol levels to male sexual dysfunction as measured by the International Index of Erectile Function. *International Journal of Impotence Research*, *21*, 207–212.

Kolettis, P. N., Purcell, M. L., Parker, W., Poston, T., & Nangia, A. K. (2015). Medical testosterone: An iatrogenic cause of male infertility and a growing problem. *Urology*, *85*, 1068–1073.

Krueger, K. E., & Papadopoulos, V. (1990). Peripheral-type benzodiazepine receptors mediate translocation of cholesterol from outer to inner mitochondrial membranes in adrenocortical cells. *The Journal of Biological Chemistry*, *265*, 15015–15022.

Lacapere, J. J., & Papadopoulos, V. (2003). Peripheral-type benzodiazepine receptor: Structure and function of a cholesterol-binding protein in steroid and bile acid biosynthesis. *Steroids*, *68*, 569–585.

Leder, B. Z., Rohrer, J. L., Rubin, S. D., Gallo, J., & Longcope, C. (2004). Effects of aromatase inhibition in elderly men with low or borderline-low serum testosterone levels. *The Journal of Clinical Endocrinology and Metabolism*, *89*, 1174–1180.

Lim, V. S., & Fang, V. S. (1976). Restoration of plasma testosterone levels in uremic men with clomiphene citrate. *The Journal of Clinical Endocrinology and Metabolism*, *43*, 1370–1377.

Lin, Y., Hou, X., Shen, W. J., Hanssen, R., Khor, V. K., Cortez, Y., et al. (2016). SNARE-mediated cholesterol movement to mitochondria supports steroidogenesis in rodent cells. *Molecular Endocrinology*, *30*, 234–247.

Liu, J., Li, H., & Papadopoulos, V. (2003). PAP7, a PBR/PKA-RIalpha-associated protein: A new element in the relay of the hormonal induction of steroidogenesis. *The Journal of Steroid Biochemistry and Molecular Biology*, *85*, 275–283.

Liu, J., Matyakhina, L., Han, Z., Sandrini, F., Bei, T., Stratakis, C. A., et al. (2003). Molecular cloning, chromosomal localization of human peripheral-type benzodiazepine receptor and PKA regulatory subunit type 1A (PRKAR1A)-associated protein PAP7, and studies in PRKAR1A mutant cells and tissues. *The FASEB Journal*, *17*, 1189–1191.

Liu, J., Rone, M. B., & Papadopoulos, V. (2006). Protein-protein interactions mediate mitochondrial cholesterol transport and steroid biosynthesis. *The Journal of Biological Chemistry*, *281*, 38879–38893.

Lo, E. M., Rodriguez, K. M., Pastuszak, A. W., & Khera, M. (2017). Alternatives to testosterone therapy: A review. *Sexual Medicine Reviews*, *6*, 106–113. https://doi.org/10.1016/j.sxmr.2017.09.004.

Luo, L., Chen, H., Trush, M. A., Show, M. D., Anway, M. D., & Zirkin, B. R. (2006). Aging and the brown Norway rat leydig cell antioxidant defense system. *Journal of Andrology*, *27*, 240–247.

Lupien, S. J., Nair, N. P., Briere, S., Maheu, F., Tu, M. T., Lemay, M., et al. (1999). Increased cortisol levels and impaired cognition in human aging: Implication for depression and dementia in later life. *Reviews in the Neurosciences*, *10*, 117–139.

McBride, J. A., Carson, C. C., 3rd, & Coward, R. M. (2016). Testosterone deficiency in the aging male. *Therapeutic Advances in Urology*, *8*, 47–60.

McCullough, A. (2015). Alternatives to testosterone replacement: Testosterone restoration. *Asian Journal of Andrology*, *17*, 201–205.

McEnery, M. W., Snowman, A. M., Trifiletti, R. R., & Snyder, S. H. (1992). Isolation of the mitochondrial benzodiazepine receptor: Association with the voltage-dependent anion channel and the adenine nucleotide carrier. *Proceedings of the National Academy of Sciences of the United States of America*, *89*, 3170–3174.

Midzak, A., Akula, N., Lecanu, L., & Papadopoulos, V. (2011). Novel androstenetriol interacts with the mitochondrial translocator protein and controls steroidogenesis. *Journal of Biological Chemistry*, *286*, 9875–9887.

Midzak, A.S., Akula, N., Rone, M.B., and Papadopoulos, V. (2015). Computational modeling and biological validation of novel non-steroidal ligands for the cholesterol recognition/interaction amino acid consensus (CRAC) motif of the mitochondrial translocator protein (TSPO). Pharmacological Research, 99C, 393–403.

Midzak, A. S., Chen, H., Papadopoulos, V., & Zirkin, B. R. (2009). Leydig cell aging and the mechanisms of reduced testosterone synthesis. *Molecular and Cellular Endocrinology, 299*, 23–31.

Midzak, A., Rammouz, G., & Papadopoulos, V. (2012). Structure-activity relationship (SAR) analysis of a family of steroids acutely controlling steroidogenesis. *Steroids, 77*, 1327–1334.

Miller, W. L. (2007). Steroidogenic acute regulatory protein (StAR), a novel mitochondrial cholesterol transporter. *Biochimica et Biophysica Acta, 1771*, 663–676.

Miller, W. L., & Bose, H. S. (2011). Early steps in steroidogenesis: Intracellular cholesterol trafficking. *Journal of Lipid Research, 52*, 2111–2135.

Mohr, B. A., Guay, A. T., O'Donnell, A. B., & McKinlay, J. B. (2005). Normal, bound and nonbound testosterone levels in normally ageing men: Results from the Massachusetts Male Ageing Study. *Clinical Endocrinology, 62*, 64–73.

Moskovic, D. J., Katz, D. J., Akhavan, A., Park, K., & Mulhall, J. P. (2012). Clomiphene citrate is safe and effective for long-term management of hypogonadism. *BJU International, 110*, 1524–1528.

Mulhall, J. P., Trost, L. W., Brannigan, R. E., Kurtz, E. G., Redmon, J. B., Chiles, K. A., et al. (2018). Evaluation and management of testosterone deficiency: AUA guideline. *The Journal of Urology, 200*, 423–432.

Nieschlag, E. (2019). Late-onset hypogonadism: A concept comes of age. *Andrology, 8*, 1506–1511. https://doi.org/10.1111/andr.12719.

Papadopoulos, V., Aghazadeh, Y., Fan, J., Campioli, E., Zirkin, B., & Midzak, A. (2015). Translocator protein-mediated pharmacology of cholesterol transport and steroidogenesis. *Molecular and Cellular Endocrinology, 408*, 90–98.

Papadopoulos, V., Baraldi, M., Guilarte, T. R., Knudsen, T. B., Lacapere, J. J., Lindemann, P., et al. (2006). Translocator protein (18kDa): New nomenclature for the peripheral-type benzodiazepine receptor based on its structure and molecular function. *Trends in Pharmacological Sciences, 27*, 402–409.

Papadopoulos, V., Berkovich, A., Krueger, K. E., Costa, E., & Guidotti, A. (1991). Diazepam binding inhibitor and its processing products stimulate mitochondrial steroid biosynthesis via an interaction with mitochondrial benzodiazepine receptors. *Endocrinology, 129*, 1481–1488.

Papadopoulos, V., Fan, J., & Zirkin, B. (2018). Translocator protein (18 kDa): An update on its function in steroidogenesis. *Journal of Neuroendocrinology, 30*. https://doi.org/10.1111/jne.12500.

Papadopoulos, V., Liu, J., & Culty, M. (2007). Is there a mitochondrial signaling complex facilitating cholesterol import? *Molecular and Cellular Endocrinology, 265–266*, 59–64.

Papadopoulos, V., Mukhin, A. G., Costa, E., & Krueger, K. E. (1990). The peripheral-type benzodiazepine receptor is functionally linked to Leydig cell steroidogenesis. *The Journal of Biological Chemistry, 265*, 3772–3779.

Parker, C. R., Jr., Slayden, S. M., Azziz, R., Crabbe, S. L., Hines, G. A., Boots, L. R., et al. (2000). Effects of aging on adrenal function in the human: Responsiveness and sensitivity of adrenal androgens and cortisol to adrenocorticotropin in premenopausal and postmenopausal women. *The Journal of Clinical Endocrinology and Metabolism, 85*, 48–54.

Payne, A. H., & Hales, D. B. (2004). Overview of steroidogenic enzymes in the pathway from cholesterol to active steroid hormones. *Endocrine Reviews, 25*, 947–970.

Payne, A. H., & Hardy, M. P. (2007). *The Leydig cell in health and disease*. Totoawa, New Jersey: Humana Press.

Peltola, V., Huhtaniemi, I., Metsa-Ketela, T., & Ahotupa, M. (1996). Induction of lipid peroxidation during steroidogenesis in the rat testis. *Endocrinology, 137*, 105–112.

Prasad, M., Kaur, J., Pawlak, K. J., Bose, M., Whittal, R. M., & Bose, H. S. (2015). Mitochondria-associated endoplasmic reticulum membrane (MAM) regulates steroidogenic activity via steroidogenic acute regulatory protein (StAR)-voltage-dependent anion channel 2 (VDAC2) interaction. *The Journal of Biological Chemistry*, *290*, 2604–2616.

Prinz, W. A. (2007). Non-vesicular sterol transport in cells. *Progress in Lipid Research*, *46*, 297–314.

Pye, S. R., Huhtaniemi, I. T., Finn, J. D., Lee, D. M., O'Neill, T. W., Tajar, A., et al. (2014). Late-onset hypogonadism and mortality in aging men. *The Journal of Clinical Endocrinology and Metabolism*, *99*, 1357–1366.

Rone, M. B., Fan, J., & Papadopoulos, V. (2009). Cholesterol transport in steroid biosynthesis: Role of protein-protein interactions and implications in disease states. *Biochimica et Biophysica Acta*, *1791*, 646–658.

Rone, M. B., Midzak, A. S., Issop, L., Rammouz, G., Jagannathan, S., Fan, J., et al. (2012). Identification of a dynamic mitochondrial protein complex driving cholesterol import, trafficking, and metabolism to steroid hormones. *Molecular Endocrinology*, *26*, 1868–1882.

Rubens, R., Dhont, M., & Vermeulen, A. (1974). Further studies on Leydig cell function in old age. *The Journal of Clinical Endocrinology and Metabolism*, *39*, 40–45.

Rupprecht, R., Papadopoulos, V., Rammes, G., Baghai, T. C., Fan, J., Akula, N., et al. (2010). Translocator protein (18 kDa) (TSPO) as a therapeutic target for neurological and psychiatric disorders. *Nature Reviews. Drug Discovery*, *9*, 971–988.

Samplaski, M. K., Loai, Y., Wong, K., Lo, K. C., Grober, E. D., & Jarvi, K. A. (2014). Testosterone use in the male infertility population: Prescribing patterns and effects on semen and hormonal parameters. *Fertility and Sterility*, *101*, 64–69.

Schultz, R., Pelto-Huikko, M., & Alho, H. (1992). Expression of diazepam binding inhibitor-like immunoreactivity in rat testis is dependent on pituitary hormones. *Endocrinology*, *130*, 3200–3206.

Shores, M. M., Smith, N. L., Forsberg, C. W., Anawalt, B. D., & Matsumoto, A. M. (2012). Testosterone treatment and mortality in men with low testosterone levels. *The Journal of Clinical Endocrinology and Metabolism*, *97*, 2050–2058.

Tajar, A., Forti, G., O'Neill, T. W., Lee, D. M., Silman, A. J., Finn, J. D., et al. (2010). Characteristics of secondary, primary, and compensated hypogonadism in aging men: Evidence from the European Male Ageing Study. *The Journal of Clinical Endocrinology and Metabolism*, *95*, 1810–1818.

Tenover, J. S., & Bremner, W. J. (1991). The effects of normal aging on the response of the pituitary-gonadal axis to chronic clomiphene administration in men. *Journal of Andrology*, *12*, 258–263.

Thompson, I. R., & Kaiser, U. B. (2014). GnRH pulse frequency-dependent differential regulation of LH and FSH gene expression. *Molecular and Cellular Endocrinology*, *385*, 28–35.

Torres-Quesada, O., Mayrhofer, J. E., & Stefan, E. (2017). The many faces of compartmentalized PKA signalosomes. *Cellular Signalling*, *37*, 1–11.

Travison, T. G., Araujo, A. B., O'Donnell, A. B., Kupelian, V., & McKinlay, J. B. (2007). A population-level decline in serum testosterone levels in American men. *The Journal of Clinical Endocrinology and Metabolism*, *92*, 196–202.

Turek, P. J., Williams, R. H., Gilbaugh, J. H., 3rd, & Lipshultz, L. I. (1995). The reversibility of anabolic steroid-induced azoospermia. *The Journal of Urology*, *153*, 1628–1630.

Urban, R. J., Veldhuis, J. D., Blizzard, R. M., & Dufau, M. L. (1988). Attenuated release of biologically active luteinizing hormone in healthy aging men. *The Journal of Clinical Investigation*, *81*, 1020–1029.

Veldhuis, J. D., & Iranmanesh, A. (2005). Short-term aromatase-enzyme blockade unmasks impaired feedback adaptations in luteinizing hormone and testosterone secretion in older men. *The Journal of Clinical Endocrinology and Metabolism*, *90*, 211–218.

Veldhuis, J. D., Urban, R. J., Lizarralde, G., Johnson, M. L., & Iranmanesh, A. (1992). Attenuation of luteinizing hormone secretory burst amplitude as a proximate basis for the hypoandrogenism of healthy aging in men. *The Journal of Clinical Endocrinology and Metabolism*, *75*, 707–713.

Venugopal, S., Martinez-Arguelles, D. B., Chebbi, S., Hullin-Matsuda, F., Kobayashi, T., & Papadopoulos, V. (2016). Plasma membrane origin of the steroidogenic pool of cholesterol used in hormone-induced acute steroid formation in leydig cells. *The Journal of Biological Chemistry*, *291*, 26109–26125.

Vigen, R., O'Donnell, C. I., Baron, A. E., Grunwald, G. K., Maddox, T. M., Bradley, S. M., et al. (2013). Association of testosterone therapy with mortality, myocardial infarction, and stroke in men with low testosterone levels. *JAMA*, *310*, 1829–1836.

Wang, C., Ilani, N., Arver, S., McLachlan, R. I., Soulis, T., & Watkinson, A. (2011). Efficacy and safety of the 2% formulation of testosterone topical solution applied to the axillae in androgen-deficient men. *Clinical Endocrinology*, *75*, 836–843.

Wu, F. C., Tajar, A., Pye, S. R., Silman, A. J., Finn, J. D., O'Neill, T. W., et al. (2008). Hypothalamic-pituitary-testicular axis disruptions in older men are differentially linked to age and modifiable risk factors: The European Male Aging Study. *The Journal of Clinical Endocrinology and Metabolism*, *93*, 2737–2745.

Xu, L., Freeman, G., Cowling, B. J., & Schooling, C. M. (2013). Testosterone therapy and cardiovascular events among men: A systematic review and meta-analysis of placebo-controlled randomized trials. *BMC Medicine*, *11*, 108.

Yanagibashi, K., Ohno, Y., Nakamichi, N., Matsui, T., Hayashida, K., Takamura, M., et al. (1989). Peripheral-type benzodiazepine receptors are involved in the regulation of cholesterol side chain cleavage in adrenocortical mitochondria. *Journal of Biochemistry*, *106*, 1026–1029.

Yen, S. S., & Laughlin, G. A. (1998). Aging and the adrenal cortex. *Experimental Gerontology*, *33*, 897–910.

Yiallouris, A., Tsioutis, C., Agapidaki, E., Zafeiri, M., Agouridis, A. P., Ntourakis, D., et al. (2019). Adrenal Aging and Its Implications on Stress Responsiveness in Humans. *Frontiers in Endocrinology (Lausanne)*, *10*, 54.

Zirkin, B. R., & Papadopoulos, V. (2018). Leydig cells: Formation, function, and regulation. *Biology of Reproduction*, *99*, 101–111.

Zirkin, B. R., & Tenover, J. L. (2012). Aging and declining testosterone: Past, present, and hopes for the future. *Journal of Andrology*, *33*, 1111–1118.

CHAPTER TWENTY-THREE

Adiponectin/AdipoRs signaling as a key player in testicular aging and associated metabolic disorders

Mayank Choubey[a,b,*], Ashutosh Ranjan[a], and Amitabh Krishna[a,*]

[a]Department of Zoology, Institute of Science, Banaras Hindu University, Varanasi, India

[b]Department of Molecular Medicine, Columbia University Irvine Medical Center, New York, NY, United States

*Corresponding authors: e-mail address: choubeymayank48@gmail.com; akrishna_ak@yahoo.co.in

Contents

Abstract

Aging undergoes serious worsening of peripheral organs and vital physiological processes including reproductive performances. Altered white adipose tissue and adipocyte functioning during aging results in ectopic lipid storage/obesity or metabolic derangements, leading to insulin resistance state. Eventually, accelerating cellular senescence thereby enhancing the high risk of age-associated metabolic alterations. Such alterations may cause derangement of numerous physiologically active obesity hormones, known as "adipokines." Specifically, adiponectin exhibits insulin sensitizing action causing anti-aging and anti-obesity effects via activation of adiponectin receptors (AdipoRs). The male reproductive physiology from reproductive mature stage to advanced senescent stage undergoes insidious detrimental changes. The mechanisms by which testicular functions decline with aging remain largely speculative. Adiponectin has also recently been shown to regulate metabolism and longevity signaling thus prolonging lifespan. Therefore, the strategy for activating adiponectin/AdipoRs signaling pathways are expected to provide a solid basis for the prevention and treatment of aging and obesity-associated reproductive dysfunctions, as well as for ensuring healthy reproductive longevity in humans.

Vitamins and Hormones, Volume 115
ISSN 0083-6729
https://doi.org/10.1016/bs.vh.2020.12.024

Abbreviations

ADIPORs	adiponectin receptors
AMPK	AMP-activated protein kinase
FSH	follicle-stimulating hormone
GnRH	gonadotropin-releasing-hormone
HPG	hypothalamic-pituitary-gonadal
HPT	hypothalamic-pituitary testicular
IR	insulin receptor
LC	Leydig cell
LH	luteinizing hormone
PPAR-α	peroxisome proliferator-activated receptor alpha
STs	seminiferous tubules
T	testosterone
T2D	type 2 diabetes
TNF-α	tumor necrosis factor-alpha
WAT	white adipose tissue

1. Introduction

Progressive medical facilities have contributed to a great extent in improving life expectancy as compared to previous generations. Although longevity may be an end in itself, the means to achieve it comprises of assorted factors. Quantitative enhancement of human lifespan coupled with qualitative aspects provides a general acknowledgment that aging should include freedom from chronic diseases and healthy physical, mental, and of course reproductive functioning (Bowling, 2007; Huijg et al., 2016; Rowe & Kahn, 1997). Reproduction is a fundamental biological feature of all living beings aimed at maintaining life through healthy progeny. Nowadays, infertility has affected a huge number of couples, and the cause is distributed to both the sexes. The reproductive potential of women is known to decline after the age of 35, but many men are unaware that their ability to contribute to conception is also adversely affected by their age too. While less is known about aging-associated male fertility, there is evidence that the older a man becomes, the more his fertility diminishes. Also, aging-associated derangement in the male reproductive tissues arises predominantly in the testis (Gunes, Hekim, Arslan, & Asci, 2016). Age-related decline in male sex hormones or testosterone (T) is a direct consequence of testicular aging. These alterations in the sex hormones in a large population of aged men may cause physiological instabilities affecting their quality of life. Aged males often retain the ability to father children, still aging influences various aspects of their gonads. The structure and function of various testicular cell types, i.e., germinal cells and somatic cells (Sertoli cells and Leydig cells) are influenced during aging. Also, the sperm quality, quantity, motility, morphology, and semen volume, are partially perturbed. Testicular metabolism and the levels of oxidative stress are altered at different rates during the different stages of the lifespan. This chapter is an amalgamation of hormonal and metabolic changes that might be responsible for decreased testicular activity during the advancement of age. Further, the present chapter summarizes the expression, physiological regulations, signaling, and the implication of obesity hormone "adiponectin" in the testicular aging processes.

2. Regulatory nexus between aging, obesity hormones, and male reproduction

The prevalence of obesity has augmented worldwide pandemic health situations (Blüher, 2019), due to its association with many metabolic disorders such as dyslipidemia, glucose intolerance, and insulin resistance resulting in the early onset of cellular aging, type 2 diabetes (T2D), and metabolism-related reproductive dysfunctions. Excessive accumulation of fat/lipid in the adipocytes of the white adipose tissue (WAT) hampers the normal metabolic processes (Scherer, 2016). Besides storing fat, mature adipocytes also synthesize and secrete numerous hormones, enzymes, cytokines, and growth factors that are associated with overall energy homeostasis of the body (Coelho, Oliveira, & Fernandes, 2013). Many of the factors that influence adipogenesis during metabolic complications are also associated with diverse body processes such as glucose homeostasis, lipid homeostasis, modulation of stress, and inflammatory responses. Further, these phenomena pave the way for major health concerns associated with the diseases such as aging, insulin resistance, obesity, T2D, cancer, and infertility (Ben-Shmuel, Rostoker, Scheinman, & LeRoith, 2016; Du-Plessis, Cabler, McAlister, Sabanegh, & Agarwal, 2010). One of the possible mechanisms underlying these metabolic complications is the altered endocrine effects of obesity-associated hormones.

Approximately ~50 different adipocyte-derived obesity hormones, known as "adipokines" are currently recognized as being secreted from the WAT. These adipokines are implicated in the modulation of several crucial physiological and metabolic processes that includes energy and glucose homeostasis, lipid metabolism, appetite control/food intake, oxidative stress, and reproduction (Coelho et al., 2013). While numerous studies have investigated metabolic diseases in-relation to reproductive functions, the underlying mechanism, particularly in males, have not yet been well elucidated (Cabler, Agarwal, Flint, & Du-Plessis, 2010; Landry, Cloutier, & Martin, 2013). Recent reports suggest that obesity has rapidly risen among Indians (Luhar, Timæus, & Jones, 2020). Also, studies have shown an association of obesity with male infertility (Du-Plessis et al., 2010). Further, there are reports that obesity affects reproductive potential in males by decreasing sperm quality together with the altered morphological and molecular structure of the germ cells in the testes (Palmer, Bakos, Fullston, & Lane, 2012). The rising prevalence of metabolic syndrome in males requires immediate attention for better public health awareness and a better understanding of the molecular mechanisms involved in the regulation of male reproductive physiology.

It is well recognized that obesity is the leading cause of hypogonadism/infertility in males (Shayeb, Harrild, Mathers, & Bhattacharya, 2011). Earlier studies have suggested the multifactorial origin of abnormal sperm parameters leading to infertility in obese males (Hammoud, Gibson, Peterson, Hamilton, & Carrell, 2006). The reduced total sperm counts were noticed in obese men due to altered reproductive hormonal profiles in comparison to the normal (Winters et al., 2006). These men have elevated estrogen and decreased production of T and Follicle-stimulating hormone (FSH) (Vermeulen, Kaufman, Deslypere, & Thomas, 1993). Other important causes of the reduction in sperm counts include direct action of adipokines at different levels of the hypothalamic-pituitary-gonadal (HPG) axis and resulting in a decline in T production, which consequently suppresses sperm production and declining fertility. Obesity is highly correlated with increased oxidative stress in the male reproductive system (Aziz et al., 2004). A study suggested that oxidative stress and lipid peroxidation could be the essential link between adiposity and infertility in obese males (Shukla et al., 2009). Obesity is known to promote oxidative stress, which in turn contributes to insulin resistance (Qatanani & Lazar, 2007). Men with T2D exhibit insulin resistance state, which is further associated with reduced T synthesis (Tsai, Matsumoto, Fujimoto, & Boyko, 2004).

3. Essential role of adipokines in regulating male reproductive physiology

Adipokines could be considered as a major link between obesity and infertility. Different studies investigated the involvement of adipokines in reproductive functions but the impact of adipokines on male reproduction/

male fertility has only recently begun to be explored. Among a huge list of adipokines, adiponectin exerts a beneficial role on sperm parameters, whereas other adipokines such as leptin, resistin, or chemerin would have a rather deleterious effect on spermatogenesis. Semen parameters seem to be impacted when leptin, resistin, and chemerin levels were increased while sperm morphology seems to be improved when adiponectin level was increased (Elfassy et al., 2018). The most studied adipokine—leptin plays a dual role with a positive effect on sperm parameters at the physiological levels and an inhibitory role at high circulating/seminal concentrations. Many semen parameters and male fertility itself are disturbed according to the semen adipokine levels, even if it is not the only interfering element. Taken together, adipokines are found in human and animal semen and most of them or their receptors are expressed in the male genital tract (Choubey et al., 2019; Choubey, Ranjan, Bora, Baltazar, & Krishna, 2019; Choubey, Ranjan, Bora, & Krishna, 2020; Ranjan, Choubey, Yada, & Krishna, 2019a; Ranjan, Choubey, Yada, & Krishna, 2019b; Singh, Choubey, Bora, & Krishna, 2018). Although the pathophysiological role of these adipokines in semen is not well elucidated, the adipokines could influence sperm functionality and could be potential biomarkers of male fertility thus requires further detailed investigation.

Earlier studies carried out in animals and humans showed localization and expression of these adipokines and their receptors in the testes. However, only a few studies have investigated the role of these adipokines on changes in reproductive activities in the case of aging/obesity/T2D males, although it has been well demonstrated that aging/obesity suppresses male fertility. Additionally, the implication of several adipokines (leptin, adiponectin, visfatin, resistin, chemerin, and nesfatin-1, etc.) in male reproductive functions have earlier been demonstrated (Table 1). While numerous studies have described the effect of adipokine, leptin on male reproduction (Tena-Sempere & Barreiro, 2002), this review emphasizes the regulatory actions of most abundant obesity hormone, the adiponectin, on the testicular functions during the aging process.

4. Adiponectin/AdipoRs: History, structure, and metabolic role

Adiponectin was discovered in 1995–96 simultaneously by four different laboratories giving different names such as AdipoQ; a 30 kDa Adipocyte complement-related protein (Acrp30); adipose most abundant

Table 1 Expression and actions of major obesity hormones "adipokines" in the regulation of male reproductive tissues.

Adipokines/receptors	Relation to obesity	Reproductive actions	References
Leptin (Ob-R)	Increases with fat accumulation	Leptin and its receptor were expressed differentially between species in peripheral reproductive tissue where it is suggested to exert direct regulatory action at different levels of the male gonadal axis that involves not only stimulatory but also inhibitory effects. Long-term administration of leptin could disturb testicular morphology and delay spermatogenesis process	Tena-Sempere and Barreiro (2002) and Kawwass, Summer, and Kallen (2015)
Adiponectin (AdipoRs)	Decreases with fat accumulation	Adiponectin and its receptors were localized in different cell types of the male reproductive tract. AdipoR2 deficiency leads to male subfertility with aspermia and atrophy of tubules, suggesting an important role of the adiponectin and its receptors in the regulation of normal reproduction	Caminos, Nogueiras, Gayta, Pineda, and Gonza (2008), Ocón-Grove, Krzysik-Walker, Maddineni, Hendricks, and Ramachandran (2008), Kawwass et al. (2015), Rak, Mellouk, Froment, and Dupont (2017), and Singh et al. (2018)
Chemerin (CMKLR1, GPR1, and CCRL2)	Increases with fat accumulation	Chemerin and its receptors are expressed in human and rat testes where it suppresses Leydig cell steroidogenesis suggesting the direct anti-gonadal actions of chemerin on testicular steroidogenesis	Li et al. (2014) and Singh et al. (2018)

Resistin	Increases with fat accumulation	Resistin is expressed particularly in testicular Leydig cells of the rat throughout the postnatal periods suggesting a direct role of resistin in testicular steroidogenesis	Nogueiras et al. (2004)
Nesfatin-1	Negatively correlated with BMI	NUCB2/nesfatin-1 protein was identified in interstitial mature Leydig cells	Ranjan et al. (2019a 2019b), and Ranjan, Choubey, Yada, and Krishna (2020)
Visfatin	Increases with visceral fat accumulation	Visfatin is expressed in the testis particularly in the Leydig cells. Visfatin increased testicular steroidogenesis from Leydig cells in vitro.	Ocon-Grove et al. (2010), Gurusubramanian and Roy (2014), and Reverchon et al. (2016)
Apelin	Increases with obesity	Infusion of apelin in male rats significantly decreases LH, serum testosterone, and number of Leydig cells	Sandal et al. (2015)
Vaspin	Increases with fat accumulation	Seminal plasma vaspin correlated negatively with ejaculate volume and positively with sperm DNA fragmentation	Thomas et al. (2013)

gene transcript 1 (apM1); and 28 kDa Gelatin binding protein (GBP28) (Hu, Liang, & Spiegelman, 1996; Maeda et al., 1996; Nakano, Tobe, Choi-Miura, Mazda, & Tomita, 1996; Scherer, Williams, Fogliano, Baldini, & Lodish, 1995). It is commonly known as adiponectin. The adiponectin gene is located on chromosome 3q27 spanning 16 kbp. The human adiponectin gene encodes a ~244 amino acid secretary peptide (~247 amino acids for mouse ortholog) of size ~30 kDa. The primary structure of adiponectin consists of four different domains, i.e., N-terminal signal peptide domain, a short variable domain, followed by the collagen-like fibrous domain, and finally a C-terminal globular domain (Ruan & Dong, 2016). The hormone adiponectin forms complex structures and circulates in the plasma as a low-molecular-weight (LMW) trimer, a medium-molecular weight (MMW) hexamer, and high-molecular-weight (HMW) multimer. All these forms of adiponectin differ in their biological activity. HMW adiponectin was shown to be the biologically active form of the hormone (Zhu et al., 2010). In addition to its complex structure, adiponectin is glycosylated, a post-translational modification that is essential to sustain its activity.

Adiponectin is the most abundantly expressed obesity hormone in the WAT and is the highest-circulating adipokines. Its circulating level ranges from 3 to 30 μg/mL in rodents as well as in humans (Arita et al., 1999; Sieminska et al., 2004). Adiponectin acts by interacting with its two isoforms of specific cell-surface receptors known as AdipoRs (AdipoR1 and AdipoR2) (Yamauchi et al., 2003). Both the receptors isoform belongs to the seven-transmembrane receptor family with an amino (N)-terminal facing inside and carboxyl (C)-terminal facing outside, thus these receptors are structurally and functionally different from those of G-protein-coupled receptors (GPCRs) (Tang et al., 2008). AdipoR1 is more firmly involved in the activation of AMP-activated protein kinase (AMPK) pathway and further regulates the inhibition of hepatic glucose production together with increased fatty-acid oxidation, whereas AdipoR2 is involved in the activation of the Peroxisome proliferator-activated receptor alpha (PPAR-α) pathways which help in increasing fatty-acid oxidation and inhibits inflammation and oxidative stress (Matsunami et al., 2010). Studies have shown the variation of these receptors in the binding affinity for various forms of adiponectin. AdipoR1 showed a higher binding affinity for globular form while AdipoR2 showed an intermediate binding affinity for both globular and full-length adiponectin. Adiponectin/AdipoRs are ubiquitously expressed (Yamauchi et al., 2003). In addition to AdipoRs, adiponectin also works through binding to receptor T-cadherin, but its role appeared to be minor (Capeau, 2007).

Adiponectin level declines in obese individuals but increased in normal and lean individuals. The study showed sex-related differences in adiponectin synthesis, with males having a lower level than females. The levels of adiponectin are significantly reduced in obese patients with T2D and infertility. The circulating concentration of adiponectin declines during metabolic disorders such as obesity, but increases during weight loss (Escobar-Morreale et al., 2006; Gavrila et al., 2003). Generally, insulin stimulates whereas tumor necrosis factor-α (TNF-α) inhibits the expression and release of adiponectin. The adiponectin hormone is known for its involvement in the control of metabolism and insulin sensitization. It also modulates inflammation by reducing the production and activity of TNF-α and interleukin-6 (IL-6). Adiponectin is also involved in the regulation of various physiological functions, such as glucose utilization, lipid synthesis, energy homeostasis, and insulin sensitivity (Kadowaki & Yamauchi, 2005; Pierre et al., 2009). Adiponectin deficiency leads to insulin resistance, glucose intolerance, and hyperlipidemia in the mice (Kubota et al., 2002; Maeda et al., 2002). Adiponectin along with other adipose-derived adipokines play major roles in the interaction between metabolism and reproduction and may be implicated in the detrimental effect of obesity on male reproductive functions.

5. Adiponectin/AdipoRs as a regulator of male reproductive physiology

Adiponectin is generally known as a "beneficial" adipokine as it improves both metabolic and reproductive profiles (Campos, Palin, Bordignon, & Murphy, 2008). In human, both hypothalamus and pituitary contain adiponectin receptors (Kubota et al., 2007; Rodriguez-Pacheco et al., 2007) and studies showed a negative impact of adiponectin on hypothalamic Gonadotropin-Releasing-Hormone (GnRH) secretion, pituitary luteinizing hormone (LH) and FSH secretion, and testicular T secretion, suggesting its important role in modulating central reproductive endocrine axis (Psilopanagioti, Papadaki, Kranioti, Alexandrides, & Varakis, 2009). Adiponectin and its two distinct receptors, i.e., AdipoR1/R2 are expressed in the HPG axis and their activation regulates the expression and/or secretion of kisspeptin, GnRH, and gonadotropins. Adiponectin/AdipoRs are expressed by different cell types of the male reproductive tract in various species like humans, rats, mice, ram, bull, and chicken (Table 2). (Caminos et al., 2008; Choubey et al., 2020; Choubey, Ranjan, Bora, Baltazar, & Krishna, 2019; Choubey, Ranjan, Bora, Baltazar, Martin, & Krishna, 2019; Civitarese et al., 2004; Kasimanickam et al., 2013; Ocón-Grove

Table 2 Studies showing localization and expression of adiponectin and its receptors in different cell types of male reproductive tissue of different species.

Species	Target tissue/cell line	Experimental evidence			References
		Adiponectin	AdipoR1	AdipoR2	
Chicken	Testis	Localized in PC and LC (IHC)	Localized mainly in PC and LC (IHC)	Localized in the SC syncytia, round spermatids, elongating spermatids, spermatozoa, and LC (IHC)	Ocón-Grove et al. (2008)
Rat	Testis	Prominent immunoreactivity detected at the cytoplasm of interstitial LC (IHC & WB)	mRNA expression mainly in ST (qPCR)	mRNA expression in adult LC (qPCR)	Caminos et al. (2008)
Mice	Testis	Localized in LC (IHC & WB)	Localized in LC (IHC & WB)	Localized in the STs particularly in the Rs (IHC & WB)	Choubey, Ranjan, Bora, Baltazar, and Krishna (2019), Choubey, Ranjan, Bora, Baltazar, Martin, and Krishna (2019), and Choubey et al. (2020)
Ram	Testis, Epididymis	Strong mRNA expression in the testis and carpus of the epididymis	Strong mRNA expression in the testis and carpus of the epididymis	Strong mRNA expression in the testis and carpus of the epididymis	Rahmanifar and Tabandeh (2012)
Human	Testicular homogenate	–	Detected adipoR1 gene expression	Detected adipoR2 gene expression highest in the testis	Civitarese et al. (2004)
Bull	Sperm	tail region of bull sperm (IF)	Sperm membrane of the equatorial region and acrosomal region (IF)	Sperm head region (IF)	Kasimanickam, Kasimanickam, Kastelic, and Stevenson (2013)

ST, Seminiferous tubules; *LC*, Leydig cells; *GC*, Germ cells; *SC*, Sertoli cells; *PC*, Peritubular cells; *IHC*, Immunohistochemistry; *WB*, Western blot; *qPCR*, Quantitative PCR; *Rs*, Round Spermatid; *RT PCR*, reverse transcriptase PCR; *IF*, immunofluorescence.

et al., 2008; Rahmanifar & Tabandeh, 2012; Wu et al., 2013). In male gonads, adiponectin regulates several functions of both somatic and germ cells, such as cellular proliferation, cell survival, apoptosis, and oxidative stress (Choubey, Ranjan, Bora, Baltazar, & Krishna, 2019). Additionally, the effect of adiponectin at the testicular level has been shown to promote the process of spermatogenesis and sperm maturation (Martin, 2014). Therefore, endocrine, paracrine, and autocrine actions of adiponectin in the testis may contribute to support male reproductive function.

Our earlier immunohistochemical studies showed the presence of AdipoRs receptors in the testis of reproductively mature mice. The AdipoR1 protein was localized in the Leydig cells while AdipoR2 protein was localized in the seminiferous tubules (STs) of the testis particularly in the round spermatids (Choubey, Ranjan, Bora, Baltazar, & Krishna, 2019). Also, chicken testis showed adiponectin and AdipoR1 expression only in peritubular cells and Leydig cells, whereas AdipoR2 is expressed in Sertoli cells and germ cells as well as in Leydig cells (Ocón-Grove et al., 2008). These findings are also of interest for men since AdipoRs mRNA has been identified in a cDNA library prepared from human testicular RNA (Bjursell et al., 2007; Civitarese et al., 2004). Also, the expression of AdipoR2 is higher than AdipoR1 in the testis. The expression of a functional AdipoR2 receptor appears to be important for the testicular functions in the mice as AdipoR2-deficient/knockout mice exhibit reduced testicular mass due to atrophy of the STs and aspermia (Bjursell et al., 2007). Moreover, Ocón-Grove and co-workers reported increased expression of AdipoRs in the testes of the chicken during sexual development (Combs et al., 2003). An earlier study suggested that adiponectin has an essential role in a variety of metabolic functions including glucose utilization, lipogenesis, immunity, and energy homeostasis (Yamauchi & Kadowaki, 2013). It is further suggested that testicular adiponectin may act as a paracrine/autocrine factor to regulate the interaction between different testicular cells. For instance, adiponectin was shown to regulate the gene expression of glucose transporters GLUT1 and GLUT3, in all ST cells in the rodent (Ceddia et al., 2005). Kasimanickam and co-workers suggested that adiponectin and its receptors contribute significantly to various fertility parameters of Holstein bulls. They found an increased serum adiponectin level and increased mRNA expression of adiponectin/AdipoRs in sperm in average and high fertility bulls compare to low fertility bulls. Adiponectin protein was abundantly localized in the lower part i.e., tail region of bull sperm, but scarce in the cytoplasm. AdipoR1 was immunolocalized mainly at the sperm membrane of the equatorial region and the acrosomal region. Furthermore, AdipoR2 was immunolocalized primarily on the sperm head region (Kasimanickam et al., 2013). A male sex steroid hormone, T synthesized predominantly by interstitial Leydig cells. The presence of AdipoRs has been demonstrated in TM3 cell line derived from mouse Leydig cell. This finding suggests an important role of adiponectin in the regulation of steroidogenesis in Leydig cells. The Leydig cells treated with pro-inflammatory cytokine was also shown to significantly down-regulated mRNA and protein expression of both AdipoR1 and AdipoR2. However, adiponectin pretreatment successfully inhibited the signaling pathway mediated by pro-inflammatory cytokines. These studies thus provided compelling evidence that adiponectin stimulates T synthesis by attenuating nuclear factor-kB (NF-kB) activation through the promotion of AMPK phosphorylation. These studies thus clearly suggested that adiponectin may be a useful target to prevent pro-inflammatory cytokine-induced dysfunction in Leydig cells through its anti-inflammatory actions (Wu et al., 2013). Pro-inflammatory cytokines such as TNF-α and IL, could lead to cytotoxicity and play inhibitory roles in the steroidogenesis of Leydig cells (Hong et al., 2004). A few studies have also examined the direct role of adiponectin on Leydig cell. The Leydig cells treated with recombinant adiponectin showed a decrease in T biosynthesis along with the reduced expression of the cholesterol transporter steroidogenic acute regulatory protein (StAR) (Pfaehler et al., 2012) whereas adiponectin also improves progesterone production through increased expression of the StAR and cholesterol side-chain cleavage (CYP11A1) steroidogenesis enzyme, suggesting that high doses of adiponectin (50, 500, or 5000 ng/mL) could promote T synthesis from the MA-10 Leydig cell line (Landry, Paré, Jean, & Martin, 2015). These studies reveal significant inconsistencies regarding the direct effects of adiponectin on testicular cells. To our knowledge, there are limited in vivo studies regarding adiponectin action on testicular functions.

The mRNA transcripts of adiponectin/AdipoRs were identified in all parts of the epididymis, i.e., caput, corpus, and cauda, as well as other reproductive parts of Iranian fat-tailed Ram with the highest expression of adiponectin and its receptors, showed by corpus epididymis as compared to caput and cauda epididymis (Rahmanifar & Tabandeh, 2012). It is well documented that sperm motility increases during their epididymal transport and they attain maximal motility in the corpus or anterior part of the epididymis (Jones, 1998; Marengo, 2008). These studies suggest that adiponectin may be a stimulatory factor responsible for the induction of sperm motility in the male reproductive tract and it may also be involved in regulating the structural and functional integrity of the epididymis.

6. Impact of aging on male reproduction

Aging or senescence is defined as a persistent decline in the fitness with the chronological age of an organism due to deterioration of functional performance expressed by an increase in mortality risk and/or decline in reproductive success (Rose, 1991). A marked change noted in the testicular activities as the male advance towards the aging process, from the neonatal and infantile period to puberty onset, adulthood, and finally senescence (Fink, 2000; Tena-Sempere & Huhtaniemi, 2003). Aging is an enormously complex, multifactorial, and unalterable process having a significant impact on male fertility (Almeida, Rato, Sousa, Alves, & Oliveira, 2017). Previous studies suggested that during aging, regressive changes occur both in spermatogenesis and testicular steroidogenesis, resulting in decreased androgen production and sperm count respectively (Perheentupa & Huhtaniemi, 2009). Numerous studies had shown a decline in the level of T as well as a decline in the quality and quantity of spermatogenesis with age (Chen et al., 2003; Kidd, Eskenazi, & Wyrobek, 2001; Kuhnert & Nieschlag, 2004; Plas, Berger, Hermann, & Pfluger, 2000). More extensive studies are required to find out the factor responsible for the onset of regressive testicular changes during aging in men.

The aged males undergo gradual hypogonadism because of a steady decline in T synthesis (Hermann, Untergasser, Rumpold, & Berger, 2000). The decreased synthesis of T may be responsible for lower sperm counts. Semen parameters, such as sperm count, motility, and viability, correlated negatively with age (Gunes et al., 2016). Recent, meta-analysis showed a decrease in sperm parameters with aging (Hellstrom et al., 2006; Johnson, Dunleavy, Gemmell, & Nakagawa, 2015). Apart from primary testicular defects, secondary mechanisms such as changes in the hypothalamic-pituitary (H-P) axis also cause reproductive aging in men. It has earlier been shown that the sensitivity of the H-P axis to negative feedback regulation by gonadal hormone changes during aging. The hormonal alterations noted during aging occurred due to decreased release of gonadotropin and this represents a primary defect on Leydig cell function (Vermeulen & Deslypere, 1985). In contrast, the aged rat also showed a decline in circulating T concentrations due to the changes in catecholaminergic transmission in the hypothalamus and consequently decline in GnRH/gonadotropin secretion, thus representing secondary rather than primary Leydig cell failure (Meites, 1988). The aging testis also showed decreased adenosine triphosphate (ATP) but increased reactive Oxygen species (ROS) production and increased apoptosis that leads to regressive changes in the testicular histology (Gunes et al., 2016). This increased ROS level in the aging male may be responsible for decreased sperm parameters (Weir & Robaire, 2007). Aging is also characterized by a progressive loss of pancreatic β-cell function and is accompanied by insulin resistance (Møller, Gormsen, Fuglsang, & Gjedsted, 2003). Moreover, it is suggested that the decline in insulin sensitivity contributes to the regressive changes in the testicular activities during aging (Pitteloud et al., 2005).

Earlier studies have reported the decreased expression of StAR protein in the testes suggesting reduced transport of cholesterol to the inner mitochondrial membrane during aging (Banerjee, Anuradha, Mukherjee, & Krishna, 2014; Chen, Huhtaniemi, & Zirkin, 1996). Studies conducted during the last few decades suggested male reproductive aging a far more complex process. Application of novel statistical approaches to the analysis of gonadotropin and T levels in serially collected blood samples from reproductively mature and elderly men revealed numerous age-related changes in the function of the hypothalamic-pituitary-testicular (HPT) endocrine axis. The aged individuals showed decreased semen quality (reduced total sperm count, concentration, and motility), and decreased in morphologically normal sperm (Cocuzza et al., 2008). Testicular histology of aged mice showed atrophied ST with depletion of germ cells and patches of vacuolization. Further, the age-related deterioration of the Sertoli cell resulted in a breakage of the blood-testes barrier, production of auto-antibodies against germ cell, and induction of an autoimmune reaction (Takano & Abe, 1987). The histomorphological study of the testis of aged rats revealed that the Leydig cell undergoes atrophic changes in size rather than a reduction in their number (Barbutska, Koeva, Bakalska, & Atanassova, 2013).

7. Adiponectin/AdipoRs in testicular aging

Over several years, theories have been suggested that testicular structure undergo regressive changes leading to hypogonadism during aging, an effective treatment that may ameliorate such age-associated testicular complications has not been reviewed in any animal model so far (Jarak et al., 2018). Earlier attempts to treat human aging by hormone replacement therapy generally failed to show desired changes (Barzilai, Huffman, Muzumdar, & Bartke, 2012). We demonstrated that an adequate expression of adiponectin and its receptors may be required for actively performing testicular functions (Choubey, Ranjan, Bora, Baltazar, & Krishna, 2019). It has also been demonstrated that the deficiency of AdipoRs in male mice may lead to subfertility due to testicular atrophy and aspermia (Bjursell et al., 2007). Recently we found that a decline in the expression of testicular adiponectin/AdipoR1/AdipoR2 proteins may be responsible for impaired testicular functions in the aged mice. Our group showed for the first time the significant variations in the expression of testicular adiponectin/AdipoRs during different maturation and aging stages starting from early postnatal, weaning, puberty, reproductively mature to late senescence period of the mice. We clearly showed changes in the expressions of testicular adiponectin/AdipoRs, which correlated significantly with the changes in testicular mass, testicular insulin receptors (IR) protein expression, and T levels during different stages of aging in the mice. The testicular adiponectin/AdipoRs protein expression enhanced gradually from the early postnatal period and attained a peak level during puberty (Fig. 1). The peak pubertal expression of adiponectin/AdipoRs persists during the reproductively mature period but declines significantly in the testis of aged mice (Choubey, Ranjan, Bora, Baltazar, Martin, & Krishna, 2019). Also, a similar pattern was observed in the avian testis, which demonstrated several folds increased expression of adiponectin mRNA from sexually immature to mature chicken, with peak expression in early adulthood but unable to depict the expression changes during senescence period (Ocón-Grove et al., 2008). Following the above findings, we can easily suggest that the upregulation of adiponectin/AdipoRs signaling may be associated with sexual maturation of the male reproductive system.

The availability of energy reserves and nutrition also leave an impact on testicular functions during aging. However, the molecular mechanisms that establish a nexus between energy reserves and male reproduction, and the signals that mediate these processes are ambiguous. Metabolism-associated hormones, such as insulin and adiponectin, play a key role in the relationship between metabolism and reproduction (Choubey, Ranjan, Bora, Baltazar, & Krishna, 2019). Adiponectin's autocrine/paracrine and endocrine signaling in the testis exerts functional metabolic effects including improvement of both insulin sensitivity in the old mice and protection from

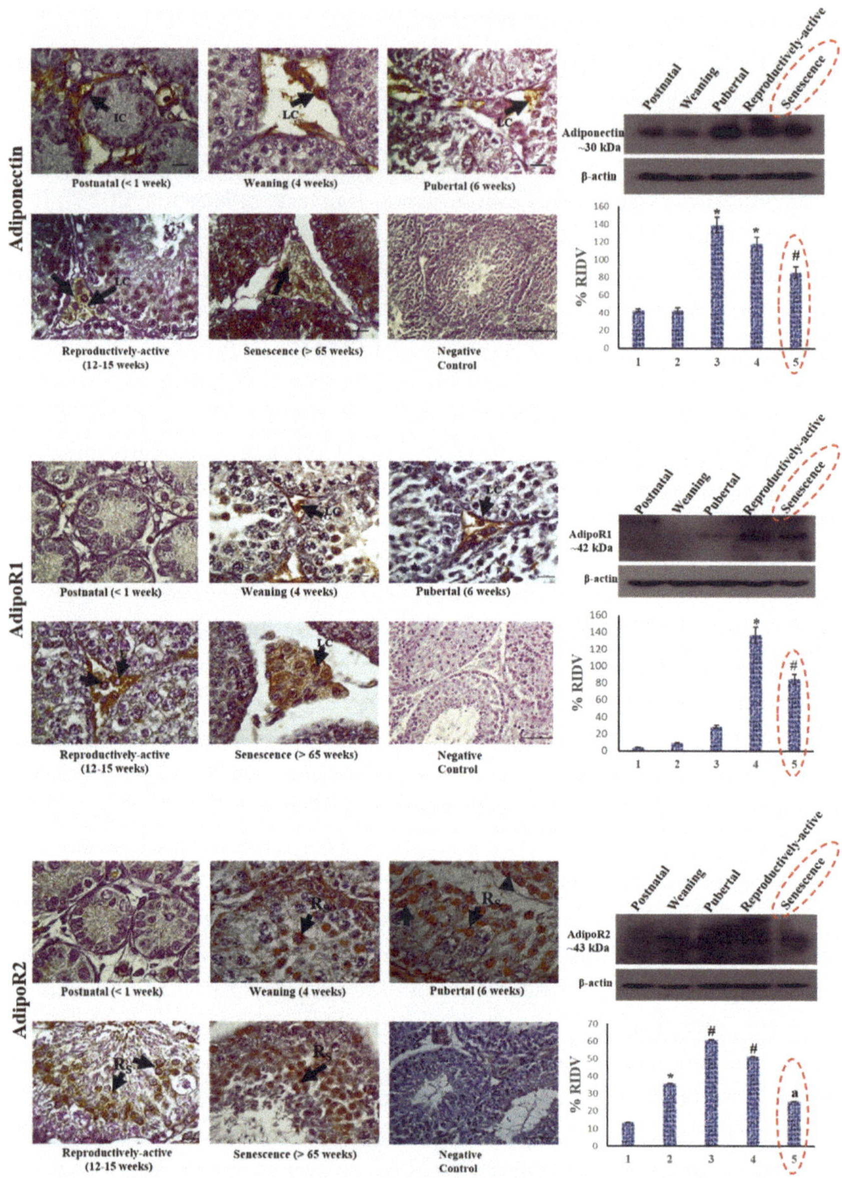

Fig. 1 Immunolocalization and Immunoblot analysis of adiponectin and its receptors (AdipoR1 and AdipoR2) in the testis of the mice during different stages of maturation (Postnatal, Weaning, Pubertal, Reproductively active) and aging (Senescence). Black arrowhead showing immunostaining of adiponectin and AdipoR1 proteins in the Leydig cells (LC) while AdipoR2 immunostaining was observed in the STs particularly in the round spermatids (Rs) (Choubey, Ranjan, Bora, Baltazar, Martin, & Krishna, 2019).

high-fat diet-induced insulin resistance in T2D mice (Choubey et al., 2020; Choubey, Ranjan, Bora, Baltazar, Martin, & Krishna, 2019). Our previous findings strongly suggest an improvement of insulin sensitivity in the aged mice after adiponectin-induced activation of testicular AdipoRs and IR pathways consecutively with phosphorylation and activation of AMPK protein (Choubey, Ranjan, Bora, Baltazar, Martin, & Krishna, 2019), a master regulator of the gonadal energy homeostasis (Bertoldo, Faure, Dupont, & Froment, 2015). The testicular AMPK activation results in increased lactate supply to the developing male germ cells by enhanced lactate dehydrogenase (LDH) activity and glucose uptake, via increased expression of testicular glucose transporter 8 (GLUT8), Monocarboxylate transporter 2 (MCT2) and MCT4 proteins of the adiponectin-treated aged mice (Choubey, Ranjan, Bora, Baltazar, Martin, & Krishna, 2019). Further, adiponectin-induced enhance expression of testicular IR protein correlated positively with the increased concentration of circulating T. This showed the involvement of insulin in the T synthesis (Bebakar, Honour, Foster, Liu, & Jacobs, 1990). Thus, the treatment of adiponectin, by up-regulating AdipoRs/AMPK signaling, increases insulin sensitivity in the testis (Choubey, Ranjan, Bora, Baltazar, Martin, & Krishna, 2019), which subsequently may stimulate T production (Schoeller, Schon, & Moley, 2012). In addition, adiponectin-treated aged mice also showed enhanced expression of certain steroidogenic marker proteins (LH-R, StAR, 3β-HSD, and AR) in the testis. These findings strongly suggest a direct involvement of adiponectin in the regulation of testicular steroidogenesis and a marked decline in its synthesis and action which may be associated with reduced testicular steroidogenesis as found during testicular aging. The aged mice showed abnormal regressive histological changes in the testis such as the enhanced vacuolization, degenerated germinal epithelium nuclei, and loss of primary spermatogenic cells in the STs. Adiponectin-treated aged mice showed an increase in tubular diameter, germinal epithelium height, improvement in testicular histoarchitecture followed by the appearance of primary spermatocytes with the fewer vacuole, increased number of type A spermatogonia (Sg), preleptotene spermatocyte (PL), pachytene spermatocytes (P) and stage 7 spermatids (S7) in the STs of the testis compared to the control aging mice (Choubey, Ranjan, Bora, Baltazar, Martin, & Krishna, 2019).

Over time, several studies suggested an imbalance between testicular ROS and antioxidant defense system as a probable cause for declined sperm counts during aging (Selvaratnam & Robaire, 2016). Recently, a study suggested an increased seminal oxidative stress which correlated with reduced sperm motility in the aging male (Gunes et al., 2016). It is very well recognized that decreased concentration of testicular anti-oxidative enzymes, superoxide dismutase (SOD), catalase (CAT), and glutathione peroxidase (GPx), while the increased concentration of testicular oxidative substance TBARS (lipid peroxidation) in the aged mice resulted in over-production of ROS. This enhanced production of ROS during aging then serves as a mechanism for decreased production of ATP and enhances apoptosis, which leads to impaired spermatogenesis (Aitken & Roman, 2008; Davalli, Mitic, Caporali, Lauriola, & D'Arca, 2016). Our laboratory findings demonstrated for the first time that treatment with adiponectin reduces the ROS formation and enhancing testicular antioxidants enzymes (SOD, CAT, and GPx) activities, thereby reducing lipid peroxidation and oxidative stress in the aged testis. This suggests that adiponectin-induced decline in oxidative stress may be responsible for the improvement in testicular functions in aging mice. Our recent finding thus portrays a possible involvement between adiponectin with oxidative stress and insulin resistance during testicular aging. Also, future studies such as siRNA-mediated adiponectin gene silencing or utilizing knockout models may be beneficial for understanding adiponectin/AdipoRs clinical potential during reproductive aging and obesity-associated reproductive dysfunction.

8. Summary

In the last 2 decades, adiponectin/AdipoRs signaling pathway has been extensively studied, and adiponectin/AdipoRs has emerged as a wonder molecule for research on metabolism and aging-related disorders. Since adiponectin/AdipoRs and its initial association with Insulin sensitivity was discovered, it is now evident that this "obesity hormone" also plays important roles in the biology of reproductive health. Adiponectin acts as an important endocrine/paracrine/autocrine factor regulating testicular physiology under different pathophysiological conditions. A marked decline in adiponectin/AdipoRs signaling during testicular aging, may cause reduced testicular steroidogenesis, (T synthesis), spermatogenesis (sperm physiology), cellular proliferation, cell survival, metabolism (glucose and lactate uptake), and enhanced testicular apoptosis and oxidative stress. Therefore, we suggest that adiponectin treatment or AdipoRs activation may be utilized as a novel therapeutic strategy for ensuring healthy reproductive longevity in patients of subfertility, precocious aging, obesity, and T2D. However, future studies needed, to understand, (a) what causes a decline in adiponectin/AdipoRs signaling during aging? (b) What causes reduced adiponectin/AdipoRs affinity during hypogonadism? (c) Whether this defective adiponectin/AdipoRs machinery may lead to precocious aging; (d), whether adiponectin/AdipoRs effects are through direct action on the gonads and/or through indirect actions involving other factors; to support our previous findings. In brief, adiponectin/AdipoRs activation may represent an important therapeutic strategy for delaying the process of reproductive aging and metabolism-associated reproductive dysfunction including infertility. Also, adiponectin treatment could be one of the promising anti-aging therapies promoting male reproductive health and of course reproductive longevity.

References

Aitken, R. J., & Roman, S. D. (2008). Antioxidant systems and oxidative stress in the testes. *Advances in Experimental Medicine and Biology*, *636*, 154–171.

Almeida, S., Rato, L., Sousa, M., Alves, M. G., & Oliveira, P. F. (2017). Fertility and sperm quality in the aging male. *Current Pharmaceutical Design*, *23*, 4429–4437.

Arita, Y., Kihara, S., Ouchi, N., Takahashi, M., Maeda, K., Miyagawa, J., et al. (1999). Paradoxical decrease of an adipose-specific protein, adiponectin, in obesity. *Biochemical and Biophysical Research Communications*, *257*, 79–83.

Aziz, N., Saleh, R. A., Sharma, R. K., Lewis-Jones, I., Esfandiari, N., Thomas, A. J., et al. (2004). Novel association between sperm reactive oxygen species production, sperm morphological defects, and the sperm deformity index. *Fertility and Sterility*, *81*, 349–354.

Banerjee, A., Anuradha, Mukherjee, K., & Krishna, A. (2014). Testicular glucose and its transporter GLUT8 as a marker of age-dependent variation and its role in steroidogenesis in mice. *Journal of Experimental Zoology Part A Ecology Genetics Physiology*, *321*, 490–502.

Barbutska, D., Koeva, Y., Bakalska, M., & Atanassova, N. (2013). Age-related changes in the steroid-producing cells of rat testis. *Scripta Scientifica Medica*, *45*, 32–35.

Barzilai, N., Huffman, D. M., Muzumdar, R. H., & Bartke, A. (2012). The critical role of metabolic pathways in aging. *Diabetes*, *61*, 1315–1322.

Bebakar, W. M., Honour, J. W., Foster, D., Liu, Y. L., & Jacobs, H. S. (1990). Regulation of testicular function by insulin and transforming growth factor-beta. *Steroids*, *55*, 266–270.

Ben-Shmuel, S., Rostoker, R., Scheinman, E. J., & LeRoith, D. (2016). Metabolic syndrome, type 2 diabetes, and cancer: Epidemiology and potential mechanisms. *Handbook of Experimental Pharmacology*, *233*, 355–372.

Bertoldo, M. J., Faure, M., Dupont, J., & Froment, P. (2015). AMPK: A master energy regulator for gonadal function. *Frontiers in Neuroscience*, *9*(235), 1–11.

Bjursell, M., Ahnmark, A., Bohlooly, Y. M., William-Olsson, L., Rhedin, M., Peng, X. R., et al. (2007). Opposing effects of adiponectin receptors 1 and 2 on energy metabolism. *Diabetes*, *56*, 583–593.

Blüher, M. (2019). Obesity: Global epidemiology and pathogenesis. *Nature Reviews Endocrinology*, *15*, 288–298.

Bowling, A. (2007). Aspirations for older age in the 21st century: What is successful aging? *The International Journal of Aging and Human Development*, *64*, 263–297.

Cabler, S., Agarwal, A., Flint, M., & Du-Plessis, S. S. (2010). Obesity: Modern man's fertility nemesis. *Asian Journal of Andrology*, *12*, 480–489.
Caminos, J. E., Nogueiras, R., Gayta, F., Pineda, R., & Gonza, C. R. (2008). Novel expression and direct effects of adiponectin in the rat testis. *Endocrinology*, *149*, 3390–3402.
Campos, D. B., Palin, M. F., Bordignon, V., & Murphy, B. D. (2008). The "beneficial" adipokines in reproduction and fertility. *International Journal of Obesity (London)*, *32*, 223–231.
Capeau, J. (2007). The story of adiponectin and its receptors AdipoR1 and R2: To follow. *Journal of Hepatology*, *47*, 736–738.
Ceddia, R. B., Somwar, R., Maida, A., Fang, X., Bikopoulos, G., & Sweeney, G. (2005). Globular adiponectin increases GLUT4 translocation and glucose uptake but reduces glycogen synthesis in rat skeletal muscle cells. *Diabetologia*, *48*, 132–139.
Chen, H., Huhtaniemi, I., & Zirkin, B. R. (1996). Depletion and repopulation of Leydig cells in the testes of aging brown Norway rats. *Endocrinology*, *137*, 3447–3452.
Chen, Z., Toth, T., Godfrey-Bailey, L., Mercedat, N., Schiff, I., & Hauser, R. (2003). Seasonal variation and age-related changes in human semen parameters. *Journal of Andrology*, *24*, 226–231.
Choubey, M., Ranjan, A., Bora, P. S., Baltazar, F., & Krishna, A. (2019). Direct actions of adiponectin on changes in reproductive, metabolic, and anti-oxidative enzymes status in the testis of adult mice. *General and Comparative Endocrinology*, *279*, 1–11.
Choubey, M., Ranjan, A., Bora, P. S., Baltazar, F., Martin, L. J., & Krishna, A. (2019). Role of adiponectin as a modulator of testicular function during aging in mice. *Biochimica et Biophysica Acta—Molecular Basis of Disease*, *1865*, 413–427.
Choubey, M., Ranjan, A., Bora, P. S., & Krishna, A. (2020). Protective role of adiponectin against testicular impairment in high-fat diet/Streptozotocin-induced type 2 diabetic mice. *Biochimie*, *168*, 41–52.
Civitarese, A. E., Jenkinson, C. P., Richardson, D., Bajaj, M., Cusi, K., Kashyap, S., et al. (2004). Adiponectin receptors gene expression and insulin sensitivity in non-diabetic Mexican Americans with or without a family history of type 2 diabetes. *Diabetologia*, *47*, 816–820.
Cocuzza, M., Athayde, K. S., Agarwal, A., Sharma, R., Pagani, R., Lucon, A. M., et al. (2008). Age-related increase of reactive oxygen species in neat semen in healthy fertile men. *Urology*, *71*, 490–494.
Coelho, M., Oliveira, T., & Fernandes, R. (2013). Biochemistry of adipose tissue: An endocrine organ. *Archives of Medical Science*, *9*, 191–200.
Combs, T. P., Berg, A. H., Rajala, M. W., Klebanov, S., Iyengar, P., Jimenez-Chillaron, J. C., et al. (2003). Sexual differentiation, pregnancy, calorie restriction, and aging affect the adipocyte-specific secretory protein adiponectin. *Diabetes*, *52*, 268–276.
Davalli, P., Mitic, T., Caporali, A., Lauriola, A., & D'Arca, D. (2016). ROS, cell senescence, and novel molecular mechanisms in aging and age-related diseases. *Oxidative Medicine and Cellular Longevity*, *3565127*, 1–18.
Du-Plessis, S. S., Cabler, S., McAlister, D. A., Sabanegh, E., & Agarwal, A. (2010). The effect of obesity on sperm disorders and male infertility. *Nature Reviews Urology*, 7, 153–161.
Elfassy, Y., Bastard, J. P., McAvoy, C., Fellahi, S., Dupont, J., & Levy, R. (2018). Adipokines in semen: Physiopathology and effects on spermatozoas. *International Journal of Endocrinology*, *3906490*, 1–11.
Escobar-Morreale, H. F., Villuendas, G., Botella-Carretero, J. I., Alvarez-Blasco, F., Sanchón, R., Luque-Ramírez, M., et al. (2006). Adiponectin and resistin in PCOS: A clinical, biochemical and molecular genetic study. *Human Reproduction*, *21*, 2257–2265.
Fink, G. (2000). Neuroendocrine regulation of pituitary function: General principles. In P. M. Conn, & M. E. Freeman (Eds.), *Neuroendocrinology in physiology and medicine* (pp. 107–134). New Jersey: Humana Press.

Gavrila, A., Chan, J. L., Yiannakouris, N., Kontogianni, M., Miller, L. C., Orlova, C., et al. (2003). Serum adiponectin levels are inversely associated with overall and central fat distribution but are not directly regulated by acute fasting or leptin administration in humans: Cross-sectional and interventional studies. *Journal of Clinical Endocrinology and Metabolism, 88*, 4823–4831.

Gunes, S., Hekim, G. N. T., Arslan, M. A., & Asci, R. (2016). Effects of aging on the male reproductive system. *Journal of Assisted Reproduction and Genetics, 33*, 441–454.

Gurusubramanian, G., & Roy, V. K. (2014). Expression of visfatin in alloxan-induced diabetic rat testis. *Acta Histochemica, 116*, 1462–1468.

Hammoud, A. O., Gibson, M., Peterson, C., Hamilton, B., & Carrell, D. (2006). Obesity and male reproductive potential. *Journal of Andrology, 27*, 619–626.

Hellstrom, W. J., Overstreet, J. W., Sikka, S. C., Denne, J., Ahuja, S., Hoover, A. M., et al. (2006). Semen and sperm reference ranges for men 45 years of age and older. *Journal of Andrology, 27*, 421–428.

Hermann, M., Untergasser, G., Rumpold, H., & Berger, P. (2000). Aging of the male reproductive system. *Experimental Gerontology, 35*, 1267–1279.

Hong, C. Y., Park, J. H., Ahn, R. S., Im, S. Y., Choi, H. S., Soh, J., et al. (2004). Molecular mechanism of suppression of testicular steroidogenesis by proinflammatory cytokine tumor necrosis factor alpha. *Molecular and Cellular Biology, 24*, 2593–2604.

Hu, E., Liang, P., & Spiegelman, B. M. (1996). AdipoQ is a novel adipose specific gene dysregulated in obesity. *Journal of Biological Chemistry, 271*, 10697–10703.

Huijg, J. M., van-Delden, A. E., van-der-Ouderaa, F. J., Westendorp, R. G., Slaets, J. P., & Lindenberg, J. (2016). Being active, engaged, and healthy: Older persons' plans and wishes to age successfully. *Journals of Gerontology Series B: Psychological Sciences and Social Sciences, 72*, 228–236.

Jarak, I., Almeida, S., Carvalho, R. A., Sousa, M., Barros, A., & Alves, M. G. (2018). Senescence and declining reproductive potential: Insight into molecular mechanisms through testicular metabolomics. *Biochimica et Biophysica Acta—Molecular Basis of Disease, 1864*, 3388–3396.

Johnson, S. L., Dunleavy, J., Gemmell, N. J., & Nakagawa, S. (2015). Consistent age-dependent declines in human semen quality: A systematic review and meta-analysis. *Ageing Research Reviews, 19*, 22–33.

Jones, R. (1998). Plasma membrane structure and remodeling during sperm maturation in the epididymis. *Journal of Reproduction and Fertility, 53*, 73–84.

Kadowaki, T., & Yamauchi, T. (2005). Adiponectin and adiponectin receptor. *Endocrine Reviews, 26*, 439–451.

Kasimanickam, V. R., Kasimanickam, R. K., Kastelic, J. P., & Stevenson, J. S. (2013). Associations of adiponectin and fertility estimates in Holstein bulls. *Theriogenology, 79*, 766–777.

Kawwass, J. F., Summer, R., & Kallen, C. B. (2015). Direct effects of leptin and adiponectin on peripheral reproductive tissues: A critical review. *Molecular Human Reproduction, 21*, 617–632.

Kidd, S. A., Eskenazi, B., & Wyrobek, A. J. (2001). Effects of male age on semen quality and fertility: A review of the literature. *Fertility and Sterility, 75*, 237–248.

Kubota, N., Terauchi, Y., Yamauchi, T., Kubota, T., Moroi, M., Matsui, J., et al. (2002). Disruption of adiponectin causes insulin resistance and neointimal formation. *Journal of Biological Chemistry, 277*, 25863–25866.

Kubota, N., Yano, W., Kubota, T., Yamauchi, T., Itoh, S., Kumagai, H., et al. (2007). Adiponectin stimulates AMP-activated protein kinase in the hypothalamus and increases food intake. *Cell Metabolism, 6*, 55–68.

Kuhnert, B., & Nieschlag, E. (2004). Reproductive functions of the aging male. *Human Reproduction Update, 10*, 327–339.

Landry, D., Cloutier, F., & Martin, L. J. (2013). Implications of leptin in neuroendocrine regulation of male reproduction. *Reproductive Biology*, *13*, 1–14.

Landry, D., Paré, A., Jean, S., & Martin, L. J. (2015). Adiponectin influences progesterone production from MA-10 Leydig cells in a dose-dependent manner. *Endocrine*, *48*, 957–967.

Li, L., Ma, P., Huang, C., Liu, Y., Zhang, Y., Gao, C., et al. (2014). Expression of chemerin and its receptors in rat testes and its action on testosterone secretion. *Journal of Endocrinology*, *220*, 155–163.

Luhar, S., Timæus, I. M., & Jones, R. (2020). Forecasting the prevalence of overweight and obesity in India to 2040. *PLoS One*, *15*(e0229438), 1–17.

Maeda, K., Okubo, K., Shimomura, I., Funahashi, T., Matsuzawa, Y., & Matsubara, K. (1996). cDNA cloning and expression of a novel adipose specific collagen-like factor, apM1 (adipose most abundant gene transcript 1). *Biochemical and Biophysical Research Communications*, *221*, 286–289.

Maeda, N., Shimomura, I., Kishida, K., Nishizawa, H., Matsuda, M., Nagaretani, H., et al. (2002). Diet-induced insulin resistance in mice lacking adiponectin/ACRP30. *Nature Medicine*, *8*, 731–737.

Marengo, S. R. (2008). Maturing the sperm: Unique mechanisms for modifying integral proteins in the sperm plasma membrane. *Animal Reproduction Science*, *105*, 52–63.

Martin, L. J. (2014). Implications of adiponectin in linking metabolism to testicular function. *Endocrine*, *46*, 16–28.

Matsunami, T., Sato, Y., Ariga, S., Sato, T., Kashimura, H., Hasegawa, Y., et al. (2010). Regulation of oxidative stress and inflammation by hepatic adiponectin receptor 2 in an animal model of nonalcoholic steatohepatitis. *International Journal of Clinical and Experimental Pathology*, *3*, 472–481.

Meites, J. (1988). Neuroendocrine biomarkers of aging in the rat. *Experimental Gerontology*, *23*, 349–358.

Møller, N., Gormsen, L., Fuglsang, J., & Gjedsted, J. (2003). Effects of ageing on insulin secretion and action. *Hormone Research*, *60*, 102–104.

Nakano, Y., Tobe, T., Choi-Miura, N. H., Mazda, T., & Tomita, M. (1996). Isolation and characterization of GBP28: A novel gelatin-binding protein purified from human plasma. *Journal of Biochemistry*, *120*, 803–812.

Nogueiras, R., Barreiro, M. L., Caminos, J. E., Gaytán, F., Suominen, J. S., Navarro, V. M., et al. (2004). Novel expression of resistin in rat testis: Functional role and regulation by nutritional status and hormonal factors. *Journal of Cell Science*, *117*, 3247–3257.

Ocón-Grove, O. M., Krzysik-Walker, S. M., Maddineni, S. R., Hendricks, G. L., & Ramachandran, R. (2008). Adiponectin and its receptors are expressed in the chicken testis: Influence of sexual maturation on testicular AdipoR1 and AdipoR2 mRNA abundance. *Reproduction*, *136*, 627–638.

Ocón-Grove, O. M., Krzysik-Walker, S. M., Maddineni, S. R., Hendricks, G. L., & Ramachandran, R. (2010). NAMPT (Visfatin) in the chicken testis: Influence of sexual maturation on cellular localization, plasma levels and gene and protein expression. *Reproduction*, *139*, 217–226.

Palmer, N. O., Bakos, H. W., Fullston, T., & Lane, M. (2012). Impact of obesity on male fertility, sperm function and molecular composition. *Spermatogenesis*, *2*, 253–263.

Perheentupa, A., & Huhtaniemi, I. (2009). Aging of the human ovary and testis. *Molecular and Cellular Endocrinology*, *299*, 2–13.

Pfaehler, A., Nanjappa, M. K., Coleman, E. S., Mansour, M., Wanders, D., Plaisance, E. P., et al. (2012). Regulation of adiponectin secretion by soy isoflavones has implication for endocrine function of the testis. *Toxicology Letters*, *209*, 78–85.

Pierre, P., Froment, P., Nègre, D., Ramé, C., Barateau, V., Chabrolle, C., et al. (2009). Role of adiponectin receptors, AdipoR1 and AdipoR2, in the steroidogenesis of the human granulosa tumor cell line, KGN. *Human Reproduction*, *24*, 2890–2901.

Pitteloud, N., Hardin, M., Dwyer, A. A., Valassi, E., Yialamas, M., Elahi, D., et al. (2005). Increasing insulin resistance is associated with a decrease in Leydig cell testosterone secretion in men. *Journal of Clinical Endocrinology and Metabolism, 90*, 2636–2641.

Plas, E., Berger, P., Hermann, M., & Pfluger, H. (2000). Effects of aging on male fertility. *Experimental Gerontology, 35*, 543–551.

Psilopanagioti, A., Papadaki, H., Kranioti, E. F., Alexandrides, T. K., & Varakis, J. N. (2009). Expression of adiponectin and adiponectin receptors in human pituitary gland and brain. *Neuroendocrinology, 89*, 38–47.

Qatanani, M., & Lazar, M. A. (2007). Mechanisms of obesity-associated insulin resistance: Many choices on the menu. *Genes & Development, 21*, 1443–1455.

Rahmanifar, F., & Tabandeh, M. R. (2012). Adiponectin and its receptors gene expression in the reproductive tract of ram. *Small Ruminant Research, 105*, 263–267.

Rak, A., Mellouk, N., Froment, P., & Dupont, J. (2017). Adiponectin and resistin: Potential metabolic signals affecting hypothalamo-pituitary gonadal axis in females and males of different species. *Reproduction, 153*, R215–R226.

Ranjan, A., Choubey, M., Yada, T., & Krishna, A. (2019a). Direct effects of neuropeptide nesfatin-1 on testicular spermatogenesis and steroidogenesis of the adult mice. *General and Comparative Endocrinology, 271*, 49–60.

Ranjan, A., Choubey, M., Yada, T., & Krishna, A. (2019b). Immunohistochemical localization and possible functions of nesfatin-1 in the testis of mice during pubertal development and sexual maturation. *Journal of Molecular Histology, 50*, 533–549.

Ranjan, A., Choubey, M., Yada, T., & Krishna, A. (2020). Nesfatin-1 ameliorates type-2 diabetes-associated reproductive dysfunction in male mice. *Journal of Endocrinological Investigation, 43*, 515–528.

Reverchon, M., Rame, C., Bunel, A., Chen, W., Froment, P., & Dupont, J. (2016). Visfatin (NAMPT) improves in vitro IGF1-induced steroidogenesis and IGF1 receptor signaling through SIRT1 in bovine granulosa cells. *Biology of Reproduction, 94*, 1–13.

Rodriguez-Pacheco, F., Martinez-Fuentes, A. J., Tovar, S., Pinilla, L., Tena-Sempere, M., Dieguez, C., et al. (2007). Regulation of pituitary cell function by adiponectin. *Endocrinology, 148*, 401–410.

Rose, M. R. (1991). *Evolutionary biology of aging* (pp. 1–221). New York: Oxford University Press.

Rowe, J. W., & Kahn, R. L. (1997). Successful aging. *The Gerontologist, 37*, 433–440.

Ruan, H., & Dong, L. Q. (2016). Adiponectin signaling and function in insulin target tissues. *Journal of Molecular Cell Biology, 8*, 101–109.

Sandal, S., Tekin, S., Seker, F. B., Beytur, A., Vardi, N., Colak, C., et al. (2015). The effects of intracerebroventricular infusion of apelin-13 on reproductive function in male rats. *Neuroscience Letters, 602*, 133–138.

Scherer, P. E. (2016). The multifaceted roles of adipose tissue-therapeutic targets for diabetes and beyond: The 2015 banting lecture. *Diabetes, 65*, 1452–1461.

Scherer, P. E., Williams, S., Fogliano, M., Baldini, G., & Lodish, H. F. (1995). A novel serum protein similar to C1q, produced exclusively in adipocytes. *Journal of Biological Chemistry, 270*, 26746–26750.

Schoeller, E. L., Schon, S., & Moley, K. H. (2012). The effects of type 1 diabetes on the hypothalamic, pituitary and testes axis. *Cell and Tissue Research, 349*, 839–847.

Selvaratnam, J. S., & Robaire, B. (2016). Effects of aging and oxidative stress on spermatozoa of superoxide-dismutase 1- and catalase-null mice. *Biology of Reproduction, 95*, 1–13.

Shayeb, A. G., Harrild, K., Mathers, E., & Bhattacharya, S. (2011). An exploration of the association between male body mass index and semen quality. *Reproductive Biomedicine Online, 23*, 717–723.

Shukla, K. K., Mahdi, A. A., Ahmad, M. K., Shankhwar, S. N., Rajender, S., Jaiswar, S. P., et al. (2009). Mucuna pruriens improves male fertility by its action on the hypothalamus-pituitarygonadal axis. *Fertility and Sterility, 92*, 1934–1940.

Sieminska, L., Marek, B., Kos-Kudla, B., Niedziolka, D., Kajdaniuk, D., Nowak, M., et al. (2004). Serum adiponectin in women with polycystic ovarian syndrome and its relation to clinical, metabolic and endocrine parameters. *Journal of Endocrinological Investigation*, *27*, 528–534.

Singh, A., Choubey, M., Bora, P., & Krishna, A. (2018). Adiponectin and chemerin: Contrary adipokines in regulating reproduction and metabolic disorders. *Reproductive Sciences*, *25*, 1462–1473.

Takano, H., & Abe, K. (1987). Age-related histologic changes in the adult mouse testis. *Archivum Histologicum Japonicum*, *50*, 533–544.

Tang, X. Y., Zhang, Q., Dai, D. Z., Ying, H. J., Wang, Q. J., & Dai, Y. (2008). Effects of strontium fructose 1,6-diphosphate on expression of apoptosis-related genes and oxidative stress in testes of diabetic rats. *International Journal of Urology*, *15*, 251–256.

Tena-Sempere, M., & Barreiro, M. L. (2002). Leptin in male reproduction: The testis paradigm. *Molecular and Cellular Endocrinology*, *188*, 9–13.

Tena-Sempere, M., & Huhtaniemi, I. (2003). Gonadotropins and gonadotropin receptors. In B. C. J. M. Fauser (Ed.), *Reproductive medicine—Molecular, cellular and genetic fundamentals* (pp. 225–244). New York: Parthenon Publishing.

Thomas, S., Kratzsch, D., Schaab, M., Scholz, M., Grunewald, S., Thiery, J., et al. (2013). Seminal plasma adipokine levels are correlated with functional characteristics of spermatozoa. *Fertility and Sterility*, *99*, 1256–1263.

Tsai, E. C., Matsumoto, A. M., Fujimoto, W. Y., & Boyko, E. J. (2004). Association of bioavailable, free, and total testosterone with insulin resistance: Influence of sex hormone-binding globulin and body fat. *Diabetes Care*, *27*, 861–868.

Vermeulen, A., & Deslypere, J. P. (1985). Testicular endocrine function in the ageing male. *Maturitas*, *7*, 273–279.

Vermeulen, A., Kaufman, J. M., Deslypere, J. P., & Thomas, G. (1993). Attenuated luteinizing hormone (LH) pulse amplitude but normal LH pulse frequency, and its relation to plasma androgens in hypogonadism of obese men. *The Journal of Clinical Endocrinology and Metabolism*, *76*, 1140–1146.

Weir, C. P., & Robaire, B. (2007). Spermatozoa have decreased antioxidant enzymatic capacity and increased reactive oxygen species production during aging in the brown Norway rat. *Journal of Andrology*, *28*, 229–240.

Winters, S. J., Wang, C., Abdlramahaman, E., Hadeed, V., Dyky, M. A., & Brufsky, A. (2006). Inhibin-B levels in healthy young adult men and prepubertal boys: Is obesity the cause for the contemporary decline in sperm count because of fewer Sertoli cells? *Journal of Andrology*, *27*, 560–564.

Wu, L., Xu, B., Fan, W., Zhu, X., Wang, G., & Zhang, A. (2013). Adiponectin protects Leydig cells against proinflammatory cytokines by suppressing the nuclear factor-κB signaling pathway. *The FEBS Journal*, *280*, 3920–3927.

Yamauchi, T., & Kadowaki, T. (2013). Adiponectin receptor as a key player in healthy longevity and obesity-related diseases. *Cell Metabolism*, *17*, 185–196.

Yamauchi, T., Kamon, J., Ito, Y., Tsuchida, A., Yokomizo, T., Kita, S., et al. (2003). Cloning of adiponectin receptors that mediate antidiabetic metabolic effects. *Nature*, *423*, 762–769.

Zhu, N., Pankow, J. S., Ballantyne, C. M., Couper, D., Hoogeveen, R. C., Pereira, M., et al. (2010). High-molecular-weight adiponectin and the risk of type 2 diabetes in the ARIC study. *Journal of Clinical Endocrinology and Metabolism*, *95*, 5097–5104.

FSC
www.fsc.org
MIX
Papier | Fördert gute Waldnutzung
FSC® C083411

Kolibri 360 GmbH
Ferdinand-Jühlke-Straße 7
99095 Erfurt, Deutschland
produktsicherheit@kolibri360.de